THE MATHEMATICS OF FINITE ELEMENTS AND APPLICATIONS V

Academic Press Rapid Manuscript Reproduction

Invited speakers at MAFELAP 1984: Standing: F. BREZZI, C. A. BREBBIA, J. R. WHITEMAN, K. W. MORTON, W. L. WENDLAND, AND A. J. BAKER: SEATED: A. R. MITCHELL, J. T. ODEN, M. F. WHEELER, O. C. ZIENKIEWICZ, AND M. A. CRISFIELD.

Based on the proceedings at a conference held at Brunel University from 1–4 May 1984

THE MATHEMATICS OF FINITE ELEMENTS AND APPLICATIONS V

MAFELAP 1984

Edited by

J. R. WHITEMAN

Department of Mathematics and Statistics
Institute of Computational Mathematics
Brunel University, Uxbridge
Middlesex, England

1985

ACADEMIC PRESS, INC.

(Harcourt Brace Jovanovich, Publishers)

London Orlando San Diego New York

Toronto Montreal Sydney Tokyo

ACADEMIC PRESS INC. (LONDON) LTD.
24–28 Oval Road
LONDON NW1 7DX

United States Edition published by
ACADEMIC PRESS, INC.
Orlando, Florida 32887

BRITISH LIBRARY CATALOGUING IN PUBLICATION DATA
MAFELAP 1984 (*Conference : Brunel University*)
The mathematics of finite elements and applications V :
MAFELAP 1984.
1. Finite element method
I. Title II. Whiteman, J.R. (John Robert)
515.3'53 TA347.F5

LIBRARY OF CONGRESS CATALOGING-IN-PUBLICATION DATA
Conference on the Mathematics of Finite Elements and
Applications (5th : 1984 : Brunel University)
The mathematics of finite elements and applications V.

Includes bibliographies and index.
1. Finite element method—Congresses. 2. Engineering mathematics—Congresses. I. Whiteman, J. R. (John Robert) II. Title.
TA347.F5C64 1984 620'.001'515353 85-47658
ISBN 0-12-747255-X (alk. paper)

PRINTED IN THE UNITED STATES OF AMERICA

85 86 87 88 9 8 7 6 5 4 3 2 1

CONTRIBUTORS

Numbers in parentheses indicate the pages on which the authors' contributions begin.

J. Aalto (105), *Institution of Mechanics, Department of General Sciences, Helsinki University of Technology, SF-02150 Espoo 15, Finland*

R. T. Ackroyd (571, 621), *UKAEA, Risley Nuclear Power Development Establishment, Risley, WA3 6AT, England*

J. E. Akin (291, 603), *Department of Mechanical Engineering and Materials Science, Rice University, Houston, Texas 77251, USA*

J. Altenbach (459), *Sektion Maschinenbau, Technische Hochschule Otto von Guericke, 3010 Magdeburg, German Democratic Republic*

M. J. Baines (421), *Department of Mathematics, University of Reading, Reading RG6 2AX, England*

A. J. Baker (391), *Department of Engineering Science and Mechanics, University of Tennessee, Knoxville, Tennessee 37996, USA*

K. J. Bathe (491), *Department of Mechanical Engineering, Massachusetts Institute of Technology, Cambridge, Massachusetts 02139, USA*

E. B. Becker (505), *TICOM, The University of Texas at Austin, Austin, Texas 78712, USA*

H. Beem (469), *Institut für Konstruktiven Ingenieurbau, Ruhr-Universität Bochum, D-4630 Bochum, Federal Republic of Germany*

H. Berger (459), *Sektion Maschinenbau, Technische Hochschule Otto von Guericke, 3010 Magdeburg, German Democratic Republic*

L. Bernspång (301), *Department of Structural Mechanics, Chalmers University of Technology, S-412 96 Göteburg, Sweden*

D. Bischoff (533), *Institut für Baumechanik und Numerische Mechanik, Universität Hannover, D-3000 Hannover 1, Federal Republic of Germany*

W. S. Blackburn (157), *Central Electricity Generating Board, Scientific Services Centre, Gravesend DA12 2RS, England*

A. Bossavit (451), *Direction des Etudes et Recherches, Electricité de France, 92141 Clamart, France*

C. A. Brebbia (229, 265), *Department of Civil Engineering, Southampton University and Computational Mechanics Institute, Southampton SO9 5NH, England*

F. Brezzi (491), *Instituto de Analisi Numerica, University of Pavia, Palazzo dell' Universita, 27100 Pavia, Italy*

J. C. Bruch, Jr. (605), *Department of Mechanical and Environmental Engineering, University of California, Santa Barbara, California 93106, USA*

C. G. Burton (123), *Materials and Engineering Research Division, Royal Aircraft Establishment, Pyestock GU14 OLS, England*

G. Caloz (431), *Départment de Mathématiques, Ecole Polytechnique Federale de Lausanne, CH-1015 Lausanne, Switzerland*

J. C. Cavendish (83), *Mathematics Department, General Motors Research Laboratories, Warren, Michigan 48090, USA*

I. Christie (415), *Department of Mathematics, West Virginia University, Morgantown, West Virginia 26506, USA*

A. W. Craig (587), *Department of Civil Engineering, University College of Swansea, Swansea SA2 8PP, Wales*

M. A. Crisfield (49), *Transport and Road Research Laboratory, Crowthorne, RG11 6AU, England*

L. Demkowicz (505), *TICOM, The University of Texas at Austin, Austin, Texas 78712, USA*

H. G. V. Der Avanessian (481), *Department of Civil Engineering, University of Surrey, Guildford GU2 5XH England*

U. Eckstein (469), *Institut für Konstruktiven Ingenieurbau, Ruhr-Universität Bochum, D-4630 Bochum, Federal Republic of Germany*

D. A. Field (83), *Mathematics Department, General Motors Research Laboratories, Warren, Michigan 48090, USA*

W. H. Frey (83), *Mathematics Department, General Motors Research Laboratories, Warren, Michigan 48090, USA*

U. Gabbert (459), *Sektion Maschinenbau, Technische Hochschule Otto von Guericke, 3010 Magdeburg, German Democratic Republic*

D. Harrison (137), *Institute of Computational Mathematics, Brunel University, Uxbridge UB8 3PH, England*

R. Harte (469), *Institut für Konstruktiven Ingenieurbau, Ruhr-Universität Bochum, D-4630 Bochum, Federal Republic of Germany*

J. Haslinger (555), *Faculty of Mathematics and Physics, Charles University, 118 00 Prague, 1, Czechoslovakia*

F. K. Hebeker (257), *Fachbereich 17, Mathematik-Informatik, Universität Gesamthochschule Paderborn, D-4790 Paderborn, Federal Republic of Germany*

T. K. Hellen (167), *Central Electricity Generating Board, Technology Planning and Research Division, Berkeley Nuclear Laboratories, Berkeley GL13 9PB England*

J. P. Hennart (309), *National University of Mexico, IIMAS-UNAM, Delegacion A. Obregon, 01000 Mexico D. F., Mexico*

R. K. Jürcke (469), *Institut für Konstruktiven Ingenieurbau, Ruhr-Universität Bochum, D-4603 Bochum, Federal Republic of Germany*

R. J. Kipp (291), *Department of Mechanical Engineering and Materials Science, Rice University, Houston, Texas 77251, USA*

W. B. Krätzig (469), *Institut für Konstruktiven Ingenieurbau, Ruhr-Universität Bochum, D-4630 Bochum, Federal Republic of Germany*

P. Lesaint (563), *Faculté des Sciences, Université de Franche-Comte, 25030 Besançon, France*

T. L. Lin (505), *TICOM, The University of Texas at Austin, Austin, Texas 78712, USA*

R. Löhner (1), *Department of Civil Engineering, University College of Swansea, Singleton Park, Swansea, SA2 8PP, Wales*

V. S. Manoranjan (175), *Department of Mathematics, University of Dundee, Dundee DD1 4HN, Scotland*

J. T. Marti (441), *Seminar für Angewandte Mathematik, Eidgenössiche Technische Hochschule, CH-8092 Zürich, Switzerland*

I. Martindale (113), *Department of Statistics and Computational Mathematics, University of Liverpool, Liverpool L69 3BX, England*

A. R. Mitchell (175), *Department of Mathematics, University of Dundee, Dundee DD1 4HN, Scotland*

K. Morgan (1), *Department of Civil Engineering, University College of Swansea, Singleton Park, Swansea SA2 8PP, Wales*

K. Moriya (283), *Department of Aeronautical Engineering, National Defense Academy, Yokosuka, Japan*

K. W. Morton (343, 421), *Oxford University Computing Laboratory, Oxford, England*

M. Nakata (367), *Exxon Production Research, Houston, Texas 77255, USA*

D. Nardini (265), *Gradjevinski Institut, University of Zagreb, 41000 Zagreb, Yugoslavia*

P. Neittannmäki (555), *Department of Mathematics, University of Jyväskylä, SF-40100 Jyväskylä 10, Finland*

J. Nittmann (157), *Etudes et Fabrication Dowell Schlumberger, Z. I. Molina la Chazotte, 42003 St. Etienne Cedex, France*

J. T. Oden (505), *TICOM, The University of Texas at Austin, Austin, Texas 78712*

P. D. Panagiotopoulos (547), *School of Technology, Aristotle University, Thessaloniki, Greece*

J. Pitkäranta (325), *Institute of Mathematics, Helsinki University of Technology, SF-02150 Espoo 15, Finland*

J. Rappaz (431), *Départment de Mathématiques, École Polytechnique Fédérale de Lausanne, CH-1015 Lausanne, Switzerland*

M. C. Rivara (595), *Department of Mathematics and Computer Science, University of Chile, Santiago, Chile*

A. B. Sabir (481), *Department of Civil and Structural Engineering, University College Cardiff, Cardiff CF2 1TA, Wales*

R. Salminen (105), *Institution of Mechanics, Department of General Science, Helsinki University of Technology, SF-02150 Espoo 15, Finland*

E.-M. Salonen (105), *Institution of Mechanics, Department of General Science, Helsinki University of Technology, SF-02150 Espoo 15, Finland*

A. Samuelsson (301), *Department of Structural Mechanics, Chalmers University of Technology, S-412 96 Göteburg, Sweden*

M. M. Sanz-Serna (415), *Departamento de Ecuaciones Funcionales, Facultad de Ciencias, Universidad de Valladolid, Valladolid, Spain*

E. Schnack (273), *Institute of Solid Mechanics, Karlsruhe University, 7500 Karlsruhe 1, Federal Republic of Germany*

B. W. Scotney (343), *Department of Mathematics, University of Ulster, Coleraine, CT52 1SA Northern Ireland*

M. S. Shephard (97), *School of Engineering, Center for Interactive Computer Graphics, Rensselaer Polytechnic Institute, Troy, New York 12181, USA*

L. W. Spradley (317), *Computational Mechanics Section, Lockheed Missiles and Space Company Inc., Huntsville Research and Engineering Center, Huntsville, Alabama 35807, USA*

R. Stenberg (325), *Institute of Mechanics, Helsinki University of Technology, 02150 Espoo 15, Finland*

G. M. Thompson (27), *GKN Technological Centre, Design Analysis Group, Birmingham, Wolverhampton, WV4 6BW, England*

T. Tiihonen (555), *Department of Mathematics, University of Jyväskylä, SF-40100 Jyväskylä 10, Finland*

R. Verfürth (335), *Mathematisches Institut, Ruhr-Universität Bochum, D-4630 Bochum, Federal Republic of Germany*

R. Wait (113), *Department of Statistics and Computational Mathematics, University of Liverpool, Liverpool L69 3BX, England*

T. J. W. Ward (123), *Institute of Computational Mathematics, Brunel University, Uxbridge UB8 3PH, England*

A. J. Wathen (421), *Department of Mathematics, University of Reading, Reading RG6 2AX, England*

A. Weiser (367), *Exxon Production Research, Houston, Texas 77255, USA*

W. L. Wendland (193), *Fachbereich Mathematik, Technische Hochschule Darmstadt, 6100 Darmstadt, Federal Republic of Germany*

Mary F. Wheeler (367), *Department of Mathematical Science, Rice University, Houston, Texas 77251, USA*

J. R. Whiteman (27, 137, 604), *Institute of Computational Mathematics, Brunel University, Uxbridge UB8 3PH, England*

N. -E. Wiberg (301), *Department of Structural Mechanics, Chalmers University of Technology, S-412 96 Göteburg, Sweden*

U. Wittek (469), *Institut für Konstruktiven Ingenieurbau, Ruhr-Universität Bochum, D-4630 Bochum, Federal Republic of Germany*

A. L. Yettram (137), *Institute of Computational Mathematics, Brunel University, Uxbridge UB8 3PH, England*

J. Z. Zhu (587), *Department of Civil Engineering, University College of Swansea, Singleton Park, SA2 8PP, Wales*

O. C. Zienkiewicz (1, 587), *Department of Civil Engineering, University College of Swansea, Singleton Park, SA2 8PP, Wales*

PREFACE

The fifth conference on The Mathematics of Finite Elements and Applications, MAFELAP 1984, was held at Brunel University in May 1984 and was organized by BICOM (the Brunel Institute of Computational Mathematics). As with previous conferences in the MAFELAP series, the intention was to bring together mathematicians and engineers, with common interests in the broad field of finite elements, to discuss finite element techniques, their theory, their application and their implementation. Following previous MAFELAP practices, topics which were considered to have gained sufficient significance in the field since the previous conference were added to the scope of the programme. On this occasion, in addition to the main themes of the field, it was decided that boundary element techniques should be included and that the finite element/computer aided design interface should again be featured.

The programme of the conference consisted of eleven invited lectures given by A. J. Baker, C. A. Brebbia, F. Brezzi, M. A. Crisfield, A. R. Mitchell, K. W. Morton, J. T. Oden, W. L. Wendland, M. F. Wheeler, J. R. Whiteman, and O. C. Zienkiewickz, thirty-five contributed papers and a similar number of poster papers. The contributed and poster papers were selected by a committee consisting of M. J. M. Bernal, D. Griffiths, D. Harrison, A. R. Mitchell, K. W. Morton, G. M. Thompson, E. H. Twizell, T. J. W. Ward, J. R. Whiteman and O. C. Zienkiewicz. The large number of papers submitted ensured that the task of selection was once again very difficult.

Comments received suggest that the fifth MAFELAP Conference was a worthy and useful successor to the previous meetings. This success was due to the contributions of the conference committee, the chairman of the sessions, the speakers and the poster session authors, as well as the unstinting efforts of the BICOM Research Fellows and the BICOM Secretary, Ms. M. E. Demmar. I express my grateful thanks to all of these people. The conference was financed in part by a grant from the United States Army European Research Office, London, which is acknowledged with great pleasure.

The complete texts of the invited papers and all but one of the contributed papers, as well as the abstracts of the poster papers, are contained in this book. All the manuscripts were prepared in camera ready form by the authors. Thus, although I have edited some of the papers quite heavily, so that major retyping has been necessary, I have made changes only in the interest of clarity. I have not sought to improve the text when the meaning is clear. My thanks go to Ms. M. E. Demmar for her invaluable help in the transforming of a set of manuscripts into book form, and to my wife who has yet again produced a detailed and valuable index for the proceedings of a MAFELAP conference.

J. R. Whiteman

CONTENTS

Abstracts of Poster Sessions

HIGH SPEED INVISCID COMPRESSIBLE FLOW BY THE FINITE ELEMENT METHOD

O.C. Zienkiewicz, R. Löhner and K. Morgan

University of Wales, Swansea, Great Britain

1. INTRODUCTION

The numerical solution of compressible flow problems has received much attention over the past thirty years due, to a large extent, to the interests of the aerospace industry. The solutions of such problems are characterised by the appearance of discontinuities, such as shock waves, in the flow field and a major topic of attention has been the development of numerical techniques which are able to adequately resolve such phenomena. A recent paper by Woodward and Collella [1] gives an excellent survey of the existing 'state of the art' and compares the performance of various widely used algorithms for certain test problems. The algorithms considered are finite difference/ finite volume based and utilise either artificial viscosity [2], linear hybridization [3] or explicit nonlinearity [4]. All the schemes considered are shown to possess certain advantages and disadvantages.

The finite element method has only recently made its appearance in this area, but it is expected that it will make a significant contribution because of the great geometrical flexibility which is inherent in the method. The characteristic Galerkin procedures of Morton [5,6] show much promise while a recent paper by Hughes [7] investigates the approach of explicit nonlinearity in a finite element context. The present authors have made some initial studies [8-10] of high speed inviscid flow problems in which they have used the finite element method and an explicit time stepping algorithm which is based on the Taylor-Galerkin schemes of Donea et al. [11-15], with an appropriate artificial viscosity [16]. In this paper, we combine this solution algorithm with an automatic mesh refinement process which is designed to produce accurate steady state solutions to problems of inviscid compressible flow in two dimensions. The results of two test problems are included which demonstrate the excellent performance characteristics of the proposed procedures.

THE MATHEMATICS OF FINITE ELEMENTS AND APPLICATIONS V

ISBN 0-12-747255-X

2. A SINGLE STEP ALGORITHM

Inviscid compressible flow of an ideal gas in two dimensions is governed by the Euler equations, which can be written in vector form as

$$\frac{\partial \underline{U}}{\partial t} + \frac{\partial \underline{F}_j}{\partial x_j} = \underline{0} \qquad j = 1,2 \qquad (2.1)$$

where

$$\underline{U} = \begin{bmatrix} \rho \\ \rho u_1 \\ \rho u_2 \\ \rho e \end{bmatrix} \qquad \underline{F}_j = \begin{bmatrix} \rho u_j \\ \rho u_1 u_j + p\,\delta_{j1} \\ \rho u_2 u_j + p\,\delta_{j2} \\ u_j(\rho e + p) \end{bmatrix} \qquad (2.2)$$

together with an equation of state

$$p = (\gamma - 1)\rho\,[e - \tfrac{1}{2} u_j u_j] \qquad (2.3)$$

Here p, ρ and e denote the pressure, density and specific internal energy of the fluid respectively while u_j is the velocity component in direction x_j of a Cartesian coordinate system. In addition, the summation convention is employed with δ_{ij} denoting the Kronecker delta.

A single step algorithm for the solution of (2.1) has been fully described elsewhere [8] and is based upon the Taylor-Galerkin methods of Donea et al. [11-15]. A brief description of the solution procedure will be given here for the sake of completeness.

Using a Taylor series expansion about time $t = t^n$ gives

$$\underline{U}^{n+1} = \underline{U}^n + \Delta t \left.\frac{\partial \underline{U}}{\partial t}\right|^n + \frac{\Delta t^2}{2} \left.\frac{\partial^2 \underline{U}}{\partial t^2}\right|^n \qquad (2.4)$$

correct to second order, where $t^{n+1} = t^n + \Delta t$ and a superscript n denotes an evaluation at time $t = t^n$. Eliminating the time derivatives via (2.1) leads to the time-stepping scheme

$$\underline{U}^{n+1} = \underline{U}^n - \Delta t \left.\frac{\partial \underline{F}_j}{\partial x_j}\right|^n + \frac{\Delta t^2}{2} \frac{\partial}{\partial x_k} \left.\left[\underline{A}_k \frac{\partial \underline{F}_j}{\partial x_j}\right]\right|^n \qquad (2.5)$$

where

$$\underline{A}^k = \frac{\partial \underline{F}_k}{\partial \underline{U}} \qquad (2.6)$$

The region Ω over which the solution is required is discretised using linear 3-noded triangular elements and a Galerkin finite element solution procedure is applied to (2.5) using approximations

$$\begin{aligned} \underline{U} &\cong \underline{U}^* = \sum_m \underline{U}_m N_m \\ \underline{F}_j &\cong \underline{F}_j^* = \sum_m \underline{F}_{jm} N_m \\ \underline{A}_k &\cong \underline{A}_k^* = \sum_e \underline{A}_{ke} P_e \end{aligned} \qquad (2.7)$$

Here N_m denotes the piecewise linear shape function associated with node m and P_e the piecewise constant shape function associated with element e. The resulting matrix equation system takes the form

$$\underline{M}(\hat{\underline{U}}^{n+1} - \hat{\underline{U}}^n) = \underline{f}^n \qquad (2.8)$$

where

$$\hat{\underline{U}}^T = (\underline{U}_1, \underline{U}_2, \ldots\ldots) \qquad (2.9)$$

and $\underline{M}$ is a standard mass matrix. An explicit solution procedure for (2.8) is adopted. For problems involving strong shocks an artificial viscosity term is included. The form used is due to Lapidus [16] and replaces the quantities calculated from (2.8) by smoothed values

$$\underline{U}_{ms}^{n+1} = \underline{U}_m^{n+1} - C_\nu \Delta t \sum_{j,e} h_e^2 \int_{\Omega_e} \left|\frac{\partial u_j}{\partial x_j}\right| \frac{\partial \underline{U}^{n+1}}{\partial x_j} \frac{\partial N_m}{\partial x_j} d\Omega \qquad (2.10)$$

where C_ν is a constant in the range $0.5 \leqq C_\nu \leqq 2$ and h_e^2 is a representative area for element e. In the representation of Donea et al. [14,15] the use of such an artificial viscous term is avoided.

3. A TWO STEP ALGORITHM

Finite difference workers have consistently avoided the use of single step explicit algorithms because of the computational expense involved in performing matrix multiplications of the form required in (2.5). They have favoured instead two-step methods which are designed to avoid this requirement. A finite element two-step algorithm may be produced which has certain features in common with the finite difference scheme of Burstein [17]. An alternative finite element two step algorithm has recently been proposed by Miner and Skop [18].

3.1 *The First Step*

Using a Taylor series expansion about time $t = t^n$ gives

$$\underline{U}^{n+\alpha} = \underline{U}^n + \alpha\Delta t \left.\frac{\partial \underline{U}}{\partial t}\right|^n \tag{3.1}$$

correct to first order, where $t^{n+\alpha} = t^n + \alpha\Delta t$, and replacing the time derivative from (2.1) produces the expression

$$\underline{U}^{n+\alpha} = \underline{U}^n - \alpha\Delta t \left.\frac{\partial \underline{F}_j}{\partial x_j}\right|^n \tag{3.2}$$

Employing triangular elements with piecewise linear and piecewise constant shape functions as previously defined, approximations are taken in the form

$$\underline{U}^n \cong \underline{U}^{*n} = \sum_m \underline{U}_m^n N_m$$

$$\underline{U}^{n+\alpha} \cong \underline{U}^{*n+\alpha} = \sum_e \underline{U}_e^{n+\alpha} P_e \tag{3.3}$$

$$\underline{F}_j^n \cong \underline{F}_j^{*n} = \sum_m \underline{F}_{jm}^n N_m$$

These approximations are substituted into (3.2) and a weighted residual statement [19] is formed using the weighting function P_E. The result is to give $\underline{U}_E^{n+\alpha}$ immediately as

$$\Delta_E \underline{U}_E^{n+\alpha} = \sum_m \underline{U}_m^n \int_{\Omega_E} N_m \, d\Omega - \alpha\Delta t \sum_m \underline{F}_{jm}^n \int_{\Omega_E} \frac{\partial N_m}{\partial x_j} d\Omega \tag{3.4}$$

where Δ_E denotes the area of element E and Ω_E denotes the surface of this element. It may be observed that $\underline{U}_E^{n+\alpha}$ can be obtained quickly as no assembly of element contributions is required to form the right hand side of (3.4).

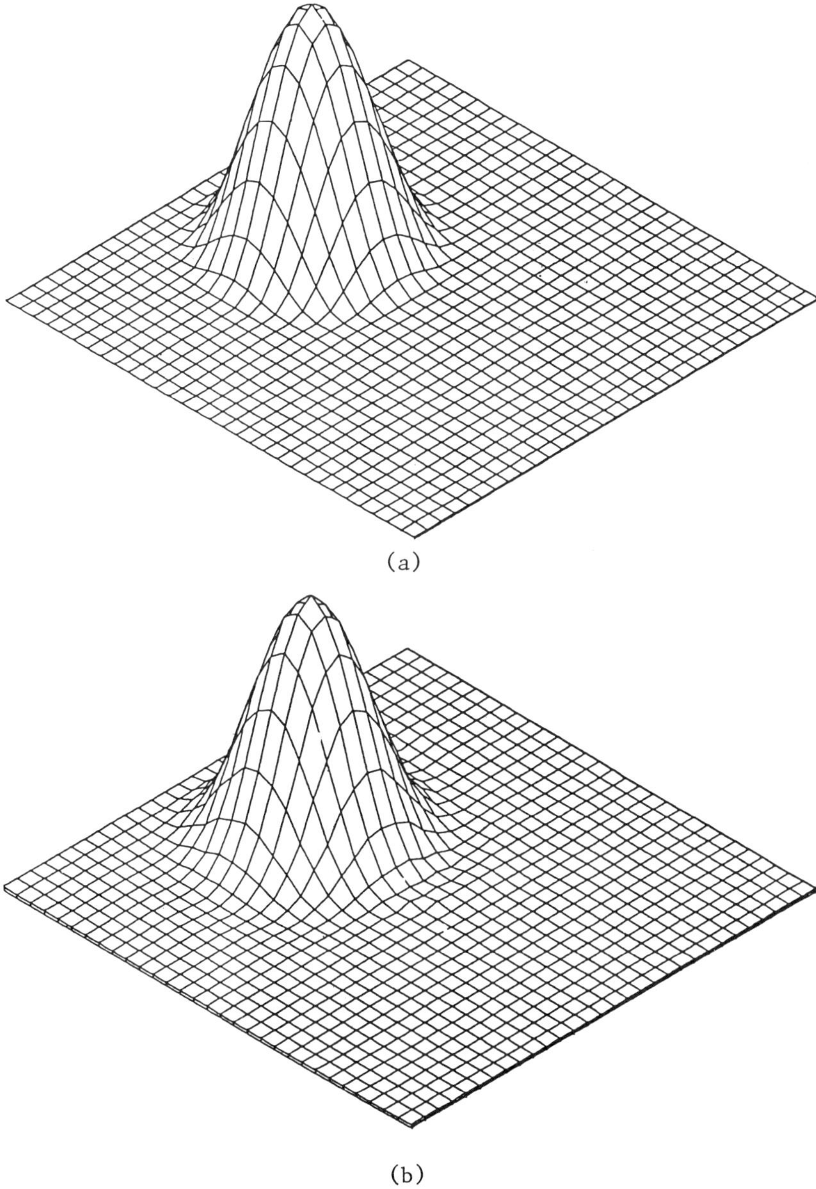

(a)

(b)

Fig.1 Pure advection in two dimensions (a) Initial concentration profile (b) Profile after one complete revolution in a rotating velocity field

3.2 The Second Step

The second step begins by making a new Taylor expansion and writing

$$\underline{U}^{n+1} = \underline{U}^{n} + \Delta t \left. \frac{\partial \underline{U}}{\partial t} \right|^{n+\alpha} \tag{3.5}$$

and again replacing the time derivative leads to

$$\underline{U}^{n+1} = \underline{U}^{n} - \Delta t \left. \frac{\partial \underline{F}_j}{\partial x_j} \right|^{n+\alpha} \tag{3.6}$$

With the approximations

$$\underline{U}^{n+1} \cong \underline{U}^{*n+1} = \sum_m \underline{U}_m^{n+1} N_m \tag{3.7}$$

$$\underline{F}_j^{n+1} \cong \underline{F}_j^{*n+\alpha} = \sum_e \underline{F}_{je}^{n+\alpha} P_e$$

an appropriated weighted residual form for (3.6) is then

$$\int_\Omega \underline{U}^{n+1} N_\ell \, d\Omega = \int_\Omega \underline{U}^{n} N_\ell \, d\Omega + \Delta t \int_\Omega \underline{F}_j^{n+\alpha} \frac{\partial N_\ell}{\partial x_j} d\Omega - \Delta t \int_\Gamma \ell_j \underline{F}_j^{n+\alpha} N_\ell \, d\Gamma \tag{3.8}$$

where $\underline{n} = (\ell_1, \ell_2)$ is the unit outward normal to the boundary Γ of Ω. Inserting (3.7) into (3.8) gives

$$\underline{M} \, (\hat{\underline{U}}^{n+1} - \hat{\underline{U}}^{n}) = \underline{g}^{n+\alpha} \tag{3.9}$$

where $\hat{\underline{U}}$ is defined in (2.9). The solution of this equation completes the second step. Again the smoothing of (2.10) is applied for problems involving strong shocks. It should be noted that for linear problems where

$$\underline{F}_k = \underline{A}_k \, \underline{U} \tag{3.10}$$

and $\underline{A}_k$ is constant, (3.8) combined with (3.9) produces an algorithm which is very similar to the single step equation system (2.8) when $\alpha = 1/2$.

This has been confirmed computationally by applying both methods to the solution of pure advection problems involving a cone-shaped profile in two dimensions. Similar problems have been studied extensively in recent years [11,20,21]. For this

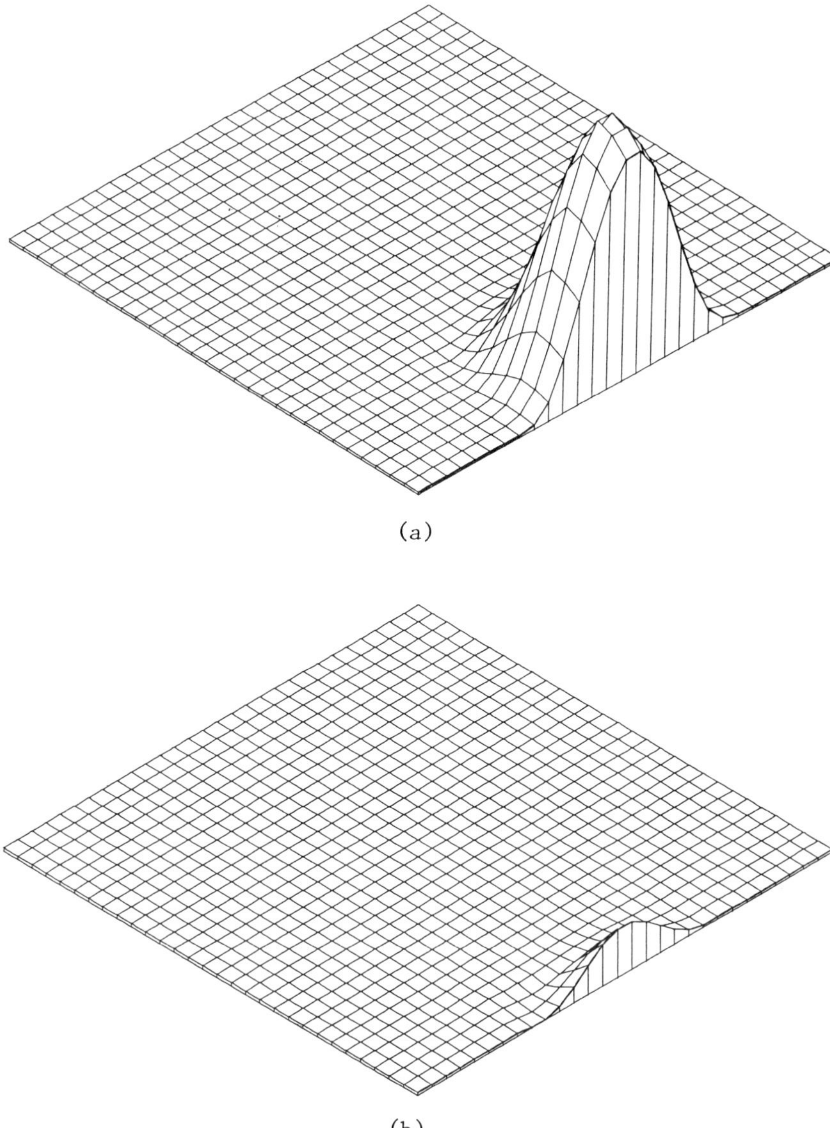

(a)

(b)

Fig.2 Pure advection in a uniform velocity field of the initial concentration shown in Fig.1(a). The concentration at two separate times is shown

problem, the governing equation is the scalar equation.

$$\frac{\partial u}{\partial t} + A_k \frac{\partial u}{\partial x_k} = 0 \qquad (3.11)$$

where (A_1, A_2) denotes a specified velocity field and U denotes the concentration. Fig.1a shows the initial profile and Fig.1b illustrates the numerical solution produced after one complete rotation in an incompressible rotating velocity field. A problem involving the transportation out of the domain of interest of an initial cone in an uniform velocity field was also investigated. Fig.2a and 2b shows the profile of the concentration as the cone leaves the region. The quality of these results is excellent and they were produced without the artificial viscosity term of (2.10) i.e. $C_\nu = 0$.

Preliminary tests show the single step and two step algorithms to be producing similar results for practical problems involving the non-linear equation set (2.1) - (2.3), with the two step method proving to be computationally more efficient.

4. DOMAIN SPLITTING

It is well-known that solutions of high speed inviscid compressible flow problems exhibit narrow regions of rapid change (e.g. shocks) which are embedded in larger regions where the solution is smooth. The economical method of producing an accurate solution is to use a discretisation which has many small elements in the region of rapid change and larger elements elsewhere. However, the small elements used might then require that a correspondingly small value of the timestep, Δt, has to be employed to ensure stability of the algorithms described in the previous section, whereas the larger elements could tolerate the use of a bigger timestep. The remedy adopted by the authors is to split the domain into regions in which different timestep sizes can be used. This method, which bears certain resemblances to the explicit procedures of Belytschko et al. [22] and Liu [23] is described in detail elsewhere [9] and will not be considered further here. It should be noted however that this procedure is performed completely automatically by the computer code at prescribed times during the analysis.

5. ADAPTIVE MESH REFINEMENT

In general, an analyst will have no a priori knowledge of the location of those areas of the solution domain in which large gradients will exist. An ideal computational algorithm would then require the ability to automatically refine the finite element mesh, in zones of high gradients, as the computation proceeds. The geometric flexibility of the linear triangular element makes it ideally suited to refinement processes of this type.

Adaptive refinement techniques are currently receiving much attention [24,25] and two general classes may be identified, viz a priori methods and a posteriori methods. Although many applications employing a priori grid refinements have been reported, at the present time this approach appears to compare unfavourably with a posteriori methods with regard to CPU time requirements [24]. For this reason, we have concentrated on the use of a posteriori techniques. Here the solution is obtained on an original mesh and, at a certain time, the mesh is refined according to some strategy. The solution then proceeds, with further refinements being made as required. The authors have investigated two alternative refinement strategies viz mesh movement and mesh enrichment.

5.1 Mesh Movement

Here the total number of nodes and elements remains constant, but the location of the nodes is changed in order to achieve a better overall distribution. Full details of the strategy adopted for moving the mesh and handling some of the subsequent problems which may arise have been given in detail elsewhere [10]. This approach alone does not appear to possess the degree of flexibility which would be required for general aerodynamic problems and so it has been replaced by mesh enrichment.

5.2 Mesh Enrichment

In this case the original mesh is held fixed, while hierarchical elements [19] or simply more elements are added. Both strategies have their merits and disadvantages. If further degrees of freedom are added hierarchically, the old shape functions are retained and the new shape functions are introduced for relative values of the unknown. Fig.3 shows the comparison of this strategy with the classical enrichment method of adding more elements. The advantages of the hierarchical technique are (i) certain matrices may be re-used, so avoiding some recomputation, (ii) multigrid solvers may be implemented naturally, (iii) no geometrical problems are encountered in the transition zone between refined and unrefined portions of the mesh. The disadvantages of the hierarchical approach are (i) more stiffness coefficients appear as all the refinement levels are interconnected, (ii) the program structure becomes very involved and, it is expected that efficient vectorisation will be virtually impossible.

The classical way of enriching grids is simply to add more elements. In this case, all the advantages of the hierarchical

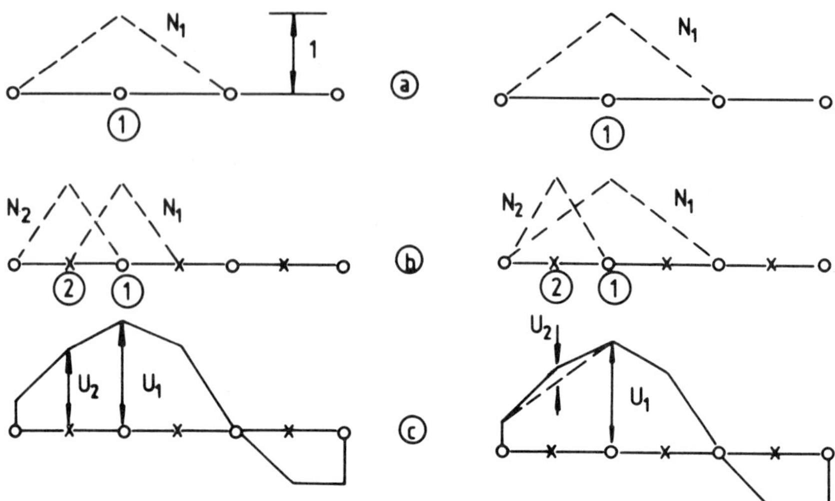

Figure 3. Comparison between standard and hierarchical refinement processes. (a) Original mesh (b) Refined mesh (c) Approximation to a function $U \cong \sum_j U_j N_j$

method are lost while the disadvantages disappear. However, when a detailed comparison was made of the two possibilities [26], it was decided that classical enrichment would be more efficient at this stage for general transient problems and so this method was implemented in the computer code.

5.3 Error Estimate

Consider a single scalar equation with unknown U and a corresponding finite element approximation U*. The aim of any adaptive mesh refinement is to attempt to minimise the highest error occurring in the elements representing the domain i.e.

$$\varepsilon = \max_e \|U-U^*\|_k^e \tag{5.1}$$

should be minimised, where $\| \ \|_k$ denotes some suitable norm. This criterion leads to the requirement that the error should be evenly distributed across the domain, which amounts to the requirement

$$\|U-U^*\|_k^e = \text{constant} \tag{5.2}$$

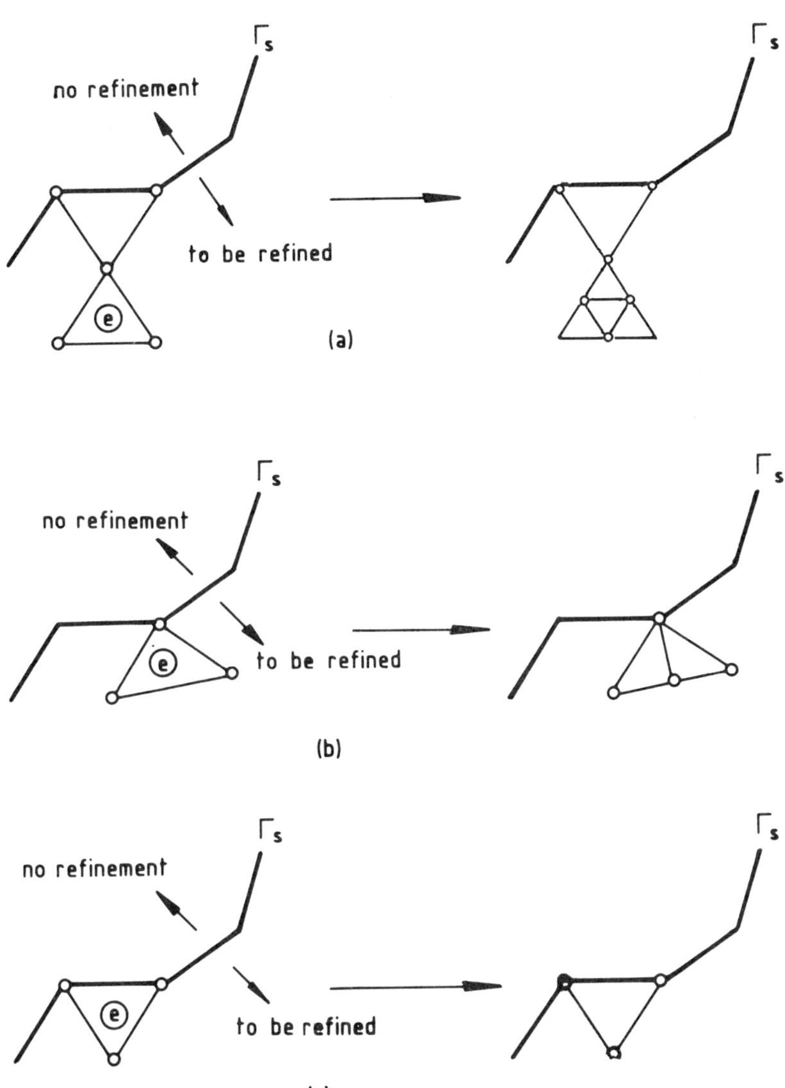

Fig.4 The refinement of element e. (a) No nodes on Γ_s (b) one node on Γ_s (c) two nodes on Γ_s

For elliptic problems, the appropriate norm appears naturally as the energy norm and a fairly complete and general theory of error estimators is available [27,28]. However, far less research has been directed towards hyperbolic problems and a detailed theory is still lacking. Here we attempt to derive a simple, yet reliable, error criterion for hyperbolic problems and linear elements.

It has been shown by Strang and Fix [29] that a natural norm for first order hyperbolic equations is the L_2 norm defined by

$$\|V\|_o^e = \left\{\iint_{\Omega_e} |V|^2 \, d\Omega\right\}^{\frac{1}{2}} \tag{5.3}$$

If the solution of hyperbolic equations is viewed as an approximation problem along characteristics then it can be shown [29] that

$$\|U-U^*\|_k^e \leqq \kappa \, h_e^{\ell-k} \, \|U\|_\ell^e \tag{5.4}$$

where h_e is a representative element length, and using the L_2 norm of (5.3) gives

$$\|U-U^*\|_o^e \leqq \kappa \, h_e^{\ell} \, \|U\|_\ell^e \tag{5.5}$$

The condition of (5.2) is now replaced by the requirement that

$$h_e^{\ell} \, \|U\|_\ell^e = \text{constant} \tag{5.6}$$

or, since the exact solution U is unknown, the practical requirement becomes

$$h_e^{\ell} \, \|U^*\|_\ell^e = \text{constant} \tag{5.7}$$

If linear elements are being employed it seems natural to choose $\ell = 1$ and to define

$$\|U^*\|_1^e = \left\{\iint_{\Omega_e} \left(\frac{\partial U^*}{\partial x_i}\right)^2 d\Omega\right\}^{\frac{1}{2}} \tag{5.8}$$

If the elements are not badly deformed then (5.7) may be approximated by the requirement that

$$\Psi^e = h_e^{\beta} \, \{\max_e U^* - \min_e U^*\} = \text{constant} \tag{5.9}$$

where β = 1/2, 1 or 3/2 for one, two or three dimensions respectively.

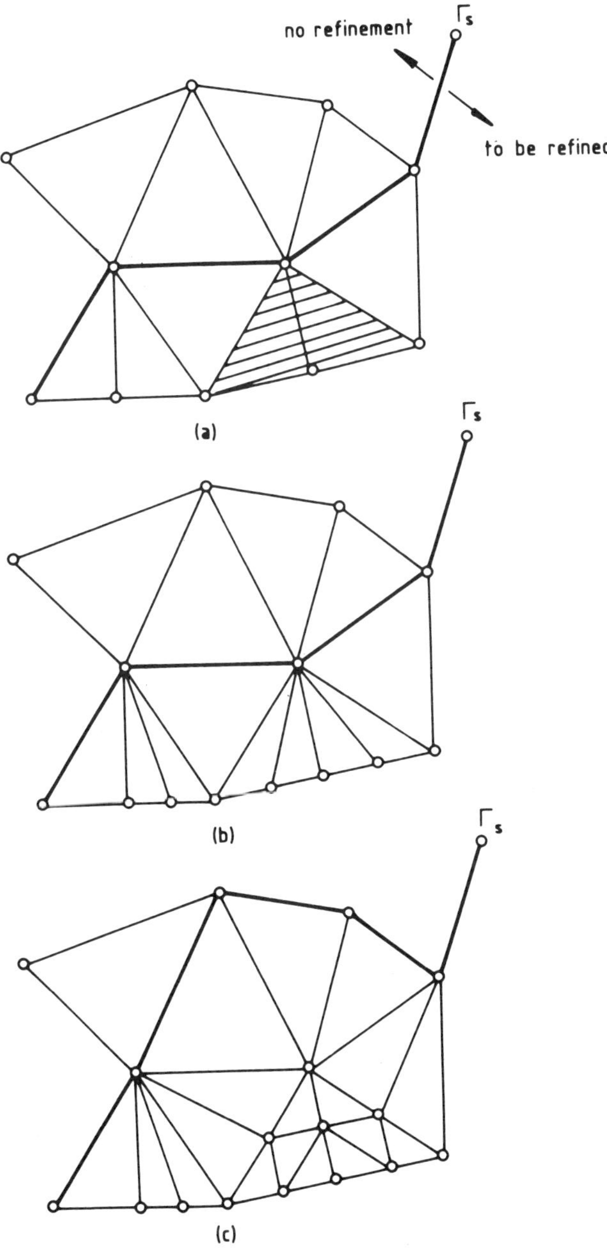

Fig.5 Avoidance of successive subdivision into two during refinement (a) The shaded element has already been subdivided at a previous step. (b) This scheme is not adopted. (c) This is the adopted scheme.

5.4 Identification of Regions to be Refined

The first step in any mesh refinement algorithm is the identification of the regions of the domain in which refinement is to be employed. For (5.9) it is clear that all elements which satisfy

$$\psi^e > \delta \max_e \psi^e \tag{5.10}$$

should be refined further. The constant δ in this equation acts as a threshold value for the refinement process. Having identified in this fashion the patches of elements that are to be refined, the boundary nodes of these patches are obtained. In order to avoid 'saw-tooth' type boundary shapes, any element which has all its nodes on these boundaries is itself included among the list of elements to be refined.

For systems of equations, a 'key-variable' has to be identified in order to employ (5.10). The choice of this variable is not obvious but the density has been chosen for the computations reported here.

5.5 Subdivision of the Elements

During the subdivision process, three different types of element can appear according to the number of nodes of the element which lie on the boundary Γ_s between the region which is to be subdivided and the region which is to remain unchanged. If the element has no nodes on Γ_s then the element is simply divided into four elements as shown in Fig.4a. If the element has one node on Γ_s then the element is subdivided into two elements as shown in Fig.4b. Finally, if the element has more than one node on Γ_s then no subdivision is performed as shown in Fig.4c. When this process has been carried out, the new nodal coordinates and the new nodal unknowns are calculated while the new element connectivities are formed. The subdivision process is completed by identifying new nodes which lie on the boundary of the physical domain and assigning to these nodes the appropriate boundary condition marker.

5.6 Mesh Quality Enhancement

In many practical situations the straightforward subdivision strategy described above will prove satisfactory but, under certain circumstances, badly deformed elements can appear. Three simple devices have thus been implemented in an attempt to improve the quality of the refined mesh and these are described below. The resulting meshes are generally free of the geometrical difficulties usually encountered when meshes are enriched by straightforward subdivision. In addition, these

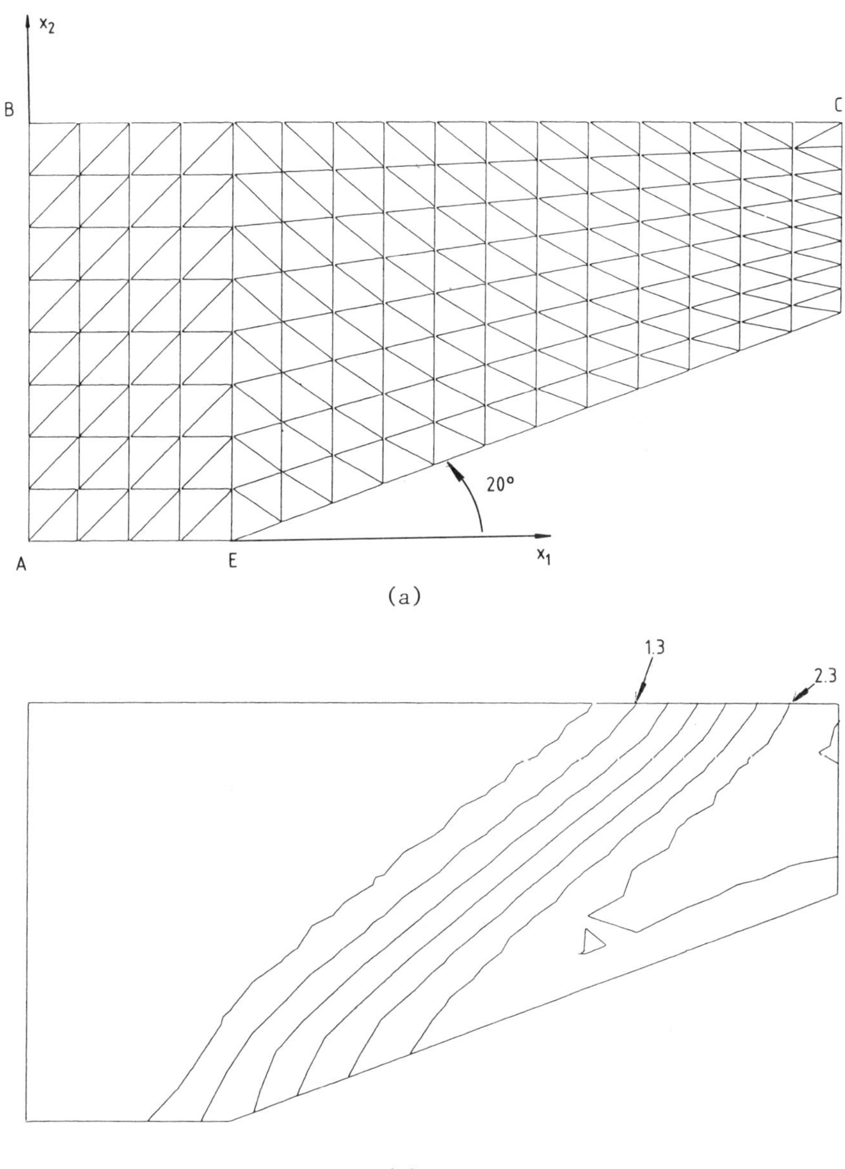

Fig.6 Supersonic flow past a wedge at Mach2. (a) Initial mesh (b) Density after 100 steps

strategies can easily be extended to three dimensions if linear tetrahedral elements are employed.

5.6.1 Smoothing With the refined grid, the sides of the elements are replaced by springs of unit stiffness. For badly deformed elements the resulting nodal forces will not be in equilibrium, whereas for regions of well-formed elements the resulting nodal forces will nearly vanish. A relaxation procedure [10] is adopted which noves the nodes until equilibrium of the nodal forces is achieved. Five steps of this process are normally employed and this ensures a local smoothing of the mesh.

5.6.2 Avoidance of Successive Subdivision into Two Badly deformed elements can appear when an element that has been initially divided into two elements as in Fig.4b is again subdivided in the same fashion. The easiest method of avoiding this second subdivision is to enlarge the region which is to be refined by adding to it all elements surrounding the element under consideration. The element under consideration can then be subdivided into eight elements, thus enhancing the regularity of the mesh. This process is illustrated in Fig.5.

5.6.3 Removal of Badly Deformed Elements If badly deformed elements are still present, after avoiding successive subdivision into two and after smoothing the mesh, then these elements are removed from the computation. This algorithm is detailed elsewhere [10] and need not be repeated here.

6. RESULTS

Results will be presented which show the application of the methods described in this paper to two separate problems viz. steady supersonic flow past a wedge and steady supersonic flow past a nose cone. As both problems are steady, the transient algorithms described previously are used as a device for integrating to steady state.

6.1 Supersonic Flow Past a Wedge

The solution domain and the initial discretisation is shown in Fig.6a. Along AB the flow variables are held fixed at the values $\rho = 1.125$, $u_1 = 1020$, $u_2 = 0$, $e = 727350$ giving a free stream Mach number of three. The boundary ED is a solid wall while 'natural' conditions are applied along BC, CD and AE. The exact solution for this problem consists of an attached plane shock at an angle of 37.5° to the x -axis with uniform flow ahead of and behind the shock.

The density distribution after 100 global timesteps is shown in Fig.6b, and, although the required features are present, the

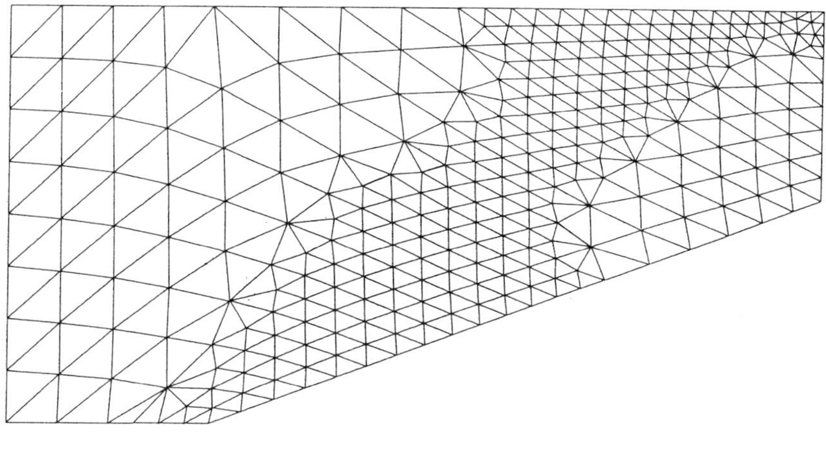

(a)

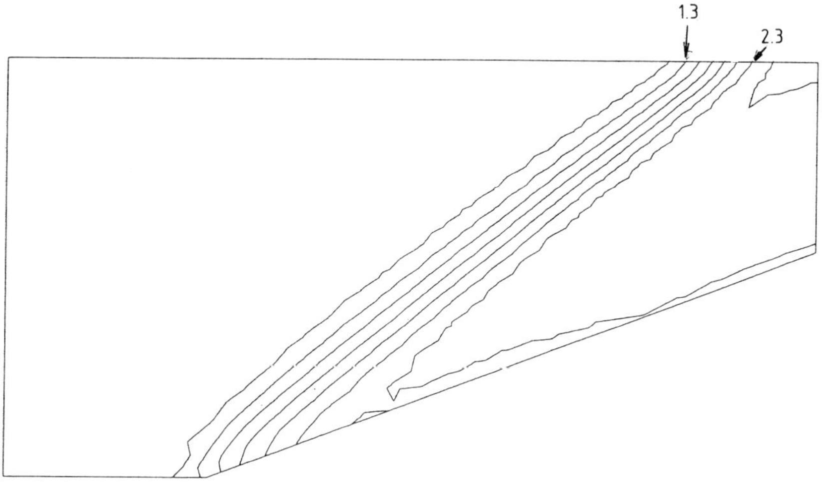

1.3

2.3

(b)

Fig.7 Supersonic flow past a wedge at Mach2. (a) First mesh refinement (b) Density after a further 100 timesteps

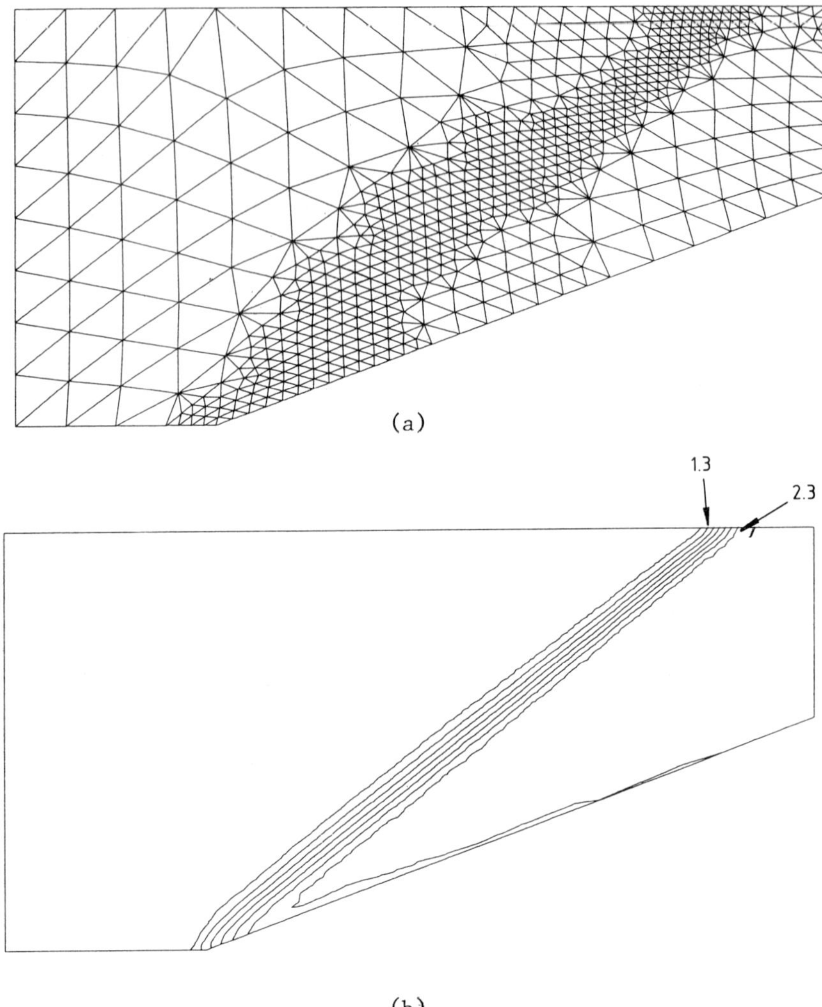

Fig.8 Supersonic flow past a wedge at Mach2. (a) Second mesh refinement (b) Density after a further 100 global timesteps

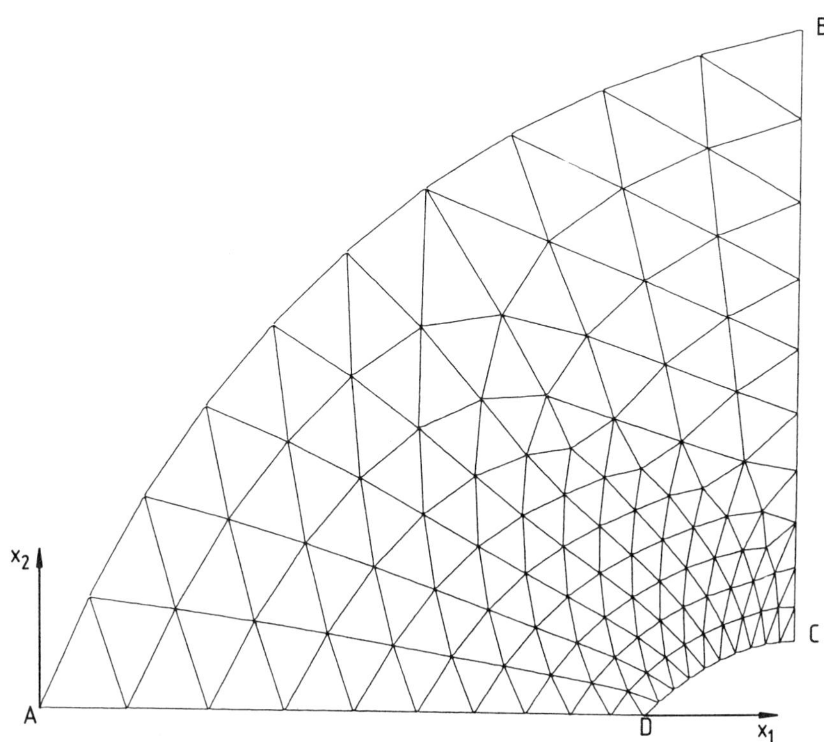

Fig.9 Supersonic flow past a nose cone - initial mesh

quality of the solution is not good because of the coarseness of the initial discretisation. The mesh is therefore automatically refined at this stage, as shown in Fig.7a, and the solution is advanced in time. After a further 100 global steps the density distribution is as illustrated in Fig.7b and the improvement in the quality of the solution is apparent. A further refinement was performed, producing the mesh of Fig.8a, which produced the density distribution shown in Fig.8b after a further 100 global steps. The quality of this solution is excellent when compared with the behaviour of the exact solution.

6.2 *Supersonic Flow Past a Nose Cone*

The solution domain and initial discretisation for this example is shown in Fig.9. Along AB the flow variables are held fixed at the values $\rho = 1$, $u_1 = 1.028$, $u_2 = 0$, $e = 1$ giving a free stream Mach number of two. The boundary CD is a solid wall and 'natural' conditions are applied over BC and DA. The initial finite element discretisation is also shown in Fig.9 and the solution was advanced 100 global timesteps on this mesh.

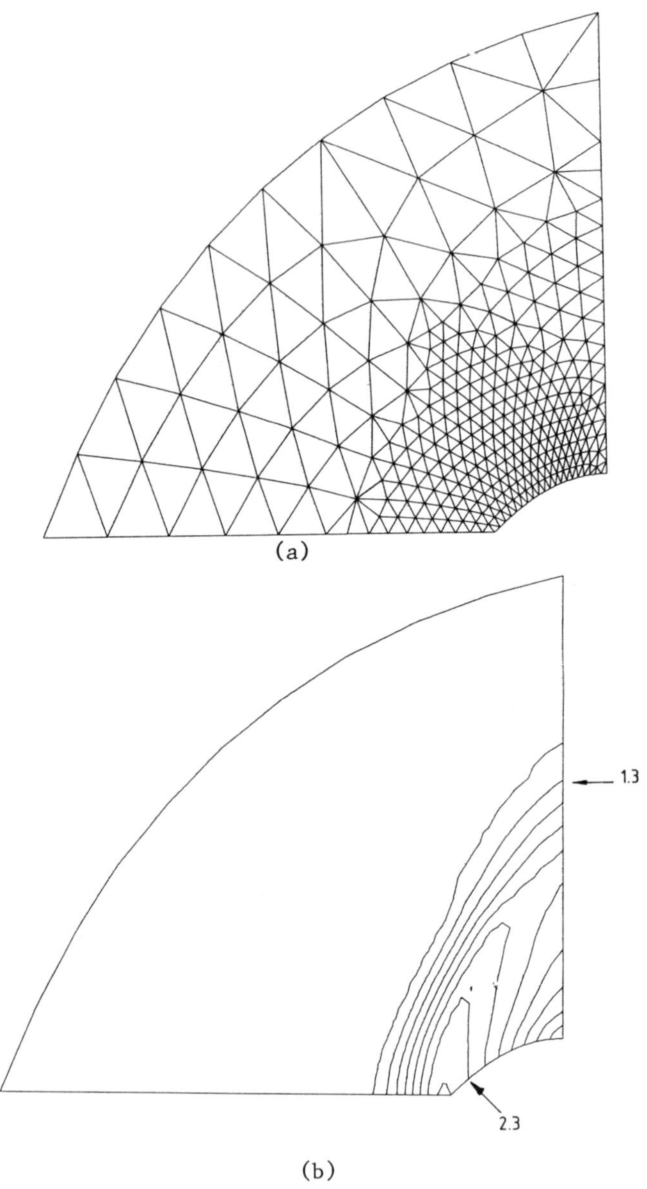

Fig.10 Supersonic flow past a nose cone (a) First mesh refinement after 100 global timesteps, (b) Density distribution after a further 100 global timesteps

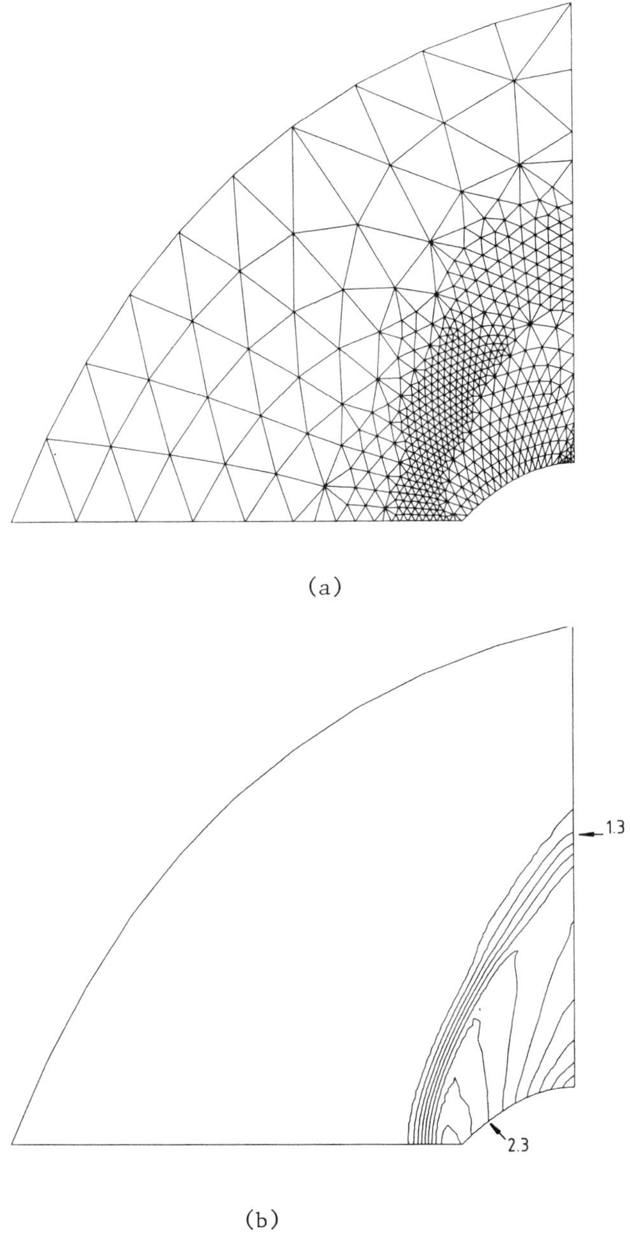

Fig.11 Supersonic flow past a nose cone (a) Second mesh refinement (b) Density distribution after a further 100 global timesteps

The mesh was then refined as shown in Fig.10a and after a further 100 global timesteps the density distribution is as illustrated in Fig.10b. The next level of refinement is displayed in Fig.11a with the corresponding density contours after a further 100 global steps appearing in Fig.11b. The final refinement is then performed as in Fig.12a, producing the density contours of Fig.12b after a further 20 global timesteps. The quality of the final solution is good, showing the detached bow shock.

7. CONCLUSIONS

Explicit solution techniques for the Euler equations have been combined with domain splitting and mesh enrichment procedures. Practical examples have been presented which confirm the viability of the procedures for the analysis of steady problems. For the analysis of true transient problems, the areas requiring refinement will traverse the whole of the domain under consideration and the refinement strategy will require suitable modification in this case [26].

8. ACKNOWLEDGEMENT

The authors would like to thank the Aerothermal Loads Branch, NASA Langley Research Center for supporting this research under Grant No. NAGW-478, and especially A.R. Wieting and K.S. Bey for guidance and encouragement. The two step computations reported here were performed by J. Peraire i Guitart and the authors are grateful for his assistance in the preparation of this paper.

9. REFERENCES

1. WOODWARD, P. and COLLELA, P, 'The numerical solution of 2D fluid flow with strong shocks' *J.Comp.Phys.* 54, 115-173 (1984).

2. VON NEUMANN, J. and RICHTMEYER, R.D., 'A method for the numerical calculation of hydrodynamic shocks', *J.Appl.Phys.* 21, 232-236 (1950).

3. HARTEN, A. and ZWAS, G., 'Self-adjusting hybrid schemes for shock computations', *J.Comp.Phys.* 6, 568-583 (1972).

4. GODUNOV, S.K., 'Finite difference methods for numerical computations of discontinuous solutions of equations of fluid dynamics', *Mat.Sb.*, 47, 271-295 (1959).

5. MORTON, K.W., 'Generalised Galerkin methods for steady and unsteady problems' in *Numerical Methods for Fluid Dynamics*, Academic Press (1982).

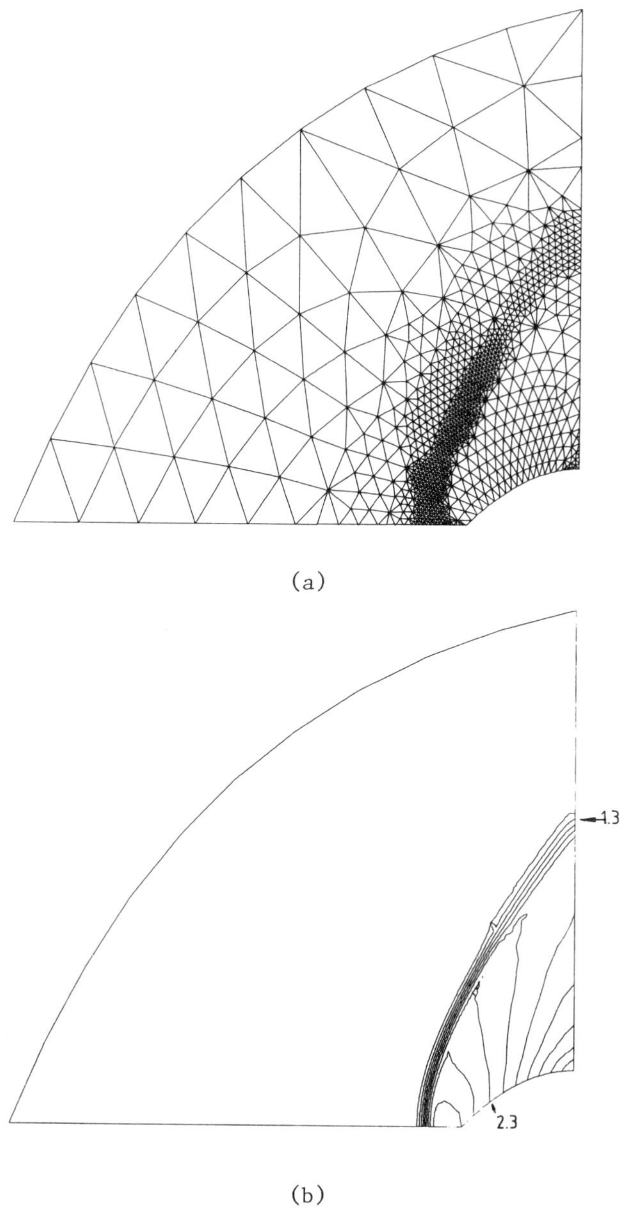

Fig.12 Supersonic flow past a nose cone (a) Final mesh (b) Density distribution after a further 20 global timesteps

6. MORTON, K.W., 'Shock capturing, fitting and recovery', *Numerical Analysis Report 5/82*, University of Reading (1982).

7. HUGHES, T.J.R., 'A shock-capturing finite element method', Preprint (1984).

8. LÖHNER, R., MORGAN, K. and ZIENKIEWICZ, O.C., 'The solution of non-linear systems of hyperbolic equations by the finite element method', accepted for publication *Int.J.Num.Meth. Fluids* (1984).

9. LÖHNER, R., MORGAN, K. and ZIENKIEWICZ, O.C., 'The use of domain splitting with an explicit hyperbolic solver', accepted for publication in *Comp.Meth.Appl.Mech.Engng.* (1984).

10. LÖHNER, R., MORGAN, K. and ZIENKIEWICZ, O.C., 'Adaptive grid refinement for the Euler and compressible Navier-Stokes equations' to appear in *Proc. Conf. Accuracy Estimates and Adaptive Refinements in Finite Element Computations*, Lisbon (1984).

11. DONEA, J., 'A Taylor-Galerkin method for convective transport problems', *Int.J.Num.Meth.Engng.* 20, 101-120 (1984).

12. DONEA, J., GIULIANI, S., LAVAL, H. and QUARTAPELLE, L., 'Time accurate solution of advection-diffusion problems by finite elements', accepted for publication in *Comp.Meth. Appl.Mech.Engng.* (1984).

13. DONEA, J., 'Recent advances in computational methods for steady and transient transport problems', accepted for publication in *Nucl.Engng.Des.* (1984).

14. SELMIN, V., DONEA, J. and QUARTAPELLE, L., 'Taylor-Galerkin method for non-linear hyperbolic equations', to appear in *Proc.Conf. Numerical Methods for Transient and Coupled Problems*, Venice (1984).

15. SELMIN, V., DONEA, J. and QUARTAPELLE, L., 'Taylor-Galerkin methods for nonlinear advection', preprint (1984).

16. LAPIDUS, A., 'A detached shock calculation by second-order finite differences', *J.Comp.Phys.*, 2, 154-177 (1967).

17. BURSTEIN, S.Z., 'Finite difference calculations for hydrodynamic flows containing discontinuities', *J.Comp. Phys.*, 2, 198-222 (1967).

18. MINER, E.W. and SKOP, R.A., 'Explicit time integration for the finite element shock wave equations', *AIAA* Paper 82 -0994 (1982).

19. ZIENKIEWICZ, O.C. and MORGAN, K., *'Finite elements and approximation'*, John Wiley, New York (1983).

20. HUGHES, T.J.R. and BROOKS, A., 'A multidimensional upwind scheme with no cross-wind diffusion', *ASME AMD*, 34, 19-36 (1979).

21. BAKER, A.J., *Finite element computational fluid mechanics'*, Hemisphere, Washington (1983).

22. BELYTSCHKO, T., YEN, H.I. and MULLEN, R., 'Mixed methods for time integration', *Comp.Meth.Appl.Mech.Engng.*, 17-18, 259-275 (1979).

23. LIU, W.K., 'Development of mixed time partition procedures for thermal analysis of structures', *Int.J.Num.Meth.Engng.*, 19, 125-140 (1983).

24. ODEN, J.T. 'Notes on grid optimization and adaptive methods for finite element methods', *TICOM* Report (1983).

25. THOMPSON, J.F. (Editor), *'Numerical grid generation'*, Elsevier, New York (1982).

26. LÖHNER, R., Ph.D. Thesis, University of Wales (1984).

27. BABUSKA, I. and RHEINBOLDT, W.C., 'A-posteriori error estimates for the finite element method', *Int.J.Num.Meth. Engng.*, 12, 1597-1615 (1978).

28. BABUSKA, I. and RHEINBOLDT, W.C., 'Computational error estimates and adaptive processes for some structural problems', *Comp.Meth.Appl.Mech.Engng.*, 34, 895-937 (1982).

29. STRANG, G. and FIX, G.J., *'An analysis of the finite element method'*, Prentice-Hall, Englewood Cliffs (1976).

FINITE ELEMENT CALCULATIONS OF PARAMETERS FOR SINGULARITIES IN PROBLEMS OF FRACTURE

J.R. Whiteman and G.M. Thompson

Institute of Computational Mathematics, Brunel University, Uxbridge, Middlesex, UB8 3PH, England

1. INTRODUCTION

In this paper singularities in problems of linear elastic fracture and of elastic-plastic fracture under strain hardening conditions are considered. The main purpose of this work is the calculation of parameters which characterise the fracture properties for stationary cracks in these contexts. The finite element method is used to treat the elastic and elastic-plastic deformation problems, thus producing approximations to the primary variables, displacements. The fracture parameters are then calculated as secondary quantities by retrieval from the calculated primary approximations. Throughout we are concerned with the obtaining of accurate approximations together, if possible, with theoretical error estimates. The marked differences from this point of view between the states-of-the-art for linear and non-linear problems are exhibited.

It should be pointed out for the material behaviour that the linear elastic model is based on the assumption that displacements and strains are small. For the elastic-plastic model it is assumed that displacements and elastic strains are small, and that strain hardening takes place. In the context of fracture the cracks are taken to be stationary.

From a mathematical point of view the linear elastic model is well established and a large number of results have been derived concerning the existence and regularity of solutions of linear elastic problems. Similarly for two-dimensional linear elastic problems in regions containing cracks the forms of singularities which occur have been derived, Grisvard [12], and again regularity results are known. The situation for three dimensions is less well understood; for example there is some debate concerning the form of the singularity in the vicinity of the

The work of both authors was supported in part by the SERC under Grant No.GR/B9679.7 and that of Whiteman additionally in part by the United States Army under Contract No.DAJA45-83-C-0055.

ISBN 0-12-747255-X

intersection of a stress-free surface with a crack front. The practical use of K (stress intensity factor) and J (path independent integral) as fracture parameters is also well established. For the finite element treatment of two-dimensional problems of this type, and to a lesser extent for similar three-dimensional problems, methods have been produced together with theoretical error bounds which are again mathematically well founded. Thus for these linear problems the situation regarding finite element techniques, both algorithmically and theoretically, is reasonably satisfactory.

As soon as nonlinear deformation behaviour is considered, the whole scene changes. Immediately problems of mathematical modelling have to be considered and aspects of engineering science have to be taken into account. Decisions have to be made concerning the choice of model and the material behaviour. With the onset of plasticity the choice is between various plasticity formulations, for which the mathematical theory is much less well understood than for the linear elastic case. In the context of nonlinear fracture mechanics (NLFM) difficulties arise due to the fact that the incremental theory of plasticity is usually employed to model the nonlinear deformation, whilst singular stress and displacement forms have been derived using the deformation theory of plasticity. Decisions have again to be made concerning the fracture criterion. Thus, although methods are described here for treating problems of NLFM using finite element methods and numerical results are presented, the mathematical basis behind the techniques is not rigorously established and for these models no theoretical error analysis exists. Nevertheless the results are thought to be useful as an indication of what can be achieved numerically.

Linear elasticity and LEF are described in Section 2, where the finite element algorithm for treating linear elasticity and for the retrieval of the fracture parameter K using the J-integral of Rice [28] is given. Error analysis results relevant to these techniques are also presented. Section 3 deals with nonlinear problems involving plasticity with strain hardening and both incremental plasticity and deformation plasticity are mentioned. The finite element algorithm for linear elasticity is extended to treat incremental plasticity, as is the retrieval algorithm to produce the fracture parameter J_p. The algorithm is finally applied to a plane strain two-dimensional problem involving a crack for which substantial yield takes place and results for J_p are presented.

2. LINEAR ELASTICITY AND LINEAR ELASTIC FRACTURE

2.1 *Linear Elasticity*

We consider first the linear elastic deformation of a three-dimensional solid defined in a region $\Omega \subset \mathbb{R}^3$ with boundary $\partial\Omega \equiv \partial\Omega_1 \cup \partial\Omega_2$. The coordinates of any point in the solid are $\underline{x} \equiv (x_1,x_2,x_3)$. In the classical formulation for the solid in equilibrium the displacement $\underline{u} = (u_1,u_2,u_3)^T$ satisfies the well-known Navier's equation

$$-\mu\Delta\underline{u} - (\lambda+\mu)\operatorname{grad}\operatorname{div}\underline{u} = \underline{f} \quad \text{in} \quad \Omega , \tag{2.1}$$

together with boundary conditions

$$\underline{u} = 0 \quad \text{on} \quad \partial\Omega_1 , \tag{2.2}$$

$$\sum_{j=1}^{3} \sigma_{ij}(\underline{u})\, n_j = g_i \quad \text{on} \quad \partial\Omega_2 , \quad 1 \leqq i \leqq 3 , \tag{2.3}$$

where λ and μ, λ, $\mu > 0$, are the Lamé constants of the material, $\sigma_{ij}(\underline{u})$ is the stress tensor, and $\underline{f} \equiv (f_1,f_2,f_3)^T \in (L_2(\Omega))^3$ and $\underline{g} \equiv (g_1,g_2,g_3)^T \in (L_2(\partial\Omega_2))^3$ are respectively prescribed volumic and surface forces.

In a Sobolev space setting let admissible displacement vectors be $\underline{v} \equiv (v_1,v_2,v_3)^T \in (H^1(\Omega))^3$ and define $V \equiv \{\underline{v} : \underline{v} \in (H^1(\Omega))^3,\ v_i\big|_{\partial\Omega_1} = 0,\ i = 1,2,3\}$. The weak form of problem (2.1) - (2.3) is

$$\text{find } \underline{u} \in V \ni a(\underline{u},\underline{v}) = F(\underline{v}) \quad \forall\ \underline{v} \in V , \tag{2.4}$$

in which

$$\begin{aligned} a(\underline{u},\underline{v}) &\equiv \int_\Omega \left\{\lambda \operatorname{div}\underline{u}\operatorname{div}\underline{v} + 2\mu \sum_{i,j=1}^{3} \varepsilon_{ij}(\underline{u})\,\varepsilon_{ij}(\underline{v})\right\} d\underline{x} \\ &= \int_\Omega \sum_{i,j=1}^{3} \sigma_{ij}(\underline{u})\,\varepsilon_{ij}(\underline{v})\, d\underline{x} , \end{aligned} \tag{2.5}$$

$$F(\underline{v}) \equiv \int_\Omega \underline{f}^T\underline{v}\, d\underline{x} + \int_{\partial\Omega_2} \sum_{i=1}^{3} g_i v_i\, ds , \tag{2.6}$$

where, for $1 \leqq i,j \leqq 3$,

$$\varepsilon_{ij}(\underline{v}) = \varepsilon_{ji}(\underline{v}) \equiv \frac{1}{2}\left(\frac{\partial v_i}{\partial x_j} + \frac{\partial v_j}{\partial x_i}\right) ,$$

$$\sigma_{ij}(\underline{v}) = \sigma_{ji}(\underline{v}) \equiv \lambda\left\{\sum_{k=1}^{3} \varepsilon_{kk}(\underline{v})\,\delta_{ij} + 2\mu\,\varepsilon_{ij}(\underline{v})\right\} .$$

The bilinear form $a(\underline{u},\underline{v})$ and the linear form $F(\underline{v})$ are continuous on V. It has also been shown that $a(\underline{u},\underline{v})$ is V-elliptic, see Ciarlet [8] and Fichera [10]. It follows that the solution $\underline{u}$ of (2.4) minimises over V the total energy functional

$$
\begin{aligned}
I[\underline{v}] &\equiv \frac{1}{2}\, a(\underline{v},\underline{v}) - F(\underline{v}) \\
&= \frac{1}{2}\int_\Omega \left\{ \lambda(\operatorname{div}\underline{v})^2 + 2\mu \sum_{i,j=1}^{3} \left(\varepsilon_{ij}(\underline{v})\right)^2 \right\} d\underline{x} - F(\underline{v}) \qquad (2.7) \\
&= \frac{1}{2}\int_\Omega \sum_{i,j=1}^{3} \sigma_{ij}(\underline{v})\, \varepsilon_{ij}(\underline{v})\, d\underline{x} - F(\underline{v}) .
\end{aligned}
$$

2.2. *Linear Elastic Fracture*

2.2.1. *Singularities, stress intensity factors and* J-*integrals*

It is well known that, under assumptions of linear elastic theory, the presence of a crack in a solid induces singular stress fields in the neighbourhood of the crack tip; see e.g. Rice [29], Westergaard [39] and Williams [44]. Following Irwin [18] three distinct stress fields can be classified; opening (Mode I), inplane sliding (Mode II), antiplane sliding (Mode III), see also Rice [28]. The important features of such stress fields can be seen from consideration of *two-dimensional* cases. For a Mode I problem, using polar coordinates (r,θ) local to the crack tip, as shown in Fig. 1, the near-tip stress field can be expressed as

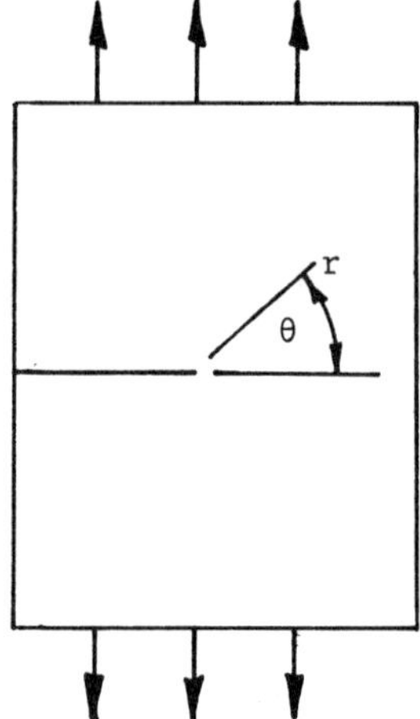

FIG. 1

$$
\begin{Bmatrix} \sigma_{11} \\ \sigma_{12} \\ \sigma_{22} \end{Bmatrix} = \frac{K_I}{(2\pi r)^{\frac{1}{2}}} \cos\theta/2 \begin{Bmatrix} 1 - \sin\theta/2 \sin 3\theta/2 \\ \sin\theta/2 \cos 3\theta/2 \\ 1 + \sin\theta/2 \sin 3\theta/2 \end{Bmatrix} , \qquad (2.8)
$$

where K_I, known as the Mode I stress intensity factor, is a measure of the magnitude of near-tip stresses.

For a similar three-dimensional linear elastic fracture problem it is known that, in the vicinity of a crack front but away from the stress-free surfaces, the stresses have the $r^{-\frac{1}{2}}$ form in planes orthogonal to the crack front. However, as has been mentioned above, in the three-dimensional setting the form of stresses near the intersection of a crack front with a stress-free surface is still not completely resolved, see Thompson and Whiteman [36] and Thompson [35].

Theoretically (2.8) would give rise to infinite stresses at the crack tip; in reality (local yielding) plasticity occurs near the tip. The fundamental assumption of linear elastic fracture mechanics (LEFM) is that this region of plasticity is small compared to the dimensions of the crack. Thus the elastic stress intensity factor K_I, which is a function of crack geometry and loading, determines whether or not the crack will propagate. Expressions similar to (2.8) exist for Mode II and Mode III stress fields; they involve the relevant stress intensity factors K_{II} and K_{III} and again exhibit the $r^{-\frac{1}{2}}$ singular form. For pure mode problems of linear elastic fracture the three stress intensity factors are *fracture criteria*, so it is important that they be calculated accurately. Propagation will occur in such problems when the stress intensity factor attains a critical value K_c (known from laboratory tests), which depends on material properties and is independent of geometry and loading. Propagation of a crack in a mixed mode problem will occur when some function of the three stress intensity factors attains an appropriate critical value, see Sih [32].

One approach for calculating the stress intensity factors as above is to relate them to the *strain energy release* rate G for the crack, where G is defined as the rate of decrease in potential energy with respect to crack length. This can be achieved by following the approach of Rice [28] who proved that, for a two-dimensional homogeneous elastic body, G is equal to the path independent integral J. Thus

$$J = G \equiv -\frac{\partial PE}{\partial L}, \tag{2.9}$$

where PE is the potential energy and L is the crack length. For a crack with flat faces parallel to the x_1-axis, see Fig. 2., J is defined as

$$J \equiv \int_\Gamma \left[W\,dx_2 - \sum_i T_i \frac{\partial u_i}{\partial x_1} ds \right], \tag{2.10}$$

in which Γ is a contour running anticlockwise from the lower to the upper crack faces enclosing the crack tip, W is the strain energy density, defined for linear elasticity as

$$W \equiv \frac{1}{2} \sum_{i,j} \sigma_{ij}\, \varepsilon_{ij}, \tag{2.11}$$

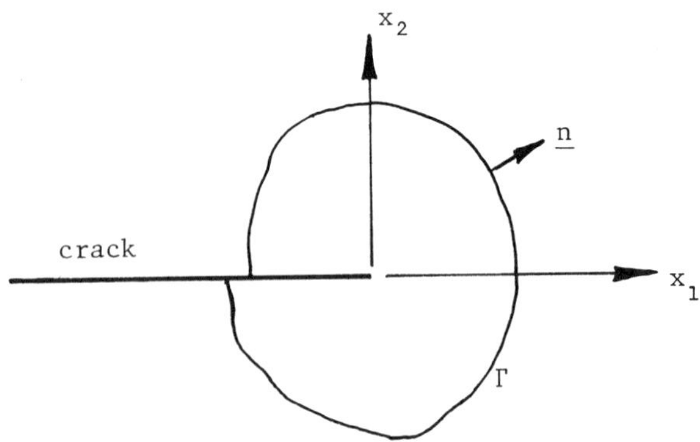

FIG. 2

T_i are tractions, defined with respect to the outward normal $\underline{n}$ to Γ as

$$T_i \equiv \sum_j \sigma_{ij} n_j \tag{2.12}$$

and ds is the increment of arc length. Rice showed further that the J-integral and the stress intensity factors are related by

$$J = \frac{H}{E}\left(K_I^2 + K_{II}^2\right) + \left(\frac{1+\nu}{E}\right) K_{III}^2 ,$$

where E is Young's modulus, ν is Poisson's ratio, $H = 1$ for plane stress and $H = (1-\nu^2)$ for plane strain. For pure Mode I fracture

$$J = \frac{H}{E} K_I^2 , \tag{2.13}$$

so that K_I can immediately be calculated from J. Clearly similar situations hold for pure Mode II or Mode III fracture. We emphasise that, as has been said above, in LEFM the stress intensity factors provide the fracture criteria, whilst the J-integral is used for their calculation.

For problems of type (2.1) - (2.3) in three dimensions containing cracks, the analysis is carried out at different points along the crack front. The J-integral at such a point is calculated in the plane through the point perpendicular to the tangent to the crack front at the point. In this case there is thus a sequence of J values, and hence K values, each associated with a point of the crack front.

Two methods are widely used for calculating J-integrals; the first is the differential stiffness method of Parks [26], which calculates J during the main (finite element) computation, and the second is the numerical integration of (2.10), which is a post-processing procedure involving the retrieval of strains and stresses from calculated values of displacements. Attention here is restricted to the second method and the retrieval is achieved using an extended version of the MODEL Finite Element Code, [13] - [15].

2.3. *Finite Element Method 1*

2.3.1. *General scheme for linear elasticity*

The finite element method can be used to treat problem (2.1)-(2.3) via the weak formulation (2.4), (or equivalently through the variational formulation (2.7)). The region Ω is discretised into elements Ω^e, such that $\Omega \equiv \bigcup_e \Omega^e$. A finite dimensional subspace $S^h \subset V$, consisting of piecewise polynomial functions defined over the partition, is chosen and the problem approximating (2.4) is that of finding

$$\underline{u}_h \in S^h \ni a(\underline{u}_h,\underline{v}_h) = F(\underline{v}_h) \qquad \forall\, \underline{v}_h \in S^h\,, \tag{2.14}$$

where h indicates element size. From the definitions (2.5) and (2.6) it follows that the equation in (2.14) can be written as

$$\int_\Omega \underline{\varepsilon}^T(\underline{v}_h)\underline{\sigma}(\underline{u}_h)\,d\underline{x} - \int_\Omega \underline{v}_h^T\,\underline{f}\,d\underline{x} - \int_{\partial\Omega_2} \underline{v}_h^T\,\underline{g}\,ds = 0 \tag{2.15}$$

and that, since $\int_\Omega = \sum_e \int_{\Omega^e}$, in an individual element the contribution to the left hand side of (2.15) is

$$\int_{\Omega^e} \underline{\varepsilon}^T(\underline{v}_h^e)\underline{\sigma}(\underline{u}_h^e)\,d\underline{x} - \int_{\Omega^e} \underline{v}_h^{eT}\,\underline{f}\,d\underline{x} - \int_{\partial\Omega_1^e} \underline{v}_h^{eT}\,\underline{g}\,ds\,, \tag{2.16}$$

where $\underline{u}_h^e \equiv \underline{u}_h(\underline{x})\big|_e$ and $\underline{v}_h^e \equiv v_h(\underline{x})\big|_e$. In each element the approximating vector $\underline{u}_h(\underline{x})$ is defined in terms of basis functions $N_i(\underline{x})$ for the n nodes of the element and, since we use only Lagrange type elements, the point evaluations $\underline{U}_i^e$ of $\underline{u}_h^e$ at the nodes. Thus

$$\underline{u}_h^e(\underline{x}) = [N]\,\underline{U}^e\,, \tag{2.17}$$

where $[N] = [\underline{N}_1(\underline{x}), \underline{N}_2(\underline{x}),\ldots, \underline{N}_n(\underline{x})]$ with $\underline{N}_j \equiv N_j[I_3]$ in which I_3 is the 3×3 unit matrix. The strains in the element can be expressed in terms of the nodal displacements as

$$\underline{\varepsilon}(\underline{u}_h^e) = [B^e]\,\underline{U}^e\,, \tag{2.18}$$

where in the matrix $[B^e]$ the elements involve derivatives of the N_i. At any point $\underline{x}$ the stresses and strains are related by a constitutive matrix [D], depending on material properties so that

$$\underline{\sigma} = [D]\,\underline{\varepsilon}\,. \tag{2.19}$$

A Galerkin method is applied to (2.14) so that in the eth element $\underline{v}_h^e$ is taken in turn to be $\underline{N}_j$, j = 1,2,...,n, and thus on substitution into (2.16) we have

$$\left\{\int_{\Omega^e}[B^e]^T[D][B^e]d\underline{x}\right\}\underline{U}^e - \int_{\Omega^e}[N]^T \underline{f}\, d\underline{x} - \int_{\partial\Omega_2^e}[N]^T\underline{g}\, ds . \quad (2.20a)$$

Amalgamation of the contributions as above from all the elements and use of (2.15) produces the system of global stiffness equations

$$[K]\ \underline{U} = \underline{F}\ , \quad (2.20b)$$

the solution $\underline{U}$ of which is the vector of the displacements at the nodal points.

2.3.2. *Error analysis and treatment of singularities*

For problem (2.4) with solution $\underline{u} \in V$ and the approximating problem (2.14) with solution $\underline{u}_h \in S^h$, where $S^h \subset V$, bounds are required on $\| \underline{u} - \underline{u}_h \|$. Since $a(\underline{u},\underline{v})$ is continuous on V and V-elliptic, it is well known, see Ciarlet [8] and Whiteman [40], that

$$\| \underline{u} - \underline{u}_h \|_V \leqq C_1 \| \underline{u} - \underline{v}_h \|_V \quad \forall\ \underline{v}_h \in S^h . \quad (2.21)$$

Further, if the space S^h of piecewise polynomial functions is defined on a quasi-uniform mesh, then

$$\| \underline{u} - \underline{v}_h \|_V \leqq C_2\ h^{\gamma} |\underline{u}|_k\ , \quad (2.22)$$

where γ depends on the order of the polynomials and on k. Since the choice of the order of the polynomials in the approximating space is to a certain extent flexible, the regularity of $\underline{u}$ which governs k is the important factor in determining γ. Equations (2.21) and (2.22) together provide a bound on the error $(\underline{u} - \underline{u}_h)$.

For problems in regions containing re-entrant corners, such as those containing cracks in the linear elastic fracture context, the regularity of the solution $\underline{u}$ is low on account of the presence of singularities with the result that the (asymptotic) rate of convergence with decreasing mesh size of $\underline{u}_h$ to $\underline{u}$ is correspondingly reduced. The practical effect of this is to reduce the accuracy of $\underline{u}_h$ in the neighbourhood of a singularity.

A considerable number of, now familiar, special finite element techniques have been proposed in recent years which, with varying degrees of success, seek to compensate for the effects of singularities on finite element solutions both theoretically by improving rates of convergence, and practically by increasing accuracy. These techniques involve such methods as the use of special singular elements around the point of singularity, Akin [1], Blackburn [7], Stern [33], Stern and Becker [34], augmentation of the test and trial function spaces with functions having the form of the singularity, Barnhill and Whiteman [3],[4], Fix, Gulatti and Wakoff [11] and Mitchell and Wait [21], and with isoparametric elements the moving of nodes in elements involving the singularity so that at the point of singularity the Jacobian of the transformation from global to local space itself becomes singular, Henshell and Shaw [16], Barsoum [5],[6], Thompson [35] and Thompson and Whiteman [36]. Extensive surveys of these elements and techniques are given by Atluri [2], Whiteman [41], Whiteman and Akin [42] and Whiteman and Schleicher [43]. O'Leary [24], specifically for the Stern-Becker singular element and in the context of a Poisson problem containing a singularity, has proved that the use of such an element produces no improvement in the *asymptotic* rate of convergence of the calculated solution. At the same time he has shown that from a practical standpoint there is a considerable gain in accuracy if this type of singular element is used. Although the results of [24] were obtained for a Poisson problem, the implications for problems of linear elastic fracture are clear.

2.3.3. *Calculation of the* J-*integral*

As has been explained in Section 2.2.1, for problems of linear elastic fracture the fundamental quantity to be approximated is the stress intensity factor, and this can be done via the J-integral. The approach adopted here is to calculate an approximation J_h to J through numerical integration of the contour integral using calculated displacements and retrieved values for strains and stresses.

The two-dimensional case of Figs. 1 and 2 is first considered and for this the integral (2.10) has to be calculated over the

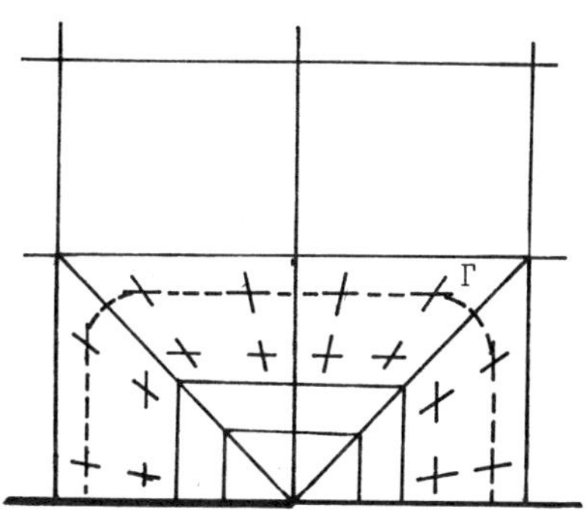

FIG. 3a

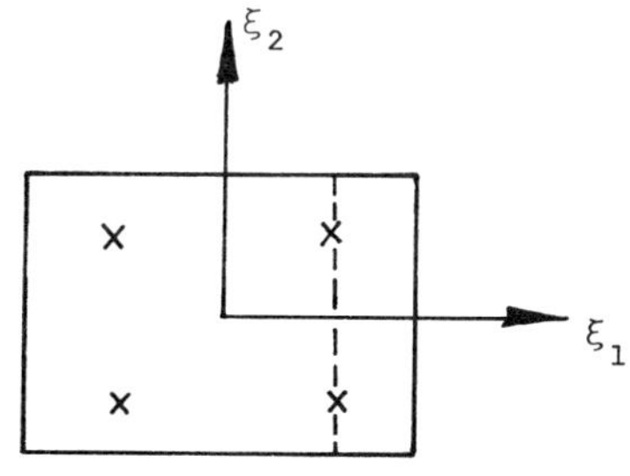

FIG. 3b

contour path Γ. Assuming that the problem domain has been partitioned into elements, as for example in Fig. 3a, and that the linear equation system (2.20) has been set up and solved, the calculated values of $\underline{U}$ at the mesh nodal points are available. A contour path Γ is chosen through Gauss points of a ring of elements Ω^e surrounding the crack tip, see Fig. 3a. Each element of the ring containing the part $\Gamma^{(e)}$ of Γ is in turn transformed onto the standard element $S \equiv \{(\xi_1,\xi_2) : -1 \leqq \xi_1,\xi_2 \leqq 1\}$ in the (ξ_1,ξ_2)-plane, with $\Gamma^{(e)}$ becoming in S either the line ξ_1 = constant or the line ξ_2 = constant through the relevant mapped Gauss points, or a *corner path* consisting in part of ξ_1 = constant and part ξ_2 = constant again through the relevant mapped Gauss points. The total calculated value J_h is the sum of the contributions $J_h^{(e)}$ of all elements involving the contour path; thus

$$J_h = \sum_e J_h^{(e)} .$$

An element with integration along a path of constant ξ_1 produces the contribution

$$J^{(e)} = \int_{-1}^{1} I_1^{(e)} \, d\xi_2 , \tag{2.23}$$

which is approximated by

$$J_h^{(e)} = \sum_j I_1^{(e)}(\xi_1^{(i)},\xi_2^{(j)})_h \, G_j , \tag{2.24}$$

where $(\xi_1^{(i)},\xi_2^{(j)})$, i = const, j = 1,2,...,NQP are the NQP Gauss points on the path and the G_j are the corresponding quadrature weights. From (2.10) - (2.12) it follows that at any of these Gauss points

$$\begin{aligned} I_1^{(e)}(.,.)_h \equiv \frac{1}{2}\Bigg[& (\sigma_{11}^e)_h \frac{\partial(u_1^e)_h}{\partial x_1} + (\sigma_{12}^e)_h\left(\frac{\partial(u_1^e)_h}{\partial x_2} + \frac{\partial(u_2^e)_h}{\partial x_1}\right) \\ & + (\sigma_{22}^e)_h \frac{\partial(u_2^e)_h}{\partial x_2}\Bigg] \frac{\partial x_2}{\partial \xi_2} \\ - \Bigg[& \Big((\sigma_{11}^e)_h n_1 + (\sigma_{12}^e)_h n_2\Big) \frac{\partial(u_1^e)_h}{\partial x_1} + \Big((\sigma_{12}^e)_h n_1 + (\sigma_{22}^e)_h n_2\Big) \frac{\partial(u_2^e)_h}{\partial x_1}\Bigg] . \\ & \left[\left(\frac{\partial x_1}{\partial \xi_2}\right)^2 + \left(\frac{\partial x_2}{\partial \xi_2}\right)^2\right]^{\frac{1}{2}} . \end{aligned} \tag{2.25}$$

In (2.25) the calculated nodal displacements $\underline{U}^e$ are used so that $\frac{\partial u_h^e}{\partial x_i} = \frac{\partial [N]}{\partial x_i} \underline{U}^e$, $\frac{\partial x_i}{\partial \xi_2}$, i = 1,2 is obtained from the Jacobian of the transformation, and the Gauss point strains and stresses are calculated (retrieved) using (2.18) and (2.19).

For integration along a path of constant ξ_2

$$J^{(e)} = \int_{-1}^{1} I_2^{(e)} \, d\xi_1 \tag{2.26}$$

and the required expressions for $J_h^{(e)}$ and $I_2^{(e)}(.,.)_h$ are obtained by replacing ξ_2 by ξ_1 in (2.24) and (2.25). For the case of corner elements

$$J_h^{(e)} = \left\{\sum_j I_1^{(e)}(\ )_h \, G_j + \sum_i I_2^{(e)}(\ )_h \, G_i\right\}\left\{\frac{1 - A(m)}{2}\right\} \tag{2.27}$$

where m, $1 \leqq m \leqq NQP$, indicates the integration path and A(m) is the local coordinate of m in the one-dimensional segment within S; e.g. with 3×3 Gauss points A(m) = ±.57735 or 0.

In practice for linear elastic fracture it is normal when calculating an approximation to J to choose at least three Γ contours and to calculate J_h for each. The approximate value is then taken as the mean of these J_h-values. In the case of pure Mode I fracture (2.13) enables an approximation K_h to K_I to be obtained from the calculated approximation to J.

For problems in three dimensions the J-integral is calculated, for points on the crack front, in planes through the points perpendicular to the tangents to the crack front. The mesh should be so designed that, for any element Ω^e through which a contour passes, after transformation the local coordinate ξ_3 is constant for such planes. For each plane the procedure for calculating J_h is exactly as above and a set of J_h values is obtained, each of which relates to a point on the crack front.

2.3.4. *Error bounds for* K_h

The error analysis of Section 2.3.2 has concerned only the calculated *primary* unknown $\underline{u}_h$. The role and importance in LEF of the stress intensity factor is now clear. In the previous section the approximation of the J-integral was described, as this provides a means of approximating the stress intensity factor. When an approximation to K is obtained via the J-integral, the computation involves numerical values of strains and stresses which are determined from calculated nodal values $\underline{U}$ and hence of $\underline{u}_h$. Any loss of accuracy in $\underline{u}_h$ due to the presence of a singularity thus affects adversely the *retrieved* quantity approximating K. Theoretical analysis of the error in this approximation is of obvious importance.

In the context of two-dimensional Poisson problems containing singularities O'Leary [24] and Schatz and Wahlbin [31] have obtained theoretical bounds for the error in approximating the coefficient of the leading singular term. For a two-dimensional Mode I LEF problem Djaoua [9] has used the J-integral approach and has proved using piecewise linear functions that $|K_I - K_h| = O(h)$. These appear to be the first results of this type.

In that the K_h is a retrieved quantity based on derivatives of the primary variable, it is to be expected that improvements on the above will be obtained by exploiting superconvergence properties of the derivatives of finite element approximations. Results on this aspect are beginning to appear, see e.g. Krizek and Neittaanmaki [19] and Levine [20].

3. ELASTO-PLASTICITY AND NONLINEAR FRACTURE

3.1. *Elasto-Plastic Deformation*

The above discussion on linear elastic fracture has assumed that the material involved is brittle. For nearly all the materials for which linear fracture mechanics is applied, nonlinear effects such as plasticity occur near the crack tip. Thus some local yielding takes place invalidating the assumptions of linear elasticity. However, provided this yield is sufficiently local, the linear elastic model and the use of the stress intensity factor are effective. For nonbrittle materials the yield is no longer local, so that it is necessary to use nonlinear fracture mechanics. For this reason we now mention *two* different theories of *plasticity*; deformation theory and incremental theory, see e.g. Washizu [38].

In the deformation theory of plasticity the characteristic assumption is that instantaneous stress and instantaneous strain are related in such a way that if either is given the other is uniquely determined. This approach thus assumes that the relationship depends only on current states of stress and not on the load history. A further assumption is that the stress-strain relations do not change during the loading process, thus restricting application to problems with monotonically increasing load. Deformation theory of plasticity is consequently indistinguishable from nonlinear elasticity. As a model of plasticity it is not in general satisfactory.

The incremental or flow theory of plasticity reflects the fact that in a zone of plasticity there is no unique relation between stress and strain; strain depends not only on the final state of stress but also on the loading history. Thus the stress-strain relations described previously must be replaced by relations between *increments* of stress and *increments* of strain and the solution procedure is based upon the technique of successively applying the loading incrementally.

The main finite element algorithm presented here for analysing elasto-plastic deformation is based upon the incremental theory of plasticity using a formulation in which the displacements are the primary unknowns. With the incremental method it is necessary to have a relation between uniaxial stress and strain. For example in Section 3.4 a power law hardening rule is assumed. We remark that both the deformation theory and incremental theory of plasticity have been treated mathematically by Necas [22] and by Necas and Hlavacek [23], using formulations in which stresses are the primary unknowns.

For problems involving cracks Hutchinson [17] and Rice and Rosengren [30], using deformation theory, have provided forms for near-tip stresses and strains; the so-called HRR singular fields. Although these have not been used in the calculations of this paper, they are presented briefly in Section 3.4, on account of the possibilities of incorporating them in the future. The route here for treating nonlinear fracture problems is to take the displacement values calculated using the incremental theory and to calculate approximations to the J_p-integral, a modified form of the J-integral.

3.2. *Finite Element Method 2: Elasto-Plastic Deformation*

Using the incremental theory of plasticity we follow Prager [27] and consider equations (2.1) - (2.3), where increments of force $d\underline{f}$ and $d\underline{g}$ are applied so that

$$-\mu \Delta d\underline{u} - (\lambda + \mu) \text{ grad div } d\underline{u} = d\underline{f}, \quad \text{in } \Omega \tag{3.1}$$

$$d\underline{u} = 0 \quad \text{on } \partial\Omega_1, \tag{3.2}$$

$$\sum_{j=1}^{3} \sigma_{ij}(\underline{u}) n_j = dg_i \quad \text{on } \partial\Omega_2, \quad 1 \leqq i \leqq 3, \tag{3.3}$$

Assuming that the increments are small, it is required to determine the stress increments $d\sigma_{ij}$ and displacement increments du_i induced on the body by the increments df_i and dg_i. The scheme for solution in the elastic-plastic case is based on that of Section 2.3.1 for each load increment. With the onset of plasticity the relation between stress and strain becomes nonlinear, so that in the incremental theory for a particular load increment the stress-strain relation involves increments of stress and strain which are related by a constitutive matrix involving current stress. Thus the matrix [D] of (2.19) must be modified appropriately when plasticity occurs. The scheme for this is now well known, see e.g. Owen and Hinton [25], so that it is not described in detail here. The actual code used in this work is again an extension of the MODEL package, in this case that developed at BICOM by Harrison and Ward, see Harrison et al. [13]-[15] and Ward and Burton [37]. It is assumed in the mathematical model that displacements and elastic strains are small, that the material is isotropic and that the plasticity is of associative type. A yield criterion (in this case the von Mises criterion) and a flow rule are chosen. Following [37], in the finite element algorithm the plastic constitutive matrix $[D_{ep}]$ is derived so that in any element

$$d\underline{\sigma} = [D_{ep}]\, d\underline{\varepsilon}\,. \tag{3.4}$$

The algorithm proceeds as follows. A suitable external force increment $d\underline{F}^{(1)}$ is applied to the structure to provide its maximum overall elastic deformation (elastic load factor) and the corresponding deflections $\underline{U}^{(1)}$ are calculated using (2.20b). Subsequently the additional incremental loadings $d\underline{F}^{(k)}$, $k = 2,3,\ldots,L$ are applied so that the kth total load is

$$\underline{F}^{(k)} = d\underline{F}^{(1)} + \sum_{q=2}^{k} d\underline{F}^{(q)} , \tag{3.5}$$

with the final load $\underline{F}^{(L)}$. After yield has taken place, in the kth load increment the $d\underline{U}^{(k)}$ are calculated in each element using either (2.20a) or

$$\int_{\Omega^e}\left\{[B^e]^T[D_{ep}][B^e]d\underline{x}\right\}d\underline{U}^{e(k)} - \int_{\Omega^e}[N]^T d\underline{f}^{(k)}d\underline{x} - \int_{\partial\Omega_2^e}[N]^T d\underline{g}^{(k)}ds . \tag{3.6}$$

Amalgamation of contributions over all the elements produces the (global) system of nonlinear equations for the load increment

$$[K(\underline{U})]\, d\underline{U}^{(k)} = d\underline{F}^{(k)} , \tag{3.7}$$

which is solved iteratively for $d\underline{U}^{(k)}$, thus producing the increment of displacement $d\underline{u}_h^{(k)}(\underline{x})$ over Ω. The elemental increments of strain and stress can then be retrieved, to produce overall $d\underline{\varepsilon}_h^{(k)}$ and $d\underline{\sigma}_h^{(k)}$. The total displacements $\underline{u}_h^{(k)}$, strains $\underline{\varepsilon}_h^{(k)}$ and stresses $\underline{\sigma}_h^{(k)}$ are then calculated. The process is repeated, incrementing up to the final load $\underline{F}^{(L)}$.

3.3. *Nonlinear Fracture Mechanics*

As has been stated earlier, a quantity of interest in NLFM is the J_p-integral which is widely employed as a guide to fracture in problems where yield has occurred. In this context the significance of J_p is analogous to that of the stress intensity factor K of LEF. The principal restriction concerning J_p arises from the fact that it is derived from deformation theory plasticity, so that theoretically no unloading may occur within the body. Further the fracture process zone must be contained well inside the zone of dominance of the HRR fields. Just as in LEF the linear elastic theory breaks down sufficiently close to the crack tip, the justification for J_p breaks down in the same way.

The J_p integral is developed from the J-integral by separating what was formerly the strain energy density term W of (2.10) and (2.11) into elastic and plastic components. Thus

$$W = W_e + W_p , \tag{3.8}$$

where the elastic component is given, in terms of the stress and elastic strain components by

$$W_e = \frac{1}{2}\sigma_{ij}(\varepsilon_{ij})_e . \tag{3.9}$$

The plastic work term W_p is defined as

$$W_p = \int_0^{\bar{\varepsilon}_p} \bar{\sigma}\, d\bar{\varepsilon}_p , \tag{3.10}$$

where $\bar{\sigma}$ and $\bar{\varepsilon}_p$ are respectively the effective stress and effective plastic strain, see Owen and Hinton [25].

3.4. *Power-Law Strain Hardening and Singular Forms*

Consider a material with *power law* strain hardening behaviour in which the effective plastic strain $\bar{\varepsilon}_p$ is given by

$$\bar{\varepsilon}_p = \frac{\bar{\sigma}}{E}\alpha\left(\frac{\bar{\sigma}}{\sigma_0}\right)^n , \tag{3.11}$$

where $\bar{\sigma}$ is the effective stress, σ_0 is the uniaxial yield stress and α and n are hardening parameters. For such a material in two dimensions using polar coordinates (r,θ) local to the crack tip, the HRR forms, [17], [30], for the near-tip stresses and plastic strains are

$$\sigma_{ij} = \sigma_0\left[\frac{EJ_p}{\alpha\sigma_0^2 M_n r}\right]^{\frac{n}{n+1}} \hat{\sigma}_{ij}(\theta,n) , \tag{3.12}$$

$$(\varepsilon_{ij})_p = \frac{\alpha\sigma_0}{E}\left[\frac{EJ_p}{\alpha\sigma_0^2 M_n r}\right]^{\frac{1}{n+1}} \hat{\varepsilon}_{ij}(\theta,n) , \tag{3.13}$$

where $\hat{\sigma}_{ij}$ and $\hat{\varepsilon}_{ij}$ are functions of θ and n, whilst M_n is a normalising constant.

Forms (3.12) and (3.13) are derived using the deformation theory of plasticity. For the reasons set out earlier, and the inadequacies of the deformation theory as indicated, the algorithm of Section 3.2 is based on the incremental theory of plasticity. As with the case of linear elastic fracture, knowledge of the form of near-tip stresses and strains is important in the nonlinear fracture case. No near-tip forms have to date been derived for the case of incremental plasticity, although clearly these would be highly desirable. Thus a situation exists in which the accepted incremental theory of plasticity does not allow for the production of near-tip singular forms, whilst those singular forms which have been derived come from deformation theory. In view of this we have chosen in our present finite element calculations not to try to incorporate special elements which would exhibit the singular behaviour of (3.12) and (3.13). Instead a simple, not theoretically controlled, form of mesh refinement local to the crack tip has been used (see Fig. 4).

In the example of Section 4 it is assumed that the material of the problem obeys a power law uniaxial stress-strain relationship of the type (3.11).

3.5. *Calculation of the* J_p*-integral*

The actual procedure for calculating approximations to J_p is very similar to that for calculating approximations to J for LEF. In the present case, for a problem with a total of L load increments a value $(J_p^{(k)})_h$, k = 1,2,...,L is calculated for each load increment k, so that a sequence of values is produced over the entire loading history.

Considering again the two-dimensional case as in Fig. 3a the contour Γ around the crack tip is chosen and $(J_p)_h^{(k)}$ is calculated by summing the contributions from all elements involving Γ; i.e.

$$(J_p)_h^{(k)} = \sum_e (J_p^e)_h^{(k)}, \tag{3.14}$$

so that for k = 2,3,...,L a sequence of approximations $(J_p)_h^{(k)}$ is obtained.

Referring back to Section 2.3.3. and in particular to (2.24), for a path of constant ξ_1 in an element the contribution $(J_p^e)_h^{(k)}$ is given by

$$(J_p^e)_h^{(k)} = \sum_j I_3^{(e)}\left(\xi_1^{(i)}, \xi_2^{(j)}\right)_h^{(k)} G_j , \tag{3.15}$$

where the $(\xi_1^{(i)}, \xi_2^{(j)})$ are the Gauss points and the G_j are the corresponding weights. If the stresses and displacements at the end of the load increment k are respectively $(\sigma_{ij}^e)_h^{(k)}$ and $(u_i^e)_h^{(k)}$, then $I_3^{(e)}(.,.)_h^{(k)}$ is given by

$$I_3^{(e)}(.,.)_h^{(k)} \equiv \frac{1}{2}\left[(\sigma_{11}^e)_h^{(k)} \frac{\partial(u_1^e)_h^{(k)}}{\partial x_1} + (\sigma_{12}^e)_h^{(k)}\left(\frac{\partial(u_1^e)_h^{(k)}}{\partial x_2} + \frac{\partial(u_2^e)_h^{(k)}}{\partial x_1}\right)\right.$$

$$\left. + (\sigma_{22}^e)^{(k)} \frac{\partial(u_2^e)_h^{(k)}}{\partial x_2} + W_p^{(k)}\right] \frac{\partial x_2}{\partial \xi_2}$$

$$- \left[\left\{(\sigma_{11}^e)_h^{(k)} n_1 + (\sigma_{12}^e)_h^{(k)} n_2\right\} \frac{\partial(u_1^e)_h^{(k)}}{\partial x_1} + \left\{(\sigma_{12}^e)_h^{(k)} n_1 + (\sigma_{22}^e)_h^{(k)} n_2\right\}\right.$$

$$\left.\frac{\partial(u_2^e)_h^{(k)}}{\partial x_1}\right]\left[\left(\frac{\partial x_1}{\partial \xi_2}\right)^2 + \left(\frac{\partial x_2}{\partial \xi_2}\right)^2\right]^{\frac{1}{2}}, \tag{3.16}$$

where $W_p^{(k)}$ is the plastic work in the increment as in (3.10).

For nonlinear fracture problems it is normal for each load increment to choose at least six contours, to calculate first the associated $(J_p)_h^{(k)}$ for each and then to derive a mean value from these. The values of $(J_p)_h^{(k)}$ together with their mean indicate the behaviour in the vicinity of the crack tip. The values of J_p vary with distance from the crack tip, reflecting the change in behaviour which occurs in this neighbourhood.

In three-dimensional problems, as for the linear elastic case, the J_P-integral is calculated for points along the crack front in planes perpendicular to the crack front. Once again the comments made previously concerning the choice of mesh hold.

4. MODE I NONLINEAR FRACTURE PROBLEM

A two-dimensional Mode I problem in the region of Fig. 1 has been analysed using the finite element method. The region has height 6.4, width 0.8 and contains a single edge crack of length 0.4. The normal stress applied to the ends is 100, whilst $E = 12500$ and $\nu = 0.25$. The principal aim is to produce, after yield has taken place, for the NLFM problem a sequence of values approximating $J_P^{(k)}$ over the load history. The problem is analysed first assuming linear elasticity and then using the methods of Sections 3.2 and 3.5 in the following way. The load is applied incrementally in six steps, $k = 1,2,\ldots,6$; step 1 involves the elastic load factor, step 2 reaches 20% of total load and each successive step adds a further 20% of total load. Linear elasticity is assumed in step 1. Subsequently either a power law stress-strain relationship as in (3.11) is assumed with $\alpha = 1$ and $n = 0.1,1,3,5,7,10$, or the material is assumed to be perfectly plastic. The finite element mesh is as in Fig. 4, which shows

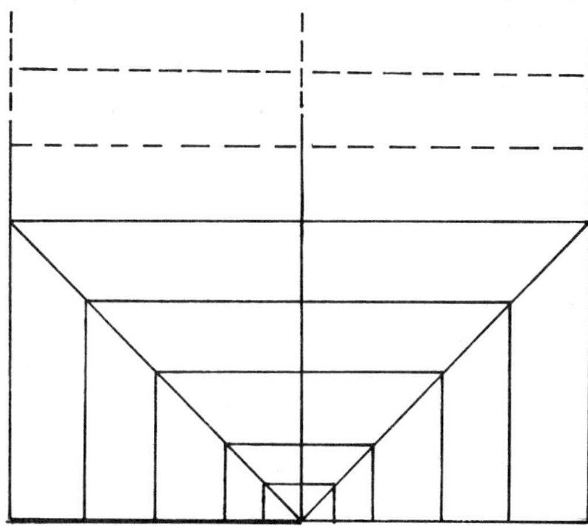

FIG. 4

only the upper half region, as symmetry has been exploited and contains 52 elements with 207 nodes. Eight node isoparametric quadratic elements are used, these being of initially triangular collapsed form around the crack tip. For linear elasticity these latter are *quarter-point* singular elements, whilst for the other cases they are of standard form.

Values of the mean $(J_P)_h^{(k)}$, $k = 1,2,\ldots,6$ are displayed in

Table 1 for the different n, together with the linear elasticity and perfect plasticity cases; the value for k = 1 is of course J_h of linear elasticity. There is considerable increase with n in mean $(J_p)_h^{(k)}$ for the higher loads, as well as divergence for perfect plasticity (the limit case $n = \infty$). The results indicate that the techniques of Section 3 can be applied to treat this type of NLFM problem.

TABLE 1

n \ k	1	2	3	4	5	6
Linear elasticity						7.542
0.1	0.0733	0.3742	1.721	4.203	7.484	11.72
1	"	0.3669	1.615	4.055	7.801	12.89
3	"	0.3648	1.639	4.456	10.78	24.90
5	"	0.3652	1.672	4.788	13.49	36.65
7	"	0.3655	1.693	5.000	15.13	44.33
10	"	0.3658	1.712	5.186	17.29	61.40
Perfect plasticity	"	0.4178	3.674	D I V E R G I N G		

No attempt has been made here to interpret these results physically, nor is any assessment made of the accuracy. We are not aware of the existence of any attempt to derive error analysis for the nonlinear case. The situation is clearly totally different to that of LEF. The discussion does, however, illustrate the marked difference between the states-of-the-art for LEF and NLF.

REFERENCES

1. AKIN, J.E., Generation of elements with singularities. *Int. J. Numer. Meth. Eng.* 10, 1249-1259 (1976).
2. ATLURI, S.N., Higher-order special and singular finite elements. Chapter 4 of A.K. Noor and W. Pilkey (eds.), *Survey of Finite Element Methods*. American Soc. Mech. Eng. (1980).
3. BARNHILL, R.E. and WHITEMAN, J.R., Error analysis of finite element methods with triangles for elliptic boundary value problems. pp.83-112 of J.R. Whiteman (ed.) *The Mathematics of Finite Elements and Applications*. Academic Press, London (1973).

4. BARNHILL, R.E. and WHITEMAN, J.R., Error analysis of Galerkin methods for Dirichlet problems containing boundary singularities. *J. Inst. Math. Applics.* 15, 121-125 (1975).
5. BARSOUM, R.S., On the use of isoparametric finite elements in linear fracture mechanics. *Int. J. Numer. Meth. Eng.* 10, 25-37 (1976).
6. BARSOUM, R.S., Triangular quarter point elements as elastic and perfectly plastic crack tip elements. *Int. J. Numer. Meth. Eng.* 12, 85-98 (1977).
7. BLACKBURN, W.S., Calculation of stress intensity factors at crack tips using special finite elements. pp.327-336 of J.R. Whiteman (ed.) *The Mathematics of Finite Elements and Applications*. Academic Press, London (1973).
8. CIARLET, P.G., *The Finite Element Method for Elliptic Problems*. North-Holland, Amsterdam (1978).
9. DJAOUA, M., Analyse Mathématique et Numerique de Quelques Problemes en Mechanique de la Rupture. *Ph.D. Thesis*, Université de Paris VI, Paris (1983).
10. FICHERA, G., Linear Elliptic Differential Systems and Eigenvalue Problems. *Lecture Notes in Mathematics 8*. Springer Verlag, Berlin (1965).
11. FIX, G.J., GULATI, S. and WAKOFF, G.I., On the use of singular functions with finite element approximation. *J. Comp. Phys.* 13, 209-228 (1973).
12. GRISVARD, P., Behaviour of the solutions of an elliptic boundary value problem in a polygonal or polyhedral domain. pp.207-274 of B. Hubbard (ed.), *Numerical Solution of Partial Differential Equations III, SYNSPADE 1975*. Academic Press, New York (1976).
13. HARRISON, D., MODEL: A general purpose modular finite element code. User manual. *Technical Report BICOM 81*, Institute of Computational Mathematics, Brunel University (1981).
14. HARRISON, D., WARD, T.J.W. and WHITEMAN, J.R., The philosophy and implementation of MODEL - A modular finite element research code. *Proc. 4th Int. Finite Element Systems Seminar*, Computational Mechanics Centre, Southampton (1983).
15. HARRISON, D., WARD, T.J.W. and WHITEMAN, J.R., Finite element analysis of plates with nonlinear properties. *Comp. Meth. Appl. Mech. Eng.* 34, 1019-1034 (1982).
16. HENSHELL, R.D. and SHAW, K.G., Crack tip finite elements are unnecessary. *Int. J. Numer. Meth. Eng.* 9, 495-507 (1975).
17. HUTCHINSON, J.W., Singular behaviour at the end of a tensile crack in a hardening material. *J. Mech. Phys. Solids* 16, 13-31 (1968).
18. IRWIN, G.R., *Structural Mechanics*. Pergamon Press, Oxford (1960).
19. KRIZEK, M. and NEITTAANMÄKI, P., Superconvergence of the finite element schemes arising from the use of averaged gradients. *Preprint 26*, Mathematics Institute, Jyvaskyla University (1984).

20. LEVINE, N., Superconvergent recovery of the gradient from piecewise linear finite element approximations. *Numerical Analysis Report 6/83*, Department of Mathematics, University of Reading (1983).
21. MITCHELL, A.R. and WAIT, R., *The Finite Element Method in Partial Differential Equations*. Wiley, London (1977).
22. NECAS, J., Variational inequalities in elasticity and plasticity with application to Signorini's problem and to the flow theory of plasticity. *ZAMM 60*, T20-T26 (1980).
23. NECAS, J. and HLAVACEK, I., *Mathematical Theory of Elastic and Elastico-Plastic Bodies: An Introduction*. Elsevier, Amsterdam (1981).
24. O'LEARY, J.R., An error analysis for singular finite elements. *TICOM Report 81-4*, Texas Institute for Computational Mechanics, University of Texas at Austin (1981).
25. OWEN, D.R.J. and HINTON, E., *Finite Elements in Plasticity: Theory and Practice*. Pineridge Press, Swansea (1980).
26. PARKS, D.M., A stiffness derivative finite element technique for determination of crack tip stress intensity factors. *Int. J. Fracture* 10, 487-502 (1974).
27. PRAGER, W., *An Introduction to Plasticity*. Addison Wesley, New York (1959).
28. RICE, J.R., A path independent integral and the approximate analysis of strain concentration by notches and cracks. *J. Appl. Mech.* 34, 379-386 (1968).
29. RICE, J.R., Mathematical analysis in the analysis of fracture. pp.191-311 of H. Liebowitz (ed.), *Fracture, Vol.2*. Academic Press, New York (1968).
30. RICE, J.R. and ROSENGREN, G.F., Plane strain deformation near a crack tip in a power law hardening material. *J. Mech. Phys. Solids* 16, 1-12 (1968).
31. SCHATZ, A. and WAHLBIN, L., Maximum norm estimates in the finite element method on plane polygonal domains. Parts I and II. *Math. Comp.* 32, 73-109 (1978) and *Math. Comp.* 33, 465-492 (1979).
32. SIH, G.C., A special theory of crack propagation. pp.XXIII - XXXIII of G.C. Sih (ed.) *Mechanics of Fracture Vol.1*. Noordhoff, Leyden (1973).
33. STERN, M., Families of consistent conforming elements with singular derivative fields. *Int. J. Numer. Meth. Eng.* 14, 409-421 (1979).
34. STERN, M. and BECKER, E., A conforming crack tip element with quadratic variation in the singular fields. *Int. J. Numer. Meth. Eng.* 12, 279-288 (1978).
35. THOMPSON, G.M., The Finite Element Solution of Fracture Problems in Two- and Three-Dimensions. *Ph.D. Thesis*, Brunel University (1983).
36. THOMPSON, G.M. and WHITEMAN, J.R., An analysis of strain representation in both singular and non-singular finite elements. (to appear).

37. WARD, T.J.W. and BURTON, C., The design and implementation of an efficient finite element code for high temperature problems. pp.123-135 of J.R. Whiteman (ed.), *The Mathematics of Finite Elements and Applications V, MAFELAP 1984*. Academic Press, London (1985).
38. WASHIZU, K., *Variational Methods in Elasticity and Plasticity*. Pergamon Press, Oxford (1968).
39. WESTERGAARD, H.M., Bearing pressures and cracks. Trans. ASME. *J. App. Mech.* 6, 49-53 (1939).
40. WHITEMAN, J.R., Some aspects of the mathematics of finite elements. pp.25-42 of J.R. Whiteman (ed.), *The Mathematics of Finite Elements and Applications II, MAFELAP 1975*. Academic Press, London (1976).
41. WHITEMAN, J.R., Problems with singularities. Sections II.6.0 and II.6.1 of H. Kardestuncer (ed.) *Finite Element Handbook*. (to appear)
42. WHITEMAN, J.R. and AKIN, J.E., Finite elements, singularities and fracture. pp.35-54 of J.R. Whiteman (ed.), *The Mathematics of Finite Elements and Applications III, MAFELAP 1978*. Academic Press, London (1979).
43. WHITEMAN, J.R. and SCHLEICHER, K-T., Introduction to the treatment of singularities in elliptic boundary value problems using finite element methods. *Technical Report BICOM 83/2*, Institute of Computational Mathematics, Brunel University (1983).
44. WILLIAMS, M.L., Stress singularities resulting from various loading conditions in angular corners of plates in extension. *J. Appl. Mech.* 24, 526-528 (1952).

NEW SOLUTION PROCEDURES FOR LINEAR AND NON-LINEAR FINITE ELEMENT ANALYSIS

M.A. Crisfield

Transport and Road Research Laboratory, Crowthorne, Berkshire, England

1. INTRODUCTION

The paper, which is divided into two parts, reviews some of the more recent solution procedures and, in addition, describes some new directions currently being explored by the author. The first part relates to linear problems and the new work involves "hierarchical preconditioning" of conjugate-gradient-like methods. The second covers non-linear analysis and concentrates on "continuation techniques" with particular emphasis on modified forms of the arc-length method. This introduction will be mainly concerned with the linear procedures.

Although most linear finite element programs use direct solution methods (such as Gauss or Cholesky [1, 2]) to solve the governing simultaneous equilibrium equations, there are reasons to believe we may yet see a resurgence of interest in iterative techniques. One reason relates to our insatiable desire to solve larger and larger problems. For example, the analysis of Statfjord B off-shore platform involved 120,000 simultaneous equations [3]. Three-dimensional analyses are now often undertaken and Hughes et al [4] have pointed out that the symmetric, banded, stiffness matrix of a 3-D finite element model of a simple cube with only twenty nodes along each edge would involve about 24,000 equations and about 29 million words of store. The detailed stress analysis of a stiffened steel box-girder bridge with a trapezoidal orthotropic deck can also involve too much storage for a direct solution. A similar situation can even occur for medium sized problems, especially if the analysis is performed on one of the new mini or micro computers that are now sufficiently cheap to be used as "individual work stations" [5]. As a further impetus to iterative solution techniques, the advent of parallel and array processors has highlighted the potential of solution methods in which similar numerical operations are performed simultaneously on different terms in a vector (or vectors).

Iterative solution techniques have, of course, been tried before. They were largely abandoned as being too unreliable although dynamic relaxation [6–9] has proved very popular when used in conjunction with finite differences. Two developments have changed this scenario. Firstly, new iterative methods have been devised (ie quasi-Newton and conjugate-gradient-like techniques [10–29]) which are less prone to rounding errors. These techniques have often been described in the mathematical programming and optimisation literatures [10–17] but have also

ISBN 0-12-747255-X

been developed and used by engineers especially for non-linear analysis [18–29]. The second development relates to better methods of preconditioning which can enormously improve the speed and robustness of the iterative procedures. While standard iterative techniques such as dynamic relaxation [6–9], conjugate-gradients [30–33] and successive over-relaxation [2] involve no matrix inversion (or factorisation), the preconditioned procedures aim to factorise a simpler matrix (or matrices) than the original stiffness matrix [34–42]. One may take direct advantage of the finite element discretisation and, at one level, factorise a set of modified element stiffness matrices [4, 43] while, if hierarchical displacement functions [44–49] are adopted, one need only factorise the structure sub-stiffness matrix associated with the lower-order displacement functions [19, 21, 22, 42]. Alternatively, one may adopt an incomplete Cholesky factorisation [34–40] with a view to either eliminating "fill-in", or else, to the inclusion of only the larger stiffness terms.

2. PART ONE: ITERATIVE SOLUTION TECHNIQUES FOR LINEAR PROBLEMS

We wish to solve the equations

$$\underline{K}\,\underline{p} - \underline{q} = \underline{0} \tag{2.1}$$

where (in structural terms) $\underline{K}$ is the stiffness matrix, $\underline{p}$ are the nodal displacements (unknown) and $\underline{q}$ are the applied loads (known). We will apply an iterative procedure of the general form:

$$\underline{p}_{i+1} = \underline{p}_i + \eta_i\,\underline{\delta}_i \tag{2.2}$$

whereby trial displacements $\underline{p}_i$ are updated to $\underline{p}_{i+1}$ using an iterative vector $\underline{\delta}_i$ and a step-length scalar η_i (which may or may not be unity). The stresses associated with the trial displacements $\underline{p}_i$ will generally not be in equilibrium with the applied loads so that the residual $\underline{g}_i$ (the symbol $\underline{g}$ is used because the vector is the gradient of the total potential energy, Φ) can be expressed as

$$\underline{g}_i = \underline{K}\,\underline{p}_i - \underline{q} \tag{2.3}$$

2.1 *Hughes' element-by-element method*

Hughes and co-workers [4, 43] use the dynamic analogy to derive an "element-by-element" solution to equation (2.1). Most iterative processes can, in some senses, be considered as "element-by-element" processes in that, with

$$\underline{K} = \sum_{i=1}^{N_{e\ell}} \underline{K}_{e,i} \quad (=,\text{ say, } \underline{K}_1 + \underline{K}_2) \tag{2.4}$$

the residual vector, $\underline{g}$, can be formed on an element-by-element basis. Standard iterative techniques apply no element or structural matrix divisions to form the iterative vector, $\underline{\delta}_i$. In contrast, Hughes and co-workers perform modified matrix divisions "at the element level". To this end, they invoke the dynamic analogy and start with the viscous dynamic equations (with no mass) at the structure level. These can be expressed in the form

$$\underline{C}\,\dot{\underline{p}} + \underline{K}\,\underline{p} = \underline{q} \tag{2.5}$$

where

$$\dot{\underline{p}} = \frac{\underline{p}_{i+1} - \underline{p}_i}{\Delta t} = \frac{\underline{\delta}_i}{\Delta t} \tag{2.6}$$

and the stiffness term $\underline{K}\ \underline{p}$ can be expressed as

$$\underline{K}\ \underline{p} = \underline{K}\,((1-\gamma)\,\underline{p}_i + \gamma\underline{p}_{i+1}) = \underline{K}\,(\underline{p}_i + \gamma\underline{\delta}_i) \tag{2.7}$$

with γ being a scalar that interpolates the interval $\underline{p}_i$ to $\underline{p}_{i+1}$.
Substitution from equations (2.3), (2.6) and (2.7) into equation (2.5) gives:

$$(\underline{C} + \gamma\Delta t\ \underline{K})\,\underline{\delta}_i = -\Delta t\,\underline{g}_i \tag{2.8}$$

and, with $\gamma > 0$, the solution involves the formation and factorisation of the structure matrix $\underline{C} + \gamma\Delta t\ \underline{K}$. Clearly, for linear problems, such a process involves (for even one iteration) as much work as the direct solution of the governing equations (2.1). However, Hughes and co-workers argue [4, 43] that an economic iterative process can be established by means of a splitting technique [50]. In particular, advantage can be taken of equation (2.4) where, for illustrative purposes, we will imagine a structure stiffness matrix composed of two element stiffness matrices $\underline{K}_1$ and $\underline{K}_2$ as indicated by the bracketed terms. We will further assume, for simplicity, that $\underline{C} = \underline{I}$ although a similar development can be made when $\underline{C}$ is some general diagonal scaling matrix (or viscous damping matrix). Using the product

$$(\underline{I} + \gamma\Delta t\,\underline{K}_1)\,(\underline{I} + \gamma\Delta t\,\underline{K}_2) = \underline{I} + \gamma\Delta t\,(\underline{K}_1 + \underline{K}_2) + (\gamma\Delta t)^2\,\underline{K}_1\,\underline{K}_2 \tag{2.9}$$

and assuming that $\gamma\Delta t$ is small, a reasonable approximation to equation (2.8) is (with $\underline{C} = \underline{I}$),

$$(\underline{I} + \gamma\Delta t\,\underline{K}_1)\,(\underline{I} + \gamma\Delta t\,\underline{K}_2)\,\underline{\delta}_i = -\Delta t\,\underline{g}_i \tag{2.10}$$

so that equation (2.10) can be solved using two "element-by-element" divisions,

$$\left.\begin{aligned} \bar{\underline{\delta}}_i &= -\Delta t\,(\underline{I} + \gamma\Delta t\,\underline{K}_1)^{-1}\,\underline{g}_i \\ \underline{\delta}_i &= (\underline{I} + \gamma\Delta t\,\underline{K}_2)^{-1}\,\bar{\underline{\delta}}_i \end{aligned}\right\} \tag{2.11}$$

While the essence of Hughes' element-by-element solution procedure [4, 43] is contained in the system of equations (2.11), Hughes and co-workers actually present a more complex two-pass system and include the diagonal scaling matrix $\underline{C}$. However no guidance appears to be given on the optimum "time" step Δt for solving linear problems. Marchuck [50] indicates that with $\gamma = 1/2$, the optimum time step of the "split" system is the same as that of the full system (equation (2.8) with $\underline{C} = \underline{I}$) and that for such a system

$$\Delta t_{opt} = \frac{2}{\lambda_{max} + \lambda_{min}} \tag{2.12}$$

Hughes et al [4] adopt a different approach and add "line-searches" [51,52] to the iterative procedure so that the $\underline{\delta}_i$ given by equation (2.8) is only a direction that will be multiplied by a scalar η_i (as in equation (2.2)). This scalar is chosen to minimise the total potential energy in the direction $\underline{\delta}_i$ using "line-search" concepts which, for linear problems, lead to the simple relationship [21, 51],

$$\eta_i = \frac{-\underline{\delta}_i^T \underline{g}_i}{\underline{\delta}_i^T \underline{K}\, \underline{\delta}_i} \tag{2.13}$$

Having introduced the scalar η_i, it follows that the Δt term on the right hand side of equation (2.8) is irrelevant and that the scalar governing the iterative direction is the total term $\gamma\Delta t$. The actual choice of γ is therefore also irrelevant (provided it is greater than zero) although Hughes et al argue that it should be unity. With line-searches being introduced, it would appear that the standard error analysis approach of [50] cannot be applied to find an optimum $\gamma\Delta t$.

It is worth noting that there is no need to use the dynamic analogy to produce a starting equation of the form of equation (2.8). Instead, by supplementing the total potential energy Φ with a Lagrangian multiplier $\lambda/2$ times a constraint on the (scaled) iterative vector $\underline{\delta}_i$, one produces

$$\Phi(\underline{p}_i + \underline{\delta}_i) = \text{const.} + \underline{g}_i^T \underline{\delta}_i + \tfrac{1}{2}\underline{\delta}_i^T \underline{K}\, \underline{\delta}_i + \frac{\lambda}{2}(\underline{\delta}_i^T \underline{C}\, \underline{\delta}_i - \Delta\ell^2) \tag{2.14}$$

which, on being made stationary, leads to

$$(\underline{C} + \lambda \underline{K})\underline{\delta}_i = -\underline{g}_i \tag{2.15}$$

This formulation has some links with non-linear "trust region" solution procedures [53–55] which will be mentioned again in the second part of the paper.

2.2 Preconditioning

If an iterative solution procedure is used to solve equation (2.1), the required number of iterations will generally be proportional to some power of the condition number of $\underline{\bar{K}}$ [56]. For structural problems, particularly those involving both in-plane and out-of-plane action, there can be a very large difference between the maximum and minimum eigenvalues of $\underline{\bar{K}}$ so that the condition number may be too large to allow an economic iterative solution. In extreme cases, the iterative process will fail to converge at all. To overcome this problem, one may (formally) aim to transform equation (2.1) into:

$$\underline{\bar{g}} = \underline{\bar{K}}\, \underline{\bar{p}} - \underline{\bar{q}} = \underline{0} \tag{2.16}$$

where

$$\underline{\bar{p}} = \underline{L}^T \underline{p} \tag{2.17}$$

$$\underline{\bar{q}} = \underline{L}^{-1} \underline{q}, \quad \underline{\bar{g}} = \underline{L}^{-1} \underline{g} \tag{2.18}$$

and

$$\underline{\bar{K}} = \underline{L}^{-1} \underline{K}\, \underline{L}^{-T} \tag{2.19}$$

where $\underline{L}$ and $\underline{L}^T$ are the Cholesky factors of some "approximate stiffness matrix", $\underline{K}_a$. ie

$$\underline{K}_a = \underline{L}\ \underline{L}^T \tag{2.20}$$

In the extreme, if the approximation is very good, $\underline{K}_a = \underline{K}$ and $\underline{\bar{K}} = \underline{I}$ which has a condition number of unity so that an iterative solution of (2.16) would be very efficient. (Of course, all the work would have been involved in forming $\underline{L}$ and $\underline{L}^T$ which process, in itself, involves a direct solution.)

In general, one should aim to find a matrix $\underline{K}_a$ that is easier to factorise than $\underline{K}$ and yet leads to a matrix $\underline{\bar{K}}$ with a lower condition number than $\underline{K}$. We will later discuss various ways in which this can be achieved but, firstly, it should be noted that there is no need to explicitly perform the transformations of equations (2.17)–(2.19). Instead, if an iterative solution procedure for equation (2.1) takes the form of equation (2.2) with

$$\underline{\delta}_i = -\alpha_i' \underline{g}_i + \beta_i' \underline{\delta}_{i-1} \tag{2.21}$$

the conceptual application of an equivalent iterative process to equation (2.16) leads to the following iterative direction for the original equations:

$$\underline{\delta}_i = -\alpha_i \underline{K}_a^{-1} \underline{g}_i + \beta_i \underline{\delta}_{i-1} \tag{2.22}$$

Equation (2.21) encompasses a number of iterative procedures (Jacobi iteration, steepest descent, conjugate-gradients, dynamic relaxation) with the specific type depending on the scalars α_i, β_i and η_i [19, 22]. Equation (2.22) defines equivalent preconditioned iterative procedures and the main problem is to define $\underline{K}_a$ and form

$$\underline{\delta}_i^* = -\underline{K}_a^{-1} \underline{g}_i \tag{2.23}$$

via some Cholesky (or Crout) factorisation of $\underline{K}_a$. The detailed application of some preconditioned conjugate-gradient-like procedures will be described later but, firstly, we will address the problem of forming $\underline{K}_a$ and/or its factors $\underline{L}$ and $\underline{L}^T$. Two options will be considered. The first involves a modification to the standard Cholesky factorisation while the second relates to the adoption of hierarchical shape functions for the finite element idealisation. In each case, the objective is to produce a reduced condition number for the implied matrix $\underline{\bar{K}}$ of equation (2.19).

2.3 Incomplete Cholesky methods

Cholesky factorisations can be applied to the upper triangle of a matrix $\underline{K}$ using either row-by-row or column-by-column techniques. We will adopt the programming convention whereby an expression on the right hand side is replaced by a term on the left hand side and will not distinguish between $\underline{K}$ and $\underline{L}$. In other words, the upper triangle $\underline{K}$ will become the factor $\underline{L}^T$ (still expressed as $\underline{K}$). The column-by-column approach then gives

$$K_{11} = K_{11}^{1/2} \tag{2.24}$$

For each column, j = 2, N

$$\left.\begin{aligned} t &= K_{ij} - \left(\sum_{\ell=1}^{i-1} K_{\ell i} K_{\ell j} \right)_{\text{if } i>1} && (2.25a) \\ K_{ij} &= t/K_{ii} && (2.25b) \end{aligned}\right\} \quad \text{for i=1, j-1}$$

$$K_{jj} = \left(K_{jj} - \sum_{\ell=1}^{j-1} K_{\ell j}^2 \right)^{1/2} \tag{2.26}$$

The alternative row-by-row procedure gives:

For each row, i = 1, N:

$$K_{ij} = K_{ij} - \left(\sum_{\ell=1}^{i-1} K_{\ell i} K_{\ell j} \right)_{\text{if } i>1} \qquad j = i+1, N \text{ (provided } i<N) \tag{2.27}$$

$$K_{ii} = \left(K_{ii} - \left(\sum_{\ell=1}^{i-1} K_{\ell i}^2 \right)_{\text{if } i>1} \right)^{1/2} \tag{2.28}$$

$$K_{ij} = K_{ij} / K_{ii} \qquad j = i+1, N \text{ (provided } i<N) \tag{2.29}$$

The column-by-column approach is easily adapted to produce a variable band-width or "sky-line" solution procedure [1]. However operation (2.25b) will often lead to "fill-in". In other words it will introduce terms into K_{ij} where originally they were zero. This does not affect a "sky-line" procedure because the "fill-in" is within the storage area, ie below the sky-line. However, particularly with a regular mesh, the required storage could be reduced to be considerably less than that of a "sky-line" procedure if one were able to avoid storing the "fill-in" terms. Strictly, this is impossible if a direct solution is attempted. However, we can adopt the spirit of the last section and neglect the "fill-in" with a view to factorising an approximate stiffness matrix $\underline{K}_a$. We must then embed the process within an iterative technique. This approach was first adopted by Meijerink and van der Vorst [40] who simply neglected the "fill-in" terms. Consequently, any K_{ij} produced by equation (2.25b) would be omitted if there were no corresponding terms in the original stiffness matrix. Such an approach leads to

$$\underline{K}_a = \underline{L}\,\underline{L}^T = \underline{K} + \underline{R} \tag{2.30}$$

where $\underline{R}$ is a residual or error matrix which does not have to be formed but is conceptually useful. For Meijerink and van der Vorst's method, the matrix $\underline{R}$ contains all the off-diagonal fill-in terms, ie

$$\underline{R} = \Sigma \begin{bmatrix} 0 & -r \\ -r & 0 \end{bmatrix} \begin{matrix} i \\ j \end{matrix} \quad (\text{columns } i, j) \tag{2.31}$$

where

$$r = t \text{ from equation (2.25a) or } K_{ij} \text{ from equation (2.27).} \tag{2.32}$$

Except in special circumstances, there is no guarantee that $\underline{K}_a$ will follow $\underline{K}$ in being positive definite. Meijerink and van der Vorst invoked such special circumstances by requiring $\underline{K}$ to be a positive definite M-matrix (with $K_{ij} < 0$ for all $i \neq j$).

If $\underline{K}$ is not positive definite, the Cholesky factorisation is impossible and, at some stage, one would be unable to take the square root of equation (2.26) (or equation (2.28)). Kershaw [39] overcame this problem by simply adding "some positive value" to any negative pivot (prior to the taking of the square root). Such a procedure will add positive terms to some of the diagonal elements of $\underline{R}$. While Kershaw presented some good numerical results, a number of questions remain unanswered. In particular, "How positive does one make the offending pivot?" and "What is the effect of a matrix $\underline{K}_a$ that is only just positive definite?"

In order to avoid such questions, Jennings and co-workers [35–38] proposed that the error matrix $\underline{R}$ should be of the form

$$\underline{R} = \Sigma \begin{bmatrix} s\,|r| & -r \\ -r & \frac{1}{s}|r| \end{bmatrix} \begin{matrix} i \\ j \end{matrix} \tag{2.33}$$

(columns i and j)

where r is given by equation (2.32) and s will be defined. Such a procedure has the major advantage that $\underline{R}$ is now positive semi-definite so that the lowest eigenvalue of $\underline{K}_a$ cannot be less than that of $\underline{K}$ and stability is assured. The constant s is chosen so that each diagonal pivot (ii and jj) experiences the same percentage change, ie

$$s = \left(\frac{K_{ii}}{K_{jj}}\right)^{\frac{1}{2}} \tag{2.34}$$

where K_{ii} and K_{jj} are the current values of the pivots at the time that the "fill-in" term is neglected. One cannot use the column-by-column approach of equations (2.24)–(2.26) to implement such a procedure because the K_{ii} term would be required for equation (2.25b) before it were fully modified. In practise, Jennings and co-workers do not avoid "fill-in" and, instead of rejecting elements because of their position, they reject terms in relation to their magnitude. In particular, K_{ij} is rejected whenever

$$|K_{ij}| < \psi\,(K_{ii} \, . \, K_{jj})^{\frac{1}{2}} \tag{2.35}$$

where ψ may lie between zero (a full Cholesky factorisation) and unity (a diagonal scaling). The terms K_{ij}, K_{ii} and K_{jj} in equation (2.35) relate to the current "reduced" values. The complete procedure is based on equations (2.27)–(2.29) and involves:

(i) Apply equation (2.27) to form the reduced K_{ij} and, at the same time, assess its magnitude using equation (2.35). If it is to be rejected, add s (equation (2.34) times $|K_{ij}|$ and 1/s times $|K_{ij}|$ to K_{ii} and K_{jj} respectively.

(ii) Having traversed row i, K_{ii} has been modified to account for rejection but further modifications are required using equation (2.28).

(iii) Re-traverse row i and apply equation (2.29) noting that the rejected K_{ij} no longer exist.

Clearly, some complicated addressings are required because the storage size (and pattern) is not known at the start. A number of integer arrays are needed for this purpose so that the potential saving in storage is only likely to be realised if integer variables of reduced word-length are available. An algorithm which includes a Fortran listing is given in [35].

Before leaving incomplete Cholesky methods, one should mention an alternative technique due to Gustafsson [34]. This procedure follows Meijerink and van der Vorst [40] and Kershaw [39] in rejecting terms according to their position (avoiding "fill-in") but is related to Jennings' method in that modifications are simultaneously made to the diagonal elements. However there are important differences. In particular, the parameter s is taken as unity and, more fundamentally, no modulus sign is adopted for the diagonal elements so that equation (2.33) is replaced by

$$\underline{R} = \Sigma \begin{bmatrix} r & -r \\ -r & r \end{bmatrix} \begin{matrix} i \\ j \end{matrix} \tag{2.36}$$

(columns i, j)

In marked contrast to Jennings procedure, the defect matrix $\underline{R}$ is negative semi-definite. As a consequence the method will only be stable for diagonally dominant matrices. The advantages of the method relate to the improved convergence characteristics when it is coupled with an iterative solution procedure (in particular, the conjugate-gradient method). It has already been noted that the iterative performance depends on the condition number of $\bar{\underline{K}}$ (equation (2.19)). For problems involving second order partial differential equations, the condition number of $\underline{K}$ will be $0(n^2)$ as $n \rightarrow \infty$ where n is the number of elements on a "typical side" (see (2.34) for further details). Gustafsson argues that the condition number of $\bar{\underline{K}}$ will also be $0(n^2)$ as $n \rightarrow \infty$ for standard incomplete Cholesky procedures. However for his technique whereby the defect matrix preserves a zero row sum (equation (2.36)), the condition number of $\bar{\underline{K}}$ is shown to be $0(n)$ as $n \rightarrow \infty$ (under certain circumstances [34]). Gustafsson also discusses methods for increasing the amount of rejection by adopting a preset storage pattern.

2.4 Hierarchical preconditioning

It has long been recognised that the adoption of local and global modes at the discretisation stage can lead to an improved conditioning of the governing equations [41, 57–60]. A special form of local-global idealisation can be provided by hierarchical displacement functions [44–49]. Using such a technique, the displacements for an element would be interpolated as

$$\underline{p} = \underline{H}_{\ell}^{T} \underline{p}_{\ell} + \underline{H}_{q}^{T} \Delta\underline{p}_{q} + \underline{H}_{c}^{T} \Delta\underline{p}_{c} \tag{2.37}$$

where $\underline{H}_{\ell}$ are the basic linear shape functions (cf global modes) while $\underline{H}_q$ and $\underline{H}_c$ include additional higher-order quadratic and cubic modes respectively (cf superimposed local modes). For the simple quadrilateral of Figure 1, the basic linear functions are

$$H_i = \tfrac{1}{4}(1 + \xi_i \xi)(1 + \eta_i \eta) \qquad i = 1, 4 \tag{2.38}$$

while the higher order terms are provided by the serendipity quadratic functions [61] which, when written in hierarchical form, are given by:

$$H_i = \tfrac{1}{2}(1 - \eta_i^2 \xi^2 - \xi_i^2 \eta^2)(1 + \xi_i \xi + \eta_i \eta) \qquad i = 5, 8 \tag{2.39}$$

The total displacement at node 5 would be given by

$$p_5 = \tfrac{1}{2}(p_1 + p_2) + \Delta p_5 \tag{2.40}$$

where Δp_5 is the variable contained in $\Delta\underline{p}_q$ (equation (2.37)).

Hierarchical formulations are becoming increasingly popular, particularly for automatic adaptive refinement [44, 46–49]. In addition, they automatically provide an ideal form of preconditioning. If the lower-order structural variables are grouped into $\underline{p}_c$ (c for coarse) while the higher-order variables are separately grouped into $\underline{p}_f$ (f for fine), the governing stiffness equations take the form:–

$$\begin{Bmatrix} \underline{q}_c \\ \underline{q}_f \end{Bmatrix} = \begin{bmatrix} \underline{K}_{cc} & \underline{K}_{cf} \\ \underline{K}_{cf}^T & \underline{K}_{ff} \end{bmatrix} \begin{Bmatrix} \underline{p}_c \\ \underline{p}_f \end{Bmatrix} \tag{2.41}$$

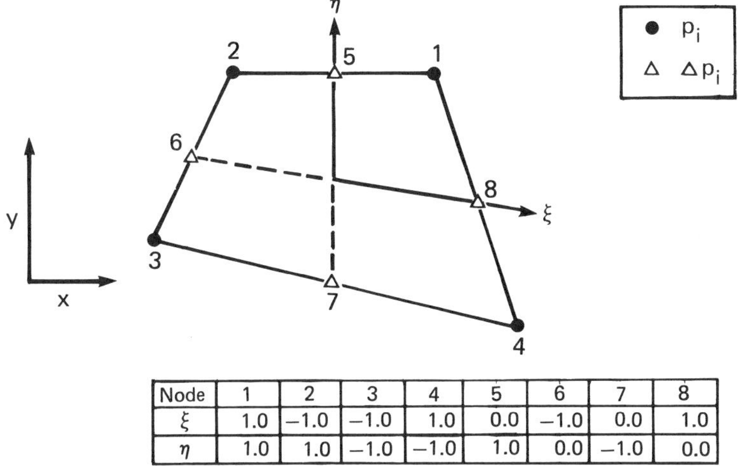

Node	1	2	3	4	5	6	7	8
ξ	1.0	−1.0	−1.0	1.0	0.0	−1.0	0.0	1.0
η	1.0	1.0	−1.0	−1.0	1.0	0.0	−1.0	0.0

Fig. 1 Node numbering and non-dimensional co-ordinates for heirarchical element

The sub-matrix $\underline{K}_{cc}$ is a structure stiffness matrix in its own right and $\underline{p}_c^T\ \underline{K}_{cc}\ \underline{p}_c$ will contain the dominant energy modes. Peano et al [46] recognised that equation (2.41) was in an ideal form for an iterative solution and applied a form of Gauss-Seidel iteration which involved factorising $\underline{K}_{cc}$ and $\underline{K}_{ff}$. Iterative solutions to equation (2.41) are also discussed by Zienkiewicz [49]. Crisfield [19, 21, 22] argued that one need only factorise $\underline{K}_{cc}$ and that an ideal pre-conditioning matrix would be provided by

$$\underline{K}_a = \begin{bmatrix} \underline{K}_{cc} & \underline{0} \\ \underline{0} & \underline{D}_{ff} \end{bmatrix} \tag{2.42}$$

where $\underline{D}_{ff}$ contains only the diagonals from $\underline{K}_{ff}$. The reasoning was largely intuitive but Axelsson and Gustafsson [42] have recently applied a precise mathematical development to arrive at the same conclusion. Crisfield used $\underline{K}_a$ of equation (2.42) in conjunction with a "secant-Newton" iterative method [19–23]. This technique can be viewed as a form of pre-conditioned conjugate gradient method (see next section). The method was applied to linear problems in [21] and to non-linear problems in [19, 22]. Axelsson and Gustafsson [42] used the preconditioned conjugate gradient method (next section) and solved problems involving second order partial differential equations. With n defining the mesh size (as before), they show that the condition number of $\underline{K}_{cc}$ will be $0(n^2)$ as $n \to \infty$ while that of $\underline{K}_{ff}$ (equation (2.41)) will be $\underline{0}(1)$. When $\underline{K}_a$ (equation (2.42)) is applied as a preconditioning matrix, the implied matrix $\underline{\bar{K}}$ of equation (2.19) has a condition number which remarkably is $0(1)$ for $n > n_o$ where n_o is sufficient to give "reasonable accuracy". (In fact, Axelsson and Gustafsson did not investigate the condition number of $\underline{\bar{K}}$ given by equation (2.19) but rather that of the related form $\underline{\bar{K}} = \underline{K}_a^{-1}\ \underline{K}$.) The number of iterations required for a preconditioned conjugate gradient method is proportional to the square root of the condition number of $\underline{\bar{K}}$ [34, 56]. Consequently there will eventually be no increase in the number of iterations as the mesh is refined. This is in marked contrast to the standard conjugate gradient method for which the number of iterations will, in theory, increase linearly with n (or the number of equations). In reality, the performance of the standard conjugate gradient method will degenerate even more rapidly. Axelsson and Gustafsson take the iterative process one stage further and apply their incomplete Cholesky procedure (last section) to the factorisation of $\underline{K}_{cc}$. Their derivation of the convergence characteristics assumed such an incomplete factorisation.

2.5 Preconditioned conjugate-gradient-like methods

The standard conjugate gradient method [2, 30–32] combines equations (2.2) and (2.13) with an iterative direction

$$\underline{\delta}_i = -\underline{g}_i + \beta_i' \underline{\delta}_i \tag{2.43}$$

where β_i' is given by

$$\beta_i' = \frac{\underline{g}_i^T \underline{g}_i}{\underline{g}_{i-1}^T \underline{g}_{i-1}} \tag{2.44}$$

for the Hestenes-Steifel formula [31] or

$$\beta_i' = \frac{\underline{g}_i^T (\underline{g}_i - \underline{g}_{i-1})}{\underline{g}_{i-1}^T \underline{g}_{i-1}} \tag{2.45}$$

for the Polak-Ribiere formula [32]. With exact arithmetic, the two methods coincide for linear problems although equation (2.45) is generally preferred for non-linear problems [33]. For linear problems,

$$\underline{g}_{i+1} - \underline{g}_i = \underline{K}\, \underline{\delta}_i \tag{2.46}$$

so that the step-length parameter η_i of equation (2.13) can be re-expressed as:

$$\eta_i = \frac{-\underline{\delta}_i^T \underline{g}_i}{\underline{\delta}_i^T (\underline{g}_{i+1}(\eta_i=1) - \underline{g}_i)} \tag{2.47}$$

and $\underline{g}_{i+1}$ can then be modified via:

$$\underline{g}_{i+1} = (1-\eta_i)\, \underline{g}_i + \eta_i\, \underline{g}_{i+1}(\eta_i = 1) \tag{2.48}$$

When preconditioning is introduced, equation (2.44) is modified to give

$$\underline{\delta}_i = -\alpha_i \underline{K}_a^{-1} \underline{g}_i + \beta_i \underline{\delta}_{i-1} = \alpha_i \overset{*}{\underline{\delta}}_i + \beta_i \underline{\delta}_{i-1} \tag{2.49}$$

where $\overset{*}{\underline{\delta}}_i$ is given by equation (2.23) and the scalar $\alpha_i = 1.0$. The preconditioned version of equation (2.45) gives

$$\beta_i = \frac{\underline{\delta}_i^{*T} \underline{g}_i}{\underline{\delta}_{i-1}^{*} \underline{g}_{i-1}} \tag{2.50}$$

while preconditioning changes equation (2.45) to

$$\beta_i = \frac{\underline{\delta}_i^{*T} (\underline{g}_i - \underline{g}_{i-1})}{\underline{\delta}_{i-1}^{*\,T} \underline{g}_{i-1}} \tag{2.51}$$

For non-linear applications, the author proposed two alternative formulae which are related to the quasi-Newton method [19, 22]. Since these techniques are already coded in the author's finite element program, one of them has been applied here to linear problems. It gives:

$$\alpha_i = \frac{-\underline{\delta}_{i-1}^{T} \underline{g}_{i-1}}{\underline{\delta}_{i-1}^{T} (\underline{g}_i - \underline{g}_{i-1})} \tag{2.52}$$

$$\beta_i = \frac{-\underline{\delta}_{i-1}^{T} \underline{g}_i}{\underline{\delta}_{i-1}^{T} (\underline{g}_i - \underline{g}_{i-1})} - \alpha_i \frac{\underline{\delta}_i^{*\,T} (\underline{g}_i - \underline{g}_{i-1})}{\eta_{i-1} \underline{\delta}_{i-1}^{T} (\underline{g}_i - \underline{g}_{i-1})} \tag{2.53}$$

For linear problems with exact arithmetic, this method coincides with the standard preconditioned conjugate-gradient method [21]. However there may be some advantage in applying equations (2.52) and (2.53) rather than $\alpha_i = 1.0$ with either equation (2.50) or equation (2.51). The former method is designed to operate (in a non-linear context) with inexact line-searches and rounding-error is bound to introduce a certain inexactness.

2.5 Applying the hierarchicaly preconditioned conjugate-gradient method

The method was applied by making modifications to an existing non-linear finite element program. As a first step, equation (2.48) was introduced in order to avoid the computation of more than one residual vector, $\underline{g}_i$ at each iteration. A straight application of a non-linear code would lead to two such calculations once exact "line-searches" were introduced. The main modification related to the hierarchical preconditioner of equation (2.42). For normal non-linear applications, $\underline{K}_a$ would be the tangent stiffness matrix at the beginning of the load increment. Luckily, hierarchical displacement functions had already been introduced (they have many advantages) and the main problem involved the partitioning of equation (2.41). In practise, this was achieved by making relatively minor modifications to an existing system for applying the "boundary conditions". This system applied prescribed displacements by condensing the active system and introducing separate vectors for the prescribed displacements and corresponding reactions. Consequently the higher-order variables, $\underline{p}_f$ were treated as "pseudo prescribed displacements" (using a preprocessor to define the appropriate mid-side nodes) and only $\underline{K}_{cc}$ was formed and factorised. Small modifications to the code produced the diagonal vector $\underline{D}_{ff}$ (see equation (2.42)) and ensured that the "reactions" became out-of-balance forces (or residuals) for the higher order variables, ie $\underline{g}_f$.

In order to test the method, progressively finer meshes were applied to a plane stress analysis of the deep cantilever shown in Figure 2. Rectangular versions of the general quadrilateral element of Figure 1 were adopted. It is not suggested that the fine meshes of Figure 2 represent a practical method for the solution of this problem but rather, that they provide a simple means of testing the solution procedure.

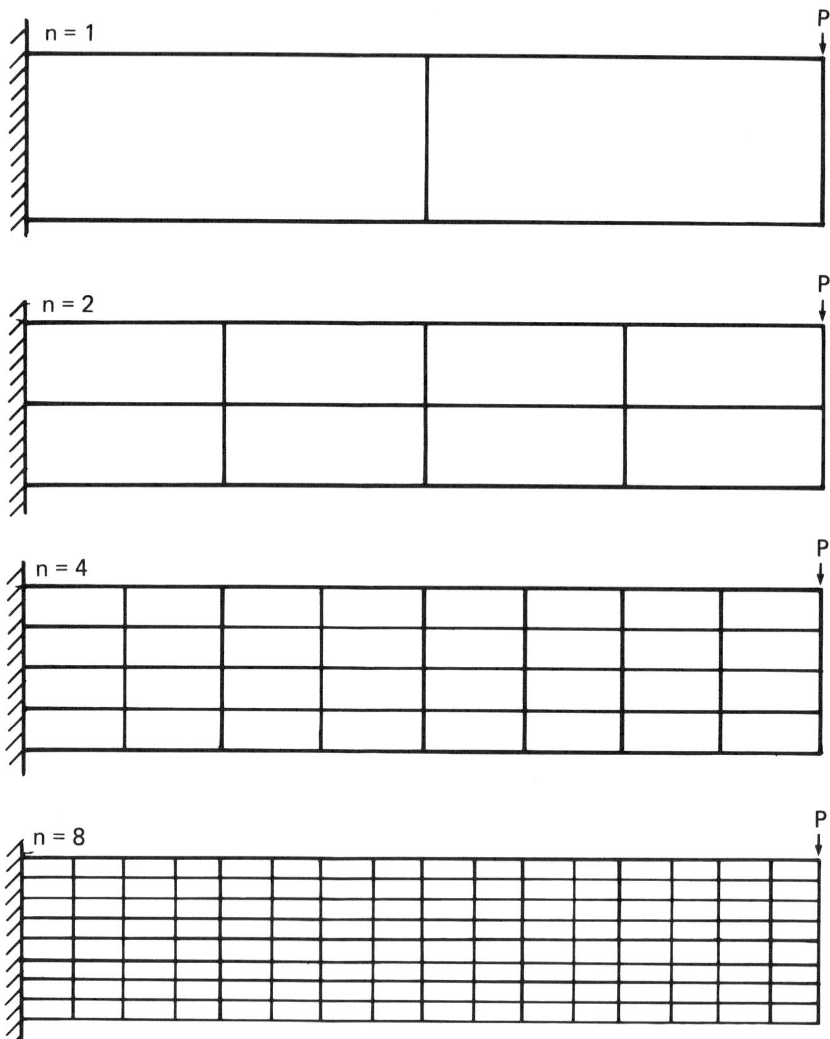

Fig. 2 Mesh divisions for cantilever

Iterations were halted once the convergence criterion,

$$\|\underline{g}\| = \sqrt{\underline{g}^T \underline{g}} < \epsilon \|\underline{q}\| \quad (\text{or } \epsilon P) \tag{2.54}$$

was satisfied. The constant ϵ was chosen to be 0.01 and at this stage, the Euclidean norm of the iterative displacements, $\|\underline{\delta}\|$ was less than 0.01 per cent of the total displacement norm $\|\underline{p}\|$. The analyses were conducted on a CDL Cyber 720 using 60 bit words and the iterative performances are recorded in Table 2 while Table 1 details the problem dimensions. Table 2 includes the starting value of ϵ (ϵ_s) that was recorded immediately after the "coarse solution".

TABLE 1

Problem dimensions for cantilever example

n	N_c	N_f	N_t	Full system			$\underline{K}_{cc}$ only		
				B_a	B_m	M	B_a	B_m	M
1	8	12	20	8	16	170	4	8	36
2	24	40	64	22	14	928	7	10	176
4	80	144	224	34	25	5624	11	14	948
8	288	544	832	58	45	37816	20	22	5804

n = mesh parameter (Figure 2); N_c = number of "coarse" variables; N_f = number of "fine" variables; N_t = total number of variables; B_a = average ½ band-width; B_m = maximum ½ band-width; M = number of matrix elements under "sky line".

TABLE 2

Required number of iterations for cantilever example

Mesh parameter n	$\underline{K}_a = \underline{I}$	$\underline{K}_a = \underline{D}$	$\underline{K}_a$ (from equation (2.42)	ϵ_s
1	334	19	14	8.65
2	Failed	54	24	11.70
4	Failed	103	27	8.96
8	Failed	Failed	27	5.31

It is clear from Table 2 that the basic conjugate-gradient method (with $\underline{K} = \underline{I}$) is totally inadequate for this problem although significant improvements were introduced once a diagonal scaling was introduced with $\underline{K} = \underline{D}$. However, even then, no solution could be obtained with the finest mesh. It should be noted that "re-starts" were applied at every N_t iterations where N_t is the total number of equations (Table 1). Such "re-starts" were recommended by Fletcher [30] and cut-out the conjugate gradient in favour of an iterative vector in the direction of the scaled steepest descent (ie with $\alpha_i = 1.0$ and $\beta_i = 0.0$).

Once hierarchical preconditioning was introduced, the iterative performance improved dramatically and conformed perfectly to the predictions of Axelsson and Gustafsson [42] in that the number of iterations became independent of n as n was increased. It is a little difficult to compare computer times because the computer system is not optimised for linear analysis. For example, the residuals, $\underline{g}$, are computed as for a non-linear analysis by calculating strains and then stresses at the "Gauss points" and hence computing the internal forces for the element via:

$$\underline{q}_i = \int \underline{B}^T \sigma \, dA \tag{2.55}$$

In addition, the times will depend on the available core-storage because the out-of-core Cholesky algorithm, which uses a "sky-line" procedure, spends more-and-more time on disc transfers as the available core-store is reduced. Nonetheless, for the adopted configuration, the hierarchically preconditioned solutions were cheaper than the diagonally preconditioned solutions for $n \geqslant 4$. But, even for the finest mesh, the solution time was twice that required for a direct solution. However this ratio is reducing rapidly with the increasing number of variables and for genuinely large problems (especially those involving three-dimensional analyses) it will clearly make sense to adopt the hierarchically preconditioned iterative approach. The direct solution routine, with the variable band-width, produces a much denser packing once the mid-side nodes are omitted (Table 2). Finally, it should be noted that the good performance of the preconditioned conjugate-gradient method relates to the conjugate-gradient technique as well as the hierarchical preconditioning. For example, no convergence was achieved with the latter technique when a straight "modified Newton-Raphson" procedure was adopted (with $\alpha_i = 1.0$ and $\beta_i = 0.0$ in equation (2.49)). This observation relates to solutions without "line-searches". Once line-searches were introduced using equations (2.47) and (2.48), convergence was achieved in 126 iterations. This performance is far worse than that of the conjugate-gradient method which took 14 iterations (Table 2) or 19 with a diagonal preconditioning.

It is hoped to apply these iterative solution procedures to folded-plated structures and, as a preliminary study, analyses were conducted on the finite element idealisation of Figure 3. This idealisation relates to a test-panel from an orthotropic steel bridge deck with trapezoidal stiffeners. Quadratic serendipity functions [61] were used for the membrane behaviour while the bending action was captured by means of a four-noded "discrete Kirchhoff" element [62], which passes the patch test. Except at plate junctions, this idealisation leads to five degrees-of-freedom at the corners (three translations and two rotations) and two at the mid-sides (the higher-order hierarchical in-plane translations). At the junctions, six degrees-of-freedom are required for the corner nodes (three translations and three rotations) and three for the mid-side nodes (higher-order hierarchical translations).

A hierarchical iterative solution was achieved by categorising the corner variables as $\underline{p}_c$ and those at the mid-sides as $\underline{p}_f$. Consequently all the bending "action" was included in the "coarse mesh idealisation". Details of the problem dimensions are given in Table 3 which shows that considerably less storage was required for the "coarse mesh stiffness matrix", $\underline{K}_{cc}$. Because the problem involves both membrane and bending action, it may be better to adopt a scaled force norm criterion instead of the direct approach of equation (2.54). In particular, one can use:

$$\|\bar{\underline{g}}\| = \sqrt{\underline{g}^T \underline{K}_a^{-1} \underline{g}} = \sqrt{-\underline{\delta}^{*T} \underline{g}} < \mu \|\bar{\underline{q}}\| \text{ or } \mu \sqrt{\underline{q} \, \underline{K}_a^{-1} \, \underline{q}} \tag{2.56}$$

Equation (2.56) introduces a similar scaling to that advocated by Peano and Riccioni [44] (and used by the author [19–22]) in which $\underline{K}_a$ would be replaced by $\underline{D}$, the diagonal stiffness matrix. The approximate stiffness matrix, $\underline{K}_a$ of equation (2.42), has been used here for convenience. Although equation (2.56) is related to an energy criterion, it differs in that $\underline{\delta}_i^*$ is used instead of the actual iterative change $\eta_i \underline{\delta}_i$. Consequently, "premature convergence" should not be triggered by a small step-length.

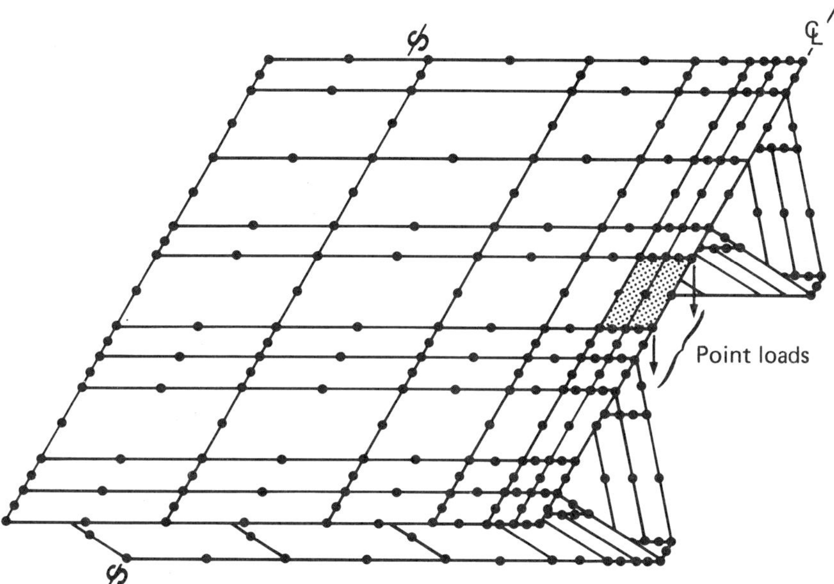

Fig. 3 Finite element idealisation for steel bridge-deck panel

Problem dimensions for folded-plate examples

N_c	N_f	N_t	Full system			$\underline{K}_{cc}$ only		
			B_a	B_m	M	B_a	B_m	M
660	535	1195	99	162	118,490	64	98	42,390

N_c = number of "coarse" variables; N_f = number of "fine variables"; N_t = total number of variables; B_a = average ½ band-width; B_m = maximum ½ band-width; M = number of matrix elements under "sky line".

Two different loadings were applied: firstly, a uniform pressure loading on the shaded patch of Figure 3 with a view to simulating a tyre loading and, secondly, two point loads applied in the same region. Using the scaled convergence criterion, μ of equation (2.56), the iterative responses were very similar (Table 4). For the patch pressure load, the 14 iterations required to reduce μ to 0.005 produced peak bending moments that were accurate (in relation to the full solution) to 3 significant figures while the equivalent axial stresses were accurate to 1–2 per cent. The iterative solution was marginally cheaper in computer time than the direct solution. However, a much more expensive solution was required (Table 4) for the point loadings when convergence was governed by an unscaled ϵ (equation (2.54)) of 0.01.

TABLE 4

Iterative performances for folded-plate examples

Loading	Start ϵ equation(2.54)	Start μ equation(2.56)	Number of iterations	End ϵ equation(2.54)	End μ equation (2.56)
Patch press.	0.18	0.05	14	0.01	0.005
Point loads	2.6	0.05	15	0.25	0.005
			35	0.05	0.001
			58	0.01	0.0002

At an earlier stage in the research, the bending action was also idealised in a hierarchical manner so that the mid-side variables included two relative rotations. The adopted bending element involved a modification [63] to the standard serendipity Mindlin formulation which eliminated "locking" and led to the passing of the "patch test". Unfortunately, the iterative performance was not as good as that already presented. The procedure was abandoned not only on this account but also because, for thin plates with straight sides, the eight-noded bending element produces displacements and bending moments that are only marginally better than those derived from the four-noded "discrete Kirchhoff" element (on an element for element basis). Nonetheless numerical experiments involving the eight-noded bending element on its own (without membrane action) indicated that an iterative solution could have possibilities especially for thick plates.

These experiments were conducted on a square clamped plate subject to a central point load. The coarse variables $\underline{p}_c$, related to the corner nodes (with w, θ_x and θ_y) while the "fine" variables, $\underline{p}_f$ related to the mid-side nodes (with $\Delta\theta_x$ and $\Delta\theta_y$). The problem dimensions and the iterative performance, which relates to a convergence factor, ϵ of (equation (2.54)) 0.01, are given in Table 5. The figures relate to a symmetric quarter of the slab. Beyond a 4 x 4 mesh, the required number of iterations reduces, assumedly because the "coarse variables", on their own, are getting closer to the exact solution. One might anticipate that, in the limit, no iterations would be required because no out-of-balance forces would be generated after the converged "coarse solution".

TABLE 5

Problem dimensions and iterative performance for a square clamped plate

					Full system			$\underline{K}_{cc}$ only		
n	N_{it}	N_c	N_f	N_t	B_a	B_m	M	B_a	B_m	M
2	9	12	8	20	9	15	180	4	8	36
4	51	40	51	96	22	34	2196	12	17	483
6	47	96	132	228	35	48	8084	18	23	1782
8	39	176	240	416	47	62	19860	24	29	4365

n = number of mesh divisions on each side for a symmetric quarter of the plate; N_{it} = required number of iterations. For other symbols see Table 3.

3. PART TWO: NON-LINEAR PROBLEMS

The earliest non-linear finite element work [64] adopted a set of linearised incremental solutions. Because these incremental solutions drift from the equilibrium curve, they were later supplemented by Newton-Raphson or modified Newton-Raphson iterations [61, 65–67]. The latter technique involves a fixed Jacobian. More recent work [18–29] has introduced a number of more mathematical iterative procedures [10–29]. However, these techniques have had to be specially adapted for structural problems. For example, quasi-Newton methods [10–12] destroyed sparsness and, although some progress was made in this area [17, 28], it was not until Matthies and Strang introduced a special vectorised version of the BFGS technique [18], that they were generally applied to structural problems [24–27]. Single-step restarted versions of these techniques were given by Crisfield [19–23] and labelled "secant-Newton methods". They can also be viewed as preconditioned conjugated gradient techniques with "inexact searches" and, as such, are related to recent mathematical papers by Shanno [14] and Buckley and Lenir [13]. The latter paper may contain useful ideas for improving the methods, particularly in relation to "re-starts" or "cut-outs". At the simplest level, the provision of "slack line searches" can significantly improve the robustness of even the basic modified Newton-Raphson procedure [68, 69].

The iterative solution procedure is usually only a part of an overall non-linear analysis because, for most structural problems, one wishes to trace the complete equilibrium path. This is particularly true of "path dependent problems" which involve plastic flow [61, 67]. In such circumstances one wishes to follow:

$$\underline{g}(\underline{p},\lambda) = \underline{f}(\underline{p}) - \lambda\,\underline{q} = \underline{0} \tag{3.1}$$

where $\underline{f}$ are the internal forces which are functions of the displacements $\underline{p}$ (and possibly the history), $\underline{q}$ is a fixed externally applied load vector and λ is the variable load level. While equation (3.1) implies proportional loading, other loadings can usually be represented using a piecewise proportional approach. In mathematical terminology, equation (3.1) represents a "continuation process" [70–74].

A related set of solution procedures involves embedding the solution to

$$\underline{g}(\underline{p}) = \underline{0} \tag{3.2}$$

(where λ is assumed fixed) in an extended equation

$$\underline{g}(\underline{p}, t) = \underline{g}(\underline{p}) - e^{-t}\underline{f}(\underline{p}_o) = \underline{0} \tag{3.3}$$

where t can be considered as a "pseudo time" parameter and as $t \to \infty$ one obtains the desired solution to equation (3.2). Such procedures are known as "homotopy methods" [75–79] and structural applications have been considered by Park [80] and Felippa [81]. They lead to solution algorithms which are akin to dynamic relaxation [6–9]. When Newton-like methods are adopted, problems are encountered if the Jacobian becomes singular. This problem has been tackled by Branin [76].

While equation (3.1) is clearly similar to equation (3.3), there are important differences. In particular λ is a real rather than a fictitious variable and, when applying equation (3.1), we generally aim to remain close to the true equilibrium path. Hence equations (3.1) are usually solved using a predictor-corrector method. The simplest predictor is the Euler tangent step

$$\Delta \underline{p} = -\Delta\lambda \frac{\partial \underline{g}}{\partial \underline{p}}^{-1} \underline{q} = -\Delta\lambda \underline{K}_T^{-1} \underline{q} = -\Delta\lambda \underline{\delta}_T \tag{3.4}$$

which was applied on its own in many early finite element solutions. However, such a procedure leads to a drift from equilibrium and equation (3.4) is usually coupled with either "modified Newton-Raphson iterations" or some of the more recent iterative procedures discussed above. An alternative approach would involve improving the predictor rather than the corrector but den Heijer and Rheinboldt argue that "high order predictions are very rarely effective" [71]. An important factor in the solution of equation (3.1) is the step-length or increment size, $\Delta\lambda$, adopted in equation (3.4). These steps should be chosen adaptively so as to reflect the current degree-of-non linearity [19, 22, 71–74].

3.1 Overcoming limit points

Some of the major problems with continuation methods relate to "limit points" at which the Jacobian, $\partial \underline{g} / \partial \underline{p}$, is singular. As will be shown in a later example, there may be a number of local maxima before the desired collapse load is reached. In structural terms, early solutions involved the introduction of artificial springs [82]. Alternatively, equilibrium iterations could be suppressed if one could predict the arrival of the limit point [73]. More recent work has concentrated on the "arc-length" method which was originally presented by Riks [83, 84] and Wempner [85]. The method is based on the observation that there may be no intersection of the equilibrium path of equation (3.1) with the plane "λ = constant" (the structure may have collapsed). As a consequence, Riks and Wempner aim to find the intersection of equation (3.1) with "s = constant" where s is the arc length defined by

$$s = \int ds \tag{3.5}$$

$$ds = \sqrt{d\underline{p}^T d\underline{p} + d\lambda^2 \beta^2 \underline{q}^T \underline{q}} \tag{3.6}$$

Related work in the mathematical literature is given in [77–79, 86, 87]. The scalar β is required in equation (3.6) because the load contribution depends on the adopted scaling between the load and displacement terms. Having introduced the arc-length s, one may attempt to directly solve

$$\underline{g}(s) = \underline{f}(\underline{p}(s)) - \lambda(s)\underline{q} = \underline{0} \tag{3.7}$$

using a higher order ODE method [86]. However, as with standard continuation techniques, it is very difficult to successfully limit the drift from equilibrium and, consequently, it may be better to adopt a predictor-corrector approach. To this end, one may replace equations (3.5) and (3.6) with an incremental constraint of the form

$$a = (\Delta\underline{p}^T \Delta\underline{p} + \Delta\lambda^2 \beta^2 \underline{q}^T \underline{q}) - \Delta\ell^2 = 0 \tag{3.8}$$

where $\Delta\ell$ is the fixed "radius" to the desired intersection (an approximation to the incremental arc-length). The vector $\Delta\underline{p}$ and scalar $\Delta\lambda$ are best thought of as incremental (not iterative) variables (see Figure 4).

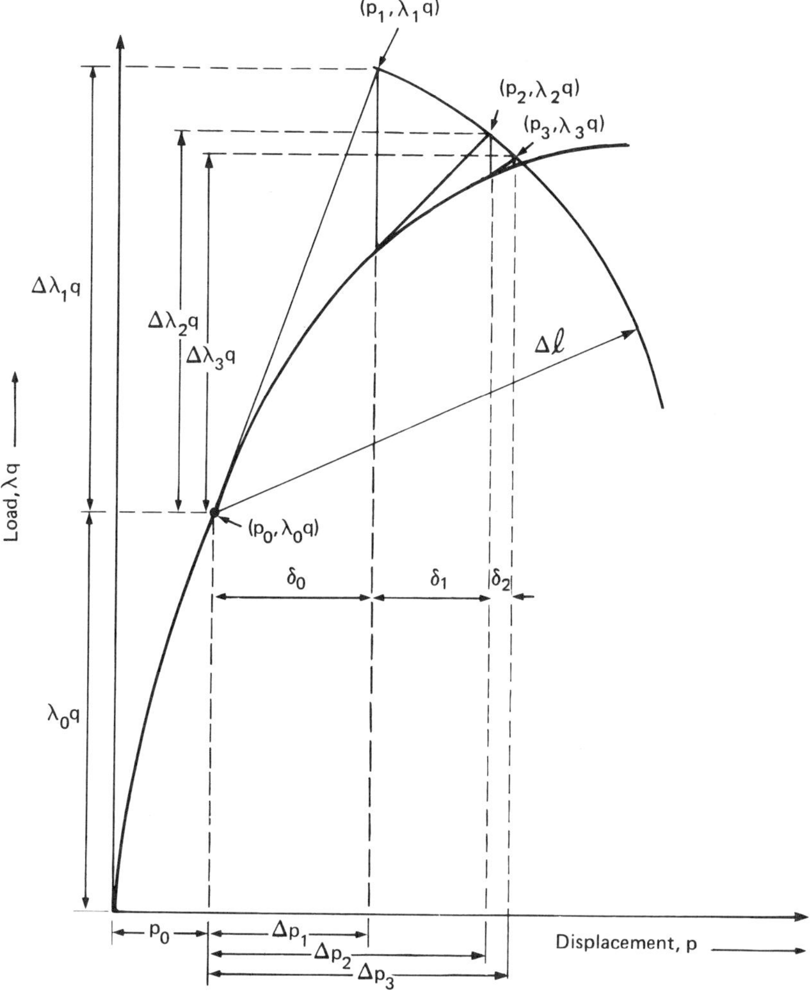

Fig. 4 Spherical arc-length procedure and notation for 1 degree-of-freedom system (with β=1)

If we combine equations (3.1) and (3.8) and apply Newton-Raphson iterations, we achieve

$$\left\{ \begin{array}{c} \underline{\delta} \\ \\ \delta\lambda \end{array} \right\} = - \begin{bmatrix} \dfrac{\partial \underline{g}}{\partial \underline{p}} & -\underline{q} \\ 2\Delta\underline{p}^T & 2\Delta\lambda\beta^2 \underline{q}^T \underline{q} \end{bmatrix}^{-1} \left\{ \begin{array}{c} \underline{g} \\ \\ a \end{array} \right\} \tag{3.9}$$

where $\underline{g}$, a and $\Delta\underline{p}$ are up-dated values from the previous iteration. In contrast to the Jacobian or stiffness matrix, $\partial\underline{g}/\partial\underline{p}$, the augmented Jacobian inside the square brackets of equation (3.9) will always be non-singular for sufficiently small increments.

If the scalar a (equation (3.8)) is assumed to be zero at the previous iteration, equation (3.9) ensures that the iterative change $(\underline{\delta}, \delta\lambda\beta\underline{q})$ is orthogonal to the incremental variables $(\Delta\underline{p}, \Delta\lambda\beta\underline{q})$ prior to the current iteration. Such a constraint has been proposed by Ramm [88, 89]. Riks and Wempner ensured orthogonality with the tangential values $(\Delta\underline{p}_1, \Delta\lambda_1 \beta \underline{q})$ (see Figure 5). In a rather different context, a similar approach was proposed earlier by Haselgrove [87].

Crisfield [19, 22, 90] directly applied the spherical constraint of equation (3.8) (Ramm also discussed this possibility [88, 89]). In order to explain how this is possible, one should firstly recognise a major limitation to the direct application of equation (3.9). For finite element applications, the Jacobian $\partial\underline{g}/\partial\underline{p}$ will generally be both symmetric and well banded while the augmented Jacobian equation (3.9) possess neither of these desirable properties. As a consequence, Crisfield [19, 22, 90] and Ramm [88, 89] adopted an "indirect solution procedure" which relates to that given by Batoz and Dhatt [91] for standard displacement-control. Using such an approach, one may specify the Newton change as:

$$\underline{\delta}_i = -\left.\frac{\partial \underline{g}}{\partial \underline{p}}\right|_i^{-1} \underline{g}(\lambda_{i+1}) = -\underline{K}_i^{-1}(\underline{f}_i(\underline{p}_i) - \lambda_{i+1}\underline{q}) = \underline{\bar{\delta}}_i + \lambda_{i+1}\underline{\delta}_{T,i} \tag{3.10}$$

where $$\underline{\bar{\delta}}_i = -\underline{K}_i^{-1}\underline{f}_i(\underline{p}_i) \tag{3.11}$$

and $$\underline{\delta}_{T,i} = \underline{K}_i^{-1}\underline{q} \tag{3.12}$$

At this stage, the load level, λ_{i+1}, is unknown but follows directly from the constraint. In particular, Ramm's linearised version [88, 89] gives

$$\lambda_{i+1} = \frac{-\Delta\underline{p}_i^T \underline{\bar{\delta}}_i + \lambda_i \Delta\lambda_i \beta^2 \underline{q}^T \underline{q}}{\Delta\underline{p}_i^T \underline{\delta}_{T,i} + \beta^2 \underline{q}^T \underline{q}\, \Delta\lambda_i} \tag{3.13}$$

where the symbol Δ indicates incremental (not iterative) values and, in relation to Figure 4,

$\underline{p}_o$ = displacement at end of previous increment, $\Delta\underline{p}_o = 0$

$\underline{f}_o(\underline{p}_o)$ = internal force vector at end of previous increment which, if there were perfect convergence at this load level, would be equal to $\lambda_o\underline{q}$, where λ_o is the load level at the end of the last increment.

For i = 0 onwards,

$$\lambda_{i+1} = \lambda_o + \Delta\lambda_{i+1} = \lambda_i + \delta\lambda_i \tag{3.14}$$

$$\underline{p}_{i+1} = \underline{p}_o + \Delta\underline{p}_{i+1} = \underline{p}_i + \eta_i\underline{\delta}_i \tag{3.15}$$

where

$$\Delta \underline{p}_{i+1} = \Delta \underline{p}_i + \eta_i \underline{\delta}_i \quad (3.16)$$

For present purposes, the step-length parameter, η_i, may be taken as unity.

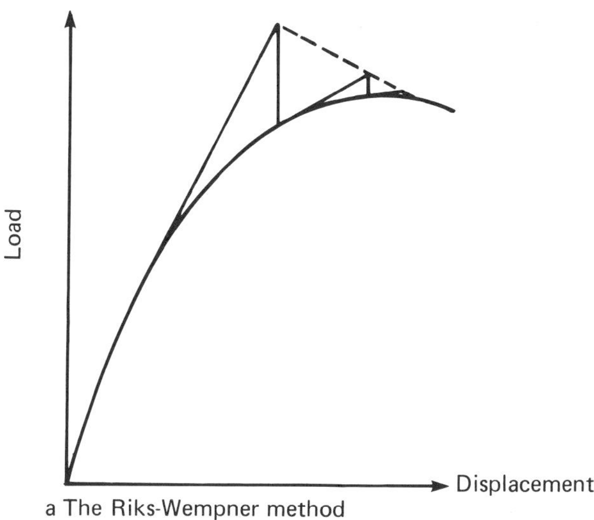

a The Riks-Wempner method

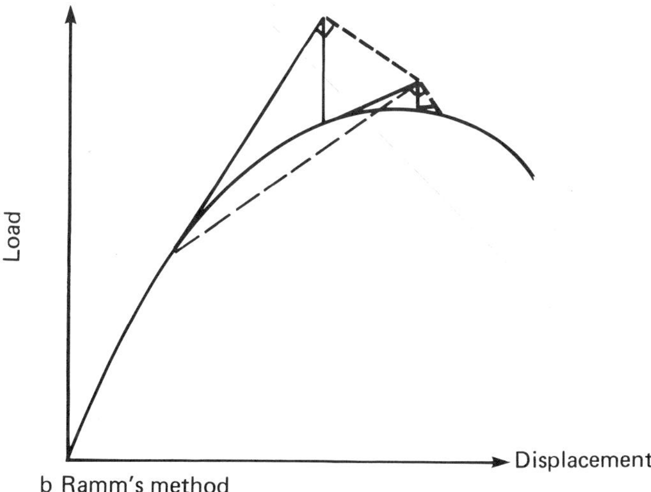

b Ramm's method

Fig. 5 Linearised arc-length methods

An "indirect" application of the Riks-Wempner method would introduce $\Delta \underline{p}_1$ and $\Delta\lambda_1$ in equation (3.13) in place of $\Delta \underline{p}_i$ and $\Delta\lambda_i$. With λ_{i+1} known, the iterative direction, $\underline{\delta}_i$ of equation (3.10) is totally defined. A theoretical disadvantage of the "indirect" procedure is that the Jacobian $\underline{K}_i$ in equation (3.10) may be singular in contrast to the augmented Jacobian of equation (3.9). However, this disadvantage is largely theoretical because, with many degrees-of-freedom, this situation does not occur in practise. It might be thought that problems would be caused by a near-singular Jacobian. However, this does not occur because the iterative deflection is constrained. In some senses, this constraint relates to that adopted with "trust regions" [53–55].

Using the "indirect approach", it is possible to apply the spherical constraint of equation (3.8) by substituting from equation (3.10). This leads to:

$$a_1\lambda^2_{i+1} + a_2\lambda_{i+1} + a_3 = 0 \qquad (3.17)$$

where

$$\left.\begin{aligned} a_1 &= \eta_i \underline{\delta}^T_{T,i}\, \underline{\delta}_{T,i} + \beta^2 \underline{q}^T \underline{q} \\ a_2 &= 2\underline{\delta}^T_{T,i}\, \Delta\underline{p}_i + 2\eta_i \underline{\delta}^T_{T,i}\, \bar{\underline{\delta}}_i - 2\lambda_{i-1}\beta^2 \underline{q}^T \underline{q} \\ a_3 &= \eta_i\, \bar{\underline{\delta}}_i^T \bar{\underline{\delta}}_i + 2\bar{\underline{\delta}}_i^T \Delta\underline{p} + \lambda_i\lambda_{i-1}\beta^2 \underline{q}^T \underline{q} \end{aligned}\right\} \qquad (3.18)$$

Equation (3.17) has two roots because the constraining sphere cuts the equilibrium path at two points. To avoid "doubling back", one should choose the root giving the largest scaled inner product $\Delta\underline{p}_i^T \Delta\underline{p}_{i+1} / (\|\Delta\underline{p}_i\| \,.\, \|\Delta\underline{p}_{i+1}\|)$ [90].

The arc-length method is best embedded in an automatic incremental (or continuation) technique whereby the radius, $\Delta\ell$ of equation (3.8) is chosen to reflect the degree of non-linearity. For example, one may choose [85, 86]

$$\Delta\ell_j = \sqrt{\frac{I_d}{I_{j-1}}}\ \Delta\ell_{j-1} \qquad (3.19)$$

where $\Delta\ell_{j-1}$ is the incremental length at the j–1 th increment, I_{j-1} is the number of iterations required at the j–1 increment and I_d is the desired number of iterations (say 4). This procedure can be applied for all increments except the first where an initial value of $\Delta\lambda_1$ will be provided by the user and from equation (3.8),

$$\Delta\ell = \sqrt{\Delta\underline{p}_1^T \Delta\underline{p}_1 + \Delta\lambda_1^2 \beta_1^2 \underline{q}^T \underline{q}} = \Delta\lambda_1 \sqrt{\underline{\delta}^T_{T,o}\, \underline{\delta}_{T,o} + \beta^2 \underline{q}^T \underline{q}} \qquad (3.20)$$

For subsequent increments, the desired "radius", $\Delta\ell$, will be known from equation (3.19) while the tangential incremental load value, $\Delta\lambda$, can be obtained from equation (3.8), so that

$$\Delta\lambda_i = \lambda_i - \lambda_o = \frac{b.\,\Delta\ell}{\sqrt{\underline{\delta}^T_{T,o}\, \underline{\delta}_{T,o} + \beta^2 \underline{q}^T \underline{q}}} \qquad (3.21)$$

where $b = \text{sign}\,(r)$ (3.22)

and r is either $r = \text{Det}\,(\underline{K})$ (3.23)

or $r = \underline{\delta}_{T,o}^T \underline{q} = \underline{\delta}_{T,o}^T \underline{K}_o \underline{\delta}_{T,o}$ (3.24)

The latter form has shown advantages for special problems involving "material instability" [69] but the former is generally more reliable.

The arc-length method has been explained in terms of internal forces, f, and total load levels, λ. In previous work [19, 22, 90] the author referred to out-of-balance forces, $\underline{g}$ and changes in load level, δλ. It is better to program in terms of the latter so that machine rounding errors can be reduced. It is easy to change from one formulation to the other, using equation (3.1).

3.2 *Modifying the arc-length method*

Perhaps the most obvious modification involves replacing $\underline{K}_i = \partial \underline{g}/\partial \underline{p}\,|_i$ with $\underline{K}_o = \partial \underline{g}/\partial \underline{p}\,|_o$, the tangent stiffness matrix at the beginning of the increment. This approximation leads to the modified Newton-Raphson method. Although this solution procedure has a slower convergence rate than the full Newton-Raphson method (particularly in the later iterations), it has the following advantages:

(i) The Jacobian need neither be reformed nor refactorised during the iterations.
(ii) The vector $\underline{\delta}_{T,i}$ of equation (3.12) becomes $\underline{\delta}_{T,o} = \underline{K}_o^{-1} \underline{q}$ which is the fixed tangent vector. Consequently, at each iteration, only $\bar{\underline{\delta}}_i$ (equation (3.11)) need be formed.
(iii) If the previous equilibrium state is in stable equilibrium, $\underline{K}_o$ will be positive definite.

The latter advantage is relevant if "line searches" are added to the arc-length method [93–95]. Such "line searches" involve a form of "damping" whereby a scalar "step-length", η_i, pre-multiplies the iterative vector, $\underline{\delta}_i$ (see equations (3.15) and (3.16)). Clasically, the "optimum" step-length is found by forcing the total potential energy to be a minimum in the direction η_i. For a stationary total potential energy,

$$\left.\frac{\partial \Phi}{\partial \eta_i}\right|_{i+1} = \left.\frac{\partial \Phi}{\partial \underline{p}}\right|_{i+1}^T \frac{\partial \underline{p}}{\partial \eta_i} = \underline{g}_{i+1} (\underline{p}_{i+1}(\eta_i))^T \underline{\delta}_i = s_j(\eta_{i,j}) = 0 \quad (3.25)$$

where $\underline{g}_{i+1}$ is the out-of-balance force vector or gradient of the total potential energy at η_i. A process of trial-and-error is usually adopted that involves interpolation and (safeguarded) extrapolation [51, 52]. A new residual or out-of-balance force vector is required at each "trial", j. Exact searches are not required, and instead, one usually aims to satisfy [51, 52, 93]

$$|\, s_j(\eta_{i,j})\,| < \psi \,|\, s_o(\eta_{i,o} = 0)\,| \quad (3.26)$$

The line-search concept can be applied even if there is no governing total potential energy. For example, plasticity or the cracking of concrete introduce a path-dependent material response with very different stress-strain relationships for "loading" and "unloading". The total potential energy may nonetheless be useful as a conceptual aid and line-searches have been found to be very beneficial for problems involving load or displacement control [68, 69].

Line-searches cannot be applied in a straightforward manner to the arc-length method because the load level varies and the iterative vector $\underline{\delta}_i$ (equation (3.10)) changes direction as the search is applied. The objective is to make the total potential energy stationary at the new load level (unknown at the start) at the end of the iteration, ie

$$\left.\frac{\partial \Phi}{\partial \eta_i}\right|_{i+1} = \underline{\delta}_i^T \underline{g}_{i+1} = (\bar{\underline{\delta}}_i + \lambda_{i+1} \underline{\delta}_T)^T (\underline{f}_{i+1}(\eta_i) - \lambda_{i+1} \underline{q})$$
$$= s_j = (\bar{\underline{\delta}}_i + \lambda_{i+1} \underline{\delta}_T)^T (\underline{f}_{i+1,j} - \lambda_{i+1} \underline{q}) = 0 \quad (3.27)$$

where $\underline{f}_{i+1,j}$ is the internal force vector at the jth estimate of η_i (ie $\eta_{i,j}$). We cannot simply apply an interpolation of the form:

$$\frac{\eta_{i,j+1}}{\eta_{i,j}} = \frac{-s_o}{s_j(\eta_{i,j}) - s_o} \tag{3.28}$$

because the resulting step-length will lead to an incremental displacement, $\Delta \underline{p}_{i+1}$ (from equation (3.16)) that violates the constraint of equation (3.8). Hence we must iterate between equations (3.8) and (3.28). Fortunately it is possible to derive an algorithm that involves previously stored "inner products" so that this iteration is very cheap and only one residual computation is required for each trial step-length, $\eta_{i,j}$ [93, 94].

It can be argued that classical line-searches are inappropriate for the arc-length method because the procedure is designed to operate with both stable and unstable equilibrium states. This latter equilibrium state will not coincide with a minimum energy configuration. However, if the modified Newton-Raphson procedure is adopted, the author has found that the iterative direction is almost invariably downhill whether the tangent stiffness matrix, $\underline{K}_o$, is positive definite or indefinite [93–95]. It appears that most uphill movement will occur in the first incremental solution, $\Delta \underline{p}_1 = \underline{\delta}_o = \Delta\lambda_o \underline{\delta}_{T,o}$ because this direction will be dominated by the lowest eigenvector of $\underline{K}_o$. The arc-length procedure ensures that subsequent iterative vectors are almost orthogonal to $\underline{\delta}_{T,o}$ and this accounts for the consequent downhill directions [93, 94].

These considerations do not apply if the full Newton-Raphson method is used and $\underline{K}_i$ is up-dated during the increment. In such circumstances, one may consider choosing a step-length, η_i, to minimise the norm of the out-of-balance forces $\| g_{i+1}(\eta_i) \|$ rather than the total potential energy [54]. Bergan [92] proposed a similar procedure for structural problems but did not couple it with a "length constraint". Indeed, variations were performed with respect to the load level, λ_i, rather than the step-length, η_i. Without the constraint, such a procedure is not iterating towards a unique point on the equilibrium curve (as would be predetermined by the intersection with the constraint). In addition, the iterations will become unbounded as the Jacobian approaches singularity. Because the constraint limits the growth of the iterative vector, one may consider the step-length, η_i, as a sophistication and consequently, the author simply abandons "line-searches" (setting η_i to unity) whenever the iterative direction is uphill (or insufficiently downhill) [95].

The modified Newton-Raphson technique is adequate for many problems (when coupled with the arc-length method). However, severe numerical difficulties can be encountered for problems involving material "unloading" [69, 93–95] especially when strain-softening is involved [69, 93–95]. In such circumstances, it can be very beneficial to up-date the Jacobian within the increment [93–95, 97]. A very low step-length (say $\eta_i < 0.1$) may be used to trigger such a computation [93–95]. However, much further work is required before a general solution strategy can be defined.

In rare circumstances, equation (3.17) will provide no real roots. The obvious solution is to reduce the length, $\Delta\ell$, because, for sufficiently small $\Delta\ell$, the spherical arc-length method will converge (quadratically [98] if the full Newton-Raphson method is adopted). If the modified Newton-Raphson method is being used with line-searches, one may abandon these searches and, instead, introduce a damping η_i into equations (3.15) and (3.16) with a view to the provision of real roots [93]. A new Jacobian may be tried instead [95]. A different solution has recently been advocated by Bathe and Dvorkin [99] who propose an alternative constraint involving a "fixed increment of external work". In particular, for the incremental (predictor) solution at $i = 0$, they propose:

$$(\lambda_o + \tfrac{1}{2}\Delta\lambda_1)\, \underline{q}^T\, \Delta\underline{p} = \Delta\lambda_1 (\lambda_o + \tfrac{1}{2}\Delta\lambda_1)\, \underline{q}^T \underline{K}_o^{-1}\, \underline{q}$$

$$= \Delta\lambda_1 (\lambda_o + \tfrac{1}{2}\Delta\lambda_1)\, \underline{q}^T \underline{\delta}_T = \Delta W \qquad (3.29)$$

where ΔW is the increment of external work which is held fixed over the increment. The scalar ΔW takes the place of $\Delta \ell$ in the spherical arc-length method. For $i \geqslant 0$, Bath and Dvorkin force the iterative external work to be zero so that

$$\tfrac{1}{2}(\lambda_i + \lambda_{i+1})\, \underline{q}^T \underline{\delta}_i = \tfrac{1}{2}(\lambda_i + \lambda_{i+1})\, \underline{q}^T (\bar{\underline{\delta}}_i + \lambda_{i+1}\, \underline{\delta}_T) = 0 \qquad (3.30)$$

which leads to a quadratic relationship for λ_{i+1} of the same form as equation (3.17) although, of course, the scalars a_1–a_3 are no longer given by equation (3.18). In contrast to equation (3.8), the constraint of equation (3.30) has the advantage that it always leads to real roots for this quadratic. Consequently, Bathe and Dvorkin advocate switching from the spherical arc-length constraint to the constant increment of external work constraint whenever the former leads to no real roots.

Bathe and Dvorkin also advocate switching to the "work constraint" whenever the solution is close to the limit point, since they argue that this procedure out-performs the arc-length method in such regions. Particularly for problems involving the cracking of reinforced concrete structures, the author's experience is that such limit points are extremely difficult to anticipate so that such a switching procedure may be difficult to implement.

An alternative modification involves "warping" the constraint surface to reflect the state of non-linearity. Padovan et al [97, 100] have proposed that the scaling parameter, in equation (3.8) should be adjusted to be proportional to Bergan's "current stiffness parameter" [73, 74]. Consequently, the stiffer the system, the more closely would the solution procedure approach load-control. Crisfield [19, 22, 90] and Ramm [88, 89] have avoided scaling problems by setting β to zero.

3.3 Applications of the arc-length method

The author has applied the arc-length method to a wide range of problems [19, 22, 90, 93–95]. A brief summary will be given here. The first example (Figure 6), which involves the analysis of stiffened diaphragm from a steel box-girder bridge, triggered the author's interest in the technique. With a view to bracketing the collapse load, different solutions were obtained with different boundary conditions [90]. When simply supported boundaries were applied, the author found it impossible to obtain a solution using conventional techniques even when displacement control was applied at the bearings. The reason related to an unexpected "snap-back" (Figure 6). The problem was overcome when the spherical arc-length method was applied although very small increments were required in the "collapse region".

The latter problem involved both geometric and material non-linearity. Simpler problems involve elastic "snap through" and "snap-back". Table 6, which is from reference [94], relates to the snap-through of a shallow elastic shell and illustrates the advantages of "slack line searches" (for increments 4 and 5). More significant advantages were obtained for problems involving material non-linearity as illustrated in Table 7, which relates to the analysis of a reinforced concrete slab. For this problem it was also necessary, on two increments, to recompute the Jacobian within the increment.

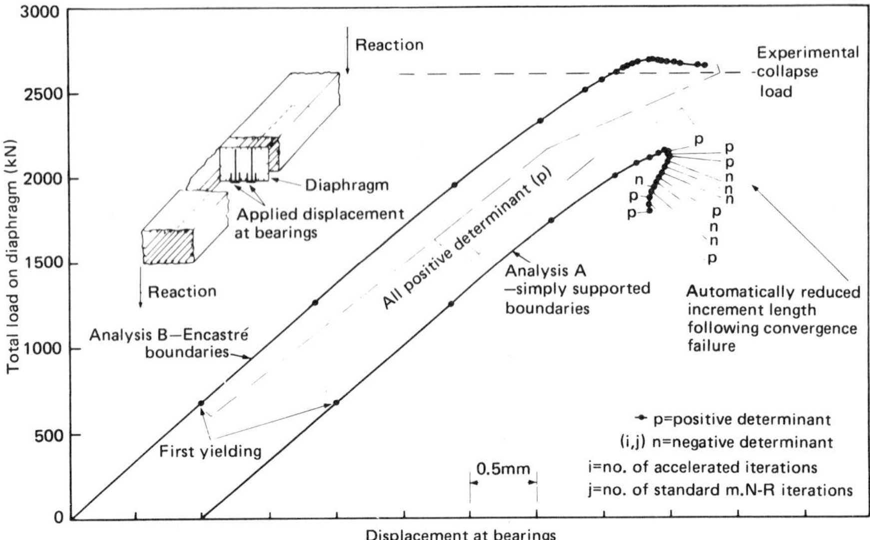

Fig. 6 Stiffened diaphragm from steel box-girder (from [90])

TABLE 6

Iterative performance for a hinged-shell (from reference [94])

Increment number, I	1	2	3	4	5
No line searches	5 (0)	4 (0)	9 (0)	22 (0)	50*
With line searches ψ(equation (3.26)) = 0.7	5 (0)	4 (0)	8 (1)	7 (2)	5 (2)
r, equation (3.24) at increment I + 1	+1	−1	−1	+1	+1

The table gives I(J) where I = number of iterations, J = number of extra residual calculations (per increment)
* Convergence not achieved and further iterations abandoned.

Finally, Figure 7 shows that many local limit points can be encountered before the final collapse load is reached. The example relates to a prestressed concrete T-beam that is described in [95]. The symbol "S" in Figure 7 indicates that a crack is shutting at a "Gauss point" while "0" indicates that it is fully open. The symbols "C" and "D" relate to different positions on the "stress-strain" curves. Full details are given in [95].

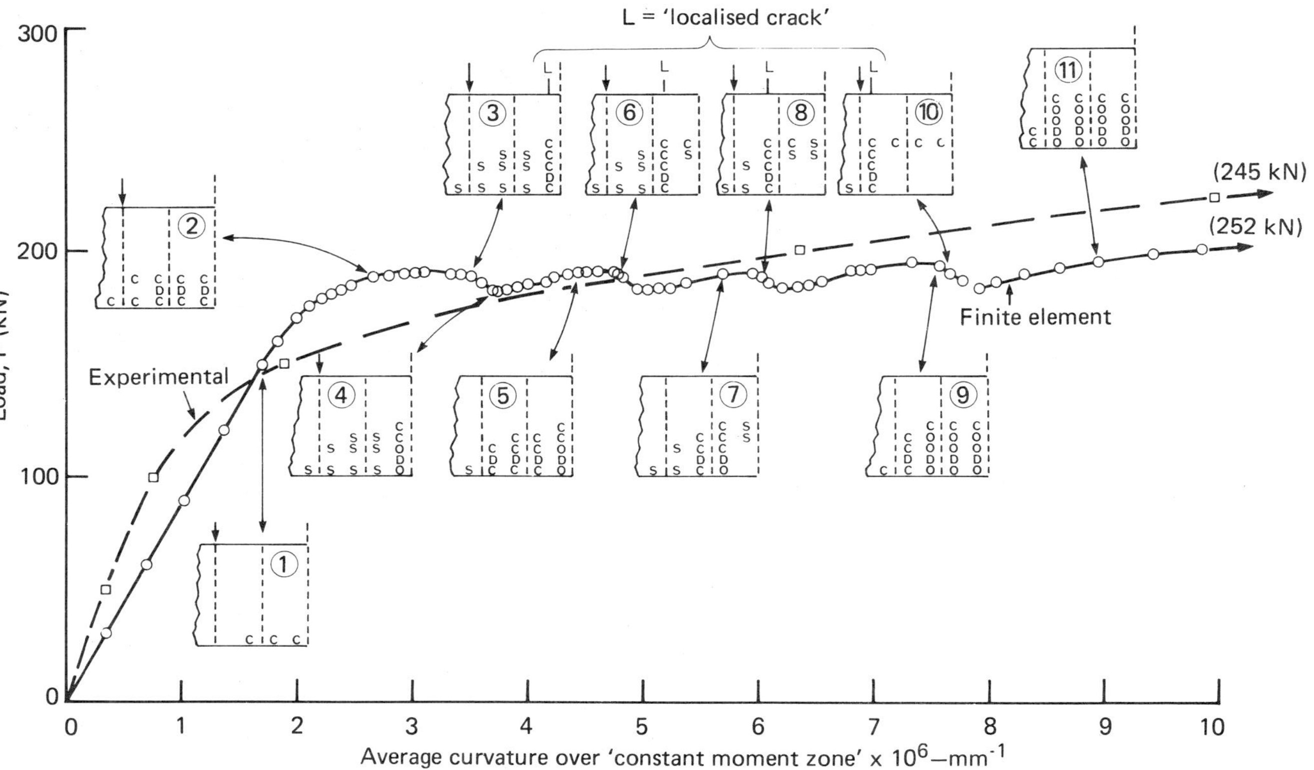

Fig. 7 Response of prestressed T-beam (from [95])

TABLE 7

Iterative performances for a reinforced concrete slab

Increment number	Line searches? No	Line searches? Yes $\psi = 0.8$
1	4	4 (0)
2	15	15 (0)
3	50‡	6 (2)
4	5	5 (0)
5	19	12 (3)
6	50‡	6 (3)§
7	50‡	11 (29)§
8	4	4 (0)
9	19	11 (2)
10	1	1 (0)
11	9	7 (2)
12	18	12 (3)
13	1	1 (0)
14	1	1 (0)

The table gives I(J), where I = number of iterations,
J = number of extra residual calculations.

‡ Convergence not achieved and further iterations abandoned.
§ One new stiffness matrix formed within the increment.

4. CONCLUSIONS

Excessive storage demands can prohibit the solution of large structural problems unless iterative methods are applied. Such techniques should be well-conditioned and robust. Two of the more important methods, which can be combined involve "incomplete Cholesky factorisation" and "hierarchical preconditioning". In this paper, the latter procedure has been applied, in conjunction with a form of conjugate gradient algorithm. Very encouraging results were obtained for two-dimensional plane stress analyses and it is anticipated that the solution procedure will become an economic proposition especially for large three-dimensional continuum problems. In comparison with other iterative methods, the technique possesses the major advantage that, beyond a certain stage, the required number of iterations does not increase as the mesh is refined.

The technique has also been applied to plate bending and folded-plated problems. The latter involve a combination of membrane and bending action and, consequently, are less amenable to iterative solution. Nonetheless, the preliminary results justify further research. Ideally, hierarchically based iterative solution procedures should be combined with adaptive idealisation techniques [44–49] whereby the relative magnitude of the out-of-balance forces after the "coarse solution" may be used to help decide which higher-order variables to add. Hierarchical iterative methods can be applied to non-linear problems.

The paper has reviewed a number of alternative solution procedures for non-linear structural analysis and has attempted to draw on both the engineering and mathematical literatures. Emphasis has been placed on the arc-length procedure and it has been shown how "line-searches" can be beneficially added to this technique. The problem of "uphill energy directions" has been discussed and it has been speculated that it may be better to minimise the out-of-balance forces rather than the total potential energy. It has been shown that, for many problems, significant advantages can be obtained by up-dating the Jacobian within an increment. It is not always necessary to go to a full Newton-Raphson procedure and further work is required on an automatic mechanism to trigger such up-dating.

5. ACKNOWLEDGEMENTS

The work described in this paper forms part of the programme of the Transport and Road Research Laboratory and is published by permission of the Director.

6. REFERENCES

1. BATHE, K-J. and WILSON, E.L., *Numerical methods in finite element analysis,* Prentice-Hall, London, (1976).
2. JENNINGS, A., *Matrix computation for engineers and scientists,* John Wiley, London, (1977).
3. HARWISS, T., Large analysis in A.S. Computas. *Finite Element News,* 6–9, (Sept. 1979).
4. HUGHES, T.J.R., LEVIT, I. and WINGET, J., An element-by-element algorithm for problems of structural and solid mechanics. *Comp. Meth. in Appl. Mech. and Engng.,* 36, 241–254, (1983).
5. HITCHINGS, D., A finite element system for in house use. *Finite Element News,* 6, 26–27, (1983).
6. CASSELL, A.C., Shells of revolution under arbitrary loading and the use of fictitious densities in dynamic relaxation. *Proc. Instn. Civ. Engrs.,* 45, 65–78, (1970).
7. FRANKEL, S.P., Convergence rates of iterative treatments for partial differential equations. *Maths. Tables Aids Comp.,* 4, 65–75, (1950).
8. OTTER, J.R.H. and DAY, A.S., Tidal computations. *The Engineer,* 209, 177–182, (1960).
9. UNDERWOOD, P.G., Dynamic relaxation – a review, Ch. 5 of T. Belytschko and T.J.R. Hughes (Eds.), *Computational methods for transient dynamic analysis,* North-Holland, Amsterdam, (1983).
10. BROYDEN, C.G., The convergence of a double-rank minimisation 2: the new algorithm. *J. Inst. Math. Appl.,* 6, 222–231, (1970).
11. DENNIS, J. Jr. and MORE, J., Quasi-Newton methods, motivation and theory. *SIAM Review,* 19, 46–84, (1977).
12. FLETCHER, R., A new approach to variable metric algorithms. *Computer J.,* 13, 317–322, (1970).
13. BUCKLEY, A. and LENIR, A., QN-like variable storage conjugate gradients. *Mathematical programming,* 27, 155–175, (1983).
14. SHANNO, D.F., Conjugate gradient methods with inexact searches. *Maths. of O.R.,* 244–256, (1978).
15. NAZARETH, L., A relationship between the BFGS and conjugate gradient algorithms and its implications for new algorithms. *SIAM J. of Numerical Analysis,* 16, 794–800, (1979).
16. NOCEDAL, J., Up-dating Quasi-Newton matrices with limited storage. *Math. of Comp.,* 35, 773–782, (1980).
17. TOINT, P.L., On sparse and symmetric matrix up-dating subject to a linear equation. *Math. Comp.,* 31, 954–961, (1977).
18. MATTHIES, H. and STRANG, G., The solution of non-linear finite element equations. *Int. J. Num. Meth. in Engng.,* 14, 1613–1626, (1979).

19. CRISFIELD, M.A., Incremental/iterative solution procedures for non-linear structural problems, pp. 261–290 of C. Taylor et al (Eds.), *Numerical methods for non-linear problems.* Pineridge, Swansea, (1980).
20. CRISFIELD, M.A., A faster modified Newton-Raphson iteration. *Comp. Meth. in Appl. Mech. and Engng.,* 20, 267–278, (1980).
21. CRISFIELD, M.A., Iterative solution procedures for linear and non-linear structural analysis. *Transport and Road Research Laboratory Report LR 900,* Crowthorne, Berkshire, (1979).
22. CRISFIELD, M.A., Solution procedures for non-linear structural problems, pp. 1–39 of E. Hinton et al (Eds.), *Recent advances in non-linear computational mechanics.* Pineridge, Swansea, (1982).
23. CRISFIELD, M.A., Alternative methods derived from the BFGS formula. *Int. J. Num. Meth. in Engng.,* 15, 1419–1420, (1980).
24. GERADIN, M., IDELSOHN, S. and HOGGE, M., Computational strategies for the solution of large non-linear problems via quasi-Newton methods. *Computers and Structures,* 13, 73–82, (1981).
25. BATHE, K.J. and CIMENTO, A.P., Some practical procedures for the solution of non-linear finite element equations. *Comp. Meth. in Appl. Mech. and Engng.,* 22, 59–86, (1980).
26. ABDEL RAHMAN, H.H., HINTON, E. and HUQ, M.M., Non-linear finite element analysis of reinforced concrete slab and slab-beam structures. pp. 493–499 of C. Taylor et al (Eds.),*Numerical methods for non-linear problems,* Pineridge, Swansea, (1980).
27. PICA, A. and HINTON, E., The quasi-Newton BFGS method in large deflection analysis of plates. pp 355–366 of C. Taylor et al (Eds.), *Numerical methods for non-linear problems.* Pineridge, Swansea, (1980).
28. KAMAT, M.P., VANDERBRINL, D.E. and WATSON, L.T., Non-linear structural analysis using quasi-Newton minimisation algorithms that exploit sparsity. pp. 1063–1075 in C. Taylor et al (Eds.),*Numerical methods for non-linear problems.* Pineridge, Swansea, (1980).
29. ELSAWAF, A.F. and IRONS, B.M., Equation solving: a retrospect. pp. 269–274 of E. Hinton et al (Eds.), *Non-linear computational mechanics.* Pineridge, Swansea, (1982).
30. FLETCHER, R. and REEVES, C.M., Function minimisation by conjugate gradients. *Computer J.,* 7, 149–254, (1964).
31. HESTENES, M. and STEIFEL, E., Method of conjugate gradients for solving linear systems. *J. of Res. of Nat. Bur. of Standards,* 49, 409–436, (1952).
32. POLAK, E. and RIBIERE, G., Note sur al convergence de methodes de directions conjugees, *Revue Francaise Inform. Rech. Operation,* 16–R1, 35–43, (1969).
33. POWELL, M.J.D., Restart procedures for the conjugate gradient method. *Computer Science and Systems Div. Report CSS 24,* AERE, Harwell, Oxford, England, (1975).
34. GUSTAFSSON, I., Stability and rate of convergence of modified incomplete Cholesky factorisation methods, *Research Report 79.02R,* Chalmers University of Technology, Gotegorg, Sweden, (1979).
35. AJIZ, M.A. and JENNINGS, A., A robust incomplete Choleski-conjugate gradient algorithm. *Int. J. for Num. Meth. in Engng.,* to be published.
36. JENNINGS, A. and MALIK, G.M., The solution of linear equations by the conjugate gradient method. *Int. J. Num. Meth. in Engng.,* 12, pp 141–158, (1978).
37. JENNINGS, A., Development of an ICCG algorithm for large sparse systems. pp 426–438 of D.J. Evans (Ed.),Preconditioning methods, Analysis and Applications. Gordon Breach, New York, (1983).
38. JENNINGS, A. and MALIK, G.M., Partial elimination, *J. Inst. Maths. Applics.,* 20, 307–316, (1977).

39. KERSHAW, D., The incomplete Cholesky conjugate gradient method for the iterative solution of systems of linear equations, *J. Comput. Phys.*, 26, pp 43–65, (1978).
40. MEIJERINK, J.A. and VAN DER VORST, H.A., An iterative solution method for linear systems of which the coefficient matrix is a symmetric M-matrix, *Math. Comp.*, 31, pp 148–162, (1977).
41. WILSON, E.L., Special numerical and computer techniques for the analysis of finite element systems. pp 3–25 of K.J. Bathe et al (Eds.), *Formulations and computational algorithms in finite element analysis.* MIT, (1977).
42. AXELSSON, O. and GUSTAFSSON, I., Preconditioning and two-level multigrid methods for arbitrary degree of approximation. *Math. of Computation,* 40, 219–242, (1983).
43. HUGHES, T.G.R., LEVIT, I. and WINGET, J., Unconditionally stable element-by-element implicit algorithms for heat conduction analysis. *ASCE, J. Engng. Mech. Div,* 109, 576–585, (1983).
44. PEANO, A. and RICCIONI, R., Automated discretisation error control in finite element analysis, pp 368–387 of J. Robinson (Ed.), *Finite elements in the commercial environment.* Robinson and Assoc., Verwood, Dorset, (1978).
45. PEANO, A., Hierarchies of conforming finite elements for plane elasticity and plate bending. *Comp. and Maths. with Appls.,* 2, 211–224, (1976).
46. PEANO, A., FANELII, R., RICCIONE, R. and SARDELLA, L., Self adaptive convergence at the crack tip of a dam buttress. pp 266–280 of A.R. Luxmore et al (Eds.), Numerical methods in fracture mechanics. Pineridge, Swansea, (1977).
47. KELLY, D.W., GAGO, J.P. de S.R., ZIENKIEWICZ, O.C. and BABUSKA, I., A posteriori error analysis and adaptive processes in the finite element method: Part 1 – Error analysis. *Int. J. Num. Meth. in Engng.,* 19, 1593–1620, (1983).
48. GAGO, J.P. de S.R., KELLY, D.W., ZIENKIEWICZ, O.C. and BABUSKA, I., Error analysis and adaptive processes in the finite element method: Part 2 – Adaptive refinement. *Int. J. for Num. Meth. in Engng.,* 19, 1621–1656, (1983).
49. ZIENKIEWICZ, O.C., The generalised finite element method – state of the art and future directions. *Report C/R/445/83,* Dept. of Civil Engng., University of Wales, Swansea, (April 1983).
50. MARCHUK, G.I., *Methods of numerical mathematics.* Applications of Mathematics, 2, Trans. by J. Ruzicka, Springer-Verlag, New York, (1975).
51. WOLFE, M.A., *Numerical methods for unconstrained optimisation – an introduction.* Van Nostran Rheinboldt Co., London, (1975).
52. GILL, P.E. and MURRAY, W., Safeguarded step-length algorithms for optimisation using descent methods. *Nat. Phys. Lab. Rep. NAL 37,* Teddington, England, (1974).
53. SORENSEN, D.C., Trust region methods for unconstrained minimisation. pp 29–38 of M.J.D. Powell (Ed.), *Non-linear optimisation.* Academic Press, London, (1981).
54. POWELL, M.J.D. A hybrid method for non-linear equations. pp 87–114 of P. Rabinowitz (Ed.), *Numerical methods for non-linear algebraic equations.* Gordon and Breach, London, (1970).
55. GOLDFELD, S.M., QUANDT, R.E. and TROTTER, H.F., Maximisation by quadratic hill-climbing. *Econometrica,* 34, 541–551, (1966).
56. AXELSSON, O., A class of iterative methods for finite element equations, *Comp. Meth. in Appl. Mech. and Engng.,* 9, 123–127, (1976).
57. MOTE, C.D. Jr., Global-local finite element. *Int. J. for Num. Meth. in Engng.,* 3, 565–574, (1971).
58. CRISFIELD, M.A. and PUTHLI, R.S., Approximations in the non-linear analysis of plated structures, pp 373–392 of P.G. Bergan et al (Eds.), *Finite elements in non-linear mechanics,* Tapir, Trondheim, (1978).

59. CRISFIELD, M.A., A combined Rayleigh-Ritz/finite element method for the non-linear analysis of stiffened plated structures. *Computers and Structures,* 8, 679–689, (1978).
60. WIBERG, N.E., Finite element method: the large system of equations and its numerical solution. *Dept. of Struct. Mech. Report 75:4,* Chalmers University of Technology, Goteborg, Sweden, (1975).
61. ZIENKIEWICZ, O.C., *The finite element method.* McGraw-Hill, London, (1977).
62. CRISFIELD, M.A., A four-noded thin-plate bending element using shear constraints – a modified version of Lyons' element. *Comp. Meth. in Appl. Mech. & Engng.,* 38, 93–120, (1983).
63. CRISFIELD, M.A., A quadratic Mindlin element using shear constraints. *Computers and Structures,* 18, 833–852, (1984).
64. TURNER, M.J., DILL, E.H., MARTIN, H.C. and MELOSH, R.J., Large deflection of structures subject to heating and external loads. *J. of Aero. Sci.,* 27, 97–106, (1960).
65. MURRAY, D.W. and WILSON, E.L., Finite element large deflection analysis of plates. *Proc. ASCE, J. of Engng. Mech. Div.,* 95, 143–165, (1969).
66. BREBBIA, C. and CONNOR, J., Geometrically non-linear finite element analysis. *Proc. ASCE, J. of Engng. Mech. Div.,* 95, 463–483, (1969).
67. OWEN, D.R.J. and HINTON, E., *Finite elements in plasticity – Theory and practice.* Pineridge, Swansea, (1980).
68. CRISFIELD, M.A., Accelerating and damping the modified Newton-Raphson method. *Computers and Structures,* 18, 395–407, (1984).
69. CRISFIELD, M.A., Accelerated solution techniques and concrete cracking. *Comp. Meth. in Appl. Mech. and Engng.,* 33, 585–607, (1982).
70. RHEINBOLDT, W.C., Numerical continuation methods for finite element applications, pp 600–631 of K-J Bathe et al (Eds.), *Formulations and computational algorithms in finite element analysis,* MIT, (1977).
71. DEN HEIJER, C. and RHEINBOLDT, W.C., On steplength algorithms for a class of continuation methods. *SIAM J. Num. Analysis,* Vol. 18, 1981, pp 925–948.
72. SCHMIDT, W.F., Adaptive step size selection for use with the continuation method. *Int. J. Num. Meth. in Engng.,* 12, 677–694, (1978).
73. BERGAN, P.G. and SOREIDE, T., Solution of large displacement and instability problems using the current stiffness parameter, pp 647–669 of P.G. Bergan et al (Eds.), *Finite elements in non-linear mechanics.* Tapir, Trondheim, (1978).
74. BERGAN, P.G., Automated incremental-iterative solution methods in structural mechanics. pp 41–62 of E. Hinton et al (Eds.), *Non-linear computational mechanics.* Pineridge, Swansea, (1982).
75. DAVIDENKO, D.F., On a new method of numerical solution of systems of non-linear equations. *Dokl. Akad. Nauk. SSSR,* 88, 601–602, (1953).
76. BRANIN, F.H. Jr., Widely convergent method for finding multiple solutions of simultaneous non-linear equations. *IBM J. Res. Dev.,* 16, 1972, 504–522.
77. KELLER, H.B., Global homotopies and Newton methods. Recent advances in numerical analysis. pp 73–94 of C. de Boor et al (Eds.), *Recent advances in numerical analysis.* Academic Press, New York, (1978).
78. GEORG, K., Numerical integration of the Davidenko equation. pp 129–161 of E.L. Allgower et al (Eds.), *Lecture notes in mathematics – Numerical solution of non-linear equations.* Springer-Verlag, Berlin, (1981).
79. ALLGOWER, E.L., A survey of homotopy methods for smooth mappings. pp 2–29 of E.L. Allgower et al (Eds.), *Lecture notes in mathematics – Numerical solution of non-linear equations.* Springer-Verlag, Berlin, (1981).
80. PARK, K.C., A family of solution algorithms for non-linear structural analysis based on relaxation equations. *Int. J. Num. Meth. in Engng.,* 18, 1337–1347, (1982).

81. FELLIPA, C.A., Dynamic relaxation and quasi-Newton methods, Proc. Second Int. Conf. on *Numerical methods for Non-linear Problems,* Barcelona, Spain, (April 1984).
82. SHARIFI, P. and POPOV, E.P., Non-linear buckling analysis of sandwich arches, *Proc. ASCE, J. of Engng. Mech. Div.,* 97, 1297–1312, (1971).
83. RIKS, E., An incremental approach to the solution of snapping and buckling problems. *Int. J. Solids and Structs.,* Vol. 15, pp 529–551, (1979).
84. RIKS, E., The application of Newton's method to the problem of elastic stability. *J. of Appl. Mech.,* 39, 1060–1066, (1972).
85. WEMPNER, G.A., Discrete approximations related to non-linear theories of structures. *Int. J. Solids and Structs.,* Vol. 7, pp 1581–1599, (1971).
86. WATSON, L.T., An algorithm that is globally convergent with probability one for a class of non-linear two-point boundary value problems. *SIAM J. Numer. Anal.,* 16, 394–401, (1979).
87. HASELGROVE, C.B., The solution of non-linear equations and of differential equations with two-point boundary conditions. *Computer Journal,* 4, pp 255–259, (1961).
88. RAMM, E., Strategies for tracing the non-linear response near limit points. pp 63–89 of W. Wunderlich (Ed.), *Non-linear finite element analysis in structural mechanics.* Springer-Verlag, Berlin, (1981).
89. RAMM, E., The Riks/Wempner approach – An extension of the displacement control method in non-linear analysis. pp 63–86 of E. Hinton et al (Eds.), *Non-linear computational mechanics.* Pineridge, Swansea, (1982).
90. CRISFIELD, M.A., A fast incremental/iterative solution procedure that handles "snap through". *Computers and Structures,* 13, 55–62, (1981).
91. BATOZ, J.L. and DHATT, G., Incremental displacement algorithms for non-linear problems. *Int. J. Num. Meth. in Engng.,* 14, 1262–1266, (1979).
92. BERGAN, P.G., Solution algorithms for non-linear structural problems, pp 13.1–13.9 of T. Harwiss et al (Eds.), *Engineering Applications of the Finite Element Method,* Computas, Hovik, Norway, (1979).
93. CRISFIELD, M.A., Variable step-lengths for non-linear structural analysis. *Transport and Road Research Laboratory Report LR 1049,* Crowthorne, Berkshire, England,(1982).
94. CRISFIELD, M.A., An arc-length method including line searches and accelerations. *Int. J. Num. Meth. in Engng.,* 19, 1209–1289, (1983).
95. CRISFIELD, M.A., Overcoming limit points with material softening and strain localisation. Proc. "Second Int. Conf. on *Numerical Methods for Non-linear Problems",* Barcelona, Spain, (April 1984).
96. PUTHLI, R.S., Inelastic post-buckling behaviour of imperfect longitudinally stiffened panels under axial load, Paper at *'The Michael R. Horne Conference, Instability and plastic collapse of steel structures',* Manchester, England, (September 1983).
97. PADOVAN, J. and TOVICHAKCHAIKUL, S., Self-adaptive predictor-corrector algorithms for static non-linear analysis. *Computers and Structures,* 15, 365–378, (1982).
98. WATSON, L.T. and HOLZER, M., Quadratic convergence of Crisfield's method. *Computers and Structures,* Vol. 17, pp 69–72, (1983).
99. BATHE, K-J. and DVORKIN, E.N., On the automatic solution of non-linear finite element equations. *Computers and Structures,* Vol. 17, pp 871–879, (1983).
100. PADOVAN, J. and ARECHAGA, T., Formal convergence characteristics of elliptically constrained incremental Newton-Raphson algorithms. *Int. J. Engng. Sci.,* 20, 1077–1097, (1982).

AUTOMATIC MESH GENERATION: A FINITE ELEMENT/COMPUTER AIDED GEOMETRIC DESIGN INTERFACE

J. C. Cavendish, D. A. Field and W. H. Frey

Mathematics Department
General Motors Research Laboratories
Warren, Michigan 48090-9055, USA

1. INTRODUCTION

Computer aided design (CAD) systems have proven extremely useful for the automation of two-dimensional design and drafting procedures. Notable success has also been achieved by three-dimensional CAD systems for the design of curves and sculptured surfaces in the automobile, aerospace and ship-building industries. Although such so-called wire-frame CAD systems are useful for the design of smooth exterior surfaces (for example, automobile sheet metal panels), they are awkward and difficult to use for the design of solid functional components such as automobile pistons, connecting rods, crankshafts, housings or other parts that are usually cast, molded or machined. Several CAD systems have recently been developed for the design of such solid objects. Among the most interesting approaches have been systems which combine (via unions, differences, and intersections) many copies of a few basic primitive solids (blocks, cylinders, cones and spheres) for the design of complex parts. Figure 1 illustrates a simple example of these operations applied to a block and a cylinder.

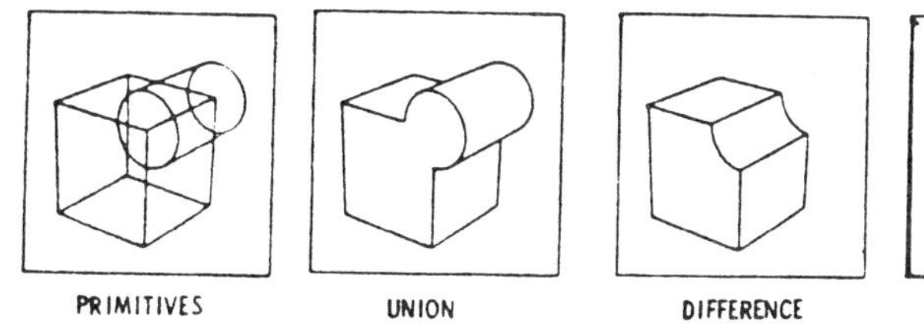

FIG. 1. Primitives and set operations.

ISBN 0-12-747255-X

A relevant question is whether or not a part designer can successfully manipulate solid primitives in a two-dimensional graphics setting to generate computer presentations of functional solid parts. Figure 2 contains examples of such functional parts designed on the solid geometric modeling system, GMSOLID, developed at the General Motors Research Laboratories [3]. GMSOLID is a computer graphics geometric modeling system that can be used to design solids by combining other simple primitive solids with set operations. The examples in Fig. 2 offer evidence that a designer can indeed effectively use such a geometric modeler to design complex automotive parts.

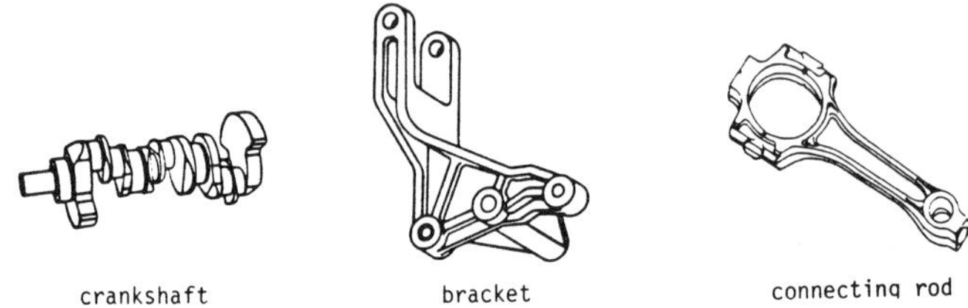

FIG. 2. Solid parts designed on GMSOLID.

With the development of solid modeling systems has come the prospect of a comprehensive CAD system which integrates the distinct functions of mechanical part design and part structural analysis. The practical construction of such a formal link between a solid modeler for part design, and finite element analysis programs for solid structures is by no means straightforward. Finite element mesh generation in solids represents the largest bottleneck in establishing this necessary link. Simply stated, it can be extremely difficult to decompose a complex solid into a valid union of finite elements. This mesh generation process becomes especially difficult when a variation in element density from region to region is required in the element idealization.

In this paper we describe an algorithm for the computer generation of tetrahedral finite element meshes for solids. As was the case of the automatic two-dimensional triangulation algorithm presented in [5] for the decomposition of planar surfaces, the proposed algorithm for solid mesh generation separates naturally into two independent modules:

1. <u>Node Point Insertion</u>. Within and on the surface of the solid, node points are first defined. This node point distribution process permits user control over local element density.

2. Three-Dimensional Triangulation. The node points are automatically connected to form a mesh of well-proportioned tetrahedral finite elements.

We remark that these two algorithms function independently of one another so that the three-dimensional triangulator can be used to triangulate points which might be entered into the structure in ways other than that described in this report. While node point insertion must precede triangulation in practice, it is more natural to discuss our triangulation method first.

The three-dimensional triangulator we shall describe makes use of two geometric constructs: the so-called Dirichlet Tessellation and the dual Delaunay Triangulation of space induced by the inserted node points. The former construct produces space-filling disjoint assemblies of convex polyhedra, and the latter produces a mesh of tetrahedra filling the convex hull of the node points. In Section 2 we develop the details of the proposed triangulation strategy using the two-dimensional analog to introduce the basic ideas, and subsequently extend these to the three-dimensional setting.

A variety of strategies may be used for defining the nodes to be triangulated. In this report, we describe a node insertion algorithm which makes use of planar cross-sections cut through the solid and on which nodes are positioned. This, of course, assumes that the geometric modeler can provide such planar cross-sections. In Section 2 we detail this approach, and in Section 3 we present examples of solid structures that were meshed using this algorithm.

We conclude this Introduction by commenting upon the apparent stigma that is associated with the use of tetrahedral elements in finite element analysis programs. Meshes are traditionally built using hexahedral elements while tetrahedral elements are reserved for regions in the solid where hexahedra cannot be conveniently fit. A principal reason why users avoid the tetrahedral element is simply the extreme difficulty of visualizing tetrahedra when manually building solid meshes. The procedure described here alleviates this problem by computerizing the process. A second objection arises from the well-known fact that linear, constant-strain tetrahedra are poor elements for analysis (that is, they are too "stiff"), requiring fine meshes to produce reasonably accurate finite element approximations. However, higher-order versions of the tetrahedron are available [1,16] and have been implemented in [11] and at the General Motors Research Laboratories and we have concluded that the quadratic tetrahedral element is competitive with the quadratic hexahedral finite element. The results generated at General Motors are being prepared for publication and pertain to analyses performed with meshes generated by the system described in this paper for problems that include torsion and bending.

2. THE PROPOSED METHOD

As was the case with the two-dimensional mesh generator presented in [5], the mesh generator proposed here involves two independent processes. First, node points must be inserted within and on the boundary of the structure to be meshed. Secondly, the node points are automatically triangulated to form a network of well-proportioned elements. The triangulation algorithm functions in both two and three-dimensional settings producing meshes of triangular and tetrahedral elements respectively. For ease of exposition, we first discuss the mesh triangulation algorithm in two dimensions.

2.1 *Mesh Triangulation Algorithm*

Our triangulation algorithm in both two and three-dimensions makes use of the Dirichlet Tessellation, a geometric construct first defined by mathematicians for theorem proving [6,14] and then rediscovered by physicists [4], geographers [8], crystallographers [7], and statisticians [9,10]. In this section we give a brief account of the Dirichlet Tessellation and direct the reader to recent articles by Sibson [13], Boyer [2], and Watson [15] for greater detail, and to Rogers [12] for a more general treatment.

Consider first the two-dimensional case. Let $p_1, p_2, \ldots p_N$ be distinct points in the plane (R^2), and define the sets V_i, $1 \leq i \leq N$, where

$$V_i = \{x: \|x - p_i\| < \|x - p_j\| \text{ for all } j \neq i\} , \qquad (2.1)$$

where $\|\cdot\|$ denotes Euclidean distance in R^2. V_i represents a region of the plane whose points are nearer to node p_i than to any other node. Thus, V_i is an open convex polygon (usually called a Voronoi polygon) whose boundaries are portions of the perpendicular bisectors of the lines joining node p_i to node p_j when V_i and V_j are contiguous. The collection of Voronoi polygons $\{V_i\}_{i=1}^N$ is called the Dirichlet Tessellation. In general, a vertex of a Voronoi polygon is shared by three neighboring polygons so that connecting the three generating points associated with such adjacent polygons forms a triangle, say T_k. The set of triangles $\{T_k\}$ is called the Delaunay Triangulation. This construct can be shown to be a triangulation of the convex hull of the node points. The Dirichlet Tessellation and dual Delaunay Triangulation is illustrated in Fig. 3 for a set of 13 node points.

An important property of the two-dimensional Delaunay Triangulation which makes it suitable for use as a finite element mesh of triangular elements is that its triangles are as close to equilateral as possible for the given set of nodes, in a sense made precise in [13]. Consequently, ill-conditioned and thin triangles are avoided wherever possible.

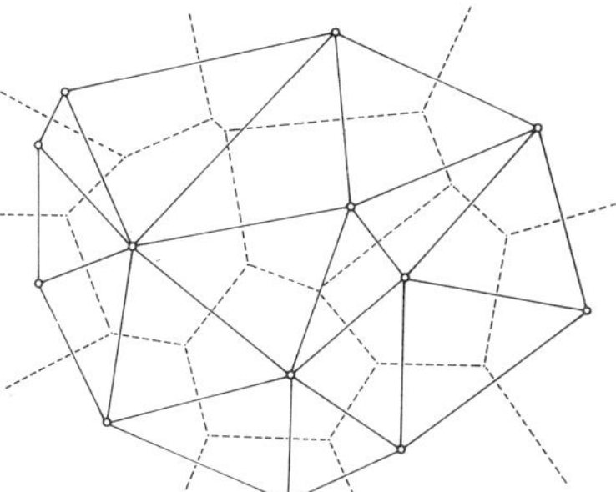

FIG. 3. Dirichlet Tessellation (dashed lines) and Delaunay Triangulation (solid lines) of 13 points.

Most investigators have been interested in the properties and uses of the Dirichlet Tessellation rather than to its dual Delaunay Triangulation. As a result, most published numerical algorithms focus on the construction of the Tessellation [2,4,7] itself. Furthermore, until recently, computation of the three-dimensional tessellation in an efficient manner remained an open question. The Delaunay Triangulation, however, is easier to compute than the Dirichlet Tessellation, and it is computed first in some algorithms which eventually compute the Tessellation. We use a modification of an algorithm presented by Watson [15]. The following terminology is helpful in describing Watson's algorithm. Three noncollinear nodes define both a triangle and a circle called the circumcircle of the triangle. The circumcircle with its interior is called a circumdisk. The radius and center of a circumcircle are called the circumcenter and circumradius, respectively, of the associated triangle.

In two dimensions, Watson's algorithm turns upon the simple observation that three given node points will form a Delaunay triangle if and only if the circumdisk defined by these nodes contains no other node points in its interior. Figure 4 illustrates this property of the Delaunay Triangulation for the node set of Fig. 3. In effect, Watson's approach is to reject from the set of all possible triangles which might be formed, those with non-empty associated circumdisks. Those triangles not rejected form the Delaunay Triangulation.

The algorithm is initialized by calculating the x,y-coordinates of three points which form a triangle T_0 that surrounds all the node points to be inserted. The circumcenter coordinates and circumradius of the circumcircle defined by T_0 are also calculated and recorded. The node points are then introduced one at a time. The algorithm operates by maintaining a list of triplets of node points which represent completed Delaunay triangles. Associated with each such triangle are the x and y-coordinates of its circumcenter and its circumradius. For each new node point entered, a search is made of all current triangles to identify those whose circumdisks contain the new point (see Fig. 5). For each such disk, the associated triangle is flagged to indicate removal. As shown in Fig. 6, the union of all such triangles forms what we call an insertion polygon containing the new node point. It can be shown that no previously inserted node is contained in the interior of the polygon and that each boundary node of the polygon may be connected to the new node by a straight line lying entirely within the polygon. Thus, a new triangulation of the region enclosed by the polygon is formed (see Fig. 7). It can also be shown that this local triangulation, when combined with the triangles outside the polygon form a new Delaunay triangulation which includes the newly added point. Repeated use of this insertion algorithm permits all node points to be entered, while ensuring that at each step the triangulation retains its Delaunay properties.

FIG. 4. Defining circumcircles for points in FIG. 3.

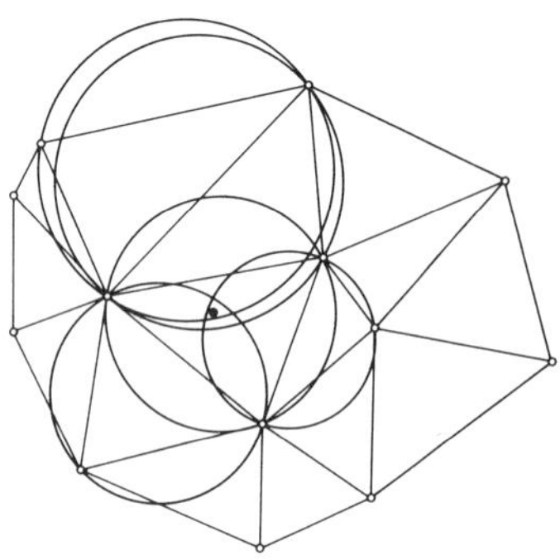

FIG. 5. Insertion of new point.

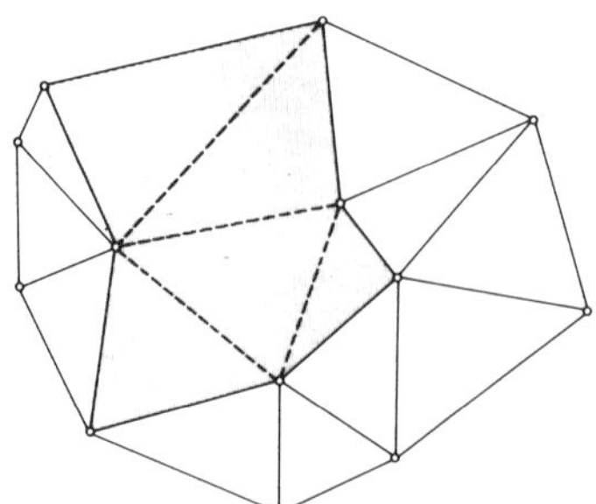

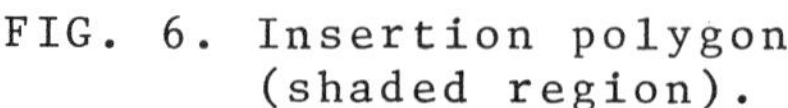

FIG. 6. Insertion polygon (shaded region).

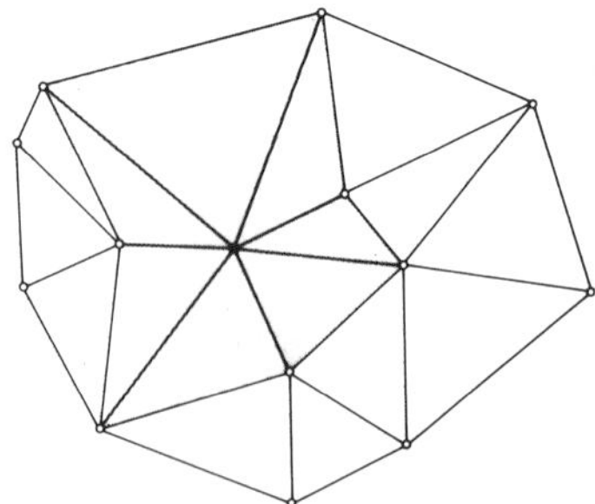

Fig. 7. Local triangulation (shaded region).

The construction just defined carries over easily to three (and higher) dimensions. In this case, a Voronoi region, V_i, is a solid polyhedron. In three-dimensions a vertex of a Voronoi polyhedron is usually shared by four neighboring polyhedra and the lines joining the four associated node points form a tetrahedron. The collection of Delaunay tetrahedra can be thought of as a three-dimensional triangulation of the node points. We conjecture that the faces of Delaunay tetrahedra are also as close as possible to equilateral triangles. Unfortunately, as we shall see later, even this would not be sufficient to insure well-proportioned tetrahedral finite elements. We shall show how the Delaunay Triangulation is modified to yield a valid assembly of tetrahedra well-suited for finite element analysis.

In three-dimensions, Watson's algorithm starts with a tetrahedron T_0 containing all points to be inserted, and new internal tetrahedra are formed as the points are entered one at a time. (Circumballs and circumspheres perform the roles of circumdisks and circumcircles respectively. The algorithm maintains a list of 4-tuples of node points and a list of the associated (x,y,z)-coordinates of their circumcenter and circumradius.) At a typical stage of the process, a new point is tested to determine which circumballs of the existing tetrahedra contain the point. The associated tetrahedra are removed, leaving an insertion polyhedron containing the new point. Edges connecting the new point to all triangular faces of the surface of the insertion polyhedron are created, defining tetrahedra which fill the insertion polyhedron. Combining these with the tetrahedra outside the insertion polyhedron produces a new Delaunay triangulation which contains the newly added point.

Several problems exist with Watson's approach. The first is the existence of so-called degenerate cases. Degenerate cases occur in practice when a newly inserted node appears to lie on the surface of a circumsphere associated with some existing tetrahedron. The problem becomes apparent whenever the distance from a newly entered nodal point to an existing circumsphere is less than ε where ε is the expected accumulated computer truncation error. When this happens, there is the danger that an incorrect or inconsistent decision will be made regarding rejection or acceptance of a given 4-tuple. This in turn produces structural inconsistencies in the triangulation (that is, overlapping tetrahedra or gaps in the mesh). A remedy for this problem which we have found useful in applications, is to slightly perturb the coordinates of a newly entered node point whenever that point is found to lie ambiguously on a circumsphere defined by some previously accepted Delaunay tetrahedron. The perturbation is just large enough and in a direction to permit correct in-out decisions to be made during the finite precision calculations. At the completion of the triangulation, all perturbed nodes are restored to their original position in the lattice of node points.

A serious problem which arises in three-dimensions, and which also requires modification of the Delaunay Triangulation occurs with the creation of tetrahedra we call "slivers". In Fig. 8a we illustrate a Delaunay tetrahedron together with its associated circumsphere. If the node point D in Fig. 8a is moved to any other position on the circumsphere, then the tetrahedron ABCD remains a valid Delaunay tetrahedron. As a worst case, suppose D were moved to point D' shown in Fig. 8b, where D' is slightly out of the plane defined by nodes A, B and C. ABCD' then defines a badly distorted Delaunay tetrahedron whose faces are well-proportioned triangles, but whose volume can be made arbitrarily small. We refer to these thin tetrahedra as slivers and we identify them in practice when the dimensionless ratio $\bar{a}$ = (radius of inscribed sphere)/(radius of circumsphere) is sufficiently small (less than 0.001). In the meshes we have generated, we have found that slivers can account for as much as 10% of the total number of tetrahedra generated.

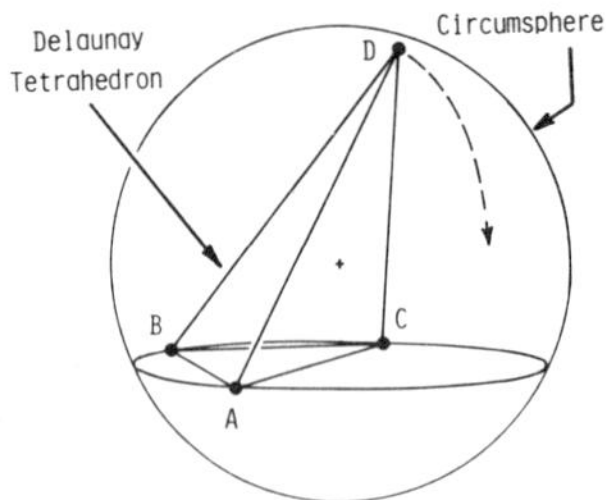

FIG. 8a. Delaunay tetrahedron and associated circumsphere

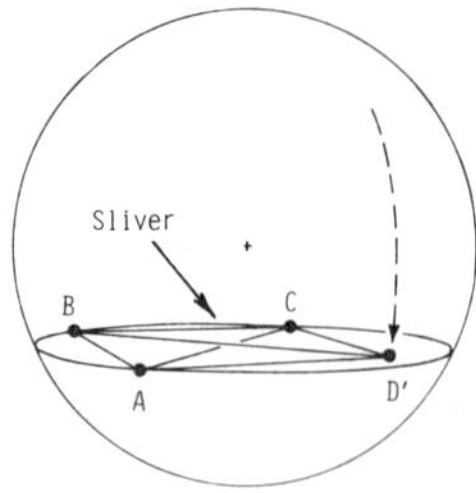

FIG. 8b. Distorted Delaunay tetrahedron

When a sliver is detected, defined say by nodes A,B,C and D, it must be removed. Our strategy is to first determine how ABCD "fits" into the triangulation. We start by determining the four tetrahedra which abut against ABCD. Consider first the case where two of these tetrahedra share a common vertex, say node E. This is the case illustrated in Fig. 9a. When this situation is detected, the sliver is removed from the collection of Delaunay tetrahedra and elements {ABDE,BCDE} are replaced by {ABCE,ACDE}.

When it is found that no two of the four surrounding tetrahedra share a common vertex, the situation is that depicted in Fig. 9b. In this case, the node point D is moved (arbitrarily) to the point (D+E)/2. The element ABCD remains an accepted tetrahedron, but the sliver has been opened up and the ratio $\bar{a}$ has been increased.

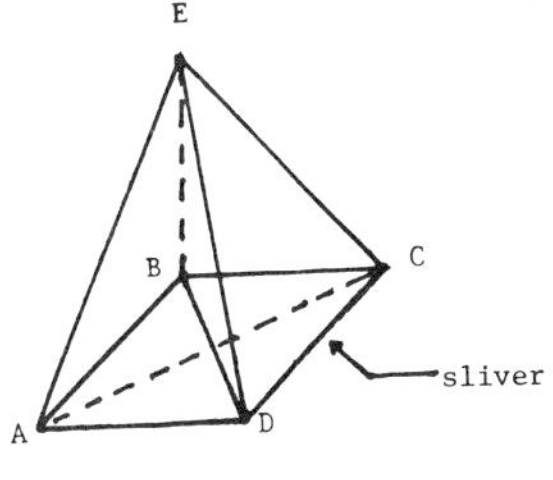

FIG. 9a. Removable sliver.

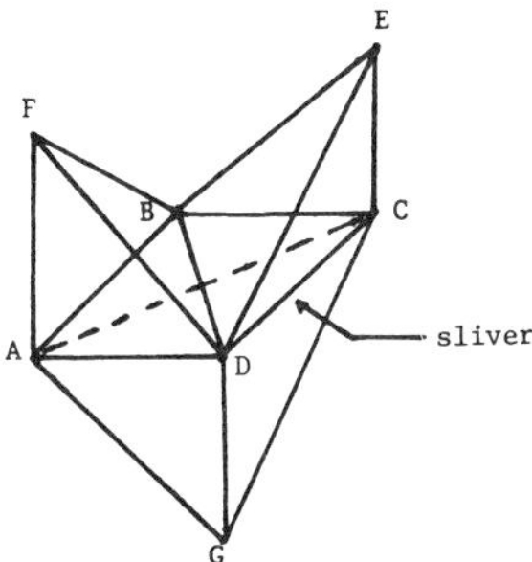

FIG. 9b Nonremovable sliver.

The procedure outlined above produces a three-dimensional triangulation of the initial tetrahedron. Therefore, the aggregate of Delaunay tetrahedra must be post-processed to determine which tetrahedra lie inside the solid and are consequently part of the solid finite element mesh. The tetrahedra associated with the interior element nodes are distinguished because they have none of the initial points, $\{p_i\}_{i=1}^{4}$, as a vertex. Of these interior tetrahedra, those that span holes or lie outside of the structure must be identified and "peeled" away. We accomplish this by simply calculating the centroid of each tetrahedron retaining only those elements whose centroids are found to lie inside the solid's bounding surfaces.

2.2 *Node Insertion Strategy*

There are a number of ways that node points can be defined within the solid. For example, the process can be done manually using digitizers to record the coordinates of the node points. In this paper we are going to assume that a facility exists to cut planes through the structure, say $P_1, P_2, \ldots, P_N$. It is within each of these cross-sections that node points will be defined. The method of node definition in planar cross-sections is that described in [5]. Figures 10, 11 and 12, taken from [5], illustrate the steps. Figure 10 shows the cross-section to be a large circle with a small interior hole. Boundary nodes have been positioned in Fig. 11 and the region covered by user-defined zones Z_1, Z_2 and Z_3. Within zone Z_i, node points are to be generated automatically so that neighboring nodes are separated by approximately r_i units. The resulting distribution of node points reflects local refinement near the small hole as shown in Fig. 12. The reader is referred to [5] for complete details of this node insertion procedure (called POINT in [5]).

In summary, node points are defined interactively a plane-at-a-time. Within each plane, the user has reasonable control over local node densities. In Fig. 13 we show a simple solid in which nodes have been inserted by the use of four slicing planes.

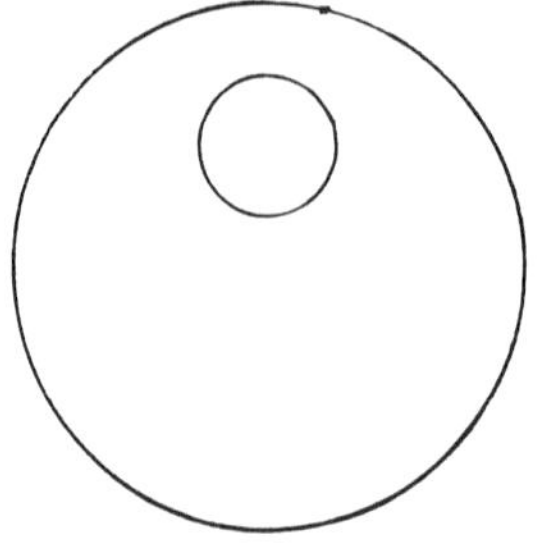

FIG. 10. Structure cross-section.

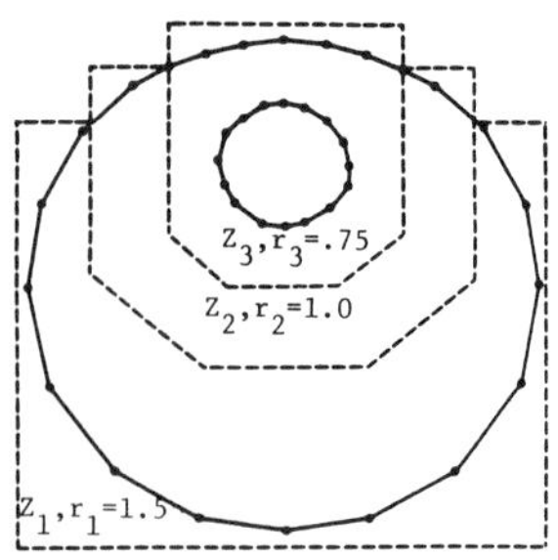

FIG. 11. Definition of node density

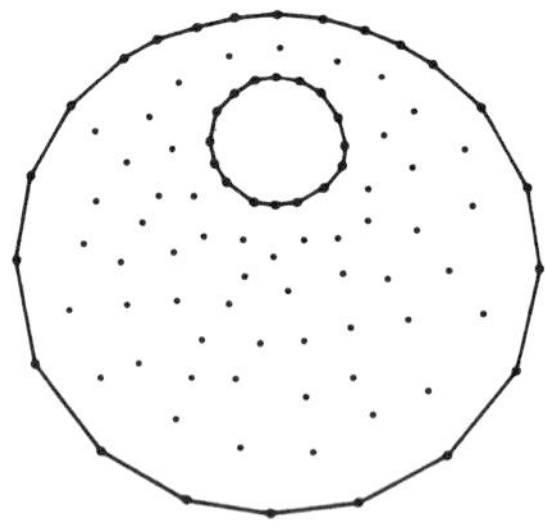

FIG. 12 Inserted nodes.

3. EXAMPLES

In this section we illustrate the results obtained using the mesh generation algorithm described here to decompose two solid structures. In all structures we have meshed, the code has terminated successfully producing a valid, well-proportioned assembly of tetrahedral elements. For display purposes, the solid meshes were interrogated and triangular faces lying on the surface of the solid were identified and recorded. Such faces are distinguished from faces interior to the solid by virtue of the fact that they are not shared by two abutting tetrahedra.

The first solid meshed is that used to illustrate the point insertion procedure (see Fig. 13). A total of 128 node points were inserted within the structure and the algorithm terminated with a mesh containing 390 tetrahedral finite elements. In Fig. 14 we display the surface triangles resulting from this decomposition.

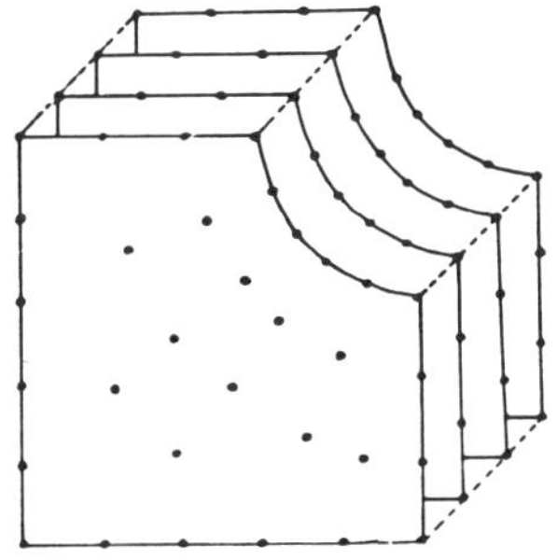

FIG. 13. Inserted nodes on four cross-sections.

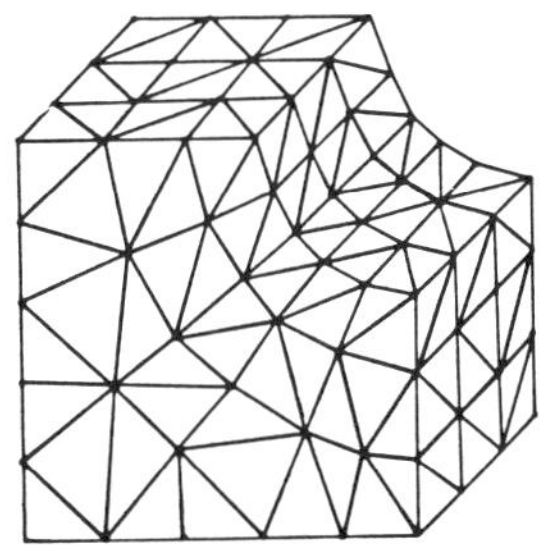

FIG. 14. Delaunay Triangulation of points in FIG. 13.

In Fig. 15a one half of an automotive bearing cap is displayed. A total of 13 cutting planes were used for node point insertion (see Fig. 15b) and 508 nodes were inserted within and on the solid. The solid mesh generator produced a total of 1896 tetrahedral elements with 696 surface triangles. The mesh is graded with local refinement around the long bolt hole in the solid. Figure 15c shows a graphics display of the resulting triangulation.

Our experience with the algorithm indicates that it is very efficient with respect to computer run time. For nodes that are entered randomly, it can be shown that the code executes in $O(n^2)$ computer time, where n denotes the number of data points. In Table 1 we tabulate the execution times required to generate the triangulations presented for several of the solids considered here. Included in this table are approximate times required to complete the node insertion phase for each of the structures. It should be noted that because of the high degree of planar symmetry in these solids, node insertion is a fairly simple task. However, even for more complex and realistic automotive parts such as crankshafts and valves, our experience has been that node insertion has not required more than 30 minutes of sit-down time at the CRT console.

We remark that the times reflect only the triangulation portion of the algorithm and not the point insertion portion. For each of these simple structures the node insertion process required less than 10 minutes of real time.

ACKNOWLEDGEMENT

We wish to express our appreciation to Fred Krull and Dave Warn of the GMR Computer Science Department for the assistance they provided in displaying the finite element meshes and models in this publication.

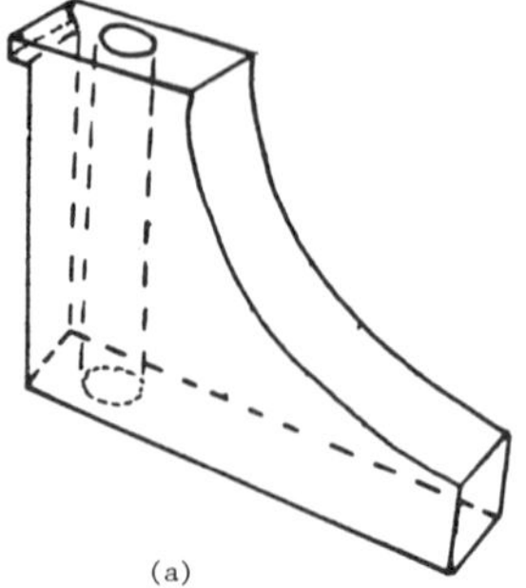

(a)

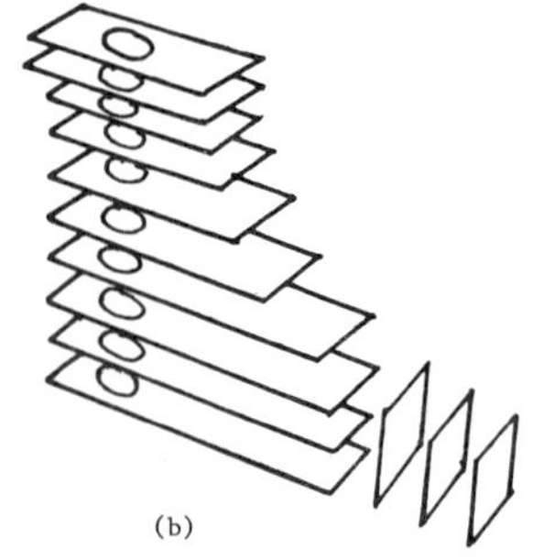

(b)

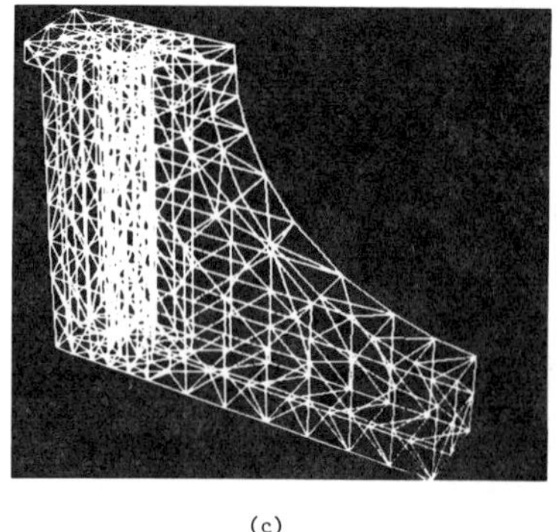

(c)

(d)

FIG. 15. (a) bearing cap, (b) cross-sections, (c)Delaunay Triangulation, (d) tetrahedral assembly.

TABLE 1

Time to Compute Delaunay Triangulation on IBM 360/3081

Structure Number	Number of Nodes	Number of Elements	CPU Time (Seconds)
1	128	390	45
2	508	1896	185

REFERENCES

1. ARGYRIS, J.H., Tetrahedron Elements with Linearly Varying Strain for the Matrix Displacement Method. *J. Roy. Aero. Soc.* 877-880 (1965).

2. BOWYER, A., Computing Dirichlet Tessellations. *The Computer Journal* 24, No. 2, (1981).

3. BOYSE, J.W. and ROSEN, J.M., 'GMSOLID' - A System for Interactive Design and Analysis of Solids. Research Publication GMR-3451, General Motors Research Laboratories, Warren, Michigan 48090 ( ).

4. BROSTOW, W. and DUSSAULT, J.P., Construction of Voronoi Polyhedra. *J. Comp. Phys.*, 29, 81-92 (1978).

5. CAVENDISH, J.C., Automatic Triangulation of Arbitrary Planar Domains for the Finite Element Method. *Int. J. Numer. Meth. Engrg.* 8, 679-696 (1974).

6. DIRICHLET, G.L., *Z. Reine Angew. Math.* 40, (1850).

7. FINNEY, J.L., A Procedure for the Construction of Voronoi Polyhedra. *J. Comp. Phys.* 32, 137-143 (1979).

8. KENDALL, D.G., Construction of Maps from 'Odd Bits of Information'. *Nature*, London 231, 158-159 (1971).

9. MILES, R.E., On the Homogeneous Planar Poisson Process. *Math. Biosci.* 6, 85-127 (1970).

10. MOLLISON, D., Spatial Contact Models for Ecological and Epidemic Spread. *J. Roy. Statist. Soc. Ser. B.* 39, 283-313 (1977).

11. NGUYEN-VAN-PHAI, Automatic Mesh Generation with Tetrahedron Elements. *Int. J. Num. Meth. Engrg.* 18, 273-289 (1982).

12. RODGERS, C.A., *Packing and Covering*, Cambridge Mathematical Tracts No. 54, Cambridge University Press.

13. SIBSON, R., Locally Equiangular Triangulations. *The Computer Journal* 21, No. 3, 243-245 (1978).

14. VORONOI, G.F., *Z. Reine Angew. Math.* 134, (1908).

15. WATSON, D.F., Computing the n-Dimensional Delaunay Tessellation with Applications to Voronoi Polytopes. *The Computer Journal* 24, No. 2, (1981).

16. ZIENKIEWICZ, O.C., *The Finite Element Method*, Third Edition, McGraw-Hill, London, (1977).

FINITE ELEMENT MODELING IN AN INTEGRATED COMPUTER AIDED DESIGN ENVIRONMENT

Mark S. Shephard

Rensselaer Polytechnic Institute, Troy, New York, U.S.A.

1. INTRODUCTION

The use of solid modeling techniques with their complete and unique geometric representations [1] allows for the automation of many time consuming engineering calculations that are carried out as part of a design process. Algorithms have been developed to automate the calculation of the various mass properties used in engineering calculations. However, the more advanced engineering analysis calculations of quantities including stresses, displacements, temperature distributions, flow properties and others have yet to be automated. Today, it is common to calculate these quantities using the finite element analysis technique. Therefore, if algorithms can be developed that reduce the time required to generate, or to even automatically define finite element models that yield an acceptable degree of accuracy, the benefits of solid modeling techniques can be more fully realized.

The purpose of this paper is to discuss techniques for generating finite element meshes for solid models. Emphasis is placed on indicating how the geometric definitions used by solid modelers can be taken advantage of in the mesh generating process. Since the geometric definitions produced by solid modelers are complete and unique, it is possible to automate any geometry controlled operation, including the subdivision of a geometry into valid finite elements.

The majority of today's finite element preprocessors employ either mapping [2] or blending function mesh generators [3,4]. These mesh generators have the advantage of allowing the creation of user controlled meshes of well shaped elements. They have the disadvantage of requiring that the domain of interest be partitioned into subdomains of fixed topologies. In two-

ISBN 0-12-747255-X

dimensions for example, the domain must be broken into a set of three and four sided subdomains. An appropriate combination of interactive graphic manipulation techniques and the ability of blending functions to mesh large subdomains allows for the efficient generation finite element meshes.

The next generation of finite element preprocessors will employ fully automatic mesh generators capable of producing valid finite element meshes for entire three-dimensional domains. These mesh generating techniques are currently being developed [5-10]. These procedures employ the complete geometric information supplied by the solid model during the generation of the mesh. In particular they employ some form of boundary representation to insure that the objects vertices, edges and surfaces are properly represented. In addition they employ interior information to insure that mesh points are in fact placed on the surface of and within the interior of the object.

The availability of dependable fully automatic mesh generators does raise the possibility of automating the entire finite element modeling process.

The next section introduces a number of geometry operators employed by mesh generating schemes to obtain geometry information from a solid model. The following two sections discuss how these operators can be employed in both blending function based interactive preprocessors and fully automatic mesh generators. The final section discusses the capabilities required to automate the entire finite element modeling process.

2. GEOMETRY COMMUNICATION OPERATORS

There are a number of possible valid mathematical representations of solid objects that can be placed in a computerized data structure [1]. However, those that are based on, or generate, a boundary representation are preferred because most mesh generators work from the surfaces to the interior. To indicate how the geometric representation employed by a solid modeler can be used in the meshing process, a set of general communication operators used to obtain specific geometric information from the solid modeler are introduced. It is possible to develop the actual algorithms required to perform these operations for the various solid modeling approaches. However, the algorithms required for specific modelers could be complex. The communication operators to be used are:

IN/OUT(IPT;ISIT,IDEN) Given a point in space, IPT, this operator is to return a TRUE or FALSE in ISIT to indicate whether the point is inside the object. In those cases where it is used, the variable IDEN returns information on the element sizes desired at that location in space.

EDGE_PLACE(NODES, WEIGHTS, EDGE; NEWNODES) Given an ordered list of points, NODES, with the first and last point at the desired location on an edge,EDGE, the operator EDGE_PLACE will return a set of point locations, NEWNODES, along the edge. Mesh gradation is controled by the parameters in WEIGHTS.

PULL_SURFACE(IPT,JSURF;NEWIPT) Given a point, IPT, near the surface, JSURF, this operater returns a new location for the point, NEWIPT, on the actual surface.

BREAK_SURFACE(EDGEPTS,JSURF;NEWEDGE) Given a set of break points, EDGEPTS, on a surface, JSURF, calculate an edge, NEWEDGE, lying on the surface going through those points.

SMOOTH_SURFACE(NODES, ELTOP, WEIGHTS, JSURF;NEW NODES) Given a set of node points, NODES, on the surface, JSURF, employ the given mesh topology information, ELTOP, and the mesh gradation parameters, WEIGHTS, to create new nodal positions, NEWNODES, that improve the gradation and shape of the resulting surface mesh.

MESH_SURFACE(EDGES, WEIGHTS, JSURF; SURFMESH) The function of this operator is to insure that the mesh generated on the surface patch, JSURE, bounded by edges, EDGES, lies on that surface. The terms WEIGHTS contains information on the desired mesh layout. This operator is less concise and well defined than the others since it implies a simultaneous knowledge of the mesh generator and the surface geometry.

3. INTERACTIVE MESH GENERATION IN A SOLID MODELING ENVIRONMENT

The minimum amount of information an interactive finite element preprocessor needs from a geometric modeler is the wireframe description of the objects boundaries. With this information the EDGE_PLACE operator can be involved to place nodes on the edges of user selected mesh patches. This discretized information is all that is needed to involve blending function mesh generators. However, the resulting mesh may violate the geometry of the object.

The additional information required to insure the proper representation of the geometry is the definition of the various surface patches bounded by the edges. With this information, the various surface operators (PULL_SURFACE, BREAK_SURFACE, SMOOTH_SURFACE and MESH_SURFACE) required to insure surface mesh points lie on the surface can be developed. This information and the associated operators, coupled with an interactive graphic preprocessor allow the user to define finite element meshes for objects. Assuming the use of blending functions or other mesh generators that are based primarily on discretized boundary information, the process of mesh generation would include the following step. The edges of mesh patches are interactively identified. Since it is not always desirable to maintain a one to one correspondence between geometric edges and the patch edges, the BREAK_SURFACE operator is needed to create additional edges to serve as patch edges. The EDGE_PLACE operator is then used to interactively place nodes, in a user specified manner, along the edges that make up to boundaries of a mesh patch.

The PULL_SURFACE and/or MESH_SURFACE operators are then used to insure that the node points on the surface of the mesh patches lie on the geometric surface. In some combinations of surface types and mesh generators, there is a priori assurance that node points are on the geometric surface, while for others a MESH_SURFACE operator can be easily developed. Other combinations may require repositioning of the nodes, one at a time, onto the actual surface by using the PULL_SURFACE operator. When the PULL_SURFACE operator is used, it is often desirable to reposition the locations of surface nodes and interior nodes to improve the "fairness" of the mesh. A number of repositioning algorithms based on Laplacian or potential functions [11] are available. These algorithms are easily applied to node points on the interior of the object, but would require a specific geometric interface for the surface nodes as indicated by the introduction of the SMOOTH_SURFACE operator.

The operators used thus far do not fully guarantee the resulting mesh is valid or properly represents the geometry of the object. Since blending function mesh generators can overspill the boundary [4], it is possible that some nodes that are suppose to be on the interior of the object are outside it. The IN/OUT operator, which is available for solid model representations, can be used to check those points that may be outside the object.

The additional geometry associativity information available in a solid representation can also be used effectively to create additional operators to make the mesh generating process more efficient. For example the selection of the numerous edges that make up a mesh patch can be sped up by allowing the user to select surface patches and to use the model data to automatically determine the edges. A number of automatic checking operations such as; whether nodes on an edge have already been used in a mesh patch and whether a patch has already been meshed can make use of the geometry database. However, most of these checks also require a separate meshing database to operate them.

Figure 1 shows a blending function mesh generated using an interactive graphic preprocessor for an object that was defined with a particular solid modeler [12].

4. AUTOMATIC MESH GENERATION

Even with the best of interactive graphic operations and the use of the geometric information supplied by a solid model, the interactive generation of finite element meshes with blending

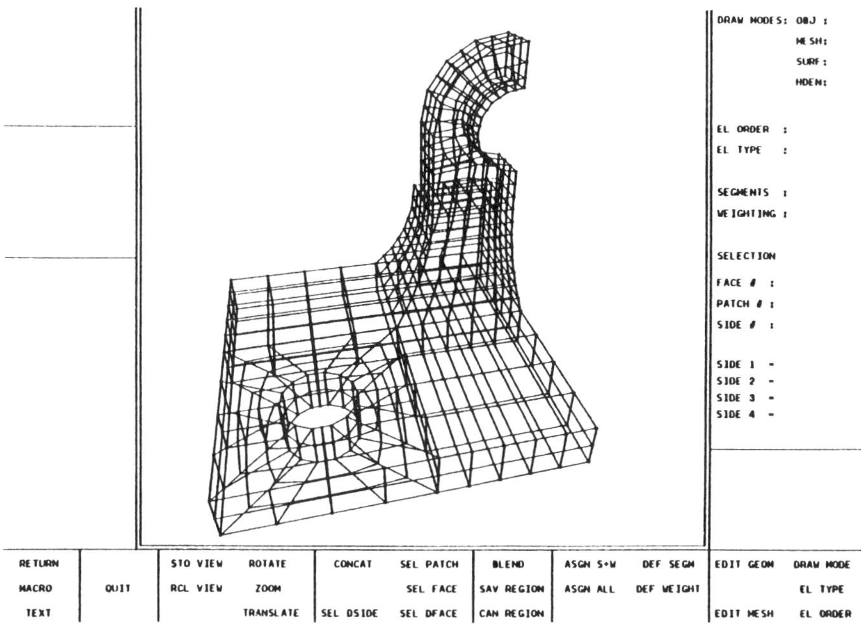

FIG. 1 Interactively Defined Mesh for a Solid Object

function mesh generators is difficult. Therefore, there is a need for fully automatic mesh generating techniques that can employ the geometric representation of the solid model to generate a finite element mesh without user intervention.

A number of algorithmic approaches are currently being developed for fully automatic mesh generation [5,10]. Although each of the approaches being taken varies with respect to the actual algorithm, a simplistic view indicates four basic algorithmic approaches, including;

1. paring elements from the object one at a time,
2. recursive object subdivision down to the element level,
3. volume triangulation of points throughout the object using tessellation operators, and
4. spatial enumeration followed mesh finalization.

In some implementation, the mesh generation process begins with a surface triangulation which is then used during subsequent steps for the definition of the solid elements.

To demonstrate the interaction between a solid model and an automatic mesh generator, a brief description of the modified-octree mesh generator [5] is given. This meshing algorithm begins by performing a spatial enumeration of an object to define its modified-octree. In a modified-octree representation, the interior of the object is represented as a set of various sized cubes, referred to as octants, with corners at specific locations in an integer coordinate system. Those octants that contain the surface of the object can have portions cut away to improve their representation of the object. The cuts are made using only a limited number of possible cutting points. A finite element mesh topology is then defined within the modified-octree representation by triangulating the individual octants in such a manner that proper matching of element faces is maintained. Computational efficiency is aided by the tree storage used [5]. Element size and mesh gradation are controlled by the size of octants in the modified-octree used in the representation of the various portions of the object. The size of octants used throughout the object is controlled by user defined parameters specified in terms of the geometric model.

The mesh is finalized by shifting the nodes from the integer coordinate system to the real coordinate system, thus yielding a proper representation of the actual geometry and an improvement in the shape of the resulting elements.

The generation of the modified-octree can be carried out by using the IN/OUT operator. The geometrically defined mesh gradation information is also communication via the IN/OUT operator. To improve the efficiency of the generation process the boundary file of the solid is employed to first define the surface modified-octree. These operations employ an extended version of the IN/OUT operator which returns information about edges and vertices within the octant in question. The definition of the interior of the modified-octree and the generation of the mesh topology requires no additional communication with the solid model.

The mesh finalization process begins by using the EDGE_PLACE operator to insure the proper representation of edges. The PULL_SURFACE and SMOOTH_SURFACE operators are then used to finalize the position of nodes on the surface of the object. The IN/OUT test can be used to insure that the repositioned interior nodes remain in the interior of the geometry.

Fig. 2 shows a mesh generated for a simple solid object using the modified-octree mesh generator.

5. AUTOMATIC FINITE ELEMENT ANALYSIS

The input to current finite element analysis codes includes a finite element mesh with the attributes of loads, material properties and boundary conditions applied to that mesh. In otherwords, the input to the finite element analysis is only a particular discretization of the original problem. The availability of fully automatic mesh generating techniques allows for the development of numerical analysis codes that accept as input the original problem description in terms of its geometric definition with all analysis attributes tied to that geometry.

As an example of such an approach, a program for the automatic tracking of cracks in two dimensional domains is currently under development [13]. The input to this program is a set of boundary curves with the problem attributes of loads, material properties and boundary conditions tied to it. The program uses

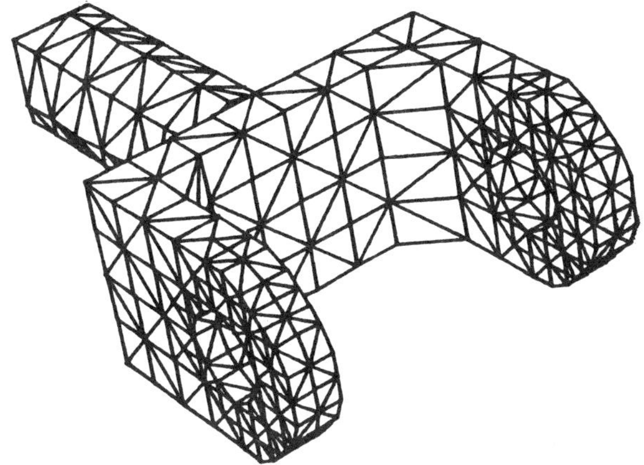

FIG. 2 Modified-Octree Mesh Example

this information to automatically mesh the geometry, with full account taken of the crack tips. The analysis is run and fracture criteria invoked to determine crack propagation direction and increment for a particular load increment. The geometry is updated to account for all crack motion. A new mesh is then automatically generated and the process continued.

Capabilities of this type are the final aim of the developers of analysis routines for an integrated computer aided design environment. However such programs must be more robust than current finite analysis codes. In particular they must be self-adaptive codes capable of controlling discretization errors to insure a desired level of accuracy.

6. CLOSING REMARKS

Developments in the areas of geometric modeling and finite element mesh generation are making the process of numerical model generation more efficient and simpler. This allows a much tighter integration of the design-analysis cycle. Although this represents a major step forward, it also presents a possible danger. The easier it is to carry out a finite element analysis, the more often such analyses will be carried out by users not well versed in the under pinnings of the analysis techniques being used. Therefore, it is not only important to continue to improve the model generation capabilities, but to also make the analysis techniques more robust and fool proof.

ACKNOWLEDGEMENT

The author would like to acknowledge General Motors Research Laboratory, International Business Machine Corporation, the National Science Foundation under grant ISP79-20240 and the RPI

CICG Industrial Associates for supporting the finite element modeling projects included in this paper. The author would also like to acknowledge Mark Yerry and Panajiotis Kotsianas for developing the programs used to generate the figures in this paper, and Diana Rogers for typing the manuscript.

REFERENCES

1. REQUICHA, A.A.G. and VOELCKER, H.B., Solid Modeling: A Historical Summary and Contemporary Assessment, *IEEE Computer Graphics and Applications*, 3, 9-24 (1982).
2. ZIENKIEWICZ, O.C. and PHILLIPS, D.V., An Automatic Mesh Generation Scheme for Plane and Curved Surfaces by Isoparametric Mapping, *Int. J. Num. Meth. Engng.*, 3, 519-528 (1971).
3. GORDON, W.J. and HALL, C.A., Construction of Curvilinear Coordinate Systems and Applications to Mesh Generation, *Int. J. Num. Meth. Engng.*, 7, 461-477 (1973).
4. HABER, R.B. and ABEL, J.F., Discrete Transfinite Mappings for the Description and Meshing of Three-Dimensional Surfaces Using Interactive Computer Grpahics, *Int. J. Num. Meth. Engng.*, 18, 41-66 (1982).
5. YERRY, M.A. and SHEPHARD, M.S., Automatic Three-Dimensional Mesh Generation by the Modified-Octree Technique, to appear, *Int. J. Num. Meth. Engng.*
6. CAVENDISH, J.C., FIELD, D.A., and FREY, W.H., An Approach to Automatic Three-Dimensional Finite Element Mesh Generation, to appear, *Int. J. Num. Meth. Engng.*
7. WOO, T.C. and THOMASMA, T., An Algorithm for Generating Solid Elements in Objects with Holes, *Computers and Structures*, 18,(2), 33-342 (1984).
8. SLUITER, M.L.C., and HANSEN, D.C., A General Purpose Automatic Mesh Generator for Shell and Solid Finite Elements, *Computers in Engineering*, 3, L.E. Hulbert, ed., Book No. G00217, ASME, 29-34 (1982)
9. WORDENWEBER, B., Volume-Triangulation, CAD Group Document No. 110, University of Cambridge, Computer Laboratory, Corn Exchange Street, Cambridge, CB2 3QG, England, (1980).
10. NGUYEN, Van-Phai, Automatic Mesh Generation with Tetrahedron Element, *Int. J. Num. Meth. Engng.*, 18, 273-289 (1982).
11. LORENSEN, W. and Grid Generation Tools for the Finite Element Analyst, *First Chautauqua on Finite Element Modeling*, J.H. Conaway, Ed., Schaffer Analysis, Inc. 99-117 (1980).
12. FITZGERALD, W., GRACER, F., and WOLFE, R., GRIN: Interactive Graphics for Modeling Solids, *IBM J. Research and Development*, Vol. 25 (4), 281-294 (1981).
13. SHEPHARD, M.S., YEHIA, N.A.B., BURD, G.S. and WEIDNER, T.J., Automatic Crack Propagation Tracking, to appear *Computer and Structures*.

A METHOD FOR DETERMINATION OF TWO DIMENSIONAL FIELD LINES

E.-M. Salonen, R. Salminen and J. Aalto

Helsinki University of Technology, Otaniemi, Finland

1. INTRODUCTION

Let us consider the following problem. A vector field

$$\underline{u}(x,y) = u_x(x,y)\underline{i} + u_y(x,y)\underline{j} \tag{1.1}$$

in a two-dimensional plane domain A is given. x and y are the rectangular Cartesian coordinates and $\underline{i}$ and $\underline{j}$ are the corresponding unit base vectors. We want to determine and plot the field line distribution corresponding to the given field.

A field line representation of a vector quantity is often needed — as for instance the streamlines in a flow problem — to clarify visually the character of the field.

The obvious way to draw the field lines is what might be called the step by step method. In its simplest form we start the plotting from a point and advance in small steps always in the direction given by the field at the point in question. Some deviations from the correct field lines usually occur and especially it may be difficult to obtain closed field lines.

We look for alternative formulations where the field lines are obtained as contour lines of a scalar function $\beta(x,y)$ to be specified later. This kind of approach has been described by Southwell in [1]. He employed the equation

$$\underline{u}\cdot\text{grad}\beta \equiv u_x \frac{\partial\beta}{\partial x} + u_y \frac{\partial\beta}{\partial y} = 0 \tag{1.2}$$

which was discretized by the finite difference method.

One advantage of being able to draw the field lines as contour lines of a scalar function β lies in the fact that when we plot a certain contour line having a given value say $\bar{\beta}$ we can easily correct the current position on the basis of the error $\beta-\bar{\beta}$ as has been shown in article [2] concerned with a method for contouring on isoparametric surfaces. In the step

ISBN 0-12-747255-X

by step approach we have no means of acting similarly.

2. THE $\gamma\beta$-METHOD

2.1 *General*

No general discussion of the necessary boundary conditions connected with equation (1.2) was given in [1] as the treatment was restricted to just one numerical example.

Equation (1.2) describes a propagation type problem. This is seen immediately as the equation can be written in the form

$$u \frac{d\beta}{d\sigma} = 0 \tag{2.1}$$

where u is the magnitude of $\underline{u}$ and σ is the arc length along a field line. Thus we have an ordinary first order differential equation on a field line. The solution is simply β = constant on a field line. To fix the constant we must give the value of β at one and only one point on each field line. To distinguish between the different field lines, a non-zero gradient of β must exist in general in the crosswise direction to the field lines. Otherwise the distribution of β can be selected arbitrarily. It is thus obvious that the construction of an automatic field line determination algorithm based on equation (1.2) seems to be rather involved, because in a way we should know in advance the field line distribution to be able to give meaningful data on β.

We shall now describe another possibility. The idea is based on the following simple observation:

> The field line distribution is determined solely by the direction of the vector field $\underline{u}$ at each point and not by its magnitude. Thus the field line distributions for vectors $\underline{u}$ and $\gamma\underline{u}$, where $\gamma(x,y)$ is an arbitrary positive scalar function, are identical. (2.2)

We now select function $\gamma(x,y)$ so that the resulting field $\gamma\underline{u}$ will be solenoidal or so-called divergence-free. This is achieved if γ satisfies the equation

$$\operatorname{div}(\gamma\underline{u}) \equiv \frac{\partial}{\partial x}(\gamma u_x) + \frac{\partial}{\partial y}(\gamma u_y) = 0 \,. \tag{2.3}$$

For the solenoidal field $\gamma\underline{u}$ we can make use of the vector potential concept in two-dimensions (or as in the fluid mechanics literature of the stream function concept):

$$\gamma u_x = \frac{\partial \beta}{\partial y}\,, \quad \gamma u_y = -\frac{\partial \beta}{\partial x}\,. \tag{2.4}$$

As is also well-known, the field lines are then contour lines of the function β to be determined from equations (2.4).

We shall call this procedure — combined with the numerical processes to be described later — the $\gamma\beta$-method as it consists of the determination of two unknown functions $\gamma(x,y)$ and $\beta(x,y)$. We are not aware of the existence of a similar procedure in the literature.

2.2 *Determination of* γ

Equation (2.3) does not determine function γ completely. Some additional information is needed. In fact the equation describes a pure convection type propagation problem. This is seen by manipulating the equation into the form

$$u \frac{d\gamma}{d\sigma} + (\operatorname{div} \underline{u})\,\gamma = 0\,. \tag{2.5}$$

Thus we have again an ordinary first order differential equation on a field line. If the value of γ is given at one and only one point on each field line, the solution is fixed.

The general solution of equation (2.3) or (2.5) can be given in the form

$$\gamma u D = c = \text{constant} \tag{2.6}$$

on a field line. Here D is the distance between the two field lines forming the field filament — an infinitesimal field tube. The absolute value of D is of no consequence; only the way D changes along a field filament matters. Result (2.6) follows directly from the well-known rule that the total flux of a solenoidal vector $(\gamma \underline{u})$ through the cross-sections of a field tube is constant. We see again that when u and D are given (as is the case since D also is determined in principle from $\underline{u}$) it is enough to give the value of γ at one point of the field line to fix the value of c and after that γ can in principle be calculated everywhere on the field line. It is also realized that if γ is positive at one point it must remain positive everywhere on the field line.

Now it first seems that we are back at the difficulty of data handling discussed in connection with equation (2.1). However, there is a way around the difficulty. With γ we have no need to develop a crosswise gradient with respect to the field lines. Instead of giving an arbitrary value of γ at one point on each field line the distribution of γ can be fixed by demanding for example that γ should differ as little as possible from a given constant value γ_o in the least squares sense. Thus we can cast the problem into an extremum value formulation: find the γ making the value of expression

$$\Pi(\gamma) = \frac{1}{2} \int (\gamma-\gamma_o)^2 dA \tag{2.7}$$

minimum under the constraint (2.3). By employing the Lagrange multiplier function $\lambda(x,y)$ we have the alternative, more convenient formulation: find the γ and λ making the value of expression

$$\Pi_L(\gamma,\lambda) = \frac{1}{2} \int (\gamma-\gamma_o)^2 dA + \int \lambda \left[\frac{\partial}{\partial x} (\gamma u_x) + \frac{\partial}{\partial y} (\gamma u_y)\right] dA \tag{2.8}$$

stationary.

From the Euler equations of functional (2.8) the result

$$c \int_0^{\ell} \frac{1}{u^2 D} d\sigma = \gamma_o \int_0^{\ell} \frac{1}{u} d\sigma \tag{2.9}$$

is obtained. Quantity ℓ is the length of the field line in question. The integrals on both sides of the equation can be evaluated in principle. Thus the equation fixes the value of the constant c and so also the particular solution for γ on the field line. No special data handling in advance is needed.

We have used the value $\gamma_o = 1$ in the numerical calculations. The selection of a different value of γ_o just means that the γ obtained with $\gamma_o = 1$ is multiplied by the new γ_o and this has no effect on the final field line distribution.

The motivation behind the use of expression (2.7) emerged from the following line of thought. If γ were a constant, the magnitude of $\underline{u}$ would be inversely proportional to the distance between the field lines drawn at equal (infinitesimal) function value intervals. (This will occur if $\underline{u}$ already happens to be solenoidal.) This well-known property is a great aid in the visual interpretation of a solenoidal field line plot. Use of expression (2.7) thus means that we try to achieve this property as far as possible even in the general non-solenoidal case.

The finite element method is used in the numerical solution. However, to avoid the discretization of λ, we have replaced expression (2.8) by its penalty method counterpart:

$$\Pi_P(\gamma) = \frac{1}{2} \int (\gamma-\gamma_o)^2 dA + \frac{\alpha}{2} \int \left[\frac{\partial}{\partial x} (\gamma u_x) + \frac{\partial}{\partial y} (\gamma u_y)\right]^2 dA. \tag{2.10}$$

α is the penalty number. We have expressed it in the form $\alpha = \alpha' A/U^2$ where α' is a dimensionless penalty number, A is the area of the domain and U is a mean value of u defined by

$U^2 = \int u^2 dA/A$.

Function γ is approximated in the normal way. The terms γu_x and γu_y are approximated using either the separate or the group representation (see for instance [3]). The group representation is computationally more economical than the separate one.

Four-noded and nine-noded Lagrangian isoparametric quadrilateral elements are used. The penalty term is underintegrated to avoid locking. Gaussian 2×2 (1×1) and 3×3 (2×2) integration points are employed corresponding to the four- and nine-noded elements. The numbers in parentheses refer to the integration rule for the penalty term.

A symmetric linear equation system for the unknown nodal values of γ is obtained in the standard way from expression (2.10) after the approximations have been substituted into it. A banded solver is used to obtain the nodal values.

The ratio $\gamma_{max}/\gamma_{min}$ of the obtained maximum and minimum values of γ give some kind of measure of the severity of the non-solenoidality of the problem.

2.3 *Determination of* β

After γ has been determined we can consider equations (2.4) to represent an overdetermined system: two equations and only one unknown function β. Thus it is natural to use the least squares approach. The least squares functional is

$$\Pi(\beta) = \frac{1}{2} \int \left[\left(\frac{\partial \beta}{\partial x} + \gamma u_y\right)^2 + \left(\frac{\partial \beta}{\partial y} - \gamma u_x\right)^2 \right] dA \qquad (2.11)$$

when equal weighting on both equations (2.4) is practised.

Function β is approximated with the same shape functions and elements which were used in the determination of γ. Gaussian 2×2 and 3×3 integration points are employed corresponding to the four- and nine-noded elements. We can approximate the data terms γu_x and γu_y again either by the separate or the group representation.

A symmetric linear equation system for the nodal values of β is obtained in the standard way. One arbitrary nodal value must be given to fix the solution. In the program the value $\beta = 0$ is used at a given node.

A finite element formulation equivalent to the use of expression (2.11) but arrived at by a somewhat different reasoning has been described in [4] in connection with the determination of the stream function in a seepage problem.

If the field $\underline{u}$ is solenoidal, we can put $\gamma = \gamma_o$ (= 1) and start directly from expression (2.11).

3. A NUMERICAL EXAMPLE

Because of the penalty formulation, double precision has been employed in the determination of γ. The value $\alpha' = 10^8$ has been used with a computer having an accuracy of 7...8 digits in single precision.

In general, the most accurate results have been achieved by employing the separate and the group representations in the determination of γ and β respectively.

A plotting routine [5] based on the one given in [2] has been used.

Results from one simple example problem are given here. More numerical results are to be found in [6]. A square domain (Fig. 1) with a radial field through point Q ($x = -0{\cdot}25$ a, $y = -0{\cdot}25$ a, a is the side length of the square) is considered. The radial component of u is taken to be a constant u_o and the circumfer-

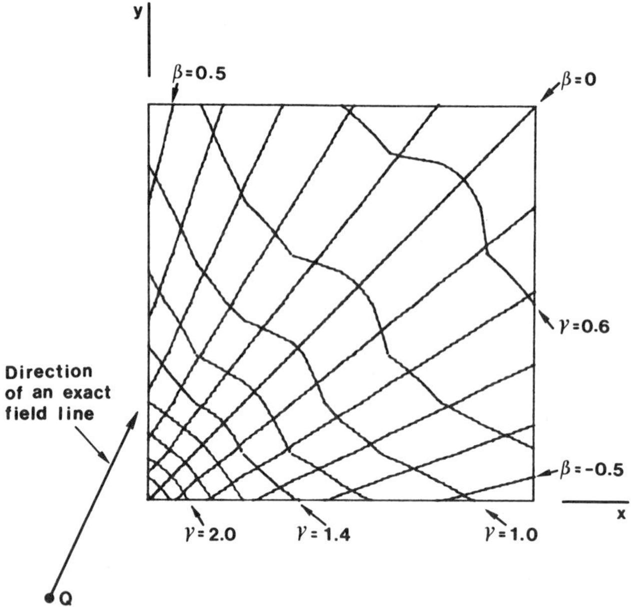

FIG. 1. Field lines (contour lines of β) and contour lines of γ.

ential component is zero. The field is non-solenoidal ($\text{div}\underline{u} = u_o/d$, where d is the distance from the point in question to point Q.) The exact field lines are straight lines through Q. A regular square $8 \times 8 = 64$ element mesh with four-noded elements is used.

It is seen that the field lines follow rather well the exact solution. The extremum values of γ were $\gamma_{max} = 2{\cdot}47$, γ_{min} =

0•52 and $\gamma_{max}/\gamma_{min}$ = 4•77. When the same problem was solved in the solenoidal case the obtained field line distribution was virtually identical to the one shown in Fig. 1.

4. CONCLUDING REMARKS

The $\gamma\beta$-method has worked quite well in several numerical test examples we have run so far when the field has not been excessively non-solenoidal.

In strongly non-solenoidal cases it is found that negative nodal values for γ are obtained. This is an indication that the discretization is inadequate. (The $\gamma\beta$-method changes the field into the opposite direction at the places where γ is negative and quite inaccurate plots are obtained.)

One possibility to proceed in such cases is to change the original field gradually. We can discard the obtained values of γ as unrealistic if they are under a certain selected critical value and modify the field only with the higher values of γ. This is allowable on the basis of statement (2.2). A new field is obtained and a new solution for γ is determined. Iteration can be continued until a reasonably solenoidal field is arrived at. The iteration process has improved the results dramatically in some difficult cases.

It may be remarked that the domain can also, if required, be divided into a certain number of subdomains and the $\gamma\beta$-method applied separately in each.

It is possible that the $\gamma\beta$-method, with further development (e.g. study of behaviour around different types [7] of singular points, study of multiply connected domains, treatment of field lines on curved surfaces by mappings etc.) could prove to be one useful addition to a graphics library.

ACKNOWLEDGEMENTS

The financial support received from the Academy of Finland is gratefully acknowledged.

REFERENCES

1. SOUTHWELL, R.V., *Relaxation Methods in Theoretical Physics. Vol. 1.* Oxford at the Clarendon Press, 164-168 (1946).

2. GRAY, W.H., AKIN, J.E., An Improved Method for Contouring on Isoparametric Surfaces. *Int. J. num. Meth. Engng.* 14, 451-472 (1979).

3. FLETCHER, C.A.J., The Group Finite Element Formulation. *Comput. Meths. Appl. Mech. Engrg.* 37, 225-243 (1983).

4. AALTO, J., Finite Element Seepage Flow Nets. *Int. J. num. and anal. Meth. Geomech.* (Accepted in 1982).

5. RÄSÄNEN, S.M. (Helsinki University of Technology), Private Communication (1980).

6. SALONEN, E.-M., SALMINEN, R., AALTO, J., Two Methods for Determination of Two-Dimensional Field Lines. *Report No. 18. Institute of Mechanics. Helsinki University of Technology.* In preparation.

7. v. KARMAN, Th., BIOT, M.A., *Mathematical Methods in Engineering*. McGraw-Hill, New York, 150-158 (1940).

FINITE ELEMENT ON THE D.A.P.

R. Wait and I. Martindale

Department of Statistics and Computational Mathematics, University of Liverpool, England.

1. INTRODUCTION - THE D.A.P.

The ICL D.A.P. (Distributed Array Processor) is a 64 x 64 array of 1-bit processors. Each processor is linked to its four nearest neighbours on a horizontal rectangular grid and it is connected vertically to its own 16K bits of store. Thus the machine can be visualised as a layer of processors sitting on top of a stack of 16K planes of store. As the processors are all very simple, they all obey the same set of instructions provided by the master control unit. The same instructions are carried out on the particular data held in the individual processor's own stack of store and hence the D.A.P. is an example of a single-instruction-multiple-data (or simd) machine. At any instant, it is possible that some of the processors may be switched off, but all the active processors have to obey the same instruction stream and the execution time is independent of the number of processors active at the time. Finite element calculations have always been viewed as forming a significant proportion of the potential market for the D.A.P. [1]. At present the D.A.P. has to be used via a host processor which in the case of the 64 x 64 D.A.P. at Q.M.C., is an ICL 2980.

2. ELEMENT MATRIX ASSEMBLY

As the processors all follow the same instruction stream they must each assemble a patch of elements that are all topologically equivalent. In the simplest case, each element would be allocated to a separate processor and the processors in excess of the number of elements remain idle. The processors are switched on and off by using logical masks. There are two ways of assigning elements to processors, one is a random allocation and the other is to map the grid onto the D.A.P. array. The random allocation is suitable for irregular grids and regions in which it is not possible to identify rows and columns of elements. It is also the most convenient form if a small amount of

ISBN 0-12-747255-X

local grid refinement is likely to be necessary. The allocation of a processor to the patch to be refined is replaced by the assignment of a number of previously idle processors, one to each of the new sub-patches. For example, Figure 1 shows how a hypothetical 4 x 4 processor array might be assigned to assemble a grid of triangular elements as the grid is successively refined. In each discretisation, each processor would assemble a single element stiffness matrix and the total time taken to assemble all the element matrices is the same for each of the grids. As the grid is refined, a progressively smaller fraction of the processor array remains idle and hence the finer grids are making a more efficient use of the machine.

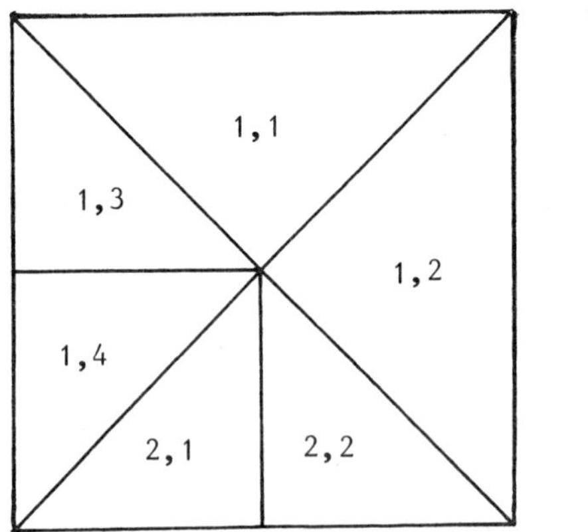

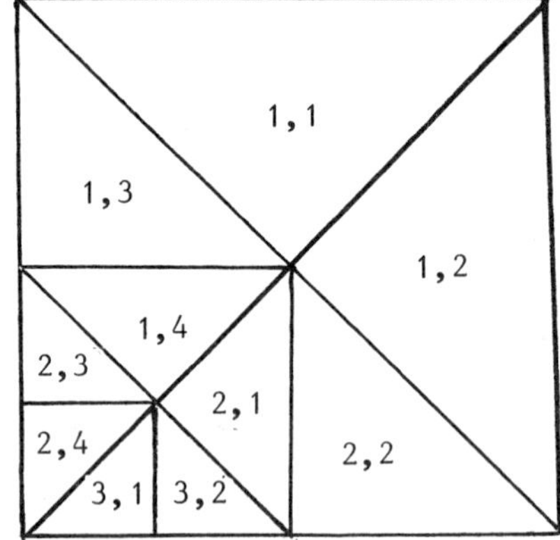

FIG. 1. Assignment of elements to a 4 x 4 D.A.P. as the grid is refined.

The alternative element assignment is to map the rows and columns of a quasi-regular grid onto the rows and columns of the processor array. With this strategy, a refinement such as illustrated in Figure 1, would be extremely difficult to implement and it would be necessary to design a grid with the processor array always in view. For example, on a 4 x 4 D.A.P. all the element matrices in the grid of Figure 2(a) would be assembled as quickly as those in Figure 2(b), the mesh is still refined in all the right places and the elements can be mapped onto the processors in a much more systematic manner.

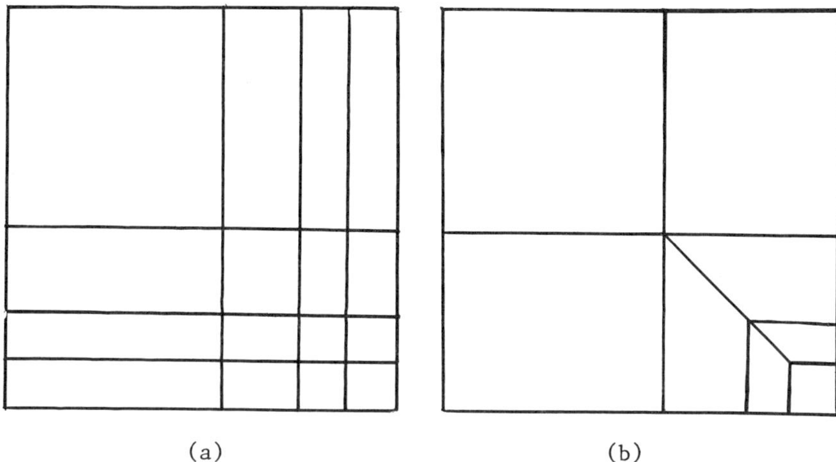

(a) (b)

FIG. 2. Graded meshes for use with a row-by-row processor allocation: (a) good (b) bad.

Assuming that each processor stores all the necessary element and nodal coordinate information relating to its own patch of elements, there is no communication between the processors during the element matrix assembly. As shown in Table 1, the timings are independent of the processor allocation scheme adopted at the element assembly but the position changes dramatically when the elements are combined into the global matrix. The times shown in Table 1 are in micro-seconds and relate to the solution of Laplace's equation on a square with Dirichlet boundary conditions. For the random allocation, the global assembly time and the time to incorporate the boundary conditions relate to a problem with 1600 nodes [2] all other timings are independent of grid size and assume one bilinear element per processor during the assembly.

TABLE 1

D.A.P. Timings

	Random	Regular
Element Assembly	250	250
Global Assembly	180	2
Incorporate Boundary Conditions	820	8
One Iteration Step	10	10

3. GLOBAL MATRIX ASSEMBLY

The equations in the global form are nodal equations, so the solution has to be based on a processor allocation that is node based rather than element based. Thus a finite element computation on the D.A.P. has a reassignment and global assembly phase for which there is no direct counterpart in a serial algorithm.

The D.A.P. equivalent of a frontal solution would involve the assembly of a substructure and the whole processor being treated in the same way as a single element on a serial machine. For the purposes of this paper, we shall restrict our attention to problems that can be solved on the D.A.P. without recourse to backing store. With existing algorithms it is reasonable to consider the allocation of up to 40 nodes per processor, that is problems with up to 160K nodes, without using any backing store.

Given that adjacent elements are assembled on different processors, it is necessary for the data to be transmitted from one processor to another in order to assemble the nodal equations. It is at this stage that the difference between the two allocation schemes has a considerable effect on the overall time for the computation. Each processor in the D.A.P. array is connected directly to its four nearest neighbours and the time taken for communication over greater distances is directly proportional to the number of nearest neighbour pairs covered by the path. Thus communication between random pairs of processors can be very expensive and the algorithm can become effectively serial at this point.

With a regular arrangement of the elements, communication is only between neighbours (including diagonally which is twice as slow) and so (see Table 1) the global assembly time is only a small fraction of the total element matrix generation time. With the (more-or-less) random assignment of elements to processors, the communication of data between processors at the global assembly stage is over large distances (in terms of the D.A.P. hardware) and the time for the global assembly dominates the whole computation.

There are two ways of reducing this communication time, one is to replace the communication by redundant calculation and the other is to use an iterative solution that only requires the residuals as these can be generated element-by-element and hence there is no need for a global matrix assembly. The redundant calculation is in the form of duplicating the element matrix assemblies so that each processor has all the element matrices necessary to assemble its nodal equations without any external communication.

4. SOLUTION OF THE NODAL EQUATIONS

It is generally accepted [3], [4], [5] that iterative methods parallelize better than direct methods for finite element calculations. The first reason is that Cholesky factorisation leads to triangular matrices and involves operations on vectors of differing lengths. Thus it is not easy to make efficient use of a processor array of a fixed size. The second reason in the case of the D.A.P. is that, although the total memory of the machine is considerable (8 megabytes), after the memory stack is partitioned between the processors there is only a relatively small amount (2 kilobytes) per processor.

Iterative methods involve operations on vectors of a constant length and there is no problem of fill-in so that only the initial non-zeros need be stored. It is possible, for the model problem, to accommodate the non-zeros of approximately 40 nodes on a single processor and so solve for a total of 160K nodes.

There is a considerable amount of literature on the parallel solution of finite element problems, particularly relating to the finite element machine at NASA Langley [3], [5]. This f.e. machine is an mimd machine (i.e. the individual processors can have independent instruction streams) with a small number (8 in August 1983) of processors, but some of the conclusions are still valid for the D.A.P. (simd with 4096 processors). We have considered a conjugate gradient solution algorithm with different forms of preconditioning as this appears to be one of the most encouraging classes of methods being investigated.

The basic algorithm for solving $\underline{A}\,\underline{x} = \underline{b}$ with a preconditioning matrix $\underline{M}$ is:

Initialize $\underline{M}\,\underline{x}_1 = \underline{b}$, $\underline{r}_1 = \underline{b} - \underline{A}\,\underline{x}_1$, $\underline{M}\,\underline{d}_1 = \underline{r}_1$

Iterate for $k = 0,1,\ldots,k_{max}$ while $\underline{r}_k^t\,\underline{r}_k > \varepsilon$

a) $\alpha_k = (\underline{r}_k^t\,\underline{M}^{-1}\,\underline{r}_k)/(\underline{d}_k^t\,\underline{A}\,\underline{d}_k)$

b) $\underline{x}_{k+1} = \underline{x}_k + \alpha_k\,\underline{d}_k$

c) $\underline{r}_{k+1} = \underline{r}_k - \alpha_k\,\underline{A}\,\underline{d}_k$

d) $\beta_k = (\underline{r}_{k+1}^t\,\underline{M}^{-1}\,\underline{r}_{k+1})/(\underline{r}_k^t\,\underline{M}^{-1}\,\underline{r}_k^t)$

e) $\underline{d}_{k+1} = \underline{M}^{-1}\,\underline{r}_{k+1} + \beta_k\,\underline{d}_k$.

It is assumed that the systems of the form $\underline{M}\,\underline{d} = \underline{r}$ can be solved cheaply. The most frequently used preconditioning schemes are either an incomplete Cholesky factorization [6] or an inner iteration such as ssor or a multigrid approximation [7], [8].In the latter cases, it is usual to take only a very small number of steps of the inner iteration [4], [7].

Both of these strategies are possible on the D.A.P., but in either case the implementation is very dependent on the allocation of nodal equations to the individual processors.

5. PROCESSOR ALLOCATION

In the element generation phase a patch of elements is assigned to each processor, but as stated earlier, the patches must all be topologically equivalent. Thus, in order to partition a general mesh into such patches, it is likely that they will be compact (almost square) patches rather than long thin rows of elements. In preparation for the solution phase, the most efficient form of global assembly will be to select a subset of the nodes from the processor patch, to be assembled on the same processor. Figure 3 gives an illustration in which four bilinear elements are assembled on each processor of a 2x2 D.A.P. and then four nodes are selected from the processor patch to be assembled.

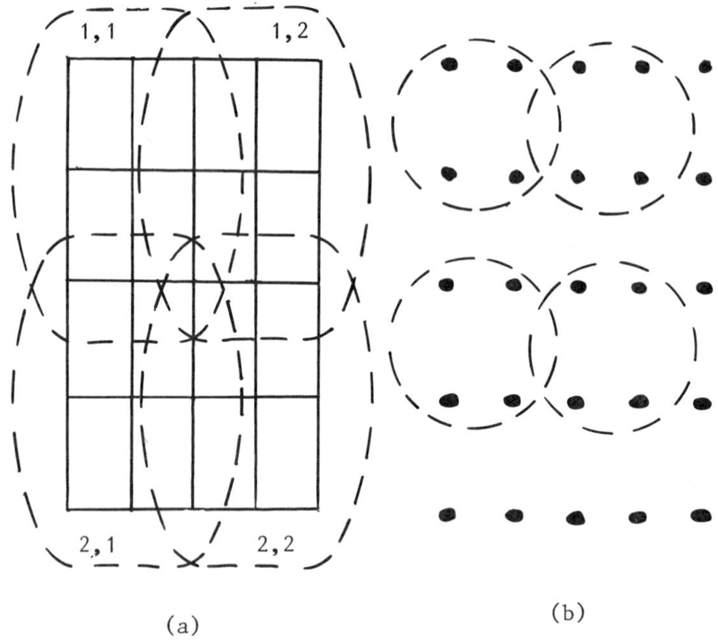

FIG. 3 Processor allocation on a 2x2 D.A.P. (a) element assignment during assembly, (b) nodal assignment during solution.

If the nodes are numbered sequentially within each processor, the example in Figure 3 leads to the node numbering

1	3	9	11
2	4	10	12
5	7	13	15
6	8	14	16

which in turn leads to a matrix $\underline{A}$ with block structure

$$\underline{A} = \underline{D} + \underline{Q} + \underline{Q}^t,$$

where the block diagonal matrix $\underline{D}$ has diagonal blocks $\underline{D}_{ij}$, that correspond to the nodes stored in processor P_{ij} alone.
In terms of D.A.P. hardware it is very convenient to use $\underline{M} = \underline{D}$ as the preconditioning matrix. If there are no more than 16 nodes per processor it is very straightforward to construct a Cholesky factorisation

$$\underline{D} = \underline{L}\,\underline{L}^t$$

by factorizing

$$\underline{D}_{ij} = \underline{L}_{ij}\,\underline{L}_{ij}^t$$

on processor P_{ij}. Similarly the systems

$$\underline{L}_{ij}\,\underline{y}_{ij} = \underline{r}_{ij}$$

$$\underline{L}_{ij}^t\,\underline{d}_{ij} = \underline{y}_{ij}$$

can be solved locally on processor P_{ij} without the need for any inter-processor communication. When larger numbers of nodes are allocated to each processor, it is necessary to solve the systems

$$\underline{M}_{ij}\,\underline{d}_{ij} = \underline{r}_{ij}$$

by, for example, ssor. Table 2 shows the results of the conjugate gradient solution of the same model problem as used in Table 1. With one node per processor, the preconditioning is simple diagonal scaling. It should be remembered that down each column, the time of a single iteration is independent of the problem size. The increase in the time taken for a single iteration as the preconditioning becomes more complex soon cancels out the gain from the reduction in the number of iterations needed for convergence.

TABLE 2

Number of Conjugate Gradient Iterations

	Nodes per Processor		
	1	4	9
Total Nodes			
900	27	22	20
3600	54	45	40
8100	82*	66	60
14400		88	80
22500			100

The stopping criterion was $\underline{r}_k^t \underline{r}_k < 10^{-7} \underline{r}_0^t \underline{r}_0$ and the * indicates an estimate using 4 nodes per processor with diagonal scaling only.

Thus, the method should be viewed as a means of extending the capability of the machine to solve larger problems, rather than a method of speeding-up convergence. As stated earlier the most efficient method for a given discretization is likely to be the method that restricts the number of idle processors i.e. spreads the work out as thinly and evenly as possible. If the data is rearranged as in Table 3, it can be seen that if the increase in the number of nodes per processor is viewed as a grid refinement over the whole region, the deterioration in the speed of convergence is approximately $h^{-\frac{1}{2}}$ that is expected from the theory of incomplete factorizations [9].

TABLE 3

Deterioration in Convergence

	Nodes per processor		
	1x1	2x2	3x3
Active D.A.P.			
20 x 20	18	30	40
30 x 30	27	45	60
40 x 40	36	59	80

REFERENCES

1. PARKINSON, D., An introduction to array processors. *Systems International* (1977).
2. DAVIES, S., *Private communication* Queen Mary College (1984).
3. STORAASLI, O.O., et al., The finite element machine: An experiment in parallel processing. *NASA Tech. Mem.* 84514, NASA, Langley (1983).
4. ADAMS, L., An M-step preconditioned conjugate gradient method for parallel computation. *NASA Contractor Rep.* 17210, NASA, Langley (1983)
5. ADAMS, L., Iterative algorithms for large sparse systems on parallel computers. *NASA Contractor Rep.* 166027, NASA, Langley (1983).
6. MANTEUFEL, T.A., The shifted incomplete Cholesky factorization. *Tech. Rep.* SAND 78 - 8226, Sandia Labs., Alberquerque (1978).
7. JACOBS, D.A., Preconditioned conjugate gradient methods for solving systems of algebraic equations. *Tech. Rep.* RD/L/N 193/80, Central Electricity Research Labs., Leatherhead (1981).
8. MARKHAM, G., The application of multigrid techniques to preconditionings for conjugate-gradient-type methods. *Tech. Rep.* TPRD/L/AP 127/M83, Central Electricity Research Labs., Leatherhead (1983).
9. AXELSSON, O., On preconditioned conjugate gradient methods. pp. 24-35 of I. S. Duff (ed.), *Conjugate gradient methods and similar techniques*. Tech. Rep. AERE - R 9636, AERE Harwell (1979).

DESIGN AND IMPLEMENTATION OF AN EFFICIENT THREE-DIMENSIONAL FINITE ELEMENT CODE FOR HIGH TEMPERATURE PROBLEMS

T.J.W. Ward* and C.G. Burton†

**Institute of Computational Mathematics, Brunel University, U.K.*

†*Royal Aircraft Establishment, Pyestock, U.K.*

1. INTRODUCTION

Experience of jet engine development over the last twenty-five years has highlighted the critical nature of the hot section. One of the most intractable problems has been the achievement of an acceptable life for these components. This has required extensive programmes of development engine testing which have had a severe impact on project timescales and costs. In order to obtain improvements in service life it is essential to be able to assess the life of a particular component at the design stage. Such assessments require the accurate determination of the stresses within the component. In principle this can be achieved using the finite element method as it is capable of modelling the complex three-dimensional geometry of modern turbine components and can incorporate a wide range of material behaviours.

The analysis would have to cater for three types of nonlinearity, those due to plasticity, creep and large displacements. In addition gas turbine components experience severe thermal loads. There are today many proprietry codes which include these properties, but it was decided that an 'in house' code should be written. It was known that, due to the complex geometry involved, the analysis would inevitably be very large (approximately 10,000 degrees of freedom). To undertake such an analysis in a realistic time would require an extremely efficient code employing the latest numerical techniques, and this could best be achieved through writing an application specific program. In addition it was vital that the code could be readily adapted to incorporate different material behaviours.

The first stage in the project was to develop an efficient three-dimensional elasto-plastic code [1]. In this paper we describe that code and particularly those factors which contribute to its efficiency. All the work described was performed on a VAX 11/780 with 5.5 Mbytes of main memory and 692 Mbytes of disc storage.

ISBN 0-12-747255-X

2. THE ELASTO-PLASTIC MATHEMATICAL MODEL

The structure to be analysed is discretised into an appropriate number of 20 node isoparametric 'brick' elements. We have assumed for the current model that displacements are small and that there are no geometric non-linearities; this means that the strain ($\underline{\varepsilon}$) - displacement ($\underline{u}_{el}$) relationship in each element is linear,

$$\underline{\varepsilon} = [B]\ \underline{u}_{el}\ , \tag{1}$$

where [B] is the strain-displacement matrix. When the structure deforms elastically the stress ($\underline{\sigma}$) - strain ($\underline{\varepsilon}$) relationship is linear

$$\underline{\sigma} = [D]\ \underline{\varepsilon}\ , \tag{2}$$

where [D] is known as the constitutive matrix. Plasticity occurs at a point within a structure when the yield criterion [3] is satisfied; that is

$$F(\underline{\sigma},\kappa) = 0\ , \tag{3}$$

where κ is the initially uniaxial yield function (which is temperature dependent) and both κ and F have to be determined experimentally. At a point of the structure which is deforming plastically the constitutive matrix is non-linear. We have adopted the incremental theory of plasticity [4], [5] which relates increments of stress ($d\underline{\sigma}$) to increments of strain ($d\underline{\varepsilon}$)

$$d\underline{\sigma} = [D_{ep}]\ d\underline{\varepsilon}\ . \tag{4}$$

The constitutive matrix $[D_{ep}]$ is stress dependent. The stresses in the structure are related to the applied forces through the principal of virtual work [2]. For the elastic deformation this gives rise to the system of equations

$$[K]\underline{u} = \underline{f}\ . \tag{5}$$

This describes the nodal deflections $\underline{u}$ of the structure in terms of the structural force vector $\underline{f}$ and the elastic stiffness matrix [K],

$$[K] = \sum_{\text{elements}} \int_{ve} [B]^T[D][B]\ dv\ . \tag{6}$$

For the plastic deformation the principal of virtual work is invoked for each load increment giving rise to the system of equations

$$[K]_T\ \Delta\underline{u} = \Delta\underline{f}\ , \tag{7}$$

where $\Delta \underline{f}$ is the increment of load, $\Delta \underline{u}$ the increment of nodal deflections,

$$[K]_T = \sum_{elements} \int_{ve} [B]^T [D_{ep}][B]\, dv \; . \tag{8}$$

The stiffness matrix $[K]_T$ is stress dependent. Equation (7) is thus nonlinear and an iterative procedure has to be adopted.

Integration within each element is performed numerically using a $2 \times 2 \times 2$ Gauss quadrature rule, so called reduced integration [2],[3]. Reduced integration is adopted since it is significantly faster than full ($3 \times 3 \times 3$ Gauss) integration, and also because there are some advantages in the accuracy of the strains and stresses which are calculated at the Gauss points of each element.

3. THE FRONTAL SOLUTION METHOD

The stiffness matrix in equations (6) and (8) is symmetric and banded, and only the components in the band in the upper half of the matrix are stored. The stiffness matrix is also positive definite and thus for Gaussian elimination [6] the equations can be reduced in any order without the necessity of searching for the largest pivot. No additional storage is required during the solution procedure. This makes the algorithm computationally very attractive, since it is both reliable and efficient. The Gaussian elimination process falls into three distinct phases, the first two are to reduce the stiffness matrix and the force vector and the third phase is back substitution. It can be shown that the number of operations required for forward reduction is of the order NB^2, N and B being the order and bandwidth of the matrix, and for back substitution and reduction of the force vector of the order of NB. In a typical turbine component we have estimated that $N = 10{,}000$, and that $B = 700$. Thus a significant saving is achieved if the stiffness matrix is unchanged for a number of right-hand sides. The most efficient method of implementing Gaussian elimination in a finite element program is the frontal solution method [7]. The alternative band solution technique is significantly less efficient, if, as in the present case, not all the stiffness matrix can be in core simultaneously. The frontal solution method, with optimal element ordering [8], is applied as follows. The stiffness matrix for the current element is first calculated (or read back from storage) and its degrees of freedom are added into the front. If the degree of freedom already exists in the front the corresponding terms of the stiffness matrix are simply added to the existing terms in the front. However, if the degree of freedom is not present in the front a new location must be allocated for it. If the new location is at either end of the front the distance between the first and last term, the frontwidth, will increase. The element degrees of freedom are then re-examined to determine which if any of them can be elminated. As the elimination proceeds the frontwidth is decreased if the degrees of freedom at either end of the front are reduced. This is desirable since the number of operations required to reduce a given degree of freedom is proportional to the square of the current frontwidth. The addition of degrees of freedom that remain in the front for a short time to the edges of the front rather than the centre will tend to promote the desired decreases in the frontwidth. This technique is used in our code, a similar technique is described by Yeo [9].

4. SOLUTION ALGORITHMS

The code contains a number of solution algorithms allowing the user to efficiently and effectively solve a range of problems with varying degrees of plasticity. For all the algorithms the solution procedure is as follows. Initially an elastic analysis is performed. The nodal centrifugal and thermal loads (see Section 5) are summed into a vector of the total load of the structure $\underline{f}$. The elastic stiffness matrices for each element of the structure are formed, and the displacements $\underline{u}$ are calculated for the system of equations

$$[K]\ \underline{u} = \underline{f}\ . \tag{9}$$

From the displacements the strains and the stresses are calculated at the Gauss points. Then using equation (3) the elastic load factor δ_1 is calculated. The elastic load factor is that proportion of the total load applied to the structure that will cause an elastic response. The elastic load is

$$\underline{f}_e = \delta_1\ \underline{f}\ . \tag{10}$$

The corresponding elastic displacements strains and stresses are determined by scaling the displacements, strains and stresses calculated from equation (9) by δ_1.

The plastic load $\underline{f}_p = \underline{f} - \underline{f}_e$ is then applied in increments. On the first iteration of each of these increments the total load $\underline{f}$ is applied to the structure. The structure is characterised by the stiffness matrix which will vary depending on the solution algorithm. Having obtained displacements corresponding to the total load these are then scaled by the load factors, δ_j, $j \geq 2$, to calculate the appropriate displacement increments, $\Delta\underline{u}_j^0$. The load factors δ_j are either calculated using the automatic loading scheme [1], [10], [11] or prescribed by the user. From the displacement increments, $\Delta\underline{u}_j^0$, the strain increments $\Delta\underline{\varepsilon}_j^0$, and stress increments $\Delta\underline{\sigma}_j^0$ are calculated at the Gauss points. At those Gauss points at which plasticity has occurred a portion of the stress increment will have had to be calculated from the strain increment using the elasto-plastic relationship between stress and strain increments, equation (4). In order that this calculation is performed as accurately as possible the elasto-plastic portion of the strain is divided into M equal portions and the calculation performed on each portion separately. M is determined in a similar manner to that suggested by Owen and Hinton [3].

At the end of the ith iteration the residual $\underline{\psi}_j^i$ is calculated from

$$\underline{\psi}_j^i = \int_V [B]^T \underline{\sigma}_j^i \, dv - \underline{f}_j , \qquad (11)$$

where [B] is the strain displacement matrix $\underline{\sigma}_j^i$ is the total stress state at the end of the current iteration, and $\underline{f}_j$ is the total load applied to the structure so far plus the reactions at the constrained nodes. The residual $\underline{\psi}_j^i$ is compared to a preset tolerance, and if convergence has not occurred then a further iteration is required and the system of equations

$$[K]_T \, \Delta\underline{u}_j^i = - \underline{\psi}_j^i \qquad (12)$$

is solved for the displacement increments $\Delta\underline{u}_j^i$. The stresses, strains and displacements are accumulated at the end of each iteration.

When using the tangent stiffness solution algorithm the stiffness matrix $[K]_T$, is recalculated at the start of each iteration. This is a second order solution method. However for large problems the reduction of the new stiffness matrix every iteration is extremely expensive in computer time. In the Hybrid Mk1 method the stiffness matrix is recalculated only on the first iteration of each load increment. This has the virtue of being significantly less demanding on computer time for large problems than the tangent stiffness method, and also models the changing stiffness of the structure provided that there is not a significant amount of plasticity within each load increment. In the Hybrid Mk2 method the stiffness matrix is recalculated only on the second iteration of each load increment. This is similar to the Hybrid Mk1 method, except that it should more accurately model the changing stiffness of the structure during a load increment, and thus allow larger load steps to be taken. The initial stiffness method uses the elastic stiffness matrix throughout. For large problems this is an extremely efficient method, but because the changing stiffness of the structure is not modelled, we would expect that this method would not converge for problems in which the stiffness of the structure changes markedly. In addition to these four solution algorithms the code allows the solution acceleration scheme suggested by Crisfield [12], [13].

5. MODELLING OF THERMAL LOAD

Thermal loads are an important contributory factor to stresses in turbine components. The stresses caused by the thermal strains will be affected by the yielding of the materials. Cast superalloys exhibit considerable temperature dependence of material properties, with the yield stress dropping by a factor of 2½ between 850 and 1000C.

In the analysis it is assumed that the structure has zero thermal strain at an ambient temperature θ_a, which is usually taken to be 20C. If the temperature at a point of the structure is different from the ambient temperature, then there will be thermal strains at this point. The size of these strains will depend on the temperature above ambient $\theta_d \equiv \theta - \theta_a$, and on the mean coefficient of linear thermal expansion, α, which will also depend on θ, $\alpha = \alpha(\theta)$. Thus

$$\underline{\varepsilon}_\theta = [\alpha\theta_d, \alpha\theta_d, \alpha\theta_d, 0, 0, 0] \tag{13}$$

where $\underline{\varepsilon}_\theta$ are the thermal strains. In general the structure will be subject to a temperature distribution and $\underline{\varepsilon}_\theta$ will vary from point to point.

The stresses at a point of the structure will be caused by the difference between the actual strains and the thermal strains. For a linear elastic analysis the constitutive relationship will take the form

$$\underline{\sigma} = [D](\underline{\varepsilon} - \underline{\varepsilon}_\theta) \ . \tag{14}$$

The equilibrium of a single element of the structure expressed in terms of the stresses is given by

$$\int_{ve} [B]^T \underline{\sigma} \, dv = \underline{f}_{el} \ . \tag{15}$$

Substituting (14) into (15) gives

$$\int_{ve} [B]^T [D](\underline{\varepsilon} - \underline{\varepsilon}_\theta) \, dv = \underline{f}_{el} \ , \tag{16}$$

or

$$\int_{ve} [B]^T [D] \, \underline{\varepsilon} \, dv = \underline{f}_{el} + \int_{ve} [B]^T [D] \, \underline{\varepsilon}_\theta \, dv \ . \tag{17}$$

Thus

$$\int_{ve} [B]^T [D][B] \, dv \, \underline{u}_{el} = \underline{f}_{el} + \int_{ve} [B]^T [D] \, \underline{\varepsilon}_\theta \, dv \ . \tag{18}$$

Summing over all the elements of the structure gives

$$[K] \, \underline{u} = \underline{f} + \underline{f}_\theta \ . \tag{19}$$

In the elasto-plastic analysis the equilibrium equation for a general load increment is given by

$$[K]\ \Delta u = \Delta \underline{f} + \Delta \underline{f}_\theta \ , \tag{20}$$

where

$$\Delta \underline{f}_\theta = \int_V [B]\ [D_{ep}]\ \Delta \underline{\varepsilon}_\theta \ dv \tag{21}$$

and $\Delta \underline{\varepsilon}_\theta$ is an increment of the thermal strain.

In our code the thermal strains resulting from the temperature distribution are calculated using equation (13). A portion of the thermal strain is applied in each load increment and the thermal load is calculated using equation (21). The size of each portion is determined by the load factor for the load increment. An alternative strategy would be to increment the temperature. However as the material properties are temperature dependent every element stiffness matrix would need to be recalculated at least once each load increment. The additional computation involved would be significant, particularly in those problems in which yielding is confined to a relatively small part of the structure.

6. TEST PROBLEMS

The code has been tested on a wide range of test problems. We present here the results for the plastic deformation of a hollow sphere of ideally plastic material obeying the Von Mises yield criterion under internal pressure and thermal loading. Analytic solutions for the above problems have been derived [4],[14],[15]. The finite element mesh used for the analysis is shown in Fig. 1, due to the symmetry only $\frac{1}{8}$ of the sphere need be considered. The mesh has 162 elements, 922 nodes and the maximum frontwidth during the reduction is 537 degrees of freedom. The inner radius, a = 100m, the outer radius b = 200m. Young's Modulus $E = 2.14 \times 10^{11}$ N/m^2, Poisson's ratio $\nu = 0.3$ and the yield stress $\sigma_Y = 2.14 \times 10^8$ N/m^2.

6.1 The Internally Pressurised Sphere

As the pressure inside the sphere is increased yielding begins at the inside surface. The pressure, p, required to cause yielding to spread to a radius, r_c, is given by

$$p = \sigma_y \left[2 \ln\left(\frac{r_c}{a}\right) + \frac{2}{3}\left(1 - \left(\frac{r_c}{b}\right)^3\right)\right] . \tag{22}$$

In the model yielding is only monitored at the element Gauss points. Initial yield will occur at the most highly stressed Gauss point at r_c = 103.0m. The analysis predicted yield at this point at 15% of the full load of 4.143 σ_y. Equation (22) gives a pressure of 15.3% of the full load, in excellent agreement with the calculated value.

In order to examine the speed and convergence of the solution algorithms the sphere was analysed using the same load history

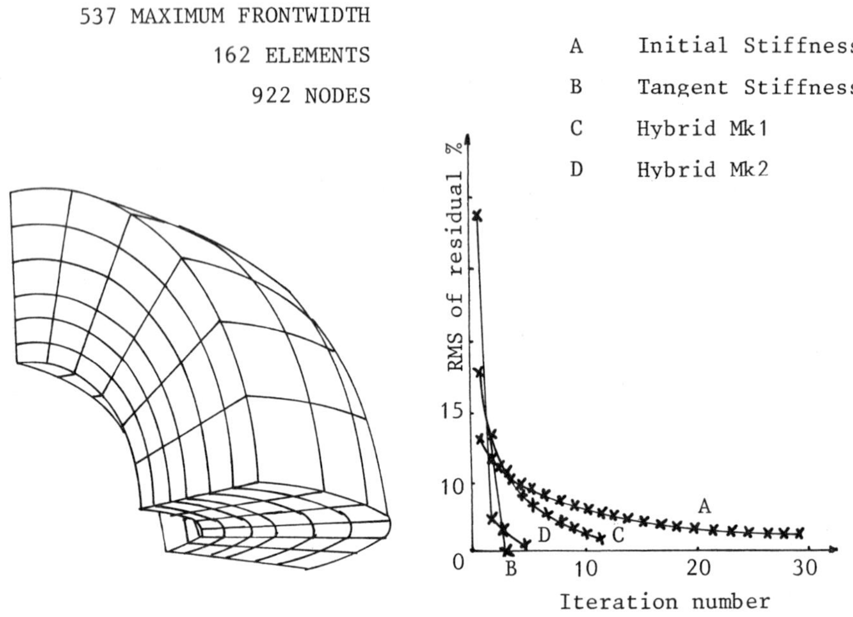

FIG.1 Finite element mesh of 1/8 sphere

FIG. 2 Comparison of convergence rates - load increment 8

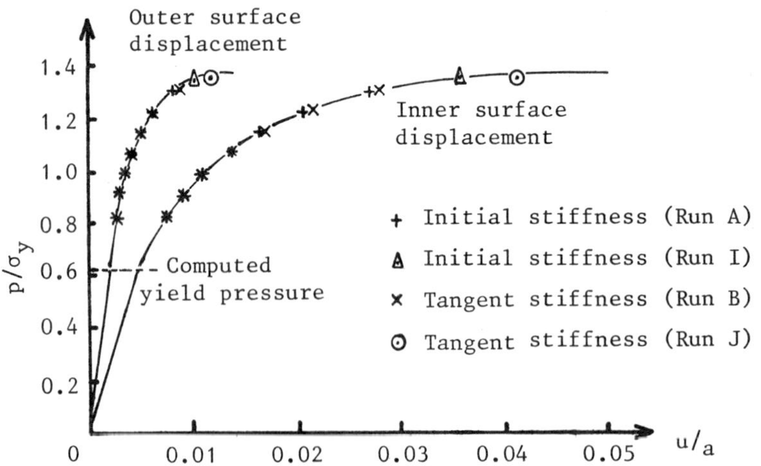

FIG. 3 Comparison of theoretical and computed load-displacement curves

for each technique. This was 8 load increments to a final load factor of 0.315. The results are summarised below.

Run	Algorithm	Total number of iterations	CPU time on VAX 11/780
A	Initial stiffness	96	5h 10m
B	Tangent stiffness	16	16h 28m
C	Hybrid Mk1	39	8h 40m
D	Hybrid Mk2	21	8h 54m
E	Hybrid Mk1+ Crisfield acceleration	-	Diverges
F	Hybrid Mk2+Crisfield acceleration	20	8h 53m
G	Initial stiffness + Crisfield acceleration	84	4h 14m
H	Hybrid Mk1+ Hybrid Mk2	24	7h 44m

Examination of intermediate timings showed that to reformulate the stiffness matrix and solve the resultant equation took 1h 1m and to perform an iteration took $2\frac{1}{2}$m, a ratio of 25:1. Thus even though the number of iterations required for the initial stiffness algorithm is large, it is actually the fastest because the stiffness matrix is only formed once. By contrast the tangent stiffness method performs relatively poorly because the stiffness matrix is reformed each iteration. The Hybrid Mk2 takes longer than the Hybrid Mk1 because it undertakes one extra stiffness matrix evaluation; in the first plastic load increment the Hybrid Mk1 algorithm uses the elastic stiffness matrix throughout. Run H shows the result of using the Hybrid Mk1 method for the first plastic load increment and the Hybrid Mk2 thereafter.

The performance of the Crisfield technique appears erratic. It may be that a more sophisticated application of the Crisfield technique in which the optimum accelerated solution is sought [11] should be incorporated in the code.

The relative convergence rates of some of the different solution algorithms are shown in Fig. 2 for a representative load increment. As expected the initial stiffness algorithm gives the worst convergence. It is clear that the number of iterations required by this algorithm could be considerably reduced if a larger convergence tolerance was used.

The initial stiffness and tangent stiffness algorithms were rerun with the whole of the post yield load up to a factor of 0.329 being applied in 1 load increment. The results are summarised below

Run	Algorithm	Total number of iterations	CPU time on VAX 11/780
I	Initial stiffness	54	3h 20m
J	Tangent stiffness	5	5h 14m

Convergence was still achieved even with such a large initial load step, and the CPU time was reduced for both algorithms.

The theoretical solution for the displacement of the interior surface of the hollow sphere is given by

$$u = \frac{2}{3}\frac{\sigma_y}{E}(1-2\nu)\left[\frac{3}{2}\frac{(1-\nu)}{(1-2\nu)}\left(\frac{r_c}{a}\right)^3 + \left(\frac{r_c}{b}\right)^3 - 1 - 3\ln\left(\frac{r_c}{a}\right)\right] a \quad (23)$$

and for the exterior surface by

$$u = \frac{\sigma_y}{E}\left(\frac{r_c}{b}\right)^3 (1-\nu) b \ . \quad (24)$$

Fig. 3 depicts these equations and the results of runs A,B,I and J.

The theoretical solution for the stresses within the sphere is given by

$$\sigma_{rr} = 2\sigma_y\left(\ln \frac{r}{a}\right) - p \ , \quad \sigma_{\theta\theta} = \sigma_y\left(1 + 2\ln \frac{r}{a}\right) - p \ , \quad (25)$$

$$\sigma_{rr} = -\frac{2\sigma_y}{3}\left(\frac{r_c}{r}\right)^3\left(1 - \frac{r^3}{b^3}\right) , \quad \sigma_{\theta\theta} = \frac{\sigma_y}{3}\left(\frac{r_c}{r}\right)^3\left(1 + \frac{2r^3}{b^3}\right) , \quad (26)$$

respectively for the plastic and elastic regions. The elasto-plastic transition radius, r_c, can be determined from

$$p = \frac{2\sigma_y}{3}\left[1 - \left(\frac{r_c}{b}\right)^3 + 3\ln\left(\frac{r_c}{a}\right)\right] . \quad (27)$$

Fig. 4 depicts the theoretical solutions and the predicted value for the final load increments of runs A and B.

In general excellent agreement is obtained between the predicted and theoretical solutions. There is no detectable difference in the magnitude of errors either between one load step or multi load step solution, or between the initial and tangent stiffness algorithm. It is clear therefore that for this problem the choice of solution strategy can be made purely on the basis of minimising the required CPU time.

6.2 *Thermal Loading*

In order to investigate the performance of the code when analysing problems involving thermal loads, the hollow sphere with an applied temperature gradient was analysed. The inner surface temperature T_i = 320C, the outer surface temperature T_0 = 20C, the mean coefficient of linear expansion $\alpha = 1.35\times10^{-5}$/C, and all material properties were assumed to be independent of temperature. The resulting steady state temperature distribution is given by

$$T = \frac{T_i a}{(b-a)}\left[\frac{b}{r} - 1\right] + T_0 \ . \quad (28)$$

The internal temperature required to cause yield to a radius r_c is given by

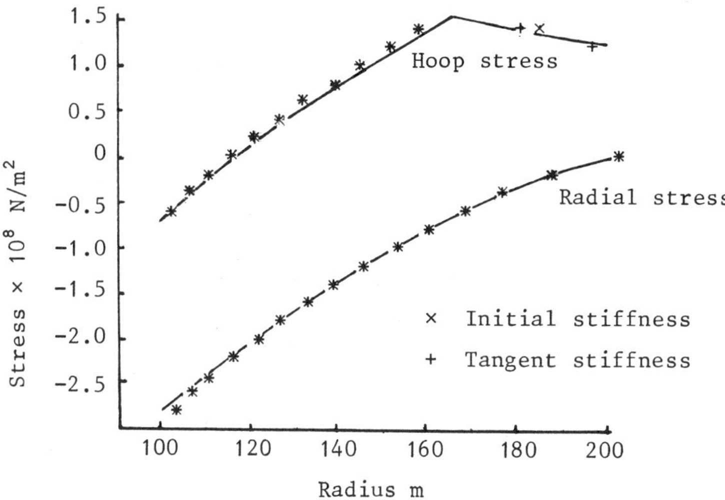

FIG. 4 Stresses after final load increment-multistep loading

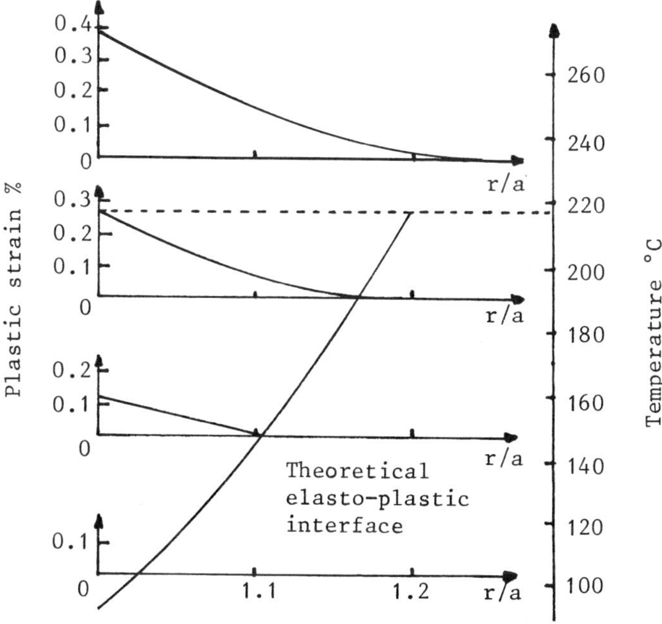

FIG. 5 Spread of plastic zone under thermal loading

$$T_i = \frac{2\sigma_y(1-\nu)(\beta-1)\left[\beta_c^3\left(3\ln\left(\frac{r_c}{a}\right)+1\right)-1\right]}{\alpha E \beta_c(\beta_c-1)(2\beta_c^2-\beta_c-1)} + T_0 \tag{29}$$

where $\beta_c = b/r_c$. Thus to cause yielding to the Gauss point radius would require an internal temperature of 102.0C or 27.3% of the total thermal load. The analysis predicted a value of 28% of the total thermal load. The analysis was performed using the initial stiffness algorithm with 5 post yield load increments. A total of 44 iterations were required and 2hr 55 min CPU time. Fig. 5 depicts the theoretical variation of the yield radius with the temperature given by equation (29) and the predicted plastic strains for load increments 1 to 4. Even though the plastic strain is only calculated at the Gauss points reasonable agreement is obtained between theoretical and predicted results.

The centre of the sphere initially yields in compression. If the internal temperature is increased eventually a second zone of yielding in tension will form. For the sphere under consideration this second zone of yielding commences at the outer surface. This will first occur when

$$\frac{\beta_c(2\beta_c^2-\beta_c-1)\ln\left(\frac{\beta_c}{\beta}\right)}{\beta_c^3\left(3\ln\left(\frac{\beta_c}{\beta}\right)-1\right)+1} = \frac{2(\beta_c-1)^2}{3(\beta_c^2-\beta_c+1)} \quad . \tag{30}$$

The solution of this equation is given approximately by $\beta_c = 1.67$, i.e. $r_c = 119.8$. Substitution of this value in equation (29) indicates that secondary yielding will occur when $T_i = 216.9$C. Although a solution was not obtained for this temperature, secondary yielding had not occurred by load increment 3 at a temperature of 190.4C but was present after load increment 4 at a temperature of 233.6C in qualitative agreement with the theoretical prediction.

REFERENCES

1. BURTON, C.G. and WARD, T.J.W., A computer suite for turbine blade stress analysis: The elasto-plastic analysis. Unpublished Ministry of Defence (Procurement Executive) Report.
2. ZIENKIEWICZ, O.C., *The Finite Element Method, 3rd edition.* McGraw Hill (1977).
3. OWEN, D.R.J. and HINTON, E., *Finite Elements in Plasticity: Theory and Practice.* Pineridge, Swansea (1980).
4. MENDELSON, A., *Plasticity Theory and Application.* Macmillan, New York (1968).
5. HILL, R., *The Mathematical Theory of Plasticity.* Oxford University Press (1950).
6. WILKINSON, J.H., *The Algebraic Eigenvalue Problem.* Oxford University Press (1965).

7. IRONS, B.M. and AHMED, S., *Techniques of Finite Elements*. Ellis Horwood, London (1980).
8. CLIFFORD, W.C., WHITWORTH, R.T. and KERSEY, A.L., Frontwidth reduction for the finite element method. *NGTE Technical Memorandum M81013*, Pyestock (1981).
9. YEO, M.F., A more efficient front solution: Allocating assembly locations by longevity considerations. *Int. J. Numer. Meth. Eng.* 7, 570-573 (1973).
10. BERGAN, P.A. and SØREIDE, T.H., Solutions of large displacement and instability problems using the current stiffness parameter. *Proc. Int. Conf. on Finite Elements in Nonlinear Solid and Structural Mechanics*, Geilo (1977).
11. HINTON, E., OWEN, D.R.J. and TAYLOR, C. (eds.) *Recent Advances in Nonlinear Computational Mechanics*. Pineridge, Swansea (1982).
12. CRISFIELD, M.A., Iterative procedures for linear and nonlinear structural analysis. *TRRL Laborary Report 900*, Crowthorne (1979).
13. CRISFIELD, M.A., A faster modified Newton-Raphson iteration. *Comp. Meth. Appl. Mech. Eng.* 20, 267-278 (1979).
14. JOHNSON, W. and MELLOR, P.B., *Engineering Plasticity*. Van Nostrand (1973).
15. VENKATRAMAN, B. and PATEL, S.A., *Structural Mechanics with Introductions to Elasticity and Plasticity*. McGraw Hill, New York (1970).

FINITE ELEMENT MODELLING OF VISCOELASTIC MATERIAL RESPONSE OF POLYMERIC STRUCTURES

D. Harrison*, J.R. Whiteman* and A.L. Yettram†

* *Institute of Computational Mathematics)* *Brunel University*
† *Department of Mechanical Engineering)*

1. INTRODUCTION

In this paper we seek to model numerically the time-dependent deformation behaviour of polymeric structures having reasonably simple geometric shapes. In particular we wish to calculate approximations $\underline{U}(\underline{x},t)$ to the displacements $\underline{u}(\underline{x},t)$. The motivation behind this work is the need eventually to be able to treat cylindrical pipes and joint components of the type found in pipe networks. The time dependence and nonlinearity of the behaviour of polymeric materials is well known, see [1]-[3], and causes such materials to be difficult to model.

In an attempt to produce an effective numerical model for this context, based on finite element techniques, an assumption of *viscoelastic* behaviour of the material has been made and various forms of viscoelasticity are considered. The work reported on here has led to an algorithm for treating this *history dependent* behaviour, based on an integral form of *constitutive law* and a classical "small strain" form of *strain-displacement law*. Several rheological configurations have been tested and applied to simple cylindrical and spherical structures, the results of which are presented in Section 4.

2. MATHEMATICAL MODEL

The historical nature of the material behaviour has been treated using a *linear* model, in which it is assumed that the constitutive relation between stress and strain is not dependent on mechanical variables (stress and strain). This takes the form of a single integral relating strain rate to stress. Similarly for simplicity a classical *small strain* (small displacement gradient) relation between strain and displacement has been adopted.

We consider a body Ω ($\in \mathbb{R}^2$ or $\mathbb{R}^3$) and for $\underline{x} \in \Omega$ relate the stress $\underline{\sigma}(\underline{x},t_j)$ at time $t = t_j$ to the rate of change with respect

ISBN 0-12-747255-X

to time ($\partial/\partial t$) of the strain $\underline{\varepsilon}(\underline{x},t)$ by

$$\underline{\sigma}(\underline{x},t_j) = \int_{-\infty}^{t_j} [D_{ve}(t_j-t)] \frac{\partial \underline{\varepsilon}(\underline{x},t)}{\partial t} dt , \qquad (2.1)$$

where $[D_{ve}(.)]$ is a viscoelastic constitutive matrix which is a function of both the shear modulus $G(t)$ and bulk modulus $K(t)$ of the material. In particular $[D_{ve}(.)]$ is constructed so that, at time $t=0$, using $G(0)$ and $K(0)$, $[D_{ve}(.)] \equiv [D_e]$, the elastic constitutive matrix, and in general

$$[D_{ve}(.)] = L^{-1}\{[D_e(G(\overline{s}),K(\overline{s}))]\} ,$$

where L^{-1} is the inverse Laplace transform, and $G(\overline{s})$ and $K(\overline{s})$ are respectively the transformed shear and bulk moduli. The matrix $[D_{ve}(.)]$ is later decomposed into two parts related respectively to the shear and bulk effects.

The lower limit of the integral of (2.1) indicates the total time history of the model. This is in practice treated by writing the constitutive law as

$$\underline{\sigma}(\underline{x},t_j) = [D_{ve}(t_j)]\underline{\varepsilon}(\underline{x},0) + \int_0^{t_j} [D_{ve}(t_j-t)] \frac{\partial \underline{\varepsilon}(\underline{x},t)}{\partial t} dt , \qquad (2.2)$$

in which, see Christensen [1], t_j represents the time level at which the deformation is currently being sought, $0 \leqq t_j \leqq t_n$, where t_n is some preset *final time* at which the calculation will be terminated.

As indicated above the viscoelastic constitutive matrix in (2.2) is decomposed into shear and bulk terms and written as

$$[D_{ve}(.)] = G^*(.)[\beta_1] + K^*(.)[\beta_2] , \qquad (2.3)$$

in which G^* and K^* are scalar material functions of the shear modulus G and bulk modulus K, and $[\beta_i]$, $i = 1,2$ are modulus transformation matrices the form of which depends on geometric properties of Ω. For axisymmetric and fully 3D problems of the type considered in Section 4, $G^* = G$ and $K^* = K$. In practice the G and K are represented as

$$G(.) = G_0 + \sum_{i=1}^{I} G_i \exp(.)/\alpha_i , \qquad (2.4)$$

$$K(.) = K_0 + \sum_{i=1}^{I} K_i \exp(.)/\gamma_i , \qquad (2.5)$$

where the G_i, K_i, α_i and γ_i are found by curve fitting to experimental uniaxial response data. The parameter I is chosen to provide an adequate fit to the data.

A linear strain-displacement law is adopted. This is implemented at the element level in the finite element technique and is thus discussed in the next section.

3. FINITE ELEMENT MODEL

The rate dependent form of (2.2) demands that the finite element algorithm shall treat the transient nature of the problem. This is done by discretising the space variables on a finite element mesh over Ω at a particular time and stepping forward through time levels t_j, $j = 1,2,\ldots,n$. Thus at time t_j a mesh of elements Ω^e, $\Omega \equiv \bigcup_e \Omega^e$, is defined and on this the approximation $\underline{U}(\underline{x},t_j)$ to the displacement $u(\underline{x},t_j)$ is defined in terms of piecewise polynomial basis functions and (unknown) point evaluations of $\underline{U}(\underline{x},t_j)$. These point values constitute the vector $\underline{U}_j$. At the element level $\underline{U}(\underline{x},t_j)\big|_{\Omega^e} = \underline{U}^e(\underline{x},t_j)$ and $\underline{U}_j\big|_{\Omega^e} \equiv \underline{U}_j^e$. The strain-displacement law at time $t = t_j$ can be written in Ω^e as

$$\underline{\varepsilon}(\underline{x},t_j) = [B(\underline{x},t_j)]\underline{U}_j^e \ , \tag{3.1}$$

where [B] is a matrix containing derivatives of the basis functions in the element.

At time level t_j the problem is assumed to be quasi-static so that a virtual work statement is taken whereby, see [6],

$$\sum_e \int_{\Omega^e} [B]^T \underline{\sigma}(\underline{x},t_j)d\Omega + \sum_e \underline{f}^e(t_j) = \underline{0} \ , \tag{3.2}$$

in which $\underline{f}^e(t_j)$ is the element force vector. Use in (3.2) of the constitutive law (2.2) and (2.3) in discrete form, together with (3.1), gives

$$\sum_e \int_{\Omega^e} [B]^T\{G^*(t_j)[\beta_1]+K^*(t_j)[\beta_2]\}[B]d\Omega \ \underline{U}^e(0)$$
$$+ \sum_e \int_0^{t_j}\int_{\Omega^e} [B]^T\{G^*(t_j-t)[\beta_1]+K^*(t_j-t)[\beta_2]\}[B]d\Omega \ \frac{\partial \underline{U}^e(t)}{\partial t}$$
$$+ \sum_e \underline{f}^e(t_j) = \underline{0} \ . \tag{3.3}$$

The time integral in (3.3) is now treated by time stepping, and for this in the interval $[t_{j-1},t_j]$ at t_j the derivative $\partial \underline{U}^e/\partial t$ is approximated by the backward difference $(\underline{U}_j^e - \underline{U}_{j-1}^e)/(t_j - t_{j-1})$.

The actual scheme is best illustrated on the *scalar analogue* of (3.3) in which the first two terms are

$$A(t_j)a_0 + \int_0^{t_j} A(t_j-t) \ \frac{\partial a(t_j)}{\partial t} \ dt \ , \tag{3.4}$$

where there is the representation, based on (2.4) and (2.5),

$$A(.) \equiv A_0 + \sum_{i=1}^{I} A_i \ \exp(.)/\lambda_i$$

in which A_0 and A_i play respectively the same role as G_0, K_0 and G_i,K_i in (2.4) and (2.5). Using the notation

$$h_i(t_j-t_{j-1}) = \frac{1}{t_j-t_{j-1}}\int_0^{t_j-t_{j-1}} \exp(-s'/\lambda_i)ds' , \qquad (3.5)$$

$$g_i(t_j) = A_i\{\exp(-t_j/\lambda_i)a(0) + \sum_{k=1}^{j-1} \exp(-(t_j-t_k)/\lambda_i)h_i(t_k-t_{k-1})\Delta a_k\} , \qquad (3.6)$$

where $\Delta a_k = a_k - a_{k-1}$ and s' is a dummy integration variable, there exists a recursive form, [4], for $g_i(t_j)$,

$$g_i(t_j)=\exp(-(t_j-t_{j-1})/\lambda_i)\{g_i(t_{j-1})+A_ih_i(t_{j-1}-t_{j-1})\Delta a_{j-1}\} ,$$

which enables (3.4) to be written as

$$\sum_{i=1}^{I} g_i(t_j) + \Delta a_j \sum_{i=1}^{I} A_ih_i(t_j-t_{j-1}) + A_0a_{j-1} . \qquad (3.7)$$

Thus (3.4) has been transformed into an expression in terms of a recursively defined variable $g_i(.)$, the unknown variable g_j at t_j, the known value a_{j-1} at t_{j-1}, the parameter A_i and the easily computed quantity $h_i(.)$.

The above scalar analysis, which has been used purely to indicate the general procedure, must now be put into the context of the vector multidimensional structural equations (3.3), recognising that the analogues of the parameters in (3.5)-(3.7) are material dependent. Thus, based on (3.7) equation (3.3) becomes at the jth time level

$$\sum_e \int_{\Omega^e} [B]^T\left\{(G_0 + \sum_{i=1}^{I} G_ih_i^G(t_j-t_{j-1}))[\beta_1]+(K_0 + \sum_{i=1}^{I} K_ih_i^K(t_j-t_{j-1}))[\beta_2]\right\} [B]d\Omega\ \Delta\underline{U}_j^e$$

$$= - \sum_e \underline{f}(t_j) - \sum_e \int_{\Omega^e} [B]^T\{G_0[\beta_1]+K_0[\beta_2]\}d\Omega\ \underline{U}_{j-1}^e$$

$$- \sum_e \int_{\Omega^e} [B]^T\left\{\sum_{i=1}^{I} g_i^G(t_j)[\beta_1] + \sum_{i=1}^{I} g_i^K(t_j)[\beta_2]\right\}d\Omega , \qquad (3.8)$$

where h_i^G, h_i^K, g_i^G, g_i^K are the forms of (3.6) and (3.7) now dependent on the shear and bulk moduli, and $\Delta\underline{U}_j^e = \underline{U}_j^e - \underline{U}_{j-1}^e$ is the increment of displacement in proceeding from the $(j-1)$th to jth time level. The algorithm for finding $\underline{U}(\underline{x},t_n)$ therefore involves solving (3.8) successively for $j = 1,2,\ldots,n$. Using matrix notation we thus have to solve

$$\sum_e \{[Q_e^G] + [Q_e^K]\}\ \Delta\underline{U}_j^e = \sum_e \underline{F}(t_j) , \quad j = 1,2,\ldots,n , \qquad (3.9)$$

where the $[Q_e]$ are viscoelastic element stiffness matrices depending on K and G and are relatively simple to calculate. The $\underline{F}$'s involve material function parameters, modulus transformation matrices and the recursive functions $g_i^G(.)$ and $g_i^K(.)$ as well as $\underline{U}_{j-1}^e$. The system (3.9) is solved successively for j = 1,2,...,n for the displacement increments $\Delta\underline{U}_j$ and the total displacement $\underline{U}_n$ at the time level t_n is accumulated.

4. NUMERICAL EXAMPLES

The scheme of Section 3 has been tested on the following two- and three-dimensional problems.

Pressurised Thick Cylinder

A thick walled cylinder in plane stress is subjected to internal pressure P(t). The two-dimensional geometry, as in Fig.1, takes account of symmetry and the inner and outer radii are initially 51.1mm and 62.5mm. The inner surface radial displacement is computed, using eight node serendipity quadrilateral elements with the number of elements, nodes and degrees of freedom, respectively N_e, N_p and N_{df}, as shown, for the case in which the internal cylinder pressure is taken as a step function at $t = 0$ with $P(t) = H(t)P_0$. The step loading is removed at $t = 8$. In keeping with normal finite element practice the pressure loading is transformed into a set of equivalent nodal loads and is then assembled into $\underline{F}(t_j)$ in (3.9).

Two material descriptions are adopted. In the first the forms of G(t) and K(t) are appropriate to a *standard linear* solid, whilst in the second they are appropriate to a *Burgers* solid; in both cases the material model exhibits a constant Poisson's ratio, a situation in which the moduli K and G are *synchronous*. The results obtained are shown in Fig. 2, together with the material functions for the models. The parameters (E_i,μ_i), i = 1,2 are those associated with the rheological analogue of the material. The interpretation of this is as shown in the figure.

Pressurised Pipe Section

A thinner section representing a pipe wall, as in Fig. 3, is taken, whilst material data in the form of *uniaxial moduli* appropriate to polymer materials are used. A loading pattern $P(t) = H(t)P_0$ is again adopted, as are material property definitions leading to a synchronous modulus. The calculated internal surface radial displacement as a function of time is shown in Fig.4 for two values of P_0.

Pressurised Spherical Cavity

A three-dimensional spherical cavity subjected to internal step pressure is now considered. The geometry of the problem and mesh details, using 20-node serendipity brick elements are given in Fig.5. Results based on two forms of material model have been obtained, one in which the shear modulus is assumed to be Maxwell-like and the bulk modulus elastic; the other in which the shear

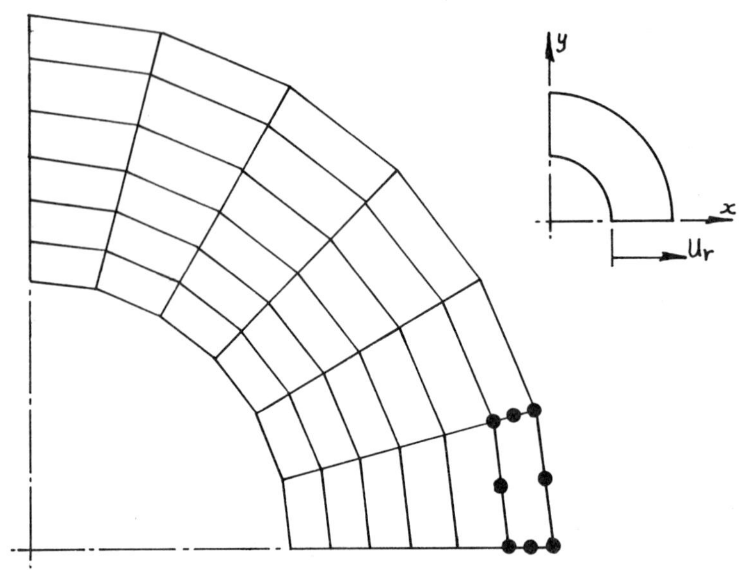

FIG. 1. Pressurised Thick Cylinder
$N_e = 36$, $N_p = 133$, $N_{df} = 266$

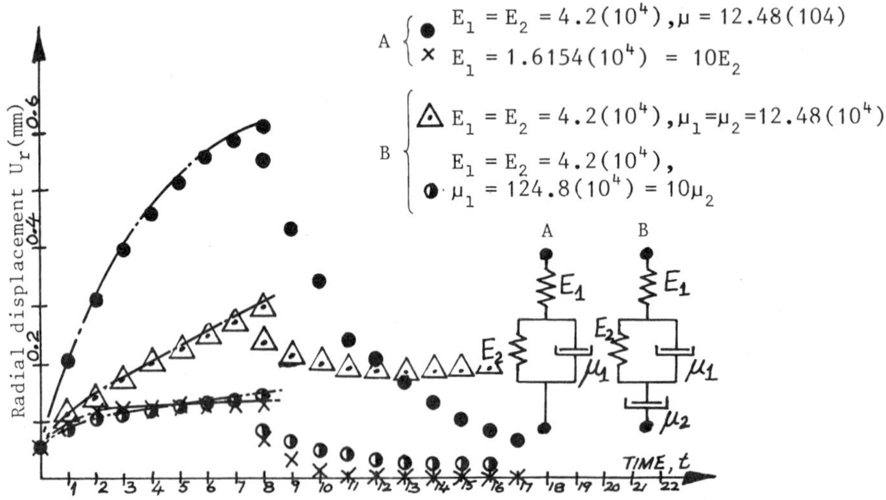

FIG. 2. Pressurised Thick Cylinder
Synchronous Modulus $\nu_0 = 0.3$. $P(t) = P_0H(t)$, $P_0 = 14N/mm^2$
A: Standard linear sold, B: Burgers solid

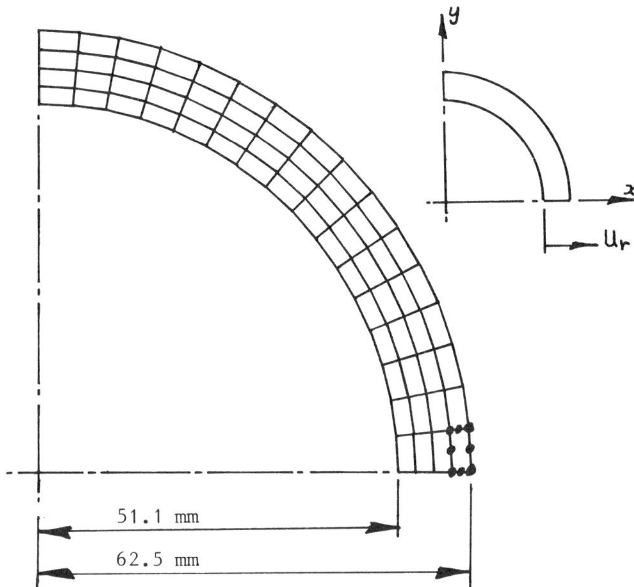

FIG. 3. Pressurised Pipe Section. Plane stress, Synchronous modulus $\nu_0 = 0.4$, Uniaxial modulus E(t), t in hours,
$E(t) = 200 + 201.9642 \exp(-t/10) - 113.0016 \exp(-t/100) + 134.0276 \exp(-t/1000) - 4.7535 \exp(-t/10000)$

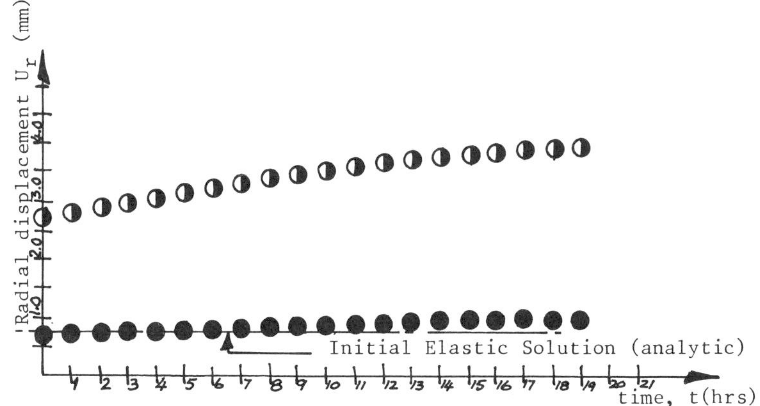

FIG. 4. Pressurised Pipe Section
● – $P_0 = 1.0\ N/mm^2$, ◐ – $P_0 = 4.0\ N/mm^2$

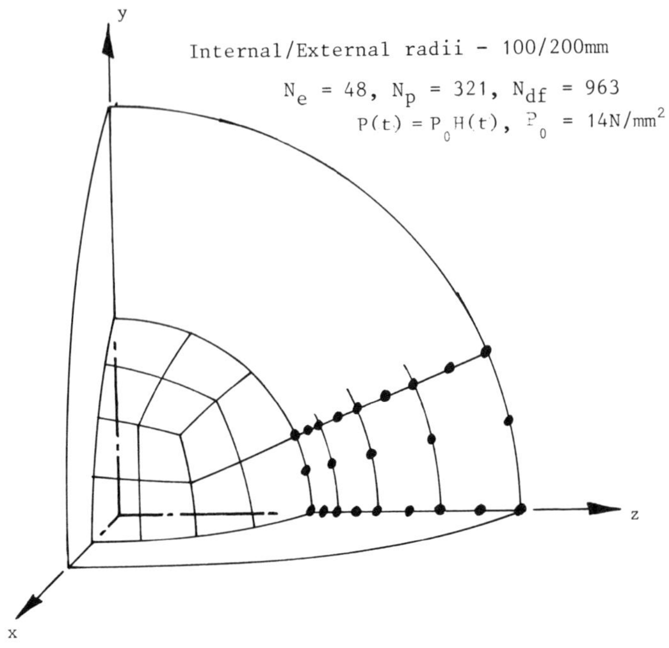

FIG. 5 Pressurised Spherical Cavity: $P(t) = P_0H(t)$, $P_0 = 14N/mm^2$

modulus is assumed to be of Burgers-type, again with an elastic bulk modulus. These forms are given in Fig. 6.

4.1 *The Results*

From the results presented we can see that the forms of response displayed by the model are entirely consistent with expectation. This correspondence is amply demonstrated in the pressurised thick cylinder example where an initial loading phase has been followed by a relaxation phase for which the applied load is zero. Whilst for the case in which the material parameters are taken as those of a standard linear solid, the displacement of the internal cylinder surface is recovered after unloading has occurred, the recovery is only partial for the Burgers-type material description. Although the results of Fig. 2 have been obtained with data describing a synchronous modulus ($\nu_0 = 0.3$), it should be noted that this is not an inherent restriction on the algorithm adopted here. We also note the ability of the model to handle the unload situation, giving rise to two distinct values of displacement; for example, at the same time level. This fully reflects an elastic unload situation.

In the results of the pressurised pipe section we note that for the case $P_0 = 1.0N/mm^2$ the initial displacement of the inner surface is somewhat underestimated. It is likely that this error

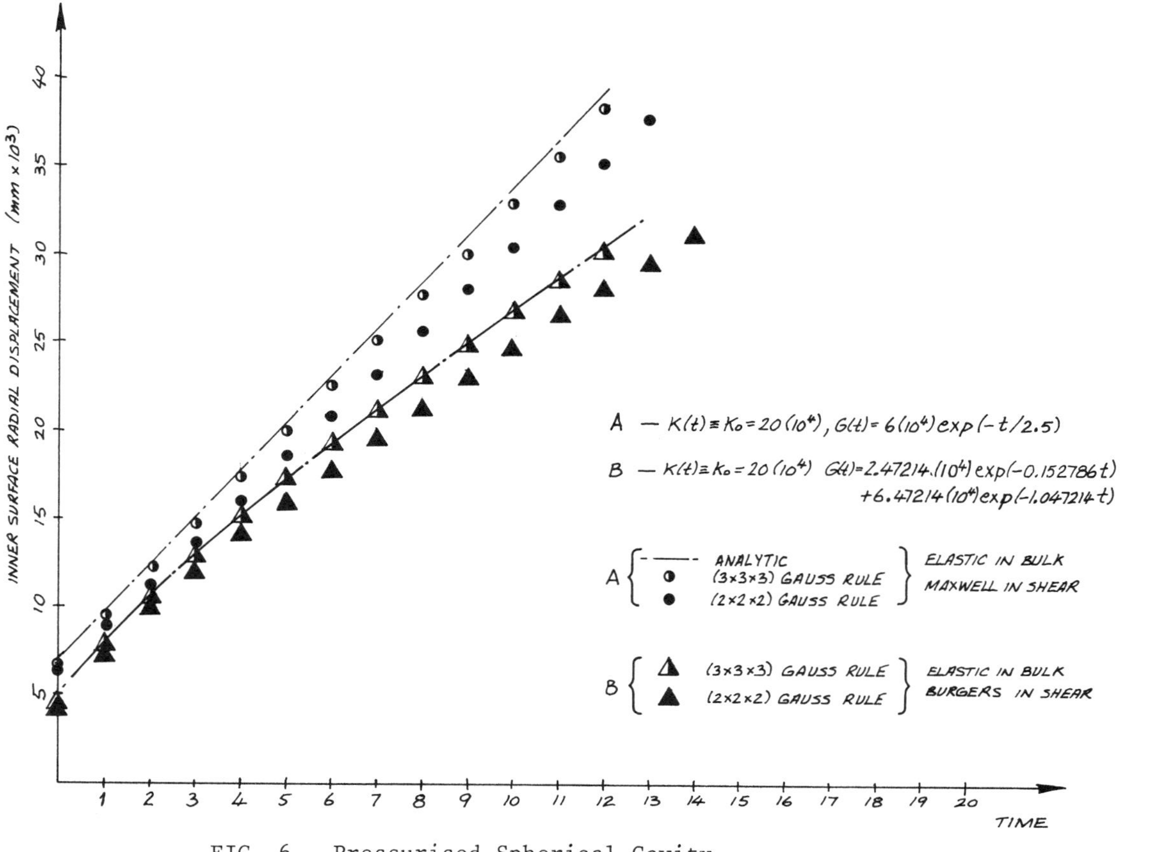

FIG. 6 Pressurised Spherical Cavity

is due to the mesh pattern adopted (Fig. 3) and the integration rule used in the general finite element process (2 × 2 Gauss). To validate this contention a parameter study could have been conducted. This has not been performed as yet. In this example we have again adopted material parameters consistent with a constant value of Poisson's Ratio. The predicted inner surface radial displacements for the two levels of applied step pressure are entirely feasible response patterns.

From the results of the internally pressurised spherical cavity it can be seen that, whilst the general patterns of internal displacement are as expected, there is a progressive error for the case where the bulk modulus,(K(t)), is assumed to be constant and the shear modulus, (G(t)), is taken to be of a Maxwell type. This error growth is seen to be crucially affected by the order of integration rule used. Thus a less marked build-up of error is noted in the case where a (3 × 3 × 3) Gauss rule is used in comparison with a (2 × 2 × 2) rule. Clearly these observations are drawn from the predictions made by the model for a given mesh pattern. We would expect that the mesh layout itself would be a contributary factor in this error build-up, but as noted above a full parametric study of these would need to be performed to enable definitive comments to be made. Also in Fig. 6 are presented the results for the case where (G(t)), the bulk modulus is taken as a Burgers form. These results are again as expected in their general form, and show, as for the Maxwell shear modulus case, an increased displacement at a given time level when the (3 × 3 × 3) integration rule is used, compared with the (2 × 2 × 2) rule.

5. COMMENTS

A number of assumptions have been built into the model of time-dependent behaviour as outlined above. Firstly a "small strain" strain-displacement law has been adopted in which second and higher order terms have been ignored. It is known that under certain loading/constraint conditions polymer structures exhibit "large" displacements confined to localised areas. To enable the model presented to cater for this type of deformation pattern a non-linear form of equation (3.1) should be incorporated. Such a development would clearly require that at each time level, at which a new displacement pattern was to be found, an iterative procedure would have to be adopted. Careful consideration would have to be made of the way in which this iterative scheme should be implemented.

In addition to the strain-displacement assumption, the scheme outlined is one in which the relationship between stress and strain is linear. From experimental observation of uniaxial specimens it is known that the behaviour of polymers can be influenced by the level of loading. This we take as implying that the level of stress in a structure influences the relationship between stress and strain since, for a *linear* stress-strain relationship, the resultant displacement pattern from one level of load can be used to extrapolate to other levels of load. Clearly, this non-linear material effect would require a form of equation (2.1) involving stress dependency. This type of extension of the linear single integral model has been studied by Knauss and Emri [3]. In addition to the mechanical nonlinearities it is well-known that temperature (and other strain inducing parameters) influence the behaviour of polymeric structures. In this case a means of extending the simple model has been identified by a number of authors, e.g. Findley et al. [2], Knauss and Emri [3], Christensen [1]. In terms of industrial application the introduction of thermal influences is highly significant since the testing procedures for polymer pipes involve assessing their behaviour at various isothermal temperatures. Standards have been adopted that take this type of testing programme into account when judging the long-term behaviour of a pipe structure under test. We therefore see the introduction of thermal effects as a most important extension to the model as presented here.

However the numerical results of Section 4 based on the models described are clearly satisfactory and indicate the power of the method. The algorithm has the potential of being developed to include these further effects.

REFERENCES

1. Christensen, R.M., *Theory of Viscoelasticity*. Academic Press, New York (1971).
2. Findley, W.N., Lai, J.S. and Onaram, K., *Creep and Relaxation Behaviour of Nonlinear Viscoelastic Materials*. North Holland, Amsterdam (1976).
3. Knauss, W.G. and Emri, I.J., Non-linear viscoelasticity based on free volume consideration. *Computers and Structures*, 13, 123-128 (1981).
4. Taylor, R.L., Pister, K.S. and Goudreau, G.L., Thermomechanical analysis of viscoelastic solids. *Int. J. Num. Meth. Eng.*, 2, 45-59 (1970).
5. Williams, J.G., *Stress Analysis of Polymers*. Longman, London (1973).
6. Zienkiewicz, O.C., *The Finite Element Method*. McGraw-Hill, New York (1977).

FINITE ELEMENTS WITH SINGULAR SHAPE FUNCTIONS FOR QUADRILATERAL AND BRICK ELEMENTS

W. S. Blackburn

Central Electricity Generating Board (SER-SSD), Canal Road, Gravesend, Kent, England

1. INTRODUCTION

In order to calculate the stress intensity factor at the tip of a crack in a linear elastic body, Tracey [1] and Blackburn [2] developed shape functions with appropriate singular derivatives at the tip and which were compatible with isoparametric elements with vertex and mid-edge nodes on the opposite side of a triangle to the tip. These were immediately incorporated into BERSAFE [3], in forms suitable for two dimensional, axisymmetric and three dimensional linear elastic materials containing a crack and have since been used with great success, the final derivation of the stress intensity factor being either by the virtual crack extension method, Hellen [4], or by extrapolation from the values of the displacements calculated at the vertex and mid-edge nodes of the faces of the crack adjacent to the tip, Hellen and Blackburn [5].

Akin [6] developed some similar shape functions with singular derivatives where the order of the singularity in the derivative is not necessarily $-\frac{1}{2}$, and which are appropriate for quadrilaterals as well as triangles. The former generalisation would be of use for sharp corners with a non-zero included angle and for certain cases where plasticity or creep dominate the strains around the tip, while the latter may be of use for an approximate analysis of a sharply kinked and/or curved crack when the stress intensity factor is required as the crack grows.

To determine which such elements should be incorporated into BERSAFE, a number of possible candidates have therefore been programmed and used with BERSAFE to analyse a test problem, viz a 50 element mesh with vertex and mid-edge nodes.

2. TEST GEOMETRY AND SHAPE FUNCTIONS

The mesh represents a symmetric quarter of a plate of 2 to 1 aspect ratio containing two parallel cracks whose length is one

ISBN 0-12-747255-X

tenth of the crack spacing and of the plate width (fig 1). There are four special square elements around the tip, the size of their sides being an eighth of the half size of the crack. The loading is applied uniformly to a free edge parallel to the crack. The spacing is sufficiently large that for unit loading and crack half size the stress intensity factor will be approximately $\sqrt{\pi}$. To test mode II, equal and opposite shear load were applied at the centres of the faces of the crack. For unit loading and crack half size the stress intensity factor will be $2/\sqrt{\pi}$.

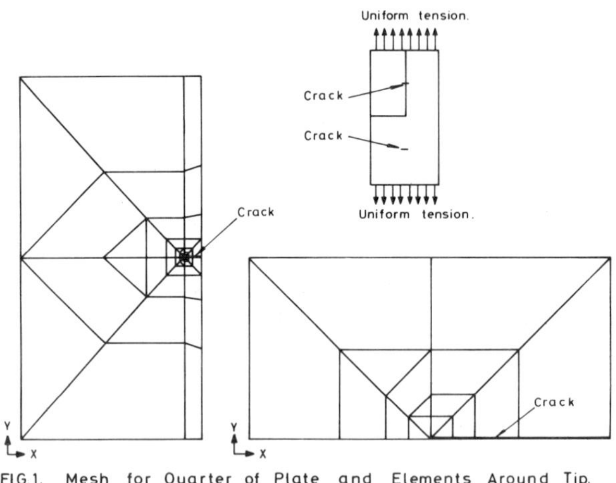

FIG.1. Mesh for Quarter of Plate and Elements Around Tip.

For a square element with sides of length 2 parallel to the ξ and η axes, with origin at its centre and with crack tip co-ordinates (-1,-1), eight shape functions are required to fit vertex and mid-edge nodes. To ensure independence of translations and rotations, terms proportional to 1, ξ and η should be incorporated. To ensure compatibility with isoparametric elements on the edges that do not pass through the tip, all terms should vary as 1, ξ and ξ^2 and as 1, η and η^2 on these edges. Terms incorporating the appropriate crack tip singularity in their gradient for a linear elastic material and which meet these conditions are $(1 + \xi)\zeta^{-\frac{1}{2}}$ and $(1 + \eta)\zeta^{-\frac{1}{2}}$ where $\zeta = \frac{1}{4}(3 + \xi + \eta - \xi\eta)$. Further terms which are zero on the edges through the tip and meet the above conditions on the opposite edges are $(1 + \xi)(1 + \eta) f(\zeta)$, $(1 + \xi)(1 + \eta)^2 g(\zeta)$ and $(1 + \xi)^2 (1 + \eta)h(\zeta)$. Because of symmetry g will be taken to be the same as h. f and g will be selected to be either both 1, or 1 and $\zeta^{-3/2}$, or $\zeta^{-3/2}$ and 1. The shape functions with singular g and unit f, and their generalisation to other orders of singularity are as follows for EP16F and EP16S elements.

$$\phi_1 = 1-(1+\tfrac{1}{2}\xi+\tfrac{1}{2}\eta)\,\{(1+\sqrt{2})\,\zeta^{-\frac{1}{2}}-2\} + \tfrac{1}{4}(1+\xi)(1+\eta)$$

$$\phi_2 = (1+\sqrt{2})(1+\xi)(\zeta^{-\frac{1}{2}}-1)$$

$$\phi_3 = \tfrac{1}{4}(1+\xi)\,\{(1+\eta)^2\,\zeta^{-3/2}-1-3\eta-2(1+\sqrt{2})(\zeta^{-\frac{1}{2}}-1)\}$$

$$\phi_4 = (1+\xi)(1+\eta)\,\{1-\tfrac{1}{2}(1+\eta)\,\zeta^{-3/2}\}$$

$$\phi_5 = \tfrac{1}{4}(1+\xi)(1+\eta)\{(2+\xi+\eta)\,\zeta^{-3/2}-3\}$$

$$\phi_6 = (1+\xi)(1+\eta)\,\{1-\tfrac{1}{2}(1+\xi)\,\zeta^{-3/2}\}$$

$$\phi_7 = \tfrac{1}{4}(1+\eta)\,\{(1+\xi)^2\,\zeta^{-3/2}-1-3\xi-2(1+\sqrt{2})\,(\zeta^{-1/2}-1)\}$$

$$\phi_8 = (1+\sqrt{2})(1+\eta)\,(\zeta^{-1/2}-1)$$

$$\phi_1 = 1-(1+\tfrac{1}{2}\xi+\tfrac{1}{2}\eta)\,\{1-\frac{1-\zeta^{\frac{-1}{n+1}}}{2^{\frac{1}{n+1}}-1}\}+\tfrac{1}{4}(1+\xi)(1+\eta)$$

$$\phi_2 = (1+\xi)\,\frac{(\zeta^{-\frac{1}{n+1}}-1)}{2^{\frac{1}{n+1}}-1}$$

$$\phi_3 = \tfrac{1}{4}(1+\xi)\,\{(1+\eta)^2\,\zeta^{-\frac{n+2}{n+1}}-1-3\eta-2\,\frac{(\zeta^{-\frac{1}{n+1}}-1)}{2^{\frac{1}{n+1}}-1}\}$$

$$\phi_4 = (1+\xi)(1+\eta)\,\{1-\tfrac{1}{2}(1+\eta)\,\zeta^{-\frac{n+2}{n+1}}\}$$

$$\phi_5 = \tfrac{1}{4}(1+\xi)(1+\eta)\{(2+\xi+\eta)\,\zeta^{-\frac{n+2}{n+1}}-3\}$$

$$\phi_6 = (1+\xi)(1+\eta)\,\{1-\tfrac{1}{2}(1+\xi)\,\zeta^{-\frac{n+2}{n+1}}\}$$

$$\phi_7 = \tfrac{1}{4}(1+\eta)\,\{(1+\xi)^2\,\zeta^{-\frac{n+2}{n+1}}-1-3\xi-\frac{2(\zeta^{-\frac{1}{n+1}}-1)}{2^{\frac{1}{n+1}}-1}\}$$

$$\phi_8 = (1+\eta)\,\frac{(\zeta^{-\frac{1}{n+1}}-1)}{2^{\frac{1}{n+1}}-1}$$

3. RESULTS:

Dimensionless stress intensity factors are presented in Table 1, i.e. stress intensity factors scaled by $\sqrt{\pi}$, the value for a small internal crack of unit half size under unit applied normal loading. The values of stress intensity factor calculated from the vertex and mid-edge nodes on the faces adjacent to the tip are presented in Table 1 with numerical integration carried out at either 2 x 2 or 3 x 3 Gauss points. It is seen that for unit loading on the top edge, there is significantly greater consistency between vertex and mid-edge results when more Gauss points are used and that unit rather than singular f improves consistency while unit or singular g makes a very small difference. When singular triangular elements were used at the crack tip instead, the corresponding results from neighbouring vertex and mid-edge nodes are .97 and .98, confirming the adequacy of the approximation to an infinite plate and the overall mesh refinement. Comparable accuracy is obtained for mode I and mode II.

TABLE 1
STRESS INTENSITY FACTOR FOR CENTRE CRACKED PLATE (2D)
(Crack size equal to tenth width)

	$f^{-2/3}$	$g^{-2/3}$	Gauss Points	Dimensionless Stress Intensity Factor Calculated at				Approx. Analytic Value	
				Vertex Mode I	Edge Mode I	Vertex Mode II	Edge Mode II	Mode I	Mode II
straight	1	1	3x3	.96	1.00			1.00	
straight	ζ	1	2x2	.93	1.17			1.00	
straight	ζ	1	3x3	.96	1.04			1.00	
straight	1	ζ	2x2	.98	1.13			1.00	
straight	1	ζ	3x3	.94	.99			1.00	
extended (1/32)	1	ζ	3x3	.97	.99			1.01	
zero size kink	ζ	1	3x3	.67	.67	.64	.20	0.625	0.5
zero size kink	1	ζ	3x3	.62	.74	.60	.58	0.625	0.5
kink size $2^{-4\frac{1}{2}}$									
parabolic edges (4)	1	ζ	3x3	.72	.74	.65	.53	0.625	0.5
parabolic edges (1)	1	ζ	3x3	.77	.74	.62	.54	0.625	0.5

Some further cases were then analysed with unit f and singular g. The first represented a crack of half size 33/32, which corresponds to a growth of 1/32 in crack length. The mesh was adjusted only at the crack tip location a and the adjacent mid-edge nodes. The mesh was thus no longer locally doubly symmetric at the tip. Little difference was found from the previous results (centre of Table 1).

Further analyses were carried out when there was a small or negligibly sized kink at the tip. These cases may be simulated by taking four quadrilaterals with curved sides through the tip. The distortion of the elements to produce the appropriate tilt will be determined using the standard isoparametric representation (i.e. parabolic sides to the elements), as the singular shape functions would imply an infinite gradient. With origin at the tip, the mid-point of that parabola with apex at the mid-point, slope m at the tip and which passes through (X,Y), is $(\frac{1}{2}X + \frac{1}{4}Y\,(Y - mX)/(X + mY),\ \frac{1}{2}Y - \frac{1}{4}X\,(Y - mX)/(X + mY))$. Two kinks were considered each at an angle of 45° so that m equals 1. The sizes of kink were zero and $\sqrt{2}/32$ times the half size of the crack. In the latter case two adjustments were made to the mid-edge nodes adjacent to the tip. They were either all adjusted to lie at the centre of parabolas which meet orthogonally at the tip (as in the case of a zero size kink) or only those on the crack face were adjusted this way, the others being kept at the centre of the straight lines to the neighbouring vertex nodes. Nemat-Nasser [7] has shown that for a small kink through a small tilt m over a proportionate length, ℓ,

$$K_I = (1 - \frac{3}{8} m^2)\, k_1 - 1\tfrac{1}{2}m\, k_2 - \frac{8}{\pi} m^2 \sqrt{(2\ell)}k_2.$$

and
$$K_{II} = \tfrac{1}{2}\, m\, k_1 + (1 - \frac{7}{8} m^2)k_2 + \frac{8}{\pi} \sqrt{(2\ell)}k_2,$$

where k_1 and k_2 are the stress intensity factors in the absence of the kink.

For the cases investigated here $k_1 = \sqrt{\pi}$, $k_2 = 0$ and $m = 1$, so that $K_I = \frac{5}{8}\sqrt{\pi}$ and $K_{II} = \frac{1}{2}\sqrt{\pi}$.

The finite element results in the final three rows of Table 1 are in satisfactory agreement, while the agreement with the result in the previous row corresponding to singular f and unit g was not so close for mode II derived from the mid-edge node.

4. SINGULAR BRICK ELEMENTS FOR THREE DIMENSIONS:

A set of shape functions for brick elements with vertex and mid edge nodes may be generated over the brick $(\xi, \eta, z = \pm 1)$ for which there is a singularity proportional to the $1/(n+1)$ power of the distance from the edge joining the 1st, 9th and 13th nodes (assuming the nodes to be arranged in the usual order, with 8 in the bottom surface, in anticlockwise order, four mid edge, and eight in the top surface). Eighteen of the twenty shape functions are presented explicitly below. The first is most conveniently derived by subtracting the sum of the second to the eighth from $\frac{1}{2}(z^2-z)$ (where z is dimensionless distance in the direction from the first to the thirteenth node) and the thirteenth by subtracting the sum of the fourteenth to the twentieth from $\frac{1}{2}(z^2+z)$. The twenty functions are linear combinations of 1, z, z^2, ξ, η, ξz, ηz and thirteen terms proportional to $\zeta^{\frac{1}{n+1}}$ (where $4\zeta = 3+\xi+\eta+\xi\eta$), the coefficients of $\zeta^{-\frac{1}{n+1}}$ being $1+\xi$, $1+\eta$, $(1+\xi)^2(1+\eta)$, $(1+\xi)(1+\eta)^2$, $(1+\xi)(1+\eta)$, $(1+\xi)z$, $(1+\eta)z$, $(1+\xi)^2(1+\eta)z$, $(1+\xi)(1+\eta)^2 z$, $(1+\xi)(1+\eta)z$, $(1+\xi)z^2$, $(1+\eta)z^2$, $(1+\xi)(1+\eta)z^2$.

$$\emptyset_2 = \tfrac{1}{2}(1-z)(1+\xi)(\zeta^{-\frac{1}{n+1}}-1)/(2^{\frac{1}{n+1}}-1)$$

$$\emptyset_3 = 1/8(1-z)(1+\xi)\{2(1-\zeta^{-\frac{1}{n+1}})/(1-2^{-\frac{1}{n+1}}) - (1-\eta)(1+\eta+z)\zeta^{-\frac{1}{n+1}}\}$$

$$\emptyset_4 = \tfrac{1}{4}(1-z)(1+\xi)(1-\eta^2)\zeta^{-\frac{1}{n+1}}$$

$$\emptyset_5 = -1/8(1-z)(1+\xi)(1+\eta)(2+z-\xi-\eta)\zeta^{-\frac{1}{n+1}}$$

$$\emptyset_6 = \tfrac{1}{4}(1-z)(1-\xi^2)(1+\eta)\zeta^{-\frac{1}{n+1}}$$

$$\emptyset_7 = 1/8(1-z)(1+\eta)\{2(1-\zeta^{-\frac{1}{n+1}})/(1-2^{-\frac{1}{n+1}})-(1-\xi)(1+\xi+z)\zeta^{-\frac{1}{n+1}}\}$$

$$\emptyset_9 = (1-z^2)(1-\zeta^{\frac{n}{n+1}})$$

$$\emptyset_{10} = \tfrac{1}{4}(1-z^2)(1+\xi)(1-\eta)\zeta^{-\frac{1}{n+1}}$$

$$\emptyset_{11} = \tfrac{1}{4}(1-z^2)(1+\xi)(1+\eta)\zeta^{-\frac{1}{n+1}}$$

$$\emptyset_{12} = \tfrac{1}{4}(1-z^2)(1-\xi)(1+\eta)\zeta^{-\frac{1}{n+1}}$$

$$\phi_{15} = 1/8(1+z)(1+\xi)\{2(1-\zeta^{-\frac{1}{n+1}})/(1-2^{-\frac{1}{n+1}})-(1-\eta)(1+\eta-z)\zeta^{-\frac{1}{n+1}}\}$$

$$\phi_{16} = \tfrac{1}{4}(1+z)(1+\xi)(1-\eta^2)\zeta^{-\frac{1}{n+1}}$$

$$\phi_{17} = 1/8(1+z)(1+\xi)(1+\eta)(\xi+\eta+z-2)$$

$$\phi_{18} = \tfrac{1}{4}(1+z)(1-\xi^2)(1+\eta)\zeta^{-\frac{1}{n+1}}$$

$$\phi_{19} = 1/8(1+z)(1+\eta)\{2(1-\zeta^{-\frac{1}{n+1}})/(1-2^{-\frac{1}{n+1}})-(1-\xi)(1+\xi-z)\zeta^{-\frac{1}{n+1}}\}$$

$$\phi_{20} = \tfrac{1}{2}(1+z)(1+\eta)(\zeta^{-\frac{1}{n+1}}-1)/(2^{\frac{1}{n+1}}-1)$$

When n equals 1, the shape functions have been used to calculate the stress intensity factor for a fully through centre cracked plate under uniform loading on the edges of the plate parallel to the crack plane. In terms of the crack size 2a, the height, width and thickness of the plate were taken to be 16a, 4a and ¼a respectively. A layer of 20 quadrilateral elements was generated to represent an eighth (because of treble symmetry) of the plate. Two cubic elements at the crack tip had sides of length 1/8a. The special shape functions were used in these two elements.

TABLE 2
STRESS INTENSITY FACTOR FOR CENTRE CRACKED PLATE (3D)
(Crack Size Equal to Half Width)

Special Element	No.of Gauss Points	Surface Vertex	Surface Mid Edge	Centre Vertex	Centre Mid Edge	2D (Bentham and Koiter)
No	13	$1.08\sqrt{\pi a}$	$.95\sqrt{\pi a}$	$1.05\sqrt{\pi a}$	$.95\sqrt{\pi a}$	$1.18\sqrt{\pi a}$
Yes	13	$1.15\sqrt{\pi a}$	$1.19\sqrt{\pi a}$	$1.12\sqrt{\pi a}$	$1.19\sqrt{\pi a}$	$1.18\sqrt{\pi a}$
Yes	2x2x2	$1.21\sqrt{\pi a}$	$1.37\sqrt{\pi a}$	$1.18\sqrt{\pi a}$	$1.36\sqrt{\pi a}$	$1.18\sqrt{\pi a}$
Yes	3x3x3	$1.15\sqrt{\pi a}$	$1.20\sqrt{\pi a}$	$1.12\sqrt{\pi a}$	$1.19\sqrt{\pi a}$	$1.18\sqrt{\pi a}$

For unit applied pressure, the stress intensity factors as determined by displacement substitution from the vertex and mid edge nodes on the crack face of the crack tip element are presented in Table 2. The value, when the height and thickness are large and small respectively when compared with the width is 1.18. This gives good agreement with the results at the mid plane when a 13 point or $3 \times 3 \times 3$ Gauss point integration rule is used. Results at the surface are about 2% higher. Use of a $2 \times 2 \times 2$ Gauss point integration rule gave higher results particularly at the mid-edge nodes. Hence it is recommended that one of the more accurate integration rules should be used.

5. CONCLUSIONS

A number of special shape functions for quadrilateral elements at a crack tip with vertex and mid-edge nodes have been tested for both a straight crack and a crack with a small kink at the tip, each in a linear elastic body under uniform loading applied on a parallel free edge. Satisfactory results have been obtained and suitable shape functions have been selected for a generalisation to non-linear situations. They have been extended also to brick elements for three dimensional analyses.

6. ACKNOWLEDGEMENTS

The author is grateful to Dr T K Hellen for many useful discussions, to Mr C H A Townley, Mr D J Lewis and Dr J N Swingler for helpful comments on the earlier draft and to the Central Electricity Generating Board for permission to publish.

7. REFERENCES

1. TRACEY, D.M., Finite elements for determination of crack tip elastic stress intensity factors. *Eng. Fract. Mech.* 3, 255-265 (1972).
2. BLACKBURN, W.S., Calculation of stress intensity factors at crack tips using special finite elements. pp.327-336 of J.R. Whiteman (ed.), *The Mathematics of Finite Elements and Applications*. Academic Press, London (1973).
3. HELLEN, T.K. and PROTHEROE, S.J., The BERSAFE finite element system. *Computer Aided Design* 6, 15-24 (1974).
4. HELLEN, T.K., On the method of virtual crack extension. *Int. J. Numer. Meth. Eng.* 9, 187-207 (1975).
5. HELLEN, T.K. and BLACKBURN, W.S., The calculation of stress intensity factors in two and three dimensions using finite elements. pp.103-120 of S.E. Benzley and E.F. Rybicki (eds.), *Computational Fracture Mechanics*. ASME, New York (1975).
6. AKIN, J.E., Generation of elements with singularities. *Int. J. Numer. Meth. Eng.* 10, 1249-1259 (1976).
7. NEMAT-NASSER, S., On stability of the growth of interacting cracks and crack kinking and curving in brittle solids. pp. 687-706 of A.R. Luxmore and D.R.J. Owen (eds.) *Numerical Methods in Fracture Mechanics*. Pineridge Press, Swansea (1980).

THE FINAL SHAPE OF ACIDIZED CHANNELS IN CHALK FORMATIONS AFTER FRACTURE CLOSURE

Johann Nittmann

Etudes et Fabrications Dowell-Schlumberger

42003 St.Etienne, France

I. INTRODUCTION

The oil industry employs special techniques to increase the production of natural oil and gas reservoirs when the output of the wells has dropped below a certain value. One of these is to decrease the flow resistance of the trapped oil (or gas) by creating long, narrow hydraulic fractures in the reservoir. This is done by pumping a viscous fluid down the well under sufficient pressure to overcome the inherent material strength and in situ stresses of the formation.

However, the creation, alone, of highly conductive fractures has little effect on the production, because the fracture fluid leaks into the permeable formation and the fracture tends to close again. Sand can be added to the fluid in order to hold the fracture walls apart and to prevent the fracture from totally closing. Another technique that is used is to etch channels in the rock along the fracture walls before fracture closure. This can be done by pumping an acid (HCl) after the fracture fluid, which as laboratory experiments show, produces a finger-like etching pattern along the fracture walls. Long channels remain after the rock is dissolved and the fracture has closed. A schematic representation of the 'frac-acid' treatment is shown in Fig.1. This technique is applied, in particular, in limestone formations.

The initial cross-sections of these channels of dissolved rock will depend on the acid-rock interaction. The final etched cross-section geometry is dependent upon the in situ stresses and the load-displacement characteristics of the chalk.

ISBN 0-12-747255-X

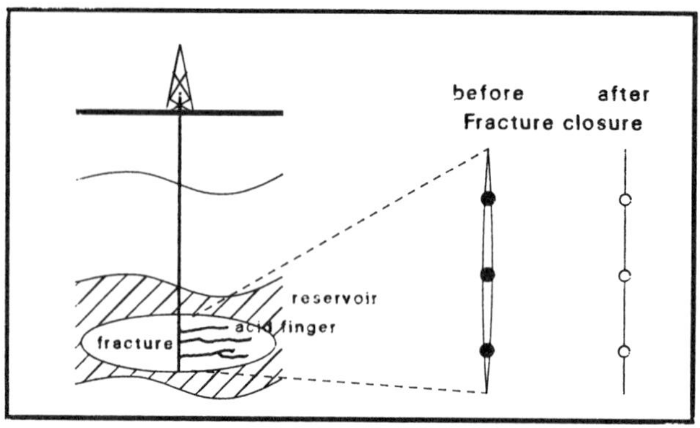

Figure 1

Our main objective was to study the effect of the confining stresses on the change of the shape and the conductivity of the channels after fracture closure. The conductivity of a channel with elliptical cross section in the x-y plans (axes a and b) is given in Equation(1)and relates the flow rate of a fluid with viscosity μ to the applied pressure drop along the z-axis, $\partial p/\partial z$:

$$q = \frac{a^3 b^3 \pi}{4\mu(a^2+b^2)} \left(-\frac{\partial p}{\partial z}\right) \tag{1}$$

Before the numerical simulations, triaxial cell tests had been performed at the Dowell-Schlumberger laboratory in St.Etienne on two types of chalk to determine their elastic properties and yielding behaviour. These studies indicated that an elasto-plastic stress-deformation relation would be justified. In addition it was found that the yield strength of the samples increases at low confining stresses and decreases at higher confining stress levels. This behaviour, which is not observed in hard rocks, is modelled by an ellipsoidal yield criterion. These results are discussed in Section 2. The finite element representation of a two-dimensional elasto-plastic structural problem is discussed in Section 3. The method employed is known as the Displacement method. In Section 4 we compare our numerical results for the conductivity reduction in a circular channel with experimental results. In addition, results of studies of elliptical channels are presented.

2. MECHANICAL PROPERTIES OF CHALK

The mechanical properties of two types of chalk (Danian and Champagne) were obtained by Fournier [3] from triaxial tests. These tests gave us not only the elastic properties, Poisson's ratio and Young's modulus, but also established the parameters associated with the elasto-plastic behaviour of the two chalks. In addition, Fournier performed a Mohr circle analysis to determine the yielding behaviour of the chalks under consideration. Fig.2 shows the stress-strain curves for different confining stresses for Champagne chalk.

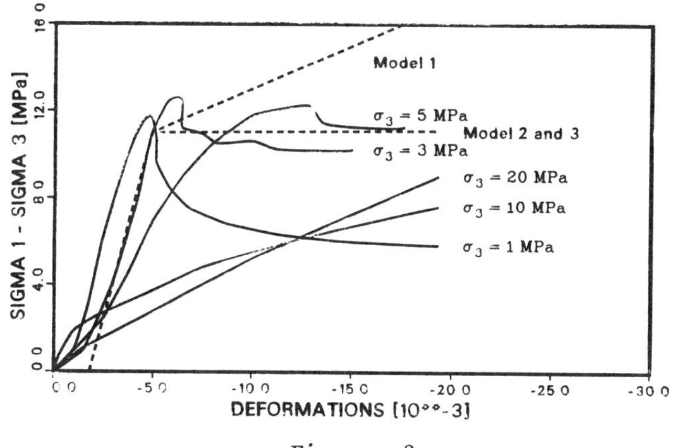

Figure 2

Fig. 2 indicates that at low confining stresses Champagne chalk exhibits, (i) brittle behaviour, (ii) an increase in yield strength with increasing confining stress. At about 5 MPa a transition from brittle to ductile behaviour appears. Above that value, it is evident that (i) a fully ductile behaviour exists and (ii) yield strength now decreases with further increases of confining stress.

This last characteristic can also be followed in Fig.3 where the corresponding Mohr circles are drawn. These where chosen to be associated with the yield point. The Mohr envelope is part of an ellipse (Models 1 and 2). If instead the point of maximum strength is chosen to represent the onset of failure, a Mohr-Coulomb type relation exists (Model 3). The results obtained with different yield criteria are compared in Section 4. Very similar mechanical properties have been found for Danian chalk. A good agreement exists between these test results and those obtained by Blanton [2].

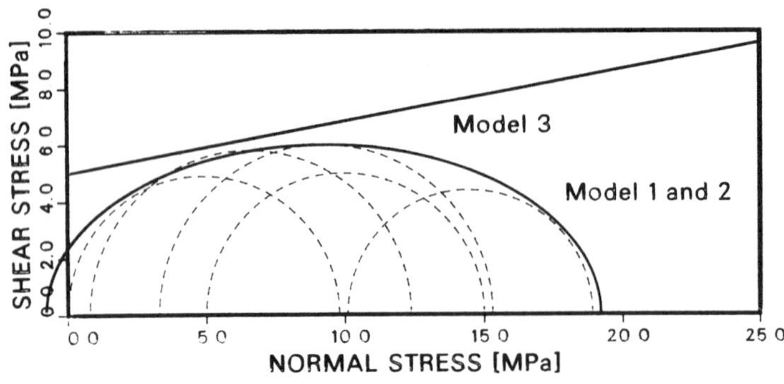

Figure 3

This yielding behaviour is very untypical for rocks. In general, harder rocks are associated with yield surfaces which are represented in the three dimensional principle stress space by open surfaces. The best examples are the Mohr-Coulomb criterion (cone) and the Griffith criterion (paraboloid) (Jaeger and Cook, [4]. The experiments have indicated, however, that a closed surface e.g. an *ellipsoid* is the best model. Ellipsoidal-type yield criteria are frequently found in soils (Mroz and Norris [5]. Similar closed yield surfaces obtained by combining a Drucker-Prager cone with an ellipsoidal cap can be found in Atkinson and Bransby [1].

The equation of an ellipsoid in the three dimensional stress space, in which the principal stresses σ_1, σ_2, and σ_3 are drawn along the axes of a cartesian coordinate system, is

$$\frac{\sigma_1{}^2}{a^2} + \frac{\sigma_2{}^2 + \sigma_3{}^2}{R^2} = 1 \quad . \tag{2}$$

The main axis is assumed to be along σ_1. The final equation for an ellipsoid whose main axis has direction cosines $(1/\sqrt{3}, 1/\sqrt{3}, 1/\sqrt{3})$ is

$$R = \left\{ \frac{\left[J_1+\sqrt{3}(c-a)\right]^2 R^2}{3a^2} + 2J_2' \right\}^{\frac{1}{2}} \tag{3}$$

with $J_1 = \sigma_1+\sigma_2+\sigma_3$ and $J_2' = 1/6\{(\sigma_1-\sigma_2)^2+(\sigma_2-\sigma_3)^2+(\sigma_3-\sigma_1)^2\}$. Apart from geometrical factors, the values for a, R, and c, are the half length of the major and minor axes and the shift of the centre of the ellipse, shown in Fig.3, along the main axis. Equation (3) has been used to represent the yielding behaviour in the following calculations.

The ellipsoidal yield criterion in the three dimensional

stress space is not identical to the yield criterion associated with an elliptic Mohr envelope in the τ_s-σ_n plane. It is, rather, a simplification of the problem, similar to the replacement of the Mohr-Coulomb criterion by the Drucker-Prager criterion.

To summarize, the triaxial cell tests have indicated that

(i) the yielding behaviour of chalk can be matched by an ellipsoidal yield criterion,

(ii) the complex post-yielding behaviour shows a change from a brittle to a ductile stress-deformation relation. This will be approximated by one elasto-plastic hardening model chosen to be independent of the confining stress levels.

3. MODEL EQUATIONS AND FINITE ELEMENT FORMULATION

3.1 Displacement Method for Elastic Problem

The governing equations for a structural problem can be obtained by either minimizing the total potential energy of the system or by using the principle of virtual work which ensures the satisfaction of the equilibrium condition.

$$\sigma_{ij,j} + f_i = 0 \; . \tag{4}$$

σ_{ij} and f_i are the stress tensor and the applied force field, respectively. By defining virtual displacements u^* and associated virtual strains ε^* one can show that an equation of the form

$$\int_V \sigma_{ij}\varepsilon^*_{ij}\,d\tau - \int_V f_i u^*_i\,d\tau - \int_\Omega t_i u^*_i\,d\omega = 0 \tag{5}$$

exists. Together with the compatibility condition and the stress-strain relation

$$\varepsilon^*_{ij} = \frac{1}{2}\,(u^*_{i,j} + u^*_{j,i}) \tag{6}$$

$$\sigma_{ij} = D_{ijkl}\,\varepsilon_{kl} \tag{7}$$

a closed system of equations is obtained for the unknown displacements. This method of approaching a numerical solution for a structural problem is known as the Displacement method (Zienkiewicz, [7]).

The geometry of our problem enables us to assume plane strain and we can write the tensors in Equation (5) in vector form.

Following the finite element representation of Equations (5)-(7) by Owen and Hinton [6], we can write the virtual displacements and strains within a particular element with n nodes as

$$\underline{u}^* = \sum_{i=1}^{n} \underline{N}_i \underline{a}^*_i \qquad \underline{\varepsilon}^* = \sum_{i=1}^{n} \underline{B}_i \underline{a}^*_i \tag{8}$$

where $\underline{N}_i$ is the matrix of the global shape functions and $\underline{B}_i$ is the global strain-displacement matrix. Substituting Equation (8) into the virtual work expression we obtain

$$\sum_{i=1}^{n} [\underline{a}_i^*]^T \{\int_V [\underline{B}_i]^T \underline{\sigma} d\tau - \int_V [\underline{N}_i]^T \underline{f} d\tau - \int_\Omega [\underline{N}_i]^T \underline{t} d\omega\} = 0 \qquad (9)$$

Since this must be true for any arbitrary set of virtual displacements $\underline{a}_i^*$ we obtain for each node i an equation of the form

$$\int_V [\underline{B}_i]^T \underline{\sigma} d\tau - \int_V [\underline{N}_i]^T \underline{f} d\tau - \int_\Omega [\underline{N}_i]^T \underline{t} d\omega = 0 \qquad (10)$$

By inserting into the stress-strain relation (7) an expression for $\underline{\varepsilon}$ similar to Equation (8), the first term of Equation (10) can be written as

$$\sum_{j=1}^{r} \underline{K}_{ij} \underline{a}_j = \int_V [\underline{B}_i]^T \underline{D} (\sum_{j=1}^{r} \underline{B}_j \underline{a}_j) \, d\tau \qquad (11)$$

where $\underline{K}_{ij}$ is a submatrix of the element stiffness matrix $\underline{K}$, and links the two nodes i and j (r nodes per element).

Equation (10) can be assembled into a set of algebraic equations of the form

$$H_{ij} . a_j = f_i \ . \qquad (12)$$

The vector components a_j are the unknown displacements, H_{ij} is the assembled stiffness matrix and f_i are the vector components of the applied loads. For a linear problem these equations can be solved directly by matrix inversion but for an elasto-plastic problem a Newton-Raphson iteration procedure is necessary.

3.2 Modification of the Element Stiffness Matrix owing to Elasto-Plastic Effects

If elasto-plastic behaviour of the material is allowed, modifications of the stress-strain relation (Equation (7)) are necessary above a certain stress value. Within the elasto-plastic region, the change of strain for an incremental change of stress, is assumed to be a sum of elastic and plastic contributions, so that

$$d\varepsilon_{ij} = d\varepsilon_{ij}^{e} + d\varepsilon_{ij}^{p} \qquad (13)$$

The elastic component can easily be obtained from Equation (7). To find an expression for the plastic component we assume that an associated theory of plasticity can be applied to model chalk. We therefore write for $d\varepsilon_{ij}^p$

$$d\varepsilon_{ij}^{p} = d\lambda \frac{\partial f}{\partial \sigma_{ij}} \tag{14}$$

This equation states that the plastic strain increment is proportional to the stress gradient of the yield function f. As the vector $\partial f/\partial \sigma_{ij}$ is directed normal to the yield surface, Equation (14) is also known as the normality condition. The plastic multiplier $d\lambda$ is a function of the steepness of the elasto-plastic part of the stress-displacement curve (E_T).

With Equation (14) a complete incremental elasto-plastic stress-strain relation can be found:

$$d\underline{\sigma} = \underline{D}_{ep} . d\underline{\varepsilon} \tag{15}$$

4. NUMERICAL COMPUTATIONS AND RESULTS

The numerical meshes for circular and elliptical holes can be seen in Fig.4. The shaded areas indicate elements for which different material properties can be assumed. 4-node isoparametric quatrilateral elements were used. Care has been taken to place the applied stress field well away from the centre of the hole. There is no length scale in the problem: our results can therefore be applied to channels of any diameter.

Three models with different material properties have been studied. These are

	E (MPa)	E_T (E)	ν	a (MPa)	c (MPa)	R (MPa)	c_{MC} (MPa)	Φ (^{o})
Model 1	3000	1/11	.3	17	2	9		
Model 2	3000	1/101	.3	17	2	9		
Model 3	3000	1/101	.3				5	9

c_{MC} and Φ are the parameters of the Mohr-Coulomb criterium - cohesion and angle of internal friction.

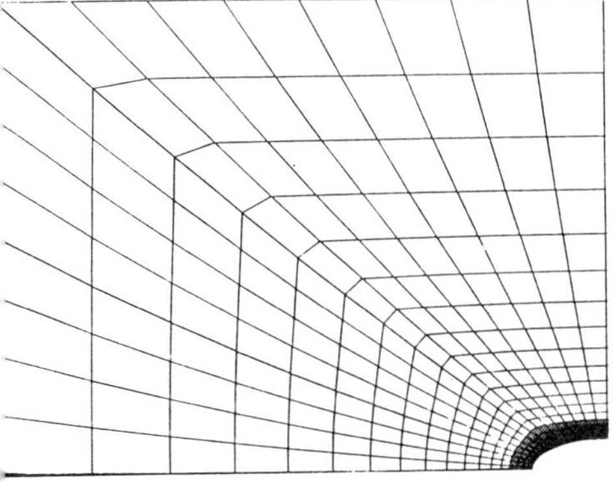
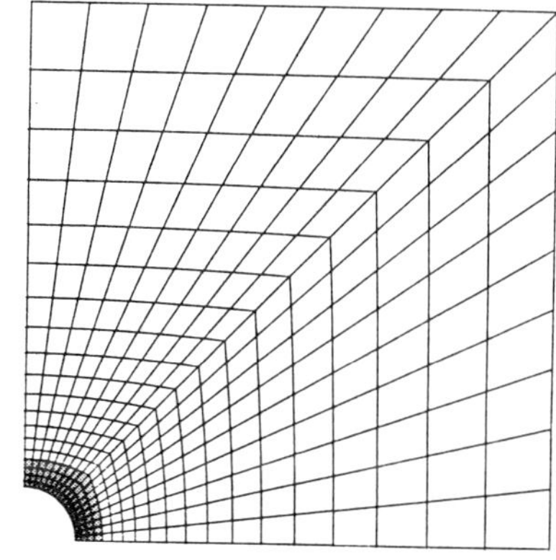

Figure 4

For the solution of the non-linear Equation (10) we were compelled to apply the initial stiffness method which converges slowly but unconditionally for small load increments.

Fig.5 compares numerical results from the three models with experimental results for a circular shaped channel. We show the change of conductivity reduction as a function of confining stresses (equal horizontal and vertical stress). We performed the experiments on cylindrical champagne chalk samples of length 8.4 cm, width 4.95 cm and with a circular hole of 1 mm radius along the main axis. We pumped oil at a constant flow rate through each sample which was positioned in a triaxial cell, and we monitored the change in pressure drop as a function of circumferential confining stress. This method did not allow us to make accurate measurements between 0 and 15 MPa owing to very small signals. However, for confining stresses between 15 and 24 MPa, at which point the conductivity was virtually zero, an almost linear decrease in conductivity was observed. The data were taken from four experiments. During the period of load increase, the chalk samples showed time dependent effects (creep). This forced us to terminate temporarily the automatic load increase until a steady state was reached.

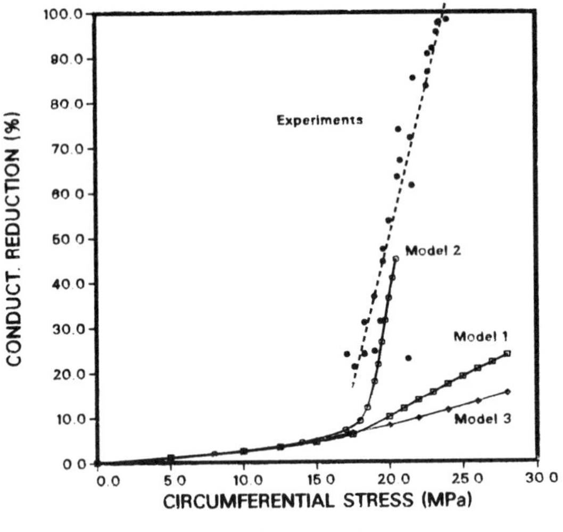

Figure 5

Comparing experimental and numerical results we came to the following conclusions.

(i) Model 2, which employs an ellipsoidal yield criterion and a post-yield behaviour characterised by small linear strain hardening is an almost perfect match to the experimental results. This implies that the complex stress-displacement behaviour of chalk can be modelled by a single elasto-plastic stress-strain relation for all confining stresses.

(ii) Model 1 underestimates the reduction.

(iii) A Mohr-Coulomb criterion as suggested by the Mohr envelope of the Mohr circles, taken at the maximum strength points, also greatly underestimates the real conductivity reduction.

After having established the correct numerical model for the material under investigation we went on to study elliptical shaped channels. We have studied channels with an axes ratio of 5/2 with the main axis parallel to the Y-axis. Figs.6-8 compare the change of shape and the associated conductivity reductions owing to equal horizontal and vertical stress (Fig.6), horizontal stress only (Fig.7) and vertical stress only (Fig.8). The amplitude of the stresses is chosen to be 13 MPa. The initial shape is indicated by circles, the final by a full line.

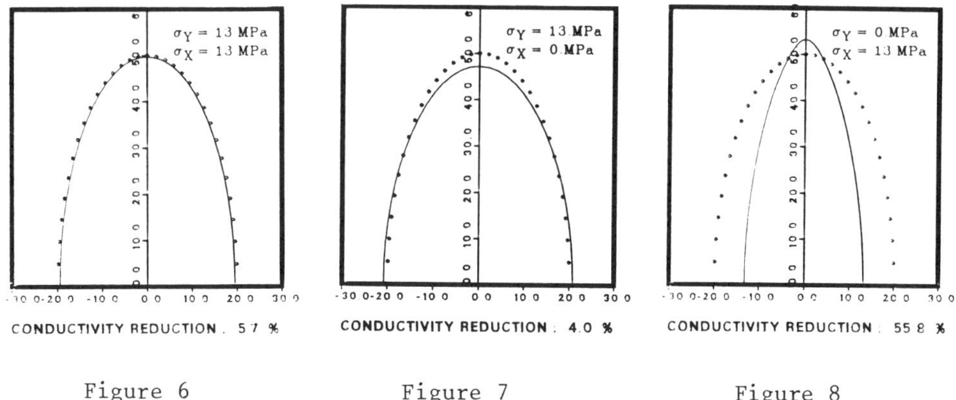

Figure 6 Figure 7 Figure 8

The interaction of acid with the rock determines the initial shape of the cross section of the channels. Our studies therefore indicate that it is extremely important to know the orientation and magnitude of the in situ confining stresses to prevent results as in Fig.8 in the oil field.

Similar numerical studies on saturated rocks are in progress. Experimental tests have indicated loss of strength when the rock is saturated with certain fluids. This and the extension of our model for the treatment of a multichannel system with two or more rocks of different mechanical properties will be considered in future work.

ACKNOWLEDGEMENT:

The author would like to thank Messieurs G. Daccord, Y. Lardin and P. Billard for their assistance during the experimental phase of this work.

REFERENCES

1. ATKINSON, J.H. and BRANSBY, P.L., *The Mechanics of Soils*. McGraw-Hill, London (1978).
2. BLANTON III, T.L., Deformation of chalk under confining pressure and pore pressure. *J. Soc. Petrol Eng.*, 43-50 (1981).
3. FOURNIER, F., Propriétés mechanique des craies de Champagne et du Danien du Danemark, *M.Sc. Thesis*, EFDS (1983).
4. JAEGER, J.C. and COOK, N.G.W., *Fundamentals of Rock Mechanics*. Chapman and Hall, London (1979).
5. MROZ, Z. and NORRIS, V.A., Elastoplastic and viscoplastic constitutive models for soils with application to cyclic loading. In G. Pande and O.C. Zienkiewicz (eds.) *Soil Mechanics - Transient and Cyclic Loads*. Wiley, London (1982).
6. OWEN, D.R.J. and HINTON, E., *Finite Elements in Plasticity*. Pineridge Press, Swansea (1980).
7. ZIENKIEWICZ, O.C., *The Finite Element Method*. McGraw-Hill, London (1977).

AN ASSESSMENT OF DIFFERENT THREE-DIMENSIONAL ELEMENT MODELS IN LINEAR ELASTIC FRACTURE MECHANICS

T K Hellen

Central Electricity Generating Board, Berkeley Nuclear Labs., Berkeley, Glos., England

1. INTRODUCTION

Fracture mechanics is an important subject in the assessment of the integrity of present day structures. In particular, in power plant it is quite feasible to operate safely with structures which contain defects arising from initial manufacture or as a result of growth through operation. Avoidance of such defects is neither possible nor necessary since their effects are rarely destructive. However, known cracks have to be analysed, and current assessment practices require accurate evaluations of stress intensity factors along such crack profiles using, often, linear elastic fracture mechanics (LEFM) theory. Comparisons with known, material and temperature dependent, critical values enable the safety of the crack to be established.

Most plant structures are of complicated geometry and so the finite element method is a powerful tool for LEFM, using various additional techniques to evaluate the stress intensity factors due to both mechanical and thermal loading. The most successful finite elements available for two and three-dimensional analysis are the quadratic displacement isoparametric elements, which exist as triangles and quadrilaterals (with three-dimensional extensions) and serendipity or Lagrangian shape functions [1]. About crack tips, special shape functions can be defined to model the known radial displacement variation, which is partly proportional to the square root of the distance from the tip. This can be done either directly [2], or by moving midside nodes adjacent to the tip half way into the tip [3,4]. For these elements, triangles or quadrilaterals are available although the former enable a better degree of refinement circumferentially about the tip.

The best element formulation to use in the tip region is not obvious and has been considered by many investigators, including other compatible models, where standard isoparametric meshes are accompanied by hybrid and other types of formulation. The present work compares some crack tip elements available in one finite element system, BERSAFE [5], so all other considerations

ISBN 0-12-747255-X

of analysis and stress intensity evaluation are common. In addition to crack tip elements, the accuracy of calculated stress intensity factors can be greatly enhanced by using the virtual crack extension method [6,7]. Such a technique is used for the present comparisons. In addition, different numerical integration rules for stiffness matrix evaluation are compared, with reduced and complete integration rules giving insight into convergence rates.

2. CRACK TIP ELEMENTS AVAILABLE

Although linear and cubic elements are available in BERSAFE, the present investigations refer only to quadratic-compatible elements because of their usual superiority. In two-dimensions, covering plane stress and plane strain (mathematical or engineering) the crack tip elements available assume triangular or quadrilateral shapes, and are shown in Figure 1 with their BERSAFE names. Thus, EP12 and EP16 are the standard 12 and 16 degree of freedom elements which are used away from the tip but, can, of course, be used there as well. As in all the quadratic elements, curved sides

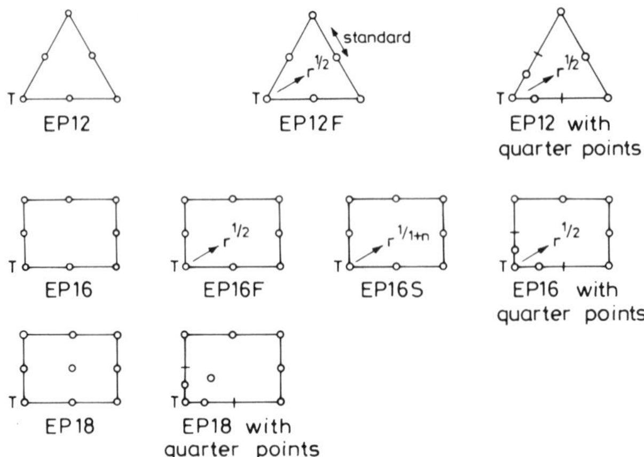

Figure 1 : Two-Dimensional Crack Tip Elements

are permitted, and the displacement functions are of the standard isoparametric form

$$u(\xi,\eta) = a_1 + a_2\xi + a_3\eta + a_4\xi\eta + a_5\xi^2 + a_6\eta^2 \qquad (1)$$

for triangles, with the extra $a_7\xi^2\eta + a_8\xi\eta^2$ for quadrilaterals, where (ξ,η) are dimensionless element coordinates.

The EP12F element incorporates the $r^{1/2}$ dependence on radial displacement, r being the distance from the tip to the point of reference in the element, and was developed, along with a linear compatible version, by Blackburn [2]. The displacement functions are of the form:

$$u(\xi,\eta) = b_1 + \frac{(b_2\xi + b_3\eta + b_4\xi\eta)}{\sqrt{(\xi+\eta)}} + b_5\xi + b_6\eta \quad (2)$$

Two new 16 degree of freedom elements exist, EP16F and EP16S [8]. The former is an extension of eqn (2), with $r^{1/2}$ dependence, and the latter incorporates a user-prescribed index n so that $u \propto r^{1/n+1}$. A technique whereby elements are degenerated by moving midside nodes to quarter positions adjacent to the crack tip is available for both triangles and quadrilaterals [3,4]. A very similar form to eqn (2) is effected for the triangle:

$$u(\xi,\eta) = c_1 + \frac{(c_2\xi + c_3\eta)}{\sqrt{(\xi+\eta)}} + \frac{c_4\xi\eta}{\xi+\eta} + c_5\xi + c_6\eta \quad (3)$$

Lagrangian elements, EP18, with the 9th centroidal node and term $a_9\xi^2\eta^2$ in eqn (1), may also be used together with the degenerated form, although here the centroidal node should be moved in sympathy to the position (1/4, 1/4). These elements may be mixed with the EP16 type elements since full edge compatibility exists. No equivalent triangle is necessary since EP12 is Lagrangian. The above elements all exist for axisymmetric structures; in BERSAFE the element names have the 'EP' replaced by 'EX', e.g. EX12 instead of EP12. The above concepts have been extended into three-dimensions, the available elements being shown in Figure 2. The standard elements are the 15 node wedge and 20 node hexahedron (brick) as EZ45 and EZ60. When used around the crack, both assume one edge lies along the crack profile, which may therefore be

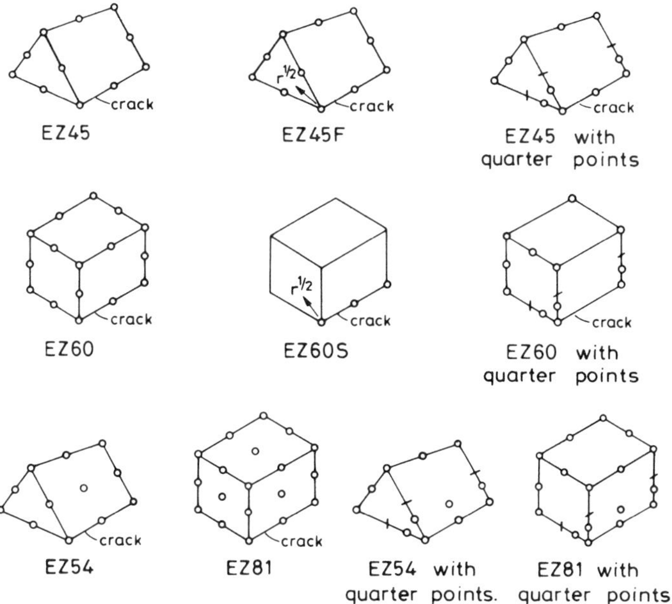

Figure 2 : Three-Dimensional Crack Tip Elements

modelled as curved, and degenerated nodes towards that profile can be define - both end faces should be so affected. EZ45F is a three-dimensional equivalent of EP12F, where $u \propto r^{1/2}$, whilst EZ60S is the equivalent of EP16S, so that $u \propto r^{1/n+1}$. EZ60F does not exist since the user merely specifies n=1 to effect the $r^{1/2}$ singularity.

The Lagrangian elements EZ54 and EZ81 are available but, unlike their two-dimensional counterparts, are no longer compatible with the other elements since the extra midface node renders incompatibility there. Quarter-point degeneracies can again be used, although consideration of moving relevant centroidal or midface nodes is required.

The main purpose in considering the Lagrangian elements around crack tips was to see if their slightly higher polynomial representation gave improved singularity behaviour, particularly when used in conjunction with the virtual crack extension method.

3. VIRTUAL CRACK EXTENSION (VCE) METHOD

This method was first described in two independent forms by Parks [6] and Hellen [7]. In application to LEFM, the former author used stiffness derivatives to evaluate the potential energy release rate due to a small, virtual, extension of crack length. The latter author used a different technique, using a property of the front solution used in BERSAFE to accumulate the strain energy during forward elimination. Provided the elements around the crack tip appear last in this sequence, their energy, plus the energy over them when the tip is moved, is available, any number of movements being allowed in one run with just a recycling over the forward elimination of those few tip elements. Thus, extensions in a circle about the tip enable an assessment of probable crack direction, in mixed mode cases, to be established based on the maximum potential energy release rate criterion, or the accuracy of the method can be assessed using a range of different extensions in one line, typically from 10^{-1} to 10^{-6} of the crack tip element size. Constant values are usually obtained over the range, so that a single result is readily obtained which in the present context is used in the comparison of different crack tip element formulations.

Hellen [9] also describes how to deal with nodal extensions along crack profiles in three-dimensions. Both vertex and midside nodes can be individually extended and the profiles may be curved. The present examples use such a scheme.

4. THREE DIMENSIONAL ANALYSES

The various types of three-dimensional crack tip element

described above have been compared in two simple test problems. The 15-noded wedge elements EZ45 have been used in a mesh of 2 layers of elements, each with 16 elements, a total of 213 nodes, representing an edge cracked plate, with crack depth a tenth the width, loaded by uniaxial tension (Mesh 1). The crack tip element size was a tenth the crack length.

The other types of element were tested in a mesh with a very similar degree of coarseness, representing this time a centre-cracked plate under uniaxial tension, the crack depth to width ratio being a half. Here, 2 layers of elements were again employed each with 20 elements, a total of 297 nodes (Mesh 2). The crack tip element size was an eighth the crack length. Planes of symmetry were used along the uncracked ligament, and, in Mesh 2, normal to the crack plane through its centre. In the half plane, either 2 hexahedral or 4 wedge shaped elements were employed for the crack tip elements.

The meshes are very coarse in order to amplify the effect of using special crack tip elements. Studies with two-dimensional equivalent meshes have been conducted but are not recorded here due to lack of space.

Several crack tip element situations were compared and are shown in Tables 1 to 3, the stress intensity factor (K) being normalised with respect to accepted alternative solutions. For the edge-cracked plate, $K = 1.19\sigma\sqrt{\pi a}$ [10] and for the centre-cracked plate $K = 1.187\sigma\sqrt{\pi a}$ [11]. The calculated K values were from the virtual crack extension method. Extensions of .001, .0001 and .00001 of the crack tip element size produced changes generally in the 5th significant digit, so only the middle result is quoted. Different numerical integration rules for stiffness evaluation are shown, NGAUS=2 denoting 2x2x2, or reduced, integration, and NGAUS=3 denoting 3x3x3 or standard integration. The 13 and 14 point rules are equivalent to 3x3x3 but use only 13 or 14 points instead of the full 27, to produce nearly equivalent accuracy with less computer time.

The results are presented along the 5 nodes comprising the crack profile: the third node is in the centre about which symmetric stress intensity factors should exist.

Table 1 shows the Mesh 1 results for three styles of the 15 noded wedge element, as ordinary (EZ45), with quarter point nodes, and the singularity form EZ45F. Generally, results within 2 or 3% of the theoretical solution are obtained with a slight increase towards the centre, node 3, as would be expected. The coarseness of the present meshes, particularly along the crack direction, precludes an accurate evaluation of the stress intensity factor profile.

The effect of reduced integration (NGAUS=2) is to raise the K values by about 5%. Often, the NGAUS=2 and 3 results bound the true result so that some insight into the state of convergence of

TABLE 1
Normalised K Results Using Mesh 1: EZ45 Type Elements

Run	Tip Element	NGAUS	Nodes Along Profile				
			Node 1	Node 2	Node 3	Node 4	Node 5
1	EZ45	2	1.031	1.045	1.054	1.045	1.031
2		3	0.983	0.997	1.002	0.997	0.983
3		13	0.984	0.999	1.002	0.997	0.982
4		14	0.983	0.997	1.002	0.997	0.983
5	EZ45+1/4pts	2	1.055	1.067	1.078	1.067	1.055
6		3	1.002	1.016	1.022	1.016	1.002
7		13	1.003	1.018	1.022	1.018	1.003
8		14	1.002	1.015	1.022	1.015	1.002
9	EZ45F	2	1.053	0.972	1.077	0.972	1.053
10		3	1.002	0.951	1.021	0.951	1.002
11		13	1.003	0.964	1.021	0.963	1.000
12		14	1.001	0.952	1.021	0.952	1.001

the mesh is offered by this difference. The other integrating rules give virtually the same results which are more accurate than NGAUS=2. Because of the asymmetric disposition of the integrating points in NGAUS=13, some slight asymmetry of calculated K values along the profile is observed, but generally the 13 and 14 point rules are quite as accurate as the full 3x3x3 rule. The singular effects about the tip clearly require these rules rather than the reduced rule despite the high quality performance of the latter in many other situations involving quadratic displacement elements.

The quarter-point and special EZ45F formulations show almost identical behaviour, as would be expected due to the similarity of eqns (2) and (3). They differ, however, by only 2% of using the standard elements around the tip due to the optimum accuracy afforded by the use of the VCE method. One important point to note is that the use of VCE with midside nodes when using EZ45F produces artificially low K values (by up to 10%), an effect that has been noted in most of our test problems. The phenomenon is not apparent with the quarter point elements, and is as yet unexplained.

The overall results are of excellent accuracy in view of the coarseness of the mesh used.

Table 2 shows the Mesh 2 results for the 20 node element as ordinary (EZ60), with quarter point nodes, and the singularity form EZ60S. The general comments of above apply, although here the relative results are slightly higher and less difference exists between reduced and full integration. The difference in K calculated using the quarter point and EZ60S versions is slightly larger than before, and again the latter element produces artificially low values at midside nodes along the profile.

TABLE 2
Normalised K Results Using Mesh 2: EZ60 Type Elements

Run	Tip Element	NGAUS	Nodes Along Profile				
			Node 1	Node 2	Node 3	Node 4	Node 5
1	EZ60	2	1.022	1.038	1.041	1.038	1.022
2		3	1.016	1.036	1.037	1.036	1.016
3		13	1.017	1.040	1.036	1.040	1.016
4		14	1.016	1.032	1.038	1.032	1.016
5	EZ60+1/4pts	2	1.038	1.007	1.062	1.007	1.038
6		3	1.035	1.031	1.054	1.031	1.035
7		13	1.035	1.072	1.055	1.077	1.032
8		14	1.035	0.997	1.053	0.997	1.035
9	EZ60S	2	1.025	0.900	1.046	0.900	1.025
10		3	1.041	0.962	1.063	0.962	1.041
11		13	1.046	1.046	1.062	1.048	1.041
12		14	1.039	0.891	1.061	0.891	1.039

Again, the overall accuracy is excellent in view of the mesh coarseness.

TABLE 3
Normalised K Results Using Mesh 2: Lagrangian Elements, NGAUS=3

Run	Tip Element	Nodes Along Profile				
		Node 1	Node 2	Node 3	Node 4	Node 5
1	EZ54	0.915	0.960	0.921	0.966	0.909
2	EZ54+1/4pts	0.977	0.976	0.996	0.964	0.980
3	EZ81	0.939	0.977	0.948	0.954	0.909
4	EZ81+1/4pts	1.184	1.071	1.167	1.030	1.128

Table 3 also concerns Mesh 2 with the Lagrangian wedge EZ54 and brick EZ81. Using only NGAUS=3, the EZ54 ordinary element around the tip gives results low by up to 9%, including midside node oscillations along the profile. A considerable improvement is observed when quarter point nodes are introduced (up to 3% low). For EZ81, the ordinary element again gives results up to 10% in error whilst the quarter point version produces larger errors. This indicates further experimentation with the VCE method is required with respect to treatment of midface and centroidal nodes. All these results show asymmetry along the crack profile.

CONCLUSIONS

The virtual crack extension method has been used in a common program with a variety of three-dimensional finite elements around the crack tip. Two very similar, coarse meshes were used to highlight differences in the various formulations. Generally, the use of serendipity-type elements with either built-in near tip displacement functions or quarter-point elements gave excellent results. Reduced integration was slightly inferior to complete rules, of which the special 13 and 14 point rules were seen to agree very well with the formal 3x3x3 (27 point) rule. Lastly, the use of Lagrangian elements gave quite erroneous results indicating further development work.

ACKNOWLEDGEMENTS

The author is grateful to Dr W S Blackburn for many useful discussions and help in performing some of the computations. The paper is published by permission of the Central Electricity Generating Board.

REFERENCES

1. IRONS, B.M.R. and AHMAD, S., Techniques of Finite Elements. *Ellis Horwood Ltd.*, Chichester, England, (1980).
2. BLACKBURN, W.S., Calculation of Stress Intensity Factors At Crack Tips Using Special Finite Elements. MAFELAP ed J. R. Whiteman, pp327-376, Academic Press, London, (1972).
3. HENSHELL, R.D. and SHAW, K.G., Crack Tip Finite Elements are Unnecessary. *Int. J. Numer. Meth. Eng.*, 9, 495-507, (1975).
4. BARSOUM, R.S., Application of Quadratic Isoparametric Finite Elements in Linear Fracture Mechanics. *Int. J. Fract.*, 10, 603-605, (1974).
5 HELLEN, T.K., *The* BERSAFE *System. A Handbook of Finite Element Systems*, ed C. A. Brebbia, CML Publications, Southampton, (1981).
6. PARKS, D. M., A Stiffness Derivative Finite Element Technique for Determination of Elastic Crack Tip Stress Intensity Factor. *Int. J. Fract*, 10, 487-502, (1974).
7. HELLEN, T. K., On the Method of Virtual Crack Extensions. *Int. J. Numer. Meth. Eng.*, 9, 187-207, (1975).
8. BLACKBURN, W. S., Finite Elements with Singular Shape Functions for Quadrilateral and Brick Elements. MAFELAP ed J. R. Whiteman, (1984).
9. HELLEN, T. K., A Substructuring Application of the Virtual Crack Extension Method. *Int. J. Numer. Meth. Eng.*, 19, 1713-1731, (1983).
10. PARIS, P. C. and Sih, G., Fracture Toughness Testing and its Applications. ASTM STP 381, 30-81, (1965).
11. ISIDA, M., Crack Tip Stress Intensity Factors for the Tension of an Eccentrically Cracked Strip. Lehigh Univ., Dept. of Mechanics, (1965).

NUMERICAL STUDIES OF BIFURCATION AND PULSE EVOLUTION IN MATHEMATICAL BIOLOGY

A. R. Mitchell and V. S. Manoranjan

University of Dundee, Dundee, Scotland

1. INTRODUCTION

Mathematical Biology is at present one of the fastest growing subjects in the applied sciences. Due to the complexity of biological systems, the subject has attracted a broad spectrum of researchers going all the way from the modern mathematical analyst (A) to the experimental biologist (B). Needless to say the gap between A and B is enormous with the former reducing the model to suit his (or her) analytical requirements and the latter resorting to brute force computer simulation on the original, usually complicated, model.

The numerical analyst has been slow to enter the fray, which is surprising considering the relative ease with which reliable numerical solutions can be obtained for many of the equations arising in mathematical biology. The purpose of this paper is to give a survey of the problems to date where numerical methods have had some success and hopefully to stimulate interest for the future.

2. THE REACTION DIFFUSION SYSTEM

Although there is often doubt concerning the mathematical model which best describes a phenomenon in areas such as biology, and physiology, where several species react with diffusion as a transport mechanism, it is generally accepted that practically all possible models lie within the general system of second order non-linear parabolic partial differential equations given by

ISBN 0-12-747255-X

$$\frac{\partial \underline{U}}{\partial t} = A\nabla^2\underline{U} + \cancel{\sum_{j=1}^{m} M_j(\underline{x},\underline{U})\frac{\partial \underline{U}}{\partial x_j}} + \underline{f}(\underline{U}) - \beta\underline{R}$$

$$\frac{\partial \underline{R}}{\partial t} = D_1\underline{U} + D_2\underline{R} \; . \tag{2.1}$$

In (2.1),
$\underline{U}(\underline{x},t)$ is an $\mathbb{R}^n$-valued function of $\underline{x} \in \mathbb{R}^m$ and $t \in \mathbb{R}^+$,
A is a diffusion matrix with non negative constant entries,
$M_j(j = 1,2,\ldots,m)$ are continuous matrix valued functions,
$\underline{f}(\underline{U})$ is a non linear function describing the reaction of the system,
D_1, D_2 are diagonal matrices,
and β is a parameter.
If (2.1) holds in a region $\Omega \times [t \geq 0]$ where Ω is in m-space with boundary $\partial\Omega$, along with (2.1) we have the initial conditions

$$\left.\begin{array}{l} \underline{U}(\underline{x},0) = \underline{U}_0(\underline{x}), \\ \underline{R}(\underline{x},0) = \underline{R}_0(\underline{x}), \end{array}\right. \quad \underline{x} \in \Omega \tag{2.2}$$

together with appropriate boundary conditions on $\partial\Omega$. Although the convective term is important in some problems it has been neglected by most authors and so we put it equal to zero in (2.1). It should be said, however, that its inclusion causes little problem to a numerical technique provided due care is taken with the "upwinding" of the first derivatives. In analytical studies, however, a large first derivative term can cause considerable problems.

3. EXAMPLES OF REACTION DIFFUSION

In order to obtain an understanding of the types of problem represented by (2.1) we look at some special cases of this system where $m = 1(\underline{x} \equiv x)$.

(1) $n = 1, \underline{U} = U, \underline{R} = R$

(a) $$\beta = 0, \quad \frac{\partial U}{\partial t} = \frac{\partial^2 U}{\partial x^2} + f(U) \tag{F.}$$

(b) $$\beta = 1, \quad \frac{\partial U}{\partial t} = \frac{\partial^2 U}{\partial x^2} + f(U) - R$$

$$\frac{\partial R}{\partial t} = b(U - dR), \quad b,d \geq 0 \, . \tag{F.N.}$$

Fisher's equation (F) with the non linear term

$$f(U) = \begin{cases} U(1-U) & (3.1a) \\ \quad\text{or} & \\ U(1-U)(U-a) \quad 0 < a \leq \frac{1}{2} & (3.1b) \end{cases}$$

appears in genetics, flame propagation etc., whereas (F.N.) with $f(U)$ given by (3.1b) governs the conduction of electrical impulses along a nerve axon.

(2) $n = 2$, $\underline{U} = [V,W]^T$, $\underline{R} = [R_1,R_2]^T$.

(a) $\beta = 0$.

(i) $A = I$, $\underline{f}(\underline{U}) = \begin{bmatrix} V(1-V) - rVW \\ -bVW \end{bmatrix}$ $r,b \geq 0$,

This is the Belousov-Zhabotinskii reaction [10] in which waves which are alternately red and blue in colour propagate away from the point at which the chemical reaction takes place.

(ii) $A = D(\text{diagonal})$, $\underline{f}(\underline{U}) = -\begin{bmatrix} \mu V + VW \\ \lambda W - VW \end{bmatrix}$, $\lambda,\mu \geq 0$.

This represents the spread of an infection within a population where V and W denote susceptible and infective individuals respectively.

(iii) $A = D$, $\underline{f}(\underline{U}) = -\begin{bmatrix} a_{11}V - a_{12}W \\ -p(t)h(V) + a_{22}W \end{bmatrix}$, $a_{11},a_{12},a_{22} > 0$.

Here the system describes the cholera-epidemic which spread in the European Mediterranean regions in 1973, where V and W describe the bacteria and the infective populations in the habitat respectively, with $p(t) = p(t+w)$, $w > 0$, a periodicity condition which accounts for seasonal fluctuations, and $h(0) = 0$, $0 < h(z') < h(z'')$ if $0 < z' < z''$.

(b) $\beta = 1$.

$$A = \begin{bmatrix} 1-\alpha & -\alpha \\ -\alpha & 1-\alpha \end{bmatrix}, \quad \underline{f}(\underline{U}) = \begin{bmatrix} F(V) - \eta(V-W) \\ F(W) + \eta(V-W) \end{bmatrix} \tag{3.2}$$

$D_1 = \epsilon I$, $D_2 = 0$. $0 \leq \alpha < \frac{1}{2}$; $\eta, \epsilon \geq 0$.

This system describes the passage of pulses on two parallel nerve fibres and constitutes the chief source of investigation in the present paper.

4. EXACT SOLUTIONS [13]

When carrying out numerical experiments on a non-linear problem it is useful to have some exact solutions available to test the numerical method in these special cases.

In (F), put

$$U(x,t) = u(x-ct) = u(\xi) \tag{4.1}$$

and so

$$u'' + cu' + f(u) = 0, \tag{4.2}$$

where $c(>0)$ is the constant speed of a wave travelling in the direction of ξ positive and a dash denotes differentiation with respect to ξ. For $f(u)$ given by (3.1a) an exact solution of (4.2) is

$$u = \left(1 + Ae^{\xi/\sqrt{6}}\right)^{-2}, \quad A > 0, \ c = 5/\sqrt{6}$$

and for $f(u)$ given by (3.1b),

$$u = \left(1 + e^{\xi/\sqrt{2}}\right)^{-1}, \quad c = \sqrt{2}(\tfrac{1}{2}-a),$$

Both of these solutions are fronts with $u(-\infty) = 1$ and $u(+\infty) = 0$ travelling in the positive ξ direction with velocity c. Another exact solution of (4.2) with $c = 0$ and $f(u)$ given by (3.1b) is

$$u = \begin{cases} 3a\{\sqrt{(2-a)(\frac{1}{2}-a)}\cosh(\sqrt{a}\xi) + (1+a)\}^{-1}, & 0 < a < \frac{1}{2} \\ \frac{1}{2} + \alpha\,\mathrm{sn}(\frac{\xi}{\sqrt{2}}(\frac{1}{2}-\alpha^2)^{\frac{1}{2}}, \alpha(\frac{1}{2}-\alpha^2)^{-\frac{1}{2}}), & a = \frac{1}{2} \end{cases} \tag{4.3}$$

where $\mathrm{sn}(G,\alpha)$ denotes the Jacobi elliptic function of argument G and $0 < \alpha < \frac{1}{2}$. The former solution in (4.3) has $u(-\infty) = u(+\infty) = 0$, whilst the latter solution is periodic. Although both of these zero speed solutions are unstable, we are actually able to use part of the former solution in (4.3) to explain some of the numerical results obtained in the section on bifurcation.

5. BIFURCATION

In boundary-value problems involving differential systems containing parameters, the number of solutions may change as the value of a parameter is altered. Such parameters are known as bifurcation parameters for the system and the positions in the parameter range where the number (or type) of solutions changes as bifurcation points. For a thorough account of bifurcation theory see Smoller [14]. In reaction-diffusion models, common bifurcation parameters are the size and shape of Ω, the region in $\underline{x}$ space involved in the problem. Some examples of this type of bifurcation will now be given.

Consider initially (F) with either the quadratic or the cubic non linear term subject to the boundary conditions

$$u(-L,t) = u(L,t) = b \ , \quad b \text{ constant} \tag{5.1}$$

and the initial condition

$$u(x,0) = u_0(x). \qquad -L \leq x \leq +L$$

For each value of L, the bifurcation parameter, Smoller and Wasserman [15] using a "time map" in phase plane analysis, obtained the exact number of steady state ($t\to\infty$) solutions of (F) subject to (5.1). In particular they show that there are at most three solutions for each value of L. Some of these solutions, of course, may be unstable. In a further analytic study, Kuo Pen Yu et al [7] showed the following upper limits (see Table 1) for the bifurcation parameter L to be sufficient for $u(x,t) \to b$ (constant solution) as $t \to \infty$.

TABLE 1

b	quadratic	cubic
0	π	$\pi/(1-a)$
a	π	π
1	π	π/a

The above critical lengths at which bifurcation occurs have been verified numerically in the sense that the constant solution has been obtained only for lengths up to and slightly beyond the values given in Table 1 [7]. In addition numerical experiments carried out for values of L considerably in excess of the values in Table 1 have produced non constant steady state solutions ($t\to\infty$) some of which are characterised by the former of the theoretical solutions (4.3).

Attempts to extend the above to higher space dimensions run into the usual geometrical difficulties with the shape of the region together with the "size" now determining possible bifurcations from the trivial solution. The problem where analysis has made some progress is that given by

$$\frac{\partial U}{\partial t} = \frac{\partial^2 U}{\partial x^2} + \frac{\partial^2 U}{\partial y^2} + U(1-U) \qquad (x,y;t) \in \Omega \times [t \geq 0]$$

$$U(x,y;0) = \phi(x,y) \qquad (x,y) \in \Omega \tag{5.2}$$

$$U(x,y;t) = 0, \qquad (x,y;t) \in \partial\Omega \times [t \geq 0]$$

with $\phi(x,y) = 0$ on $\partial\Omega$. The trivial solution $U = 0$ is perturbed and the linearised form of the steady state $(t \to \infty)$ perturbation $\delta(x,y)$ is given by

$$\nabla^2 \delta + f'(0)\delta = 0 \tag{5.3}$$

where $\nabla^2 \equiv \frac{\partial^2}{\partial x^2} + \frac{\partial^2}{\partial y^2}$ and a dash denotes differentiation with respect to U. If λ_1 is the first eigenvalue of

$$\nabla^2 \delta + \lambda\delta = 0$$

on the specified domain with zero Dirichlet boundary conditions, then if

$$\lambda_1 < f'(0) = 1, \tag{5.4}$$

the trivial solution is unstable. Examples of shapes where λ_1 is known in closed form are

(i) Rectangle of side lengths a and b,

$$\lambda_1 = \pi^2 \left(\frac{1}{a^2} + \frac{1}{b^2} \right)$$

(ii) Circle of radius r,

$$\lambda_1 = p_1^2 / r^2$$

where p_1 is the first zero of J_0, the Bessel function of the first kind and order 0. Thus the trivial solution is <u>unstable</u> for a square of side a if $a > \sqrt{2}\pi \approx 4.4$ and for a circle of radius r if $r > p_1 \approx 2.4$. First eigenvalues for more complicated shapes have been obtained approximately by Murray and Sperb [12]. Analytical studies showing the existence of non constant stable solutions for $a > \sqrt{2}\pi$ and $r > p_1$ in the cases of the square and circle respectively have been carried out by Kuo Pen Yu et al [8] and numerical studies [1] involving the time dependent problems confirm these results in the case of the square. In fact the numerical experiments were carried out for $a = 4.0, 4.8$ for a variety of initial conditions and as $t \to \infty$ the trivial solution is obtained for $a = 4.0$ and a non constant solution for $a = 4.8$.

This example of bifurcation applies to the spruce budworm problem, where a budworm population $U(x,y)$ infests a finite area $\Omega(x,y)$ of spruce forest at $t = 0$, and the lethal condition $U = 0$, imposed on the boundary $\partial\Omega$, either by spraying or removing the spruce trees, is required to wipe out the budworm population in Ω as $t \to \infty$. This of course, from the analysis, only happens if Ω is sufficiently small and (5.4) is not satisfied.

6. NUMERICAL METHOD

Since the principle aim of this paper is the numerical solution of the system of equations modelling the transmission of pulses along parallel nerve fibres, we return to (2.1) and (3.2) and describe our numerical method for the solution of this system. Our numerical experience in the past has shown that the non linear cubic term $F(V)$ can be replaced by a linear caricature given by the formula

$$F(V) = V(1-V)(V-a) \approx -V + H(V-a), \tag{6.1}$$

without having a significant effect on the results. Consequently we adopt the linear simplification (6.1) together with a similar replacement for $F(W)$, where H is the Heaviside function. This linearisation although essential for analytical studies is not required for numerical work.

We now return to (2.1) suitably modified to incorporate the values given by (3.2). The weak solution after application of the Extrapolated Crank-Nicolson method [5] leads to

$$\left[\frac{1}{k}(V^{n+1}-V^n),\psi_j) + \tfrac{1}{2}((1-\alpha)\frac{\partial}{\partial x}(V^{n+1}+V^n),\frac{\partial\psi_j}{\partial x}) - \tfrac{1}{2}(\alpha\frac{\partial}{\partial x}(W^{n+1}+W^n),\frac{\partial\psi_j}{\partial x}\right]$$
$$= (F(\tfrac{3}{2}V^n-\tfrac{1}{2}V^{n-1}),\psi_j) - (R_1^{n+1},\psi_j) - \eta((V^{n+1}-W^{n+1}),\psi_j) \tag{6.2}$$

and

$$\left[\frac{1}{k}(W^{n+1}-W^n),\psi_j) + \tfrac{1}{2}((1-\alpha)\frac{\partial}{\partial x}(W^{n+1}+W^n),\frac{\partial\psi_j}{\partial x}) - \tfrac{1}{2}(\alpha\frac{\partial}{\partial x}(V^{n+1}+V^n),\frac{\partial\psi_j}{\partial x}\right]$$
$$= (F(\tfrac{3}{2}W^n-\tfrac{1}{2}W^{n-1}),\psi_j) - (R_2^{n+1},\psi_j) - \eta((W^{n+1}-V^{n+1}),\psi_j), \tag{6.3}$$

followed by

$$\frac{1}{k}(R_1^{n+1}-R_1^n) = \tfrac{1}{2}\,\epsilon\,[(V^{n+1}+V^n) - b(R_1^{n+1}+R_1^n)] \tag{6.4}$$

and

$$\frac{1}{k}(R_2^{n+1}-R_2^n) = \tfrac{1}{2}\,\epsilon\,[(W^{n+1}+W^n) - b(R_2^{n+1}+R_2^n)] \tag{6.5}$$

where (,) denotes the L_2 inner product, $\psi_j \forall j$ are test functions, $t = nk$, $n = 0,1,2,\ldots$, and integration by parts, neglecting boundary terms, has been carried out for second order derivatives in space. In order to start the calculation in time $F(\frac{3}{2}V^n - \frac{1}{2}V^{n-1})$ and $F(\frac{3}{2}W^n - \frac{1}{2}W^{n-1})$ are replaced by $F(V^n)$ and $F(W^n)$ respectively for $n = 0$. Discretisation in space is carried out by putting

$$V = \sum_i V_i(t)\phi_i(x), \quad \text{and} \quad W = \sum_i W_i(t)\phi_i(x)$$

where $\{\phi_i(x)\}$ are the standard piecewise linear trial functions, each with a support of 2h. In all calculations, the test functions $\{\psi_j(x)\}$ are also taken to be standard piecewise linears. Neglecting boundary terms in integration by parts is permissible since we only carry out initial value problem calculations. Also using the caricature instead of the original cubic non linear function prevents us having to use product approximation [2] on the non-linear term. This of course may be a mixed blessing.

7. PROPAGATION OF NERVE PULSES

The nervous system is made up of cells which communicate with each other by means of electrical signals. Consequently the conduction of electrical impulses in a nerve axon is one of the most important problems in neurobiology. The most acceptable mathematical model for this phenomenon is due to Hodgkin and Huxley [6] and can be written in the general form

$$\begin{aligned} \frac{\partial u}{\partial t} &= \frac{\partial^2 u}{\partial x^2} - I(u,\underline{w}) \\ \frac{\partial \underline{w}}{\partial t} &= P(u)\underline{w} + \underline{q}(u) \end{aligned} \qquad \text{(HH)}$$

where u is the electrical potential across an axon and $\underline{w}$ is a three dimensional vector function. The 3×3 matrix P and the vector function $\underline{q}$ depend nonlinearly on u, and $I(u,\underline{w})$ is linear in u and nonlinear in $\underline{w}$. The exact form of (HH) with suitable details is given in [6].

Reliable numerical results for the full Hodgkin-Huxley equations are very difficult to obtain mainly due to the complicated nonlinear functions I, P, and $\underline{q}$. The interested reader should consult [3]. Fortunately a simpler formulation which reproduces most of the features of the original system is (2.1), which in the forms (F.N.) and (3.2), deals with pulse transmission along a single and double nerve respectively. Details of numerical and analytical results for the single fibre are given in [11], the principal result being that for (F.N.) with $f(U)$ given by (3.1b), a train of pulses travels along a semi infinite nerve axon $0 \le x < \infty$ for certain values of the parameters a, b, d and I where $\frac{\partial u}{\partial x} = -\frac{1}{2}I$ at $x = 0$ for $t > 0$, and $I > 0$. Also the speed of the pulse train depends on a, b, and d but not on the current I.

In the case of pulse transmission on two parallel nerve fibres the problem is much more complicated. A typical question to be answered now is "does a pulse in one fibre induce a potential in the neighbouring fibre?" Laboratory experiments give evidence of such "crosstalk". The two fibres are assumed to be identical and to be enclosed by a sheath, the respective volumes inside the fibres and between the fibres and the sheath being filled with fluids of different conductivities. The sheath radius, fibre radius, and fibre separation distance influence the parameters α, η and ϵ in the system

$$\frac{\partial}{\partial t}\begin{bmatrix} V \\ W \end{bmatrix} = \begin{bmatrix} 1-\alpha & -\alpha \\ -\alpha & 1-\alpha \end{bmatrix} \frac{\partial^2}{\partial x^2}\begin{bmatrix} V \\ W \end{bmatrix} + \begin{bmatrix} F(V) \\ F(W) \end{bmatrix} + \eta\begin{bmatrix} -1 & +1 \\ +1 & -1 \end{bmatrix}\begin{bmatrix} V \\ W \end{bmatrix} - \begin{bmatrix} R_1 \\ R_2 \end{bmatrix}$$

$$\frac{\partial}{\partial t}\begin{bmatrix} R_1 \\ R_2 \end{bmatrix} = \epsilon\begin{bmatrix} V \\ W \end{bmatrix}, \tag{7.1}$$

which is (2.1) rewritten to incorporate (3.2). From experiments it appears that the distance apart of the fibres, provided it is not too large, does not have an effect on the interaction between the fibres. The parameters α and η are coupling parameters for the components V and W and we note that $\alpha = \eta = 0$, $\epsilon = b$ in (7.1) restores the single fibre case represented by (F.N.) with $d = 0$. The matrix $\begin{bmatrix} 1-\alpha & -\alpha \\ -\alpha & 1-\alpha \end{bmatrix}$ is positive semi-definite leading to the restricted range $0 \le \alpha \le \frac{1}{2}$ for the coupling parameter α. In fact from experimental work it appears that α is close to zero.

The principal analytic attempt to deal with pulse transmission in coupled fibres is due to Luzader [9]. For α-coupling ($\alpha \neq 0$, $\eta = 0$) on two coupled fibres for example, he uses the expansions

$$u^{(1)} = u_0 + \alpha u_1(\delta)$$

$$u^{(2)} = u_0 + \alpha u_1(-\delta)$$

for the respective pulse velocities, where u_0 is the velocity of a pulse in a single uncoupled fibre, and δ is the distance apart of the pulses ($\delta > 0$ means that pulse 2 is ahead of pulse 1). For pulses travelling at the same speed

$$u^{(1)} = u^{(2)} = u,$$

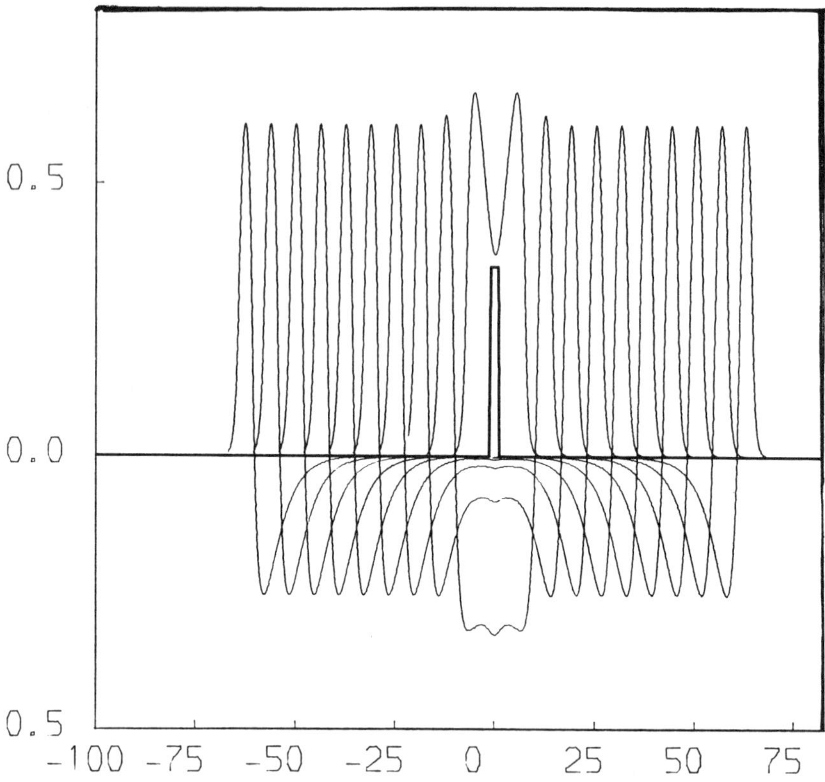

FIG. 1. Pulses in active fibre, α-coupling, box initial condition.

and so the dependent variables in (7.1) become functions of $x - ut$ leading to a system of ordinary differential equations. The special values of δ at which $u^{(1)} = u^{(2)}$ are zeros of the function

$$D_\alpha(\delta) = \frac{u_1(\delta) - u_2(\delta)}{u_0}$$

and such a phenomenon is known as "pulse locking". Using the caricature for the cubic nonlinear function and (7.1) transformed to a system of ordinary differential equations, Luzader found two stable states for δ, one of which is $\delta = 0$, and one unstable state. Some confirmation of these results for "pulse locking" is given in [4] where (7.1) in partial differential form is solved numerically using a "Hopscotch" finite difference method. The initial condition consisted of one pulse on each

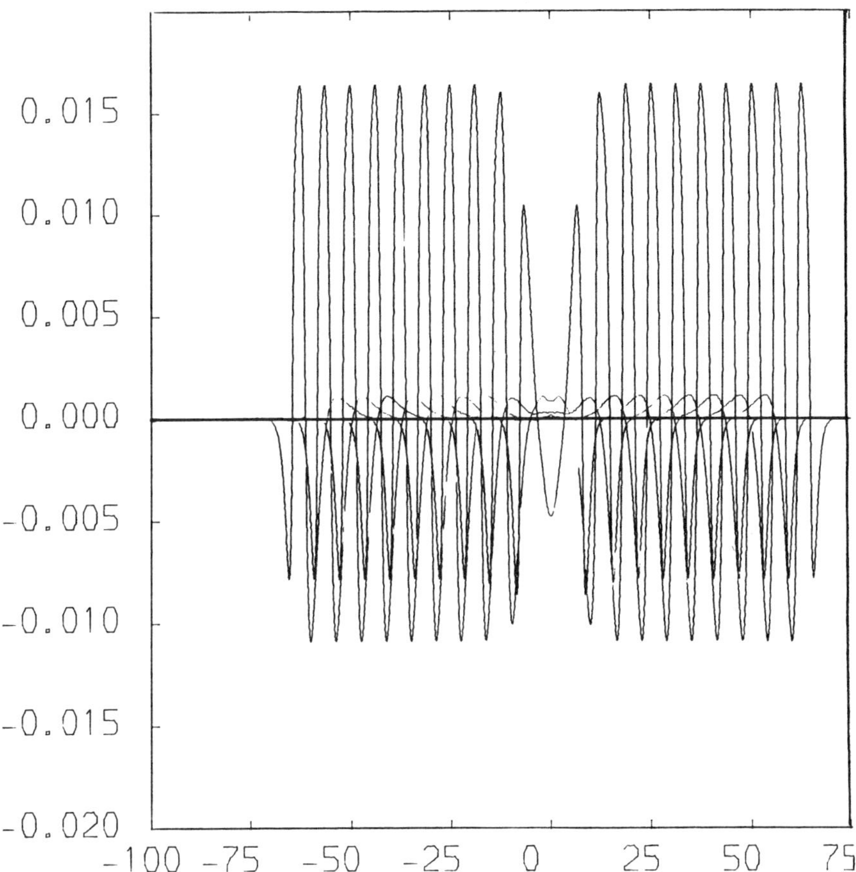

FIG. 2. Pulses in inactive fibre, α-coupling, box initial condition.

fibre separated by a distance δ_0. Each initial pulse is the exact solution to the uncoupled single fibre problem.

8. NUMERICAL RESULTS FOR TWO COUPLED FIBRES

We now solve (7.1) numerically using the method described in Section 6. The space discretisation has been carried out over elements of length h and the time levels are $t = nk$, $n = 0,1,2...$ The parameters involved are α, η, ϵ, and a, the first two being coupling parameters and a being crucial in the caricature representation of the cubic nonlinear function F. Only the pure initial value problem is considered with the initial condition in the active fibre being either a rectangular

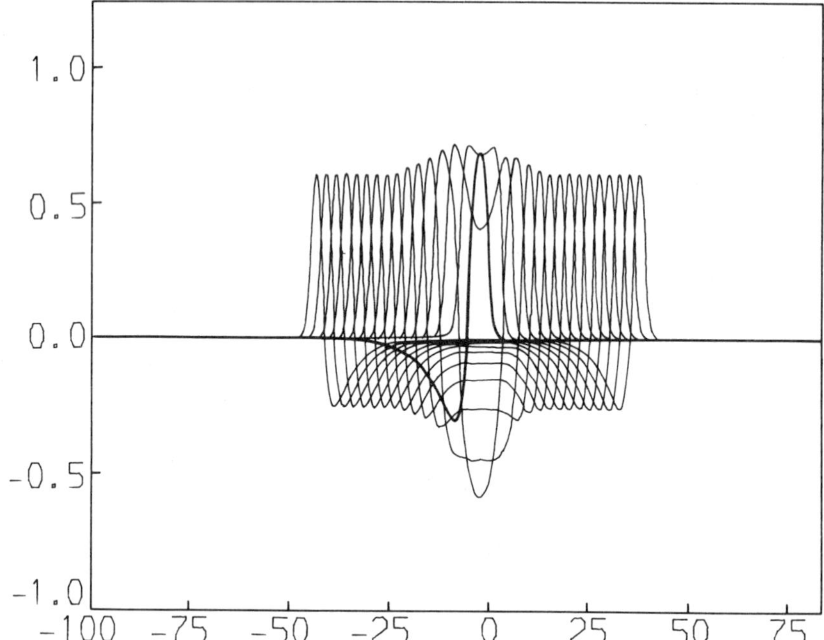

FIG. 3. Pulses in active fibre, α-coupling, pulse initial condition.

box

$$V = \begin{cases} 0 & -\infty < x < -x_1 \\ H \text{ (constant)} & -x_1 \le x \le +x_1 \\ 0 & +x_1 < x < +\infty. \end{cases}$$

or the pulse from the single fibre [11], together with the recovery variable $R_1 = 0$. In the inactive fibre $W = R_2 = 0$. We look at two distinct cases;

(i) α-coupling ($\alpha \neq 0$, $\eta = 0$). The parameter values are

$$\alpha = 0.1,\ \eta = 0,\ \epsilon = 0.1,\ a = 0.3.$$

and the grid sizes are

$$h = 0.25,\ k = 0.05.$$

The principal results from the numerical calculations are as follows;

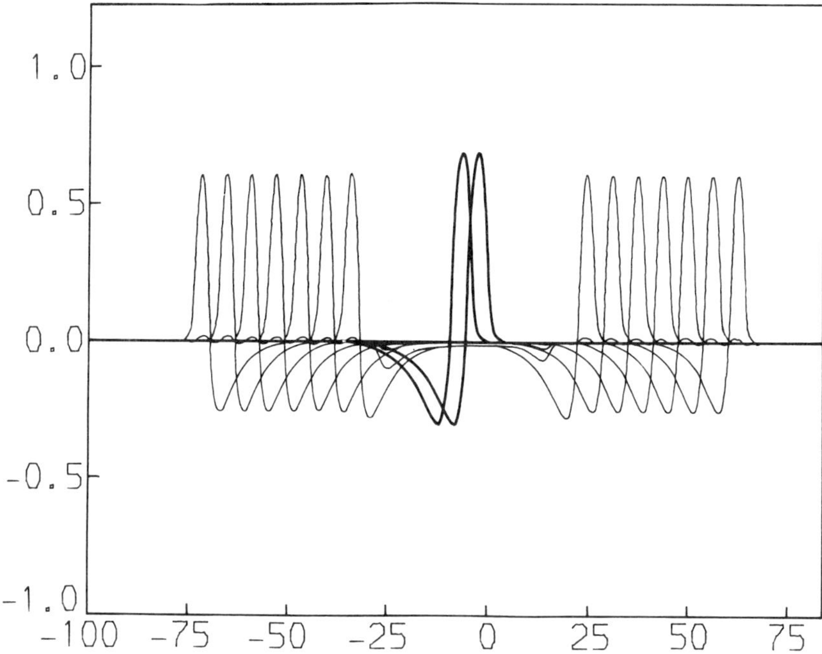

FIG. 4. Pulses in active fibre, α-coupling, twin pulse initial condition.

For a rectangular box initial condition of breadth 2.5 ($x_1 = 1.25$) in the active fibre, provided $H \geq 0.35$, (threshold value) there is an induced potential in the "inactive" fibre. The pulses in the active and inactive fibres are shown in Fig. 1 and Fig. 2 respectively for $H = 0.35$. Note the difference in shape of the potentials in the two fibres. Irrespective of the value of $H(\geq 0.35)$ the pulse speed is 0.500 for large t. Numerical experiments were also carried out with the pulse in the single fibre as the initial condition and the picture for increasing time shown in Fig. 3. The single pulse with a velocity of 0.713 breaks up into two pulses one in each direction with an asymptotic $(t \to \infty)$ speed of 0.500. A very small potential induced in the inactive fibre is not shown. Finally two initial pulses one in each fibre originally pointing in the same direction but with a phase difference of 4 units become two pulses travelling in opposite directions again with a velocity of 0.500. This is illustrated in Fig. 4. A strange result which we note for the present but which requires further investigation is that rectangular box initial conditions in both fibres with $H = 0.5$ reduce to zero after a moderate time interval.

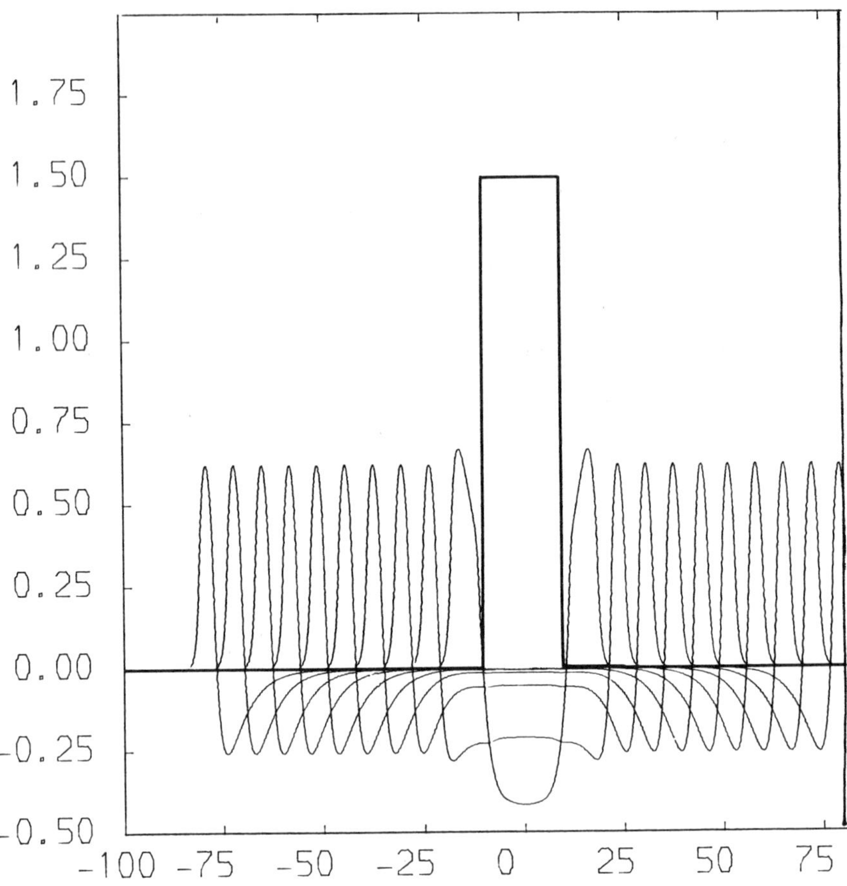

FIG. 5. Pulses in active fibre, η-coupling, box initial condition.

(ii) η-coupling ($\alpha = 0$, $\eta \neq 0$). The parameter values are

$$\alpha = 0, \ \eta = 0.005, \ \epsilon = 0.1, \ a = 0.3$$

and the grid sizes are

$$h = 0.25, \ k = 0.05.$$

A rectangular box initial condition of breadth 20 units ($x_1 = 10$) and height $H = 1.5$ produces pulses in the active fibre (Fig. 5) and an induced potential in the inactive fibre (Fig. 6). For large t, the constant speed attained by the pulses is 0.556 and the phase shift between the pulses 0.04.

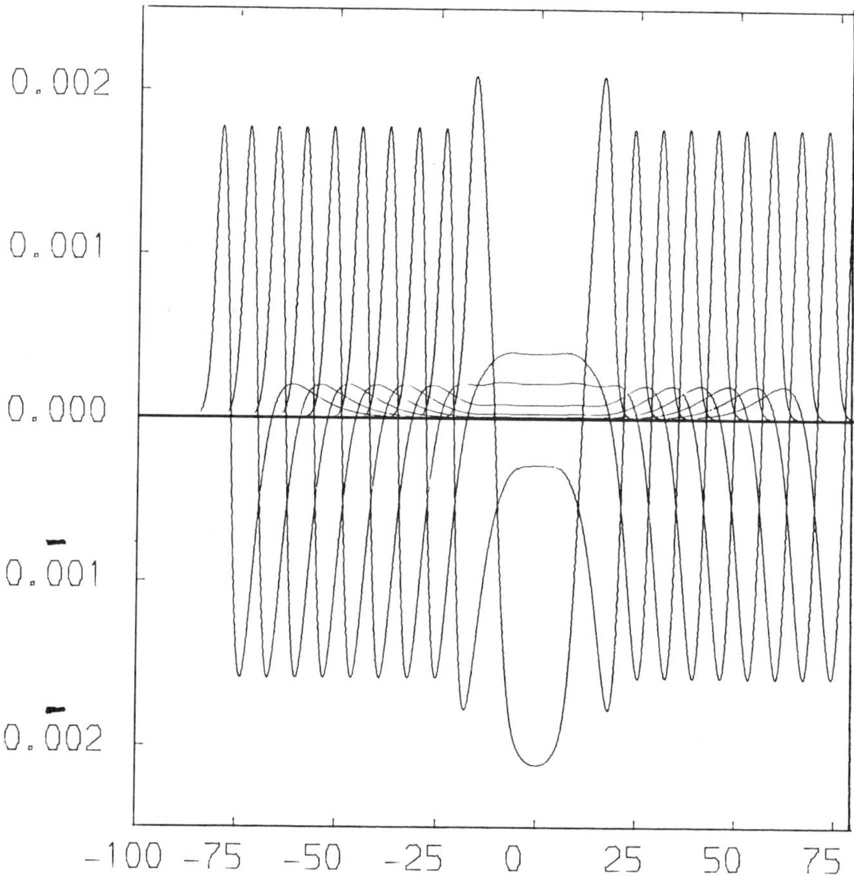

FIG. 6. Pulses in inactive fibre, η-coupling, box initial condition.

Raising the coupling parameter to $\eta = 0.055$ with all the other conditions similar failed to produce a sustained pulse on the active fibre and hence no induced pulse on the inactive fibre. More numerical work is required to fully explain the η-coupling.

ACKNOWLEDGEMENTS

The authors acknowledge considerable assistance given by Professors B.D. Sleeman and Kuo Pen Yu during the preparation of this paper. Financial support by the Science and Engineering Research Council is also acknowledged by both authors. Finally grateful thanks are due to Miss L. Simpson for her expert typing of the manuscript.

REFERENCES

1. BROWN, A.M., The Numerical Study of a Class of Reaction-Diffusion Equations. M.Sc. Thesis, University of Dundee (1983).

2. CHRISTIE, I., GRIFFITHS, D.F., MITCHELL, A.R., and SANZ-SERNA, J.M., Product Approximation for Non-linear Problems in the Finite Element Method. *IMAJNA* 1, 253-266 (1981).

3. COOLEY, J.W., and DODGE, F.A., Digital Computer Solutions for Excitation and Propagation of the Nerve Impulses. *Biophys. J.* 8, 583-599 (1966).

4. EILBECK, J.C., LUZADER, S.D., and SCOTT, A.C., Pulse Evolution on Coupled Nerve Fibres. *Bull of Math. Biology* 43, 389-400 (1981).

5. EWING, R.E., Time Stepping Galerkin Methods for Nonlinear Sobolev Partial Differential Equations. *SIAM J. Numer. Anal.* 15, 1115-1150 (1978).

6. HODGKIN, A.L., and HUXLEY, A.F., A Qualitative Description of Membrane Current and its Application to Conduction and Excitation in Nerves. *J. Physiol.* 117, 500-544 (1952).

7. KUO PEN YU, MANORANJAN, V.S., MITCHELL, A.R., and SLEEMAN, B.D., Bifurcation Studies in Reaction-Diffusion. *(To appear in Journal of Computational and Applied Mathematics)* (1984).

8. KUO PEN YU, MITCHELL, A.R., and SLEEMAN, B.D., Spatial Patterning of the Spruce Budworm in a Circular Region. University of Dundee Report DE 83:5 (1983).

9. LUZADER, S.D., Neurophysics of Parallel Nerve Fibres. Ph.D. Thesis, University of Wisconsin (1979).

10. MANORANJAN, V.S., and MITCHELL, A.R., A Numerical Study of the Belousov-Zhabotinskii Reaction using Galerkin Finite Element Methods. *J. Math. Biology* 16, 251-260 (1983).

11. MITCHELL, A.R., and MANORANJAN, V.S., Finite Element Studies of Reaction-Diffusion. pp.17-36 of J.R. Whiteman (Ed.), *The Mathematics of Finite Elements and Applications*. Academic Press, London (1981).

12. MURRAY, J.D., and SPERB, R.P., Minimum Domains for Spatial Patterns in a Class of Reaction-Diffusion Equations. (Preprint) (1981).

13. SLEEMAN, B.D., and TUMA, E., On Exact Solutions of a Class of Reaction Diffusion Equations. *(To appear in I.M.A. J. of Applied Mathematics)* (1984).

14. SMOLLER, J., *Shock Waves and Reaction-Diffusion Equations*. Springer Verlag, New York (1983).

15. SMOLLER, J., and WASSERMAN, A., Global Bifurcation of Steady-State Solutions. *J. Diff. Equs.* 39, 269-290 (1981).

ON SOME MATHEMATICAL ASPECTS OF BOUNDARY ELEMENT METHODS FOR ELLIPTIC PROBLEMS

W.L. Wendland

Fachbereich Mathematik, Technische Hochschule Darmstadt, B.R.D.

1. INTRODUCTION

During the last decade, boundary element methods have become well established as engineering computational methods for solving boundary value problems, in addition to finite difference and finite element methods. In [58] one finds a review of more than 80 boundary element codes showing the fast growth of corresponding software. For the numerous engineering applications see e.g. [16], [17]. The mathematical numerical analysis of these methods, however, is by no means sufficiently far developed yet, although boundary integral equations have been used for boundary value problems as much as 100 years ago [71]. One of the reasons for this lies in the fact that boundary integral equations underlying BEM form a significantly larger class than classical Fredholm equations. Unifying mathematical analysis is only possible in the framework of the theory of pseudodifferential operators. This has been developed during the last two decades but is not as yet everyday mathematics (see e.g. [73], [93]).

I try here to give a survey of the mathematical analysis of boundary element methods and confine myself to elliptic boundary value problems. This lecture is a continuation of [98]. However, since 1982 some of the concepts have become clearer and more rigorous.

The conversion of an engineering boundary value problem into an economical computer program is by no means a straightforward task and requires many decisions concerning the choice of elementary numerical manipulations. I try to analyze this scheme and to formulate some of the corresponding mathematical questions. Such an attempt cannot of course be completely satisfactory, but I hope to clarify at least some aspects for helping to improve the efficiency of boundary element methods.

The above-mentioned scheme begins with a given interior or

THE MATHEMATICS OF FINITE ELEMENTS AND APPLICATIONS V

ISBN 0-12-747255-X

exterior boundary value problem consisting of differential equations and boundary conditions. The boundary element methods require the explicit knowledge of the fundamental solution of the differential equations. Then representations of the solution in terms of boundary potentials lead, in connection with the boundary conditions, to boundary integral equations which live on the boundary manifold only. The corresponding reduction of the dimension by one corresponds to the sparsity of finite element matrices. This step already offers many different possibilities as we shall exemplify for elastostatic problems. A systematical analysis of corresponding advantages or disadvantages is not yet available. In [20], [22], however, we show for rather general boundary value problems how to choose boundary integral equations whose variational formulation resembles the energy bilinear forms of the original boundary value problem. The variational formulations provide the basic properties for further analysis as coercivity, continuity and sometimes symmetry. The approximate solution of the boundary integral equations in connection with finite elements on the boundary manifold creates a boundary element method. Again, different choices of boundary elements and different discretizations such as point collocation, Galerkin's method or least squares offer a whole variety of methods. For two-dimensional problems, the corresponding asymptotic error analysis is now relatively complete whereas for the real life three-dimensional problems the convergence properties of point collocations are not yet analyzed at all. In addition to asymptotic error estimates such an analysis also gives estimates for the conditioning of the discrete equations to be solved on the computer. For building up the influence matrix and right hand sides of the linear equations it is necessary to perform numerical integrations very extensively. This contributes the major part of the computational costs. Here I combine the required accuracy with the asymptotic errors of the theoretical discretizations to obtain criterions for an optimal choice of the numerical integration. However, a systematic analysis and development of higher dimensional efficient numerical quadratures involving the singularities of the kernels and the boundary elements is not available yet. I present a survey.

Another costly part of the BEM method is the solution of the large linear systems of equations. The development and use of efficient fast solvers in boundary element methods is yet to be done. Lastly, the boundary element methods require a postprocessing for the evalutation of the originally sought solution of the boundary value problem. Also here a systematic analysis is not yet available. Nevertheless I hope that this lecture helps to clarify some directions of the work in BEM.

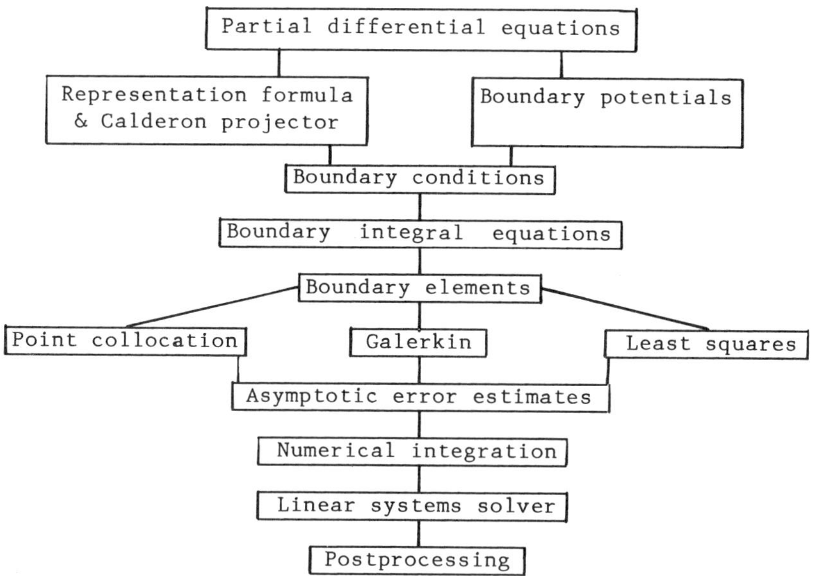

Scheme of boundary element methods

2. FORMULATION OF BOUNDARY INTEGRAL EQUATIONS

As we shall see, to one boundary value problem one can find (arbitrarily) many different boundary integral equations which have in general also different mapping properties implying different advantages and disadvantages of the resulting numerical schemes. The two most common formulations are either based on special potential representations - this is the classical approach - or on the so-called "direct method". Up to now the only systematic study of the different boundary integral equation formulations can be found in [20],[22] where the connection between the variational formulations of the boundary value problem and the boundary integral equations is the basic concept.

For simplicity let us consider the example of the traction problem of homogeneous isotropic elastic bodies. Let $\Omega \subset \mathbb{R}^n$, $n = 2$ or 3, denote a given bounded domain having sufficiently smooth boundary Γ describing the elastic body and let $\vec{u}(x)$ denote the displacement field, $x \in \Omega$ (or $x \in \Omega^c = \mathbb{R}^n \setminus \overline{\Omega}$ for exterior problems). Then $\vec{u}$ is sought as a solution of the governing Navier differential equations

$$L\vec{u} = \mu\Delta\vec{u} + (\lambda+\mu)\,\mathrm{grad}\,\mathrm{div}\,\vec{u} = \vec{\Phi} \quad \text{in} \quad \Omega \quad (\text{or } \Omega^c) \,. \tag{2.1}$$

Here $\mu > 0$ and $\lambda > -\frac{2}{n}\mu$ are the Lamé constants and $\vec{\Phi}$ is a given body force. In a preliminary step let us compute some particular solution of (2.1), e.g. by

$$\vec{u}_p(x) = \int \underset{\sim}{F}(y,x)\vec{\Phi}(y)dy \tag{2.2}$$

where

$$\underset{\sim}{F}(y,x) = \frac{\lambda + 3\mu}{4\pi(n-1)(\lambda+2\mu)} \left\{\gamma(y,x)I + \frac{\lambda+\mu}{\lambda+3\mu} \frac{(x-y)(x-y)^T}{|x-y|^n}\right\} \tag{2.3}$$

denotes the *fundamental solution* to (2.1) with I the identity matrix and

$$\gamma(y,x) = \begin{cases} -\log|x-y| & \text{for } n = 2 , \\ \frac{1}{|x-y|} & \text{for } n = 3 . \end{cases} \tag{2.4}$$

Hence, further on we restrict ourselves to the case $\vec{\Phi} = \vec{0}$, i.e. no body forces.

All boundary element methods are based on the representation of the desired solution $\vec{u}(x)$ in terms of potentials generated by boundary charges supported by Γ . Here let us particularly use Betti's formula which relates for the interior problems the displacement by

$$\vec{u}(x) = \int_\Gamma \{\underset{\sim}{F}(y,x)\vec{t}(y) - \underset{\sim}{T}(y,x)\vec{\phi}(y)\}ds_y \quad \text{for } x \in \Omega \tag{2.5}$$

with the *Cauchy data* , i.e. the pair of boundary functions

$$\vec{\phi}(y) := \vec{u}_{|\Gamma} , \quad \vec{t}(y) := T[\vec{u}]_{|\Gamma} \quad \text{for } y \in \Gamma ; \tag{2.6}$$

here the boundary displacement $\vec{\phi}$ and the boundary traction $\vec{t}$, where T denotes the traction operator on Γ ,

$$T[\vec{u}] = \lambda\vec{\nu}(y)\text{div } \vec{u} + 2\mu \frac{\partial\vec{u}}{\partial\nu} + \mu\vec{\nu}(y) \wedge \text{curl } \vec{u}(y)_{|\Gamma} . \tag{2.7}$$

$\vec{\nu}(y)$ is the exterior normal vector at $y \in \Gamma$, ds_y the surface element and

$$\underset{\sim}{T}(y,x) = (T_{(y)}[\underset{\sim}{F}(y,x)])^T .$$

The representation formula (2.5) yields with the jump relations for elastic potentials on Γ [48], [63] the relations

$$\vec{u}(x) = \frac{1}{2}\vec{\phi}(x) - \int_{\Gamma\setminus\{x\}} \underset{\sim}{T}(y,x)\vec{\phi}(y)ds_y + \int_{\Gamma\setminus\{x\}} \underset{\sim}{F}(y,x)\vec{t}(y)ds_y , \tag{2.8}$$

$$T[\vec{u}](x) = -T_{(x)}\Big[\int_{\Gamma\setminus\{x\}} \underset{\sim}{T}(y,x)\vec{\phi}(y)ds_y\Big] + \frac{1}{2}\vec{t}(x) + \int_{\Gamma\setminus\{x\}} \underset{\sim}{T}^T(x,y) \cdot \vec{t}(y)ds_y . \tag{2.9}$$

The matrix of operators on the right hand side defines the "Calderon projector" P for solutions of the homogeneous Navier equations (2.1) which reduces to the identity if $\vec{\phi}$ and $\vec{t}$ correspond to the Cauchy data (2.6) of some solution $\vec{u}$ of $L\vec{u} = \vec{0}$. In short, we write (2.8), (2.9) as

$$\begin{pmatrix} \vec{u} \\ T[\vec{u}] \end{pmatrix}(x) = P \begin{pmatrix} \vec{\phi} \\ \vec{t} \end{pmatrix}(x) \quad for \quad x \in \Gamma . \tag{2.10}$$

The pair of functions $\begin{pmatrix} \vec{\phi} \\ \vec{t} \end{pmatrix}$ *defines Cauchy data for some solution of* (2.1) *with* $\vec{\Phi} = \vec{0}$ *if and only if*

$$\begin{pmatrix} \vec{\phi} \\ \vec{t} \end{pmatrix}(x) = P \begin{pmatrix} \vec{\phi} \\ \vec{t} \end{pmatrix}(x) \quad for \quad x \in \Gamma . \tag{2.11}$$

Therefore, if one part of the Cauchy data or, more generally, a combination of them is given by a boundary condition, then (2.11) provides us with two or by combination with a whole family of boundary integral equations for the determination of the missing parts of the Cauchy data.

In the particular case of *given tractions* on Γ ,

$$T[\vec{u}](y) = \vec{\psi}(y) , \qquad y \in \Gamma , \tag{2.12}$$

the missing Cauchy datum is $\vec{u}(y)$ for $y \in \Gamma$.

The choice of the first equation in (2.11), i.e. (2.8), corresponds to the *direct method* and provides the boundary integral equation with side condition for $\vec{u}(y)$ (and $\vec{\omega} = \vec{0}$) ,

$$\begin{aligned} A_1\vec{u}(x) + M(x)\vec{\omega} &:= \vec{u}(x) + \varepsilon\, 2 \int_{\Gamma\setminus\{x\}} \underset{\sim}{T}(y,x)\vec{u}(y)ds_y + M(x)\vec{\omega} \\ &= \vec{f}_1(x) := \varepsilon \int_{\Gamma\setminus\{x\}} \underset{\sim}{E}(y,x)\vec{\psi}(y)ds_y , \end{aligned} \tag{2.13}$$

$$\Lambda\vec{u} := \int_\Gamma M^T(y)\vec{u}(y)ds_y = \vec{0} , \tag{2.14}$$

where $\varepsilon = 1$ for the interior boundary value problem (and $\varepsilon = -1$ for the exterior problem [45]). $M(x)$ is the rigid motion matrix

$$M(x) = \begin{pmatrix} 1, 0, -x_2, 0, 0, x_3 \\ 0, 1, x_1, 0, -x_3, 0 \\ 0, 0, 0, 1, x_2, -x_1 \end{pmatrix} \quad \text{for } n = 3 , \tag{2.15}$$

for $n = 2$ skip the last three columns and the last row. If for the interior problem , $\varepsilon = 1$, the given tractions $\vec{\psi}$ in (2.12) satisfy the equilibrium conditions of vanishing total forces and total momentums, then $\vec{\omega} = \vec{0}$ in (2.13) for any solution $\vec{u}$ which is uniquely determined by (2.14) [45] . (For exterior problems see also [45]).

Here the boundary integral equation is a *singular integral equation* with Cauchy kernel for $n = 2$ and with Giraud-Mikhlin kernel for $n = 3$ (see e.g. [62]) . In the framework of pseudodifferential operators, A_1 in (2.13) is a strongly elliptic pseudodifferential operator of order zero [73],[100],[102]. Note that the kernel $2\,\underset{\sim}{T}(y,x)$ is *not* symmetric, therefore the discretization of (2.13) can never become symmetric. The system (2.13), (2.14) is always uniquely solvable in any Sobolev space of boundary fields $H^\sigma(\Gamma)$ if $\vec{f}_1$ is given in $H^\sigma(\Gamma)$ or $\vec{\psi} \in H^{\sigma-1}(\Gamma)$, respectively, $\sigma \in \mathbb{R}$.

If we choose the second equation of the Calderon projector in (2.11), i.e. (2.9) then we obtain a *different* boundary integral equation for $\vec{u}$, namely

$$A_2\vec{u} + M\vec{\omega} := \int_{\Gamma\setminus\{x\}} T_{(x)}[(T_{(y)}[\underset{\sim}{F}(y,x)])^T](\vec{u}(y)-\vec{u}(x))ds_y + M(x)\vec{\omega}$$

$$= \vec{f}_2(x) := \frac{\varepsilon}{2}\vec{\psi}(x) + \int_{\Gamma\setminus\{x\}} \underset{\sim}{T}^T(x,y)\vec{\psi}(y)ds_y\,, \qquad (2.16)$$

$$\Lambda\vec{u} = \int_\Gamma M^T(y)\vec{u}(y)ds_y = \vec{0} \qquad (2.17)$$

with $\varepsilon = 1$ for the interior (and $\varepsilon = -1$ for the exterior) problem. Again, $\vec{u}$ is uniquely determined by (2.16), (2.17) and $\vec{\omega} = \vec{0}$ if $\vec{\psi}$ satisfies the equilibrium conditions [45]. This boundary integral equation has a *hypersingular nonintegrable kernel*. However, (2.16) is already regularized and exists in the sense of a Cauchy principal value for differentiable $\vec{u}$ and has been obtained from (2.9) since there the first term of the right hand side vanishes for constant $\vec{\phi}$. A_2 is a strongly elliptic pseudodifferential operator of order one [45], [102]. Note that the operators in (2.16), (2.17) are symmetric. Hence, the Galerkin procedure will provide symmetric (positive definite) systems of discrete equations. Moreover, the corresponding bilinear form agrees with the energy form of the generated fields [20], [22].

The direct method with the boundary integral equations (2.13), (2.14) is the most popular formulation underlying engineering computations [17], [23], [46], [79]. The use of the hypersingular equation (2.16) is more recent, [64], [69], [94] (for more references see [102]). Clearly, any combination of the equations in (2.11) (and also any combination of tangential derivatives of (2.8) with (2.9)) would yield further boundary integral equations for $\vec{u}_{|\Gamma}$ (or its tangential derivatives). Hence one has a large variety available and can choose formulations which have additional features.

A further variety of formulations stems from the classical approach via potentials which yield either singular integral equations or Fredholm integral equations of the second kind, see e.g. [17], [46], [48], [62], [63].

If the boundary condition (2.12) is replaced by different boundary conditions one can find a third type of integral equation. E.g. for given displacement

$$\vec{u}(y) = \vec{\phi}(y) \quad \text{for} \quad y \in \Gamma \,, \tag{2.18}$$

the choice of the first equation in (2.11), i.e. the direct method, (2.8) yields for the unknown boundary traction $\vec{t}(y)$ a *Fredholm integral equation of the first kind* with the weakly singular kernel $\underset{\sim}{E}$, i.e.

$$A_3\vec{t} + M\vec{\omega} := \int_{\Gamma\setminus\{x\}} \underset{\sim}{E}(y,x)\vec{t}(y)ds_y + M(x)\vec{\omega} = \vec{f}_3(x)$$

$$:= -\frac{\varepsilon}{2}\vec{\phi}(x) + \int_{\Gamma\setminus\{x\}} \underset{\sim}{T}(y,x)\vec{\phi}(y)ds_y \,, \tag{2.19}$$

$$\Lambda\vec{t} = \int_\Gamma M^T(y)\vec{t}(y)ds_y = \vec{0} \tag{2.20}$$

with $\varepsilon = 1$ for interior (and $\varepsilon = -1$ for exterior) problems. Here A_3 defines a strongly elliptic pseudodifferential operator on Γ of order -1 . For a detailed analysis and extensive references see [42] in case $n = 2$, for three-dimensional problems see e.g. [67], [68].

The choice of the second equation in (2.11), i.e. (2.9) yields again a singular integral equation for $\vec{t}_{|\Gamma}$ which has a singular integral operator adjoint to that in (2.13),[48],[63].

3. BOUNDARY ELEMENTS

For the numerical treatment of the boundary integral equations one needs a discretization of the desired boundary functions. In the boundary element method the trial functions are chosen as finite elements on Γ . To this end we assume that Γ is given by local parameter representations such that a family of regular partitions in the parameter domain is mapped onto corresponding partitions of Γ . On the partitions of the parameter domain we use a regular $S_h^{k,m}$ family of finite elements in the sense of Babuška and Aziz [12] $(m,k \in \mathbb{N}_o\ ,\ m < k)$. The parameter representation of Γ then transplants the finite elements onto the boundary Γ defining a family of boundary elements. For the computations this approach requires that the parameter representations are fully available in the computer. In particular in three-dimensional problems this can become rather involved. Therefore often an additional approximation of the boundary surface Γ is used leading to isoparametric boundary elements similar to shell elements as investigated by Nedelec [67] .

3.1 *No approximation of* Γ

<u>$n = 2$:</u> We assume that Γ is a system of mutually disjoint smooth (e.g. C^∞) Jordan curves where each can be represented by a regular parameter representation

$$\Gamma : \vec{x} = \vec{X}_j(t) \quad \text{for} \quad t \in \mathbb{R} , \ j = 1,\ldots,L , \tag{3.1}$$

with 1-periodic smooth functions $\vec{X}_j(t)$. Then we may identify functions on Γ with L-vector valued 1-periodic functions on $\mathbb{R}$ and the boundary integral equations become systems of such for vector-valued 1-periodic functions $\vec{u}(t)$. On $\mathbb{R}$ we choose a family of 1-periodic grid partitions $0 = t_o < t_1 < \ldots < t_N = 1$,

$$\Pi_h = \{t_i\}_{i=-\infty}^{\infty} , \quad t_{i+N} = t_i + 1 , \ h = \max\{t_{i+1} - t_i\} , \tag{3.2}$$

$0 < h \to 0$. Here $S_h^{k,m}$ denotes the family of 1-periodic splines of degree $k-1$ subordinate to the partitions Π_h and belonging to $C^{m-1} \subset H^s(\Gamma)$, $s < m + 1/2 < k$. The spaces of trial functions H_h then are defined by

$$\vec{u}_h(x) = \vec{\Phi}_h(\vec{X}_\ell^{-1}(x)) \Bigg/ \left| \frac{d\vec{X}_\ell}{dt} \right| \tag{3.3}$$

for $x \in \Gamma_\ell$ and with $\vec{\Phi}_h \in (S_h^{k,m})^p$. By $\vec{\mu_j}$, $j = 1,\ldots,N$ let us denote a basis of H_h with functions $\vec{\mu}_j$ having smallest supports.

<u>$n = 3$:</u> Let Γ be given by a finite union of covering local regular parameter representations

$$\Gamma : \vec{x} = \vec{X}_\ell(t) , \quad t \in \mathcal{U}_\ell \subset \mathbb{R}^2 , \quad \ell = 1,\ldots,M \tag{3.4}$$

over two-dimensional parameter regions $\mathcal{U}_\ell$ such that every $\mathcal{U}_\ell$ has polygonal boundary and such that mutual interior overlapping on Γ is excluded [67] .

On every $\mathcal{U}_\ell$ there is given a family of triangulations Π_h with maximal meshsize h and a subordinated system $S_h^{k,m}$ of finite elements [12] satisfying additional compatibility conditions across the boundaries of $\mathcal{U}_\ell$ such that H_h defined by (3.3) satisfies $H_h \in H^s$, $0 \le m \le k-1$ with $s \le m$.

3.2 Additional approximation of Γ *by* Γ_h

According to [53] for $n = 2$ and [67] for $n = 3$ we now also approximate $\Gamma \in C^{\kappa+2}$, $\kappa \in \mathbb{N}_o$ by a Lagrangian system $S_h^{\kappa+1,1}$ of finite elements defined on Π_h and associated with a unisolvent set of grid points $\{p_{\ell i}\}$ in the triangle T_ℓ such that the interpolation problem $\psi_h(p_{\ell i}) = \psi(p_{\ell i})$ for $\psi_h \in S_h^{\kappa+1,1}$ is uniquely solvable. Then Γ_h is defined by the interpolation of $\vec{X}_\ell(t)$ in (3.4) with elements in $S_h^{\kappa+1,1}$ and is a continuous smooth surface given by

$$\vec{y} = \vec{X}_{h\ell}(t) \;, \quad t \in \mathcal{U}_\ell \;, \quad \ell = 1,\dots,M \;. \tag{3.5}$$

Now choose a $S_h^{k,m}$ system as in the previous Section 3.1 with $0 \le m \le k-1 \le \kappa$ and $m \le 1$. Then we define on Γ_h isoparametric elements by

$$\vec{\chi}_h(y) := \vec{\Phi}_h \circ \vec{X}_{h\ell}^{-1}(y) \Big/ \left| \frac{d\vec{X}_{h\ell}}{dt} \right| \tag{3.6}$$

for $y \in \Gamma_h$, $\vec{\Phi}_h \in (S_h^{k,m})^p$. For their transplantation onto Γ define the mapping $y = \Psi^{-1}(x)$ by the nearest hit of the rays through $x \in \Gamma$ in the $\pm\vec{\nu}(x)$-directions with Γ_h, $y \in \Gamma_y$. Then

$$\vec{u}_h(x) := \left[\vec{\Phi}_h \circ \vec{X}_{h\ell}^{-1} \Big/ \left| \frac{d\vec{X}_h}{dt} \right| \right] \circ \Psi^{-1}(x) \tag{3.7}$$

defines the *boundary element space* H_h.

All these boundary element spaces have the following APPROXIMATION PROPERTY: [7], [12], [28], [67] .

Let $-\infty < \sigma \le s \le k$. *Then to any* $\vec{u} \in H^s(\Gamma)$, *the usual Sobolev-Slobodetski space on* Γ, *and* H_h *there exists* $\vec{v}_h \in H_h$ *such that*

$$\| \vec{u} - \vec{v}_h \|_{H^\sigma(\Gamma)} \le c h^{s-\sigma} \| \vec{u} \|_{H^s(\Gamma)} \tag{3.8}$$

where the constant c *is independent of* $\vec{u}, \vec{v}_h$ *and* h *provided*

$$\sigma < m + \tfrac{1}{2} \text{ for (3.3) and } n=2 \;; \quad \sigma < \min\{\tfrac{3}{2}, m + \tfrac{1}{2}\} \text{ for (3.7) and } n=2 \;, \tag{3.9}$$

$$\sigma \le m \text{ for (3.3) and } n=3 \;; \quad \sigma \le \min\{1,m\} \text{ for (3.7) and } n=3 \;.$$

If we assume in addition that the partitions Π_h satisfy regularity assumptions then H_h also provide the INVERSE ASSUMPTION [7], [12], [28], [67] :

For any σ *and* s *with* $s \le \sigma$ *and* σ *satisfying* (3.9) *there exists a constant* c *such that the inequality*

$$\| \vec{v}_h \|_{H^\sigma(\Gamma)} \le c h^{s-\sigma} \| \vec{v}_h \|_{H^s(\Gamma)} \tag{3.10}$$

holds for all $\vec{v}_h \in H_h$.

4. DISCRETIZATIONS AND BOUNDARY ELEMENT METHODS

For the boundary element methods, the boundary integral equations of the form

$$A\vec{u}(x) + M(x)\vec{\omega} = \vec{f}(x) \;, \quad \Lambda\vec{u} = \int_\Gamma \lambda(y)\vec{u}(y)ds_y = \vec{\beta} \tag{4.1}$$

are to be approximated where $\vec{u}$ is the unknown and $\vec{f}$ a given p-vector valued function, $\vec{\omega} \in \mathbb{R}^{\tilde{q}}$ an unknown and $\vec{\beta} \in \mathbb{R}^q$ a given constant vector. The three methods mostly used are *point collocation, Galerkin's method* and the *least squares method*. For their formulation let $\vec{\mu}_1(x),\dots,\vec{\mu}_N(x)$ be a basis of H_h for fixed h and set

$$\vec{u}_h(x) := \sum_{j=1}^{N} \gamma_j \vec{\mu}_j(x) \tag{4.2}$$

with constants γ_j to be determined.

4.1 The point collocation

With an appropriate set of collocation points $\{x_k\} \subset \Gamma$, $k = 1,\dots,N_1$ with $p\cdot N_1 + \tilde{q} = q + N$, the collocation equations read as

$$\begin{aligned} &\sum_{j=1}^{N} \gamma_j A\vec{\mu}_j(x_k) + M(x_k)\vec{\omega}_h = \vec{f}(x_k)\ , \qquad k = 1,\dots,N_1\ , \\ &\sum_{j=1}^{N} \gamma_j \int_\Gamma \lambda(y)\vec{\mu}_j(y)ds_y = \vec{\beta} \end{aligned} \tag{4.3}$$

forming a quadratic system of linear equations for the unknowns γ_j and $\vec{\omega}_h$. As we shall see, the appropriate choice of collocation points x_k depends decisively on the mapping properties of A and is by no means trivial. Nevertheless this method is most popular in boundary element computations [17], [46]. Note that the coefficient matrix $(A\vec{\mu}_j(x_k))$ usually *cannot be symmetric*.

4.2. The Galerkin method

Here (4.1) is approximated by the system of linear Galerkin equations

$$\begin{aligned} &\sum_{j=1}^{N} \gamma_j (A\vec{\mu}_j,\vec{\mu}_k)_{L^2(\Gamma)} + (M\vec{\omega}_h,\vec{\mu}_k)_{L^2(\Gamma)} = (\vec{f},\vec{\mu}_k)_{L^2(\Gamma)}\ , \quad k=1,\dots,N\ , \\ &\sum_{j=1}^{N} \gamma_j \int_\Gamma \lambda(y)\vec{\mu}_j(y)ds_y = \vec{\beta}\ . \end{aligned} \tag{4.4}$$

Note that for a selfadjoint system (4.1) the coefficient matrix in (4.4) becomes symmetric. This is the case for (2.16), (2.17) and for (2.19), (2.20), but *not* for (2.13) and (2.14). For (4.4), being a quadratic system of linear equations, we need $q = \tilde{q}$ which restricts the generality of (4.1). We shall see that for the boundary element method A must even be *strongly elliptic*. Although the numerical implementation of Galerkin's method seems to require more computational effort due to double integration we shall see later that it is just a modified collocation method and indeed is used effectively in numerical computations [33], [34], [37], [41], [42], [53], [68], [69],[70] .

4.3 The least squares method

The least squares method leads to the system of *symmetric, linear, positive definite equations* if we suppose uniqueness for (4.1),

$$\sum_{j=1}^{N} \gamma_j\{(A\vec{\mu}_j, A\vec{\mu}_k)_{L^2(\Gamma)} + \Lambda\vec{\mu}_j \cdot \Lambda\vec{\mu}_k\} + (M\vec{\omega}_h, A\vec{\mu}_k)_{L^2} = (\vec{f}, A\vec{\mu}_k)_{L^2} + \vec{\beta}\cdot\Lambda\vec{\mu}_k\ , \quad k = 1,\ldots,N\ ,$$

$$\sum_{j=1}^{N} \gamma_j(A\mu_j, M^T)_{L^2} + (M\vec{\omega}_h, M^T)_{L^2} = (\vec{f}, M^T)_{L^2}\ . \qquad (4.5)$$

As we shall see, (4.5) can be used under the weakest conditions on A but has on the other hand in general the worst conditioning of the three methods. Nevertheless also least squares are used in practical computations [82] .

5. ASYMPTOTIC ESTIMATES FOR THE METHOD ERRORS AND REQUIREMENTS FOR THE BOUNDARY INTEGRAL EQUATIONS

Up to now, all rigorous asymptotic error estimates for the three boundary element methods of Section 4 have been based on *variational methods* for the boundary integral equations on Γ . (Only in the very restricted case of Fredholm integral equations of the second kind with weakly singular kernels different techniques are available [3], [9].) For two-dimensional problems, $n = 2$, these asymptotic error estimates based on variational methods are now rather complete whereas for $n = 3$ and the point collocation, results are only available for the already mentioned class of Fredholm integral equations of the second kind with weakly singular kernels (see [11], [78]) .

Let us suppose that A in (4.1) is an *elliptic pseudo-differential operator of* order 2α on Γ [87] and that (4.1) is uniquely solvable. Then (4.1) defines a continuous, bijective mapping $(H^{\sigma+\alpha}(\Gamma))^p \times \mathbb{R}^q \to (H^{\sigma-\alpha}(\Gamma))^p \times \mathbb{R}^{\tilde{q}}$ and we have the a-priori estimate

$$\|\vec{u}\|_{H^{\sigma+\alpha}} + |\vec{\omega}| \le c\{\ \|A\vec{u} + M\vec{\omega}\|_{H^{\sigma-\alpha}} + |\Lambda\vec{u}|\} \le c'\{\ \|\vec{u}\|_{H^{\sigma+\alpha}} + |\vec{\omega}|\}\ . \qquad (5.1)$$

However, ellipticity is sufficient only for the convergence of the least squares method whereas for the point collocation and the Galerkin method it is necessary to require for A to belong to the much smaller proper subclass of *strongly elliptic* pseudodifferential operators of order 2α ; i.e. to A there exists a regular matrix-valued function $\Theta(x)$ (which is real for real equations) and a pseudodifferential operator of order $2\alpha-\varepsilon$ with some $\varepsilon > 0$ and a positive constant γ_0 such that coerciveness in the form of Garding's inequality holds, i.e.

$$\mathrm{Re}((\Theta A + C)\vec{v},\vec{v})_{L^2(\Gamma)} \geq \gamma \|\vec{v}\|^2_{H^\alpha(\Gamma)} \quad \text{for all } \vec{v}\in H^\alpha(\Gamma) . \tag{5.2}$$

Strong ellipticity is a consequence of positive definiteness of the principal symbol of ΘA (see [73]) and has been proved for many boundary equations of mathematical physics (see e.g. [1], [7], [21], [30], [45], [48], [62], [88], [89], [96], [99], [100], [101], [102]). Direct proofs of (5.2) without using the symbol can be found in [20], [22], [43], [68].

We now shall summarize the asymptotic estimates of the method errors which are all of optimal order; i.e. they are of the form

$$\| e \|_\sigma := \| \vec{u}-\vec{u}_h \|_{H^\sigma(\Gamma)} + |\vec{\omega}-\vec{\omega}_h| \leq ch^{s-\sigma} \| \vec{u} \|_{H^s(\Gamma)} . \tag{5.3}$$

More precisely: For the $\sigma \leq s$ restricted below there exist $c, h_0 > 0$ such that for every $h > 0$, with $h \leq h_0$ the respective equations of Section 4 are uniquely solvable. The corresponding approximate solutions $\vec{u}_h$, $\vec{\omega}_h$ satisfy the estimate (5.3) where c is independent of h, $\vec{u}$, $\vec{\omega}$, $\vec{u}_h$, $\vec{\omega}_h$.

For the *point collocation* method we restrict ourselves to $n = 2$ and to *naive collocation* for "smoothest" splines, i.e. $k = m+1$, where

$$x_k = \begin{cases} \vec{X}(t_k) & \text{for odd } m \text{ and} \\ \vec{X}(\frac{1}{2}(t_{k+1} + t_k)) & \text{for even } m . \end{cases} \tag{5.4}$$

An extension to more general choices of collocation points has been presented in [86] which converges with optimal order for a slightly larger class of equations than with strongly elliptic A .

First results for uniform Π_h , $m = 1$ and strongly elliptic Cauchy singular integral equations (in Hölder spaces) have been obtained in [76] by using explicit eigenvalue estimates for circulant matrices. For arbitrary Π_h , odd m and the general class of strongly elliptic pseudodifferential equations variational methods have been used in [7]. With Fourier series, the case of even m , regular Π_h and "constant coefficients" was treated in [81] which has been generalized to more general equations in [75] and to arbitrary equations and arbitrary m in [8] . The necessity of strong ellipticity was proved in [85] based on [76] .

For $n = 3$ the question of convergence of point collocation (for strongly elliptic A) is still open.

The stability and convergence of *Galerkin's method* for strongly elliptic A in the energy space $H^{\alpha}(\Gamma)$ follows from Gårding's inequality (5.2) by standard arguments [61],[39],[43] and for $2\alpha-k \le \sigma \le \alpha$ by using the Aubin-Nitsche trick [44] (see also [68]). For $n=2$ these results have been slightly extended in [8] and the *necessity* of strong ellipticity for the optimal order convergence of the spline Galerkin method was shown in [85].

For the *least squares method*, the optimal order convergence has been proved in [90] (see also [65], [66], [77]) . Here, the operator A^*A is always strongly elliptic and of order 4α . Therefore the asymptotic estimates from Galerkin's method with A^*A provide here corresponding estimates and in special cases also the necessity of ellipticity for optimal order convergence.

We collect these results in the following Table 1 .

n = 2	n = 2 and 3	n = 2 and 3
Naive point collocation	Galerkin	Least squares
Strong ellipticity is sufficient (and necessary)	Strong ellipticity is sufficient (and necessary)	Ellipticity is sufficient (and necessary)
$k = m + 1$,	$0 \le m < k$,	$0 \le m < k$,
$2\alpha \le \sigma \le s \le k$,	$2\alpha-k \le \sigma \le s \le k$,	$4\alpha-k \le \sigma \le s \le k$,
m odd: $2\alpha < m$,	$\alpha < m+1/2$ for n=2 , $\alpha \le m$ for n=3 ,	$2\alpha < m+1/2$ for n=2 , $2\alpha \le m$ for n=3 ,
$\sigma \le \alpha + \frac{k}{2} \le s$ for arbitrary Π_h.	$\sigma \le \alpha \le s$ for arbitrary Π_h ,	$\sigma \le 2\alpha \le s$ for arbitrary Π_h ,
$2\alpha < m + \frac{1}{2} \wedge 2\alpha + \frac{1}{2} < s$ for uniform Π_h .	$\sigma \le m$ for Π_h with (3.10)	$\sigma \le m$ for Π_h with (3.10).
m even: $2\alpha < m + \frac{1}{2}$		
uniform Π_h and $\sigma < m + \frac{1}{2} \wedge 2\alpha + \frac{1}{2} < s$	$\sigma < m + \frac{1}{2}$ for n=2 and uniform Π_h .	$\sigma < m + \frac{1}{2}$ for n=2 and uniform Π_h .
Highest order: $\|e\|_{2\alpha} \le ch^{k-2\alpha}\|\vec{u}\|_k$	$\|e\|_{2\alpha-k} \le ch^{2k-2\alpha}\|\vec{u}\|_k$	$\|e\|_{4\alpha-k} \le ch^{2k-4\alpha}\|\vec{u}\|_k$

Table 1

Remark 5.1: For $n = 2$, $k = m+1$ and $\alpha < 0$ let us summarize some of the restrictions on the Sobolev indices for (5.3) in the following Figure 1 .

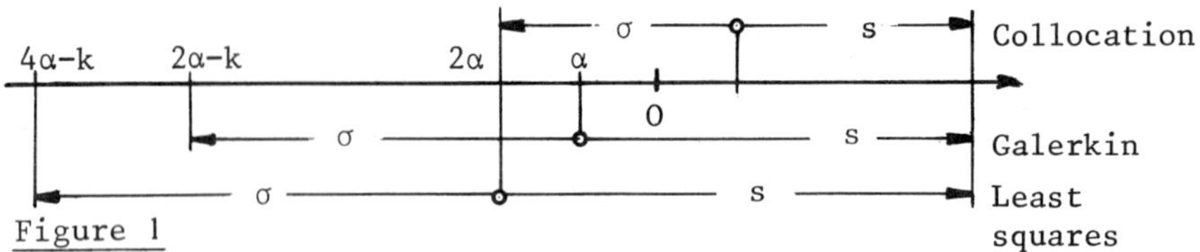

Figure 1

Hence, to obtain the *same* highest order ℓ_o of convergence with each of the three methods the corresponding degrees of splines m_c, m_G, m_{LS} must satisfy

$$m_c = 2m_G + 1 = 2m_{LS} + 1 - 2\alpha = \ell_o - 1 + 2\alpha \ . \tag{5.5}$$

On the other hand, for Galerkin's method and least squares one needs double integration to compute the coefficients of the discrete equations in contrast to point collocation. This effect will also be reflected by the necessary accuracy of the corresponding numerical integrations so that - after all - the computing times and corresponding costs are essentially the same for all three methods in the case $n = 2$.

Remark 5.2: The trigonometric-spline Galerkin-Petrov methods in [4], [5] even converge exponentially.

6. NUMERICAL INTEGRATION [1)]

For the implementation of any of the boundary element methods in Section 4 one meeds to compute numerically the coefficients a_{jk} and $\vec{f}_k$ given by

	Point collocation	Galerkin	Least squares	
$a_{jk} =$	$A\vec{\mu}_j(x_k)$	$(A\vec{\mu}_j, \vec{\mu}_k)_{L^2}$	$(A\vec{\mu}_j, A\vec{\mu}_k)_{L^2}$	(6.1)
$\vec{f}_k =$	$\vec{f}(x_k)$	$(\vec{f}, \vec{\mu}_k)_{L^2}$	$(\vec{f}, A\vec{\mu}_k)_{L^2}$	(6.2)

and $(\lambda, \vec{\mu}_j)_{L^2}$, $(M, \vec{\mu}_k)_{L^2}$, $(M, A\vec{\mu}_k)_{L^2}$, respectively. For accurate numerical integrations the method errors (5.3) dominate. But in view of 50% computing time for accurate computations of the above weights in most of the boundary element packages it is ridiculous to perform the numerical integrations more accurately than necessary. Hence, the error of the solution due to numerical integration should be of the *same order* as the method error in (5.3). Therefore we first indicate how to relate the accuracy of the weights with the methodical asymptotic errors. Then we survey currently available numerical integrations.

1) I want to thank Mr. Schwab for providing me with a rather thorough survey of two-dimensional numerical integration.

6.1 The precision of weights and asymptotic errors

For estimating the influence of *erroneous weights* $\tilde{a}_{jk}$, $\tilde{\vec{f}}_k$ we use the *inverse assumption*, hence confine us to *quasiuniform families* Π_h and obtain stability of the discrete equations from (5.3) which is used in the second Strang lemma [31, p. 144]. Let us begin with stability.

<u>Lemma 6.1:</u> *Let* Π_h *be quasiuniform such that the inverse assumption* (3.10) *holds. Then*

$$\|\vec{u}_h\|_{L^2} + |\vec{\omega}_h| \le ch^{\beta}\{\|\vec{f}_h\|_{L^2} + |\vec{\beta}|\} \quad \text{and}$$

$$\|\vec{f}_h\|_{L^2} + |\vec{\beta}| \le c'h^{\beta'}\{\|\vec{u}_h\|_{L^2} + |\vec{\omega}_h|\} \tag{6.3}$$

for $\vec{f}_h \in H_h$ *and the corresponding solution of* (4.3), (4.4) *or* (4.5), *respectively, provided the assumptions in Section 5 are fulfilled, where*

$$\beta = \begin{cases} \min\{0,2\alpha\} \\ \min\{0,4\alpha\} \end{cases} \text{and } \beta' = \begin{cases} -\max\{0,2\alpha\} & \text{for point collocation or Galerkin ,} \\ -\max\{0,4\alpha\} & \text{for least squares .} \end{cases}$$

The proof can be found for $n = 2$ in [7]and can easily be extended from [43], [44] to our slightly more general case. We omit the details. Note that (6.3) implies a conditioning of $O(h^{-2|\alpha|})$ for the discrete equations from point collocation and Galerkin's method and $O(h^{-4|\alpha|})$ for least squares which is for $\alpha \neq 0$ the price to pay for the wide applicability of least squares. (For $n = 2$, $\alpha = -1/2$ and an ordinary computer with e.g. 10 digits, the round-off errors exceed the systematic errors in [41] for $N \geq 400$ with Galerkin's method. Hence least squares become unstable already for $N \geq 20$.)

<u>Theorem 6.2:</u> (See [42] and [97].) *Let* Π_h *be quasiuniform and the assumptions of Section 5 be fulfilled. Let the coefficients* a_{jk} *and* f_k *of the discrete equations in Section 4 and their numerically integrated counterparts* $\tilde{a}_{jk}$ *and* $\tilde{f}_k$ *satisfy estimates*

$$|a_{jk} - \tilde{a}_{jk}| \le ch^{\rho} \; , \; |f_k - \tilde{f}_k| \le c_f h^{\rho-(n-1)} \; . \tag{6.4}$$

Then the solutions $\tilde{\vec{u}}_h$, $\tilde{\vec{\omega}}_h$ *of the numerically integrated equations corresponding to* (4.3), (4.4) *or* (4.5),*respectively, satisfy asymptotic estimates*

$$\|\vec{u}_h - \tilde{\vec{u}}_h\|_{H^{\sigma}(\Gamma)} + |\vec{\omega}_h - \vec{\omega}_h| \le ch^{r}\{\|\vec{u}_h\|_{L^2} + c_f\} \tag{6.5}$$

where

$$r = \rho + \min\{0,-\sigma\} - (n-1) + \begin{cases} \min\{0,2\alpha\} & \textit{for point collocation}, \\ \min\{0,2\alpha\}-(n-1) & \textit{for Galerkin}, \\ \min\{0,4\alpha\}-(n-1) & \textit{for least squares}, \end{cases} \tag{6.6}$$

provided $r > 0$. *The right hand side of* (6.5) *can further be estimated by using*

$$\|\vec{u}_h\|_{L^2} \le c \begin{cases} \|\vec{u}\|_{H^{\max\{0,2\alpha\}}} & \textit{for point collocation} \\ \|\vec{u}\|_{H^{\max\{0,2\alpha-k\}}} & \textit{for Galerkin's method}, \\ \|\vec{u}\|_{H^{\max\{0,4\alpha-k\}}} & \textit{for least squares}. \end{cases} \tag{6.7}$$

Consequently, the complete systematic error of the fully discretized equations is given by

$$\|\tilde{\vec{u}}_h - \vec{u}\|_\sigma + |\tilde{\vec{\omega}} - \vec{\omega}| \le ch^{s-\sigma}\|\vec{u}\|_s + ch^r\{\|\vec{u}\|_\tau + c_f\} \tag{6.8}$$

with τ given by (6.7). This suggests the choice $r=s-\sigma$ for best efficiency. For balancing e.g. L_2-*errors* we need

$$\sigma = \begin{cases} \max\{0,2\alpha\} & \text{for point collocation} \\ \max\{0,2\alpha-k\} & \text{for Galerkin's method} \\ \max\{0,4\alpha-k\} & \text{for least squares} \end{cases} \tag{6.9}$$

in view of Section 5. Here (6.9) implies requirements

$$\rho \overset{(\ge)}{=} \begin{cases} k+1-2\alpha & \text{for point collocation and } n=2, \\ k+2(n-1)-\min\{0,2\alpha\}-\min\{0,k-2\alpha\} & \text{for Galerkin's method}, \\ k+2(n-1)-\min\{0,4\alpha\}-\min\{0,k-4\alpha\} & \text{for least squares}. \end{cases} \tag{6.10}$$

In order to recover the *highest orders* of convergence one needs

$$\rho \overset{(\ge)}{=} \begin{cases} k+1-4\alpha & \text{for point collocation and } n=2, \\ 2k+2(n-1)-2\alpha-\min\{0,2\alpha\} - \min\{0,k-2\alpha\} & \text{for Galerkin's method}, \\ 2k+2(n-1)-4\alpha-\min\{0,4\alpha\} - \min\{0,k-4\alpha\} & \text{for least squares}. \end{cases} \tag{6.11}$$

Remark 6.1: For $n=2$, with ℓ_o as in Remark 5.1 and without the approximation of Γ we need for the same highest order of accuracy

$$\rho \ge \ell_o + \begin{cases} 1-2\alpha & \text{for point collocation}, \\ 2-\min\{0,2\alpha\}-\min\{0,k-2\alpha\} & \text{for Galerkin's method}, \\ 2-\min\{0,4\alpha\}-\min\{0,k-4\alpha\} & \text{for least squares}. \end{cases}$$

This shows that for the *same* asymptotic *accuracy* the three methods require *different degree splines* (5.5) but *essentially the same accuracy of numerical integration.* For more details see [6].

In general, the errors in (6.4) consist of two contributions, one due to numerical integration and the other due to the approximation of Γ by Γ_h. Estimates of the latter are based on the following lemma.

Lemma 6.3 [18], [34], [67] : *Let* $\Gamma \in C^{\kappa+2}$. *Then there exist positive constants* c *independent of* Π_h *such that for every patch* $T_j \subset \mathcal{U}$ *and* $0 \le \beta \le \kappa$

$$\| \vec{X}_{\ell h} - \vec{X}_\ell \|_{C^\beta(T_j)} + \| \psi o \vec{X}_{\ell h} - \vec{X}_{\ell h} \|_{C^\beta(T_j)} \le c h^{\kappa+1-\beta} \| \vec{X}_\ell \|_{C^{\kappa+2}(T_j)} . \tag{6.12}$$

Moreover,

$$|J_h(y) - J o \psi(y)| + \big| |y-y'|^2 - |\psi(y)-\psi(y')|^2 \big| \, |y-y'|^{-2} \le c h^{\kappa+1} \| \vec{X}_\ell \|_{C^{\kappa+2}}$$

and

$$c^{-1} \le |\psi(y) - \psi(y')| \, |y-y'|^{-1} \le c .$$

Here J and J_h denote the Jacobians of $\vec{X}_\ell$ and $\vec{X}_{\ell h}$, respectively. These estimates can be used in the kernel expansions of A with respect to the local coordinates on Γ and Γ_h, respectively. However, some geometrical quantities converge faster than indicated by (6.12) and direct comparison of the explicitly given operators on Γ and Γ_h yield sharper results. Then one obtains asymptotic estimates of the form (6.5) with $\vec{u}_h$, $\vec{\omega}_h$ being now the solutions of the equations in Section 4 on Γ_h instead of Γ provided quasiuniform Π_h and the assumptions of Section 5 are satisfied, where

$r = \kappa + \min\{0,-\sigma\} + \min\{0,2\alpha\} + \varepsilon$ with

- $\varepsilon = 1$ for collocation and Galerkin's method and Fredholm equations of the second kind, $2\alpha=0$, $n=2,3$ [33], [100],
- $\varepsilon = 1$ for equations as (2.19), (2.20) and Galerkin's method, $2\alpha = -1$, $n = 2$ [53], $n = 3$ [67],
- $\varepsilon = 0$ for hypersingular equations as (2.16), (2.17) and Galerkin's method, $2\alpha = 1$, $n = 2,3$ [34] .

To the comparison of the operators in the parameter domain we shall come back later on.

Once the exponent ρ in (6.4) is fixed we must choose also *numerical integration procedures* accordingly. This is by no means trivial and a *systematic* investigation and development of efficient methods - in particular with respect to the singularities of the kernels - is yet to be done.

Let us begin with the simplest case of *smooth integrals* since usually the kernels can be split into the principal parts with the singularities in some standard form and smooth remaining kernels.

6.2. Numerical integration of smooth integrands

In view of (3.3) and (3.6), respectively, all the wieghts in (6.1) and (6.2) on Γ or on Γ_h involve integrals solely *in the paramter domain* which are of the form

$$\int_{T_\ell} F(t)\tilde{\mu}_j(t)dt = h^{n-1} \int_{\tilde{T}} F(\tilde{\psi}_\ell(\tilde{t}))\tilde{\mu}_j(\tilde{\psi}_\ell(\tilde{t}))\tilde{J}_\ell d\tilde{t} \tag{6.13}$$

with smooth F. Here $\tilde{T}$ denotes *a reference element* and $\tilde{\psi}_\ell : \tilde{T} \to T_\ell$ is the affine linear mapping with constant Jacobian $\tilde{J}_\ell$ satisfying

$$\frac{\partial\tilde{\psi}_\ell}{\partial\tilde{t}} = h^{(n-1)}\tilde{J}_\ell\ , \quad |\tilde{J}_\ell| + |\tilde{J}_\ell^{-1}| \le c \tag{6.14}$$

with c independent of h for Π_h quasiregular. $\tilde{\mu}_j$ is a (piecewise) polynomial on T_ℓ. Although straightforward numerical quadrature of (6.14) - mostly Gaussian quadrature - is used in most programs [15], [17], [47], [50] the weight $\tilde{\mu}_j$ in (6.13) reduces the order of accuracy.

<u>Lemma 6.4 [51, Lemma 33]:</u> *Let* $\tilde{d}$ *be the degree of precision of the numerical integration formula used for* (6.13). *Let* $\tilde{\mu}_j$ *be a (piecewise) polynomial of degree* $\tilde{k}$. *Then the error satisfies estimates*

$$|\tilde{F}_{j\ell} - \int_{T_\ell} F(t)\tilde{\mu}_j(t)dt| \begin{cases} \le c_1 h^{\tilde{d}+n-\tilde{k}} \| F^{(\tilde{d}+1-\tilde{k})} \|_{W^o_\infty(T_\ell)}\ , \\ \le c_2 h^{\tilde{d}+\frac{(n+1)}{2}-\tilde{k}} \| F^{(\tilde{d}+1-\tilde{k})} \|_{L^2(T_\ell)}\ . \end{cases}$$

Note that $\tilde{\mu}_{j|T_\ell}$ is a special polynomial on $\tilde{T}$ with $k-1 \le \tilde{k}$. For Gaussian quadrature $\tilde{d}+1 = 2\cdot$(number of node points). The proof follows easily with the Taylor formula applied to $F\circ\psi$, exact integration of polynomials of degree $\tilde{d}$ and (6.14).

If we apply product integrations to (6.1), (6.2) then we have in (6.4)

	Point collocation	Galerkin	Least squares	
$\rho =$	$\tilde{d} + n - \tilde{k}$	$\tilde{d} + 2n-1-\tilde{k}$	$\tilde{d} + 2n-1-\tilde{k}$	(6.15)

If one wants to avoid this loss of order $\tilde{k}$ it is advisable to use *weighted integration formulas* with the non-negative weight functions $\tilde{\mu}_{j}|_{T_\ell}$. If the computations are to be performed only for one or two different partitions then (weighted) Gaussian quadrature will provide good accuracy and efficiency. If, however, one wants to perform *mesh refinements* then Gaussian quadratures lead to a great increase of computing times. Hence, for mesh refinement one should use integration formulas with nodes on a regular grid which forms a subset of the refined grid and then perform the necessary function evaluations only at the *new* grid points, reusing the already computed values from the coarse grid. The numerical experiments in [41], [42] revealed very high efficiency of such grid point integrations.

n = 2 : Here the integrations over T_ℓ are one-dimensional and in [35] one finds how to obtain weights and nodes for weighted Gaussian formulas. (See also [92]).

Weighted grid point formulas have been developed in [6], [41], [42], [96] for splines μ_j and uniform families of partitions of the real axis. Here $k = m+1$, $\tilde{\mu}_j(t) = \mu(\frac{t}{h} - j+1)$ and the grid point formulas are of the form

$$\int F\tilde{\mu}_j(t)dt \approx h \sum_{\ell=-M}^{M} F(\vec{X}((j+\frac{m-1}{2} + \ell\cdot\gamma)h))$$

with $\tilde{d} = 2M+1$, $b_\ell = b_{-\ell}$ and $\gamma = 1$ or $\frac{1}{2}, \frac{1}{4}, \ldots$, respectively.

	Point collocation	Galerkin	Least squares
ρ =	2M + 3	2M + 4	2M + 4

For $\gamma = 1$ and $\frac{1}{2}$, the weights b_ℓ can be found in [6, Table 3] and [96, p. 258]. Note that for $\gamma\cdot M \geq m$ some of the b_ℓ become negative which causes large round off errors due to cancellation of digits. Hence $\gamma\cdot M < m$ with positive b_ℓ is preferable.

n = 3 : Here the integrations over T_ℓ are two-dimensional. Formula (6.13) shows that we need an integration formula

$$\int_{\tilde{T}} \tilde{\mu}(\tilde{t})\Phi(\tilde{t})dt = \sum_{i=1}^{K} w_i\Phi(\tilde{t}_i) + \text{remainder} \tag{6.16}$$

where $\tilde{\mu}(\tilde{t})$ is one of given nonnegative weight functions $\tilde{\mu}_j(\psi_\ell(\tilde{t}))$ on the reference element $\tilde{T}$. For (6.16) we require exact integration *for all polynomials of degree* $\leq\tilde{d}$. Hence $\{\tilde{t}_i\}$ must be a set of unisolvent points for these polynomials. If the node points $\tilde{t}_i$ are *given* and $K \geq (1/2)\tilde{d}(\tilde{d}-1)$ then (6.16) becomes just a system of $(1/2)\tilde{d}(\tilde{d}-1)$ linear equations for the w_i .

Gridpoint integration: If the node points correspond to the grid belonging to the boundary elements or its refinement then also here we can find grid point integration formulas. (For

$\tilde{d} = 2$ see [10].) These can be computed for the reference element and the given μ and then are ready to be used in the actual boundary element computations. For finding the w_i we need the exact values of the moments, i.e. the left hand integrals in (6.16) for Φ running through a basis of the polynomials of degree $\le \tilde{d}$. These can be obtained via extrapolation based on [55]. In general, positive weights w_i can only be found if $\tilde{K}$ is chosen large enough. For larger ρ, i.e. larger $\tilde{d}$, this can result in a rather fine grid on Γ or Γ_h connected with Π_h.

Optimal and Gaussian formulas: The exact integration of (6.16) for all polynomials Φ of degree $\le \tilde{d}$ can also be used to find the smallest number and location of all node points t_i together with the weights – or of *some* of the node points and *given remaining* nodes. In [91] it was shown that in the first case the nodes belong to the intersection points of the zero sets of the orthogonal polynomials subject to $\tilde{T}$ and μ. The equations for w_i and t_i are nonlinear. In general, their solvability properties are not yet clear. However, for $\mu \equiv 1$ and several shapes of $\tilde{T}$ integration formulas (6.16) have been obtained in [13], [29], [57], [103]. For more general μ, also our experiments in combination with extrapolation methods for finding the moments, i.e. the left hand sides of (6.16) have been rather encouraging.

If in addition Γ *is approximated by* Γ_h then the additional error of the smooth remainders can be estimated easily.

Lemma 6.5: *Let* F *depend smoothly on* $\vec{X}_\ell$ *and its derivatives up to order* β *and let* F_h *be defined by replacing* $\vec{X}_\ell$ *in* F *by* $\psi \circ \vec{X}_{\ell h}$ *or just by* $\vec{X}_{\ell h}$. *Then we have for the integrals (in the parameter domain)*

$$\left|\int_{T_\ell} F_h(t)\tilde{\mu}_j(t)dt - \int_{T_\ell} F(t)\tilde{\mu}_j(t)\right| \le ch^{n+\kappa-\beta} .$$

Hence the contribution to (6.4) *due to* Γ_h *instead of* Γ *from the smooth remainders is of order*

$$\rho = n+\kappa-\beta + \begin{cases} 0 & \textit{for collocation} \\ n-1 & \textit{for Galerkin's method and least squares.} \end{cases}$$

6.3 *Computation of the singular principal parts*

As in the example of Section 3, the principal parts of the boundary integral operators involved are given by *pseudodifferential operators of integer orders* [20], [22, Lemma 3.1] defined by compositions of different operators with integral operators having singular kernels. Therefore their corresponding numerical evaluation for the weights (6.1) is nontrivial. On one hand their computation requires special care whereas on the other hand higher order formulas are in opposition to higher singularities of the higher derivatives of the singular integrand. Therefore the singular integrals should either be computed analytically or by special integration formulas involving singular weights. At present there is a big need for efficient higher order integration formulas, in particular in case $n = 3$.

n = 2 : Since for two-dimensional problems the operators in A are given on the *one-dimensional* boundary Γ or, in parameter representation on $\mathbb{R}$, they can always be written in the form

$$A\left\{v/\left|\frac{d\vec{X}}{dt}\right|\right\} = \sum_{\beta=-L}^{2\alpha} A_\beta v(\tau) + \int_{\mathbb{R}^{n-1}} k(\tau,t)v(t)dt \quad \text{for } v \in C_0^\infty \tag{6.18}$$

where $L \geq -2\alpha$ is any chosen integer and the kernel function k is in C^{L-1} if $L \geq 1$. The operators A_β are given by

$$\begin{aligned}(A_\beta v)(\tau) &= a_\beta(\tau)\, D^\beta v(\tau) - b_\beta(\tau) \int_{\mathbb{R}} \frac{D^\beta v(t)}{t-\tau}\, dt \quad \text{for } \beta \geq 0 ,\\ &= a_\beta D^\beta v(\tau) + b_\beta(\tau) \int_{\mathbb{R}} \log|t-\tau|\, D^{\beta+1} v(t)dt\end{aligned} \tag{6.19}$$

where $D = \frac{d}{dt}$ and

$$(A_\beta v)(\tau) = \int_{\mathbb{R}} |t-\tau|^{-\beta-1} \{p_1(\tau,\text{sign}(t-\tau)) + p_2(\tau,\text{sign}(t-\tau)) \cdot \log|t-\tau|\}\, v(t)dt \quad \text{for } \beta \leq -1 . \tag{6.20}$$

The coefficients a_β, b_β, $p_1(\tau,+1),\ldots,p_2(\tau,-1)$ are all smooth functions. (This special form can be obtained from [2] and the fact that for $\beta = 0$ all pseudodifferential operators are given by (6.19). Note that (6.20) can also be written as (6.19) if D^{-1} denotes any right inverse to D .) Inserting (6.18), (6.19) into (6.1) on Γ or on Γ_h , respectively, results in integrals of the form

$$\int_{\text{supp}(\tilde{\mu}_j)} \log|t-\tau_k|\, D^{\beta+1}\tilde{\mu}_j(t)dt \quad \text{for point collocation ,} \tag{6.21}$$

$$\int_{\text{supp}(\tilde{\mu}_k)} \int_{\text{supp}(\tilde{\mu}_j)} \log|t-\tau|\, c(\tau)\tilde{\mu}_k(\tau)(D^{\beta+1}\tilde{\mu}_j(t))dtd\tau \quad \text{for Galerkin,} \tag{6.22}$$

$$\int_{\tau\in I} \int_{\text{supp}(\tilde{\mu}_{jk})} \int_{\text{supp}(\tilde{\mu}_j)} \log|t_1-\tau|\log|t_2-\tau|\, c(\tau)(D^{\beta_1+1}\tilde{\mu}_j(t_1)) \cdot (D^{\beta_2+1}\tilde{\mu}_k(t_2))dt_1dt_2d\tau$$

plus terms of the form (6.22) for the least squares. (6.23)

Here β can be any (also negative) integer. Since here $\tilde{\mu}_j = \mu_j |\frac{d\vec{X}}{dt}|$ is a piecewise polynomial, all the integrations over t can be performed analytically, [17], [53]. However, the resulting expressions must be formulated very carefully since they behave like finite difference formulas on the grid points in $\text{supp}(\mu_j)$ causing cancellation of digits (see [41], [42]). If $c(\tau)$ is a piecewise polynomial then also the exterior integrations can be performed analytically.

A different way is to introduce new coordinates

$$\xi = t-\tau \ , \quad \eta = t+\tau \ ; \quad t = \frac{1}{2}(\xi+\eta) \ , \quad \tau = \frac{1}{2}(\eta-\xi) \ .$$

Then (6.22) becomes

$$\frac{1}{2} \sum_{0\le\ell,p\le m} \int_{t_{j-1}-t_{k+m}}^{t_{j+m}-t_{k-1}} \log|\xi| \left\{ \int_{\max\{2t_{j+\ell-1}-\xi,\ 2t_{k+p-1}+\xi\}}^{\min\{2t_{j+\ell}+\xi,\ 2t_{k+p}-\xi\}} \cdot \right.$$

$$\left. \cdot\ c(\tfrac{1}{2}(\eta-\xi))D^{\beta+1}\tilde{\mu}_j(\tfrac{1}{2}(\xi+\eta))\tilde{\mu}_k(\tfrac{1}{2}(\eta-\xi))d\eta \right\} d\xi \ . \tag{6.24}$$

The outer integration can be done by Gaussian quadrature with the weight $\log|\xi|$ and at the corresponding ξ-nodes, the inner integrations can be performed with ordinary Gaussian formulas (see [41], [42]).

The third possibility is to perform the outer integrations in (6.1) for Galerkin's method and for least squares as in Section 6.2 . Then one again needs (6.2) where τ_k are now the node points corresponding to the outer integration.

The computational expense for (6.21)-(6.23) reduces significantly for splines on *uniform partitions* with $t_j = j\cdot h$, $h = \frac{1}{N}$, $N \in \mathbb{N}$, $\tau_k = h\cdot(k + \frac{m-1}{2})$ and

$$\tilde{\mu}_j(t) = \mu(\tfrac{t}{h} - j + 1) \quad \text{for} \quad t_{j-1} \le t \le t_{j-1}+1 \ , \tag{6.25}$$

where μ is a given fixed reference B-spline on $[0,m+1] \subset \mathbb{R}$. This is not a severe restriction, since the choice of the parametrization $\vec{X}(t)$ gives enough freedom for a nonuniform grid on Γ . Now (6.21) e.g. becomes

$$h^{-\beta} \int_0^{m+1} [\log h + \log|t'+(j-k) - \tfrac{m+1}{2}|]\ \mu^{(\beta+1)}(t')dt'$$

$$= h^{-\beta} [\log h \int_0^{m+1} \mu^{(\beta+1)}dt' + W_{(j-k)}] \ . \tag{6.26}$$

Hence, the elements of (6.21) form a *Toeplitz matrix* whose entries are defined via (6.26) with weights W_ρ , $\rho \in \mathbb{Z}$ which are independent of Γ and h . These can be computed once for all and then are ready to be used.

If we insert (6.25) into (6.22) and (6.24) then for *constant* $c(\tau)$ and Galerkin's procedure we again find Toeplitz matrices defined by similar weights W_ρ . (For further details see [41], [42],[96].)

Remark 6.2: The representation (6.18), (6.19) includes also the operators with hypersingular kernels as A_2 in (2.16) with $\beta = 2\alpha = 1$, $a_1 = 0$, $b_1 = \frac{\mu(\lambda+\mu)}{\pi(\lambda+2\mu)} I$, I the unit matrix.

Remark 6.3: Instead of using the logarithmic potential of (6.19), (6.20) in (6.20)-(6.23) one could use also the Hilbert transform i.e. the first version of (6.19). Then the corresponding formulas would contain $D^\beta\tilde{\mu}_j$ instead of $D^{\beta+1}\tilde{\mu}_j$. For the Hilbert transform, Piessens in [74] and Elliott and Paget in [26], [27] give numerical quadrature formulas. In [17] Appendix A it is suggested to use part finie integral formulas given by Kutt [49]. In both cases, however, the formulas contain *negative weights* implying *cancellation of digits* and require therefore further improvements. By contrast, our approach yields positive weights and the expressions $D^{\beta+1}\tilde{\mu}_j(t)$ are explicitly given by the B-spline $\tilde{\mu}_j$.

Remark 6.4: In the case of *uniform partitions* with (6.25) it follows for the influence matrices (6.1) that they can be decomposed into a sum of a Toeplitz matrix and a matrix defined by the discretization of an operator of order less than 2α . (For point collocation and Galerkin's scheme this was indicated before. For the least squares method this can be shown by using periodic functions and Fourier expansions.)

n = 3 : Since A is a classical integer order pseudodifferential operator on the two-dimensional surface Γ or Γ_h it also admits the form (6.18) in local coordinates with $t = (t_1,t_2)$, $\tau = (\tau_1,\tau_2)$, $t-\tau = r(\cos\phi, \sin\phi)$ in the parameter domain where

$$(A_\beta v)(\tau) = \sum_{\gamma_1+\gamma_2=\beta} a_\gamma(\tau)D^\gamma v(\tau) + \lim_{\varepsilon\to 0+} \sum_{\gamma_1+\gamma_2=\beta} \int_{\mathcal{U}\cap r\geq\varepsilon} \frac{1}{r}\Phi_\gamma(\tau,\phi)D^\gamma_{(t)}v(t)\,dr\,d\phi \quad \text{for } \beta \geq 0 \qquad (6.27)$$

and for $v\in C_0^\infty(\mathcal{U})$. Here $D^\gamma = D_1^{\gamma_1} D_2^{\gamma_2}$, $D_j = \frac{\partial}{\partial t_j}$ or $\frac{\partial}{\partial \tau_j}$, respectively, $\gamma_j \in \mathbb{N}_0$, and the characteristics Φ_γ satisfy

$$\int_{\phi=0}^{2\pi} \Phi_\gamma(\tau,\phi)\,d\phi = 0 \ . \qquad (6.28)$$

Moreover,

$$(A_\beta v)(\tau) = \int_{\mathcal{U}} r^{-1-\beta}\{\Phi_{1\beta}(\tau,\phi) + (1+\beta)\log r\,\Phi_{2\beta}(\tau,\phi)\}\,v\,dr\,d\phi \quad \text{for } \beta \leq -1 \ . \qquad (6.29)$$

The coefficients $a_\gamma(\tau)$ and the characteristics $\Phi_{\ell\beta}(\tau,\phi)$ are smooth in $\mathcal{U}$ and $\mathcal{U}\times\mathbb{R}$ and 2π-periodic in ϕ, respectively. The above representations follow e.g. from [87, Section III], [62]. For $v \in \mathring{H}^\beta(\mathcal{U})$, (6.27) becomes after integration by parts with respect to ϕ,

$$(A_\beta v)(\tau) = \sum_{\gamma_1+\gamma_2=\beta} a_\gamma(\tau)D^\gamma v(\tau) + \lim_{\varepsilon\to 0+} \sum_{\lambda=0}^{\beta} \int_{\mathcal{U}\cap r\geq\varepsilon} r^{-\lambda-1} \cdot \\ \cdot\ \underset{\sim}{\Phi}_{\lambda\beta}(\tau,\phi)\left(\frac{\partial}{\partial r}\right)^{\beta-\lambda} v\, dr d\phi\ ,$$

$$\text{where} \quad \sum_{\lambda=\beta-j}^{\beta} \int_0^{2\pi} \frac{j!}{(\lambda+j-\beta)!}\ \underset{\sim}{\Phi}_{\lambda,\beta}(\tau,\phi)e^{ij\phi}d\phi = 0$$

for $j = -\beta,\ldots,\beta$, or, after another integration by parts with respect to r, also for $\beta \geq 0$,

$$(A_\beta v)(\tau) = \sum_{\gamma_1+\gamma_2=\beta} a_\gamma(\tau)D^\gamma v(\tau) + \text{part finie} \int_{\mathcal{U}} r^{-\beta-1}\ \Phi_{1\beta}(\tau,\phi) \\ \cdot\ v\, dr d\phi\ . \tag{6.31}$$

Hence, we need sufficiently accurate numerical integrations with $v = \tilde{\mu}_j \in S_h^{k,m}$ for (6.27) and (6.29) or, if $m \geq \beta$, for (6.30) or (6.31) in case of collocation or for the corresponding iterated integrals in case of Galerkin's or least squares methods. Note that $\tilde{\mu}_j$ is a polynomial of degree $\tilde{k}$ on each of the triangles $T_\ell \subset \operatorname{supp}(\tilde{\mu}_j)$ in the parameter domain. τ can either belong to $\overline{T}_\ell$ or be outside the domain of integration. Let us consider first the case $\tau \in \overline{T}_\ell$ when the integrals are singular.

6.3.1 The use of standard routines for the weakly singular integrals. In [80] one finds a general standard routine to evaluate two-dimensional weakly singular integrals including the cases (6.27) and (6.29). However, the experiments, Examples 22,24,27, Table 6.5 [80] show that for each of the approximately h^{-2} many singular integrals in (6.1) we need *several thousand* function evaluations. This shows that we cannot use standard routines as a black box to compute the matrix elements in (6.1).

6.3.2 Exptrapolation techniques. Lyness in [54], [56] developed for weakly singular integrals, (6.29) for $\beta \leq -1$, an extrapolation procedure with T_ℓ a square or a simplex and τ a corner point by using meshrefinement. Using a harmonic refinement one has the least number of necessary node points. For $\Phi_{1,-1} = 1$ and T_ℓ a triangle we needed 200 to 2 000 node points for accuracies of 4 to 10 digits, respectively (with double precision in the inner loops.) For the Bulirsch refine-

the corresponding number of node points was 2 000 to 20 000 . Therefore these techniques are also far too costly be to be used in a boundary element program as a standard routine.

6.3.3 Special cases. For some special operators in boundary element methods, e.g. in classical potential theory and elasticity, one finds special integrations which are either analytical or partly analytical. For (6.29) with $\beta = -1$, $\Phi_{1,-1} \equiv 1$ and special μ_j on triangles T_ℓ we find partly analytical numerical integration formulas in [24] and for the operators of elasticity of Section 2 in [94]. If τ is a corner point of T_ℓ then triangular coordinates transform the integral into one with regular integrand [17, Section 6.9], [52]. However, this regularization does not work properly for non-constant, i.e. oscillatory $\Phi_{1,-1}$ as e.g. for the *classical double layer potential* with

$$\Phi_{1,-1} = -\frac{1}{2}(b_{11}(\tau)\cos^2\phi + 2b_{12}(\tau)\cos\phi\sin\phi + b_{22}(\tau)\sin^2\phi) \tag{6.32}$$

where $((b_{jk}))$ is the second fundamental tensor. (See [59, Chap. 2] where one can also find $\Phi_{1,\beta}$ for $\beta = -2,-3,\ldots$.) For this special operator see also [10], [95].

6.3.4 Numerical integration with part finie integrals. In [17, p. 451 ff.] it is suggested to use for (6.32) Kutt's formulas with respect to r and then formulas as Gaussian quadrature with respect to ϕ yielding

$$\text{part finie} \int_{T_\ell} r^{-\beta-1}\Phi_{1\beta}(\tau,\phi)\tilde{\mu}_j dr d\phi$$

$$\approx \sum_{i=1}^{I} w_i \Phi_{1\beta}(\tau,\phi_i) \text{ part finie} \int_{0<r\leq\rho(\phi_i)} r^{-\beta-1}\tilde{\mu}_j(r\cos\phi_i, r\sin\phi_i)dr. \tag{6.33}$$

As before, $\tilde{\mu}_j$ is a polynomial of degree $\leq \tilde{k}$ and the boundary of T_ℓ is given by the *piecewise* smooth representation $r = \rho(\phi)$. To date there are two severe disadvantages:

a) For T_ℓ a square with τ the midpoint we have

$$\rho(\phi) = h \min\left\{\frac{1}{|\sin\phi|}, \frac{1}{|\cos\phi|}\right\}$$

and the analytic continuation of the outer integrand in (6.34) has essential singularities in the complex ϕ-plane near $\kappa\frac{\pi}{4}$, $\kappa \in \mathbb{Z}$ implying very slow convergence for increasing I .

b) Kutt's formulas contain large negative weights and also nodes with negative r . Besides the cancellation of digits due to the oscillation of $\Phi_{1\beta}$, the negative weights yield an additional cancellation. (Moreover, the papers by Kutt, loc. cit. 7 and 8 in [17, p. 454] are not at all accessible.)

However, this approach is rather general and, thus, should be further improved. Although for each single integration method the degree of precision is known, a systematic error analysis is also yet to be done.

6.3.5 Approximation by Γ_h . Since the characteristics Φ_γ in (6.27) and (6.29) are defined by the derivatives $D^\lambda\vec{X}(\tau)$ one can prove the following general Lemma. In particular cases, however, it is possible to obtain sharper results as in [10], [33], [34], [67], [100].

Lemma 6.6: *Let* Φ_γ *be defined by* $D^\mu\vec{X}(\tau)$ *for* $|\mu| \le \lambda$ *and* $\Phi_{\gamma h}$ *correspondingly. Let* $|\gamma| = \beta$ *and* $\Gamma \in C^{\kappa+2}$ *. Then for the corresponding weights we have*

$$|a_{jk}-\tilde{a}_{jk}| \le ch^{\kappa+1-\lambda-\beta+\varepsilon}|\log h| \qquad (6.34)$$

where $\varepsilon = 0$ *for collocation and* $\varepsilon = 2$ *for Galerkin's and least squares methods.*

Proof: Let us consider only collocation and $\beta \ge 0$. The remaining cases follow in the same manner.

$$\Big|\int_{T_\ell} \frac{1}{r}(\Phi_\gamma-\Phi_{\gamma h})D^\gamma\tilde{\mu}_j \, drd\phi\Big|$$

$$\le \Big|\int_0^{2\pi}\int_{r=0}^{\delta} \frac{1}{r}(\phi_\gamma-\phi_{\gamma h})D^\gamma\tilde{\mu}_j \, drd\phi\Big| + \int_0^{2\pi}\int_\delta^h \frac{1}{r}\,|\Phi_\gamma-\Phi_{\gamma h}|\,|D^\gamma\tilde{\mu}_j|drd\phi \,.$$

The first term can be estimated as in [62, p. 226] and the second with Lemma 6.3 yielding

$$|a_{jk}-\tilde{a}_{jk}| \le c\delta\|D^{\beta+1}\tilde{\mu}_j\|_{C^0(T_\ell)} + ch^{\kappa+1-\lambda}\log\frac{h}{\delta}\,\|\tilde{\mu}_j\|_{C^\beta(T_\ell)}.$$

The choice $\delta = h^{\kappa+2-\lambda}$ and $\|\tilde{\mu}_j\|_{C^{\beta'}(T_\ell)} \le ch^{-\beta'}$ yields (6.34).

6.3.6 Nearly singular integrals. For $\tau \notin \overline{T}_\ell$, the integrands in (6.27)-(6.31) are all smooth. However, if τ is near to T_ℓ then the singularity of the integral kernel outside the region of integration still destroys the accuracy. Therefore in [10] the integrals are computed exactly. In [25], [47], [72] a combination of recursive mesh refinement and product Gauss integration is used with an empirical error indicator. This method could be improved by extrapolation as in Section 6.3.2. But it seems that these methods are responsible for the long computing times and high costs - there is a big need for efficient integration formulas. This is yet to be done.

7. CONCLUDING REMARKS

Once the coefficients and right hand sides of the discrete linear equations are established we find a fully populated $(N+q) \times (N+q)$ matrix. Albeit the a_{jk} in (6.1) look rather different, all three discretizations can be viewed as Galerkin-Petrov methods with same trial but different test functions as delta functions for (4.3), the trial functions for (4.4) and the image functions $A\vec{\mu}_k$ in (4.5). (For a further modification see [4], [5].) Correspondingly, also the numerical weights $\tilde{a}_{jk}$ are obtained in very similar procedures due to outer numerical integrations:

	Point collocation	Galerkin's method	least squares
$\tilde{a}_{jk} \sim$	$A\vec{\mu}_j(x_k)$	$\sum_{\ell=1}^{P} w_{k\ell} A\vec{\mu}_j(x_{k\ell})$	$\sum_{\ell=1}^{Q} w_\ell A\vec{\mu}_j(x_\ell) \cdot A\vec{\mu}_k(x_\ell)$

Here $w_{k\ell}$ and w_ℓ are weights and $x_{k\ell}$ and x_ℓ are nodes of appropriate numerical integrations, respectively. Hence, Galerkin's method and least squares can be considered as *modified collocation methods*. Correspondingly, the computation of $A\vec{\mu}_j(x_p)$ at appropriate nodes in the first place and their combination to the Galerkin or least squares weights saves computing time, see [15], [41], [42], [51], [84] .

Since in three-dimensional problems N rather quickly becomes several thousands, effective linear systems solvers are necessary. Here symmetry and positive definiteness are advantageous. For the moment general fast solvers are not yet available. In two dimensions the matrix $\tilde{a}_{jk}$ is a sum of a Toeplitz matrix and a discrete approximation of a compact mapping. Hence the combination of fast Toeplitz solvers [38] and multigrid methods for Fredholm integral equations of the second kind [36] provides fast solution methods as in [83] for some of our equations. For the general case our asymptotic error estimates in Section 5 are related to the behaviour of the spectra of A_h for $h \to 0$ which underlies the general multigrid methods in [60]. Another (related) possibility for an effective solution of the linear systems is the lumping technique in [14] .

The substructuring of Ω produces a matrix as in Figure 2 having a band-like structure [17], [72] allowing a more efficient solution. This structure is very well suited for vector processing [40] which seems to work particularly well with BEM.

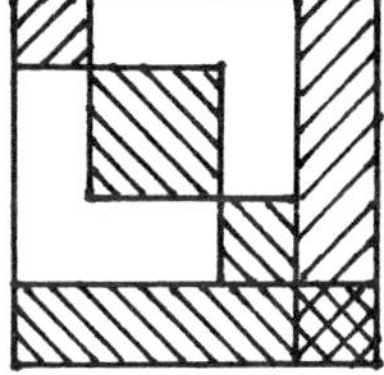

Figure 2

The postprocessing for the computation of the desired solution and its derivatives at any given point x via the representing boundary potentials as e.g.

$$D^{\delta}\vec{u}(x) = \int_{\Gamma} \{D^{\delta}_{(x)}\underset{\sim}{E}(y,x)\vec{t}(y) - D^{\delta}_{(x)}\underset{\sim}{T}(y,x)\vec{\phi}(y)\}ds_y \qquad (7.1)$$

works extremely well as long as x is not near the boundary. Here we have even pointwise *highest order* approximations as in the Table in Section 5. But if x approaches Γ then the kernels in (7.1) start oscillating (due to the jump relation) and for acceptable accurate results one needs a very accurate numerical integration - in contrary to the advantages of the BEM. This is very much related to 6.3.6. Here we need (asymptotic) expansions near Γ to simplify the computations and to make the BEM more efficient.

As we could see from the preceding remarks, there is no uniform recipe for the BEM. If we choose the numerically most stable singular integral equations (2.13) where $2\alpha = 0$ then the discrete equations are symmetric only for the least squares method which is most costly with respect to numerical integration. If we choose the hypersingular equations (2.17) or the first kind equations (2.19), respectively, then the conditioning of order h^{-1} will be worse (Theorem 6.1) due to $2\alpha = 1$ or -1, respectively. If we choose collocation then we have less trouble with numerical integration but need higher order elements than for Galerkin's method (Remark 5.1). But this situation is very familiar from finite element methods and our asymptotic error analysis is also basically the same as for finite elements. If we compare the two methods then BEM is advantageous with respect to quantities on Γ and for *exterior* problems whereas FEM can handle also variable coefficients and nonlinear problems i.e. the methods are somehow complementary. The complexity and costs of both methods are essentially the same (see Table 2).

	FEM	BEM
Number of grid points	h^{-n}	h^{1-n}
Storage of the influence matrix	h^{2-2n}	h^{2-2n}
Sparse elimination [32]	h^{2-3n}	
Gaußian elimination		h^{3-3n}
Fast solvers	$h^{2-2n}\|\log h\|$	
Postprocessing per point	——	h^{-n+1}

Table 2 : Comparison of BEM and FEM

In view of the petite différence a marriage of FEM and BEM will probably create a new generation of numerical methods superior to each , [104].

REFERENCES

1. AGRANOVIČ, M.S., Spectral properties of diffraction problems. In: N.N. Voitovic, B.Z. Katzenellenbaum, A.N. Sivov, *The General Method of Natural Vibrations in Diffraction Theory* (Russian), Izdat. Nauka, Moscow, (1977).
2. AGRANOVIČ, M.S., Spectral properties of elliptic pseudo-differential operators on closed curves. *Funct. Anal.* 13, 54-56 (1979).
3. ANSELONE, P.M., *Collectively Compact Operator Approximation Theory*. Prentice Hall, Englewood Cliffs, N.J. (1971).
4. ARNOLD, D.N., A spline-trigonometric Galerkin method and an exponentially convergent boundary integral method. *Math. Comp.* 41, 383-397 (1983).
5. ARNOLD, D.N., The effect of the test functions on the convergence of spline projection methods for singular integral equations. pp. 1-4 of A. Gerasoulis and R. Vichnevetsky (Ed.), *Numerical Solution of Singular Integral Equations*, Publ. IMACS (1984).
6. ARNOLD, D.N. and WENDLAND, W.L., Collocation versus Galerkin procedures for boundary integral methods. pp. 18-33 of C.A. Brebbis (Ed.). *Boundary Element Methods in Engineering*. Springer-Verlag, Berlin, (1982).
7. ARNOLD, D.N, and WENDLAND, W.L., On the asymptotic convergence of collocation methods. *Math. Comp.*, 41, 349-381 (1983).
8. ARNOLD, D.N. and WENDLAND, W.L., The convergence of spline collocation for strongly elliptic equations on curves. In preparation.
9. ATKINSON, K.E., *A Survey of Numerical Methods for the Solution of Fredholm Integral Equations of the Second Kind*. SIAM, Philadelphia, Penns. (1976).
10. ATKINSON, K.E., Piecewise polynomial collocation for integral equations on surfaces in three dimensions. To appear.
11. ATKINSON, K., GRAHAM, I. and SLOAN, I., Piecewise continuous collocation for integral equations. *SIAM J. Numer. Anal.*, 20, 172-186 (1983).
12. BABUŠKA, I. and AZIZ, A.K., Survey lectures on the mathematical foundations of the finite element method. pp. 3-359 of A.K. Aziz (Ed.), *The Mathematical Foundation of the Finite Element Method with Applications to Partial Differential Equations*, Academic Press, New York, (1972).
13. BECKER, T., *Iterative Konstruktion von Kubaturformeln mit geraden Exaktheitsgraden über 2-dimensionalen, symmetrischen Integrationsbereichen*. Bericht 6/82, Fachgebiet Leichtbau, Prof. Dr.-Ing. J. Wissmann, Technische Hochschule Darmstadt, (1982).

14. BEER, G. and MEEK, J.L., On the boundary element analysis of excavation in rock. pp. 623-634 of C.A. Brebbia, T. Futagami, M. Tanaka (Ed.), *Boundary Elements*. Springer-Verlag, Berlin (1983).
15. BOLTEUS, L. and TULLBERG, O., BEMSTAT-A new type of boundary element program for two-dimensional elasticity problems. pp. 518-537 of C.A. Brebbia (Ed.), *Boundary Element Methods*, Springer-Verlag, Berlin (1981).
16. BREBBIA, C.A., The boundary element method in engineering practice, *Engineering Analysis*, 1, 3-12 (1984).
17. BREBBIA, C.A., TELLES, J.C.F. and WROBEL, L.C., *Boundary Element Techniques*. Springer-Verlag, Berlin (1984).
18. CIARLET, P.G. and RAVIART, P.A., General Lagrange and Hermite interplation in $\mathbb{R}^n$ with applications to finite element methods. *Arch. Rational Mech. Anal.*, 46, 177-199 (1972).
19. COLTON, D. and KRESS, R., *Integral Equation Methods in Scattering Theory*. John Wiley & Sons, New York (1983).
20. COSTABEL, M., *Starke Elliptizität von Randintegraloperatoren erster Art*. Habilitationsschrift, Technische Hochschule Darmstadt (1984).
21. COSTABEL, M., STEPHAN, E., A direct boundary integral equation method for transmission problems. *J. Math. Analysis Appl.*, to appear.
22. COSTABEL, M. and WENDLAND, W.L., Strong ellipticity of boundary integral operators, in preparation.
23. CRUSE, T.A., Application of the boundary integral equation method to three-dimensional stress analysis. *Comp. Struct.* 3, 309-369 (1973).
24. DJAOUA, M., *Méthode d'éléments finis pour la résolution d'un problème extérieur dans* $\mathbb{R}^3$. Centre de Math. Appl., Ecole Polytechniques, Rapp. Int. 3, Palaiseau, France (1975).
25. DREXLER, W., *Ein Beitrag zur Lösung rotationssymmetrischer, thermoelastischer Kerbprobleme mittels der Randintegralgleichungsmethode*. Doctoral Thesis, Maschinenwesen, Techn. Univ. München (1982).
26. ELLIOTT, D. and PAGET, D.F., On the convergence of a quadrature rule for evaluating certain Cauchy principal value integrals. *Numer. Math.* 23, 311-319 (1975).
27. ELLIOTT, D. and PAGET, D.F., On the convergence of a quadrature rule for evaluating certain Cauchy principal value integrals: An addendum. *Num. Math.* 25, 287-289 (1976).
28. ELSCHNER, J. and SCHMIDT, G., *On spline interpolation in periodic Sobolev spaces*. P-Math-01/83, Akad. Wiss. DDR, Inst. Math., DDR 1080 Berlin (1983).
29. ENGELS, H., *Numerical Quadrature and Cubature*. Academic Press, London (1980).

30. FISCHER, T., An integral equation procedure for the exterior three-dimensional viscous flow. *Int. Equs. and Operator Theory* 5, 490-505 (1982).
31. FIX, G.J. and STRANG, G., *An Analysis of the Finite Element Method.* Prentice Hall, Englewood Cliffs, N.J. (1973).
32. GEORGE, A., Block elimination on finite element systems. In D.J. Rose and R.A. Willoughby (Ed.), *Sparse Matrices and their Applications,* Plenum Press, New York (1972).
33. GIROIRE, J., *Integral equation methods for exterior problems for the Helmholtz equation.* Rapp. Int. 40, Centre Math. Appl., Ecole Polytechnique, Palaiseau, France (1978).
34. GIROIRE, J. and NEDELEC, J.C., Numerical solution of an exterior Neumann problem using a double layer potential, *Math. Comp.* 34, 973-990 (1978).
35. GOLUB, G.H. and WELSH, J.H., Calculation of Gauss quadrature rules. *Math. Comp.* 23, 221-230 (1969).
36. HACKBUSCH, W., Die schnelle Auflösung der Fredholmschen Integralgleichungen zweiter Art. *Beiträge zur Numerischen Mathematik* 9 , 47-62 (1981).
37. HA DUONG, T., La méthode de Schenck pour la résolution numérique du problème de radiation acoustique. *E.D.F., Bull. Dir. Etudes Rech., Ser. C, Math. Inf. et Math. Appl.* 2, 15-50 (1979).
38. HEINIG, G. and ROST, K., *Algebraic Methods for Toeplitz-like Matrices and Operators.* Akademie-Verlag, Berlin (1984).
39. HILDEBRANDT, St. and WIENHOLTZ, E., Constructive proofs of representation theorems in separable Hilbert space. *Comm. Pure Appl. Math.* 17, 369-373 (1964).
40. HODOUS, M.F., KATNIK, R.B., BOZEK, D.G. and KLINE, K.A., Vector processing applied to boundary element algorithms on the CDC CYBER-205. In: *Vector and Parallel Computing in Scientific Applications,* Pluralis, Paris (1983).
41. HSIAO, G.C., KOPP, P. and WENDLAND, W.L., A Galerkin collocation method for some integral equations of the first kind. *Computing* 25, 89-130 (1980).
42. HSIAO, G.C., KOPP, P., WENDLAND, W.L., Some applications of a Galerkin collocation method for integral equations of the first kind. *Math. Meth. Appl. Sci.* 6, 280-325 (1984).
43. HSIAO, G.C. and WENDLAND, W.L., A finite element method for some integral equations of the first kind. *J. Math. Anal. Appl.* 58, 449-481 (1977).
44. HSIAO, G.C. and WENDLAND, W.L., The Aubin-Nitsche lemma for integral equations. *J. Integral Equations,* 3 299-315 (1981).
45. HSIAO, G.C. and WENDLAND, W.L., On a boundary integral method for some exterior problems in elasticity. To appear in: *Dokl. Akad. Nauk SSSR,* special issue ded. Academician V.D. Kupradze to his 80th birthday.

46. JASWON, M.A. and SYMM, G.T., *Integral Equation Methods in Potential Theory and Elastostatics*. Academic Press, London (1977).
47. KUHN, G. and MÖHRMANN, W., Boundary element method in elastostatics: theory and application. *Applied Math. Modelling*, 7, 97-105 (1983).
48. KUPRADZE, V.D., GEGELIA, T.G., BASHELEISHVILI, M.O. and BURCHULADZE, T.V., *Three-Dimensional Problems of the Mathematical Theory of Elasticity and Thermoelasticity*, North-Holland, Amsterdam (1979).
49. KUTT, H.R., The numerical evaluation of principal value integrals by finite part integration. *Numer. Math.* 24, 205-210 (1975).
50. LACHAT, J. C. and WATSON, J.O., Effective numerical treatment of boundary integral equations: A formulation for three-dimensional elastostatics. *Int. J. Numer. Meth. Eng.* 10, 991-1005 (1976).
51. LAMP, U., SCHLEICHER, T., STEPHAN, E. and WENDLAND, W.L., Galerkin collocation for an improved boundary element method for a plane mixed boundary vlaue problem. *Computing*, to appear (1984).
52. LEAN, M.H. and WEXLER, A., Electromagnetic scattering from arbitrary shapes with the boundary element method, *IEEE Trans. Magnetics*, to appear.
53. LE ROUX, M.N., *Résolution Numérique du Problème du Potential dans le Plan par une Méthode Variationelle d'Elements Finis*. Doctoral Thesis, L'Univ. de Rennes, Ser. A, No. 347, Ser. 38 (1974).
54. LYNESS, J.N., An error functional expansion for N-dimensional quadrature with an integrand function singular at a point, *Math. Comp.* 30, 1-23 (1976).
55. LYNESS, J.N., Quadrature on a Simplex: Part 2. A representation for the error functional. *SIAM J. Num. Anal.* 15, 870-887 (1978).
56. LYNESS, J.N., Quadrature error functional expansions for the simplex when the integrand function has singularities at the vertices. *Math. Comp.* 34, 213-225 (1980).
57. LYNESS, J.N. and JESPERSEN, J., Moderate degree symmetric quadrature rules for the triangle. *J. Inst. Maths. Appl.* 15, 19-32 (1975).
58. MACKERLE, J. and ANDERSSON, T., Boundary element software in engineering. *Advances in Eng. Software* 6, 66-102 (1984).
59. MARTENSEN, E., *Potentialtheorie*. Teubner, Stuttgart (1968).
60. MC CORMICK, S.F., Multigrid methods for variation problems: The V-cycle, *Math. and Comp. in Simulation* 25, 63-65 (1983).
61. MICHLIN, S.G., *Variationsmethoden der Mathematischen Physik*. Akademie-Verlag, Berlin (1962).
62. MICHLIN, S.G. and PRÖSSDORF, S., *Singuläre Integraloperatoren*, Akademie-Verlag, Berlin (1980).

63. MUSKHELISHVILI, N.I., *Some Basic Problems of the Mathematical Theory of Elasticity*. Noordhoff, Groningen (1963).
64. MUSTOE, G.W., VOLAIT, F. and ZIENKIEWICZ, O.C., A symmetric direct boundary integral equation method for two-dimensional elastostatics. *Res. Mechanica* 4, 57-82 (1982).
65. NASHED, M.Z., On moment-discretization and least-squares solutions of linear integral equations of the first kind, *J. Math. Anal. Appl.* 53, 359-366 (1976).
66. NATTERER, F., *Discretising ill-posed problems*. Problemi non Ben Posti ed Inversi, n. 9, Publ. dell' Inst. di Analisi Globale e Appl., Conous. Naz. delle Ric., Firenze (1983).
67. NEDELEC, J.C. Curved finite element methods for the solution of singular integral equations on surfaces in $\mathbb{R}^3$. *Comp. Math. Appl. Mech. Eng.* 8, 61-80 (1976).
68. NEDELEC, J.C., *Approximation des Equations Intégrales en Mécanique et en Physique*. Lecture Notes, Centre Math. Appl., Ecole Polytechnique, Palaiseau, France (1977).
69. NEDELEC, J.C., Integral equations with non integrable kernels. *Int. Equs. and Operator Theory* 5, 562-572 (1982).
70. NEDELEC, J.C. and PLANCHARD, J., Une méthode variationelle d'éléments finis pour la résolution numérique d'un problème extérieur dans $\mathbb{R}^3$. *R.A.I.R.O.* 7 R3, 105-129 (1973).
71. NEUMANN, C., *Untersuchungen über das logarithmische und Newtonsche Potential*, Teubner, Leipzig (1877).
72. NEUREITER, W., *Boundary-Element-Programmrealisierung zur Lösung von zwei- und dreidimensionalen thermoelastischen Problemen mit Volumenkräften*. Doctoral Thesis, Maschinenwesen, Techn. Univ. München (1982).
73. PETERSEN, B.E., *Introduction to the Fourier Transform and Pseudo-Differential Operators*. Pitman, London (1983).
74. PIESSENS, R., Numerical evaluation of Cauchy principal values of integrals. *BIT* 10, 476-480 (1970).
75. PRÖSSDORF, S., Ein Lokalisationsprinzip in der Theorie der Splinapproximationen und einige Anwendungen. *Math. Nachr.*, to appear.
76. PRÖSSDORF, S. and SCHMIDT, G., A finite element collocation method for singular integral equations. *Math. Nachr.* 100, 33-60 (1981).
77. RICHTER, G.R., Numerical solution of integral equations of the first kind with nonsmooth kernels. *SIAM J. Numer. Anal.* 17, 511-522 (1978).
78. RICHTER, G.R., Superconvergence of piecewise polynomial Galerkin approximations for Fredholm integral equations of the second kind. *Numer. Math.* 31, 63-70 (1978).
79. RIZZO, F.J., An integral equation approach to boundary value problems of classical elastostatics. *Quart. Appl. Math.* 25, 83-95 (1967).

80. ROBINSON, I. and DE DONCKER, E., Algorithm 45. Automatic computation of improper integrals over a bounded or unbounded planar domain. *Computing* 27, 253-284 (1981).
81. SARANEN, J. and WENDLAND, W.L., On the asymptotic convergence of collocation methods with spline functions of even degree, *Math. Comp.*, to appear (1984).
82. SARKAR , T.K. and RAO, S.M., An iterative method for electrostatic problems. *IEEE Trans. AP.* 30, 611-616 (1982).
83. SCHIPPERS, H., Multigrid methods for boundary integral equations. *Numer. Math.*, to appear.
84. SCHLEICHER, T., *Die Randelementmethode für gemischte Randwertaufgaben auf Eckengebieten unter Berücksichtigung von Singulärfunktionen.* Diplom Thesis, Fachber. Mathematik, Techn. Hochschule Darmstadt (1983).
85. SCHMIDT, G., The convergence of Galerkin and collocation methods with splines for pseudo-differential equations on curves. *Zeitschr. Angew. Analysis*, to appear.
86. SCHMIDT, G., On spline collocation methods for boundary integral equations in the plane. *Math. Meth. in the Appl. Sci.*, to appear.
87. SEELEY, R., Topics in pseudo-differential operators. pp. 169-305 of L. Nirenberg (Ed.), *Pseudo-Differential Operators*, C.I.M.E., Edizione Cremonese, Roma (1969).
88. STEPHAN, E., Solution procedures for interface problems in acoustics and electromagnetics. pp. 291-348 of P. Filippi (Ed.), *Theoretical Acoustics and Numerical Techniques*, CISM Courses 277, Springer-Verlag, Wien, New York (1983).
89. STEPHAN, E., *Boundary Integral Equations for Mixed Boundary Value Problems, Screen and Transmission Problems in* $\mathbb{R}^3$. Habilitationsschrift, Technische Hochschule Darmstadt (1984).
90. STEPHAN, E. and WENDLAND, W.L., Remarks to Galerkin and least squares methods with finite elements for general elliptic problems. *Lect. Notes Math.*, Springer, Berlin 564, 461-471 and *Manuscripta Geodaetica* 1, 93-123 (1976).
91. STROUD, A.H., *Approximate Calculation of Multiple Integrals.* Prentice-Hall, Englewood Cliffs, N.J. (1971).
92. STROUD, A.H. and SECREST, D., *Gaussian Quadrature Formulas.* Prentice-Hall, Englewood Cliffs, N.J. (1966).
93. TAYLOR, M.E., *Pseudodifferential Operators*, Princeton Univ. Press, Princeton, N.J. (1981).
94. WATSON, J.O., Advanced implementation of the boundary element method for two- and three-dimensional elastostatics. pp. 31-63 of P.K. Banerjee and R. Butterfield (Ed.), *Developments in Boundary Element Methods - 1* , Appl. Science Publ. LTD, London (1979).
95. WENDLAND, W.L., *Lösung der ersten und zweiten Randwertaufgaben des Innen- und Außengebietes für die Potentialgleichung im* $\mathbb{R}^3$ *durch Randbelegungen*, Doctoral Thesis, Techn. Univ. Berlin D-83 (1965).

96. WENDLAND, W.L., On Galerkin collocation methods for integral equations of elliptic boundary value problems. pp. 244-275 of J. Albrecht, L. Collatz (Ed.), *Numerical Treatment of Integral Equations*, Intern. Ser. Num. Math., Birkhäuser, Basel 53 (1980).
97. WENDLAND, W.L., Asymptotic accuracy and convergence. pp. 289-313 of C.A. Brebbia (Ed.), *Progress in Boundary Element Methods*, Pentech Press, London (1981).
98. WENDLAND, W.L., On the asymptotic convergence of some boundary element methods. pp. 281-312 of J.R. Whiteman (Ed.), *The Mathematics of Finite Elements and Applications IV*, Academic Press, London (1982).
99. WENDLAND, W.L., On applications and the convergence of boundary integral methods. pp. 463-476 of C.T.H. Baker and G.F. Miller (Ed.), *Treatment of Integral Equations by Numerical Methods*, Academic Press, London (1982).
100. WENDLAND, W.L., Boundary element methods and their asymptotic convergence. pp. 135-216 of P. Filippi (Ed.), *Theoretical Acoustics and Numerical Techniques*, CISM Courses 277, Springer-Verlag, Wien, New York (1983).
101. WENDLAND, W.L., Asymptotic error analysis for boundary integral methods and applications. pp. 119-132 of D. Lascaux (Ed.), *Numerical Methods in Engineering*, Pluralis, Paris (1983).
102. WENDLAND, W.L., Asymptotic accuracy and convergence for point collocation methods. Chap. 9 of C.A. Brebbia (Ed.), *Progress in Boundary Element Methods* 4, Springer-Verlag, Berlin, to appear.
103. WISSMANN, J., *Mehrdimensionale Interpolation, orthogonale Polynome und numerische Integration.* Bericht 3/81, Fachgebiet Leichtbau, Prof. Dr.-Ing. Wissmann, Technische Hochschule Darmstadt (1981).
104. ZIENKIEWICZ, O.C., KELLY, D.W. and BETTES, P., Marriage à la mode. The best of both worlds (finite elements and boundary integrals). pp. 81-107 of R. Glowinski, E.Y. Rodin, O.C. Zienkiewicz (Ed.), *Energy Methods in Finite Element Analysis*, J. Wiley, Chichester (1979).

THE SOLUTION OF TIME DEPENDENT PROBLEMS
USING BOUNDARY ELEMENTS

C.A. Brebbia

Southampton University and Computational Mechanics Institute
Southampton, England

1. INTRODUCTION

The present paper studies the solution of parabolic and hyperbolic time dependent problems using boundary elements [1], [2]. It starts by reviewing the different ways in which the time dependent diffusion equation can be solved, including i) Laplace transformation, ii) combination of finite differences and boundary elements and iii) boundary elements in time and space. The latter formulation employs time and space dependent fundamental solutions and is the one that is generally recommended. Two schemes are then presented for the solution, one based on a domain discretization to integrate the initial conditions and the other consisting of only boundary integrals. The second approach is the one that is most attractive to the user as it requires considerably less data than the first scheme.

The paper then concentrates on the study of the hyperbolic transient wave equation for which only the second approach is recommended. In this way the boundary element solution of the problem retains its intrinsic elegance as neither cells nor their associated domain integrals are needed. Unfortunately the problem becomes more complex, especially for the two dimensional case and its extension to study equations such as those governing elastodynamics is rather difficult to implement. Consequently at this stage a simplification started to be investigated and a new approach was proposed.

The original approach developed for the solution of the transient or steady state wave equation consists of an approximation which allows the dynamics problem to be formulated in terms of an inertial or mass matrix. Using simple trial functions this matrix can be expressed in terms of the boundary values only, without having to compute any internal integrations or defining any cells. The approach also eliminates the need to

ISBN 0-12-747255-X

use fundamental solutions which are frequency dependent and permits to find the fundamental frequencies of the system - if required - as the solution of an algebraic eigenvalue problem. For time dependent problems on the other hand, the problem can be expressed in terms of mass matrices - plus damping terms if necessary - and the resulting system of differential equations in time can be integrated using any of the available time marching schemes.

2. TIME DEPENDENT DIFFUSION PROBLEMS

In this section we will study the boundary element solution of the diffusion equation,

$$\nabla^2 u(x,t) - \frac{1}{\alpha}\frac{\partial u(x,t)}{\partial t} = 0 \qquad \text{in } \Omega \tag{2.1}$$

using boundary conditions of the following two types

$$\begin{aligned} &\text{Essential conditions} \quad u(x,t) = \bar{u}(x,t) \quad \text{on } \Gamma_1 \\ &\text{Natural conditions} \quad q(x,t) = \frac{\partial u(x,t)}{\partial n(x)} = \bar{q}(x,t) \quad \text{on } \Gamma_2 \end{aligned} \tag{2.2}$$

The coefficient α depends on the material properties for the type of physical problem under consideration and will be considered constant in time and space in what follows.

The solution of equation (2.1) requires the specification of some initial conditions at time $t = t_o$, i.e.

$$u(x,t) = u_o(x,t_o) \qquad \text{in } \Omega \tag{2.3}$$

Our aim now is to recast the problem represented by equation (2.1) to (2.3) into an integral form. Three different formulations have been used toward this end in the literature;

i) Transformation of the problem by removing its time dependency using the Laplace transform.
ii) Use of a finite difference approximation to discretize the time dependent term in equation (2.1).
iii) Employing time dependent fundamental solutions to discretize the problem in time as well as in space.

The first approach was proposed as early as 1970 by Rizzo and Shippy [3] who applied the direct boundary element formulation in conjunction with Laplace transform. The solution was found for a series of real, positive values of the transform parameter. Then a numerical transform inversion was employed to compute the physical variables in the actual time domain. Using this approach the time dependence of the problem is temporarily removed and the problem transformed into the solution of a sequence of elliptic problems.

The Laplace transform of equation (2.1) can be written as,

$$\nabla^2 U(x,\lambda) - \frac{\lambda}{\alpha} U(x,\lambda) + \frac{1}{\alpha} u_o = 0 \tag{2.4}$$

where λ is the transform parameter. The boundary conditions can now be written as,

$$\begin{aligned} &\text{Essential} \quad U(x,\lambda) = \bar{U}(x,\lambda) = \frac{\bar{u}}{\lambda} \quad \text{on } \Gamma_1 \\ &\text{Natural} \quad U(x,\lambda) = \bar{Q}(x,\lambda) = \frac{\bar{q}}{\lambda} \quad \text{on } \Gamma_2 \end{aligned} \tag{2.5}$$

The governing integral expression can be found by using weighted residuals [1],[2], defining a function U^* which represents the weighting function, i.e.

$$\begin{aligned} &\int_\Omega \{\nabla^2 U(x,\lambda) - \frac{\lambda}{\alpha} U(x,\lambda) + \frac{1}{\alpha} u_o\} U^*(\xi,x,\lambda) d\Omega(x) = \\ &= \int_{\Gamma_2} \{Q(x,\lambda) - \bar{Q}(x,\lambda)\} U^*(\xi,x,\lambda) d\Gamma(x) \\ &\quad - \int_{\Gamma_1} \{U(x,\lambda) - \bar{Q}(x,\lambda)\} U^*(\xi,x,\lambda) d\Gamma(x) \end{aligned} \tag{2.6}$$

Integrating by parts twice one obtains,

$$\begin{aligned} &\int_\Omega \{\nabla^2 U^*(\xi,x,\lambda) - \frac{\lambda}{\alpha} U^*(\xi,x,\lambda)\} U(x,\lambda) d\Omega(x) \\ &\quad + \frac{1}{\alpha} \int u_o U^*(\xi,x,\lambda) d\Omega(x) = \\ &= - \int_\Gamma Q(x,\lambda) U^*(\xi,x,\lambda) d\Gamma(x) + \int_\Gamma U(x,\lambda) Q^*(\xi,x,\lambda) d\Gamma(x) \end{aligned} \tag{2.7}$$

where $\Gamma = \Gamma_1 + \Gamma_2$ is the total boundary, i.e. we do not differentiate for the moment between those parts of the boundary where one or the other type of boundary condition has been applied.

The weighting function required to transform equation (2.7) into a boundary integral form is the fundamental solution of the Helmholtz equation, i.e.

$$\nabla^2 U^*(\xi,x,\lambda) - \frac{\lambda}{\alpha} U^*(\xi,x,\lambda) + \Delta(\xi,x,\lambda) = 0 \qquad (2.8)$$

where Δ represents a Dirac delta function applied at the source point ξ. The resulting U^* is well known and can be expressed in terms of Bessel's functions.

Applying (2.8) to equation (2.7) one can write,

$$U(\xi) + \alpha \int_\Gamma U(x,\lambda) Q^*(\xi,x,\lambda) d\Gamma(x) = \qquad (2.9)$$
$$= \alpha \int_\Gamma U^*(\xi,x,\lambda) Q(x,\lambda) d\Gamma(x) + \int_\Omega u_o U^*(\xi,x,\lambda) d\Omega(x)$$

Taking the point to the boundary it is easy to prove that the following equation holds,

$$c(\xi)U(\xi) + \alpha \int_\Gamma U(x,\lambda) Q^*(\xi,x,\lambda) d\Gamma(x) =$$
$$= \int_\Gamma U^*(\xi,x,\lambda) Q(x,\lambda) d\Gamma(x) \qquad (2.10)$$
$$+ \int_\Omega u_o U^*(\xi,x,\lambda) d\Omega(x)$$

where the c coefficient is $\frac{1}{2}$ for smooth boundaries and proportional to the solid angles for corner points.

The above approach requires the back transformation of the results from the 'λ' plane to the actual time 't' space. This back transformation is the most difficult part of the technique as it needs certain knowledge of the behaviour of the solution. For comparatively simple cases, series representations are used, fitting the coefficients in the λ plane and inverting its analytical forms to return to the time space. Durbin [4] has also proposed using the Fast Fourier transform for more general problems. The values of λ selected in these processes are in principle totally arbitrary. In practice however, they are related to the sampling times t_i in order to avoid numerical problems, i.e. $\lambda_i \cong 1/(\alpha t_i)$.

The second approach to the solution of equation (2.1) is basically a coupling between boundary element and finite difference methods. It assumes that the time derivative in equation (2.1) can be approximated in a finite difference form for a sufficiently small interval Δt, i.e.

$$\frac{\partial u(x,t)}{\partial t} = \frac{u(x,t+\Delta t) - u(x,t)}{\Delta t} \tag{2.11}$$

Equation (2.1) can now be written as,

$$\nabla^2 u(x,t+\Delta t) - \frac{1}{\alpha\Delta t} u(x,t+\Delta t) + \frac{1}{\alpha\Delta t} u(x,t) = 0 \tag{2.12}$$

The fundamental solution corresponding to the adjoint of this equation is of the same form as in the previous case, i.e. as for the Helmholtz formula.

The boundary integral formulation can also be obtained using weighted residual considerations. It gives,

$$\begin{aligned} c(\xi)u(\xi,t+\Delta t) &+ \alpha \int_\Gamma u(x,t+\Delta t)q^*(\xi,x,\Delta t)d\Gamma(x) \\ &= \alpha \int_\Gamma q(x,t+\Delta t)u^*(\xi,x,\Delta t)d\Gamma(x) \\ &\quad + \frac{1}{\Delta t} \int_\Omega u(x,t)u^*(\xi,x,\Delta t)d\Omega(x) \end{aligned} \tag{2.13}$$

The solution can start at the initial time $t = t_o$ with u values given by u_o initial conditions and then proceed on time, by solving equation (2.13) numerically. Values of u at time $t = t_o+\Delta t$ can then be computed at a sufficiently large number of internal points to be used as the starting values for the second time step. The procedure can then continue in this manner.

Numerical results obtained using this scheme indicate that small time steps are needed in order to achieve accurate results.

A more promising approach is the use of a time dependent fundamental solution, which starts by considering that the weighting is carried out in time and space as follows,

$$\int_{t_o}^{t_F}\int_{\Omega}\left\{\nabla^2 u(x,t) - \frac{1}{\alpha}\frac{\partial u(x,t)}{\partial t}\right\} u^*(\xi,x,t_F,t)\,d\Omega(x)\,dt$$
$$= \int_{t_o}^{t_F}\int_{\Gamma}\{q(x,t) - \bar{q}(x,t)\}u^*(\xi,x,t_F,t)\,d\Gamma(x)\,dt \tag{2.14}$$
$$- \int_{t_o}^{t_F}\int_{\Gamma}\{u(x,t) - \bar{u}(x,t)\}q^*(\xi,x,t_F,t)\,d\Gamma(x)\,dt$$

Notice that using this approach u, q, u^* and q^* need to be functions of time and space.

Integrating by parts equation (2.14) gives,

$$\int_{t_o}^{t_F}\int_{\Omega}\left\{\nabla^2 u^*(\xi,x,t_F,t) + \frac{1}{\alpha}\frac{\partial u^*(\xi,x,t_F,t)}{\partial t}\right\} u(x,t)\,d\Omega(x)\,dt \tag{2.15}$$
$$- \frac{1}{\alpha}\left|\int_{\Omega} u(x,t)u^*(\xi,x,t_F,t)\,d\Omega(x)\right|_{t=t_o}^{t=t_F}$$
$$= - \int_{t_o}^{t_F}\int_{\Gamma} q(x,t)u^*(\xi,x,t_F,t)\,d\Gamma(x)$$
$$+ \int_{t_o}^{t_F}\int_{\Gamma} u(x,t)q^*(\xi,x,t_F,t)\,d\Gamma(x)$$

The time dependent fundamental solution which satisfies the adjoint equation, i.e.

$$\nabla^2 u^*(\xi,x,t_F,t) + \frac{1}{\alpha}\frac{\partial u^*(\xi,x,t_F,t)}{\partial t} + \Delta(\xi,x)\Delta(t_F,t) = 0 \tag{2.16}$$

is a very simple exponential solution of the type

$$u^* = \frac{1}{(4\pi\alpha\tau)^{d/2}}\exp\left\{-\frac{r^2}{4\alpha\tau}\right\} \tag{2.17}$$

where $\tau = t_F - t$ and d is the number of spatial dimensions of the problem - i.e. d = 3 for three dimensional problems, d = 2 for two dimensions. The solution is identically zero for $t > t_F$ due to the causality condition.

Solution u^* represents a source applied at a point ξ, and at time t_F. In order to avoid ending the integrations in equation (2.15) exactly at the peak of a Dirac delta function, one can add to the upper limit of the integral an arbitrarily small quantity ε, hence the causality condition will apply and the 'final' conditions term at $t = t_F$ will be identically zero. The limit of equation (2.15) then yields,

$$u(\xi,t_F) + \alpha \int_{t_o}^{t_F} \int_\Gamma u(x,t) q^*(\xi,x,t_F,t) d\Gamma(x) dt = \int_{t_o}^{t_F} \int_\Gamma q(x,t) u^*(\xi,x,t_F,t) d\Gamma(x) dt + \int_\Omega u_o(x,t_o) u^*(\xi,x,t_F,t_o) d\Omega(x) \tag{2.18}$$

where $q^*(\xi,x,t_F,t) = \partial u^*(\xi,x,t_F,t)/\partial n(x)$.

Taking the point ξ to the boundary and accounting for the jump on the left hand side integral, the following boundary integral equation is obtained,

$$c(\xi)u(\xi,t_F) + \alpha \int_{t_o}^{t_F} \int_\Gamma u(x,t) q^*(\xi,x,t_F,t) d\Gamma(x) dt = \alpha \int_{t_o}^{t_F} \int_\Gamma q(x,t) u^*(\xi,x,t_F,t) d\Gamma(x) dt + \int_\Omega u_o(x,t) u^*(\xi,x,t_F,t_o) d\Omega(x) \tag{2.19}$$

where $c(\xi)$ is as previously a function of the solid angle at the point ξ.

Since the time variation of function u and q is not known a priori, a time-stepping technique has to be introduced for the numerical solution of equation (2.19). This technique should not be confused with the previous finite difference one. Here, the fundamental solution itself is time dependent and as a result large time steps can generally be adopted.

Two different time marching schemes can be employed for the numerical solution (see Figure 1),

i) Scheme 1 This first scheme treats each time step as a new problem and so at the end of each of them the function u needs to be computed at a sufficient large number of internal points in order to use these values as pseudo-initial values for the next step.

ii) Scheme 2 This other scheme starts the process always at time t_o and so in spite of the increasing number of intermediate steps, does not require any values of u at internal points. Furthermore if the u_o field (at $t = t_o$) satisfies Laplace's equation even the domain integral in (2.19) can be transformed into an equivalent boundary integral.

In both cases functions u and q are discretized in time as well as in space, generally using constant, linear or quadratic interpolation functions.

The main difference between the two schemes lies in the way in which the values of the variables up to the actual instant of time are taken into account in order to solve for a new instant of time. In scheme 1 they are accounted for in the domain integral,for instance the values at t_{n-1} are initial values, to compute the new values at t_n, etc. In scheme 2 instead their variation is considered through a summation of boundary integrals. Scheme 1 has been employed in references [5] to [7] and [8]. Scheme 2 was suggested by Thaler and Mueller [9] in conjunction with the indirect formulation as early as 1970 and was subsequently used in references [10][11] and [12]. Let us consider both schemes in more detail.

Scheme 1 The initial values of u at time t_o are supposed to be specified over the domain Ω and this domain is subdivided into a series of cells (figure 2). The initial conditions at t_o or any other steps are taken into account through a numerical integration over Ω and in order to do so the values of u at a certain number of internal points are needed. At the end of each step, the values of u at the previously selected internal points are recomputed to be used as initial values for the next step. In this case we are applying equation (2.19) from an initial time $t_o = t_{n-1}$ to a final time $t_F = t_n$. If the time step is constant the boundary element matrices generated by this solution can be used throughout all the marching process. Notice that these coefficients are functions of the fundamental solution which is practically always applied for a constant time step, i.e. from time t_{n-1} to t_n. The numerical procedure and necessary algorithm to find these coefficients and the corres-

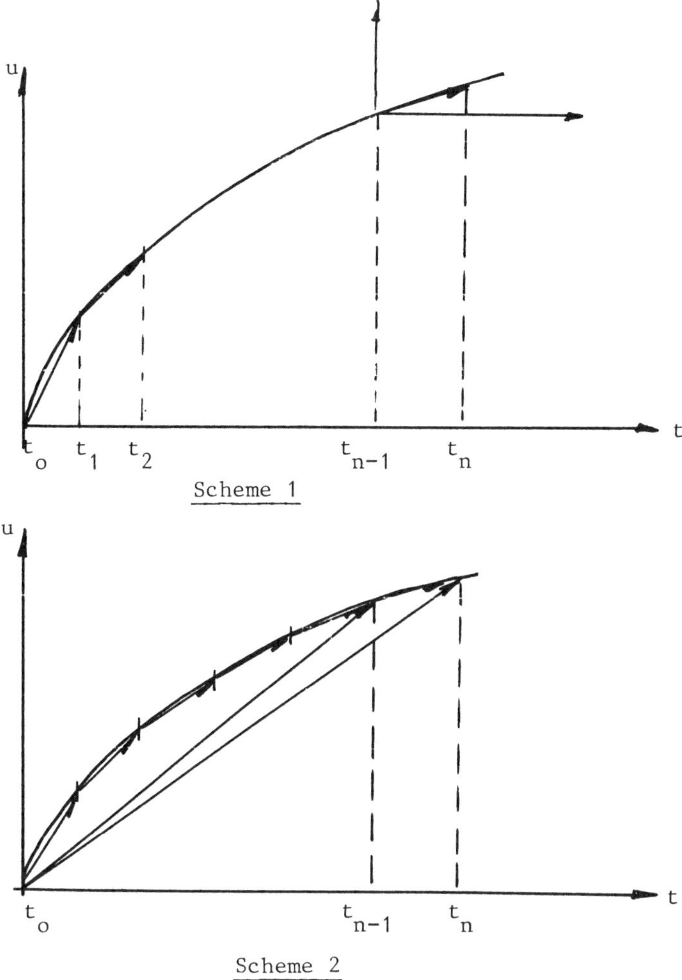

Figure 1 The Two Different Schemes used for the Time and Space Dependent Boundary Element Formulation

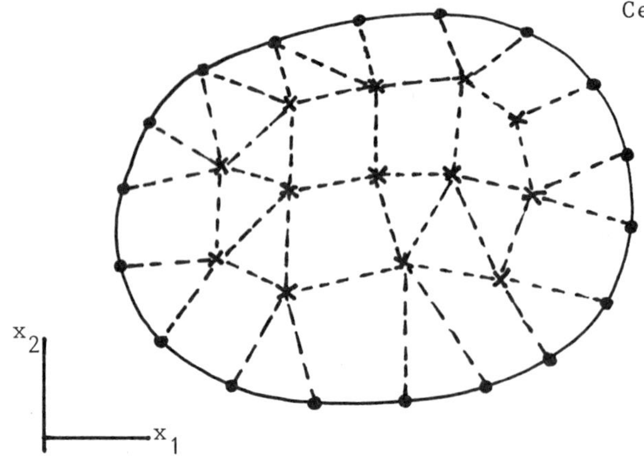

Figure 2 Scheme 1 Continuum discretized into boundary elements and internal cells

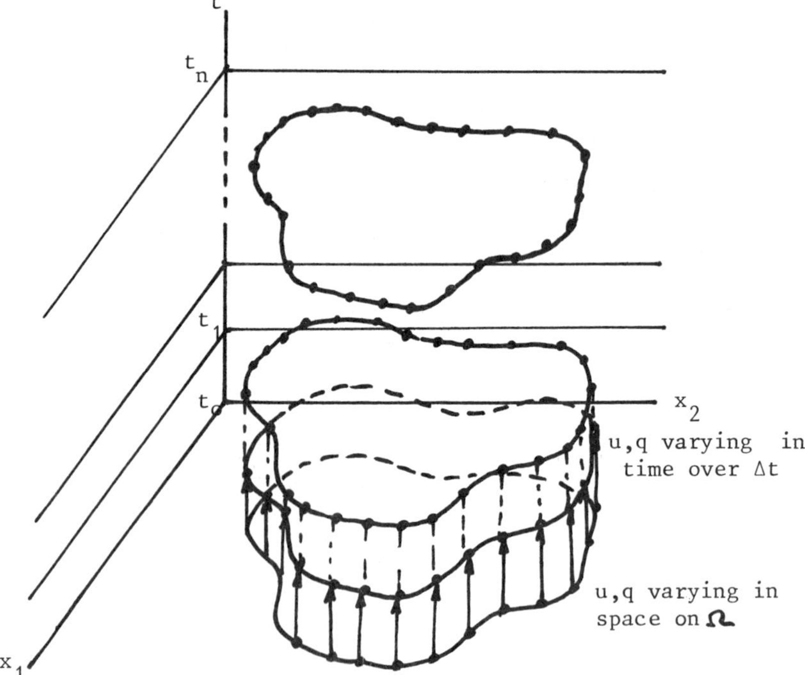

Figure 3 Scheme 2 Time and space discretization of the boundary only

ponding matrices are discussed in great detail in reference [2].

Scheme 2 In this approach the values of u at internal points need not be recomputed at the end of each time step. A domain integral accounting for the initial conditions at time $t = t_o$, is required if $u_o \neq 0$, but if $\nabla^2 u_o = 0$ even the domain integrals can be transformed into equivalent boundary integrals. As this condition is easy to satisfy in practice, the dimensionality of the problems is effectively reduced during the integration. In this case we are applying equation (2.19) from the actual initial time t_o to a final time which varies as $t_1, t_2 \ldots t_{n-1}, t_n$ (figure 1). The boundary integrals need however to be discretized in time to represent u and q from t_o to an arbitrary time t_n. Hence the number of boundary integrals increases as time progresses and a selective numerical integration scheme is required to make the program more efficient [2]. This scheme is such that if a constant time step is adopted throughout the analysis only two new matrices need to be evaluated for each step, making it possible for all the others to be kept in disc storage. The approach is graphically represented in figure 3.

The above schemes yield similar numerical results for the same time step and space discretization and are totally equivalent from the results point of view. With regard to stability condition, we note that the boundary element formulation is implicit in character and thus relatively free from stability problems. In fact a mathematical proof of uniform convergence and stability of the boundary element method as applied to linear two-dimensional transient conduction problems has been reported in reference [13].

Example 1

The example presented in figure 4 studies the heat conduction on a prolate spheroid initially at zero temperature and subject to a unit surface temperature at $t = 0$. The problem is axisymmetric and the relevant theory is given in great detail in reference [14]. A parametric representation of points in the surface in the R-Z plane (figure 4) may be written as,

$$\begin{aligned} R &= L_1 \cos \phi \\ Z &= L_2 \sin \phi \end{aligned} \qquad (a)$$

where the ϕ angle is as indicated in the figure.

The discretization adopted is shown in the figure and the numerical values assumed for this analysis were,

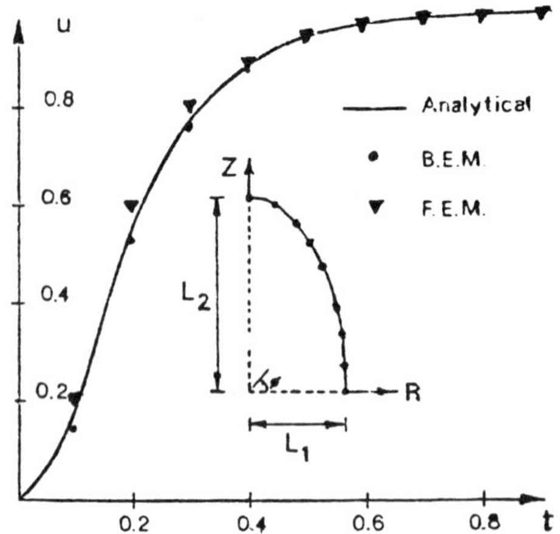

Figure 4 Temperature at the Centre of the Prolate Spheroid

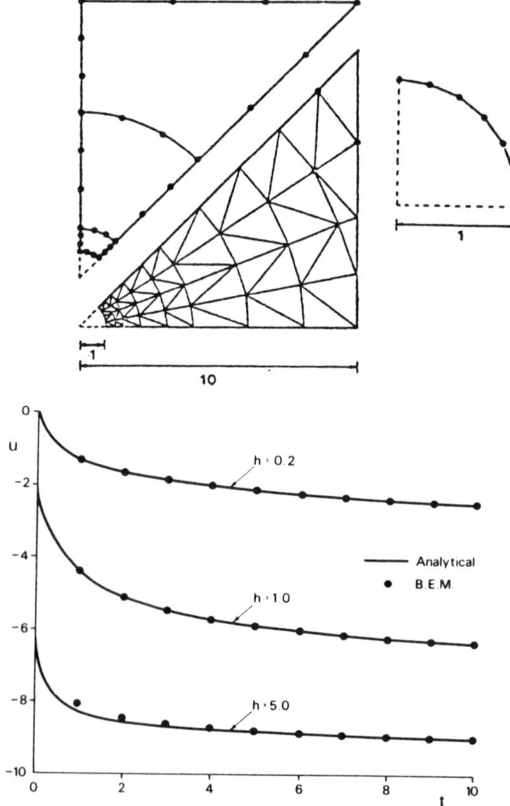

Figure 5 Hole in an Infinite Domain

$\alpha = 1, \quad L_1 = 1 \quad , \quad L_2 = 2$

Results for the temperature at the centre point (R = Z = 0) are compared in figure 4 with an analytical solution [14] and a finite element solution [16] obtained using parabolic three-dimensional isoparametric elements. The finite element analysis was performed with a time step value $\Delta t = 0.025$ whereas the boundary element solution was found with a time step $\Delta t = 0.050$.

Example 2

For problems involving regions extending to infinity, boundary element solutions can be much more economical than finite element ones. In order to demonstrate this, let us consider a circular opening in an infinite plane region with initial temperature $u_0 = 10$, subject to the heat transfer condition $q + hu = \gamma$ (figure 5). The radius of the hole is unity, its ambient temperature equals zero and the material properties of the medium are assumed to be unity for simplicity (i.e. $\alpha = 1$).

The variation of the surface temperature with time is presented in figure 5 for various values of the heat transfer coefficient h and compared against the analytical solution given in [17]. The agreement between both solutions is excellent. A time step $\Delta t = 0.5$ was adopted for the boundary element analysis using stepwise constant functions in time and linear in space. Because of symmetry only one quarter of the interface needed to be discretized employing only 7 nodes.

The problem has also been studied using finite elements [15] but in this case the infinite region needs to be limited by a finite, non-conducting boundary. In order to achieve the same degree of accuracy as in Boundary Elements, the Finite Element solution was carried out using a time step $\Delta t = 0.05$, i.e. ten times smaller. The domain was discretized using 70 triangular elements or 3 cubic isoparametric elements.

Notice that the BE solution in this case has been obtained using the scheme 2 which is ideally suited to problems extending to infinity. If scheme 1 was employed a domain discretization would be required, in order to integrate over the domain.

3. SCALAR WAVE PROPAGATION EQUATION

This section deals with the boundary element solution of the transient scalar wave equation, i.e.

$$\nabla^2 u(x,t) - \frac{1}{c^2}\frac{\partial^2 u(x,t)}{\partial t^2} = 0 \tag{3.1}$$

with the boundary conditions,

$$\begin{aligned} &\text{Essential} \quad u(x,t) = \bar{u} \quad \text{on } \Gamma_1 \\ &\text{Natural} \quad q(x,t) = \bar{q} \quad \text{on } \Gamma_2 \end{aligned} \tag{3.2}$$

and the following initial conditions

$$\begin{aligned} u(x,t) &= u_o(x,t_o) \quad \text{in } \Omega \\ \dot{u}(x,t) &= \frac{\partial u(x,t)}{\partial t} = \left[\frac{\partial u(x,t)}{\partial t}\right]_o \quad \text{in } \Omega \end{aligned} \tag{3.3}$$

The constant c represents here the wave propagation speed or celerity. The integral formulation for this problem goes back to Kirchhoff [18] - for three dimensional cases - and can be found in different ways in Baker and Copson [19], Lamb [20] and Morse and Feshbach [21]. Numerical boundary integral type of formulations to solve this equation have been presented by Friedman and Shaw [22], Shaw [23], Groenenboom [24] and Mansur and Brebbia [25],[26].

For harmonic waves - i.e. assuming a time dependency of the form $e^{-i\omega t}$ - the problem can be solved in the frequency domain where it reduces to the solution of the Helmholtz equation, i.e.

$$\nabla^2 u(x) + \kappa^2 u(x) = 0 \tag{3.4}$$

in which $\kappa = \omega/c$ is the wave number.

In what follows we will only consider the time and space boundary element solution of equations (3.1) to (3.3). These formulae can be recast into an integral equation for the u and q using weighted residuals in the same manner as it was done in Section 2. One can start by distributing in time as well as in space, i.e.

$$\int_{t_o}^{t_F}\int_{\Omega}\left\{\nabla^2 u(x,t) - \frac{1}{c^2}\frac{\partial^2 u(x,t)}{\partial t^2}\right\} u^*(\xi,x,t_F,t)\,d\Omega(x)\,dt$$
$$= \int_{t_o}^{t_F}\int_{\Gamma_2}\{q(x,t) - \bar{q}(x,t)\}u^*(\xi,x,t_F,t)\,d\Gamma(x)\,dt$$

$$- \int_{t_o}^{t_F} \int_{\Gamma_1} \{u(x,t) - \bar{u}(x,t)\} q^*(\xi,x,t_F,t) d\Gamma(x) dt \qquad (3.5)$$

Integrating by parts twice the Laplacian with respect to space and twice the time derivatives with respect to time we obtain,

$$\int_{t_o}^{t_F} \int_{\Omega} \left\{ \nabla^2 u^*(\xi,x,t_F,t) - \frac{1}{c^2} \frac{\partial^2 u^*(\xi,x,t_F,t)}{\partial t^2} \right\} u(x,t) d\Omega(x) dt$$

$$+ \frac{1}{c^2} \left[\int_{\Omega} \left\{ u(x,t) \frac{\partial u^*(\xi,x,t_F,t)}{\partial t} - \frac{\partial u(x,t)}{\partial t} u^*(\xi,x,t_F,t) \right\} d\Omega \right]_{t=t_o}^{t=t_F}$$

$$= - \int_{t_o}^{t_F} \int_{\Gamma} q(x,t) u^*(\xi,x,t_F,t) d\Gamma(x) dt + \qquad (3.6)$$

$$+ \int_{t_o}^{t_F} \int_{\Gamma} u(x,t) q^*(\xi,x,t_F,t) d\Gamma(x) dt$$

The fundamental solution u^* satisfies,

$$\nabla^2 u^*(\xi,x,t_F,t) - \frac{1}{c^2} \frac{\partial^2 u^*(\xi,x,t_F,t)}{\partial t^2} + \Delta(\xi,x)\Delta(t_F,t) = 0 \qquad (3.7)$$

The fundamental solution for three dimensional problems is given in terms of a Dirac delta function and a 'retarded' time and has some interesting properties which will be shortly discussed. For the two dimensional case instead the solution is based on a Heaviside type function and this produces a more complex formulation.

We will assume that the u^* solution is known and that it obeys the causality condition, i.e.

$$u^*(\xi,x,t_F,t) = 0 \qquad \text{for } c(t_F - t) < |x - \xi| \qquad (3.8)$$

In order to avoid ending the integration process right at the peak of a Dirac delta function we can add to the upper limit of the time integrals an arbitrarily small quantity ε. Then we have that when the $\varepsilon \to 0$ equation (3.6) becomes,

$$c(\xi)u(\xi,t_F) + \int_{t_o}^{t_F}\int_\Gamma u(x,t)q^*(\xi,x,t_F,t)d\Gamma(x)dt$$

$$= \int_{t_o}^{t_F}\int_\Gamma q(x,t)u^*(\xi,x,t_F,t)d\Gamma(x)dt \tag{3.9}$$

$$- \frac{1}{c^2}\int_\Omega \left\{u_o(x,t_o)\left[\frac{\partial u^*(\xi,x,t_F,t)}{\partial t}\right]_o - \left[\frac{\partial u(x,t)}{\partial t}\right]\right.$$

$$\left. \times\ u^*(\xi,x,t_F,t_o)\right\}d\Omega$$

where the c coefficient is equal to one if the point is internal or is proportional to the solid angle for points on the boundary.

Equation (3.9) is considerably simplified for the three dimensional case. Due to the finiteness of the wave propagation velocity the influence between fields at two points separated in space will not be instantaneous. The time lag, which depends on the distance between the source and field points is called the 'retardation'. The fundamental solution for three dimensions is

$$u^* = \frac{\Delta(t,t_R)}{4\pi r} \tag{3.10}$$

where r is the distance between the point ξ and x and $t_R = t_F - r/c$ is the 'retardation'. The normal derivative of u^* along Γ is given by

$$q^* = \frac{\partial u^*}{\partial n} = -\frac{1}{4\pi r}\left[\frac{\Delta(t,t_R)}{r} - \frac{1}{c}\frac{\partial\Delta(t,t_R)}{\partial t}\right] n_r \tag{3.11}$$

$n_r = \partial r/\partial n.$

The time integrals in (3.9) can then be evaluated analytically for this particular case. Equation (3.9) now becomes,

$$c(\xi)u(\xi,t_F) = \frac{1}{4\pi r}\int\left\{q(x,t_R)+n_r\left(\frac{1}{r}u(x,t_R)+\frac{1}{c}\left[\frac{\partial u(x,t)}{\partial t}\right]_{t=t_R}\right)\right\}d\Gamma$$

$$+ \frac{1}{4\pi}\left\{t\ N_o + \frac{\partial}{\partial t}[t\ M_o]\right\} \tag{3.12}$$

where M_o and N_o are respectively the mean values of u_o and $(\partial u/\partial t)_o$ over a spherical surface with centre at ξ and variable radius ct. Equation (3.12) is the well known Kirchhoff or Huyghens integral expression. A special feature of this formulation is that no time integration is required. The same unfortunately does not apply for the two dimensional case

The fundamental solution for the two dimensional transient scalar wave equation is of the form,

$$u^* = \frac{c}{2\pi[c^2(t_F-t)^2-r^2]^{1/2}} H[c(t_F-t)-r] \tag{3.13}$$

where H is the Heaviside function. In this case the influence of a source function at a point on the potential at another is no longer restricted to the value of the retarded time but has to be integrated from the initial time t_o up to the actual time t_F. Thus the marching procedures of the type discussed in section 2 can be used for the numerical solution.

The following example has been solved using the type of Scheme 2 mentioned earlier. In this case internal cells are not required even for the comparatively complex two dimensional case.

Example 3

The subject of this study is a square membrane with an initial velocity $\bar{v}_o$ prescribed over the domain Ω_o as shown in figure 6 and zero displacements over all the boundary.

The boundary was discretized into 32 elements and Ω_o was divided into four cells (figure 6). Analytical [28] and BEM results for displacements at point (a/2, a/2) and the normal derivative of displacements at the point (a, a/2) were compared.

The values of u and q for β = 0.2 are plotted in figure 7 where β is the Courant number, i.e.

$$\beta = \frac{c\Delta t}{\ell}$$

ℓ: length of an element.

The results agree well with those due to the analytical solution. Notice that the value of ct has been plotted in the horizontal axis to follow the wave propagation in terms of the characteristic size length a.

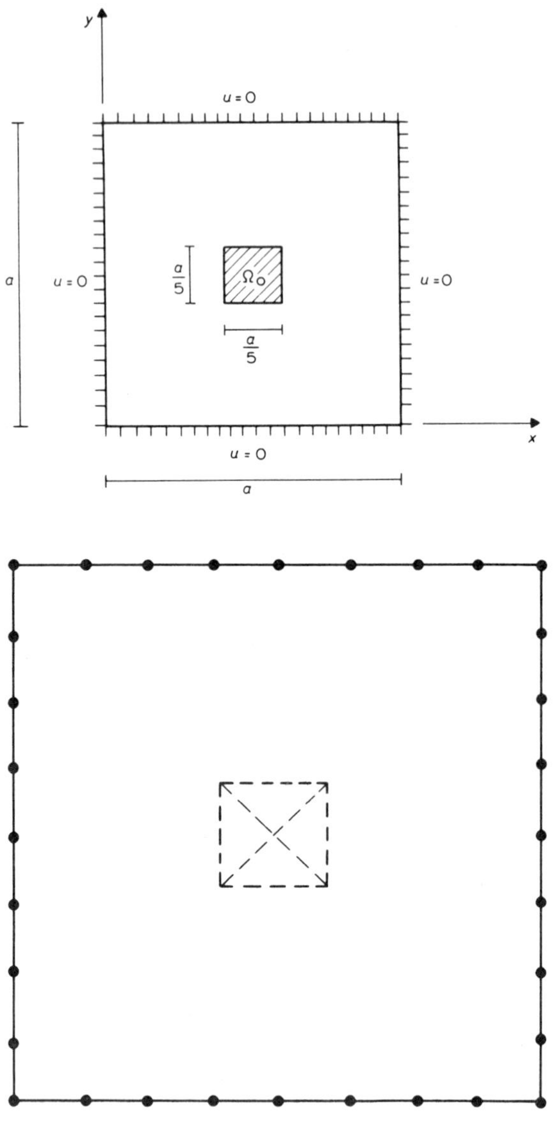

Figure 6 Geometry Definition and Mesh Discretization for the Membrane Analysis

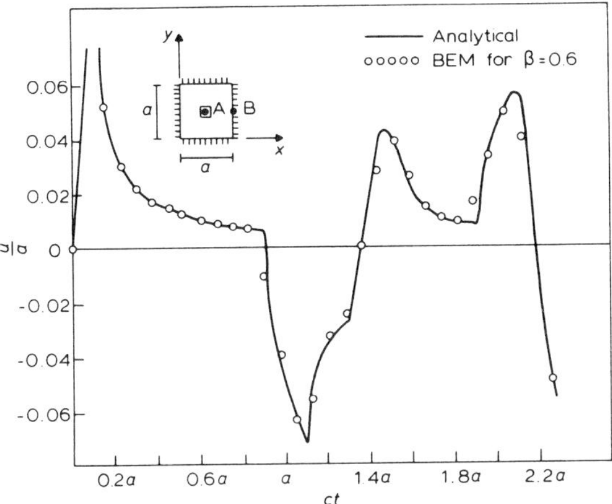

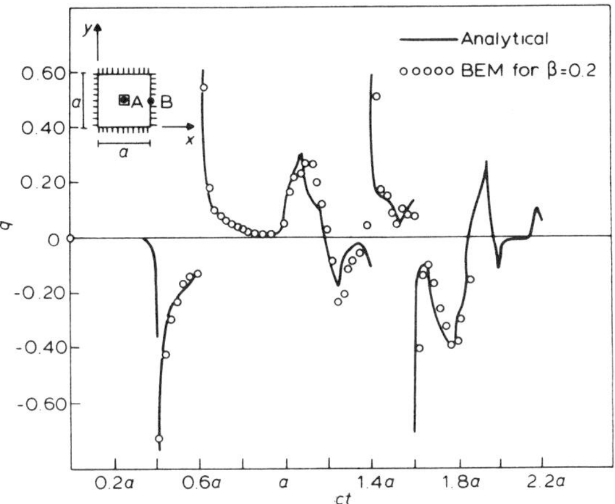

Figure 7 Displacements and Normal Derivatives for the Membrane

Different values of β and number of boundary elements were tried to investigate the accuracy of the solution. In general, as in all numerical solutions of hyperbolic problems, the choice of time intervals and boundary discretizations is critical and the causality condition ($\beta \leq 1$) should be respected. This and other examples [26] demonstrate nevertheless the remarkably good accuracy that can be obtained using boundary elements for hyperbolic problems and validate the use of the technique for the transient scalar wave equation.

4. A NEW APPROACH FOR THE SOLUTION OF THE WAVE EQUATION

The formulation of hyperbolic problems specially when trying to reduce them to a boundary only problem, becomes extremely complicated and it is difficult to extend it to problems such as elastodynamics. The formulation is however of great practical interest if it allows the representation of the problem without having to define any internal cells. A new and simple way of presenting a boundary only formulation for hyperbolic problems was needed, and one that could also solve algebraic eigenvalue problems which are important in many practical applications but cannot be easily obtained with the usual frequency dependent fundamental solutions.

Because of this Nardini and Brebbia decided to investigate the formulation of dynamic problems in terms of a mass or inertial matrix such that it could then be reduced to the boundary without having to perform any internal integrations. They started in 1982 [27] by proposing a new approach to free vibrations using boundary elements. Their approach eliminated the need of using fundamental solutions which were themselves frequency dependent, producing therefore a non-algebraic eigenvalue corresponding to equation (33) - For these cases integration over the boundary had to be carried out for each trial frequency, which made the procedure very uneconomic in practice. Nardini and Brebbia's approach instead reduces the problem to a boundary problem using simple fundamental solutions which are frequency independent. The main advantage of the approach is that the boundary integrals need to be computed only once as they are frequency independent. The procedure has been extended to solving transient problems in which case the standard time marching schemes can be applied [29].

To understand the new approach let us start by considering the wave equation (3.1) with the corresponding boundary condition (3.2) and propose the following weighted residual statement.

$$\int_{\Omega} \left\{ \nabla^2 u(x,t) - \frac{1}{c^2} \frac{\partial^2 u(x,t)}{\partial t^2} \right\} u^*(\xi,x) d\Omega(x) = \tag{4.1}$$

$$= \int_{\Gamma_2} \{q(x,t) - \bar{q}(x,t)\} u^*(\xi,x) d\Gamma(x)$$

$$- \int_{\Gamma_1} \{u(x,t) - \bar{u}(x,t)\} q^*(\xi,x) d\Gamma(x)$$

Notice that we are not integrating on time as previously and consequently the initial conditions do not need to be introduced in the problem at this stage.

If we integrate equation (4.1) twice in space, one obtains

$$\int_{\Omega} \{\nabla^2 u^*(\xi,x)\} u(x,t) d\Omega(x) - \frac{1}{c^2} \int_{\Omega} \frac{\partial^2 u(x,t)}{\partial t^2} u^*(\xi,x) d\Omega(x)$$

$$= - \int_{\Gamma} q(x,t) u^*(\xi,x) d\Gamma(x) + \int_{\Gamma} u(x,t) q^*(\xi,x) d\Gamma(x) \tag{4.2}$$

Let us now apply the simple fundamental solution for the time independent Laplacian equation, i.e.

$$\nabla^2 u^*(\xi,x) + \Delta(\xi,x) = 0 \tag{4.3}$$

Applying equation (4.3) into (4.2) one obtains

$$c(\xi) u(\xi,x) + \frac{1}{c^2} \int_{\Omega} \frac{\partial^2 u(x,t)}{\partial t^2} u^*(\xi,x) d\Omega(x) \tag{4.4}$$

$$= \int_{\Gamma} q(x,t) u^*(\xi,x) d\Gamma(x) - \int_{\Gamma} u(x,t) q^*(\xi,x) d\Gamma(x)$$

Notice that the integral on the right hand side of equation (4.4) is a domain integral. In order to formulate the problem in terms of boundary values only a suitable approximation to the potential u inside the domain will now be proposed.

A class of functions denoted by f^j will be chosen such that the potential u (or its acceleration $\ddot{u}$) will be approximated using a set of unknown coefficients α_j i.e.

$$\text{and}\quad \begin{aligned} u(x,t) &= \alpha_j(t)\, f^j(x) \\ u(x,t) &= \ddot{\alpha}_j(t)\, f^j(x) \end{aligned} \tag{4.5}$$

Hence the domain integral in (4.4) becomes

$$\int_\Omega \ddot{u}(x,t)u^*(\xi,x)d\Omega(x) = \ddot{\alpha}_j(t) \int_\Omega f^j(x)u^*(\xi,x)d\Omega(x) \tag{4.6}$$

Notice that the resulting integrals now contain only known functions. Our objective is to transform them into an equivalent boundary integral. The similarity between the form of (4.6) and the domain term obtained for the right hand side of a Poisson's equation indicates that the functions f^j can be treated as a right hand side term of some Poisson's equation, such that the following equation is satisfied

$$\nabla^2 \psi^j = f^j \tag{4.7}$$

The pseudo potential field represented by ψ can also be associated with a set of boundary fluxes η such that,

$$\eta(x) = \frac{\partial \psi(x)}{\partial n} \tag{4.8}$$

Hence we can now write the domain term (4.6) in an integral form if we carry out the product of the pseudo field ψ,η and the fundamental solution field u^*,q^*. This gives,

$$\begin{aligned} &\ddot{\alpha}_j(t) \int_\Omega f^j(x)u^*(\xi,x)d\Omega(x) = \\ &= \ddot{\alpha}_j(t) \left\{ -c(\xi)\psi^j(\xi) + \int_\Gamma u^*(\xi,x)\eta(x)d\Gamma(x) \right. \\ &\qquad \left. - \int_\Gamma q^*(\xi,x)\psi(x)d\Gamma(x) \right\} . \end{aligned} \tag{4.9}$$

This idea, simple but ingenious, allows us to write (4.4) in terms of boundary integrals only, i.e.

$$
\begin{aligned}
& c(\xi)u(\xi,t) - \int_\Gamma u^*(\xi,x)q(x,t)d\Gamma(x) \\
& \quad + \int_\Gamma u(x,t)q^*(\xi,x)d\Gamma(x) + \\
& + \ddot{\alpha}_j(t)\frac{1}{c^2}\left\{-c(\xi)\psi^j(\xi) + \int_\Gamma u^*(\xi,x)\eta(x)d\Gamma(x)\right. \\
& \qquad \left. - \int_\Gamma q^*(\xi,x)\psi(x)d\Gamma(x)\right\} = 0
\end{aligned}
\tag{4.10}
$$

This equation contains the unknown potential and fluxes on the boundary as well as the unknown coefficients $\ddot{\alpha}_j$. The boundary can now be divided into a series of elements of a given type as usual and the potential and fluxes can be approximated in terms of the nodal values using simple space dependent element shape functions. Notice that if we assume the same variation for η and ψ as for **q** and u the influence matrices for the terms on the left or right hand side of (4.10) are the same. The terms on the right hand side are associated with inertial effects.

The final system of equations can be written in matrix form as [29],

$$
[H]\{U\} - [G]\{Q\} + \frac{1}{c^2}(-[H][\psi] + [G][\eta])\{\ddot{\alpha}\} = 0 \tag{4.11}
$$

The relationship between $\{\ddot{\alpha}\}$ and $\{\ddot{U}\}$ can be established applying equation (4.5) at all boundary points, i.e.

$$
\{\ddot{U}\} = [F]\{\ddot{\alpha}\} \tag{4.12}
$$

The elements of the matrix [F] are simply the values of the functions $f^j(x)$ at the nodal points. If the functions f^j are linearly independent and their number is chosen to be equal to the number of nodes, then the matrix [F] is square and regular and possesses an inverse, $[E] = [F]^{-1}$, i.e.

$$
\{\ddot{\alpha}\} = [E]\{\ddot{U}\} \tag{4.13}
$$

Substituting (4.13) into (4.11) we obtain

$$[H]\{U\} - [G]\{Q\} + [M]\{\ddot{U}\} = \{0\} \tag{4.14}$$

where

$$[M] = \frac{1}{c^2}\{-[H][\psi] + [G][\eta]\}[E] \tag{4.15}$$

Notice that for the harmonic wave equation

$$\{\ddot{U}\} = -\omega^2\{U\} \tag{4.16}$$

and equation (4.14) becomes

$$[H]\{U\} - [G]\{Q\} = \omega^2[M]\{U\} \tag{4.17}$$

After applying the boundary conditions those equations can be expressed as an algebraic eigenvalue problem with the ω^2 as unknowns.

The above procedure has been explained in detail in references [27] and [29] where some particular solution for f^j and the resulting ψ and η fields are presented.

5. CONCLUSIONS

This paper has shown how the Boundary Element method can be applied to solve time dependent problems of parabolic and hyperbolic type. The study concentrates on the application of the technique using boundary values only, without the need to carry out integrations in the domain. This has been achieved in two ways, i) by using fundamental solutions which are time as well as space dependent, ii) by using simple stationary fundamental solutions and converting the remaining domain - or inertial - integral using a boundary formulation.

The applications shown in this paper and others presented in some of the references point out that the Boundary Element method can be an efficient tool for the solution of time dependent problems, in particular by producing accurate solutions for comparatively large time steps and by reducing the problem to a boundary only formulation.

REFERENCES

1. BREBBIA, C.A. *The Boundary Element Method for Engineers* Pentech Press, London & Wiley, NY, 1978, Second Edition.

2. BREBBIA, C.A., TELLES, J.L. & WROBEL,L.C. *Boundary Element Techniques - Theory and Applications in Engineering* Springer Verlag, Berlin & N.Y,(1984.)

3. RIZZO, F.Y. & SHIPPY, D.J. A Method of Solution for Certain Problems of Transient Heat Conduction. *AIAA Journal*, 8, No. 11, 2004-2009.

4. DURBIN, F. Numerical Inversion of Laplace Transform: An Efficient Improvement to Dubner and Abate's Method. *Computer J.*, 17, 371-376 (1974).

5. CHANG, Y.P., KANG, C.S. and CHEN, D.U. The Use of Fundamental Green's Functions for the Solution of Problems of Heat Conduction in Anisotropic Media. *Itn. J. Heat and Mass Transfer* 16, 1905-1918 (1973).

6. SHAW, R.P. An Integral Equation Approach to Diffusion. *Int. J. of Heat and Mass Transfer* 17, 693-699 (1974).

7. WROBEL, L.C. and BREBBIA, C.A. The Boundary Element Method for Steady State and Transient Heat Conduction, in *Numerical Methods in Thermal Problems*. R.W. Lewis and K. Morgan (Eds) Pineridge Press, Swansea (1979).

8. DUBOIS, M. AND BUYSSE, M. Transient Heat Transfer Analysis by the Boundary Integral Equation Method in *New Developments in Boundary Element Methods*, C.A. Brebbia (Ed), CML Publications, Southampton (1980) Reprinted 1983.

9. THALER, R.H. and MUELLER, W.K. A New Computational Method for Transient Heat Conduction in Arbitrarily Shaped Regions. *Fourth Int. Heat Transfer Conf.* U. Grigull & E. Hahne (Eds) Elsevier Publishing Co. Amsterdam (1970).

10. LIGGETT, J.A. and LIU, P.L.F. Unsteady Flow in Confined Aquifers: A Comparison of Two Boundary Integral Methods. *Water Resources Research* 15, No.4, 861-866 (1979).

11. WROBEL, L.C. and BREBBIA, C.A. Time Dependent Potential Problems, in *Progress in Boundary Element Methods*, 1, C.A. Brebbia (Ed), Pentech Press, London, Wiley, NY (1981).

12. ONISHI, K. and KUZOKI, T. Boundary Element Method in Transient Heat Transfer Problems. *The Bulletin of the Institute for Advanced Research of Fukuoka Univ, No.52, Fukuoka (1981).*

13. ONISHI, K. Convergence in the Boundary Element Method for Heat Equation. *TRU Mathematics* 16, No.2, (1981).

14. WROBEL, L.C. and BREBBIA, C.A. A Formulation of the Boundary Element Method for Axisymmetric Transient Heat Conduction. *Int. J. Heat and Mass Transfer* 24, No.5, 843-850, (1981).

15. ZIENKIEWICZ, O.C. and PAREKH, C.J. Transient Field Problems: Two-dimensional and Three-dimensional Analysis by Isoparametric Finite Elements. *Int. J. Num. Methods in Engng* 2, No.1, 61-71 (1970).

16. HAJI, A., SHEIK, I. and SPARROW, E.M. Transient Heat Conduction in a Prolate Spheroidal Solid. *J. Heat Transfer Trans. ASME* 88C 331-333 (1966).

17. CARSLAW, H.S. and JAEGER, J.C. *Conduction of Heat in Solids*. 2nd Edition, Clarendon Press, Oxford (1959).

18. KIRCHHOFF, G. *Zur Theorie der Liechtstrahlen*. Berliner Ber., 641 (1982).

19. BAKER, B.B. and COPSON, E.T. *The Mathematical Theory of Hyghen's Principle*. 2nd Ed. Oxford University Press, Oxford (1953).

20. LAMB, H. *Hydrodynamics*. 6th Ed. Cambridge University Press, Cambridge (1932).

21. MORSE, P.M. and FESHBACH, H. *Methods of Theoretical Physics*. McGraw Hill, NY (1953).

22. FRIEDMAN, M.B. and SHAW, R.P. Diffraction of a Plane Shock Wave by an Arbitrary Rigid Cylindrical Obstacle. *Trans. ASME, J. Appl. Mech.* 29, 40-46 (1962).

23. SHAW, R.P. An Outer Boundary Integral Equation applied to Transient Wave Scattering in an Inhomogeneous Medium. *Trans. ASME, J. Appl. Mech.* 42, 147-152 (1975).

24. GROENENBOOM, P.H.L. Wave Propagation Phenomena, in *Progress in Boundary Elements*, 2, C.A. Brebbia (Ed) Pentech Press, London and Springer Verlag, N.Y. (1983).

25. MANSUR, W.J. and BREBBIA, C.A. Formulation of the Boundary Element Method for Transient Problems governed by the Scalar Wave Equation. *Appl. Math. Modelling* 6, 307-311, August 1982.

26. MANSUR, W.J. and BREBBIA, C.A. Numerical Implementation of the Boundary Element Method for Two Dimensional Transient Scalar Wave Propagation Problems. *Appl. Math. Modelling* 6, 299-306, August (1982).

27. NARDINI, D. and BREBBIA, C.A. A New Approach to Free Vibration Analysis using Boundary Elements. in *Proc. of the 4th Int. Conference on Boundary Element Methods*, Southampton, September 1982. C.A. Brebbia (Ed) Springer Verlag, Berlin & N.Y., (1982).

28. MORSE, P.M. and INGARD, K.V. *Theoretical Aeronautics.* McGraw-Hill, London (1968).

29. BREBBIA, C.A. and NARDINI, D. Dynamic Analysis in Solid Mechanics by an Alternative Boundary Element Procedure. *Int. J. of Soil Dynamics and Earthquake Engineering* 2, No. 4, October (1983).

A BOUNDARY ELEMENT METHOD FOR STOKES EQUATIONS IN 3-D EXTERIOR DOMAINS

F.K. Hebeker

Universität-GHS, Paderborn

1. INTRODUCTION

The steady-state slow viscous incompressible fluid flow past a moving body in a body-fixed frame may be described mathematically by the 3-D exterior Stokes boundary value problem (with normed viscosity $\nu = 1$):

$$-\Delta \underline{v} + \nabla p = \underline{f} \ , \quad \operatorname{div} \underline{v} = 0 \quad \text{in } \Omega' \ ,$$

$$\underline{v}_{|\partial\Omega} = \underline{g} \ , \qquad \underline{v} = \underline{v}_\infty \quad \text{at infinity} \ , \tag{1.1}$$

for the velocity field $\underline{v}$ and the pressure function p of the flow in a domain $\Omega' = \mathbb{R}^3 - \overline{\Omega}$ exterior to a smoothly bounded ($\partial\Omega \in C^\infty$) rigid body Ω, where $\underline{v}_\infty$ is the uniform onflow at infinity. Boundary element methods have recently been applied to solve 3-D slow viscous flow problems in domains with smooth, but otherwise arbitrary, shape; see Hsiao-Wendland-Fischer [3], Nedelec-Zhu [14], and Hebeker [5], [6]. Apart from the general advantages of such methods over domain-type methods, we point out that boundary element methods yield approximate flow fields which automatically satisfy the incompressibility constraint. Conversely, the disadvantage of difficulty in treating nonhomogeneities should be noted.

In the present paper a collocation-type boundary element method is described for discretizing integral equations of the second kind, corresponding to Stokes flow. This is found to be particularly efficient when the resulting linear algebraic system is solved iteratively with a multigrid method. Finally, results of numerical test computations are discussed, and in particular are compared with the Stokes formula of the drag of a sphere.

THE MATHEMATICS OF FINITE ELEMENTS AND APPLICATIONS V

ISBN 0-12-747255-X

2. THE BOUNDARY INTEGRAL EQUATIONS

The Stokes differential equations form a generalized elliptic system in the sense of Agmon-Douglis-Nirenberg. However a simple fundamental matrix $\underline{\Gamma}$,

$$\Gamma_{ij}(x) = -\frac{1}{8\pi}\left(\frac{\delta_{ij}}{|x|} + \frac{x_i x_j}{|x|^3}\right), \qquad i,j = 1,2,3 , \tag{2.1}$$

exists, and the ith column $\underline{\Gamma}^{(i)}$ of $\underline{\Gamma}$ solves Stokes equations with pressure $-x_i/(4\pi|x|^3)$. With the interior normal vector $\underline{n}$ w.r.t. Ω' the stress vector of a flow is

$$\underline{t}(\underline{u},p) = -p\underline{n} + \Sigma(\nabla u_j)n_j + \frac{\partial u}{\partial n},$$

which has adjoint $\underline{t}'(\underline{u},p) = \underline{t}(\underline{u},-p)$. The potential theory which follows can be found in [10].

By means of the Stokes volume potential $\underline{Uf}$, where

$$(Uf)_i(x) = \int_{\Omega'} \underline{\Gamma}^{(i)}(x-z)\cdot\underline{f}(z)\,dz,$$

problem (1.1) may be reduced to the homogeneous form

$$\begin{aligned} -\Delta\underline{u} + \nabla q = \underline{0}, \quad & \operatorname{div}\underline{u} = 0 \quad \text{in } \Omega', \\ \underline{u}|_{\partial\Omega} = \underline{a}, \quad & \underline{u} = 0 \quad \text{at infinity}, \end{aligned} \tag{2.2}$$

where

$$\underline{u} = \underline{v} + \underline{Uf} - \underline{v}_\infty \quad \text{and} \quad \underline{a} = \underline{g} + \underline{Uf}|_{\partial\Omega} - \underline{v}_\infty .$$

The classical representation formula for regular solutions of the homogeneous Stokes equations is

$$\begin{aligned} \underline{u} &= \underline{V}\{\underline{t}(\underline{u},q)|_{\partial\Omega}\} - \underline{W}\{\underline{u}|_{\partial\Omega}\}, \\ q & \quad \text{analogous}, \end{aligned} \tag{2.3}$$

with the Stokes simple-layer potential $\underline{V\phi}$

$$(V\phi)_i(x) = \int_{\partial\Omega} \underline{\Gamma}^{(i)}(x-z)\cdot\underline{\phi}(z)\,do_z, \tag{2.4}$$

and the Stokes double-layer potential $\underline{W\phi}$

$$(W\phi)_i(x) = \int_{\partial\Omega} \underline{t}'_z(\underline{\Gamma}(x-z))\cdot\underline{\phi}(z)\,do_z. \tag{2.5}$$

Both surface potentials solve the homogeneous Stokes equations in $\mathbb{R}^3 - \partial\Omega$ with corresponding pressure potentials; see [10]. The potential $\underline{V\phi}$ is continuous in the whole space, but $\underline{W\phi}$ jumps when passing through the boundary $\partial\Omega$:

$$\underline{W}^{+} - \underline{W} = \underline{W} - \underline{W}^{-} = -\frac{1}{2}\underline{\phi} \quad \text{on } \partial\Omega \ , \tag{2.6}$$

where $\underline{W}^{+}$, $\underline{W}^{-}$ denote respectively the interior (w.r.t Ω') or exterior limits as the boundary is approached.

The classical potential representation for solving (2.2) is stated in terms of a double-layer potential ($\underline{u} = \underline{W}\underline{\phi}$). However, for this, due to the jump relation (2.6), we are led to an integral equation of the second kind in which eigensolutions arise. To circumvent the serious practical difficulties due to this, we propose (extending an idea of Leis, Brakhage and P. Werner for the case of Helmholtz' equation) the *combined* representation

$$\underline{u} = (\underline{W} + \eta\underline{V})\underline{\phi} \ , \quad q \text{ analogous} \ , \tag{2.7}$$

where η is a free parameter. The unknown surface source vector $\underline{\phi}$ is determined by the jump relation (2.6) and the continuity of $\underline{V}$. In view of the boundary data this produces the *boundary integral equation system of the second kind*

$$(\underline{I} - 2\underline{W} - 2\eta\underline{V})\underline{\phi} = -2\underline{a} \text{ on } \partial\Omega \tag{2.8}$$

($\underline{I}$ = unit matrix). The following theorem (see [5],[8]) arises:

Theorem 1: Choose $\eta > 0$. Then for any continuous boundary data $\underline{a}$ the system (2.8) has a unique continuous solution $\underline{\phi}$. The corresponding potential (2.7) is the unique classical solution of (2.2) and the drag $\underline{D}$ of the body is

$$\underline{D} = \eta \int_{\partial\Omega} \underline{\phi} \, do \ . \tag{2.9}$$

Formula (2.9), which is important for applications in fluid dynamics, was brought to our attention by Dr. T. Fischer (Darmstadt).

3. THE BOUNDARY ELEMENT METHOD

How is (2.7) to be discretised? Perhaps the most simple and efficient way of doing this is to use a collocation-type boundary element procedure. As coordinate functions we use globally continuous and piecewise bilinear polynomials on the parameter space of the boundary. The integrals (2.7), (2.8) are evaluated using a simple midpoint-rule, and this introduces the major part of the discretization error. Gauss quadrature formulae have been implemented in this context by Zhu [14], but no mathematical convergence theory seems to date to have been derived.

Denoting the grid-size by h we are led to a linear algebraic system which can be written formally as

$$(\underline{I}_h - 2\underline{W}_h - 2\ \underline{V}_h)\underline{\phi}_h = -2\underline{a}_h \ . \tag{3.1}$$

The matrix of that system is nonsparse but relatively small, as only unknowns located on the boundary $\partial\Omega$ are involved. The following convergence theorem has been proved, see [8].

Theorem 2: Choose $\eta > 0$. Then, for sufficiently small h and for any continuous boundary data $\underline{a}$, the linear algebraic system (3.1) has a unique (discrete) solution. The sequence $(\underline{\phi}_h)$ of approximate densities converges almost linearly:

$$\sup_{\partial\Omega} |\underline{\phi}_h - \underline{\phi}| \leq \text{const. } h \log \frac{1}{h} . \tag{3.2}$$

The corresponding sequence $(\underline{u}_h)$ of approximate potentials converges in the same manner:

$$\sup_{Q} |\underline{u}_h - \underline{u}| \leq \text{const. } h \log \frac{1}{h} , \tag{3.3}$$

uniformly in any domain $Q \subset \Omega'$ with positive distance from $\partial\Omega$. Finally, for the approximate drag $\underline{D}_h = \eta \int \underline{\phi}_h$ we have

$$|\underline{D}_h - \underline{D}| \leq \text{const. } h \log \frac{1}{h} . \tag{3.4}$$

4. THE ITERATIVE PROCEDURE

Large systems like (3.1) are usually solved iteratively. For this we propose here a two-level multi-grid method. This consists of an alternating sequence of just one Gauss-Jacobi iterative step on some fine grid, followed by direct solution of the defect equations on the double-sized grid. As a starting value one may take the exact solution of the linear system corresponding to (3.1) on the coarse grid. The details of the method are given in [7] and [8]. For general multigrid methods we refer to [4].

Geometric convergence

$$\sup_{\partial\Omega} |\underline{\phi}_h^{(k)} - \underline{\phi}_h| \leq (\text{const. } h \log \frac{1}{h})^k \sup_{\partial\Omega} |\underline{\phi}_h^{(o)} - \underline{\phi}_h| ,$$

can be proved for the iterates $\underline{\phi}_h^{(k)}$, and related convergence results hold for the corresponding approximate potentials and drags. Note that in contrast to classical iterative procedures the convergence rate is *improved* when the grid is refined.

5. SOME NUMERICAL RESULTS

Our test calculations are performed on the specific example of flow exterior to a unit ball. However, we point out that no symmetry properties of the flows have been utilised. The polar coordinates are used only as a global coordinate frame. Hence the method and the results can be considered to be typical of 3-D flows in more general flow regions.

In Fig. 1 the influence of the free parameter η in the representation (2.7) on the numerical accuracy is studied. Here

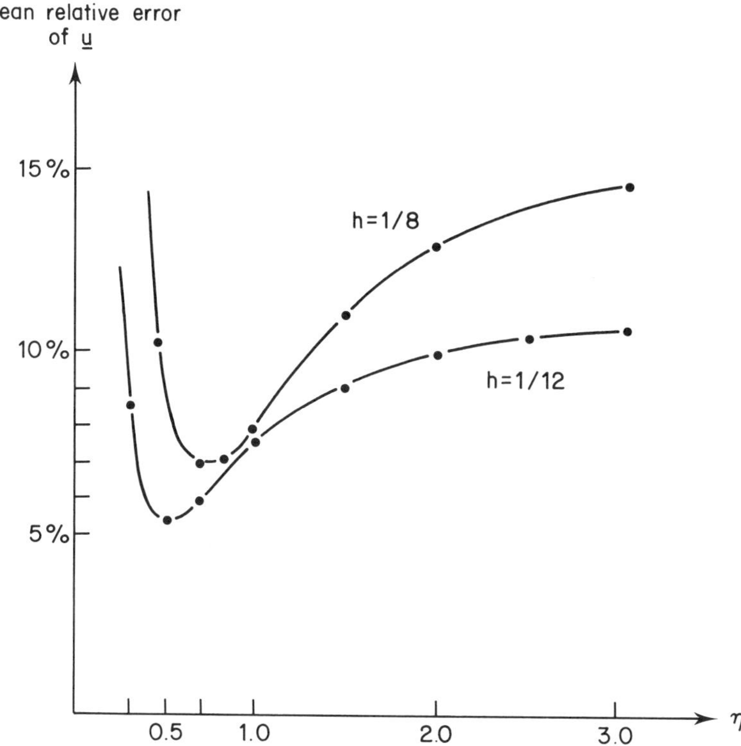

FIG.1. Influence of η.

the system (3.1) is solved by Gauss elimination on some coarser grids with mesh-sizes $h = 1/8$ and $h = 1/12$. In a test example with known solution it turns out that the size of η influences the numerical results markedly, and hence η may be regarded as a *conditioning parameter*.

The efficiency of the iterative procedure is studied at $\eta = 0.5$ on a fine grid with mesh-size $h = 1/16$. The nonsparse matrix here is of size 867×867 and the computations are performed with simple precision. As a measure for the accuracy of an iterate $\underline{X}$ in Fig. 2 the maximum modulus of the defect

$$(\underline{I}_h - 2\underline{W}_h - 2\eta\underline{V}_h)\underline{X} + 2\underline{a}_h$$

no. of cycle	two-grid method	Gauß-Jacobi
1	.24	1.5
2	.38 -1	.76
3	.69 -2	.38
4	.77 -3	.19
5	.10 -3	.10
6	.13 -4	.50 -1
.	.	.
.	.	.
15	.	.11 -3
.	.	.
.	.	.
20	.	.40 -5

FIG.2. Defects of iterates.

is computed, and is stated in the form .45 −3 (≡.00045). As a starting value we choose $\phi_h \equiv 0$; it is found that one two-grid cycle yields the same accuracy as three Gauss-Jacobi iterative steps (Gauss elimination is much more expensive). This relative superiority of the two-grid method has been confirmed in several test series with different starting values (see [7],[8]).

In the final example our procedure is tested against Stokes well-known formula for the drag of a sphere (e.g. [11]), where the uniform onflow $\underline{v}_\infty = (1,0,0)$ is parallel to the x-axis. With some two-grid cycles we obtain an approximate value D_x^h for the x-component D_x of the drag:

D_x^h	D_x	Relative Error
19.45	18.85	3.2%

These numerical results give good insight into the efficiency of our boundary element method in a full 3-D viscous flow problem.

ACKNOWLEDGEMENTS

The author wishes to express since thanks to Professor Dr. W.L. Wendland for some helpful discussions and suggestions. Further, partial support of this work by the Deutsche Forschungsgemeinschaft under its special program "Finite Approximationen in der Strömungsmechanik" is gratefully acknowledged.

REFERENCES

1. BREBBIA, C.A. et al. (eds.) *Boundary Element Methods.* Berlin (1983).

2. FAXEN, H., Fredholm'sche Integralgleichungen zu der Hydrodynamik zäher Flüssigkeiten. *Ark. Mat. Astr. Fys.* 21A, 14, pp.1-40 (1929).

3. FISCHER, T.M., An integral equation procedure for the exterior 3-D slow viscous flow. *Integral Equ. Op. Th.* 5, 490-505 (1982).

4. HACKBUSCH, W. and TROTTENBERG, U., *Multigrid Methods.* Springer Verlag, Berlin (1982).

5. HEBEKER, F.K., A theorem of Faxen and the boundary integral method for 3-D viscous incompressible fluid flow. *Preprint*, University of Paderborn (1982).

6. HEBEKER, F.K., A boundary integral approach to compute the 3-D Oseen's flow past a moving body. pp.124-130 of M.Pandolfi and R. Piva (eds.), *Numerical Methods in Fluid Mechanics.* Braunschweig (1984).

7. HEBEKER, F.K., On a multigrid method to solve the integral equations of 3-D Stokes' flow. Conference at Kiel, January 27-29 (1984).

8. HEBEKER, F.K., Zur Randelemente-Methode in der 3-D viskosen Strömungsmechanik. *Preprint*, University of Paderborn (1984).

9. HSIAO, G.C. and WENDLAND, W.L., The Aubin-Nitsche lemma for integral equations. *J. Integral Equ.* 3, 299-315 (1981).

10. LADYZHENSKAJA, O.A., *The Mathematical Theory of Viscous Incompressible Flow.* Gordon and Breach, New York (1969).

11. SCHLICHTING, H., *Grenzschicht-Theorie*, 8. ed., Karlsruhe (1982).

12. WENDLAND, W.L., Die Behandlung von Randwertaufgaben im $\mathbb{R}^3$ mit Hilfe von Einfach- und Doppelschichtpotentialen. *Num. Math.* 11, 380-404 (1968).

13. WHITEMAN, J.R., (ed.) *The Mathematics of Finite Elements and Applications IV, MAFELAP 1981.* Academic Press, London (1982).

14. ZHU, J., A boundary integral equation method for the stationary Stokes' problem in 3-D. pp.283-292 of [1].

CREATION OF A MASS MATRIX IN ELASTODYNAMICS BY THE BOUNDARY ELEMENT METHOD

D. Nardini

University of Zagreb, Yugoslavia

C. A. Brebbia

University of Southampton, England

1. INTRODUCTION

The aim of this paper is to present an alternative formulation of equations of motion, using the boundary integral approach. The two existing boundary integral methods, based on impulse[1], and harmonic [2] responses, although being applicable to the general case of transient elastodynamics, have disadvantages in practical applications. Namely, in those formulations, the coefficients of the governing equations are either time-dependent or frequency-dependent, causing serious computational difficulties. Nevertheless, these solutions have given some valuable results in the field of wave propagation [3].

If the problem of elastodynamics does not involve the shock-type disturbances, it is shown here that the solution procedure can be formulated by a set of linear differential equations, with the unknowns only at the boundary, and possibly at a limited number of internal points. In this way, the problem is defined only by a small number of unknowns, which makes the method most favourable from the computational point.

2. FORMULATION OF THE ELASTODYNAMIC PROBLEM

An elastic homogeneous body V is considered to be enclosed by the boundary S. The external influences are confined to the boundary in form of the prescribed displacements $\bar{u}$ on S_1 and prescribed tractions $\bar{p}$ on S_2 part of the boundary.

The equilibrium conditions of an infinitesimal volume are given by the well-known tensor expression

ISBN 0-12-747255-X

$$\sigma_{ij,j} + b_i = 0 \tag{2.1}$$

with σ_{ji} being the Cartesian stress tensor, and b_i the body force vector.

Confining the external influences to the boundary reduces the body force only to the inertial part:

$$b_i = - \rho \ddot{u}_i \tag{2.2}$$

where ρ is the mass density, and $\ddot{u}_i$ is the acceleration vector. With this substitution, equation (2.1) becomes

$$\sigma_{ij,j} - \rho \ddot{u}_i = 0 \tag{2.3}$$

The boundary conditions are given by

$$u_i = \bar{u}_i \ (t) \qquad \text{on } S1 \tag{2.4}$$

$$p_i = \bar{p}_i \ (t) \qquad \text{on } S2 \tag{2.5}$$

In order not to unnecessarily obscure the presentation, the initial conditions at time zero will be taken as homogeneous.

3. THE BOUNDARY INTEGRAL FORMULATION

A usual way of deriving the direct boundary integral expression is by utilising the generalised Maxwell-Betti reciprocal theorem, including two independent displacement modes. The first displacement mode represents the unknown solution, while the second one is usually a singular solution of the governing differential equation in the unbounded domain, also reffered to as the fundamental solution.

The fundamental solution to equation (2.3) can either be expressed in the time domain (impulse response), or frequency domain (harmonic response). The disadvantages of these procedures have been noted in the introduction.

The alternative approach, described here, will use the fundamental solution of the statical part of the governing differential equation, which will be denoted by U_{ki}. By applying the Maxwell-Betti theorem, one arrives at the following integral expression

$$\rho \int_V \ddot{u}_i \, U_{ki} \, dV = - C_{ki} U_i + \int_S U_{ki} P_i dS - \int_S P_{ki} U_i dS \tag{3.1}$$

where P_{ki} represents tractions due to the displacement field U_{ki} while C_{ki} are constants dependent on the position of the source in respect to the boundary, which is well known from the statical formulation [4].

The penalty for using only the partial fundamental solution is the retention of the domain integral in equation (3.1), which is due to inertia. In order to be able to express this integral in terms of the boundary integrals, an approximation to the acceleration $\ddot{u}_i$ in the domain will be made in the following manner.

Time dependent displacements will be represented as a sum of m coordinate functions $f^j(\xi)$, multiplied by unknown time functions $\ddot{\alpha}_i^j(t)$:

$$u_i(\xi,t) = \alpha_i^j(t)\, f^j(\xi) \qquad (3.2)$$

with summation on j=1 to m implied here and subsequently. Accelerations are derived from this expression as

$$\ddot{u}_i = \ddot{\alpha}_i^j(t)\, f^j(\xi) \qquad (3.3)$$

With this substitution, the domain integral of equation (3.1) is given by

$$\int_V \ddot{u}_i U_{ki} dV = \ddot{\alpha}_i^j \int_V f^j U_{ki} dV = \ddot{\alpha}_l^j \int_V \delta_{li}\, f^j U_{ki} dV \qquad (3.4)$$

The resulting integral can be transformed into the equivalent boundary integrals [5] in the following form

$$\int_V \delta_{li} f^j dV = C_{ki}\,\psi_{li}^j - \int_S U_{ki}\,\eta_{li}^j dS + \int_S P_{ki}\,\psi_{li}^j dS \qquad (3.5)$$

where ψ_{li}^j is the solution to the elastostatic problem given by

$$\sigma_{lim,m}^j + \delta_{li} f^j = 0 \qquad (3.6)$$

in the unbounded domain, and η_{li}^j are the corresponding tractions. Using expressions (3.4) and (3.5), equation (3.1) becomes

$$C_{ki}U_i + \int_S P_{ki}U_i dS - \int_S U_{ki}p_i dS +$$
$$\rho(C_{ki}\,\psi_{li}^j + \int_S P_{ki}\,\psi_{li}^j dS - \int_S U_{ki}\,\eta_{li}^j dS)\,\ddot{\alpha}_l^j = 0 \qquad (3.7)$$

From this boundary integral expression, a suitable numerical solution procedure can be derived.

4. THE BOUNDARY ELEMENT SOLUTION

Discretising the boundary into elements, equation (3.7) is approximated by a set of equations in the form

$$\underline{C}_n\underline{U}_n + \underline{h}_{ne}\underline{U}_e - \underline{g}_{ne}\underline{p}_e +$$
$$\rho(\underline{C}_n\,\underline{\psi}_n^j + \underline{h}_{ne}\,\underline{\psi}_e^j - \underline{g}_{ne}\,\underline{\eta}_e^j)\,\underline{\ddot{\alpha}}^j = \underline{0} \qquad (4.1)$$

with sum on e over elements, and j over interpolation functions. Submatrices $\underline{h}$ *and* $\underline{g}$ *are well known from elastostatics, while elements of* $\underline{\psi}$ *and* $\underline{\eta}$ *are assembled from the values of the corresponding functions at boundary nodes. This system of equations is written in the assembled matrix from as*

$$\underline{H}\underline{u} - \underline{G}\underline{p} + \rho\ (\underline{H}\ \underline{\psi} - \underline{G}\ \underline{\eta})\ \ddot{\underline{\alpha}} = \underline{0} \tag{4.2}$$

The transformation from the coordinate space $\underline{\alpha}$ *to the displacement unknowns* $\underline{u}$ *is established using the relationship given by equation (3.2). When it is applied to each boundary node in turn, a linear transformation is obtained in the form*

$$\underline{u} = \underline{F}\ \underline{\alpha} \tag{4.3}$$

with coefficients of $\underline{F}$ *being the values of functions* $f^j(\xi)$ *at the nodes. If the set* f^j *is chosen to be linearly independent, the matrix* $\underline{F}$ *is square and regular, and the inverse transformation to (4.3) can be derived as*

$$\underline{\alpha} = \underline{E}\underline{u} \tag{4.4}$$

with $\underline{E} = \underline{F}^{-1}$*. Using this substitution, equation (4.2) becomes*

$$\underline{H}\underline{u} + \underline{M}\ddot{\underline{u}} = \underline{G}\underline{p} \tag{4.5}$$

where the generalised mass matrix is given by

$$\underline{M} = \rho\,(\underline{H}\ \underline{\psi} - \underline{G}\ \underline{\eta})\ E \tag{4.6}$$

Equation (4.5) represents a system of linear differential equations with constant coefficients, the solution to which can be obtained by direct time integration.

5. *TYPES OF ANALYSIS*

The system of differential equations (4.5) is applicable to transient analysis of both types of external influences, given in form of support displacements $\bar{u}$ *at* S_1*, or boundary tractions* $\bar{p}$ *at* S_2*. In order to be able to distinguish known from unknown quantities in the solution process, the system of equations is partitioned, retaining only the unknown variables at the left side. This results in the following system of equations*

$$\hat{\underline{H}}\underline{u}_2 + \hat{\underline{M}}\ddot{\underline{u}}_2 = \underline{G}\underline{p}_2 + \hat{\hat{\underline{H}}}\underline{u}_1 + \hat{\hat{\underline{M}}}\ddot{\underline{u}}_1 \tag{5.1}$$

where the modified matrices reflect the partitioning, whilst indices 1 and 2 refer to S_1 *and* S_2 *respectively. There are three distinct types of dynamical problems which will be discussed separately.*

5.1 *Eigenvalue analysis*

The free vibration case arises in the absence of the external disturbances, therefore equation (5.1) is reduced to

$$\hat{\underline{H}}\underline{u}_2 + \hat{\underline{M}}\ddot{\underline{u}}_2 = \underline{0} \qquad (5.2)$$

Solution to this equation is harmonic, therefore

$$\ddot{\underline{u}}_2 = -\omega^2 \underline{u}_2 \qquad (5.3)$$

where ω is the natural circular frequency. With this substitution, equation (5.2) becomes

$$\hat{\underline{H}}\,\underline{u}_2 = \omega^2 \hat{\underline{M}}\underline{u}_2 \qquad (5.4)$$

which represents the generalised algebraic eigenvalue problem.

5.2 *Forced response analyses*

When supports (S_1 part of the boundary) are fixed, equation (5.1) is reduced to

$$\hat{\underline{H}}\underline{u}_2 + \hat{\underline{M}}\ddot{\underline{u}}_2 = \hat{\underline{G}}\underline{p}_2 \qquad (5.5)$$

Tractions are given functions of time, and this system of equations can be solved by a direct time integration procedure.

5.3 *Support excitations*

If all points belonging to S_1 are translating as a rigid body, the resulting volume forces in the relative coordinate system are given by

$$\underline{b}_i = -\rho\,\underline{a}_i \qquad (5.6)$$

where a_i is the support acceleration in the i-th direction. Taking this into consideration, the general dynamic equation (4.5) is augmented to

$$\underline{H}\underline{u} + \underline{M}\ddot{\underline{u}} = \underline{G}\underline{p} + \underline{b} \qquad (5.7)$$

Elements of vector $\underline{b}$ are generally derived as

$$\underline{b}_k = (\rho \int_V \underline{U}_{ki} dV)\, \underline{a}_i \qquad (5.8)$$

There are various ways to express this domain integral by the equivalent boundary integrals, so that the volume integration need not be performed. Taking into the account that in this case $\underline{u}_1 = 0$ and $\underline{p}_2 = 0$, equation (5.7) yields

$$\underline{\hat{H}}\,\underline{u}_2 + \underline{\hat{M}}\,\underline{u}_2 = \underline{\hat{b}}$$

which again may be solved by directly integrating in time.

6. APPLICATIONS

A computer program based on the outlined method has been developed for the case of plane elastodynamics. The boundary elements employed are three-noded, quadratic. A number of examples have been worked out, and the results were compared to the ones obtained by the finite element method.

6.1 Free vibrations of a hollow square

The first four modes of free vibrations were computed for the plane structure shown in Fig. 1, ($E/\rho = 10^5$ and $\nu = 0{,}25$). The number of boundary elements employed was 24, and the resulting natural frequencies are compared to the ones obtained by a finite element model consisting of 180 nodes.

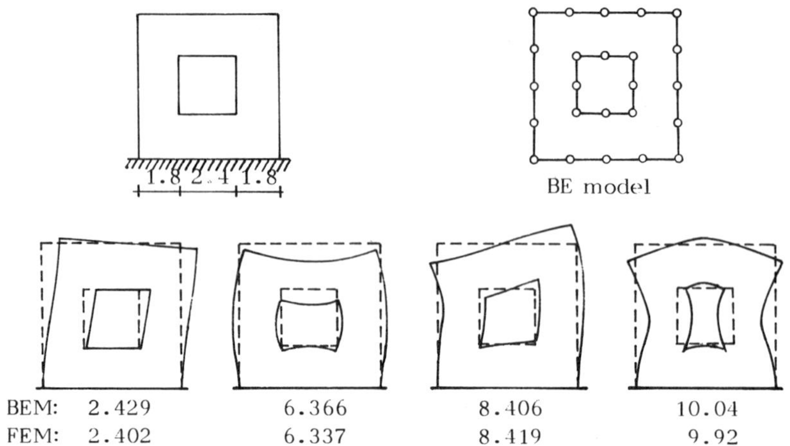

FIG. 1. Free vibration modes and frequencies of a hollow square

6.2 Forced vibrations of a thich arch

The arch shown in Fig. 2 ($E = 10^5$, $\rho = 1$, $\nu = 0{,}25$) is subjected to a suddenly applied side pressure of intensity, varying along the arch, according to the cosine. Horizontal displacement of point A was plotted versus time as obtained by the present method on a 22 element model, and was compared to the finite element results (114 node model).

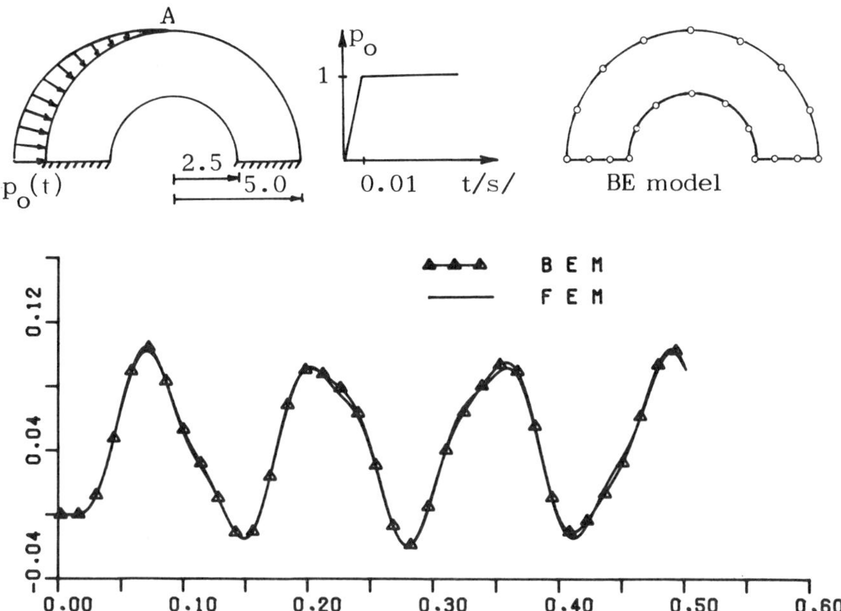

FIG. 2. Forced vibrations of a thick arch

6.3 Support excitations of a trapezoidal dam

A dam-like structure shown in Fig. 3 was subjected to harmonic excitations with the forcing frequency of 16 Hz, which is the mean of the first two skew-symmetric natural frequencies of the structure (f1 = 11.04 Hz, f2 = 20,72 Hz). Results of horizontal displacements of the crest point, obtained by the boundary element model consisting of only 12 elements were compared to the FEM results (124 node model).

7. CONCLUSION

The newly developed method based on boundary integrals is shown to be efficient and accurate in application to various types of problems. The main advantage in comparison to other approaches is that the integration involved is confined only to the boundary, and that the general transient dynamic problem is reduced to a set of linear differential equations with constant coefficients. It was demonstrated that this method can be applied to various types of analyses, ranging from free vibrations to the problem of support excitations.

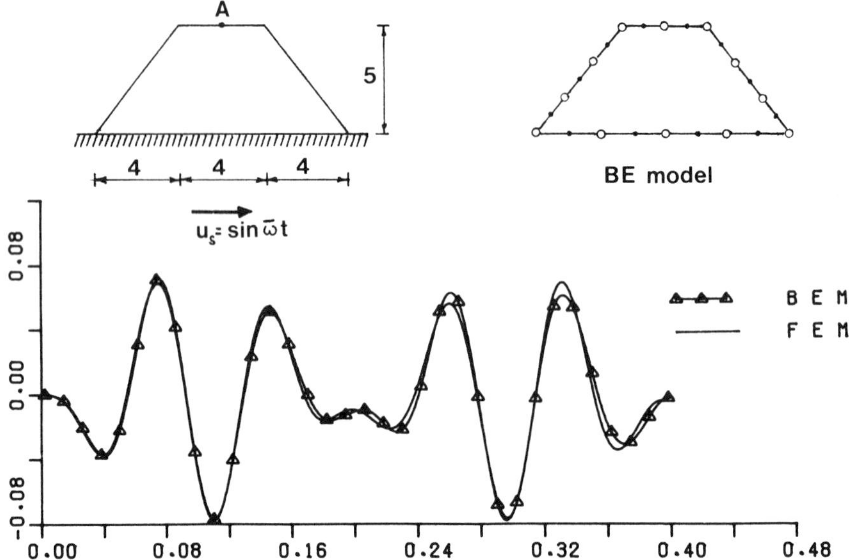

FIG.3. Transient response of a dam subjected to support excitations

REFERENCES

1. COLE, D.M. et al., A Numerical Boundary Integral Equation Method For Elastodynamics I. *Bull. Seism. Soc. Amer.*, 68, (1978)

2. CRUSE, T.A. and RIZZO, F.J., A Direct Formulation and Numerical Solution of the General Transient Elastodynamic Problem. *J. Math. Anal. Appl.* 22 (1968)

3. MANSUR, W. and BREBBIA, C.A., Formulation of the Boundary Element Method for Transient Problems Governed by the Scalar Wave Equation, *App. Math. Modell.*, Vol.6, 307-311, (1982).

4. BREBBIA, C.A., *The Boundary Element Method for Engineers.* Pentech Press, London, (1978).

5. NARDINI, D. and BREBBIA, C.A., Transient Dynamic Analysis by the Boundary Element Method. *Proceedings Fifth Boundary Element Conference*, Ed. C.A.Brebbia, Springer Verlag, Berlin, (1983).

STRESS ANALYSIS WITH A COMBINATION OF HSM AND BEM

E.Schnack

Institute of Solid Mechanics/Karlsruhe University/F.R. Germany

1. INTRODUCTION

An elastic body V is defined in the interior of a smooth closed curve or surface in $\mathbb{R}^n$, n = 2 or 3. The material of V is homogeneous, isotropic and linear elastic. The deformation is thus small so that the strain tensor e_{km} can be defined in terms of the displacement vector $\underline{u}$ as

$$e_{km} = \frac{1}{2} (u_{k,m} + u_{m,k}) \text{ in } V \ . \tag{1}$$

The displacement field should be compatible, so that the incompatibility tensor disappears, i.e.

$$i_{rs} = f_{(e_{km})} = 0 \in V \ . \tag{2}$$

The continuum should also be in equilibrium, and it is assumed that there are no volume forces and no residual stresses. The stress tensor τ_{km} then satisfies

$$\tau_{km,k} = 0 \in V \ . \tag{3}$$

The stresses are related to the strains through the material law,

$$\tau_{km} = 2G\left(e_{km} + \frac{\nu}{1-2\nu} \delta_{km} e_{qq} \right) , \tag{4}$$

where G is the shear modulus and ν the Poisson's ratio. Three kinds of boundary conditions may be considered and produce respectively the Dirichlet problem, the Neumann problem, or a combination of both. The Dirichlet problem, formulated with the predetermined displacement vector $\underline{\tilde{u}}$ has boundary condition

THE MATHEMATICS OF FINITE
ELEMENTS AND APPLICATIONS V

ISBN 0-12-747255-X

$$\underline{u}\big|_{\partial V_K} = \tilde{\underline{u}} \ . \tag{5}$$

With the external normal vector $\underline{n}$ to ∂V and the unit vector $\underline{e}$, the Dirichlet problem leads for the predetermined traction vector $\tilde{\underline{t}}$ to

$$\underline{t} := \tau_{km}\, n_k\, \underline{e}_m\big|_{\partial V_S} = \tilde{\underline{t}} \ . \tag{6}$$

In the numerical process for solving the elasticity problem the domain V has to be discretised into elements. For this purpose two domains, V_1 and V_2 with $V = V_1 \cup V_2$, are defined. In V_1 there is high stress concentration, whilst in V_2 there are only small stress gradients, see Fig. 1.

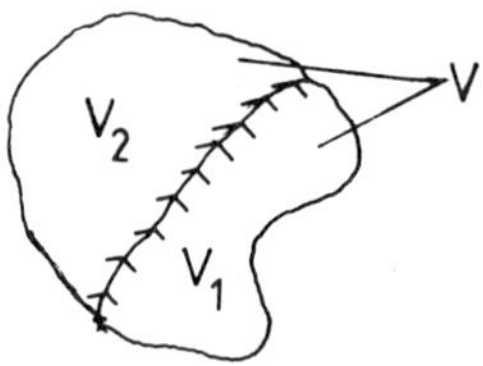

FIG. 1. Domain V with the domains V_1 and V_2

A discretization is effected in V_2 with classical displacement elements. For this triangle elements with six nodes and quadratic displacement distributions can be used. For V_1 special hybrid elements will be developed, which render a precise approximation of high function gradients possible with low CPU-time. This has already been undertaken successfully for plane problems [1-3]. The use of mixed methods combining BEM and the classical FEM has already been proposed, [5-7]. In the three-dimensional context hybrid elements have so far been developed only for axi-symmetric problems, in which the traction condition (equn. (6)) cannot be fulfilled exactly [4]. It will be shown here, that by application of the BEM [8] an *efficient* hybrid element (H-element) can be developed by satisfying (6) exactly.

2. BASIC EQUATIONS FOR THE HYBRID ELEMENT (H)

The hybrid elements with their volumes V_n and their surfaces ∂V_n (see Fig. 2) of the domain V_1 have to satisfy the compatibility and equilibrium conditions. Because in combination with elements of domain V_2 these conditions cannot be satisfied exactly, only a *weak* formulation of that in [1] will be required. The generalized compatibility equation has the form

$$\int_{\partial V_n} \underline{t}^T\, (\underline{u}_V - \underline{u}_{\partial V})\, ds = 0 \ , \tag{7}$$

where $\underline{u}_V$ is the displacement vector of the stress field ((1) and

(4)) inside of H, see Fig. 2. The displacement distribution for the surface ∂V_n is $\underline{u}_{\partial V}$ and is independent from $\underline{u}_V$, so that an inter-element compatibility can be reached exactly for the classical displacement elements. The generalized equilibrium condition has the form:

$$\int_{\partial V_S} \underline{\tilde{t}}^T \underline{u}_{\partial V} \, ds - \int_{\partial V_n} \underline{t}^T \underline{u}_V \, ds = 0 . \tag{8}$$

The predetermined traction vector $\underline{\tilde{t}}$ (see (6)) acts on ∂V_S of ∂V_n.

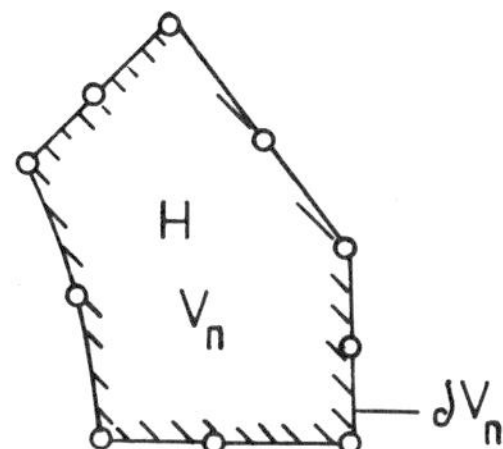

FIG. 2. Hybrid element

The displacement vector $\underline{u}_V$ must satisfy (2), whilst $\underline{t}$ satisfies

$$\int_{\partial V_n} \underline{t} \, ds = \underline{0} \tag{9}$$

and

$$\int_{\partial V_n} \underline{r} \times \underline{t} \, ds = \underline{0} , \tag{10}$$

where $\underline{r}$ is the position vector to the origin of the co-ordinate system. It can be seen from (7) and (8) that a formulation of the stress distribution τ_{km} inside of V_n is not necessary, by contrast to the classical hybrid stress method [1].

In the next section we explain how the trial and test functions for the traction and displacement vectors can be formulated using (7) and (8).

3. TRIAL FUNCTIONS

Fig. 3 shows an actual discretization of the domain V with finite elements for a plane problem.

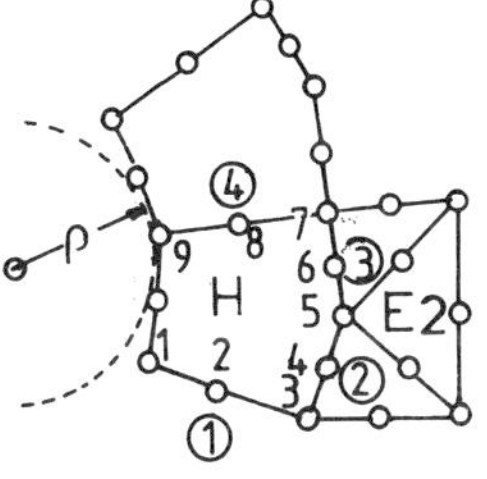

FIG. 3. H: hybrid element of V_1, see Fig. 1
E2: displacement elements of V_2, see Fig. 1.

The notch contour of the stress concentration domain V_1 is approximated by circular arcs, see fig. 3. The notch boundary may be loadfree, so that from (6) and (8)

$$\underline{\tilde{t}} = \underline{0} \in \partial V_S. \tag{11}$$

The residual boundaries of the element H consist of straight lines, so that compatible connection with the classical elements having quadratic displacement distributions of type E2 is possible, see Fig. 4.

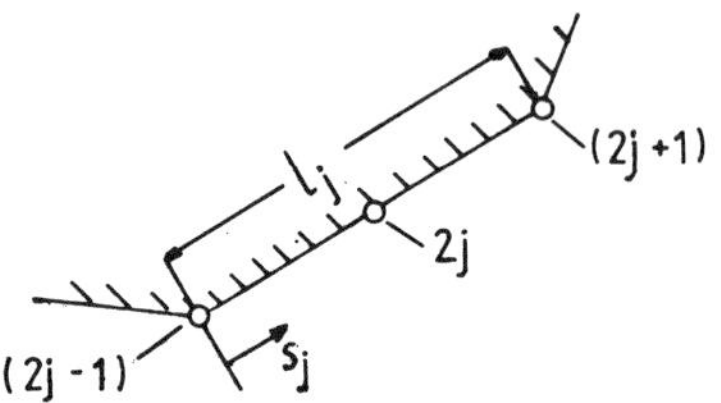

FIG.4. Side j

With the co-ordinate

$$\xi_j = s_j / l_j \tag{12}$$

the quadratic displacement distribution can be formulated for plane bodies with the nodal displacements u_{xk}, u_{yk} as follows:

$$\underline{u}_{\partial V} = \begin{bmatrix} u_x \\ u_y \end{bmatrix}_{\partial V} = \begin{bmatrix} A_{(\xi_j)} & B_{(\xi_j)} & C_{(\xi_j)} & 0 & 0 & 0 \\ 0 & 0 & 0 & A_{(\xi_j)} & B_{(\xi_j)} & C_{(\xi_j)} \end{bmatrix} \begin{bmatrix} u_{x\,2j-1} \\ u_{x\,2j} \\ u_{x\,2j+1} \\ u_{y\,2j-1} \\ u_{y\,2j} \\ u_{y\,2j+1} \end{bmatrix} \tag{13}$$

For A, B, C one has

$$\left.\begin{aligned} A_{(\xi_j)} &= 1 - 3\xi_j + 2\xi_j^2 \\ B_{(\xi_j)} &= 4\xi_j\,(1 - \xi_j) \\ C_{(\xi_j)} &= \xi_j\,(2\xi_j - 1) \end{aligned}\right\} . \tag{14}$$

By taking the compatibility condition into account (n_s is the number of straight-lined sides)

$$\underline{u}_{\partial V(\xi_{j-1}=1)} = \underline{u}_{\partial V(\xi_j=0)}, \qquad j \in \{2, n_s\} \tag{15}$$

one obtains

$$\underline{u}_{\partial V} = \underline{N}\,\underline{d} \in \partial V_n \tag{16}$$

with

$$\underline{d}^T = \{u_{x1}\ u_{x2} \ldots u_{xn} \mid u_{y1}\ u_{y2} \ldots u_{yn}\}\ , \tag{17}$$

n being the number of nodes.

In the same way the traction function must be formulated

$$\underline{t} = \begin{bmatrix} t_x \\ t_y \end{bmatrix} = \underline{N}\,\underline{f} \in \partial V_n \tag{18}$$

with the same form function as in equ. (16). The traction node vector has the form:

$$\underline{f}^T = \{f_{x1}\ f_{x2} \ldots f_{xn} \mid f_{y1}\ f_{y2} \ldots f_{yn}\}\ . \tag{19}$$

Because there should be no volume forces, no residual stresses, and no single forces acting on the element, the traction field is in equilibrium, provided that (9) and (10) have been satisfied. Because ∂V_S is loadfree, see (11), one finds for plane problems that

$$\left.\begin{aligned} &\sum_{j=1}^{n_S} \int_{\partial V_j} t_x\, l_j\, d\xi_j = 0 \\ &\sum_{j=1}^{n_S} \int_{\partial V_j} t_y\, l_j\, d\xi_j = 0 \\ &\sum_{j=1}^{n_S} \int_{\partial V_j} (x t_y - y t_x)\, l_j\, d\xi_j = 0 \end{aligned}\right\}\ . \tag{20}$$

Equation (20) is re-written in matrix form as

$$\underline{G}\,\underline{f} = \underline{0}\ , \tag{21}$$

where the matrix $\underline{G}$ has the structures

$$\underline{G} = \left[\begin{array}{c|c} \underline{G}^T & \underline{0}^T \\ \hline \underline{0}^T & \underline{G}^T \\ \hline \underline{E}^T & \underline{M}^T \end{array}\right]\ , \tag{22}$$

the submatrices being row vectors of the order n.

The three dependent elements are combined to form $\underline{f}_S$, whilst the others constitute $\underline{f}_0$ and, using (22), one obtains

$$[\underline{G}_0\ \underline{G}_s] \begin{bmatrix} \underline{f} \\ \underline{f}_s \end{bmatrix} = \underline{0} \quad . \tag{23}$$

Equation (18) is split similarly so that

$$\underline{t} = [\underline{N}_0\ \underline{N}_s] \begin{bmatrix} \underline{f}_0 \\ \underline{f}_s \end{bmatrix} \cdot \tag{24}$$

If the determinant $|\underline{G}_s| \neq 0$, $\underline{t}$ can be written as a function of $\underline{f}_0$ only; i.e.

$$\underline{t} = \underbrace{(\underline{N}_0 - \underline{N}_s \bullet \underline{G}_s^{-1} \bullet \underline{G}_0)}_{\underline{R}}\ \underline{f}_0 \ . \tag{25}$$

The dependent parameters $\underline{f}_s$ cannot be chosen freely, because $|\underline{G}_s|$ must be different from zero. Since

$$\underline{f}_s^T = \{f_{x2} f_{x4}\ f_{y2}\} \tag{26}$$

it follows that

$$|\underline{G}_s| = \frac{8}{27}\, l_1^2\ l_2\ (y_{(4)} - y_{(2)}) \ , \tag{27}$$

so that the consequence must be: $y_{(4)} \neq y_{(2)}$, which can always be satisfied for the problems treated here.

In order to be able to construct a stiffness matrix from (7) and (8), one must have the vector $\underline{u}_V$. This $\underline{u}_V$ represents the boundary displacement field of the hybrid element as a function of the previously formulated traction vector, (25). In this case it is advantageous to determine the displacement vector numerically by means of a suitable integral equation of the BEM.

From the many boundary element methods, [9], the <u>direct method</u> is chosen to solve the Neumann problem

$$\underline{W}_{(q)}\underline{u}_{V(q)} + \int_{\partial V_n} \left\langle \underline{T}_{(q,s)}\ \underline{u}_{V(s)} - \underline{U}_{(q,s)}\ \underline{t}_{(s)} \right\rangle ds = \underline{0},\ q \in \partial V_n \tag{28}$$

The elements of $\underline{W}$ are

$$w_{ij} = \delta_{ij} \left\langle (\alpha_2 - 2\mu\alpha_1)(\Theta_E - \Theta_A) + (\delta_{i2} - \delta_{i1})\ \mu\alpha_1 \ \cdot \right. \tag{29}$$

$$\left. \cdot\ (\sin 2\Theta_E - \sin 2\Theta_A) \right\rangle + (\delta_{ij} - 1)\ 2\mu\alpha_1\ (\sin^2\Theta_E - \sin^2\Theta_A)$$

for corners without singularities. The coefficients α_1 and α_2 can be defined by the Lamé constants λ and μ:

$$\alpha_1 = -\frac{\lambda+\mu}{\lambda+3\mu}\ , \qquad \alpha_2 = \frac{2\mu^2}{\lambda+3\mu} \ . \tag{30}$$

The definition of the angles can be seen from fig. 5.

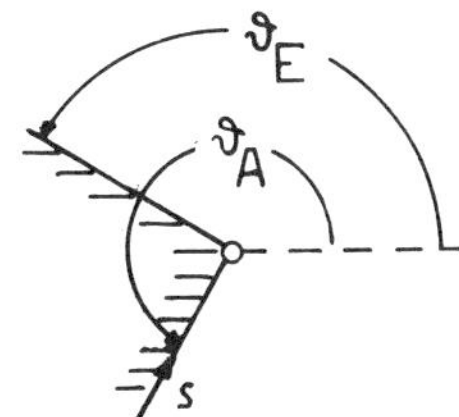

FIG.5. Corners of the hybrid element

The fundamental matrices $\underline{T}$ and $\underline{U}$ are well known from the Kelvin solution [10].

For the numerical implementation, the boundary ∂V_n of the hybrid element will be split up into intervals (Fig. 6), and constant functions will be formulated piecewise for ∂V_j, so that

$$\left.\begin{aligned} \underline{\tilde{t}}_j &= \underline{\tilde{R}}_j\, \underline{f}_0 \\ \text{and} \quad \underline{\tilde{u}}_{Vj} &= \underline{\tilde{L}}_j\, \underline{f}_0 \end{aligned}\right\} . \quad \text{(The number of collocation points is m.)} \tag{31}$$

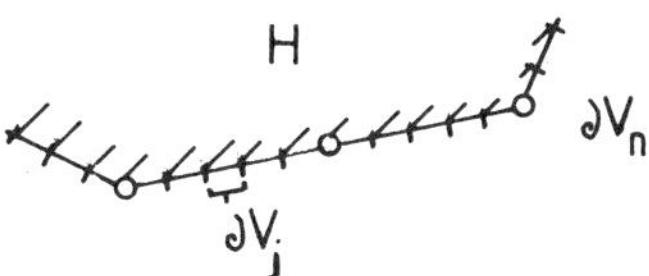

FIG.6.Discretization of ∂V_n

In discretized form (28) leads to

$$\underline{W}_i\, \underline{u}_{Vi} + \sum_{j=1}^{m} \left\langle \int_{\partial V_j} \underline{T}_{(i,s)}\, ds\, \underline{u}_{Vj} - \int_{\partial V_j} \underline{U}_{(i,s)}\, ds\, \underline{t}_j \right\rangle = \underline{0} \quad \text{for } i=1,\ m. \tag{32}$$

From this results the linear equation system

$$\underline{A}\, \underline{\tilde{u}} = \underline{B}\, \underline{\tilde{t}}. \tag{33}$$

The definitions of $\underline{\tilde{u}}$ and $\underline{\tilde{t}}$ can be seen from:

$$\underline{\tilde{u}}^T = \{\tilde{u}_{V1}^T \ \ldots\ \tilde{u}_{Vj}^T \ \ldots\ \tilde{u}_{Vm}^T\} \tag{34}$$

and

$$\underline{\tilde{t}}^T = \{\underline{\tilde{t}}_1^T \ \ldots.\ \underline{\tilde{t}}_j^T \ \ldots\ \underline{\tilde{t}}_m^T\}. \tag{35}$$

The insertion of (31), (34), (35) into (33) leads to

$$(\underline{A}\, \underline{\tilde{L}} - \underline{B}\, \underline{\tilde{R}})\, \underline{f}_0 = \underline{0}. \tag{36}$$

In consideration of the rigid motions, by exchange of columns in $\underline{A}$ and $\underline{B}$, the matrix $\underline{A}$ is regular and leads to

$$\underline{\tilde{L}} = \underline{A}^{-1} \underline{B}\, \underline{\tilde{R}}. \tag{37}$$

In the last step it is necessary to produce the stiffness matrix for the hybrid element. Supposed that $\underline{u}_V$ is produced exactly from the integral equation. It can be written:

$$\underline{u}_V = \underline{L}\, \underline{f}_0 . \tag{38}$$

so that for (7) one can set up

$$\underline{f}_0^T \left\langle \underbrace{\int_{\partial V_n} \underline{R}^T \underline{L}\, dA}_{\underline{H}}\, \underline{f}_0 - \underbrace{\int_{\partial V_n} \underline{R}^T \underline{N}\, dA}_{\underline{S}}\, \underline{d} \right\rangle = \underline{O}. \tag{39}$$

Because $\underline{\tilde{L}}$ can only be produced as an approximation to the exact matrix $\underline{L}$, it can only lead to an approximation of the matrix $\underline{H}$

$$\underline{\tilde{H}} = \sum_{j=1}^{m} \underline{\tilde{R}}_j^T\, \underline{\tilde{L}}_j\, \Delta s_j . \tag{40}$$

Numerical experiments show that, [11], $\underline{\tilde{H}}$, as an approximation to $\underline{H}$ is computed as a symmetric and positive definite matrix with suitable interpolation nodes. So the stiffness matrix resulting from the equilibrium condition (8) is symmetric and positive definite [1]

$$\underline{k} = \underline{S}^T\, \underline{\tilde{H}}^{-1}\, \underline{S}. \tag{41}$$

The matrix $\underline{S}$ from (39) and (41) can be analytically integrated. In combination with the stiffness matrices of the classical displacement elements, in consideration of the rigid motions, the matrix $\underline{k}$ becomes a symmetric, positive definite stiffness matrix with band structure.

As a first numerical test [11], the tensile-loaded plate was computed, in the course of which the displacement amplitudes were to be determined more precisely than 1%. Trial examples for two and three dimensional problems with stress-concentrations based on the presented algorithm are in progress.

3. CONCLUSION

The afore-constructed algorithm contains the advantages of the hybrid assumed stress method, i.e. the structure is computed neither too weakly nor too stiffly. Compared with the classic hybrid stress method, the stress formulations can be formulated more easily. It is easier now, to satisfy exactly the traction condition for three-dimensional problems. The notch parameters are included in the trial functions, because the displacement field is computed with the BEM depending on the boundary structure. Further, by contrast to the BEM, symmetric positive definite band matrices result from the combination with the classical FEM.

4. REFERENCES

1. SCHNACK, E., Singularities of Cracks with Generalized Finite Element Methods. p. 16, *Singularities and Constructive Methods for their Treatment*, Mathematisches Forschungsinstitut Oberwolfach (21-25 Nov. 1983).
2. SCHNACK, E. a. WOLF, M., Application of Displacement and Hybrid-Stress Methods to Plane Notch and Crack Problems. *Int. J. Num. Meth. Engng.* 12, No. 6, 963-975 (1978).
3. DRUMM, R., Zur effektiven FEM-Analyse ebener Spannungskonzentrationsprobleme. *Fortschr.Ber. VDI-Z, Reihe 18*, 13 (1983).
4. SCHNACK, E., Beitrag zur Berechnung rotationssymmetrischer Spannungskonzentrationsprobleme mit der FEM. *Diss. Techn. Univ. München* (1973).
5. BART, J., Kombination eines Integralgleichungsverfahrens mit der Methode der finiten Elemente zur Berechnung ebener Spannungskonzentrationsprobleme. *Diss. Techn. Univ. München* (1974).
6. ZIENKIEWICZ, O.C., KELLY, D.W., a. BETTESS, P., The Coupling of Finite Element and Boundary Solution Procedures. *Int. J. Num. Meth. Engng.* 11, 355-375 (1977).
7. HISTAKE, H.T., ITO, T, a. UEDA, H., Three Dimensional Coupling Method of Boundary and Finite Element Methods. Osaka University, Japan. *5th Int. Conf. on Boundary Elements, Hiroshima* (1983).
8. BREBBIA, C.A., *The Boundary Element Method for Engineers*. Pentech Press, London:Plymouth (1978).
9. HSIAO, G.C. a. WENDLAND, W.L., On a Boundary Integral Method for some Exterior Problems in Elasticity. *Preprint 769, FB Math., Techn. Hochschule Darmstadt* (1983).
10. RIZZO, F.J., An Integral Equation Approach to Boundary Value Problems of Classical Elastostatics.*Quart. Appl. Math.* 25, 83-95 (1967).
11. CHENG, K.C., Ein gemischtes Verfahren auf der Basis von BEM/FEM. *Diplomarbeit/Master's Thesis, Institute of Solid Mechanics*, Karlsruhe University (1983).

INFINITE ELEMENTS FOR THE ANALYSIS OF UNBOUNDED DOMAIN PROBLEMS

K. MORIYA

National Defense Academy, Yokosuka, Japan

1. INTRODUCTION

Finite element analysts often encounter the problems with infinite domains. The domain is divided into the interior region of prime interest and the remaining exterior region. In order to deal with the exterior region within FEM context, various infinite elements have been proposed to date[1]-[6] and each has certain merits and drawbacks.

Several years ago, the author developed a two-node semi-infinite element based on the assumed displacement hybrid finite element model for the analysis of two dimensional potential and wave problems[4],[5]. The element has only two nodes which are common to those of interior elements and requires no interpolation nodes in the exterior region. Therefore, both the number of the total degrees of freedom and the bandwidth of the global stiffness matrix remain the same as those of the interior problem alone. All the integrations for the direction extending to infinity, which are required to generate the element stiffness matrix, can be performed exactly even for wave problems. Better still, the element stiffness matrix can be expressed explicitly.

The element is problem oriented and it requires a general or at least an asymptotic solution of the governing equation. In this paper, the infinite elements are developed for the analyses of 2-D, 3-D and axisymmetric problems in elastostatics. The application to steady-state wave problems is also presented.

2. FINITE ELEMENT FORMULATION

In the interior region the following conventional variational functional is used

THE MATHEMATICS OF FINITE
ELEMENTS AND APPLICATIONS V

ISBN 0-12-747255-X

$$\Pi_I = \sum_m [\iiint_{\partial V_m} A(u_i)dV - \iint_{(\partial V_\sigma)_m} \bar{T}_i u_i dS] \tag{2.1}$$

where $\sum_m$ denotes summation over all interior elements; $(\ )_m$, the quantity which belongs to the *m*th element; V, volume; A, the strain energy function; u_i, displacement component; ∂V_σ, the boundary where the traction T_i is prescribed; $(\bar{\ })$, the prescribed quantity.

The functional for the exterior region can be written as[7],

$$\Pi_E = \sum_n [\iint_{\partial V_n} T_i \tilde{u}_i dS - \frac{1}{2}\iint_{\partial V_n} T_i u_i dS - \iint_{(\partial V_\sigma)_n} \bar{T}_i u_i dS] \tag{2.2}$$

where T_i denotes the boundary traction derived from the displacement field u_i which satisfies both the equilibrium and compatibility conditions; $\tilde{u}_i$, displacement assumed only on element boundaries.

The total functional can be written as $\Pi = \Pi_I + \Pi_E$ and the vanishing of $\delta\Pi$ for arbitrary variations of u_i and $\tilde{u}_i$ gives the following Euler equations on the interelement boundaries between infinite elements and those between interior and infinite elements.

$$u_i = \tilde{u}_i \ , \ T_i^{(a)} + T_i^{(b)} = 0 \tag{2.3}$$

where the superscripts a and b denote the two adjacent elements. Hereafter, only the functional Π_E will be considered and the third term in (2.2) is neglected for simplicity.

The functional Π_E can be written in a matrix form as

$$\Pi_E = \sum_n [\iint_{\partial V_n} \underline{T}^T \underline{\tilde{u}} \ dS - \frac{1}{2}\iint_{\partial V_n} \underline{T}^T \underline{u} \ dS] \tag{2.4}$$

where $\underline{T} = \{\sigma_{1j}\nu_j, \sigma_{2j}\nu_j, \ldots\}$, $\underline{u} = \{u,v,w\}$, $\underline{\tilde{u}} = \{\tilde{u},\tilde{v},\tilde{w}\}$.

The matrices $\underline{T}$ and $\underline{u}$ can be written in the form

$$\underline{T} = \underline{\underline{R}}\underline{\beta} \ , \ \underline{u} = \underline{\underline{U}}\underline{\beta} \tag{2.5}$$

where $\underline{\beta} = \{\beta_1, \beta_2, \ldots, \beta_{N\beta}\}$, $N\beta$ is a total number of parameters β_j.

The displacement $\underline{\tilde{u}}$ can be assumed in terms of the generalized nodal displacement $\underline{q}$ as

$$\underline{\tilde{u}} = \underline{\underline{L}}\underline{q} \tag{2.6}$$

where $\underline{L}$ is the interpolation function matrix defined on each element face.

Substitution of (2.5) and (2.6) into (2.4) yields an equation of the form

$$\Pi_E = \sum_n [\underline{\beta}^T \underline{\underline{G}}\underline{q} - \frac{1}{2}\underline{\beta}^T \underline{\underline{H}}\underline{\beta}] \tag{2.7}$$

where $\underline{G}$ and $\underline{H}$ are given by

$$\underline{G} = \iint_{\partial V_n} \underline{R}^T \underline{L} dS \quad , \quad \underline{H} = \frac{1}{2} \iint_{\partial V_n} (\underline{R}^T \underline{U} + \underline{U}^T \underline{R}) dS \tag{2.8}$$

The stationary condition of Π_E with respect to $\underline{\beta}$ which is independently assumed from one element to the other yields

$$\underline{\beta} = \underline{H}^{-1} \underline{G} \underline{q} \tag{2.9}$$

Substituting (2.9) back into (2.7), the functional Π_E can be expressed

$$\Pi_E = \Sigma \frac{1}{2} \underline{q}^T \underline{K} \underline{q} \tag{2.10}$$

where $\underline{K} = \underline{G}^T \underline{H}^{-1} \underline{G}$ is the stiffness matrix of the present infinite element.

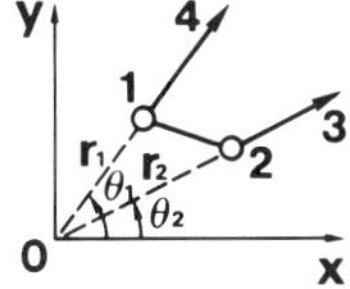

FIG. 1. 2-D element.

3. ELEMENT MATRICES FOR ELASTOSTATICS

3.1 Two-Dimensional Infinite Element

The geometrical shape of the two-dimensional infinite element is a sectorial strip with two straight edges extending in the radial direction away from the origin as shown in Fig. 1. Nodes 1 and 2 are real nodes which are common to those of interior elements. The other two nodes (3 and 4) are imaginary nodes placed at an infinite distance from the origin and they do not appear in the actual computation.

Stresses and displacements are expressed in terms of complex potentials $\psi(z)$ and $\chi(z)$ by

$$\sigma_x + \sigma_y = 4\mathrm{Re}[\psi'(z)], \quad \sigma_y - \sigma_x + 2i\tau_{xy} = 2[\bar{z}\psi''(z) + \chi''(z)]$$
$$2G(u - iv) = \mu\bar{\psi}(\bar{z}) - \bar{z}\psi'(z) - \chi'(z) \tag{3.1}$$

where $z = x + iy = re^{i\theta}$, $\mu = 3 - 4\nu$, Re means the real part of the function following it, G denotes the shear rigidity and ν is the Poisson's ratio.

ψ and χ can be represented by polynomials consisting of only negative powers of z as

$$\psi(z) = \sum_{j=1}^{N} (\beta_j + i\beta_{N+j}) z^{-j}$$
$$\chi'(z) = \sum_{j=1}^{N} (\beta_{2N+j} + i\beta_{3N+j}) z^{-j} \tag{3.2}$$

in which N is a positive integer and β's are real constants.

Substituting (3.2) into (3.1), stresses and displacements are expressed in terms of β's.

The boundary displacement $\underline{\tilde{u}}$ is assumed on segment 1-2 so that it coincides with $\underline{u}$ of the neighboring interior element. On segments 2-3 and 1-4, which extend towards infinity, $\underline{\tilde{u}}$ is assumed so as to simulate the correct far-field behavior of displacements. Since the dominant behavior of displacements indicates $1/r$ variation, $\underline{\tilde{u}}$ is assumed on segment 2-3 as

$$[\tilde{u},\tilde{v}]^T = \begin{bmatrix} 0 & 0 & r_2/r & 0 \\ 0 & 0 & 0 & r_2/r \end{bmatrix} [\underline{q}_1^T \ , \ \underline{q}_2^T]^T \tag{3.3}$$

where $\underline{q}_i^T = [u,v]_i$. $\underline{\tilde{u}}$ is assumed on segment 1-4 in the same way.

In order to generate the element stiffness matrix, two line integrals in (2.8) need be evaluated along the element contour. On the edges extending to infinity, these integrals can be evaluated explicitly. For example, on segment 2-3, the (i,j) component of $\underline{H}$ is given by

$$\underline{H}(i,j) = [r_2^{-(i+j)}/4G(i+j)]\{[\mu(i^2+j^2) + 2ij]\sin(i+j+2)\theta_2 + (i-j)(\mu+ij)\sin(i-j)\theta_2\} \tag{3.4}$$

When the total number of β_j is four (N=1), $\underline{H}$ and $\underline{G}$ are calculated on segments 2-3 and 1-4 as

$$\underline{H} = \sum_{j=1}^{2} \frac{\mu+1}{8G} \frac{(-1)^j}{r_j^2} \begin{bmatrix} 2\sin 4\theta_j & -2\cos 4\theta_j & -\sin 2\theta_j & \cos 2\theta_j \\ & -2\sin 4\theta_j & \cos 2\theta_j & \sin 2\theta_j \\ \text{S.Y.M.} & & 0 & 0 \\ & & & 0 \end{bmatrix}$$

$$\underline{G} = [\underline{g}_1 \quad \underline{g}_2] \tag{3.5}$$

where

$$\underline{g}_j = \frac{(-1)^j}{2r_j} \begin{bmatrix} \sin 3\theta_j + \sin\theta_j & -\cos 3\theta_j + \cos\theta_j \\ -\cos 3\theta_j - \cos\theta_j & -\sin 3\theta_j + \sin\theta_j \\ -\sin\theta_j & \cos\theta_j \\ \cos\theta_j & \sin\theta_j \end{bmatrix}$$

3.2 Axisymmetric and Three-Dimensional Infinite Elements

In this section, the axisymmetric and three-dimensional infinite elements for elastic half space problems are derived.

For the axisymmetric infinite element shown in Fig. 1, where x is the axis of symmetry, y is the radial coordinate and the domain occupies $x \geq 0$, the Boussinesq's point load solution is used for the stress and displacement assumption. For example, σ_x and u_x are assumed as

$$\sigma_x = (3\cos^3\theta/r^2)\beta_1, \quad 2Gru_x = -(\cos^2\theta + 2-2\nu)\beta_1 \tag{3.6}$$

where $r = \sqrt{x^2 + y^2}$.

For the three-dimensional infinite element, the geometrical shape is a truncated pyramid with straight edges extending towards infinity from a common vertex at the origin as shown in Fig. 2. Again the Boussinesq's solution is used for the stress and displacement assumption.

In order to evaluate $\underline{G}$ and $\underline{H}$ on lateral faces (1-2-6-5 etc.), i.e. to integrate the two surface integrals in (2.8), Gauss-Legendre quadrature is used for the finite direction, but they are analytically evaluated for the infinite direction. When the Poisson's ratio is 0.5, all the surface tractions are identically zero on lateral faces and there is no need to carry out the integrations in (2.8) over these faces. Even if the Poisson's ratio is other than 0.5, the integrations can be omitted on lateral faces for practical purposes. Although the inconsistency of the stresses and displacements occurs on these faces, numerical results show that the finite element solutions with excellent accuracy are obtained. Since for the axisymmetric and three-dimensional infinite elements the total number of parameter β's is one, the inverse of matrix $\underline{H}$ is directly obtained from $\underline{H}^{-1} = 1/\underline{H}$.

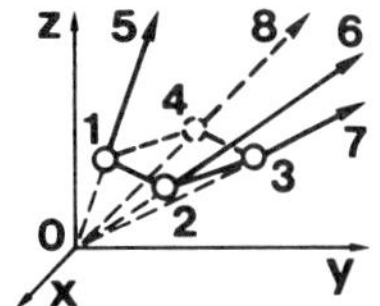

FIG. 2. 3-D element.

4. APPLICATION TO STEADY-STATE WAVE PROBLEMS

The derivation of the hybrid infinite elements for the wave radiation problems governed by Helmholtz's equation follows the same line as that for elastostatics given in the preceding sections except that the wave potential ϕ and its normal derivative $\partial\phi/\partial\nu$ take the place of u_i and T_i, respectively.

For the two-dimensional infinite element, ϕ can be written as

$$\phi = \underline{U\beta} = [H_0(kr), H_1(kr)\cos\theta, \ldots]\{\beta_1, \beta_2, \ldots, \beta_{2N+1}\} \qquad (4.1)$$

where k denotes the wave number. The interpolation function matrix $\underline{L}$ is assumed so that ϕ varies linearly on segment 1-2 in Fig. 1, while on segments 2-3 and 1-4, in order to simulate the basic wave shape and its decay, $\underline{L}$ is assumed as follows.

$$\underline{L}_{2-3} = [0, \sqrt{r_2/r}\, e^{ik(r-r_2)}], \quad \underline{L}_{1-4} = [\sqrt{r_1/r}\, e^{ik(r-r_1)}, 0] \qquad (4.2)$$

where the subscript denotes the node number. On these segments, the integrations in (2.8) can be performed explicitly. For example, on segment 2-3, matrix $\underline{H}$ is calculated as follows.

$$\underline{H}(1,1) = 0, \quad \underline{H}(2,2) = \frac{1}{2} \sin 2\theta_2 \int_{r_2}^{\infty} H_1(kr)^2 \frac{dr}{r}$$
$$= \frac{1}{4} \sin 2\theta_2 [H_0(kr_2)^2 + H_1(kr_2)^2], \quad \underline{H}(1,2) = \frac{1}{2} \sin \theta_2 \times$$
$$\int_{r_2}^{\infty} H_0(kr) H_1(kr) \frac{dr}{r} = \frac{1}{2} \sin \theta_2 \{kr_2 [H_2(kr_2) H_0(kr_2)$$
$$- H_1(kr_2)^2] - H_1(kr_2) H_0(kr_2)\} \qquad (4.3)$$

Matrix $\underline{G}$ is calculated in a similar manner.

For the axisymmetric infinite element, ϕ is assumed as

$$\phi = \sum_{j=0}^{N-1} \beta_{j+1}\, h_j(kr) P_j(\cos\theta) \tag{4.4}$$

When segment 1-2 is assumed to be a circular arc of radius a and N = 1 is used, the stiffness matrix of the axisymmetric infinite element can be written explicitly as follows,

$$\underline{K} = -\frac{2\pi a(1-ika)}{\cos\theta_1-\cos\theta_2}\begin{bmatrix} f^2 & -fg \\ -fg & g^2 \end{bmatrix} \tag{4.5}$$

where $f = (\sin\theta_1-\sin\theta_2)/(\theta_1-\theta_2) - \cos\theta_1$

$g = (\sin\theta_1-\sin\theta_2)/(\theta_1-\theta_2) - \cos\theta_2$

For the three-dimensional infinite element, ϕ is assumed as

$$\phi = \sum_{m=0}^{M}\sum_{n=0}^{N} P_n^m(\cos\theta) h_n^{(2)}(kr)(\alpha_{mn}\cos m\chi + \alpha'_{mn}\sin m\chi)\;(m \leq n) \tag{4.6}$$

If only the first term of (4.6) is used, the normal derivative $\partial\phi/\partial\nu$ is zero over all lateral faces and there is no need to calculate the improper integrals in (2.8) to evaluate the element stiffness matrix.

5. NUMERICAL EXAMPLES

The first is the tunnel with a circular cross section of radius a (=2) bored in a medium which is subjected to a unit horizontal in situ stress. The finite element mesh is shown in Fig. 3, where the region is discretized by conventional linear elements and the present infinite elements with N=1 (4-β).

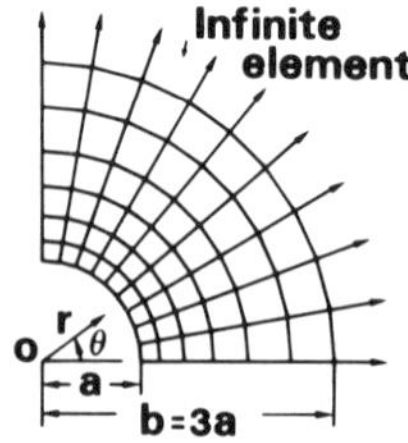

FIG. 3. Tunnel.

FIG. 4. Radial disp.

TABLE 1

Maximum Radial Displacements

Mesh (b/a)	3	4	6	8	10	Theory
Max disp.	3.57	3.56	3.55	3.55	3.55	3.64

In a plane strain condition with E=1 and ν=0.3, the calculated radial displacements at the periphery are plotted in Fig. 4 along with the exact solutions. The maximum radial displacements are also listed in Table 1 for several meshes which are generated by successively attaching the outer layers of interior elements to the mesh shown in Fig. 3.

Next, the axisymmetric problem of a circular distributed load (p=1) acting on an elastic half space is solved using the linear model in Fig. 5. The outermost contour of the interior region is varied from b=5 to b=1 by removing the outer layers of interior elements from the mesh shown in Fig. 5, where b denotes the distance between the origin and the outer boundary of

TABLE 2

Surface Deflections

r	0	1	2	3	5
Theory	1.9800	1.2605	0.5121	0.3348	0.1990
Mesh 1 (b=5)	1.9668	1.1844	0.5080	0.3399	0.2081
Mesh 2 (b=4)	1.9703	1.1890	0.5151	0.3500	
Mesh 3 (b=3)	1.9763	1.1991	0.5361	0.3726	
Mesh 4 (b=2)	1.9914	1.2439	0.6553		
Mesh 5 (b=1)	1.9217	2.7792			

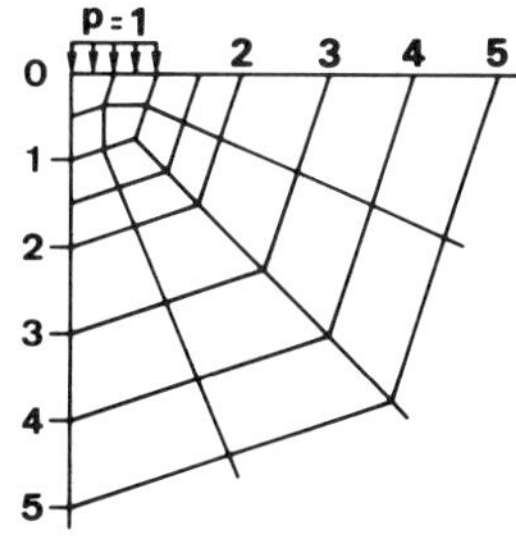

FIG. 5. Distributed load.

the interior region. The material properties are characterized by E=1 and ν=0.1. The computed surface deflections are listed in Table 2. It is seen that mesh 5, in which the infinite element nodes are located only a unit distance from the origin, still gives the satisfactory solution.

The Boussinesq's problem is solved using the three-dimensional linear models in Fig. 6. The interface between the interior region and the exterior one is placed on the surface of the cube of side length b. The size of the interior region is varied from b=1 to b=4. The computed surface deflection corresponding to E=1 and ν=0.1 is listed in Table 3.

TABLE 3

Surface Deflections

r	0	1	2	3	4
Theory	∞	0.3151	0.1575	0.1050	0.07878
Mesh 1 (b=4)	3.2022	0.3209	0.1559	0.1045	0.07793
Mesh 2 (b=3)	3.2026	0.3213	0.1559	0.1033	
Mesh 3 (b=2)	3.2039	0.3223	0.1537		
Mesh 4 (b=1)	3.2242	0.3412			

The final example is the acoustic radiation from a rigid square piston of side length 2a (=2) mounted in a rigid infinite baffle. The boundary of the interior region is placed on the surface of the cube of side b. Two different quadratic finite element meshes are used to model one quadrant of the domain: coarse (b=4/3) and fine (b=6/3). The coarse mesh is shown in Fig. 7. The percentage error of the computed radiation resistance is shown in Fig. 8, in which k and S represent the wave number and the surface area of the piston, respectively.

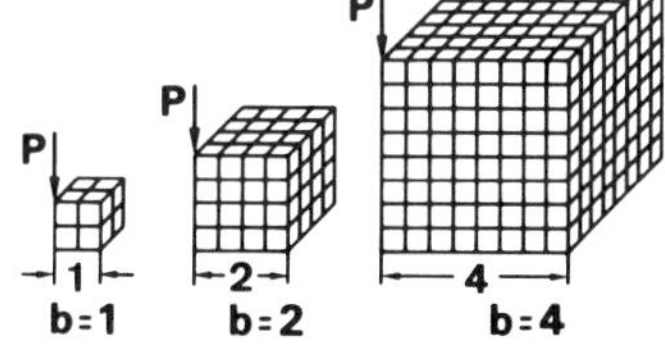

FIG. 6. Boussinesq's problem.

In every example problem, it is observed that the finite element solutions obtained using relatively small interior meshes agree well with the corresponding theoretical solutions.

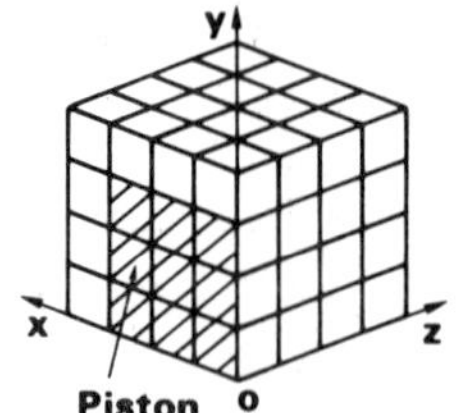

FIG. 7. Acoustic radiation.

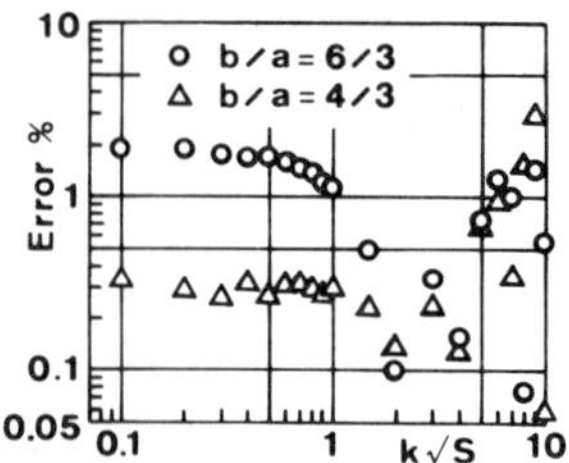

FIG. 8. Percentage errors. ($N\beta=1$)

6. CONCLUSIONS

Simple and efficient infinite elements have been developed based on the assumed displacement hybrid model for 2-D, 3-D and axisymmetrical infinite domain problems in elastostatics. The application to steady state wave problems is also presented. The present infinite elements are problem oriented and require the general solution or at least an asymptotic solution of the governing equation. The element has the following features: 1) Since the element nodes are common to those of interior elements, the number of the total degrees of freedom and the bandwidth of the global stiffness matrix remain the same as those of the bounded problem of the same interior structure. 2) All the integrations for the direction extending to infinity can be done exactly. When the total number of internal parameters β is chosen to be one, the element stiffness matrix can be generated without the calculation of improper integrals nor the matrix inversion and it can be expressed explicitly.

REFERENCES

1. MEI, C.C. and CHEN, H.S., A Hybrid Element Method for Steady Linearized Free-Surface Flows. *Int. J. Num. Meth. Eng.* 10, 1153-1175 (1976).
2. BETTES, P., Infinite Elements. *Int. J. Num. Meth. Eng.* 11, 53-64 (1977).
3. LYNN, P.P. and HADID, H.A., Infinite Element with $1/r^n$ Type Decay. *Int. J. Num. Meth. Eng.* 17, 347-355 (1981).
4. MORIYA, K., Two Node Semi-Infinite Element for the Analysis of Exterior Field Problems. pp.181-192, *Theoretical and Applied Mechanics* 30. Univ. of Tokyo Press, Tokyo (1980).
5. MORIYA, K., Two Node Hybrid Infinite Element for the Analysis of Steady-State Wave Problems. pp.49-61, *Theoretical and Applied Mechanics* 31. Univ. of Tokyo Press, Tokyo (1981).
6. BEER, G. and MEEK, J.L., Infinite Domain Elements. *Int. J. Num. Meth. Eng.* 17, 43-52 (1983).
7. TONG, P., PIAN, T.H.H. and LASRY, S.J., A Hybrid-Element Approach to Crack Problems in Plane Elasticity. *Int. J. Num. Meth. Eng.* 7, 297-308 (1973).

COMBINING ISOPARAMETRIC AND HIERARCHICAL ELEMENT PROCEDURES

J.E. Akin, R.J. Kipp

Department of Mechanical Engineering and Material Science, Rice University, Houston, Texas, U.S.A.

1. Introduction

The use of hierarchical elements has been made popular by Szabo [5], Babuska [3], Zienkiewicz [2], and others. Progress has been made in the use of hierarchical formulations for self-adaptive solutions. The architecture of these programs is significantly different from the conventional finite element systems. This paper reports on initial studies into how hierarchical elements can best be used to enhance existing non-adaptive finite element programs. A set of model equations is utilized to evaluate methods for implementing the hierarchical error indicators. This allows the use of the large investment in existing finite element programs and provides additional output to assist the analysts. Since existing programs utilize isoparametric elements the model study uses hierarchical extensions of isoparametric elements rather than the complete hierarchical family. The exact solution is compared with the isoparametric finite element solution and the hierarchical extension of the isoparametric formulation.

Of great concern to the finite element analyst is the accuracy of the computed result. It can be reduced by, among other factors, the discretization error; i.e., the error caused by attempting to model a continuous system by a system made up of discrete elements. Obviously, increasing the order of the (polynomial) interpolation functions and/or the density of the mesh will increase the number of degrees of freedom and tend to reduce this error, but will typically require complete reassembly and recalculation of the solutions. The resulting additional computational effort (and cost) will be significant and not necessarily warranted. It would be of great benefit to the analyst to have an indication of the improvement in accuracy

THE MATHEMATICS OF FINITE ELEMENTS AND APPLICATIONS V

ISBN 0-12-747255-X

which would result by the addition of a specified number of degrees of freedom without having to expend the additional computational effort as described above. The informed analyst could then decide if the addition of more degrees of freedom and recalculation of the solution was warranted and, if so, could confidently specify the number and location of additional degrees of freedom.

It bears mentioning that "self-adaptive" finite element codes are being developed which automatically evaluate a measure of error and increase the number of degrees of freedom as needed, making use of hierarchical finite element procedures. These codes will certainly be useful, but the usefulness of the large number of non-self-adaptive codes, both at present and in the future, would be greatly enhanced if similar hierarchical finite element procedures were adapted to them, giving them the capability to calculate and display error indicators and/or estimates to the user. With the aforementioned in mind, it is the goal of this work to develop and demonstrate hierarchical finite element procedures which can be adapted to existing isoparametric procedures and enhance their usefulness.

In reviewing the typical literature on hierarchical finite element approaches, eg [2]-[10], one may get some impressions that are not always true. For example, many people believe that these procedures always lead to an improved condition number for the matrix system to be solved. That is not true. Also the example model problems are often simplified and not typical of common engineering problems. Therefore when searching for a typical model problem our goal was to select one that would be somewhat more realistic.

The work discussed in this investigation is carried out in one dimension. It is demonstrated that a lower-order isoparametric formulation with a higher-order hierarchical term added results in a solution equivalent to the complete higher-order solution. A technique is demonstrated by which a correct higher-order solution can be iteratively calculated using the initial lower-order stiffness matrix and without assembly of the higher-order stiffness terms. Also shown is a method for calculating a useful element-level error indicator.

The one-dimensional model problem used is of a form which yields non-zero off-diagonal hierarchical stiffness terms; a situation typical in cases of higher dimension. Consequently, this work provides a basis for extension of the concepts into higher dimensions, where the benefit of the hierarchical enhancement will be most significant.

2. Model Equation

A second order ordinary differential equation is selected as the model equation:

$$A \frac{d^2u}{dx^2} + Bu = C \quad , \quad x\varepsilon[0,1]. \tag{2.1}$$

Depending on the values of the boundary conditions and constant coefficients (A,B, & C) this equation can illustrate significantly different behavior. For example, if A approaches zero so A<<B one has a classical singular perturbation problem that has a sharp boundary layer solution near one end of the solution domain. But for other cases like A>B one can have a typical thermal problem of heat conduction and convection along a bar [1]. Then the solution tends to be relatively smooth with no boundary layer effects.

In the first case we select

$$\begin{aligned} A &= -e^2 = -0.01 \\ B &= C = 1 \\ u'(0) &= u(1) = 0 \end{aligned} \tag{2.2}$$

which yields a sharp corner in the solution near x = 1. The exact solution is

$$u(x) = 1 - (\text{Cosh } (x/e))/(\text{Cosh } (1/e)). \tag{2.3}$$

In the second case we associate A and B with the conductive and convective heat transfer contributions, respectively, in a bar. To compare with previous numerical results [1], we select

$$\begin{aligned} C &= 0 \\ m^2 &= B/A = 0.8 \\ u(0) &= 30 \\ u'(1) &= 0 . \end{aligned} \tag{2.4}$$

The corresponding exact solution is

$$u(x) = (\text{Cosh}(m(1-x)))/\text{Cosh}(m) . \tag{2.5}$$

The standard isoparametric solutions to these problems are well known [10],[6].

Assume that in a typical element the approximation to u(x) is given by an interpolation

$$u = v = \underline{n}^T\underline{a} \tag{2.6}$$

where a_j are coefficients (degrees of freedom) to be determined and the n_j are the corresponding interpolation functions. Then we know that the Galerkin weak form approximation yields

$$\underline{K}\ \underline{a} = \underline{f} \tag{2.7}$$

where the typical element contributions are

$$k^e_{ij} = \int_{L^e} (B^e n_i n_j - A^e n'_i n'_j)\, dx \qquad (2.8)$$

$$f^e_i = \int_{L^e} C^e n_i \, dx \quad . \qquad (2.9)$$

In the above relation the a_j may represent the approximate nodal values of u or they may represent derivatives of u.

3. Hierarchical Partitions

If one selects the common linear interpolation function so that there are two degrees of freedom per element the element matrices are

$$\underline{K}^e = \begin{bmatrix} (Bh/3 - A/h) & (Bh/6 + A/h) \\ (Bh/6 + A/h) & (Bh/3 - A/h) \end{bmatrix} \qquad (3.1)$$

and $$\underline{f}^{eT} = [Ch/2 \quad Ch/2] \qquad (3.2)$$

where h denotes the element length. If we were to increase the isoparametric interpolation from linear to quadratic then K^e becomes 3x3 and contains coefficients totally different from (3.1). Consequently, in an isoparametric element calculation, increasing the order of the interpolation functions would result in a completely different global system of simultaneous equations to be solved, requiring the relatively large amount of computer time for complete re-solution of the simultaneous equations. Consider now a hierarchical quadratic interpolation function

$$n_H(r) = n_3(r) = (r^2 - r)/2 \qquad (3.3)$$

to be used with the second derivative at the midpoint and where $0 \leq r \leq 1$ is the local element coordinate. Adding this to the previous linear approximations gives element matrices that can be partitioned into a form

$$\left[\begin{array}{c|c} K_{oo} & K_{oH} \\ \hline K^T_{oH} & K_{HH} \end{array}\right]^e \quad \text{and} \quad \left\{\begin{array}{c} f_o \\ \hline f_H \end{array}\right\}^e \qquad (3.4)$$

where $\underline{K}_{oo}$ and $\underline{f}_o$ are the original lower order terms in (3.1) and (3.2) and where the new hierarchical contributions are

$$\underline{K}^{eT}_{oH} = [-1\ -1]Bh^3/24$$

$$\underline{K}^e_{HH} = Bh^5/120 - Ah^3/12,\quad \underline{f}^e_H = -ch^3/12\ . \tag{3.5}$$

As a result of the hierarchical functions and this partition, the vector $\underline{a}_o$ continues to represent the approximated value of the function u at the original nodal points. The new hierarchical vector $\underline{a}_H$ represents approximated values of the derivative(s) of the function u. For more commonly selected model problems the coupling, $\underline{K}_{oH}$, between the original and hierarchical terms is identically zero. This can improve the condition number of the resulting system equations. However, for higher dimensional spaces the coupling terms are almost always non-zero, as is the case in our model equation.

4. Matrix Conditioning

For certain problems, a hierarchical formulation may lead to improved elemental and global stiffness matrix conditioning and consequently improved efficiency of the solution of the global system. This, however, is not the general case.

Consider the model problem (2.1). For consistency with the work of Peano et al. [6], the local coordinate system $-1 \leq s \leq 1$ shall be used. The integral equation for the stiffness matrix is given by (2.8). Using linear shape functions $n_1(s)=(1-s)/2$, $n_2(s)=(1+s)/2$ and a quadratic hierarchical function $n_H(s)=(s^2-1)/2$, associated with d^2u/ds^2, the hierarchical form of (2.8) becomes

$$\underline{K}^e = \frac{A}{30}\begin{bmatrix} -15 & 15 & 0 \\ 15 & -15 & 0 \\ 0 & 0 & -20 \end{bmatrix} + \frac{B}{30}\begin{bmatrix} 20 & 10 & -10 \\ 10 & 20 & -10 \\ -10 & -10 & 8 \end{bmatrix}\ . \tag{4.1}$$

By way of comparison, if one utilizes the standard isoparametric quadratic shape function yields (2.8) in the form

$$\underline{K}^e = \frac{A}{30}\begin{bmatrix} -35 & 40 & -5 \\ 40 & -80 & 40 \\ -5 & 40 & -35 \end{bmatrix} + \frac{B}{30}\begin{bmatrix} 8 & 4 & -2 \\ 4 & 32 & 4 \\ -2 & 4 & 8 \end{bmatrix}\ . \tag{4.2}$$

If coefficients are selected in the model problem to be A = -1, B = 0 (used by Zienkiewicz, et al.[9]), the hierarchical formulation has a matrix condition number of 1.5 while the isoparametric form has a value of 4.0. Clearly the hierarchical formulation has better conditioning. However, for A = -.01, B = 1, as investigated here, the hierarchical formulation has a condition number of 29.7 while the isoparametric form has a condition number of 6.2. In this second case, the hierarchical formulation is more poorly conditioned.

It is worthwhile to note that as A becomes smaller, the function of u develops a sharper corner near x = 1 and becomes more difficult to approximate accurately. Correspondingly, the conditioning of the hierarchical formulation becomes poorer in relation to the complete quadratic isoparametric form. Actually, one can show that the consideration of hierarchical condition numbers depends on the choice of the local interpolation space (-1 to +1 vs 0 to 1), and whether the hierarchical degrees of freedom are local derivatives or global derivatives.

5. Architectural Considerations

As mentioned earlier the common numerically integrated isoparametric finite element codes have a software architecture that is significantly different from that utilized in self-adaptive programs. Since many isoparametric codes exist one should consider possible modifications to take advantage of the known advantages, such as error indicators, of hierarchical element enhancements.

The major cost of a finite element solution is that associated with the factorization of the stiffness matrix. The forward and backward substitution using the forcing vector is relatively cheap by comparison. Many codes have existing procedures for repeating the substitutions for several force vectors, such as a time integration. This feature is used here.

Basically, the hierarchical components of the element level stiffness matrix and force vector are used with the initial nonhierarchical displacement values to calculate an updated approximation of the force vector at the element level. The force vector is then reassembled and the system is resolved for new displacement values. The method is repeated until convergence is obtained.

To illustrate the concept, consider the system of equations

$$\begin{bmatrix} \underline{K}_{oo} & \underline{K}_{oH} \\ \underline{K}_{oH}^{T} & \underline{K}_{HH} \end{bmatrix} \begin{Bmatrix} \underline{a}_{o} \\ \underline{a}_{H} \end{Bmatrix} = \begin{Bmatrix} \underline{f}_{o} \\ \underline{f}_{H} \end{Bmatrix} . \qquad (5.1)$$

If the original d.o.f., $\underline{a}_o$ have already been computed at the system level by the conventional formulation

$$\underline{K}_{oo}\, \underline{a}_o = \underline{f}_o \qquad (5.2)$$

then we can write an approximation for $\underline{a}_H$ at the element level as

$$\underline{a}_H^e \cong \underline{K}_{HH}^{e^{-1}} (\underline{f}_H^e - \underline{K}_{oH}^{e^T} \underline{a}_o) . \tag{5.3}$$

This is an approximation because the $\underline{a}_o$ is a global (system) level value while the other components are calculated at the element level only. From (5.1)

$$\underline{a}_o \cong \underline{K}_{oo}^{e^{-1}} (\underline{f}_o^e - \underline{K}_{oH}^e \underline{a}_H^e) \tag{5.4}$$

can be written. At the element level, $\underline{K}_{oo}^e$ may be singular. However, $\underline{K}_{oo}$ will not be singular at the system level. The value in parenthesis in (5.4) can be calculated at the element level using the approximated value of $\underline{a}_H^e$ from (5.3). This can be written

$$\underline{f}_o^{e*} = \underline{f}_o^e - \underline{K}_{oH}^e \underline{a}_H^e . \tag{5.5}$$

Here $\underline{f}_o^{e*}$ can be thought of operationally as an updated force vector and can be assembled into the global force vector and a new, updated value for $\underline{a}_o$ can be obtained by forward and back substitution. This sequence of operations is repeated until convergence is obtained. Use of this method with the one dimensional model problems has shown convergence to the appropriate higher order solutions.

6. Error Indicator

An error indicator first proposed by Peano et al. [5] was evaluated for its usefulness in isoparametric routines enhanced by hierarchical elements. This indicator was selected because it is computationally convenient to solve and is not limited to any particular element type or dimension space. The indicator is based on evaluation of the gradient of system potential energy with respect to higher order degrees of freedom.

If the derivative of the energy is taken with respect to a hierarchical degree of freedom a_H with the value of a_H set to zero, the following is obtained:

$$\frac{\partial u}{\partial a_H} = \underline{K}_{oH} \underline{a}_o - f_H . \tag{6.1}$$

Thus, the gradient of the total potential energy with respect to hierarchical degrees of freedom can be determined from the present value of the lower order degrees of freedom. For use as an error indicator, the gradient should be normalized in energy by dividing by the square root of the pivot, $\underline{K}_{HH}$. The indicator thus becomes

$$I = (\underline{K}_{oH}\ \underline{a}_o - f_H)/K_{HH}^{1/2} \quad . \tag{6.2}$$

Peano states that this indicator can be used to determine which hierarchical degrees of freedom to include in the solution, and which should be left out.

For the selected model equations the actual error in the energy norm was computed for comparison by numerical integration. It is

$$\| e \|_E^2 = \int [A(u' - u_h')^2 + B\ (u - u_h)^2]\ dx \tag{6.3}$$

where u is the exact solution from (2.3) or (2.5), and u_h is the finite element approximation.

7. Sample Results

Before attempting the hierarchical enhancements of the isoparametric formulation two groups of problems were solved to provide benchmark results. One group used elements based on Lagrangian interpolation; while the second group utilized a hierarchical family of interpolation equations. Results were obtained for linear, through pentic polynomials. Relatively crude meshes were employed with up to four elements of equal size.

The iterative technique discussed previously was successfully applied to isoparametric solutions to yield hierarchical enhancements that were consistent with standard higher order solutions. It also provided a useful evaluation of the error indicator proposed by Peano. The following observations are true for all cases considered: (1) Upon convergence, the hierarchical interpolation enhancement of the isoparametric form yields displacement values ($\underline{a}_o$) equivalent to the complete higher-order solution of the same order. (2) The final values of the hierarchically calculated derivatives ($\underline{a}_H$) are equivalent to the same order derivatives at the corresponding points of the complete higher-order solution. (3) The actual error of the energy norm is reduced in going from the isoparametric solution to the higher order hierarchical formulation. (4) The initial error indicator provides a reasonable element by element indication of the relative amount by which the square of the energy error norm will be reduced by adding the next higher-order hierarchical enhancement. In addition, both the initial and final error indicator values appear to provide qualitative indications of where the variation in the problem function is the greatest; and therefore, which elements are likely to have the highest error. (5) No problems were experienced in obtaining convergence.

Typical numerical results are shown in Tables 1&2. Table 1 shows nodal values of u that are exact, Lagrangian cubic, Lagrangian quadratic with cubic hierarchical solved as (a)a full system and (b)approximate element iteration solutions, and finally, the unenhanced quadratic results. Table 2 shows data on the estimated hierarchical derivatives, the element error indicator, and the actual element error. Similar results were obtained for larger numbers of elements. Even when the number of hierarchical degrees of freedom were increased to about 90% of the original d.o.f. the element iteration still gave accurate results for the added d.o.f..

TABLE 1

Nodal Comparisons for $A=-1/100$, $B=C=1$, $u'(0)=0$, $u(1)=0$

6x	u, exact	u_h:(1)	u_h:(2)	u_h:(3)	u_h:(4)	u_h:(5)
0	0.99991	0.99992	0.99992	0.99992	0.99981	0.99971
1	0.99975	0.99978	0.99978	0.99978	0.99977	0.99947
2	0.99873	0.99881	0.99881	0.99881	0.99886	0.99721
3	0.99326	0.99389	0.99389	0.99389	0.99391	0.99051
4	0.96432	0.96550	0.96550	0.96550	0.96555	0.94728
5	0.81113	0.82305	0.82305	0.82304	0.82306	0.81994
6	0	0	0	0	0	0

(1)Cubic Lagrangian, (2)Quadratic Lagrangian plus cubic hierarchical, (3)Eight element iterations of (2) with local hierarchical derivative d.o.f., (4)Eight element iterations of (2) with global hierarchical derivative d.o.f., (5)Quadratic Langrangian before element iteration.

TABLE 2

Element Comparisons of Cubic Hierarchical Enhancement of u_h:(5)

El No.	Before Iteration u_h'''	Before Iteration I	Before Iteration $\|e\|^2$	After Eight Iterations u_h'''	After Eight Iterations I	After Eight Iterations $\|e\|^2$	Cubic u_h'''
1	0.07	3.46D-10	2.57D-7	-0.26	5.01D-9	1.00D-9	-.026
2	-11.85	1.01D-5	3.98D-4	-7.90	4.51D-6	3.97D-7	-7.90
3	-225.40	3.67D-3	3.89D-3	-229.05	3.79D-3	2.22D-4	-229.06

Notes: D-n = 10^{-n} ; ( )' = d ( )/dx

8. Conclusions

Enhancement by an element iteration procedure was accurate, quick to converge, and economical. The use of the error indicator at an element level seems useful. The element iteration can be sensitive to the element matrix condition number.

References

1. AKIN, J.E., *Application and Implementation of Finite Element Methods*, Academic Press, London, (1982).
2. BABUSKA, I. and W.C. RHEINBOLDT, On Reliability and Optimality of the Finite Element Method, *Computers and Structures*, vol. 10, p. 87-94, (1979).
3. BABUSKA, I, B.A. SZABO, I.N. KATZ, The P-version of the Finite Element Method, *SIAM J. Num. Anal.*, 18, p. 515-546, (1981).
4. KELLY, D.W., J. GAGO, O.C. ZIENKIEWICZ, I. BABUSKA, A-posteriori Error Analysis and Adaptive Processes in the Finite Element Method - Parts I and II, *IJNME*, vol. 19, no. 11, (1983).
5. PEANO, A., M. FANELLI, R. RICCIONI, L. SARDELLA, Self-Adaptive Convergence at the Crack Tip of a Dam Buttress, *Proceedings First Int. Conf. Numerical Methods in Fracture Mechanics*, Swansea U. K., p. 268-280, (1978).
6. PEANO, A., A. PASINI, R. RICCIONI, L. SARDELLA, Adaptive Approximations in Finite Element Structural Analysis, *Computers & Structures*, vol. 10, p. 333-342, (1979).
7. SZABO, B.A., Some Recent Developments in Finite Element Analysis, *Comp. and Maths. with Appls.*, vol. 5, p. 99-115, (1979).
8. ZIENKIEWICZ, O.C., *The Finite Element Method*, (Third Edition), McGraw-Hill, London, (1977).
9. ZIENKIEWICZ, O.C., J.P. De S. R. GAGO, D.W. KELLY, The Hierarchical Concept in Finite Element Analysis, *Computers & Structures*, vol. 16, no. 1-4, p. 53-65, (1983).
10. ZIENKIEWICZ, O.C., D.W. KELLY, J. GAGO, I. BABUSKA, Hierarchical Finite Element Approaches, Error Estimates and Adaptive Refinement, *The Mathematics of Finite Elements and Applications IV*, Academic Press, p. 313-346, (1982).

ON THE IMPROVEMENT OF THE NUMERICAL ACCURACY OF FEM-SOLUTIONS

N.-E. Wiberg, A. Samuelsson, and L. Bernspång
Department of Structural Mechanics,
Chalmers University of Technology, Göteburg, Sweden.

1. INTRODUCTION

In the finite element method applied to elastomechanics the state of deformations and stresses of each element is approximated by a combination of a finite number of functions. The element analysis gives equations that are constraints between chosen parameters at the element boundaries while interelement boundary conditions give rise to a set of equations in force, displacement, deformation, or mixed form. In all cases a large sparse equation system is obtained. This may, however, have quite different character with respect to numerical stability depending on which method is chosen.

The displacement method with a standard basis quite often gives an ill-conditioned set of equations. The numerical performance may be improved with another basis having relative displacements as variables. With the hierarchical p-version approach a simple displacement field over a basic mesh gives the main behaviour and some additional relative displacements improve the solution without destroying the condition number. A third way of using the displacement method is to establish non-singular element matrices, factorize them and then establish the triangular factors of the final system.

With force or mixed variables (diacoptics) the condition number mostly is considerably reduced.

Iterative solutions by such methods as conjugate-gradient and viscous relaxation can be implemented in a way that avoids the assemblage of the total matrix thus reducing the round-off errors. Calculations making comparisons are performed here on an IBM 3081.

2. DISPLACEMENT METHOD

The solution of the Navier differential equation for linear elasticity $Lu + f = 0$ defined on a region Ω with boundary conditions, by the standard finite element method is based on a finite element mesh, a variational, weak Galerkin formulation, and an

ISBN 0-12-747255-X

approximation of the field variable u. The field variable may be written $u = \phi \underline{\tilde{u}}$ where the row matrix $\underline{\phi}$ contains the basis for the shape functions and $\underline{\tilde{u}}$ the nodal values. The Galerkin procedure gives a matrix equation of the form

$$\underline{S}\,\underline{\tilde{u}} = \underline{f} \tag{2.1}$$

where $\underline{S}$ is a positive definite matrix and $\underline{f}$ a load vector.

It has been found, see [1], that the error $\delta\underline{\tilde{u}}$ in the solution vector $\underline{\tilde{u}}$ for a disturbed system

$$(\underline{S} + \delta\underline{S})(\underline{\tilde{u}} + \delta\underline{\tilde{u}}) = \underline{f} + \delta\underline{f} \tag{2.2}$$

is bounded according to

$$\frac{\|\delta\underline{\tilde{u}}\|}{\|\underline{\tilde{u}}\|} \leq \frac{\kappa(\underline{S})}{1 - \kappa(\underline{S})\,\|\delta\underline{S}\|/\|\underline{S}\|}\left(\frac{\|\delta\underline{S}\|}{\|\underline{S}\|} + \frac{\|\delta\underline{f}\|}{\|\underline{f}\|}\right) \tag{2.3}$$

where $\|\delta\underline{S}\| < \|\underline{S}\|$ and the spectral condition number $\kappa(\underline{S})$ is defined as $\kappa(\underline{S}) = \|\underline{S}\| \cdot \|\underline{S}^{-1}\| = \lambda_n/\lambda_1$ where $\|\cdot\|$ is the Euclidean norm for vectors and the spectral norm for matrices and λ_1 is the lowest and λ_n the highest eigenvalue for the matrix $\underline{S}$. Thus large condition numbers increase the numerical errors

A calculation of a cantilever with elements based on a bilinear approximation of u with a mesh 10×100 is shown in Fig. 1, [2]. The result in Fig. 1 shows that the displacement method solution in pure single precision is reliable up to a slenderness ratio of $\psi = L/h = 4$; then the errors rapidly increase due to increase in the condition number, especially if single precision accumulation is used. In Fig. 1 p_c is the calculated value and $p_b = PL^3/3EI$ is the cantilever displacement according to technical beam theory. Observe that the bilinear approximation for the given mesh gives very accurate solution with full double precision also for high ψ-ratios, except for approximation errors.

An iterative improvement $\underline{u}_2$ of the first solution $\underline{\tilde{u}} = \underline{\tilde{u}}_1$ is achieved by a double precision computation, [2], of the residual $\underline{r}_1 = \underline{f} - \underline{S}\underline{\tilde{u}}_1$ followed by a single precision solution

$$\underline{\tilde{u}}_2 = \underline{\tilde{u}}_1 + \Delta\underline{\tilde{u}}_1, \quad \underline{S}\Delta\underline{\tilde{u}}_1 = \underline{r}_1 \tag{2.4}$$

in which the already factorized matrix $\underline{S}$ is used. With this iteration reliable results are obtained for the slenderness ratio up to $\psi = 5$ for single precision accumulation and up to $\psi = 7$ for double precision accumulation at the factorization of S. A single refinement often gives the main effect.

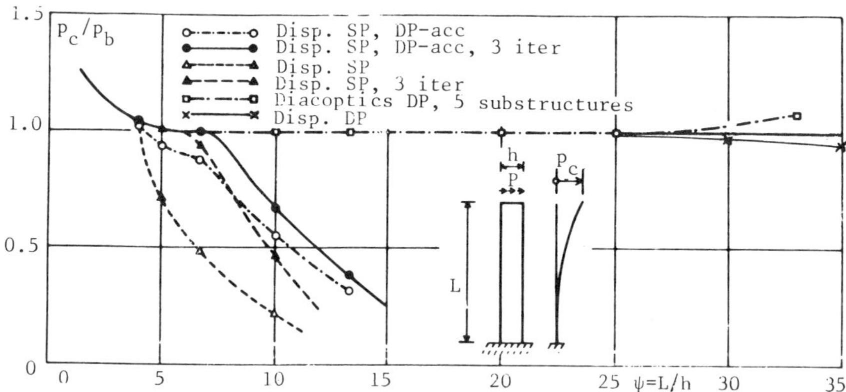

FIG. 1. Cantilever top displacements

3. CHANGE OF BASIS ON ELEMENT LEVEL

3.1 Relative displacements, deformation method

By introduction of a new set of variables $\tilde{\underline{u}}^r$ by $\tilde{\underline{u}} = \underline{H}\tilde{\underline{u}}^r$, where $\underline{H}$ is a non-singular matrix, a new set of equations is established according to

$$\underline{H}^t \underline{S}\,\underline{H}\,\tilde{\underline{u}}^r = \underline{H}^t \underline{f} \tag{3.1}$$

If the variables $\tilde{\underline{u}}^r$ are relative displacements (deformations) the equations (3.1) can physically be understood as equilibrium equations for substructures. The stiffness matrix $\underline{H}^t\underline{S}\,\underline{H}$ then generally gets much better condition number than $\underline{S}$. A very simple example is the numerical improvement obtained for a tall building with relative sway $\tilde{\underline{u}}^r$ as variables instead of absolute sway $\tilde{\underline{u}}$.

3.2 Hierarchical p-formulation

With a hierarchical p-formulation the field of displacements is approximated as $u = \underline{\phi}_b\tilde{u}_b + \underline{\phi}_h\tilde{u}_h$ where the row matrix $\underline{\phi}_b$ contains the basis for primary shape functions defined for a finite element mesh and $\underline{\phi}_h$ a hierarchical basis of additional shape functions of higher order defined on the same mesh. These are mostly chosen as Legendre polynomials or trigonometric functions. The Galerkin procedure gives a matrix equation of the form

$$\underline{S}\,\underline{u} = \underline{f}\ , \quad \begin{bmatrix} \underline{S}_{bb} & \underline{S}_{bh} \\ \underline{S}_{hb} & \underline{S}_{hh} \end{bmatrix} \begin{bmatrix} \tilde{\underline{u}}_b \\ \tilde{\underline{u}}_h \end{bmatrix} = \begin{bmatrix} \underline{f}_b \\ \underline{f}_h \end{bmatrix} \tag{3.2}$$

Here $\underline{\tilde{u}}_h$ represents relative displacements.

It has been shown, [3], that the hierarchical formulation gives a much better conditioned system than the conventional one, especially if the functions are normalized in energy norm. In many common cases the condition number is bounded for $p \to \infty$, see [3]. Calculated condition numbers for a cantilever structure for some standard and hierarchical functions are shown in Fig. 2.

The eight node serendipity element and the 4-node bilinear element + one hierarchical polynomial for each side give the same approximation. We observe that the numerical accuracy of the system is mainly dependent on the condition number of the basic system. The method with hierarchical functions can be regarded as a systematic way of choosing a basis for the displacement field that gives a stable procedure. Besides, (3.2) is well suited for a block iterative solution as the first equation with $\underline{\tilde{u}}_h$ equal to zero gives a good first approximation.

	4-node bilinear	8-node serendipity	4-node bilinear + 1 hierarchical polynomial for each side	4-node bilinear + 1 hierarchical sine-function for each side
κ	$33.6 \cdot 10^3$	$313.7 \cdot 10^3$	$58.8 \cdot 10^3$	$56.7 \cdot 10^3$

FIG. 2. Condition number for cantilever structure L/b = 5.0, mesh 3×15

4. CHANGE OF BASIS ON SYSTEM LEVEL

4.1 Displacement method, preconditioning

A change of basis directly on a displacement formulation is called preconditioning. It is used mainly for iterative methods, see below.

4.2 Force method, Diacoptics

The element forces N can be written as $\underline{N} = \underline{N}_p + \underline{Z}\underline{R}$ where $\underline{N}_p$ is a particular solution, $\underline{R}$ statically indeterminant forces, and $\underline{Z}$ a transformation matrix. If $\underline{F}$ is a flexibility matrix, and if a suitable basis for $\underline{R}$ is constructed a very accurate solution can be obtained by a force solution.

$$\underline{Z}^t\underline{F}\underline{Z}\underline{R} = -\underline{Z}^t\underline{F}\underline{N}_p = \underline{r} \qquad (4.1)$$

If we split the structure in parts X_0 and X_1 it is possible to combine the force- (in X_0) and the displacement- (in X_1) methods into a mixed method called diacoptics [2], Fig. 3a. The matrix Z in (4.1) obtains then a block form shown in Fig. 3c.

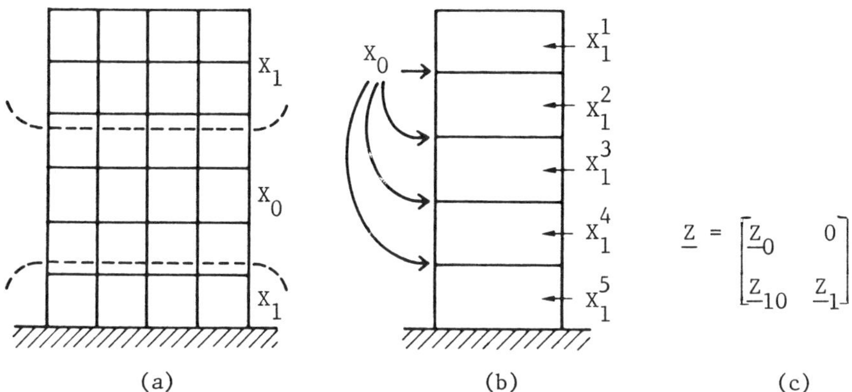

FIG. 3. Structure with arbitrary dissection

By a study of the governing equations a final system of equations will be found as

$$\begin{bmatrix} \underline{F}_{00} & -\underline{A}_{10}^t \\ \underline{A}_{10} & \underline{S}_{11} \end{bmatrix} \begin{bmatrix} \underline{R} \\ \underline{u}^r \end{bmatrix} = \begin{bmatrix} \underline{r} \\ \underline{f} \end{bmatrix}, \quad \underline{F}_{00} = \underline{Z}_0^t \underline{F} \underline{Z}_0 \qquad (4.2)$$

where $\underline{u}^r$ are relative displacements in a substructure. In special applications the substructure X_0 does not contain any structural elements but only coupling forces, as in Fig. 3b. This means that $\underline{F}_{00} = \underline{0}$.

In the example, Fig. 1, the cantilever is subdivided into five equal substructures, see Fig. 3b, each having a mesh of 10×20. From the results shown in Fig. 1 it is found that diacoptics gives a good result in single precision calculations even for fairly large ψ-values. For ψ-values larger than 25 the result begins to be inaccurate, which depends on the fact that the displacement method solution begins to be incorrect for each substructure ($\psi_{sub} = 5$). A greater number of substructures will enable ψ-values much greater than 25 to be treated.

5. SOLUTION TECHNIQUES

5.1 Direct method, element factorization

The factorization of the assembled stiffness matrix $\underline{S} = \underline{U}^t\underline{U}$ in the standard displacement method may give large numerical errors. It has been shown that a direct factorization of the differential stiffness matrix $\underline{S}_d = \underline{U}_d^t\underline{U}_d$ (or of the non-singular element stiffness matrix $\underline{S}_n = \underline{U}_n^t\underline{U}_n$) reduces the numerical errors [4]:

$$\underline{S} = \underline{U}^t\underline{U} = \underline{B}^t\underline{U}_d^t\underline{U}_d\underline{B} \tag{5.1}$$

The triangular factor U may be obtained by a QR-transformation of the rectangular matrix $\underline{U}_d\underline{B}$. The numerical error of this procedure is reduced by the factor $1/\sqrt{\kappa(\underline{S})}$ compared to standard displacement method. This method applied to the example in Fig. 1 shows a stable performance in single precision up to about $\psi = 100$ and is almost as good as ordinary double precision. If the cantilever is given a large stiffness at the top the condition number grows and so do the numerical errors. Fig. 4a shows the effect on the stability of different methods if the stiffness at the top is increased by a factor 10^m.

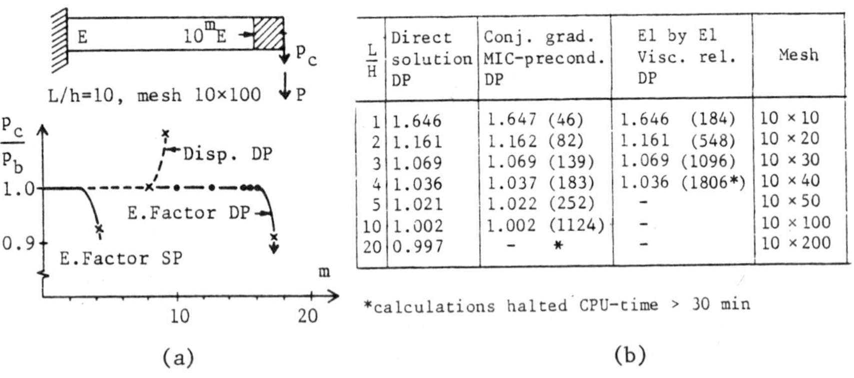

L/H	Direct solution DP	Conj. grad. MIC-precond. DP	El by El Visc. rel. DP	Mesh
1	1.646	1.647 (46)	1.646 (184)	10 × 10
2	1.161	1.162 (82)	1.161 (548)	10 × 20
3	1.069	1.069 (139)	1.069 (1096)	10 × 30
4	1.036	1.037 (183)	1.036 (1806*)	10 × 40
5	1.021	1.022 (252)	-	10 × 50
10	1.002	1.002 (1124)	-	10 × 100
20	0.997	- *	-	10 × 200

*calculations halted CPU-time > 30 min

(b)

FIG. 4. Cantilever calculations, (a) p_c/p_b for stiffened cantilever, Element factorization, (b) p_c/p_b for uniform beam, Fig. 1. Iterative methods with different number of square elements. Number of iterations within parenthesis

5.2 Iterative methods; conjugate gradient with preconditioning, viscous relaxation with splitting

Solution of $\underline{S}\underline{\tilde{u}} = \underline{f}$ iteratively by the preconditioned conjugate gradient method is made according to

$$\underline{r}^{(i)} = \underline{S}\underline{\tilde{u}}^{(i)} - \underline{f}; \qquad \underline{d}^{(i)} = -\underline{C}^{-1}\underline{r}^{(i)} + \beta_i\underline{d}^{(i-1)} \tag{5.2}$$
$$\underline{\tilde{u}}^{(i+1)} = \underline{\tilde{u}}^{(i)} + \lambda_i\underline{d}^{(i)}, \quad i = 0,1,2,\ldots$$

where β_i, λ_i are determined from minimization of corresponding error functionals and $\underline{C}$ is a well chosen preconditioning matrix.

The convergence depends on the condition number $\kappa = \kappa(\underline{C}^{-1}\underline{S})$ which is less than $\underline{\kappa}(\underline{S})$.

Many proposals for selection of an efficient C have been made. In [5] a modified incomplete Cholesky factorization (MIC) in a multigrid manner with kept row sums is applied and in [6] and [7] element-by-element factorization techniques are utilized. A calculation of our beam, Fig. 1, with the MIC-matrix was not very promising, see Fig. 4b. In [6] it is pointed out that the element-by-element preconditioning method is not very successful for structural elements, such as beams, plates and shells.

Viscous relaxation with line search - that is (5.2) with $\beta_i = 0$, see [8], [9], [10], [11], has been successfully applied to many problems, also in a split element-by-element version. For our beam problem, Fig. 1, this method was not efficient, see Fig. 4b. However, a combination of a p-version hierarchical basis and viscous relaxation with the linear approximation giving the preconditioning matrix seems to be useful, see [11] and [12].

We have observed that in all iterative procedures it is essential to calculate the residual with a minimum of rounding errors. A simple and efficient way is to calculate it from relative displacements, see [5].

7. SUMMARY AND CONCLUSIONS

Some devices for the improvement of the numerical accuracy have been discussed. Change of basis at element level, change of basis

Device for improvement		Cond. number compared to pure displ. method	Operation count	Storage requirement	Numerical Accuracy	Coding complexity
Change of basis element level	Relative displacements	Reduced κ(S)	Same	Same	Better	Higher
	Hierarchical functions	Reduced κ(S)	Same	Mostly reduced	Better	Higher
Change of basis system level	Diacoptics	Reduced κ(S)	Mostly reduced	Same	Better	Higher
Solution techniques	Conventional substructuring	Same	Mostly reduced	Mostly reduced	Same	Higher
	Element factorization	Error reduced by $1/\sqrt{\kappa(S)}$	Increased	Same	Better	Same
	Precond. CG, MIC	$\kappa = \kappa(C^{-1}S)$	Problem dependent	Reduced	Problem dependent	Lower
	Viscous relaxation, Splitting	?	Problem dependent	Reduced	Problem dependent	Lower

FIG. 5. On the improvement of numerical accuracy and efficiency for some methods compared to conventional displacement method with Gaussian elimination

at system level and different direct and iterative solution techniques have been compared with the conventional displacement method with Gaussian elimination. A general picture of the merits of the different methods is given in Fig. 5.

8. REFERENCES

1. WILKINSON, J.H., Rounding Errors in Algebraic Processes, H.M. Stationary Office, London (1963).
2. WIBERG, N.-E., Matrix Structural Analysis with Mixed Variables, *Int. J. Num. Meth. Engng.* 8, 167-194 (1974).
3. WIBERG, N.-E., MÖLLER, P., and SAMUELSSON, A., A Study of the Hierarchical p-version of the Finite Element Method for Static Problems. Int. Conf. on Accuracy Estimates and Adaptive Refinement in Finite Element Computations, Lisbon, June (1984).
4. WILHELMY, V., On the Element Stiffness Factor Formulation, *Comp. Meth. in Appl. Mech. and Engng.* 11, 75-95 (1977).
5. AXELSSON, O., GUSTAFSSON, I., A Preconditioned Conjugate Gradient Method for Finite Element Equations, which is Stable For Rounding Errors, *Information Processing* 80, 1980.
6. HUGHES, T., RAEFSKY, A., MULLER, A., WINGET, J., LEVIT, I., A progress Report on EBE Solution Procedures in Solid Mechanics. 2nd Intern. Conference on Numerical Methods for Nonlinear Problems, Barcelona (1984).
7. GUSTAFSSON, I., LINDSKOG, G., A Preconditioning technique based on element matrix factorization. Report 84.01R, Chalmers Univ. of Techn., Dep. Comp. Sci. (1984).
8. HUGHES, T., LEVIT, I., WINGET, J., Element-by-Element Solution Algorithm for Problems of Structural and Solid Mechanics, *Comp. Meth. Appl. Mech. and Engng.* 36, 241-254 (1983).
9. KÄRRHOLM, G., A Method of Iteration Applied to Beams Resting on Springs, Transactions of Chalmers Univ. of Tech., Publ. 199, Göteborg (1958).
10. MARCHUK, G.I., Numerical Methods in Weather Prediction, Academic Press, INC, New York, 88-99 (1974).
11. ZIENKIEWICZ, O.C., Iterative Methods and Hierarchical Approaches. A Prospect for the Future of the Finite Element Method. ASEA centenniel lecture (1984).
12. AXELSSON, O., GUSTAFSSON, I., Preconditioning and Two-Level Multigrid Methods of Arbitrary Degree of Approximation. *Math. of Comp.* 40, 219-242 (1983).

A GENERAL FINITE ELEMENT FRAMEWORK
FOR NODAL METHODS

J.P. HENNART

National University of Mexico, Mexico D.F., Mexico

1. INTRODUCTION

Modern coarse-mesh or nodal methods have been developed during the 1970's in numerical nuclear reactor calculation. (See for instance, the recent review paper by J. Dorning [1]). Broadly speaking, nodal methods are fast and accurate methods which combine many attractive features of the finite element method (f.e.m.) as well as of the finite difference method (f. d.m.). With the f.e.m., they have in common the fact that the unknown function is approximated by a piecewise continuous, usually polynomial function over a given coarse mesh. Since in the original versions the final equations are normally derived from physical considerations, the relationship with known finite element schemes was somewhat obscure until recently and the tendency in the past has been to consider them as quite distinct methods. With the f.d.m., they share the fact that the resulting algebraic systems are in principle quite sparse and well structured : this is true of course for meshes not too irregular, like unions of rectangles, for which standard, i.e. conforming, finite elements would lead to much more coupled systems of equations. Modern nodal schemes can thus be viewed as fast solvers, to which techniques of vectorization can easily be applied. They are perfectly well suited to all physical situations which have been traditionally solved by finite difference techniques, over domains not too irregular, for which homogenized piecewise constant properties are available. This is typically the case of numerical reactor calculation at the subassembly level, after some homogenization procedure has been carried out. The subsequent coarse mesh calculation result must normally go through some dehomogenization procedure if fine structure effects are to be seen. There are many other situations of particular interest for which pre- and postprocessing are not needed, but only a coarse mesh calculation : let us mention for instance groundwater hydrology, oil reservoir simulation, air pollution modelling problems, etc. : here the physical parameters such as porosity, permeability, diffusivity, etc. are usually not known with great accuracy and over each node, a constant or mean

ISBN 0-12-747255-X

value only is available.

In Section 2, we recall some of the simplest original nodal schemes and explain why they are not quite satisfactory, in the sense that they do not climb in order correctly. Section 3 introduces a more general family of nodal schemes, which do not exhibit that defect, as shown by some simple numerical examples. In Section 4, practical implementation aspects are discussed for different basic situations, as well as relationships with some other discretization techniques.

2. RECALLING SOME EARLY NODAL SCHEMES

Let us consider the typical second order elliptic equation

$$Lu \equiv -\operatorname{div}(p \operatorname{grad} u) + qu = f, \quad \text{on } \Omega \subset \mathbb{R}^n \qquad (2.1)$$

subject for instance to homogeneous Dirichlet conditions on the boundary Γ of Ω. As a rule, p and q are assumed to be piecewise constant, and in particular constant over each "cell", "block", "element" or "node" of some discretization Ω_h of Ω. For the sake of simplicity, we shall in the following restrict ourselves to two-dimensional situations (n=2). The nodes will be supposed rectangular and always referred to the unit square $C \equiv [-1, +1] \times [-1, +1]$ by an affine mapping. Let us introduce the following elementary spaces of polynomials.

$$\mathcal{Q}_{k,\ell} \equiv \left\{ x^a\, y^b, \quad a \le k, \; b \le \ell \right\} \qquad (2.2a)$$

and

$$\mathcal{P}_k \equiv \left\{ x^a\, y^b, \quad a+b \le k \right\} \qquad (2.2b)$$

as well as the linear forms associated with the cell

$$m_C^{ij} = \int_C P_i(x)\, P_j(y)\, u(x,y)\, dxdy \,/\, N_i \cdot N_j, \qquad (2.3a)$$

and with its faces, as for instance

$$m_L^i = \int_{-1}^{+1} P_i(y)\, u(-1,y)\, dy / N_i \quad , \qquad (2.3b)$$

where the subindex L stands for "left". Similarly, m_R^i, m_B^i and m_T^i can be defined with appropriate modifications, R, B and T meaning "right", "bottom" and "top" respectively. In (2.3), $N_i = 2/(2i+1)$ is a convenient normalization factor while P_i is the normalized Legendre polynomial over $[-1, +1]$.

With this notation, it is now easy to describe (in a slightly modified way) the so-called sum or Σ nodal schemes introduced in [2,3]. They consist of looking for finite dimensional approximations $u_h \in V_h$ of u, the restrictions of which to C for a nodal scheme of type (k), k=0,1,.., are finite dimensional spaces of polynomials S_k with degrees of freedom belonging to sets D_k where

$$S_k = \mathcal{Q}_{k+2,0} \cup \mathcal{Q}_{0,k+2}, \quad \dim S_k = 2k+5 \tag{2.4a}$$

$$D_k = \left\{ m_L^0, m_R^0, m_B^0, m_T^0, m_C^{i0}, \ i = 0,..k, \ m_C^{0j}, \ j=1,...,k \right\}, \quad \text{card } D_k = 2k+5 \tag{2.4b}$$

It is easy to check that any polynomial in S_k is uniquely determined by its moments in D_k : in other words, D_k is S_k-*unisolvent*.

In [2,3], the final nodal equations are derived on the basis of physical arguments by expressing that (eventually weighted) balance equations are satisfied over the cell for all weights $w \in S_{k-2}$ (2k+1 equations) and that the mean values of the normal flux $j_h \cdot n = -p \text{grad}\, u_h \cdot n$ are continuous through the faces of the nodes (4 equations), the mean values of u_h being already continuous by construction.

With the f.e.m., a weak form of (2.1) is first considered after multiplication by a test function v and integration by parts, namely $u \in V$ is looked for such that

$$a(u,v) = f(v), \ \forall v \in V, \tag{2.5}$$

where $a(\cdot,\cdot)$ and $f(\cdot)$ are standard bilinear and linear forms, while V is some infinite dimensional space, in our case $H_1(\Omega)$. In standard, i.e. *conforming*, f.e. situations, a finite dimensional subspace V_h of V is considered and a discretized form of the above problem consists of finding $u_h \in V_h$ such that

$$a(u_h, v_h) = f(v_h) \qquad \forall v_h \in V_h. \tag{2.7}$$

If (2.4) is used to describe V_h, $V_h \not\subset V$ since only the mean value of u_h (and not u_h itself) is continuous through the faces of C. We are thus in a *nonconforming* situation : (2.7) can still be used but $a(\cdot,\cdot)$ has to be replaced by $a_h(\cdot,\cdot)$ where

$$a_h(u,v) = \sum_{C \in \Omega_h} \int_C (p \text{ grad } u \cdot \text{grad } v + quv)\, dx \tag{2.8}$$

The convergence of this approximate solution, obtained by the so-called m.n.m. (for mathematical nodal method), is guaranteed by the mean value continuity through the faces of the nodes, which ensures that the "patch test" [4] is passed at the lowest order.

In a recent paper [5], we proved that the p.n.m. (for physical nodal method) described before is in fact a m.n.m. where the matrix elements are not calculated exactly but only approximately by using some numerical quadrature of the Radau type in a nonstandard way. The equivalence of some of the early nodal schemes with a nonstandard nonconforming f.e.m. was thus shown for the first time, leading by the way to the first complete numerical analysis of the corresponding schemes. It can be shown in particular that in norm H_1 the error is of $O(h)$ *independently of* k, as confirmed by numerical results given in [5]. The reasons for this are that $\mathscr{P}_2 \not\subset S_k$, $\forall$ k <u>and</u> also that with only mean values on each face the patch test cannot be passed at higher orders.

3. A GENERAL FAMILY OF NODAL SCHEMES

A family of nodal schemes "climbing" correctly in order would ideally exhibit errors in norm H_1 of $O(h^{k+1})$ if the solution were smooth enough. This can be achieved if $\mathscr{P}_{k+1} \subset S_k$, $\forall$ k <u>and</u> with moments up to order k on the faces so that higher order patch tests can be passed. Such a general family of nodal schemes was developed in [6]: for a nodal scheme of type (k), V_h is described by

$$S_k = \mathscr{Q}_{k+2,k} \cup \mathscr{Q}_{k,k+2}, \quad \dim S_k = (k+1)(k+5) \tag{3.1a}$$

$$D_k = \Big\{ m_L^i,\ m_R^i,\ m_B^i,\ m_T^i,\ i = 0,\ldots,k\ ; \quad m_c^{ij},\ i,j=0,\ldots,k \Big\}, \quad \text{card } D_k = (k+1)(k+5) \tag{3.1b}$$

In [6], it is shown that D_k is S_k-*unisolvent*. Moreover, the basis functions associated with the boundary and cell parameters are given in particularly compact formulae using the Legendre polynomials. For instance, associated to m_L^i we have

$$\mu_L^i(x,y) \propto (P_{k+1}(x) - P_{k+2}(x)) \cdot P_i(y), \tag{3.2}$$

which is equal to zero at the (k+2) right Radau abscissae.

Some sample calculations have been run to check the theoretical convergence orders in the case of smooth analytical solutions like $(1-x^2)(1-y^2)$ (Problem P1) and $(1-x^4)(1-y^4)$ (Problem P2) over C. The corresponding discrete L_2 errors are reported in Table 1 for different meshes and are clearly of order k+2, consistent with order k+1 in norm H_1.

TABLE 1
Discrete L_2 Errors for Sample Calculations

Problem & Nodal Scheme Type :	P1 k=0	P1 k=1	P1 k=2	P2 k=2
Mesh 2×2	5.33(-1)	4.55(-2)	0.	1.01(-1)
Mesh 4×4	1.83(-1)	4.81(-3)	0.	4.58(-3)
Mesh 6×6	-	-	0.	8.92(-4)
Mesh 8×8	5.34(-2)	5.20(-4)	0.	-

These results have been produced by the m.n.m. In [6], where by the way many more details can be found, it is also proved that the corresponding p.n.m. can be obtained from the m.n.m. by numerical quadratures of the Radau type.

4. PRACTICAL IMPLEMENTATION PROBLEMS

If we denote by H,V and C the sets of horizontal parameters (m_L^i, m_R^i), vertical parameters (m_B^i, m_T^i) and cell parameters (m_C^{ij}) respectively, it is easy to verify that the elementary mass and stiffness matrices couple the sets H and V only through C. The use of particular quadrature formulae can make this decoupling even stronger. Anyhow, these properties directly lead to alternating direction implementations, which is not the case with the standard f.e.m. (see however [7]).

Dimensionally reduced nodal methods can be obtained in a more direct way, by using *transverse integration*. Assuming that V_h restricted to a given node C is described by (3.1), we consider successively the k+1 normalized moments of Eq. (2.1) over each dimension (x or y) within C. The result is a set of 2(k+1) ordinary differential equations in the corresponding transverse moments : the (k+1) moments $u_y^\ell(x)$, $\ell=0,\ldots,k$ with respect to y for instance satisfy the following 1D equations

$$-p \frac{d^2u_y^\ell(x)}{dx^2} + q\, u_y^\ell(x) = \hat{f}_y^\ell(x), \quad \ell=0,\ldots,k, \tag{4.1}$$

where only the second member depends on ℓ and is an effective source term including the ℓ'th moment of f with respect to y, as well as a *transverse leakage term*, arising from the cell boundary values of the y moment of the differential operator with respect to y. There is also a compatibility condition, namely,

$$\left[u_y^\ell(x)\right]_x^k = \left[u_x^k(y)\right]_y^\ell \quad , \forall\, C \in \Omega_h, \tag{4.2}$$

which is automatically satisfied as a unique representation of u is provided over each cell. In the general n dimensional case, the transverse moments would be taken n times separately over (n-1) (all but one) dimensions, resulting in n(k+1) 1D equations per node.

The final transverse integrated nodal equations are then solved iteratively : at iteration (i), the (k+1) 1D equations in x (4.1) are solved assuming that the transverse leakage terms in the y direction are known from iteration (i-1). The same is then done for the (k+1) 1D equations in y assuming that the transverse leakage terms in the x direction are known form iteration (i) and so on. The corresponding iteration will eventually converge and several acceleration procedures are available for these alternating direction like methods.

The fast solver aspect of this algorithm is clear since the x or y 1D equations are all similar and with different second members only. Increasing the order k is thus not very expensive, since assembly and factorization is done once for all. Moreover, vectorization possibilities are obvious.

The numerical solution of each of the equations (4.1) is quite naturally obtained by a standard Galerkin approach using the same set of parameters as above, namely end-point values and in-cell moments : in other words, a nodal method in 1D is used, in either the mathematical or physical version, and it turns out to be conforming. For a nodal scheme of type (k), the standard choice for V_h in x or y and for any of the (k+1) corresponding transverse moments is described by

$$S_k = \mathscr{P}_{k+2},\ \dim S_k = k+3 \tag{4.3a}$$

$$D_k = \left\{ m_L,\ m_R,\ m_C^i,\ i=0,..,k \right\},\ \mathrm{card}\ D_k = k+3 \tag{4.3b}$$

Another choice of S_k leading to *analytical nodal methods* consists of the direct sum of the 2 dimensional null space of $L\cdot \equiv -p d^2\cdot/dx^2 + q\cdot$ and of the (k+1) dimensional space of solutions of $Lu = \mathscr{P}_k$.

In the basic elliptic situation described by Eq. (2.1) it is worth mentioning that it is always possible to replace the basis functions associated to m_L and m_R, namely u_L and u_R, by these functions plus a linear combination of the cell basis functions u_C^i, where the coefficients of the linear combination are chosen such that the particular combination of the mass and stiffness matrices over a given node used for the assembly of the final algebraic system only couples the end-point parameters and is otherwise diagonal. As a result, the numerical solution of the transverse integrated nodal equations boils down to solving sequences of tridiagonal algebraic systems for the end-point values as well as local (i.e. node by node) 1×1 algebraic equations for each of the in-cell values. In parabolic situations like $u_t+Lu = f$, the same is true except that the local combination of mass and stiffness matrices is different and now depends on time integration parameters, such as the time step and an eventual θ parameter.

In the above cases, the techniques of quasidiagonalization of the final matrix at the node level (and not of the mass and stiffness matrices separately) do in fact improve upon well-known static condensation techniques (which in fact block-triangularize the nodal matrix) or recent techniques used in multidimensional situations to separate the cell parameters from the boundary ones [8] (where the resulting equations for the boundary values turn out to be much more coupled). The fact that the resulting tridiagonal algebraic systems couple the end-point values is quite natural. One might be tempted however to obtain similar (block) tridiagonal systems for the in-cell values (reminiscent of mesh-centered finite differences) : this is possible only approximately, for instance through the use of adhoc quadrature rules (see e.g.[9, Chapters 3&4]). The interesting result in that case is that the average value of p used in the final formula is the harmonic average value of p used in one, as it should be for problems with rough coefficients, at least in 1D. As a matter of fact, there is a close relationship between the transverse integrated nodal methods and the standard mixed f.e.m. which also picks up the harmonic average [10]. For more details, see Section 6 of [6].

Finally of few words about hyperbolic problems : after transverse integration has been performed, the result is again a set of ordinary differential equations, which are not of the boundary value but instead of the initial value type. In the case of the linear 2D neutron transport equation for instance and when the standard discrete ordinates method is used for the angular discretization, these equations should be integrated from left to right and from bottom to top if an angular direction in the first quadrant is considered. On the other hand, all the the nice properties mentioned above are still valid [9, Chapter 12] and piecewise continuous integration techniques, either of the polynomial or analytical type can again be used [9,Chapter 11]

REFERENCES

1. DORNING, J.J., Modern coarse-mesh methods. A development of the '70's. pp. 3.1-3.31 of *Computational Methods in Nuclear Engineering*. American Nuclear Society, Williamsburg, Virginia (1979).

2. LANGENBUCH. S., MAURER. W., and WERNER. W., Coarse-mesh flux-expansion method for the analysis of space-time effects in large light water reactor cores. *Nucl. Sci. & Engng.* 63, 437-456(1977).

3. LANGENBUCH. S., MAURER, W., and WERNER, W., High-order schemes for neutron kinetics calculations, based on a local polynomial approximation. *Nucl. Sci. & Engng.* 64, 508-516 (1977).

4. STRANG, G., and FIX, G.J., *An Analysis of the Finite Element Method.* Prentice Hall, Englewood Cliffs, New Jersey (1973).

5. FEDON MAGNAUD, C., HENNART, J.P., and LAUTARD, J.J., On the relationship between some nodal schemes and the finite element method in static diffusion calculations. pp. 987-1000 of *Advances in Reactors Computations.* American Nuclear Society, Salt Lake City, Utah (1983).

6. HENNART. J.P., A general family of nodal schemes. *Com. Tec. Serie NA.* 354, 63 p., IIMAS-UNAM (1983).

7. HENNART, J.P., On the implementation of finite elements for parabolic evolution problems. pp. 385-394 of J.R. Whiteman (ed.) *The Mathematics of Finite Elements and Applications IV. MAFELAP 1981.* Academic Press, London (1982).

8. GREENSTADT, J., The cell discretization algorithm for elliptic partial differential equations. *SIAM J. Sci. Stat. Comput.* 3, 261-288 (1982).

9. HENNART, J.P., et al, Advances in numerical reactor calculations. In Preparation.

10. BABUSKA, I., and OSBORN, J.E., Generalized finite element methods : Their performance and their relation to mixed methods. *SIAM J. Num. Anal.* 20, 510-536 (1983).

11. HENNART, J.P., Piecewise continuous discretization techniques for initial value problems with applications to kinetics and transport. *Transport Theory and Statistical Physics* 12, 233-250 (1983).

ACKNOWLEDGEMENT

This research was supported in part by the IBM Corporation Scientific Centers in Palo Alto, California and Mexico City, by the INRIA in France and by the Mexican CONACYT.

IMPLEMENTING FINITE ELEMENT METHODS ON A SUPERCOMPUTER

Lawrence W. Spradley

Lockheed-Huntsville Research & Engineering Center
Huntsville, Alabama 35806 USA

1. INTRODUCTION

The art of computational mechanics is the construction of an appropriate numerical equivalent of the conservation equations since there is no systematic procedure which will yield the optimum approximation for a general system. The classical works use finite difference techniques, but the concept of the finite element has become increasingly popular. Scientists with extensive backgrounds in finite difference solutions defend their approach against the finite element advocates. Likewise, finite element proponents claim that the other side did not understand their method. The resulting arguments have produced many papers and few solutions. An occasional attempt has even been made to relate the two approaches. In Ref. 1, the author presents a consistent approach which clarifies the situation and, as it turns out, demonstrates that both approaches have merit and that each side can learn a great deal from the other.

Regardless of the numerical approach chosen, all three-dimensional computational models share two common difficulties - large storage requirements and long computer run times. The modern "supercomputers" have helped the situation somewhat with their relatively large memories and increased processing speeds. Machines such as the CRAY and CDC CYBER 200 series are now used extensively for fluid dynamic simulations. To get maximum efficiency from these supercomputers, the numerical algorithm must be designed specifically for the machine architecture. Reworking older codes for the new generation of machines has achieved only limited success. However, a code has been built for the CYBER 200 machines which shows significant improvement in processing speed due to vectorization of the algorithm and coding.

This paper summarizes a numerical algorithm that has been implemented on CDC CYBER 203/205 machines for solution of

ISBN 0-12-747255-X

three-dimensional viscous fluid flows. Extensive calculations have been done and experience gained in using finite element methods on supercomputers (1-8). This paper will share these experiences with the reader in an attempt to guide future activities in implementing finite element methods on supercomputers. Example performance numbers are given but, due to space limitations, actual problem solutions will only be referenced to the published literature.

2. THE GENERAL INTERPOLANTS METHOD (GIM) CODE

The General Interpolants Method (GIM) was introduced by Lockheed in 1977 as a consistent approach for deriving families of numerical schemes, explicit and implicit, finite difference and element, from a single point of departure. A set of shape functions is chosen to approximate flow variables in the differential equations over a small element. A weighted integral of the approximated differential equation is set to zero producing a very general finite element/discrete difference relation. The GIM computer code (1-8) solves the multi dimensional Navier-Stokes equations for complex, arbitrary shaped geometric domains with flow fields which may be elliptic, parabolic, or hyperbolic in nature in a steady or unsteady state. The fundamental difference between this approach and the overwhelming majority of methods in current use is that local coordinates are used solely for discretizing the grid, and the actual differencing is performed in the Cartesian physical domain.

The General Interpolants Method has now been extended by Stalnaker (9) to formulate a consistent numerical algorithm for solving very general conservation law equations including time-dependent spatial coordinates. The algorithm is termed Progressive Assembly of Generalized Elements (PAGE). The approach differs from the original GIM in a number of aspects. The full domain equations are formed from the local element equation by a progressive (dynamic) assembly which allows for time-dependent element approximations. The PAGE algorithm also treats certain geometric terms differently than GIM to produce a more consistent formulation. Apart from the geometric versatility of the PAGE approach, other less apparent advantages arise from this formulation in the physical domain. First, the formulation admits any discrete differencing algorithm for which appropriate weight and shape functions can be developed. Second, the question of the form of the conservation law is rendered moot; since, in the Cartesian physical space the strong and chain-rule conservation-law-form are degenerate. Third, the differencing algorithm and the metric information are developed in a completely consistent fashion which does not allow geometrically induced errors to manifest themselves through indiscriminate differencing of the metric terms. Finally, the formulation admits a large variety of approximating shape functions and there is no restriction that the shape functions be time-independent.

The GIM code has been exercised extensively for the past five years. Flow fields have been computed for both two- and three-dimensional problems and range from inviscid Euler solutions to Navier-Stokes solutions with turbulence modeling. The codes for the CYBER 200 machines are actually a series of solvers rather than a single code. Versions currently in use include the following: (1) elliptic module - solves time dependent or asymptotic steady state problems which include both upstream and downstream boundary conditions; (2) quasi-parabolic (or quasi elliptic) - solves a parabolic form of the viscous equations (with unsteady terms included) from an up-stream boundary condition; (3) hyperbolic solver - integrates the inviscid Euler equations for supersonic, inviscid flow by a spatial marching technique; and (4) parabolized Navier-Stokes module - currently under development for solving a parabolic set of viscous equations by a spatial marching technique starting from an upstream boundary.

Problems which have been solved with the GIM code and published in the literature are summarized below:

- Two-and three-dimensional inviscid flow in a converging-diverging nozzle
- Two-dimensional viscous flow in a solid rocket motor submerged nozzle
- Shock-boundary layer interaction in a two-dimensional duct including separation of the boundary layer
- Three-dimensional turbulent shear layers
- Hypersonic aerodynamics on three-dimensional projectiles
- Flow in an aircraft storage cavity
- Natural convection in cavity geometries at varying levels of gravity
- Chemically reacting flow of hydrogen and air in a two-dimensional duct with finite rate reactions
- External aerodynamics on wing-fuselage configurations
- Viscous interaction flow of a sonic jet firing perpendicular to an oncoming supersonic stream
- Shock-on-shock interaction (inviscid) for high speed external flow
- Viscous, three-dimensional hot gas flow in complex internal networks of turbine engines.

Nodal counts used thus far range from a few hundred for two-dimensional inviscid flow to several hundred thousand for three-dimensional parabolized viscous problems. Time iteration counts have ranged from a few hundred steps to several tens of thousands. Run times vary from a few seconds up to an hour each. Most large problems are run for a specified number of steps, plotted and studied, and then restarted until steady state convergence is proclaimed. References 1-8 contain details of these calculations.

3. CODE VECTORIZATION

The GIM and PAGE codes are operational on the CYBER 203 and 205 machines. The U.S. Air Force Parabolized Navier-Stokes (PNS) code is operational on the CRAY 1S. Most of the subsequent discussion deals with experiences on the CYBER machines with the GIM code.

A very efficient code will result if an algorithm and code layout are designed for a specific machine architecture. Taking advantage of specific computer features allows a higher degree of vectorization than a general purpose code written to be machine independent. An attempt to vectorize an older scalar code can actually produce poor results. The Air Force PNS code structure was not amenable to vectorization on the CRAY. It is now being reworked extensively to realize a speed improvement.

The GIM code has been run for example problems using both 64 bit and 32 bit word lengths. No significant loss in accuracy has been seen from the half word size. A factor of approximately 2 is seen in speed improvement and, of course, available storage is doubled. The CYBER 205 system allows both full and half word FORTRAN.

Explicit algorithms naturally vectorize to a higher extent than do implicit ones. Most implicit schemes either require data that are not available until the end of a cycle or involve some type of recursive operation. Operations which can be performed on contiguous data strings will achieve the highest degree of vectorization on a memory-to-memory machine such as the CYBER 205. Matrix operations for solving simultaneous systems of linear algebraic equations have presented a surprising result. Direct methods such as Gauss elimination are slower in solving a system than iterative methods such as Gauss-Seidel. The iterative methods do not require recursion and hence can be highly vectorized.

On CYBER 200 machines it is important to use vector lengths that are large. For sequential operations involving only a few datum, scalar coding may be more efficient because of the vector "startup" time. Vector lengths equal to the number of mesh points in the problem result in very efficient pipelining on the CYBER 203/205. The time consuming operation in GIM is multiplication of a matrix by a column vector:

$$\dot{U}_i = \sum_{j=1}^{M} A_{ij} E_j \qquad i=1,2,\ldots,N$$

Summation is on j over all node points having an influence on point i. The total nodal connectivity M for each mesh point is at most 27 for three dimensions, but is typically eight for one-sided differences. The value of N, however, is quite large, i.e., total number of nodes in the problem. This can easily be several thousand for two dimensions and tens of thousands for three dimensions. The vectorization procedure for the CYBER 200 machines is to use N as the vector length and reverse the order of summation and multiplication. One part of a sum is computed for each node, i, followed by a second part and continued until all M components for node i are computed. The scalar method is to perform the matrix product in the normal manner. The vectorized scheme thus proceeds down diagonals and then across rows. The result is that long vector lengths are used in the pipeline multiply process.

On supercomputers that have a large vector-to-scalar speed ratio, avoiding scalar operations is vital (10). For example, consider a code that is 90% vectorized. If the vector to scalar speed ratio is 5, then the 10% scalar operation count takes about 35% of the total time. If the speed ratio goes up to 20, then the 10% scalar operation requires almost 70% of the total time. Scalar operations should be avoided on the CYBER equipment when at all possible. Boundary conditions are usually the most difficult item to vectorize since they require an arithmetic operation sequence which is different from the interior grid points. The use of "Gather/Scatter" techniques on the CYBER has proved to be useful. In this approach, the boundary node points are "gathered" into a contiguous data base, boundary conditions are applied in vector operations, and then the arrays are "scattered" back into their original arrays. Another scalar-type operation to avoid is use of "conditional" statements, such as IF and GO TO, inside large index loops. This hinders the automatic vectorization on both the CRAY and CYBER machines. However, the use of "bit" control vectors, and associated vector functions on the CYBER systems allows the user to create very powerful conditional programming constructs.

Perhaps even more important than the speed increase on supercomputers is the relatively large amount of high speed memory compared to the Univac 1100, IBM 360, and CDC 7600 machines. Most of the CRAY 1 and CYBER 203 systems have 1 or 2 million 64 bit words of high speed access memory. The scalar equipment usually has from 32K to 200K of high speed core plus several hundred thousand of lower speed access storage. The CYBER 200 series supercomputer uses the virtual memory concept while the CRAY designers choose to require the user to write input/output onto tape or solid state storage devices. The virtual memory system moves "pages" of data into and out of central memory as the code needs it. This concept allows the programmer to write coding as if the amount of memory were infinite and can leave the I/O to the virtual software. This concept works well only if the data base in a code is organized into pages which can be accessed in sequential manner. Otherwise, "memory thrashing" can occur which dominates the machine and runs the cost of processing up exponentially. A well organized data base is thus essential for economic reasons. It is often more efficient to recalculate certain data than to store it and "page out" of memory.

A series of parametric cases has been run with the GIM code to establish a central processor time versus problem size as compared with the GIM CDC 7600 and Univac 1108 scalar algorithms. The computation time for the fully vectorized CYBER 203 code, expressed in seconds per iteration per grid point, is 2.2×10^{-5} for two dimensions and 3.3×10^{-5} for three dimensions. These numbers represent a factor of approximately 20 improvement over a scalar CDC 7600 code and almost 200 times faster than the same scalar code on a Univac 1100 machine. The CYBER 205 code has shown a factor of approximately 3 improvement in speed over the 203 version. Special tailoring to the CYBER 205 should eventually produce a factor of 5 increase over the CYBER 203. Table 1 summarizes the performance numbers and gives an example.

TABLE 1

Performance Comparison of GIM Code on Various Computers

System	2-D	3-D	Example 10,000 Node, 500 Steps, 3-D
Univac 1108	3.7×10^{-3}	5.9×10^{-3}	8 hours
CDC 7600	3.6×10^{-4}	5.8×10^{-4}	48 minutes
CYBER 203	2.2×10^{-5}	3.3×10^{-5}	3 minutes
CYBER 205	5.5×10^{-6}	8.3×10^{-6}	1 minute

NOTE: Values are Central Processor seconds per grid point per iteration.

The CRAY system appears to be better suited to users who want to write codes in standard (scalar) FORTRAN and have the software perform vectorization. However, the CYBER 200 machines allow the programmer specialist more versatility through use of vector language and to thus realize more speed gain from the vectorization. This author thus far has no experience with the new Japanese machines.

4. FUTURE TRENDS

The future of this type of computation depends on advances in computer hardware and in algorithm development. The Japanese machines (especially Hitachi and Fujitsu), the CRAY 2, the CYBER 2xx/GF-10, and the National Aeronautics and Space Administration's NAS facility, appear to be major advances (11). High speed memory is a major item. Some estimates indicate that several hundred million 64 bit words of high speed memory will be required to store the quantity of data needed to simulate high Reynolds number viscous flows. Access to this data must be rapid or the machine becomes input/output bound and the central processor speed advantage is lost. Sustained computing speeds of 10 gigaflops will soon be a reality (1 gigaflop = 1,000,000,000 floating point operations per second). This speed regime will allow three-dimensional viscous flows to be computed economically if turbulence is treated with a "model." Actual simulation, to the eddy level, of turbulent flows will require at least another order of magnitude increase (100 gf) in central processor speed. Color graphics and even three dimensional visualization equipment will be vital to interpreting the large volume of data.

These hardware requirements are based on numerical algorithms which exist today; advances in solution techniques will (perhaps) alter the hardware requirements. Most current algorithms rely on elements which are spatially fixed with respect to time. The approximating functions for the elements are chosen a priori and never changed. Future algorithms for computing realistic flow fields must contain moving, adaptive grid methods for locating elements in positions to reduce discretization errors. Adaptive, time dependent, element approximation functions will also be necessary to maintain a given error level without requiring unrealistically large numbers of grid points. Polynomials, whose orders change to reduce discretization errors, will most likely be used in early developments due to speed and ease of integration. Moving mesh schemes, adaptive polonomials, and method of adding and removing mesh points will all eventually be needed in a robust algorithm.

Fully coupled solutions including the flow of a fluid over a solid surface, the response of that solid surface, and the resulting interaction have been accomplished only in very approximate means. The fluid-solid-thermal coupled interaction problems will provide the computational analyst with challenges far into the future.

5. ACKNOWLEDGMENTS

This paper was prepared under a Lockheed Independent Research Program. The methods, machine implementation and computations described herein are the result of a team effort by the Computational Mechanics Group at Lockheed's Huntsville, Alabama, facility.

6. REFERENCES

1. PROZAN, R.J., L.W. SPRADLEY, P.G. ANDERSON, and M.L. PEARSON, "The General Interpolants Method," AIAA Paper 77-642, Albuquerque, New Mexico, (June 1977).
2. SPRADLEY, L.W., J.F. STALNAKER, and A.W. RATLIFF, "Solutions of the Three-Dimensional Navier-Stokes Equations on a Vector Processor," AIAA J., 19, No. 10 (October 1981).
3. SPRADLEY, L.W., J.F. STALNAKER, and K.E. XIQUES, "Quasi-Parabolic Technique for Spatial Marching Navier-Stokes Computations," AIAA J., 20, No. 1 (January 1982).
4. XIQUES, K.E., E.G. RAWLINSON, J.F. STALNAKER, and L.W. SPRADLEY, "Computation of Three-Dimensional Inviscid Flow over Hypersonic Missile Configurations Using the GIM Code," AIAA Paper 82-0248 (January 1982).
5. ANDERSON, P.G., and L.W. SPRADLEY, "Finite Difference Grid Generation by Multivariate Blending Function Interpolation," NASA CP-2166, Langley Research Center, Hampton, Va. (October 1980).
6. RAWLINSON, E.G., and L.W. SPRADLEY, "Computation of Two-Dimensional Jet Interaction Flow Field," AIAA Paper 83-1546 Montreal, Canada, (June 1983).
7. DRESSLER, R.F., S.J. ROBERTSON, and L.W. SPRADLEY, "Effect of Rayleigh Accelerations Applied to an Initially Moving Fluid," Materials Processing in the Reduced Gravity Environment of Space, Ed. by G.E. Rendone, Elsevier Science Publishing Company, Inc. (1982).
8. STALNAKER, J.F., and K.E. XIQUES, "High Speed Aerodynamic Applications of the GIM Code," AIAA Paper 83-1801 Boston, Mass., (July 1983).
9. STALNAKER, J.F., "A Unifying Principle for Discrete Models of the Conservation Laws of Continuum Mechanics with Applications to Solution-Adaptive Techniques," International Conference on Accuracy Estimates and Adaptive Refinements in Finite Element Computations, Lisbon, Portugal (June 1984).
10. KUMAR, AJAY, D.H. RUDY, J.P. DRUMMOND, and J.E. HARRIS, "Experiences with Explicit Finite Difference Schemes for Complex Fluid Dynamics Problems on STAR-100 and CYBER 203 Computers," CYBER 205 Applications Symposium, Fort Collins, Colo. (August 1982).
11. ELSON, B.M., "Computational Fluid Dynamics," Aviation Week and Space Technology, 50-57, (1983).

ERROR BOUNDS FOR THE APPROXIMATION OF THE STOKES PROBLEM USING BILINEAR/CONSTANT ELEMENTS ON IRREGULAR QUADRILATERAL MESHES

J. Pitkäranta and R. Stenberg

Institute of Mathematics, Helsinki University of Technology

1. INTRODUCTION

One of the most popular methods to numerically solve the Stokes equations in fluid mechanics is to use a mixed finite element method where the velocities are approximated with continuous isoparametric bilinear elements on quadrilateral meshes whereas a piecewise constant approximation is used for the pressure. After eliminating the pressure by simple perturbation techniques one obtains a positive definite system for the velocities alone. Another way to obtain the same method is to apply penalty techniques with reduced/selective integration (cf. [6], [8]). In numerical computations this method has been shown to give excellent results for the computed velocities and also for the pressure provided that the latter has been "smoothed" in an appropriate way (cf. e.g. [6]). From a theoretical point of view this success has been considered somewhat surprising since it is well known that the method is not uniformly stable in the sense of Babuška [1] and Brezzi [3]. For rectangular meshes, however, it has been possible to analyze the method, cf. [7]. The analysis of [7] relies on a weak Babuška-Brezzi-type stability condition together with a careful consistency estimate and shows that the method converges with the optimal rate provided the exact solution is smooth enough. In this note we will extend and improve the analysis of [7]. We will show that the method in fact converges (after a pressure smoothing) for a very general class of meshes and that this happens without any extra smoothness assumptions on the exact solution. The fact that the extra smoothness assumption of [7] is not required was also observed by Boland and Nicolaides [2] in the case of a rectangular grid.

2. NOTATION AND PRELIMINARIES

Let Ω be a polygonal domain in $\mathbb{R}^2$ with boundary Γ. The

ISBN 0-12-747255-X

problem under consideration consists of the stationary Stokes equations for an incompressible viscous fluid:

$$\begin{aligned} -\nu\Delta u + \nabla p &= f \quad \text{in } \Omega, \\ \operatorname{div} u &= 0 \quad \text{in } \Omega, \\ u &= 0 \quad \text{on } \Gamma, \end{aligned} \tag{2.1}$$

where u is the fluid velocity, p is the pressure, f is the body force and $\nu > 0$ is the kinematic viscosity.

We denote by $|\cdot|_{s,T}$ and $\|\cdot\|_{s,T}$, respectively, the semi-norm and norm of the Sobolev space $[H^s(T)]^\alpha$ where s and α are integers. As usual $H_0^1(T)$ denotes the subspace of $H^1(T)$ consisting of functions with vanishing trace on ∂T. We will also introduce the space

$$L_0^2(T) = \{p \in L^2(T) \mid \int_T p\, dx = 0\}.$$

The inner product in $[L^2(T)]^\alpha$, for integral α, is denoted by $(\cdot,\cdot)_T$. The subscript T will be dropped if $T = \Omega$. As usual we will denote by C and C_j positive constants, possibly different at different occurences, which are independent of the mesh parameter h.

In variational form (2.1) reads: Find $u \in [H_0^1(\Omega)]^2$ and $p \in L_0^2(\Omega)$ such that

$$\begin{aligned} \nu(\nabla u,\nabla v) - (\operatorname{div} v,p) &= (f,v) \quad \forall v \in [H_0^1(\Omega)]^2, \\ (\operatorname{div} u,\mu) &= 0 \quad \forall \mu \in L_0^2(\Omega). \end{aligned} \tag{2.2}$$

In the finite approximation of (2.2) the spaces $[H_0^1(\Omega)]^2$ and $L_0^2(\Omega)$ are replaced by the finite dimensional subspaces V_h and P_h, respectively. Below we define the subspaces as

$$V_h = \{v \in [H_0^1(\Omega)]^2 \mid v|_K \in [Q_1(K)]^2 \quad \forall K \in C_h\}$$

and

$$P_h = \{p \in L_0^2(\Omega) \mid p|_K \text{ is constant} \quad \forall K \in C_h\},$$

where C_h stands for a partitioning of Ω into convex quadrilaterals and $Q_1(K)$ is the space of (isoparametrically) transformed bilinear functions [4]. As usual, the mesh parameter h is defined as $h = \max_{K \in C_h} h_K$, where h_K denotes the diameter of K.

We now specify our assumptions on the partitioning C_h. First, we assume that C_h is a refinement of a coarser partitioning C_{2h}, obtained by subdividing each $\tilde{K} \in C_{2h}$ into four quadrilaterals by joining the midpoints of the opposite sides of $\tilde{K}$ by straight lines. Second, we assume that C_{2h} is also a similar refinement of a still coarser partitioning C_{4h}. Third, regarding C_{4h}, we merely assume that C_{4h} is regular. By this we mean that there are the constants $\sigma > 1$ and $0 < \gamma < 1$ independent of h such that

$$h_K \leq \sigma \rho_K, \quad |\cos \theta_{iK}| \leq \gamma; \; i = 1,2,3,4, \quad \forall K \in C_{4h},$$

where h_K, ρ_K and θ_{iK} are respectively the diameter of K, the diameter of the largest circle contained in K, and the angles of K.

Below we refer to the quadrilaterals of C_{2h} or C_{4h} as "macroelements" and denote them by M. We also introduce the subspace.

$$V_{2h} = \{v \in [H_0^1(\Omega)]^2 \mid v_{|M} \in [Q_1(M)]^2 \quad \forall M \in C_{2h}\}$$

where $Q_1(M)$ is as above. The space P_h will be written as the sum of three subspaces. The unit square $\hat{K}$ is partitioned into subdomains $\hat{K}_{ij} = \{(x_1,x_2) \in \hat{K} \mid \frac{(i-1)}{2} \leq x_1 \leq \frac{i}{2}, \frac{(j-1)}{2} \leq x_2 \leq \frac{j}{2}\}$, $i,j = 1,2$ and on $\hat{K}$ we define the function η through

$$\eta_{|\hat{K}_{ij}} = (-1)^{i+j}, \quad i,j = 1,2.$$

We then define the subspaces

$$P_{h1} = \{p \in P_h \mid p_{|M} \text{ is constant } \forall M \in C_{2h}\},$$

$$P_{h3} = \{p \in P_h \mid p_{|M} = c_M \eta o F_M^{-1}, \; c_M \in \mathbb{R}, \forall M \in C_{2h}\}$$

where F_M is the bilinear mapping of $\hat{K}$ onto M. The orthogonal complement of P_h with respect to $P_{h1} \oplus P_{h3}$ is denoted by P_{h2}. Finally we introduce a "pressure smoothing operator" $\pi: P_h \to P_{h1} \oplus P_{h2}$. Every $p \in P_h$ can be written uniquely as $p = \sum_{i=1}^{3} p_i$, $p_i \in P_{hi}$. The filtered pressure πp is then defined as $\pi p = p_1 + p_2$.

3. ERROR ANALYSIS

Let us start with a consistency estimate which is crucial for the analysis in this paper.

Lemma 3.1. For each $v \in V_{2h}$ and $p \in P_{h3}$ we have

$$(\text{div } v, p) = 0.$$

Proof. Consider a macroelement $M \in C_{2h}$ with nodes x^i, $i = 1,2,3,4$, as in the figure below and suppose $p_M = \eta \circ F_M^{-1}$ takes the values ± 1 as in the figure.

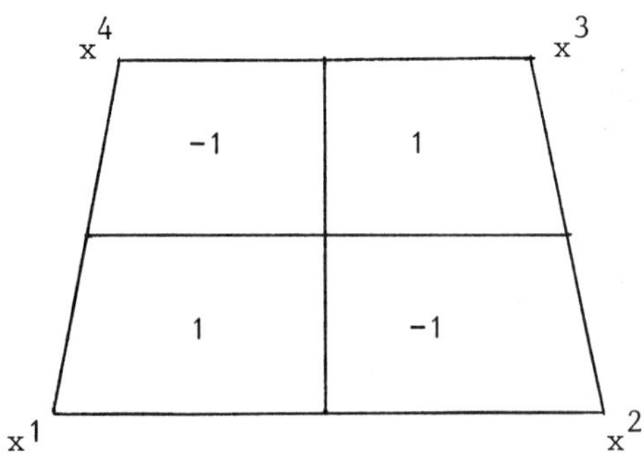

Denote by $v^i = v(x^i)$, $i = 1,2,3,4$, the degrees of freedom of $v \in V_{2h|M}$ and write

$$x^i x^j = x^i - x^j, \quad i,j = 1,2,3,4.$$

It will further be convenient to use the (scalar valued) vector product in $\mathbb{R}^2$, i.e. if $a = (a_1,a_2)$ and $b = (b_1,b_2)$ we define

$$a \wedge b = a_1 b_2 - a_2 b_1.$$

Using Green's formula and integrating over the sides in M one obtains

$$(\operatorname{div} v, p_M)_M = \frac{1}{4}\Bigg[(v^1-v^4) \wedge x^4x^1$$

$$+ 2\left(-\left(\frac{v^1+v^2}{2}\right) + \left(\frac{v^3+v^4}{2}\right)\right) \wedge \left(\frac{x^4x^1+x^3x^2}{2}\right)$$

$$+ (v^2-v^3) \wedge x^3x^2 + (-v^1+v^2) \wedge x^2x^1$$

$$+ \left(\left(\frac{v^1+v^4}{2}\right) - \left(\frac{v^2+v^3}{2}\right)\right) \wedge \left(\frac{x^2x^1+x^3x^4}{2}\right) + (-v^4+v^3) \wedge x^3x^4\Bigg]$$

$$= \frac{1}{8}\,[(v^1-v^4-v^2+v^3) \wedge (x^4x^1-x^3x^2)$$

$$+ (-v^1+v^2+v^4-v^3) \wedge (x^2x^1-x^3x^4)] = 0,$$

since by the definition of x^ix^j, $i,j = 1,2,3,4$ we have $x^4x^1-x^3x^2 = x^2x^1-x^3x^4$.

Since $p_{|M} = c_Mp_M$, $c_M \in \mathbb{R}$, for every $p \in P_{h3}$, the assertion is proved. □

Next we will turn to the stability estimate, the proof of which will only be sketched since the arguments are very similar to those given in [9].

Lemma 3.2. There is a constant $C > 0$ such that

$$\sup_{\substack{u\in V_h \\ u\neq 0}} \frac{(\operatorname{div} u,p)}{|u|_1} \geq C\|\pi p\|_0 \qquad \forall p \in P_h.$$

Proof. Consider a macroelement $M \in C_{4h}$ and define

$$V_{0,M} = \{v \in [H_0^1(M)]^2 \mid v_{|K} \in [Q_1(K)]^2 \quad \forall K \subset M,\ K \in C_h\}$$

and

$$N_M = \{p \in P_{h|M} \mid (\operatorname{div} v,p)_M = 0 \quad \forall v \in V_{0,M}\}.$$

A straightforward calculation shows that $N_M = \{c_1\psi_1^M + c_2\psi_3^M,\ c_1,c_2 \in \mathbb{R}\}$, where ψ_1^M is constant on M and ψ_3^M takes the values ± 1 in a chessboard - like manner on the subrectangles of M. Let $\tilde{P}_{hi} = \{p \in Q_h \mid p_{|M} = c_M\psi_i^M,\ c_M \in \mathbb{R}\}$, $i = 1,3$, and let $\tilde{P}_{h2}$ be the orthogonal complement of P_h to $\tilde{P}_{h1} \oplus \tilde{P}_{h3}$. By the same arguments as those leading to the macroelement principle introduced in [9] (cf. Lemma 3.1 and Lemma 3.2 of

[9]) one now concludes that for every $p \in P_h$, $p = \sum_{i=1}^{3} \tilde{p}_i$, $\tilde{p}_i \in \tilde{P}_{hi}$, there is a $v \in V_h$ such that $v_{|M} \in V_{0,M}$ $\forall M \in C_{4h}$ and

$$(\mathrm{div}\ v,p) \geq C_1 \|\tilde{p}_2\|_0^2 \tag{3.1}$$

and

$$|v|_1 \leq \|\tilde{p}_2\|_0 . \tag{3.2}$$

By the same reasoning as in Lemma 3.3 of [9] one can also show that for every $\tilde{p}_1 \in \tilde{P}_{h1}$ there is a $g \in V_{2h}$ such that

$$(\mathrm{div}\ g,\tilde{p}_1) = \|\tilde{p}_1\|_0^2 \tag{3.3}$$

and

$$|g|_1 \leq C_2 \|\tilde{p}_1\| . \tag{3.4}$$

Since $g \in V_{2h}$ we have by Lemma 3.1

$$(\mathrm{div}\ g,\tilde{p}_3) = 0. \tag{3.5}$$

Let now $p \in P_h$ be arbitrary and write $p = \sum_{i=1}^{3} \tilde{p}_i$. Define $z = v + \frac{2C_1}{(1+C_2^2)} g$, where v, g, C_1 and C_2 are as above. A straightforward calculation, using the relations (3.1) to (3.5), then gives (cf. the proof of Theorem 3.1 in [9])

$$\frac{(\mathrm{div}\ z,p)}{|z|_1} \geq C(\|\tilde{p}_1\|_0 + \|\tilde{p}_2\|_0) \geq C\|\pi p\|_0 . \qquad \square$$

As a final preparation for our error estimate we will introduce a seminorm on P_h defined through

$$|p|_h = \sup_{\substack{v \in V_h \\ v \neq 0}} \frac{(\mathrm{div}\ v,p)}{|v|_1} \qquad \forall p \in P_h .$$

The following estimate is an immediate consequence of Lemma 3.2 and the definition of the seminorm $|\cdot|_h$

$$\|p\|_0 \geq |p|_h \geq C\|\pi p\|_0 \qquad \forall p \in P_h . \tag{3.6}$$

We are now ready to prove

Theorem 3.1. Let (u,p) be the solution to (2.1) and let $(u_h,p_h) \in V_h \times P_h$ be its finite element approximation defined as above. Then we have the error estimate

$$|u-u_h|_1 + \|p-\pi p_h\|_0 \leq Ch(|u|_2 + |p|_1),$$

provided $u \in [H^2(\Omega)]^2$ and $p \in H^1(\Omega)$.

Moreover, if Ω is a convex region, we have the additional estimate

$$\|u-u_h\|_0 \leq Ch^2(|u|_2 + |p|_1).$$

Proof. Let $\tilde{u} \in V_h$ be the interpolant to u and let $\tilde{p}$ be the L^2-projection of p onto P_h. By the general theory of Babuška [1] and Brezzi [3] (cf. also [7]) one concludes that there exists $v \in V_h$ and $\mu \in P_h$ such that

$$|v|_1 + |\mu|_h \leq C,$$

and

$$|u_h-\tilde{u}|_1 + |p_h-\tilde{p}|_h \leq C\{|(\nabla u-\tilde{u}),\nabla v) \mid (\operatorname{div} v,p-\tilde{p})| + \quad (3.7)$$
$$+ \ (\operatorname{div}(u-\tilde{u}),\mu)|\}.$$

By standard interpolation theory [4] the first two terms on the right hand side of (3.7) can be estimated as

$$|(\nabla(u-\tilde{u}),\nabla v)| \leq |u-\tilde{u}|_1|v|_1 \leq Ch|u|_2, \quad (3.8)$$

$$|(\operatorname{div} v,p-\tilde{p})| \leq |v|_1\|p-\tilde{p}\|_0 \leq Ch|p|_1. \quad (3.9)$$

To estimate the third term we write $\mu = \pi\mu + (I-\pi)\mu$ so as to obtain

$$|(\operatorname{div}(u-\tilde{u}),\mu)| \leq |(\operatorname{div}(u-\tilde{u}),\pi\mu)| + \quad (3.10)$$
$$+ \ |(\operatorname{div}(u-\tilde{u}),(I-\pi)\mu)|$$

For the first term on the right hand side of (3.10) we obtain, using the estimate (3.6),

$$|\operatorname{div}(u-\tilde{u}),\pi\mu)| \leq |u-\tilde{u}|_1\|\pi\mu\|_0 \leq Ch|u|_2. \quad (3.11)$$

To estimate the second term on the right hand side of (3.10) we introduce the interpolant $\approx{u} \in V_{2h}$ to u. Since $\operatorname{div} u = 0$,

$(I-\pi)\mu \in P_{h3}$ and $\tilde{\tilde{u}} \in V_{2h}$ we have by Lemma 3.1

$$|(\operatorname{div}(u-\tilde{u}),(I-\pi\mu))| = |(\operatorname{div}(\tilde{u}-\tilde{\tilde{u}}),(I-\pi)\mu)| \le \tag{3.12}$$
$$\le |\tilde{u}-\tilde{\tilde{u}}|_1(|\mu|_h + \|\pi\mu\|_0) \le C|\tilde{u}-\tilde{\tilde{u}}|_1|\mu|_h \le C|\tilde{u}-\tilde{\tilde{u}}|_1 \le Ch|u|_2.$$

Here the last inequality is a consequence of standard interpolation error estimates. Upon collecting the estimates (3.7) through (3.12) we obtain

$$|u_h-\tilde{u}|_1 + |p_h-\tilde{p}|_h \le Ch(|u|_2 + |p|_1). \tag{3.13}$$

The asserted estimate for $|u-u_h|_1$ now follows using the triangle inequality. To obtain the estimate for $\|p-\pi p_h\|_0$ we use (3.6) and get

$$\|\pi p_h - \pi\tilde{p}\|_0 \le Ch(|u|_2 + |p|_1). \tag{3.14}$$

The asserted estimate is now obtained upon applying the triangle inequality together with the estimate

$$\|p-\pi\tilde{p}\|_0 \le Ch|p|_1. \tag{3.15}$$

In order to obtain the L^2-estimate for the velocity we first note that if we replace $\tilde{p}$ by $\pi\tilde{p}$ in (3.7) through (3.12) we obtain

$$|p_h-\pi\tilde{p}|_h \le Ch(|u|_2 + |p|_1). \tag{3.16}$$

From (3.16), (3.6) and (3.14) we then obtain

$$|(I-\pi)p_h|_h \le |p_h-\pi\tilde{p}|_h + |\pi\tilde{p}-\pi p_h|_h \tag{3.17}$$
$$\le |p_h-\pi\tilde{p}|_h + \|\pi\tilde{p}-\pi p_h\|_0 \le Ch(|u|_2 + |p|_1).$$

We now proceed using the Aubin-Nitsche trick. Let $(z,\lambda) \in [H_0^1(\Omega)]^2 \times L_0^2(\Omega)$ be the solution to the problem

$$\nu(\nabla z,\nabla v) - (\operatorname{div}\lambda,v) = (u-u_h,v) \quad \forall v \in [H_0^1(\Omega)]^2,$$
$$(\operatorname{div} z,\mu) = 0 \qquad \forall \mu \in L_0^2(\Omega).$$

In the usual way (cf. e.g. [5]) we then obtain

$$\|u-u_h\|_0^2 \le Ch\|u-u_h\|_1(|z|_2 + |\lambda|_1) + |(\operatorname{div}(z-\tilde{z}),p-p_h)|, \tag{3.18}$$

where $\tilde{z} \in V_h$ is the interpolant to z. To estimate the second term on the right hand side of (3.18) we repeat the arguments used in proving the estimates (3.10) through (3.12). Let $\tilde{\tilde{z}}$ be V_{2h}-interpolant to z. Since $\operatorname{div} z = 0$ and $\tilde{\tilde{z}} \in V_{2h}$ we obtain

$$|(\operatorname{div}(z-\tilde{z}),p-p_h)| \leq |(\operatorname{div}(z-\tilde{z}),p-\pi p_h)| + \tag{3.19}$$
$$+ |(\operatorname{div}(\tilde{z}-\tilde{\tilde{z}}),(I-\pi)p_h)| \leq Ch|z|_2(\|p-\pi p_h\|_0 +$$
$$+ |(I-\pi)p_h|_h).$$

The asserted estimate now follows upon combining (3.18), (3.19) and (3.17) and using the regularity estimate

$$|z|_2 + |\lambda|_1 \leq C\|u-u_h\|_0. \qquad \square$$

Remark. We could have simplified the above analysis by choosing the interpolants $\tilde{u}$ and $\tilde{z}$ in V_{2h}, i.e. setting $\tilde{u} = \tilde{\tilde{u}}$, $\tilde{z} = \tilde{\tilde{z}}$. The reason for not doing this was to show that the error constants in the final estimates are not substantially larger than what they would be if the method were uniformly stable in the classical sense - a fact that has also been confirmed by numerous numerical calculations.

REFERENCES

1. I. Babuška, Error bounds for finite element methods, *Numer. Math.* 16, 322-333 (1971).

2. J.M. Boland and R.A. Nicolaides, Stable and semistable low order finite elements for viscous flows, Preprint, Univ. of Connecticut (1984).

3. F. Brezzi, On the existence, uniqueness and approximation of saddle-point problems arising from Lagrange multipliers, *RAIRO* 8-R2, 129-151 (1974).

4. P.G. Ciarlet, *The Finite Element Method for Elliptic Problems*, North Holland (1978).

5. V. Girault and P.-R. Raviart, *Finite Element Approximation of the Navier-Stokes Equations*, Lecture notes in Mathematics, Springer, Berlin (1979).

6. T.J.R. Hughes, W.K. Liu and A. Brooks, Finite element analysis of incompressible viscous flows by the penalty function formulation, *J. Comp. Phys.* 30, 1-60 (1979).

7. C. Johnson and J. Pitkäranta, Analysis of some mixed finite element methods related to reduced integration, *Math. Comp.* 38, 375-400 (1982).

8. D. Malkus and T.J.R. Hughes, Mixed finite element methods - reduced and selective integration techiques: a unification of concepts, *Comp. Meth. Appl. Mech. Eng.* 15, 63-81 (1978).

9. R. Stenberg, Analysis of mixed finite element methods for the Stokes problem: a unified approach, *Math. Comp.* 42, 9-23 (1984).

MIXED FINITE ELEMENT APPROXIMATION OF A FLUID FLOW PROBLEM

R. Verfürth

Mathematisches Institut, Ruhr-Universität Bochum, Bochum, West Germany

1. INTRODUCTION

We consider the stationary Navier-Stokes equations

$$-\nu\Delta\underline{u} + \nabla p + (\underline{u}\nabla)\underline{u} = \underline{f} \quad , \quad \operatorname{div}\underline{u} = 0 \tag{1.1}$$

in a bounded domain $\Omega\subset\mathbb{R}^3$ with the mixed boundary conditions

$$\underline{u}\cdot\underline{n} = \underline{n}\cdot\underline{\underline{T}}(\underline{u},p)\cdot\underline{\tau}_k = 0 \quad , \; k=1,2 \; , \tag{1.2}$$

on $\Gamma:=\partial\Omega$. Here, $\underline{n}$ denotes the outward normal of Ω, $\underline{\tau}_k$, $k=1,2$, are orthonormal vectors spanning the tangent plane and

$$\underline{\underline{T}}(\underline{u},p)_{ij} := -p\delta_{ij} + \nu\left\{\frac{\partial u_i}{\partial x_j} + \frac{\partial u_j}{\partial x_i}\right\} \; , \; 1\leq i,j\leq 3 \; ,$$

is the stress tensor. The boundary Γ has to be of class C^3. To simplify the notation we assume that Ω is convex. Problem (1.1), (1.2) arises as a subproblem in fluid flow governed by surface tension where Ω is not known a priori. Instead it is determined by the additional condition that the mean curvature of Γ is proportional to the normal stress $\underline{n}\cdot\underline{\underline{T}}(\underline{u},p)\cdot\underline{n}$ [2,3,9].

We consider a nonconforming mixed finite element discretization of (1.1), (1.2). For sufficiently small data $\nu^{-2}\underline{f}$ the continuous and discrete problem have unique solutions. We obtain $O(h^{1/2})$ error estimates for the H^1-norm of the velocities and the L^2-norm of the pressure and $O(h)$ error estimates for the L^2-norm of the velocities. The suboptimality of these error

ISBN 0-12-747255-X

estimates is due to the nonconformity. However, this cannot be avoided as is shown in the last section by a Babuška-type paradox.

2. FINITE ELEMENT DISCRETIZATION

Denote by $H^k(\Omega)$, $k \geq 0$, and $L^2(\Omega) := H^0(\Omega)$ the Sobolev and Lebesgue spaces equipped with the usual norm $\|.\|_k$. The inner product of $L^2(\Omega)$ is denoted by $(.,.)_0$. Let S be the space of rotations transforming Ω into itself. Put

$$X := \{\underline{u} \in H^1(\Omega)^3 : \underline{u} \cdot \underline{n} = 0 \text{ on } \Gamma\}/S \quad ,$$

$$M := \{p \in L^2(\Omega) : (p,1)_0 = 0\}$$

and denote by $\underline{\underline{D}}(\underline{u})$ the deformation tensor

$$\underline{\underline{D}}(\underline{u})_{ij} := \frac{\partial u_i}{\partial x_j} + \frac{\partial u_j}{\partial x_i} \quad , \quad 1 \leq i,j \leq 3 .$$

We introduce the following three bilinear resp. trilinear forms

$$a(\underline{u},\underline{v}) := \frac{\nu}{2} \int_\Omega \underline{\underline{D}}(\underline{u}):\underline{\underline{D}}(\underline{v}) \, dx \quad ,$$

$$b(\underline{u},p) := - \int_\Omega p \operatorname{div} \underline{u} \, dx \quad ,$$

$$N(\underline{u},\underline{v},\underline{w}) := \int_\Omega [(\underline{u}\nabla)\underline{v}]\underline{w} \, dx \quad , \quad \forall \ \underline{u},\underline{v},\underline{w} \in H^1(\Omega)^3, \ p \in L^2(\Omega) .$$

The weak formulation of (1.1), (1.2) to which we will refer as *Problem (N)* then is to find $(\underline{u},p) \in X \times M$ such that

$$a(\underline{u},\underline{v}) + b(\underline{v},p) + N(\underline{u},\underline{u},\underline{v}) = (\underline{f},\underline{v})_0 \quad \forall \ \underline{v} \in X$$

$$b(\underline{u},q) = 0 \quad \forall \ q \in M.$$

The corresponding linear problem without the term $N(\underline{u},\underline{u},\underline{v})$ will be refered to as *Problem (S)*. Problem (S) always has a unique solution and the regularity estimate

$$\|\underline{u}\|_2 + \|p\|_1 \leq c\|\underline{f}\|_0 \tag{2.1}$$

holds [2,3,9]. If the data $\nu^{-2}\|\underline{f}\|_0$ are sufficiently small, the same is true for Problem (N) [2,3].

Let $\Omega_h \subset \bar{\Omega}$ be a family of polyhedrons with sides of length O(h) such that all vertices of Ω_h lie on Γ. We divide each Ω_h into tetrahedrons with sides of length O(h) such that the resulting triangulation T_h satisfies the usual regularity assumptions [6].Moreover, we assume that each face of Ω_h is the face of one $T \in T_h$. Denote by S_h^r the space of continuous finite elements of polynomial degree r corresponding to T_h and put

$$X_h := \{\underline{u} \in (S_h^2)^3 : \underline{u} \cdot \underline{n} = 0 \text{ at all vertices of } \Omega_h\}/S,$$

$$M_h := \{p \in S_h^1 : \int_{\Omega_h} p\, dx = 0\}.$$

Note that $X_h \subset H^1(\Omega_h)^3$, but $X_h \not\subset X$ and $\underline{u} \cdot \underline{n}_h \neq 0$ on Γ_h where $\underline{n}_h$ is the normal to Ω_h.

The norm of $H^k(\Omega_h)$ is denoted by $\|.\|_{k,h}$; the inner product of $L^2(\Omega_h)$ is denoted by $(.,.)_{o,h}$. The discrete analogs of a,b and N are given by

$$a_h(\underline{u},\underline{v}) := \frac{\nu}{2} \int_{\Omega_h} \underline{\underline{D}}(\underline{u}) : \underline{\underline{D}}(\underline{v})\, dx \quad ,$$

$$b_h(\underline{u},p) := -\int_{\Omega_h} p \operatorname{div} \underline{u}\, dx \quad ,$$

$$N_h(\underline{u},\underline{v},\underline{w}) := \frac{1}{2} \int_{\Omega_h} \{[(\underline{u}\nabla)\underline{v}]\underline{w} - [(\underline{u}\nabla)\underline{w}]\underline{v}\}\, dx$$

$$\forall\ \underline{u},\underline{v},\underline{w} \in H^1(\Omega_h)^3,\ p \in L^2(\Omega_h) .$$

The discretization of Problem (N) to which we will refer as *Problem* (N_h) then is to find $(\underline{u}_h,p_h) \in X_h \times M_h$ such that

$$a_h(\underline{u}_h,\underline{v}_h) + b_h(\underline{v}_h,p_h) + N_h(\underline{u}_h,\underline{u}_h,\underline{v}_h) = (\underline{f},\underline{v}_h)_o$$

$$b_h(\underline{u}_h,q_h) = 0$$

$$\forall\ \underline{v}_h \in X_h\ ,\ q_h \in M_h .$$

The corresponding linear problem will be refered to as *Problem* (S_h).

In the sequel $c, c_o, c_1, \ldots$ denote various constants independent of h which may have different values depending on the context.

3. ERROR ESTIMATES FOR THE LINEAR PROBLEM

Recall that $\Gamma = \partial\Omega$, $\Gamma_h = \partial\Omega_h$ and that $\underline{n}$ and $\underline{n}_h$ denote the outward normal to Ω and Ω_h. The following Lemma is proved in [11].

LEMMA 3.1: There is an $h_o>0$ such that the following boundary estimates hold for all $0<h\le h_o$:

$$\|\underline{u}\cdot\underline{n}_h\|_{L^2(\Gamma_h)} \le ch^{1/2}\|\underline{u}\|_{1,h} \quad , \forall\ \underline{u}\in X_h ,$$

$$\|\underline{u}\cdot\underline{n}_h\|_{L^2(\Gamma_h)} \le ch\|\underline{u}\|_1 \quad , \forall\ \underline{u}\in X . \quad \square$$

Since X_h contains all continuous, piecewise quadratic finite elements which vanish on Γ_h, we conclude from [4,10] that

$$\inf_{p\in M_h\setminus\{0\}} \sup_{\underline{u}\in X_h\setminus\{0\}} \frac{b_h(\underline{u},p)}{\|\underline{u}\|_{1,h}\|p\|_{o,h}} \ge \beta > 0 \quad (3.1)$$

holds with a constant β independent of h. The inequalities of Korn and Poincaré-Friedrichs [9] and Lemma 3.1 imply the existence of an $h_o>0$ and an $\alpha>0$ independent of h such that (cf. [11])

$$a_h(\underline{u},\underline{u}) \ge \alpha\nu\|\underline{u}\|^2_{1,h} \qquad \forall\ \underline{u}\in X_h ,\ 0<h\le h_o \ .(3.2)$$

The continuity of a_h, b_h, equs. (3.1), (3.2) and standard results on mixed problems [5] imply the unique solvability of Problem (S_h) for all $0<h\le h_o$.

THEOREM 3.2: Let $(\underline{u},p)\in X\times M$ and $(\underline{u}_h,p_h)\in X_h\times M_h$, $0<h\le h_o$, be the unique solutions of Problem (S) and (S_h) resp. Then the error estimate

$$\|\underline{u}-\underline{u}_h\|_{o,h} + h^{1/2}\|\underline{u}-\underline{u}_h\|_{1,h} + h^{1/2}\|p-p_h\|_{o,h} \le ch\|\underline{f}\|_o$$

holds.

Proof: The functions $\underline{u}\in H^2(\Omega)^3$, $p\in H^1(\Omega)$ (cf. (2.1)) are weak solutions of

$$-\nu\Delta\underline{u} + \nabla p = \underline{f} \quad , \quad \operatorname{div}\underline{u} = 0 \quad \text{in } \Omega,$$
$$\underline{u}\cdot\underline{n} = \underline{n}\cdot\underline{\underline{T}}(\underline{u},p)\cdot\underline{\tau}_k = 0 \quad \text{on } \Gamma,\ k=1,2. \quad (3.3)$$

Multiplying (3.3) with $\underline{v}_h\in X_h$, $q_h\in M_h$, integrating over Ω_h, using partial integration and subtracting the defining equations for $\underline{u}_h, p_h$, we obtain [11]

$$a_h(\underline{u}-\underline{u}_h,\underline{v}_h) + b_h(\underline{v}_h,p-p_h) = \int_{\partial\Omega_h} \underline{n}_h\cdot\underline{\underline{T}}(\underline{u},p)\cdot\underline{v}_h \quad \forall\ v_h\in X_h$$
$$b_h(\underline{u}-\underline{u}_h,q_h) = 0 \qquad \forall\ q_h\in M_h . \quad (3.4)$$

The inhomogeneous right hand side is due to the nonconformity. With similar arguments as for standard mixed finite element approximations [5,10,11] we conclude from (3.4)

$$\begin{aligned} &\|\underline{u}-\underline{u}_h\|_{1,h} + \|p-p_h\|_{o,h} \\ &\leq c \inf_{\underline{v}_h\in X_h} \|\underline{u}-\underline{v}_h\|_{1,h} + c \inf_{q_h\in M_h} \|p-q_h\|_{o,h} \\ &+ c \sup_{\underline{v}_h\in X_h\setminus\{0\}} \frac{1}{\|\underline{v}_h\|_{1,h}} \int_{\Gamma_h} |\underline{n}_h\cdot\underline{\underline{T}}(\underline{u},p)\cdot\underline{v}_h| \ . \end{aligned} \tag{3.5}$$

Since the interpolate of $\underline{u}$ lies in X_h the first two terms on the right hand side of (3.5) can be bounded by $ch\|\underline{f}\|_o$. To estimate the third term, let $\underline{v}_h\in X_h$. From Lemma 3.1 and equ. (2.1) we obtain

$$\begin{aligned} &\int_{\Gamma_h} |\underline{n}_h\cdot\underline{\underline{T}}(\underline{u},p)\cdot\underline{v}_h| \\ &\leq ch^{1/2}\|\underline{v}_h\|_{1,h}\|\underline{f}\|_o + c\|\underline{v}_h\|_{1,h} \sum_{k=1}^{2} \|\underline{n}_h\cdot\underline{\underline{D}}(\underline{u})\cdot\underline{\tau}_{hk}\|_{L^2(\Gamma_h)} . \end{aligned}$$

Here, $\underline{\tau}_{hk}$, k=1,2, are orthonormal vectors spanning the tangent space of Ω_h. Since $\underline{n}\cdot\underline{\underline{D}}(\underline{u})\cdot\underline{\tau}_k = 0$ on Γ, we conclude [11]

$$\sum_{k=1,2} \|\underline{n}_h\cdot\underline{\underline{D}}(\underline{u})\cdot\underline{\tau}_{hk}\|_{L^2(\Gamma_h)} \leq ch\|\underline{f}\|_o \ . \tag{3.6}$$

Finally, the L^2-estimate for the velocities follows by a standard duality argument. □

Note that the error estimate of Theorem 3.2 can only be improved by requiring $\underline{v}\cdot\underline{n}_h = 0$ on Γ_h for all $\underline{v}\in X_h$. Example 5.1 shows that this assumption is not appropriate for our problem.

4. ERROR ESTIMATES FOR THE NONLINEAR PROBLEM

THEOREM 4.1: There is an $h_o>0$ and a constant $K>0$ which does not depend on h such that Problems (N) and (N_h), $0<h\leq h_o$, have unique solutions $(\underline{u},p)\in X\times M$ and $(\underline{u}_h,p_h)\in X_h\times M_h$ resp. provided $\nu^{-2}\|\underline{f}\|_o < K$. Moreover, the error estimate

$$\|\underline{u}-\underline{u}_h\|_{o,h} + h^{1/2}\|\underline{u}-\underline{u}_h\|_{1,h} + h^{1/2}\|p-p_h\|_{o,h} \leq ch\|\underline{f}\|_o$$

holds. □

We only sketch the crucial steps of the proof here. For more details we refer to [11]. The existence of a unique solution to Problems (N) and (N_h) resp. for small data $\nu^{-2}\|f\|_0$ follows from a standard fixed point argument as in [8] and Chap. IV of [7]. Together with Theorem 3.2 it also yields the H^1-error estimate for the velocities. The proof of the error estimate for the pressure is similar to that in the linear case. The L^2-error estimate for the velocities follows by a duality argument. This requires the solution of a Stokes problem with additional first order terms and with boundary conditions (1.2). Due to the nonconformity, we have to estimate additional boundary integrals of the form

$$\int_{\Omega_h} \underline{n}_h \cdot \underline{\underline{T}}(\underline{v},q)\cdot(\underline{u}-\underline{u}_h) \quad \text{and} \quad \int_{\Omega_h} (\underline{u}\cdot\underline{n}_h)\underline{v}\cdot(\underline{u}-\underline{u}_h) \ .$$

This is done using Lemma 3.1 and equ. (3.6).

5. A BABUŠKA-TYPE PARADOX

We have seen in §3 that the error estimates of Theorems 3.2 and 4.1 could only be improved by requiring $\underline{u}\cdot\underline{n}_h = 0$ on Γ_h for all $\underline{u}\in X_h$. The following example shows that this assumption is not appropriate for the problem under consideration.

EXAMPLE 5.1: Let $\Omega\subset\mathbb{R}^2$ be the interior of the unit circle and $w\in C^\infty([0,1])$ be a function with

$$w(r) = \begin{cases} 0 & , \ 0\le r\le \frac{1}{3} \ , \\ \frac{3}{2}r & , \ \frac{2}{3}\le r\le 1 \ . \end{cases}$$

Put $\underline{u}(x,y) := w(r)\ (-\sin\phi,\cos\phi)^t$ where r,ϕ are polar coordinates in Ω. Obviously, $\underline{u}\in C^\infty(\bar\Omega)^2$, div $\underline{u} = 0$ in Ω and $\underline{u}\cdot\underline{n} = \underline{n}\cdot\underline{\underline{D}}(\underline{u}) = 0$ on Γ.
Using the same notation as in §2 put

$$\tilde{X}_h := \{\underline{v}\in(S_h^2)^2 : \underline{v}\cdot\underline{n}_h = 0 \text{ on } \Gamma_h\} \ .$$

Assume that h is small enough such that $|\underline{u}|\ge 1$ on Γ_h. Denote by $(\underline{u}_h,p_h)\in\tilde{X}_h\times M_h$ the unique solution of

$$\begin{aligned} a_h(\underline{u}_h,\underline{v}_h) + b_h(\underline{v}_h,p_h) &= (-\Delta\underline{u},\underline{v}_h)_{0,h} & \forall\ \underline{v}_h\in\tilde{X}_h \\ b_h(\underline{u}_h,q_h) &= 0 & \forall\ q_h\in M_h \ . \end{aligned} \qquad (5.1)$$

The continuity of $\underline{u}_h$ then implies $\underline{u}_h(p) = 0$ for all vertices p of Ω_h and therefore

$$\|\underline{u}_h\|_{L^2(\Gamma_h)^2} \leq ch^{1/2}\|\underline{u}_h\|_{1,h} \leq c'h^{1/2}\|\underline{u}\|_1 .$$

Hence, we have

$$\|\underline{u}-\underline{u}_h\|_{1,h} \geq c_o\|\underline{u}-\underline{u}_h\|_{L^2(\Gamma_h)^2} \geq c_1 - c_2h^{1/2}\|\underline{u}\|_1 .$$

Thus the solutions of the discrete problems (5.1) do not converge to the solution of the corresponding continuous problem. Instead, the $\underline{u}_h$ converge to the weak solution of the Stokes problem in Ω with right hand side $-\Delta\underline{u}$ and *homogeneous Dirichlet boundary conditions* (cf. [11]). □

Finally, let us remark that a similar result is well known as Babuška-paradox in plane elasticity [1]. The above example, however, seems to be new in fluid dynamics. It shows that problems with mixed boundary conditions behave essentially different from those with Dirichlet boundary conditions.

REFERENCES

1. BABUŠKA, I., The theory of small changes in the domain of existence in the theory of partial differential equations and its applications. In: *Differential equations and their applications*. Academic Press, New York (1963)
2. BEMELMANS, J., Gleichgewichtsfiguren zäher Flüssigkeiten mit Oberflächenspannung. *Analysis* (1981)
3. BEMELMANS, J., Liquid drops in a viscous fluid under the influence of gravity and surface tension. *Manuscripta Math.* 36, 105-123 (1981)
4. BERCOVIER, M., PIRONNEAU, O., Error estimates for finite element method solution of theStokes problem in the primitive variables. *Numer. Math.* 33, 211-224 (1979)
5. BREZZI, F., On the existence, uniqueness and approximation of saddle-point problems arising from Lagrangian multipliers. *RAIRO Anal. Numér.* 8 (R-2), 129-151 (1974)
6. CIARLET, Ph. G., *The finite element method for elliptic problems*. North Holland, New York (1980)
7. GIRAULT, V., RAVIART, P.-A., *Finite element approximation of the Navier-Stokes equations*.Springer, Berlin (1979)
8. LETALLEC, P., A mixed finite element approximation

of the Navier-Stokes equations. *Numer. Math.* 35, 381-404 (1980)

9. SOLONNIKOV, V.A., SČADILOV, V.E., On a boundary value problem for a stationary system of Navier-Stokes equations. *Proc. Steklov Inst. Math.* 125, 186-199 (1973)
10. VERFÜRTH, R., Error estimates for a mixed finite element approximation for the Stokes equations. *RAIRO* (to appear)
11. VERFÜRTH, R., Mixed finite element approximation of Navier-Stokes equations with mixed boundary conditions. (submitted)

PETROV-GALERKIN METHODS AND DIFFUSION-CONVECTION PROBLEMS IN 2D

*K. W. Morton and †B. W. Scotney

*Oxford University Computing Laboratory, 8-11 Keble Road, Oxford
†University of Reading, Whiteknights, Reading
United Kingdom

1. INTRODUCTION

One of the most attractive features of the finite element method has always been its capacity for yielding optimal approximations in some appropriate energy norm: intimately related to this are the superconvergence phenomena, which are often the practical means by which one is able to exploit these optimal approximation properties. As compared with finite difference methods, one is able to get away from estimating truncation errors in terms of inaccessible higher derivatives of the unknown function and, even more difficult, estimating the effect of these on the eventual approximation error. As a result one can remove the emphasis on asymptotic orders of accuracy as the mesh size tends to zero and confidently use coarse meshes where the smoothness of the solution or the sought-after information warrants it.

This is familiar enough in problems where a quadratic extremal principle leads naturally to the Galerkin approximation: that is, where the associated differential equation is linear, elliptic and self-adjoint. However, in more general problems optimal approximation properties are much more difficult to achieve. Considerable attention has been focussed on this area in recent years, in particular on the use of Petrov-Galerkin and other generalised Galerkin methods. The deficiencies of the simple Galerkin approach are similar to those encountered with central difference schemes and in various situations show themselves as:-

- (i) greatly reduced stability in unsteady problems ;
- (ii) divergence of standard iterative procedures for solving the linear algebraic equations in steady problems;
- (iii) much reduced accuracy in both cases.

ISBN 0-12-747255-X

Rapid progress has been made in overcoming these for non-self-adjoint elliptic problems and for hyperbolic problems. We shall concentrate on the former here but shall also comment on some links with the latter.

It was Zienkiewicz who in MAFELAP 1975 [21] first brought to the attention of many of us the diffusion-convection problems which not only form an important class of practical problems in their own right but also provide an admirable testing ground for the methods needed for general non-self-adjoint problems. A typical problem has the form:-

$$-\underline{\nabla}.(a\underline{\nabla}u-\underline{b}u) = f \text{ in } \Omega \quad , \tag{1.1a}$$

$$u = g \text{ on } \Gamma_D \ , \quad \partial u/\partial n = 0 \text{ on } \Gamma_N, \tag{1.1b}$$

where a is a diffusion coefficient and $\underline{b}=(b_1,b_2)^T$ is a convective velocity which we shall assume is incompressible, i.e.

$$\underline{\nabla}.\underline{b} = 0. \tag{1.2}$$

We shall assume Ω is a bounded region of the plane with boundary $\Gamma_D \cup \Gamma_N$ and $\partial/\partial n$ is in the outward normal direction.

With the standard notation $H^1(\Omega)$ for the Sobolev space of functions with first derivatives square integrable, we define the following subspaces:-

$$H^1_E := \{v \varepsilon H^1(\Omega) \,|\, v = g \text{ on } \Gamma_D\} \tag{1.3a}$$

$$H^1_{E_0} := \{v \varepsilon H^1(\Omega) \,|\, v = 0 \text{ on } \Gamma_D\} \quad . \tag{1.3b}$$

Then (1.1) can be written in weak form as

$$B(u,w) = (f,w) \quad \forall w \varepsilon H^1_{E_0} \ , \tag{1.4}$$

where the unsymmetric bilinear form $B(\cdot,\cdot)$ is defined as

$$B(v,w) := (a\underline{\nabla}v,\underline{\nabla}w) + (\underline{\nabla}.(\underline{b}v),w), \tag{1.5}$$

and $(\cdot,\cdot)$ denotes the L^2 inner product over Ω. We shall assume that $a\varepsilon C^0(\bar{\Omega})$ is strictly positive, that $\underline{b}\varepsilon[H^1(\Omega)]^2$ and that $g\varepsilon H^{\frac{1}{2}}(\Gamma_D)$. Then $B(\cdot,\cdot)$ is clearly continuous over $H^1_{E_0} \times H^1_{E_0}$ and is coercive with

$$B(v,v) = (a\underline{\nabla}v,\underline{\nabla}v) + \tfrac{1}{2}\int_{\Gamma_N}(\underline{b}.\underline{n})v^2 ds \tag{1.6}$$

if we assume, as we shall, that $\underline{b}.\underline{n} \geq 0$ on Γ_N, i.e. that u is prescribed on all inflow boundaries. It follows from the Lax-Milgram theorem, see for instance Ciarlet [8], that (1.4) has a unique solution for all $f \varepsilon L^2(\Omega)$ and that this is also the solution of (1.1).

We assume that Ω is polygonal and that on a suitable triangulation, with maximal triangle diameter h, we construct a finite element approximation space $S^h \subset H^1(\Omega)$ of the form

$$S^h := \{V(\underline{x}) = \Sigma_{(j)} V_j \phi_j(\underline{x})\}. \qquad (1.7)$$

We suppose that the Dirichlet data g is the restriction to Γ_D of a function $G \in H^1(\Omega)$ and then define, corresponding to (1.3),

$$S_0^h := S^h \cap H^1_{E_0}, \quad S_E^h := S_0^h \oplus \{G\}. \qquad (1.8)$$

For a Petrov-Galerkin approximation, S_E^h forms the trial space but we need also a test space T_0^h with basis functions $\psi_j(\underline{x})$ and the same dimension as S_0^h:-

$$T_0^h := \{W(\underline{x}) = \Sigma_j W_j \psi_j(\underline{x}) \,|\, W(\underline{x}) = 0 \text{ on } \Gamma_D\}. \qquad (1.9)$$

Thus we seek $U \in S_E^h$ such that

$$B(U,W) = (f,W) \quad \forall W \in T_0^h . \qquad (1.10)$$

The special case $T_0^h \equiv S_0^h$ gives the Galerkin approximation with all the difficulties associated with central differencing. From the viewpoint of seeking an optimal approximation, we then obtain only

$$||u-U||_{B_1} \leq [1+\text{const}.h(|\underline{b}|/a)_{max}] \inf_{V \in S_E^h} ||u-V||_{B_1}, \qquad (1.11)$$

where $h(|\underline{b}|/a)_{max}$ corresponds to a mesh Peclet number, see Morton [16]: here we have used the norm

$$||v||^2_{B_1} \equiv B_1(v,v), \quad B_1(v,w) := (a\underline{\nabla}v, \underline{\nabla}w), \qquad (1.12)$$

which is essentially the symmetric part of B(v,w), but the result would hold in any equivalent norm.

Just as with difference methods, most improvements in the choice of T_0^h have been based on some form of upwinding. That which arises naturally from the problem is due to Hughes & Brooks [13,14]: though originally motivated by enhancing the diffusion only in the direction of $\underline{b}$, and hence often called the streamline diffusion method, it is now generally treated as a Petrov-Galerkin method with basis functions ψ_j given in terms of the trial space basis functions ϕ_j as

$$\psi_j = \phi_j + (\alpha h/2|\underline{b}|)\underline{b}.\underline{\nabla}\phi_j \qquad (1.13)$$

with some locally-defined parameter α. In one dimension a piecewise linear trial space has the familiar hat-shaped basis functions ϕ_j: clearly then the Ψ_j of (1.13) is increased on the upwind side of the central node and decreased on the downwind side. The result will be a discontinuous test function to which the simple standard theory does not apply, but element-by-element evaluation of the inner products leads to a straightforward extension. Alternative test functions with similar properties abound in the literature: the proceedings of the special ASME conference on this topic, edited by Hughes [12] and including the survey [10], and the report on the IAHR workshop on the topic, prepared by Smith & Hutton [19], give a good indication of the variety of methods presently in use.

In this paper, we shall first show how the correct choice of test space T_0^h can lead to an optimal approximation to (1.4) and how from this viewpoint upwinded test functions reduce the factor in (1.11) to a value near unity which is independent of mesh Péclet number. Then in section 3 we shall present a new mixed method developed from that presented by Barrett & Morton [5] at the last MAFELAP conference: these seek approximations which are optimal in a norm differing from that of (1.12) in that they tend to the L^2 norm as the mesh Péclet number tends to infinity. Hence we can trace the limit of the procedure to that appropriate to hyperbolic problems and in particular to the characteristic Galerkin methods that have been developed for these in recent years. Finally in section 4 we present comparative numerical results for this method and standard upwind methods both for the CEGB test problem reported on in [19] and a modification which has a strong tangential boundary layer.

2. OPTIMAL AND NEAR-OPTIMAL APPROXIMATIONS

2.1 Theoretical Framework

Suppose we equip $H^1_{E_0}$ with any inner product $B_m(\cdot,\cdot)$ and corresponding norm $\|\cdot\|_{B_m}$ with respect to which $B(\cdot,\cdot)$ is therefore continuous and coercive. Then for fixed $w\in H^1_{E_0}$, $B(v,w)$ is a bounded linear functional over $v\in H^1_{E_0}$. Thus by the Riesz representation theorem and by the linear dependence on w there exists a bounded linear operator $R_m: H^1_{E_0}\to H^1_{E_0}$ such that

$$B(v,w) \equiv B_m(v,R_m w) \quad \forall v,w\in H^1_{E_0}. \tag{2.1}$$

The error u-U in the Petrov-Galerkin approximation (1.10) corresponding to a conforming test space $T_0^h\subset H^1_{E_0}$ clearly satisfies

$$B(u-U,W) = 0 \quad \forall W\in T_0^h : \tag{2.2a}$$

so this can be written instead as the orthogonality relation

$$B_m(u-U,R_m W) = 0 \quad \forall W\in T_0^h . \tag{2.2b}$$

From the coercivity of $B(\cdot,\cdot)$ it follows that R_m is invertible on $H^1_{E_0}$ and hence there exists a test space, spanned by $\{\psi_i^*\}$

say, such that

$$\mathrm{span}\{R_m\psi_i^*\} = \mathrm{span}\{\phi_i\} = S_0^h \quad . \tag{2.3}$$

If we denote by U_m^* the corresponding Petrov-Galerkin approximation obtained with this test space, we have

$$B_m(u-U_m^*,\phi_i) = 0 \quad \forall\phi_i\in S_0^h \tag{2.4}$$

and hence the optimal approximation property

$$\|u-U_m^*\|_{B_m} = \inf_{V\in S_E^h}\|u-V\|_{B_m} \quad . \tag{2.5}$$

Thus we are assured that the Petrov-Galerkin approach is capable of yielding what we seek.

In practice, of course, R_m is not very easy to deal with: in some one-dimensional cases explicit expressions for R_m and for ψ_i^* have been given by Barrett & Morton [6] but these show that R_m is a non-local operator and serve to emphasise the difficulties of a direct approach to solving (2.3) in two-dimensional problems. However, the above framework gives a valuable basis for an error analysis of more practical Petrov-Galerkin methods and also leads to the more indirect approaches which will be exploited in the next section.

As regards error bounds, the following result is established in Morton [16] and in [6]. Suppose the closeness with which S_0^h can be approximated by $R_mT_0^h$ is described by a constant $\Delta_m \equiv \Delta_m(h)$:-

$$\inf_{W\in T_0^h}\|V-R_mW\|_{B_m} \le \Delta_m\|V\|_{B_m} \quad \forall V\in S_0^h \quad . \tag{2.6}$$

Then if $\Delta_m<1$, the Petrov-Galerkin approximation U is uniquely defined by (1.10) and satisfies the error bound

$$\|u-U\|_{B_m} \le (1-\Delta_m^2)^{-\frac{1}{2}} \inf_{V\in S_E^h}\|u-V\|_{B_m} \quad . \tag{2.7}$$

This gives a sharper result than the generalised Lax-Milgram lemma of Babuska & Aziz [2]. In addition one can go some way towards setting up an explicit calculation for Δ_m. In the form given by Scotney [18] and discussed in [6], let us define the NxN matrices A, B and C, where $N = \dim S_0^h$, having the entries

$$A_{ij} = B_m(R_m\psi_j,R_m\psi_i) \tag{2.8a}$$

$$B_{ij} = B_m(\phi_j,R_m\psi_i) \equiv B(\phi_j,\psi_i) \tag{2.8b}$$

$$C_{ij} = B_m(\phi_j,\phi_i) \quad . \tag{2.8c}$$

Then we find that

$$1 - \Delta_m^2 = \inf_{V\varepsilon S_0^h}\left[\frac{\underline{V}^T B^T A^{-1} B\underline{V}}{\underline{V}^T C\underline{V}}\right], \tag{2.9}$$

where $\underline{V}$ is the N-vector of nodal parameters for V.

Note that in calculating (2.9) knowledge of the explicit form of R_m is needed only in the calculation of A. Even this can be considerably reduced and sometimes completely eliminated. Let us consider the case of the norm given by (1.12), which we shall regard as the particular case of the above formulae in which we put m=1. Then the defining relation for R_1 becomes

$$(\underline{\nabla}v, a\underline{\nabla}w+\underline{b}w) \equiv (\underline{\nabla}v, a\underline{\nabla}(R_1 w)) \quad \forall v, w\varepsilon H^1_{E_0}, \tag{2.10}$$

where we have used $\underline{\nabla}.\underline{b} = 0$ to write

$$(\underline{\nabla}.(\underline{b}v), w) = (\underline{b}.\underline{\nabla}v, w) = (\underline{\nabla}v, \underline{b}w) . \tag{2.11}$$

Then we obtain from (1.12), using (2.10) twice,

$$\begin{aligned} B_1(R_1\psi_j, R_1\psi_i) &= (\underline{\nabla}(R_1\psi_j), a\underline{\nabla}(R_1\psi_i)) \\ &= (\underline{\nabla}(R_1\psi_j), a\underline{\nabla}\psi_i+\underline{b}\psi_i) \\ &= (a\underline{\nabla}\psi_j+\underline{b}\psi_j, \underline{\nabla}\psi_i)+(\underline{\nabla}(R_1\psi_j), \underline{b}\psi_i). \end{aligned} \tag{2.12}$$

This can be put in a more symmetric form by first noting that

$$(\underline{b}v, \underline{\nabla}w) + (\underline{\nabla}v, \underline{b}w) = \int_{\Gamma_N}(\underline{b}.\underline{n})vw\,ds \quad \forall v, w\varepsilon H^1_{E_0} . \tag{2.13}$$

Thence we obtain

$$\begin{aligned} B_1(R_1\psi_j, R_1\psi_i) &= (a\underline{\nabla}\psi_j, \underline{\nabla}\psi_i) + \tfrac{1}{2}\int_{\Gamma_N}(\underline{b}.\underline{n})\psi_j\psi_i\,ds \\ &+ \tfrac{1}{2}(\underline{b}.\underline{\nabla}(R_1\psi_j), \psi_i) + \tfrac{1}{2}(\underline{b}.\underline{\nabla}(R_1\psi_i), \psi_j). \end{aligned} \tag{2.14}$$

To reduce this still further we will make some simplifying assumptions.

We suppose that a is constant, that we have only Dirichlet boundary conditions, and that Ω is convex. Then $\underline{b}\psi_i \varepsilon [H_0^1(\Omega)]^2$ can be decomposed into

$$(\underline{b}/a)\psi_i = \underline{\nabla}q_i + \underline{\nabla}\times\chi_i \tag{2.15}$$

where $\underline{\nabla}\times v := (\partial v/\partial y, -\partial v/\partial x)^T$ and $q_i, \chi_i \varepsilon H_0^1(\Omega)$. Then applying the decomposition to the last term of (2.12) we obtain

$$\begin{aligned} B_1(R_1\psi_j, R_1\psi_i) &= (a\underline{\nabla}(R_1\psi_j), \underline{\nabla}(\psi_i+q_i) + \underline{\nabla}\times\chi_i) \\ &= (a\underline{\nabla}\psi_j+\underline{b}\psi_j, \underline{\nabla}\psi_i+(\underline{b}/a)\psi_i)+(a\underline{\nabla}(R_1\psi_j)-a\underline{\nabla}\psi_j-\underline{b}\psi_j, \underline{\nabla}\times\chi_i). \end{aligned} \tag{2.16}$$

Finally, application of Green's theorem to the last term in (2.16), using the assumption that a is constant and the corresponding decomposition for $\underline{b}\psi_j$, gives the symmetric form

$$B_1(R_1\psi_j, R_1\psi_i) = (a\underline{\nabla}\psi_j + \underline{b}\psi_j, \underline{\nabla}\psi_i + (\underline{b}/a)\psi_i) - (a\underline{\nabla}\times\chi_j, \underline{\nabla}\times\chi_i) \qquad (2.17)$$

which does not require knowledge of the operator R_1 at all.

2.2 Analysis in one dimension for $B_1(\cdot,\cdot)$

In one dimension, with b constant to conform with (1.2), (2.10) can be solved explicitly to give

$$(R_1 w)(x) = w(x) + b\int_0^x \frac{w(t)-\gamma\bar{w}}{a(t)}dt, \qquad (2.18)$$

where $\gamma = 0$ or 1 according to whether a Neumann or Dirichlet boundary condition is applied at the outflow end of the interval: in (2.18) $\bar{w}$ denotes the average of w(x) with the weighting $a^{-1}(x)$ so that $R_1 w$ satisfies the outflow boundary condition when it is of Dirichlet type. It is then easy to see that

$$B_1(R_1\psi_j, R_1\psi_i) = \int a[\psi_j' + b(\psi_j - \gamma\bar{\psi}_j)/a][\psi_i' + b(\psi_i - \gamma\bar{\psi}_i)/a]dx, \qquad (2.19)$$

which can readily be evaluated. Scotney [18] has used this to calculate (2.9) for a number of commonly used test spaces corresponding to a piecewise linear trial space and where Ω is the unit interval and a is also constant. As shown by Barrett & Morton [6], in this case the exponential test functions used by Hemker [11] form an ideal test space satisfying the relations (2.3) and therefore giving $\Delta_1 = 0$. For the quadratic conforming test space used by Christie et al [7] and Heinrich et al [9] the error factors are given in Table 1: this space has basis functions

$$\psi_i(x) := \phi_i(x) + \alpha\sigma_i(x) \qquad (2.20a)$$

with

$$\sigma_i(x) := \begin{cases} 3(x-x_{i-1})(x_i-x)/h^2, & x_{i-1} \le x \le x_i \\ -3(x_{i+1}-x)(x-x_i)/h^2, & x_i \le x \le x_{i+1} \end{cases} \qquad (2.20b)$$

where $\phi_i(x)$ is the piecewise linear hat function and α a free parameter. Table 1 also contains similar results for the non-conforming test space of Hughes & Brooks [13,14] given by (1.13), where the computations have to be carried out element-by-element and the above theory does not strictly apply - though see below.

For both of these test spaces the free parameter (α in (1.13) and in (2.20a)) has been chosen to reproduce the difference operator due to Allen & Southwell [1], as does the Hemker test space: that is, it is set equal to the well-known exponential fitting factor

$$\alpha = \xi := \coth(bh/2a) - (2a/bh), \tag{2.21}$$

to be substituted in the scheme

$$-(a/h)[1+\tfrac{1}{2}\alpha bh/a]\delta^2 U_i + b\Delta_0 U_i = (f,\psi_i) \tag{2.22a}$$

$$\text{or } -(a/h)\delta^2 U_i + b[(1-\alpha)\Delta_0 + \alpha\Delta_-]U_i = (f,\psi_i) \tag{2.22b}$$

using the standard difference notation $\delta^2 U_i := U_{i+1} - 2U_i + U_{i-1}$, $\Delta_0 U_i := \frac{1}{2}(U_{i+1} - U_{i-1})$ and $\Delta_- U_i := U_i - U_{i-1}$. The corresponding columns headed HHMZ and HB of the table demonstrate clearly how effective both spaces are in eliminating the dependence of the error factor on the mesh Péclet number, as compared with the divergence of this same factor exhibited for the simple Galerkin method.

The norm $\|\cdot\|_{B_1}$ in this one-dimensional constant coefficient case leads to a best fit solution U_i^* which is exact at the nodes. Indeed Hemker derived his test functions as the Green's functions for the adjoint problem at the nodes to achieve just this for any trial space. Thus these test spaces will give very good nodal accuracy when the error norm for the best fit $\|u-U^*\|_{B_1}$ is small: when it is not, of course even a small increase can lead to considerable inaccuracy at the nodes, as may happen with a sharp boundary layer caused by Dirichlet boundary conditions.

TABLE 1

bh/A	Error factor $(1-\Delta_1^2)^{-\frac{1}{2}}$				
	GALERKIN	HHMZ	HHMZ*	HB	SHB
2	1.1547	1.0060	1.0060	1.0924	1.0178
5	1.7559	1.0468	1.0428	1.1509	1.0597
50	14.468	1.2022	1.0945	1.1547	1.1406
500	144.34	1.2344	1.0954	1.1547	1.1532
10^5	28868.	1.2383	1.0954	1.1547	1.1546

Error factors in (2.7) given by (2.9) for m=1:HHMZ as in [9] using (2.20) with (2.21); HHMZ* with (2.23); HB as in [13,14] using (1.13) with (2.21); SHB is the "super" Hughes & Brooks as in [18].

The table also contains two other results of some interest. The discontinuous streamline diffusion test space can be modified to make it conforming: Scotney does this by modifying the square pulses generated by $\phi_j(x)$ to have sloping sides over an interval εh but the same area so that the Allen & Southwell operator continues to be generated. Then ε can be chosen to minimise the error constant. The results for this conforming test space are shown in the table as SHB, for "super Hughes & Brooks" as they are generally better than the corresponding ones for the streamline diffusion method. More importantly, they are rigorous bounds and the computation of the optimal ε shows that it tends to zero as the mesh Péclet number tends to infinity: thus the streamline diffusion method can be regarded as the limit of this more accurate conforming method, which yields the same difference operator and differs from it only in the sampling of any source terms f.

The second result relates to the column headed HHMZ*. In this the free parameter multiplying the quadratic terms in the test functions (2.20) used by Heinrich et al.[9] has been chosen to minimise the error factor. It turns out to be given by

$$\alpha_{opt} = \frac{5bh}{36a} \cdot \frac{1 + 12(a/bh)^2}{1 + 10(a/bh)^2} \tag{2.23}$$

so that in the limit of infinite mesh Péclet number, instead of having an equal weighting of $\phi_i(x)$ and $\sigma_i(x)$ in (2.20a) as would happen with the conventional choice (2.21), the quadratic term $\sigma_i(x)$ is wholly dominant. This means that the corresponding difference scheme, instead of tending to an upwind approximation to bu' as would be given by (2.22) as $\alpha \to 1$, tends to the second difference approximating bu": clearly this is because the oscillatory quadratic $\sigma_i(x)$ yields in this limit an approximation to f' on the right hand side.

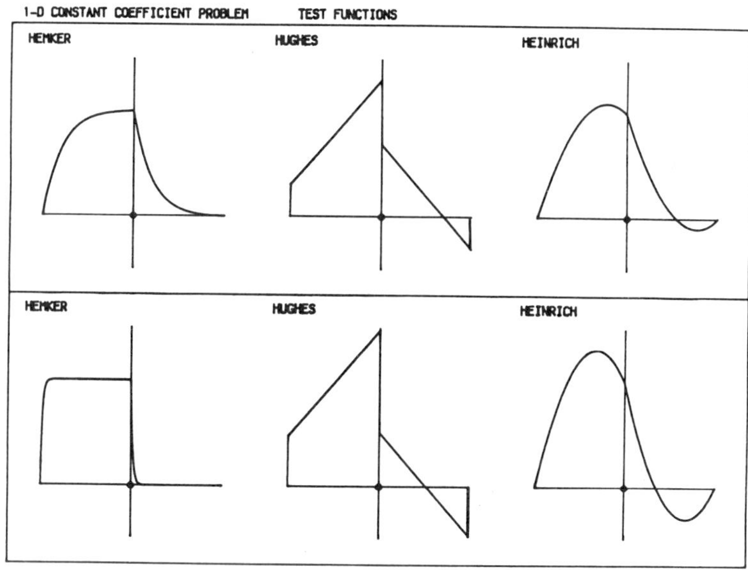

FIG. 1. Test functions for mesh Péclet number 5 (top) and 50 (bottom).

We see from this table both the limitation of being too closely guided by finite difference arguments and the shortcomings of the two test spaces (1.13) and (2.20). The Allen & Southwell scheme will give exact nodal values with a constant source term no matter what test space is used to obtain it. But it will only give good nodal accuracy for an arbitrary source term if the test functions approximate the Hemker test functions. It is clear from Fig.1 where the three test functions of Hemker, Hughes (et al.) and Heinrich (et al.) are plotted, that the last two are not very good approximations to the first when bh/a becomes large. Mixing in some proportion of the characteristic function for the upwind interval is clearly indicated and this, of course, is all that will be needed in the limit when one is approximating only $bu'=f$. Interestingly, Griffiths (private communication) has also noted that this test function is needed when the equation is generalised to $-au''+bu'+cu = f$ or when a time derivative or derivatives with respect to a second space coordinate are simulated in the analysis.

3. A NEW MIXED METHOD

The existence of ideal test spaces given by (2.3) prompts an alternative approach to the diffusion convection problem which has some advantages over a direct Petrov-Galerkin formulation. Since by (1.8) the Petrov-Galerkin approximation, given by (1.10), is such that we have $U-G \in H^1_{E_0}$, we can write (1.10) as

$$B(U-G,W) = (f,W) - B(G,W) \qquad \forall W \in T_0^h, \tag{3.1a}$$

$$\text{i.e. } B_m(U-G,R_m W) = (f,W) - B(G,W) \quad \forall W \in T_0^h. \tag{3.1b}$$

Now let us suppose that, for each basis function ϕ_i, we obtain ψ_i^* such that $R_m\psi_i^* = \phi_i$ and take $T_0^h = \text{span}\{\psi_i^*\}$. Then (3.1b) will give the best fit approximation U_m^* defined by (2.4) in the form

$$B_m(U_m^*-G,\phi_i) = (f,\psi_i^*) - B(G,\psi_i^*) \tag{3.2a}$$

$$\text{i.e. } B_m(U_m^*-G,\phi_i) = (f,R_m^{-1}\phi_i) - B(G,R_m^{-1}\phi_i) \quad \forall \phi_i \in S_0^h \,. \tag{3.2b}$$

The significant points here are that the system of equations for U_m^*-G is now symmetric and that $\psi_i^*=R_m^{-1}\phi_i$ only appears in the right hand side. Thus approximations to U_m^* can be formed by approximating $R_m^{-1}\phi_i$ in the terms on the right.

This is the basis of the approximate symmetrization approach developed by Barrett & Morton [3,4,5,6]. It should be noted that the relationship $R_m\psi_i^*=\phi_i$ is quite onerous to satisfy and means that the ψ_i^* are non-local. Thus even for m=1 and the one-dimensional model problem the ψ_i^* are linear combinations of the Hemker test functions. This is why even in one dimension these authors only approximately symmetrize - though see Rheinhardt [17] for an equivalent completely symmetric approach, which is closely related to the two-dimensional method described at the end of this section.

For two-dimensional problems Barrett & Morton in [6] proposed a mixed method approach to (3.2) and it is this that we shall develop here. The aim is to treat separately the convective features of the flow and the interaction of diffusion with convection. For this purpose we assume the flow has no recirculating regions and referring to the problem statement of (1.1), introduce a function $h \in H^1(\Omega)$ given by the convection equation

$$\underline{b}.\underline{\nabla}h = f \text{ in } \Omega,\ h = g \text{ on } \Gamma_D \cap \Gamma_- , \tag{3.3}$$

where Γ_- is the inflow boundary, on which $\underline{b}.\underline{n}<0$. Then we introduce the flux function

$$\underline{v} = \underline{b}(u-h) - a\underline{\nabla}u \tag{3.4}$$

so that the diffusion convection equation (1.1a) reduces to

$$\underline{\nabla}.\underline{v} = 0 . \tag{3.5}$$

This enables a stream function ψ to be introduced such that

$$\underline{v} = \underline{\nabla}\times\psi := (\partial\psi/\partial y, -\partial\psi/\partial x)^T \tag{3.6a}$$

$$\underline{b}.\underline{v} = -\underline{p}.\underline{\nabla}\psi,\ \underline{p}.\underline{v} = \underline{b}.\underline{\nabla}\psi = -a\underline{p}.\underline{\nabla}u, \tag{3.6b}$$

where $\underline{p} = (b_2,-b_1)^T$ is perpendicular to $\underline{b}$.

In common with Barrett & Morton we shall use as the main part of our symmetric form

$$B_2(v,w) := (a\underline{\nabla}v,\underline{\nabla}w) + ((b^2/a)v,w) \quad \forall v,w\in H^1(\Omega) . \tag{3.7}$$

It is easily checked that for $w\in H^1_{E_0}$, this has the alternative form

$$B_2(v,w) = (a\underline{\nabla}v-\underline{b}v, \underline{\nabla}w-(\underline{b}/a)w) + \int_{\Gamma_N} (\underline{b}.\underline{n})vw\,ds \quad \forall v\in H^1(\Omega),\ w\in H^1_{E_0} . \tag{3.8}$$

Thus if we factor the diffusion convection operator into $L_1^*L_2$, where L_1^* is the adjoint of $L_1 := a^{\frac{1}{2}}\underline{\nabla}$ and $L_2 := a^{\frac{1}{2}}\underline{\nabla}-(\underline{b}/a^{\frac{1}{2}})$, then $B_1(\cdot,\cdot)$ is the symmetric form based on L_1 and $B_2(\cdot,\cdot)$ is the natural alternative based on L_2. We have already seen that in one dimension R_1 has a simple form and in two dimensions its defining relation (2.10) enables considerable simplication to be effected in the expressions for $B_1(R_1\psi_j, R_1\psi_i)$: on the other hand it is seen in [6] that R_1^{-1} involves rather awkward integrals over sharply decaying exponentials. Just the opposite holds for the bilinear form $B_2(\cdot,\cdot)$: while the corresponding R_2 involves the awkward exponential factors making the error analysis for a standard Petrov-Galerkin method difficult, the form of R_2^{-1} is simpler making the use of (3.25) more attractive in this case. Indeed in the one dimensional Dirichlet case, corresponding to (2.18) we obtain

$$(R_2^{-1}w)(x) = w(x) + b\int_x^1 \frac{w(t) - C\bar{w}e^{-\lambda(t)}}{a(t)}\,dt, \qquad (3.9)$$

where $\lambda(x) = \int_0^x (b/a)dt$ and C is determined by the boundary condition at x=0: the similarity of form to (2.18) is clear.

Thus for the solution u of the diffusion convection problem (1.1) we obtain an equation of the form (3.2) with m=2 as follows:-

$$B_2(u,w) := (a\underline{\nabla}u,\underline{\nabla}w)+((b^2/a)u,w)=B(u,w)+((b^2/a)u-\underline{b}.\underline{\nabla}u,w)$$

$$= (f,w) + ((\underline{b}/a).(\underline{v}+h\underline{b}),w) \quad \forall w\varepsilon H^1_{E_0} \; ;$$

that is,

$$B_2(u,w) = (f+(b^2/a)h+(\underline{b}/a).\underline{v},w) \quad \forall w\varepsilon H^1_{E_0} \; , \qquad (3.10)$$

in terms of the quantities h and $\underline{v}$ introduced in (3.3)-(3.6).

The identification with (3.2) is seen most readily in special cases when only one form of inhomogeneous data is present. Suppose, for instance, g=0; then

$$B_2(u,w) = (f,w) + ((\underline{b}/a).(\underline{b}u-a\underline{\nabla}u),w) \equiv (f,R_2^{-1}w) \quad \forall w\varepsilon H^1_{E_0} \qquad (3.11)$$

in which the similarity of the form of R_2^{-1} with that of (3.9) is evident as u is wholly determined from $\underline{\nabla}.(\underline{b}u-a\underline{\nabla}u) = f$ and homogeneous boundary conditions. On the other hand, suppose $f \equiv 0$ and that tangential boundary conditions are compatible with the Dirichlet data on the inflow boundary: then $\underline{b}.\underline{\nabla}h = 0$ and we can take $G \equiv h$ so that we get

$$B_2(u-G,w) = -B(G,w) - [B_2(G,w)-B(G,w)] + ((\underline{b}/a).(\underline{b}u-a\underline{\nabla}u),w)$$

$$= -B(G,w) - (\underline{b}/a).[\underline{b}(G-u)-a\underline{\nabla}(G-u)],w)$$

$$= -B(G,R_2^{-1}w) \qquad \forall w\varepsilon H^1_{E_0} \; , \qquad (3.12)$$

with u now being linearly dependent on G.

The relations (3.10) however are the starting point for our approximation method. The flux function $\underline{v}$ will be determined from its stream function ψ and $\underline{b}.\underline{v}$ replaced by $\underline{p}.\underline{\nabla}\psi$. The equation for ψ is obtained from (3.4) and (3.6a) as

$$\underline{\nabla}.(a^{-1}\underline{\nabla}\psi) = \underline{\nabla}\times(\underline{v}/a) = \underline{\nabla}\times[(\underline{b}/a)(u-h)] = -\underline{\nabla}.[(f/a)(u-h)]. \qquad (3.13)$$

In weak form this becomes then for $w\varepsilon H^1(\Omega)$

$$(a^{-1}\underline{\nabla}\psi,\underline{\nabla}w) = \int_\Gamma[\frac{\partial\psi}{\partial n} + (\underline{p}.\underline{n})(u-h)](w/a)ds-(a^{-1}(u-h),\underline{p}.\underline{\nabla}w). \quad (3.14)$$

On the boundary Γ, from (3.4) and (3.6) we obtain

$$\frac{\partial\psi}{\partial n} = -(\underline{p}.\underline{n})(u-h) + a\frac{\partial u}{\partial s}, \quad (3.15)$$

where $\partial/\partial s$ is the tangential derivative, in an anti-clockwise direction for convex Ω. Thus (3.14) simplifies to

$$(a^{-1}\underline{\nabla}\psi,\underline{\nabla}w) = \int_\Gamma \frac{\partial u}{\partial s} wds - (a^{-1}(u-h),\underline{p}.\underline{\nabla}w). \quad (3.16)$$

The essence of the method is then to iterate between (3.16) for ψ and (3.10) for u.

However, there is one further ingredient to add. In many problems there may be sharp gradients of u normal to the velocity field $\underline{b}$ and it is the maintenance of these in the presence of small diffusion coefficients which is of great interest. There may therefore be some advantage in exploiting the freedom that we have in choosing our norm to emphasise these features. Fortunately, relations (3.6b) show how this may be achieved: it follows from these that

$$(a\underline{p}.\underline{\nabla}u,\underline{p}.\underline{\nabla}w) = -(\underline{b}.\underline{\nabla}\psi,\underline{p}.\underline{\nabla}w) \qquad \forall w\varepsilon H^1(\Omega). \quad (3.17)$$

This can be combined with (3.10) in some proportion and hence we introduce the more general bilinear form

$$B_\gamma(v,w) := B_2(v,w) + \gamma^2(a\underline{p}.\underline{\nabla}v,\underline{p}.\underline{\nabla}w) \qquad v,w\varepsilon H^1(\Omega), \quad (3.18)$$

which we shall use in our mixed method.

The method has been implemented on rectangular elements using bilinear basis functions ϕ_i. It is assumed that the streamlines for the velocity field $\underline{b}$ are available, at least as straight line segment approximations across the elements. Then an approximation H to h is obtained by integrating the ordinary differential equations (3.3) along each (approximate) streamline: in many practical cases, of course, f=0 except possibly in a few confined regions. The spacing of the streamlines needs to be sufficiently fine that H can be evaluated at the 2x2 Gauss points used in the quadratures for (3.10) and (3.16). The bilinear elements are used for both U and Ψ. Then using superscripts to denote successive iterates and starting with $U^{(0)} = G$, we obtain the pair of equations for $\ell = 1,2,\ldots$

$$(a^{-1}\underline{\nabla}\Psi^{(\ell)},\underline{\nabla}\phi_i) = \int_\Gamma \frac{\partial U^{(\ell-1)}}{\partial s}\phi_i ds - (a^{-1}(U^{(\ell-1)}-H),\underline{p}.\underline{\nabla}\phi_i) \quad \forall\phi_i\varepsilon S^h, \quad (3.19)$$

$$B_\gamma(U^{(\ell)},\phi_i) = (f+(b^2/a)H-(\underline{p}/a).\nabla\Psi^{(\ell)},\phi_i)-\gamma^2(\underline{b}.\underline{\nabla}\Psi^{(\ell)},\underline{p}.\underline{\nabla}\phi_i)\forall\phi_i\varepsilon S_0^h. \quad (3.20)$$

Note that each system of equations is symmetric, though they are of different order because all the boundary conditions for Ψ are of Neumann type, the arbitrary additive constant being of no significance. In practice it has been found that convergence takes place after three or four iterations.

4. RESULTS FOR THE CEGB TEST PROBLEMS

Considerable stimulus for the development of the method presented in the previous section has been provided by the test problems prepared by research workers at the CEGB for an IAHR workshop held there in 1981. The main problem concerned the convection of a sharply varying temperature profile by a completely reversed flow and the results submitted for this are summarised and reviewed in Smith & Hutton [19]. The second problem was harder and involved a tangential boundary layer. We have slightly modified the latter and compared various standard finite difference and finite element methods with our new method for each of them.

The domain Ω for these problems is a rectangle $-1 \leq x \leq 1$, $0 \leq y \leq 1$ and the velocity field is specified analytically by a stream function $(1-x^2)(1-y^2)$ as

$$\underline{b} = (2y(1-x^2), -2x(1-y^2))^T \ . \tag{4.1}$$

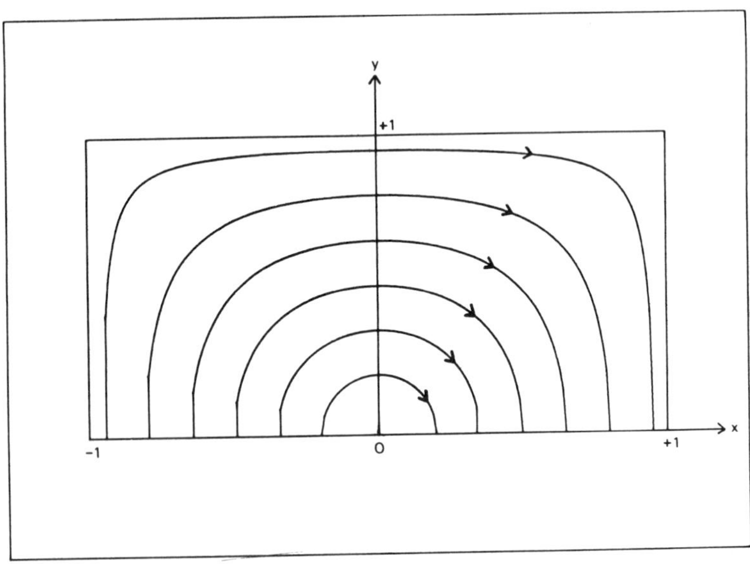

FIG. 2. Test problem flow field.

Streamlines, and boundary conditions for the second (modified) problem, are sketched in Fig. 2. In the first problem, the data on the inlet boundary $-1 \leq x \leq 0$, $y = 0$ are given by

$$u(0,x) := g(x) := 1 + \tanh[(2x+1)\alpha] \qquad (4.2)$$

with $\alpha = 10$ in all cases considered: on the tangential boundaries $x = -1$, $y = 1$ and $x = 1$ the compatible Dirichlet condition $g \equiv 1-\tanh\alpha$ is imposed which for $\alpha = 10$ is for all practical purposes equivalent to $g \equiv 0$. On the outflow boundary $0 \leq x \leq 1$, $y = 0$ no condition is imposed for the Petrov Galerkin methods, corresponding to a homogeneous Neumann condition.

The main test in the first problem is for the output profile for various values of the diffusion coefficient a, and thus for the Péclet number. The standard mesh used as one of the means of comparing methods in [19] was $\Delta x = \Delta y = 0.1$: this is the only mesh for which we shall give results. Values of a that we have used are 10^{-6}, 2×10^{-3}, 10^{-2} and 10^{-1} giving mesh Péclet numbers, with max $\{b\} = 2$, of 2×10^5, 100, 20 and 2 respectively. To cover all of these cases effectively on such a coarse mesh is quite a severe test of any method: the velocity field has very high curvature near the origin so that inaccuracies due to cross-wind diffusion are very easily introduced; thus some methods may be designed to deal well with high Péclet numbers but then fail to pick up changes as a is increased; while methods which concentrate too heavily on avoiding non-physical oscillations will smear the profile for small values of a.

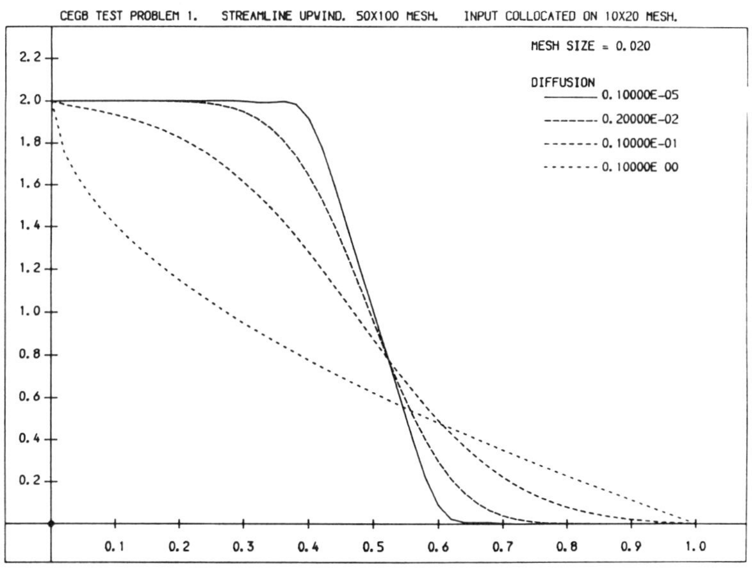

FIG. 3. Fine mesh accurate solution for first problem.

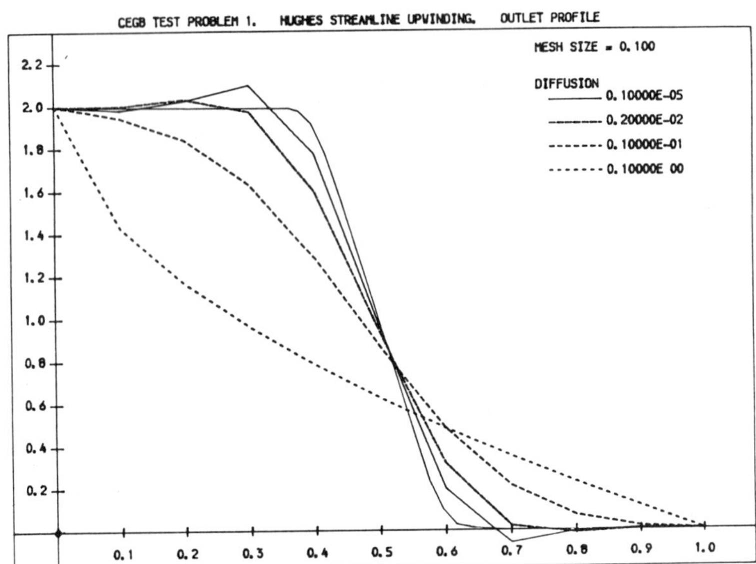

FIG. 4. Standard mesh solution (plus fine mesh for a = 10^{-6}) for streamline diffusion method.

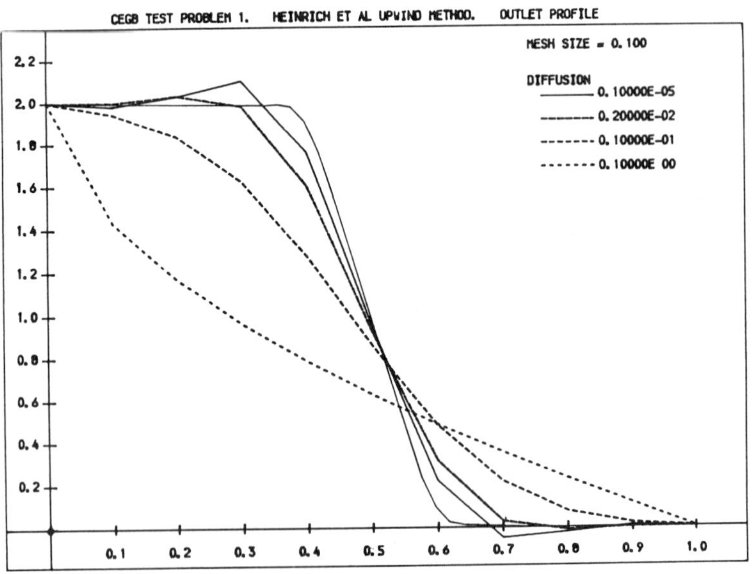

FIG. 5. Standard mesh solution (plus fine mesh for a = 10^{-6}) for Heinrich et al. method.

The input profile itself is not well represented on this mesh so it should be noted that we have used the same data for all methods, namely grid-point values of (4.2) with linear interpolation where required. The accurate solution used for comparison is obtained using this same data but with a mesh of $\Delta x = \Delta y = 0.02$: calculated with the streamline diffusion method, it is given as Fig. 3 and the result for the highest Péclet number is reproduced on each of the subsequent figures. The first of these, Fig. 4, gives the streamline diffusion results on the standard mesh. They are clearly very good for all but the very highest Péclet number. Even in that case there is only about a five percent overshoot and undershoot and the profile has not lost much of its sharpness. Similar results are given in Fig. 5 for the Heinrich et al. [9] upwind method and indeed the results are remarkably similar.

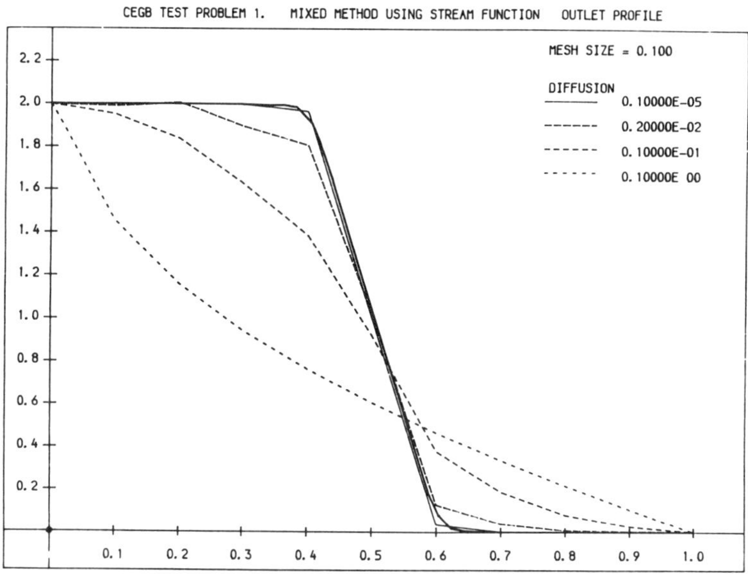

FIG. 6. Standard mesh solution (plus fine mesh for $a = 10^{-6}$) for new mixed method.

In Fig. 6 we give the corresponding results for the new mixed method. As might be expected from the construction of the method, these are very much better for the highest Péclet number, being as good then as one could hope to get on this mesh. They are not quite as good as those for the two earlier methods in the next case, where the mesh Péclet number is 100, but are comparable for the last two cases, of mesh Péclet numbers 20 and 2. The value of γ^2 used for these results was 60; thus since most of the variation of u is normal to the field lines, the norm here is very comparable with the $\|\cdot\|_{B_1}$ norm we have assumed appropriate for the other two methods and therefore the results are strictly comparable.

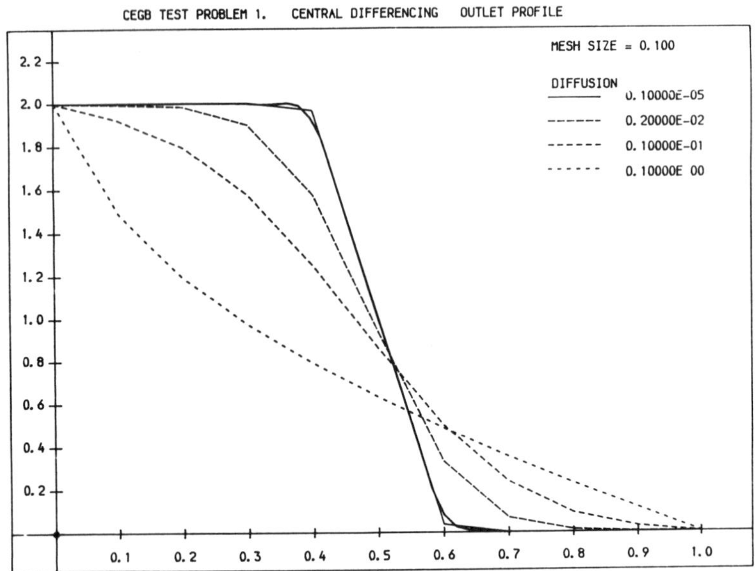

FIG. 7. Standard mesh solution (plus fine mesh for $a = 10^{-6}$) for central differencing.

Finally for this problem we give three sets of finite difference results. In Fig. 7 we give those obtained with central differences: they are remarkably good! However, as noted by Smith & Hutton [19], there is a symmetry condition operating to eliminate mesh oscillations just at the outlet while only one element in from the boundary there is a 20% overshoot. (In all the earlier cases we have checked that the output profile is not particularly flattering and that there is not a deterioration of the accuracy away from the outlet boundary.) Then in Fig. 8 we give results for a scheme used by Spalding [20] in which the differences in each direction at a mesh point are switched between central and fully upwind according to whether $b_i h/a$ is less than or greater than two. Clearly it gives good results (corresponding to the central difference scheme) for the most diffusive case but is hopelessly inaccurate for all other cases. Thirdly, Fig. 9 gives results for a two dimensional version of the Allen & Southwell scheme studied by Kellogg [15]: rather disappointingly this is very little better than the simple switch between central and upwind differences.

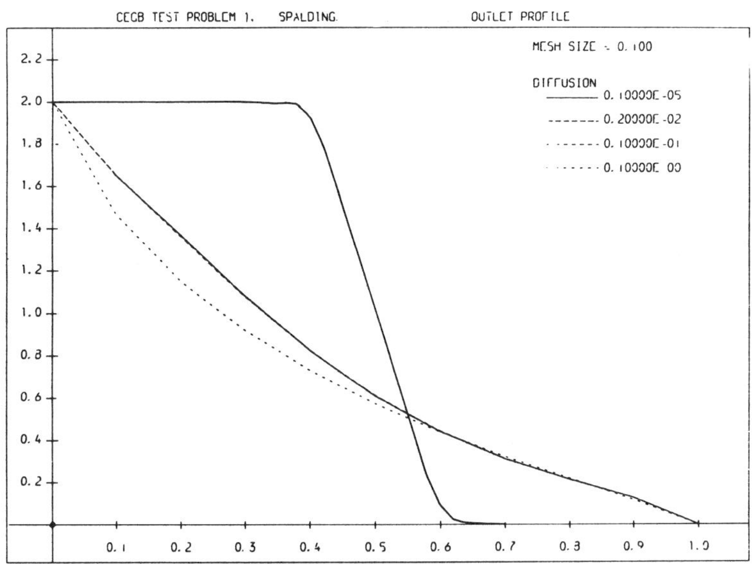

FIG. 8. Standard mesh solution (plus fine mesh for a = 10^{-6}) for upwind differencing.

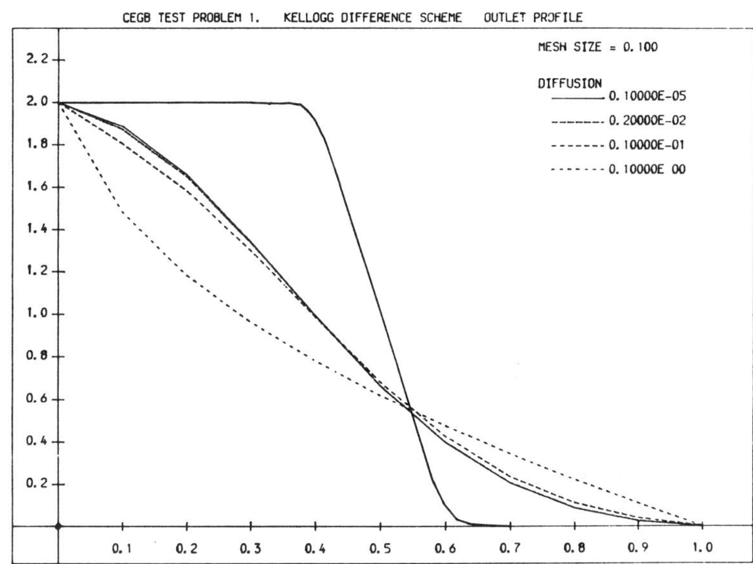

FIG. 9. Standard mesh solution (plus fine mesh for a = 10^{-6}) for an Allen & Southwell scheme.

The second CEGB problem has been modified to have zero Dirichlet data on $x = -1$, most of $y = 1$ and at the inflow $-1 \leq x \leq 0$, $y = 0$, while on $x = 1$ we impose

$$u(1,y) = g(1,y) := 100: \tag{4.3}$$

to ensure regularity of the data, g is then brought down linearly from $g(1,1) = 100$ to $g(0.9,1) = 0$. This gives a sharp boundary layer starting from the singular point $x = 1$, $y = 1$ which broadens as it is carried down to the outlet at $y = 0$. The results are therefore presented, for each of the three finite element methods, as plots of this boundary layer at the three stations $y = 0.9$, $y = 0.5$ and $y = 0$. These are given in Figs. 10-13 for the four Péclet numbers.

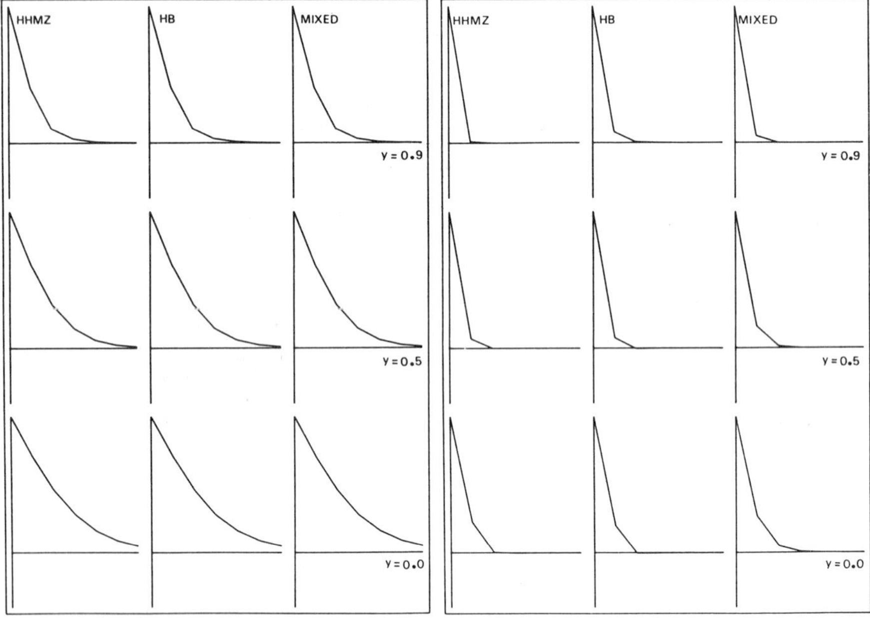

FIGS. 10,11. Boundary layers in second problem for $a=10^{-1}, 10^{-2}$.

There is good agreement between the three methods at the lower Péclet numbers but considerable difference at the higher values. The two methods based on the $B_1(\cdot,\cdot)$ form should be strictly comparable and aim to give a positive, monotone boundary layer: clearly the upwinding method has been more successful than the streamline diffusion. The results for the new method however are with $\gamma^2 = 0$ so they are based on the $B_2(\cdot,\cdot)$ form. Thus at higher Péclet numbers they should be close to L^2 best fits, and they indeed exhibit the typical oscillatory behaviour associated with this property.

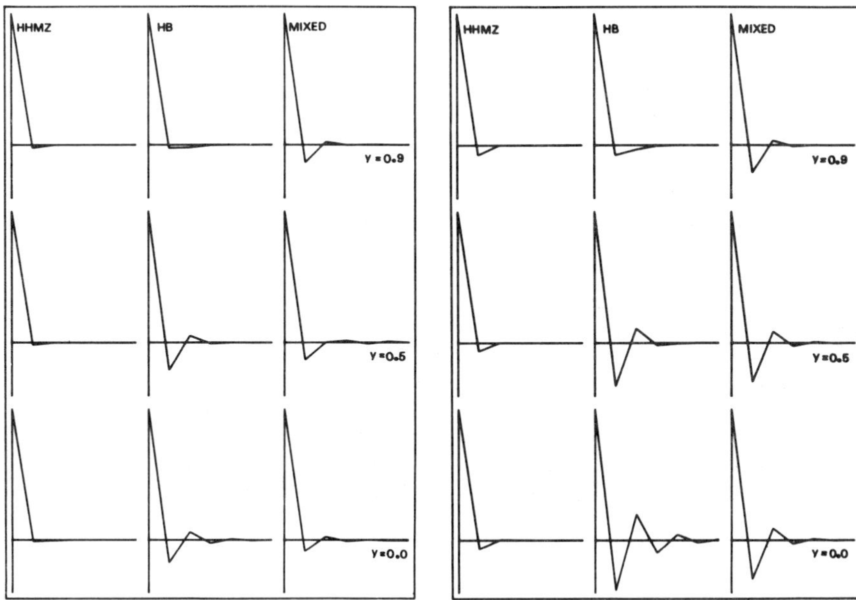

FIGS. 12,13. Boundary layers in second problem for $a=2x10^{-3}, 10^{-6}$.

If one makes the assumption that the boundary layer is exponential in form, one can attempt to recover its key parameters by exploiting the best fit property aimed for with the new method. Suppose we do this locally at each y level, $y_j = j/10$, $j = 0,1,\ldots,9$: that is, we assume a form

$$\tilde{u}_j(x,y) = a_j e^{b_j(x-1)} + c_j \tag{4.4}$$

with a_j, b_j, c_j chosen so that $\tilde{u}_j(1,y) = 100$ and also so that

$$B_2(\tilde{u}_j - U, \phi_{jk}) = 0 \quad k = K-1,\ K-2, \tag{4.5}$$

where k = K corresponds to the right hand boundary. Then we find very good agreement with an asymptotic analysis of the boundary layer broadening [18]: for example, with the mesh Péclet number 20, matching the results to a half width of 0.0569 at y = 0, the asymptotic analysis gives 0.0430 at y = 0.5 compared with the recovered value from (4.5) of 0.0405. This is an example of sub-gridscale recovery using a priori assumptions regarding the solution.

5. CONCLUSIONS

For the particular class of problems studied here, steady linear diffusion-convection, we have seen that the Petrov-Galerkin formulation can be very powerful. It has a firm theoretical foundation which can lead to sharp error estimates; a considerable variety of ways in which it can be applied have been demonstrated; and the results for quite severe test problems are impressive. Not surprisingly, the longer established methods are now widely used in practice.

Several other classes of problems which require generalisations of the basic Galerkin idea are of current interest. Nonlinear hyperbolic equations form a particularly active area of development at the moment and there are clear links to the present topic. Petrov-Galerkin methods, though useful, are not there so effective as other generalisations such as characteristic Galerkin methods. Consider for example the scalar conservation law [with the possibility of added diffusion]

$$\partial_t u + \partial_x f(u) \; [-\partial_x (a\partial_x u)] = 0. \tag{5.1}$$

Then a typical characteristic Galerkin method using the L^2 norm would take the form

$$\left(\frac{U^n - U^{n-1}}{\Delta t}, \phi_i\right) + (\partial_x f(U^n), \Phi_i^n)[+(a\partial_x U^n, \partial_x \phi_i)] = 0, \tag{5.2}$$

where U^n denotes the approximation at time $n\Delta t$ obtained with basis functions ϕ_i and

$$\Phi_i^n(x) = \frac{1}{b\Delta t}\int_{x-b\Delta t}^{x} \phi_i(z)dz\ , \tag{5.3}$$

where $b = b(u) = \partial f/\partial u$. The treatment of the first two terms is now well established, while that of the last term which forms the link with the present study, may not be the most appropriate. Already however the form of (5.3) indicates possible links with formulae such as those leading to (3.9).

6. REFERENCES

1. ALLEN, D.N. de G. & SOUTHWELL, R., Relaxation methods applied to determine the motion, in two dimensions, of a viscous fluid past a fixed cylinder. *Quart. J. Mech. and Appl. Math.* VIII, 129-145 (1955).
2. BABUSKA, I. & AZIZ, A. K., Survey lectures on the mathematical foundations of the finite element method. pp.3-363 of A. K. Aziz (Ed.), *The Mathematical Foundations of the Finite Element Method with Applications to Partial Differential Equations.* Academic Press, New York (1972).

3. BARRETT, J. W. & MORTON, K. W., Optimal finite element solutions to diffusion-convection problems in one dimension. *Int. J. of Num. Meth. Engng.* 15, 1457-1474 (1980).
4. BARRETT, J. W. & MORTON, K. W., Optimal Petrov-Galerkin methods through approximate symmetrization. *IMA J. Numer. Anal.* 1, 439-468 (1981).
5. BARRETT, J. W. & MORTON, K. W., Optimal finite element approximation for diffusion-convection problems. pp.403-411 of J. R. Whiteman (Ed.), *The Mathematics of Finite Elements and Applications 1981*. Academic Press, London (1982).
6. BARRETT, J. W. & MORTON, K. W., Approximate symmetrization and Petrov-Galerkin methods for diffusion-convection problems. *Comp. Meth. in Appl. Mech. Eng.*, to appear.
7. CHRISTIE, I., GRIFFITHS, D. F., MITCHELL, A. R. & ZIENKIEWICZ, O. C., Finite element methods for second order differential equations with significant first derivatives. *Int. J. Num. Meth. Engng.* 10, 1389-1396 (1976).
8. CIARLET, P. G., *The Finite Element Method for Elliptic Problems*. North-Holland, Amsterdam (1978).
9. HEINRICH, J. C., HUYAKORN, P. S., MITCHELL, A. R. & ZIENKIEWICZ, O. C., An upwind finite element scheme for two-dimensional convective transport equations. *Int. J. Num. Meth. Engng.* 11, 131-143 (1977).
10. HEINRICH, J. C. & ZIENKIEWICZ, O. C., The finite element method and 'upwinding' techniques in the numerical solution of convection dominated flow problems. pp.105-136 of T. J. R. Hughes,(Ed.), *Finite Element Methods for Convection Dominated Flows*. AMD, 34, Am. Soc. Mech. Eng., New York (1979).
11. HEMKER, P. W., *A numerical study of stiff two-point boundary problems*. Thesis, Mathematisch Centrum, Amsterdam (1977).
12. HUGHES, T. J. R. (Ed.), *Finite Element Methods for Convection Dominated Flows*. AMD, 34, Am. Soc. Mech. Eng., New York (1979).
13. HUGHES, T. J. R. & BROOKS, A. N., A multi-dimensional upwind scheme with no crosswind diffusion, pp.19-35 of T. J. R. Hughes, (Ed.), *Finite Element Methods for Convection Dominated Flows*. AMD, 34, Am. Soc. Mech. Eng., New York (1979).
14. HUGHES, T. J. R. & BROOKS, A. N., A theoretical framework for Petrov-Galerkin methods with discontinuous weighting functions: application to the streamline-upwind procedure. pp. 47-65 of R. H. Gallagher, D. H. Norrie, J. T. Oden & O. C. Zienkiewicz (Eds.), *Finite Elements in Fluids*, 4, J. Wiley & Sons, New York (1982).
15. KELLOGG, R. B., Analysis of a difference approximation for a singular perturbation problem in two dimensions. pp.113-117 of J. J. H. Miller (Ed.), *BAILI*. Boole Press, Dublin (1980).

16. MORTON, K. W., Finite element methods for non-self-adjoint problems. pp.113-148 of P. R. Turner (Ed.), *Proc. SERC Summer School, 1981*. Lect. Notes in Maths. 965, Springer-Verlag, Berlin (1982).

17. RHEINHARDT, H. J., A-posteriori analyses and adaptive finite element methods for singularly perturbed convection-diffusion equations. *Math. Meth. in Appl. Sci.* 4, 529-547 (1982).

18. SCOTNEY, B. W., Error analysis and numerical experiments for Petrov-Galerkin methods. *Univ. of Reading Num. Anal. Report* 11/82.

19. SMITH, R. M. & HUTTON, A. G., The numerical treatment of convection - a performance comparison of current methods. *Jnl. Num. Methods Heat Transfer*, 5, 439-461 (1982).

20. SPALDING, D. B., A novel finite difference formulation for differential expressions involving both first and second derivatives. *Int. J. Num. Meth. Engng.* 4, 551-559 (1972).

21. ZIENKIEWICZ, O. C., GALLAGHER, R. H. & HOOD, P., Newtonian and non-Newtonian viscous imcompressible flow, temperature induced flows: finite element solutions. pp.235-267 of J. R. Whiteman (Ed.), *The Mathematics of Finite Elements and Applications*. Academic Press, London (1975).

SOME SUPERCONVERGENCE RESULTS FOR MIXED FINITE ELEMENT METHODS FOR ELLIPTIC PROBLEMS ON RECTANGULAR DOMAINS

Mie Nakata,* Alan Weiser,**
and Mary Fanett Wheeler*

*Rice University, Houston, Texas USA
**Exxon Production Research, Houston, Texas USA

1. INTRODUCTION

In this paper we consider mixed finite element methods for approximating the pair $(u;p)$ satisfying the first order system

$$u = -a\cdot\nabla p, \qquad x \in \Omega, \tag{1.1}$$

$$\nabla\cdot u = f, \qquad x \in \Omega, \tag{1.2}$$

$$u\cdot\nu = 0, \qquad x \in \partial\Omega, \tag{1.3}$$

where $\Omega = (0,1)^2$ with boundary $\partial\Omega$ and ν is the outward normal vector on $\partial\Omega$. We assume that the coefficients $a^i \in W^1_\infty(\Omega)$ satisfy $a_0 \le a^i(x,y) \le a_1$, for positive constants a_0 and a_1, and $f \in L_2(\Omega)$.

We shall restrict our attention to the Raviart-Thomas approximating spaces. Optimal L_2 error estimates are derived for these spaces in Brezzi [1], Falk and Osborn [6], and Raviart and Thomas [9]. In this paper, we derive a discrete L_2 superconvergence result for these spaces, obtaining an additional power of h. Previous numerical results exhibiting this superconvergence can be found in [4, 5].

Russell and one of the authors [10] have shown that the standard five point block-centered finite difference method can be viewed as a mixed finite element method using the lowest order Raviart-Thomas space $(r = 0)$ and special numerical quadrature formulae. In this paper, we establish an $\mathcal{O}(h^2)$

ISBN 0-12-747255-X

discrete L_2 superconvergence result for this scheme. Manteuffel and White [8] have indicated that a similar result may be obtained via matrix arguments. Since the local truncation error for this scheme is only $\mathcal{O}(1)$ for non-uniform meshes, this scheme may be termed "supra-convergent" [7].

In this paper, no quasi-uniformity assumptions are made on the mesh. Extensions to other boundary conditions, and to rectangular parallelepipeds in $\mathbb{R}^n$, are straightforward.

The remainder of this paper is organized as follows. In Section 2, we formulate the mixed finite element method and establish notation. In Section 3, we derive discrete L_2 superconvergence results for these methods for any $r \geq 0$, assuming exact integration is used. We treat the block-centered finite difference scheme $(r = 0)$ in Section 4. In Section 5, collaborating numerical results are presented for the scheme in Section 4.

2. FORMULATION OF THE MIXED FINITE ELEMENT METHOD

Denote by $(z,w) = \int_\Omega z\cdot w dx$, the standard L_2 inner product, where $\cdot$ is multiplication (dot product) if z is a scalar (vector) function. Let $H(\text{div};\Omega)$ be the set of vector functions $v \in (L_2(\Omega))^2$ such that $\nabla\cdot v \in L_2(\Omega)$, and let

$$V = H(\text{div};\Omega) \cap \{v \mid v\cdot\nu = 0 \quad \text{on} \quad \partial\Omega\}.$$

Let $D = L_2(\Omega)$ and

$$W_p^s(\Omega) = \left\{ f:\Omega \to \mathbb{R} \;\middle|\; \frac{\partial^m\partial^n}{\partial x^m \partial y^n} f \in L_p(\Omega), \right.$$

$$\left. \text{for all integers } m \geq 0 \text{ and } n \geq 0 \text{ such that } m + n \leq s \right\}$$

$$\| f \|^p_{W_p^s(\Omega)} = \sum_{\substack{m,n \\ m+n\leq s}} \left\| \frac{\partial^m \partial^n f}{\partial x^m \partial y^n} \right\|^p_{L_p(\Omega)} .$$

Let Δ_x and Δ_y be partitions of $[0,1]$;

$$\Delta_x : 0 = x_0 < x_1 < \ldots < x_{N_x} = 1,$$

and

$$\Delta_y : 0 = y_0 < y_1 < \ldots < y_{N_y} = 1.$$

Let $h_{i-1} = x_i - x_{i-1}$, $h'_{j-1} = y_j - y_{j-1}$ and $h = \max_{i\,j}\{h_i, h'_j\}$. Set $V_h^r = V_h^{r,1} \times V_h^{r,2}$ with $V_h^{r,1} = \mathring{M}_0^{r+1}(\Delta_x) \otimes M_{-1}^r(\Delta_y)$ and $V_h^{r,2} = M_{-1}^r(\Delta_x) \otimes \mathring{M}_0^{r+1}(\Delta_y)$, and $D_h^r = M_{-1}^r(\Delta_x) \otimes M_{-1}^r(\Delta_y)$, where

$$M_t^s(\Delta_x) = \{v \in C^t(I) \,|\, v_{|I_i} \in P_s(I_i),\ \forall i = 1, \ldots, N_x\},$$

with I_i the ith subinterval and $P_s(I_i)$ the set of all polynomials of degree less than or equal to s. We denote by $C^t(I)$ the set of functions defined on I that have t continuous derivatives and by $C^{-1}(I)$ the set of functions defined on I. Let $\mathring{M}_t^s(\Delta_x) = \{v \in M_t^s(\Delta_x) \,|\, v(0) = v(1) = 0\}$. The spaces V_h^r and D_h^r are usually referred to as Raviart-Thomas spaces [9], and we note that $\operatorname{div} V_h^r \subset D_h^r$. We let $\{v_i^x\}$, $i = 1, \ldots, N_x(r+1) - 1$, be a basis of $\mathring{M}_0^{r+1}(\Delta_x)$. We choose the basis $\{w_{jk}^y\}$, $j = 1, \ldots, N_y$, $k = 1, \ldots, r+1$ of $M_{-1}^r(\Delta_y)$ such that w_{jk}^y is a polynomial of degree r on $[y_{j-1}, y_j]$ satisfying $w_{jk}^y(y_{jm}) = \delta_{km}$, where $\{y_{jk}\}$ are the $r+1$ Gauss points on $[y_{j-1}, y_j]$ for the $(r+1)$-point Gauss quadrature formula, and $w_{jk}^y = 0$ on $I \setminus [y_{j-1}, y_j]$. Similarly, we denote by $\{v_j^y\}$, $j = 1, \ldots, N_y(r+1) - 1$ a basis of $\mathring{M}_0^{r+1}(\Delta_y)$, and we choose the basis $\{w_{i\ell}^x\}$, $i = 1, \ldots, N_x$, $\ell = 1, \ldots, r+1$ of $M_{-1}^r(\Delta_x)$ like $\{w_{jk}^y\}$.

For a function $f: \Omega \to \mathbb{R}$ we define

$$|||\, f \,|||_x^2 = \sum_{i=1}^{N_x} \sum_{j=1}^{N_y} \sum_{\ell=1}^{r+2} \sum_{k=1}^{r+1} f(\bar{x}_{i\ell}, y_{jk})^2 h_{i-1} h'_{j-1},$$

$$|||\, f \,|||_y^2 = \sum_{i=1}^{N_x} \sum_{j=1}^{N_y} \sum_{\ell=1}^{r+1} \sum_{k=1}^{r+2} f(x_{i\ell}, \bar{y}_{jk})^2 h_{i-1} h'_{j-1},$$

and

$$|||\, f \,|||^2 = \sum_{i=1}^{N_x} \sum_{j=1}^{N_y} \sum_{\ell=1}^{r+1} \sum_{k=1}^{r+1} f(x_{i\ell}, y_{jk})^2 h_{i-1} h'_{j-1},$$

where $\{x_{i\ell}\}$ $(\{\bar{x}_{i\ell}\})$, $\{y_{jk}\}$ $(\{\bar{y}_{jk}\})$ are the $r+1$ $(r+2)$ Gauss points on $[x_{i-1}, x_i]$ $([y_{j-1}, y_j])$, respectively.

Now, we formulate the mixed finite element method. In (1.1), multiplying a^i by $(a^i)^{-1} v^i$, $(v^1, v^2) \in V$, and integrating by parts, we obtain

$$((a^1)^{-1}u^1,v^1) + ((a^2)^{-1}u^2,v^2) = (-\nabla p,v) = (p,\mathrm{div}\ v) \quad (2.1)$$

Multiplying (1.2) by $w \in D$ and integrating the result, we note that

$$(\mathrm{div}\ u,w) = (f,w), \qquad w \in D. \quad (2.2)$$

The first order system (2.1)-(2.2) is a weak form for the solution pair $(u;p)$, and motivates the definition of the mixed finite element method.

For a sequence of mesh parameters $h > 0$, we choose finite dimensional subspaces $V_h^r \subset V$ and $D_h^r \subset D$, and seek the solution pair $(U;P) \in V_h^r \times D_h^r$ satisfying

$$((a^1)^{-1}U^1,v^1) + ((a^2)^{-1}U^2,v^2) - (P,\mathrm{div}\ v) = 0,$$

$$v = (v^1,v^2) \in V_h^r, \quad (2.3)$$

$$(\mathrm{div}\ U,w) = (f,w), \qquad w \in D_h^r. \quad (2.4)$$

In Sections 3 and 4 we will denote by C a generic constant.

3. SUPERCONVERGENCE RESULTS FOR THE RAVIART-THOMAS SPACES

The main result of this section is the following superconvergence result. For simplicity we assume that $a^1 = a^2 = a$.

THEOREM 3.1. Let $(u;p)$ satisfy (2.1)-(2.2) and assume $p \in W_3^{r+2}(\Omega)$ and $u^i \in W_\infty^{r+2}(\Omega)$ for $i = 1,2$. Further, assume that $(p-P,1) = 0$. Let $(U;P)$ satisfy (2.3) and (2.4). Then there exists a constant $C > 0$, independent of h, such that

$$|||p-P||| + |||u^1-U^1|||_x + |||u^2-U^2|||_y \leqslant C\, h^{r+2} \quad (3.1)$$

We note that (3.1) shows that the approximation $(U;P)$ is close to $(u;p)$ at certain Gauss points. If the $|||\cdot|||$ seminorms were replaced by L_2 norms, the error estimate would be $O(h^{r+1})$. The latter result can be found in Raviart-Thomas [9], Brezzi [1] and Falk and Osborn [6].

In order to prove Theorem 3.1 we shall need to define several projections. Let $\hat{P} \in D_h^r$ satisfy

$$(p - \hat{P},\ d) = 0, \quad d \in D_h^r, \quad (3.2)$$

and $\hat{U}_a^i \in V_h^{r,i}$ satisfy

$$(a^{-1}(u^i-\hat{U}_a^i),v) = 0, \qquad v \in V_h^{r,i}, \ i = 1,2. \tag{3.3}$$

We call these projections L_2-projections. For convenience, we set $\hat{U}^i = \hat{U}_a^i$ for $a \equiv 1$. Let $\tilde{U}^1 \in V_h^r$ satisfy

$$((u^1-\tilde{U}^1)_x,v_x) = 0, \qquad v \in V_h^{r,1}, \tag{3.4}$$

and let $\tilde{U}^2 \in V_h^{r,2}$ satisfy

$$((u^2-\tilde{U}^2)_y,v_y) = 0, \qquad v \in V_h^{r,2}. \tag{3.5}$$

We call these H_0^1-projections.

Using (2.1) we observe that

$$(a^{-1}\hat{U}_a,v) - (\hat{P}, \text{div } v) = 0, \qquad v \in V_h^r. \tag{3.6}$$

One can easily show that

$$(\text{div } \tilde{U}, w) = (\text{div } u, w) = (f,w), \qquad w \in D_h^r,$$

or

$$(\text{div } \tilde{U}, w) = (\text{div } U, w), \qquad w \in D_h^r, \tag{3.7}$$

since $\text{div } V_h^r \subset D_h^r$, and $\tilde{U}\cdot\nu = 0$ and $u\cdot\nu = 0$ on $\partial\Omega$. Subtracting (3.6) from (2.3) and setting $v = U - \tilde{U}$, we obtain

$$(a^{-1}(U-\hat{U}_a), U-\tilde{U}) = 0,$$

since $\text{div}(U-\tilde{U}) = 0$.

Thus,

$$(a^{-1}(U-\hat{U}_a), U-\hat{U}_a) = (a^{-1}(U-\hat{U}_a), \tilde{U}-\hat{U}_a),$$

or

$$\begin{aligned}\| U-\hat{U}_a \|_{L_2(\Omega)} &\leqslant C \| \hat{U}_a-\tilde{U}\|_{L_2(\Omega)} \\ &\leqslant C(\| \hat{U}-\tilde{U}\|_{L_2(\Omega)} + \| \hat{U}_a-\hat{U}\|_{L_2(\Omega)}).\end{aligned} \tag{3.8}$$

We bound the two terms on the right-hand side of (3.8) in the next two lemmas.

LEMMA 3.1. Let $\hat{U}$ and $\tilde{U}$ be defined by (3.3) and (3.4)-(3.5). Then there exists a constant $C > 0$, independent of h, such that

$$\| \tilde{U}-\hat{U}\|_{L_2(\Omega)} \leqslant C\, h^{r+2}.$$

PROOF. It suffices to prove this lemma componentwise. So we will establish it for $\tilde{U}^1-\hat{U}^1$.

Define the functions z_{jk} on I by

$$z_{jk}(x) = \frac{1}{h'_{j-1}} \int_{y_{j-1}}^{y_j} u^1(x,y) w^y_{jk}(y)\,dy, \quad j = 1,2, \ldots, N_y, \quad k = 1,2, \ldots, r+1.$$

Note that $z_{jk} \in W^{r+2}_{\infty}(I)$.

Let $\hat{Z}_{jk}$ be the L_2-projection of z_{jk} onto $\mathring{M}^{r+1}_0(\Delta_x)$.

Let $\tilde{Z}_{jk}$ be the H^1_0-projection of z_{jk} onto $\mathring{M}^{r+1}_0(\Delta_x)$.

Then it can be computed easily that

$$\hat{Z}_{jk}(x) = \frac{1}{h'_{j-1}} \int_{y_{j-1}}^{y_j} \hat{U}^1(x,y) w^y_{jk}(y)\,dy,$$

and

$$\tilde{Z}_{jk}(x) = \frac{1}{h'_{j-1}} \int_{y_{j-1}}^{y_j} \tilde{U}^1(x,y) w^y_{jk}(y)\,dy.$$

Note that

$$\hat{Z}_{jk}(x) = A_k \hat{U}^1(x,y_{jk}),$$

and

$$\tilde{Z}_{jk}(x) = A_k \tilde{U}^1(x,y_{jk}),$$

where A_k is the k-th weight for the (r+1)-point Gauss quadrature formula.

Thus,

$$\begin{aligned} &\| \tilde{U}^1-\hat{U}^1 \|_{L_2(\Omega)} \\ &\leqslant C ||| \tilde{U}^1-\hat{U}^1 |||_x \\ &= C \left\{ \sum_i \sum_\ell \sum_j \sum_k h_{i-1} h'_{j-1} (\tilde{Z}_{jk}-\hat{Z}_{jk})^2(\bar{x}_{i\ell}) \right\}^{\frac{1}{2}} \\ &\leqslant C \max_{jk} \| \tilde{Z}_{jk}-\hat{Z}_{jk} \|_{L_\infty(I)} \end{aligned}$$

$$\leqslant C \max_{jk} \left[\| \tilde{Z}_{jk} - z_{jk} \|_{L_\infty(I)} + \| \hat{Z}_{jk} - z_{jk} \|_{L_\infty(I)} \right]$$

$$\leqslant C\, h^{r+2}.$$

The last inequality follows from Wheeler [13] for the H_0^1-projection, and from Douglas, Dupont, and Wahlbin [2] and Dupont [3] for the L_2-projection.

LEMMA 3.2. Let $\hat{U} = \hat{U}_1$ and $\hat{U}_a$ be defined by (3.3). Then there exists a constant C, independent of h, such that $\| \hat{U} - \hat{U}_a \|_{L_2(\Omega)} \leqslant C\, h^{r+2}$.

PROOF. It suffices to prove the result for the first component. By (3.3) we have

$$(a^{-1}(\hat{U}^1 - \hat{U}_a^1), \hat{U}^1 - \hat{U}_a^1)$$

$$= (\hat{U}^1 - u^1,\ a^{-1}(\hat{U}^1 - \hat{U}_a^1))$$

$$= (\hat{U}^1 - u^1,\ a^{-1}(\hat{U}^1 - \hat{U}_a^1) - \psi), \qquad \psi \in V_h^{r,1}$$

$$\leqslant \| \hat{U}^1 - u^1 \|_{L_2(\Omega)} \cdot \inf_{\psi \in V_h^{r,1}} \| a^{-1}(\hat{U}^1 - \hat{U}_a^1) - \psi \|_{L_2(\Omega)}.$$

Thus,

$$\| \hat{U}^1 - \hat{U}_a^1 \|_{L_2(\Omega)}^2 \leqslant C \| \hat{U}^1 - u^1 \|_{L_2(\Omega)} \cdot \inf_{\psi \in V_h^{r,1}} \| a^{-1}(\hat{U}^1 - \hat{U}_a^1) - \psi \|_{L_2(\Omega)}.$$

Since $\| \hat{U}^1 - u^1 \|_{L_2(\Omega)} \leqslant C\, h^{r+1}$ by approximation theory, it suffices to show that

$$\inf_{\psi \in V_h^{r,1}} \| a^{-1}(\hat{U}^1 - \hat{U}_a^1) - \psi \|_{L_2(\Omega)} \leqslant C\, h \| \hat{U}^1 - \hat{U}_a^1 \|_{L_2(\Omega)}.$$

Now,

$$\inf_{\psi \in V_h^{r,1}} \| a^{-1}(\hat{U}^1 - \hat{U}_a^1) - \psi \|_{L_2(\Omega)}$$

$$= \inf_{\psi\ V_h^{r,1}} \| (a^{-1} - \alpha^*)(\hat{U}^1 - \hat{U}_a^1) + \alpha^*(\hat{U}^1 - \hat{U}_a^1) - \psi \|_{L_2(\Omega)}$$

$$\leq \|(a^{-1}-\alpha^*)(\hat{U}^1-\hat{U}_a^1)\|_{L_2(\Omega)} + \inf_{\psi \in V_h^{r,1}} \|\alpha^*(\hat{U}^1-\hat{U}_a^1)-\psi\|_{L_2(\Omega)}$$

$$\leq \|a^{-1}-\alpha^*\|_{L_\infty(\Omega)} \|\hat{U}^1-\hat{U}_a^1\|_{L_2(\Omega)} + \inf_{\psi \in V_h^{r,1}} \|\alpha^*(\hat{U}^1-\hat{U}_a^1)-\psi\|_{L_2(\Omega)},$$

for any $\alpha^* \in M_0^1(\Delta_x) \otimes M_{-1}^0(\Delta_y)$.

Since α^* can be chosen from $M_0^1(\Delta_x) \otimes M_{-1}^0(\Delta_y)$ such that

$$\|a^{-1}-\alpha^*\|_{L_\infty(\Omega)} \leq C\, h \|a^{-1}\|_{W_\infty^1(\Omega)},$$

the following lemma completes the proof of Lemma 3.2.

LEMMA 3.3. Let $z \in V_h^{r,1}$ and $\alpha^* \in M_0^1(\Delta_x) \otimes M_{-1}^0(\Delta_y)$. Then

$$\inf_{\psi \in V_h^{r,1}} \|\alpha^* z-\psi\|_{L_2(\Omega)} \leq C(\|\alpha^*\|_{W_\infty^1}) h \|z\|_{L_2(\Omega)} .$$

PROOF. Let $F = \alpha^* z$. Then for $\tilde{F} \in V_h^{r,1}$,

$$\inf_{\psi \in V_h^{r,1}} \|F-\psi\|_{L_2(\Omega)} \leq \|F-\tilde{F}\|_{L_2(\Omega)},$$

where $\tilde{F}(x_\ell,y_{jk}) = F(x_\ell,y_{jk})$, $\ell = 0,1, \ldots, N_x$, $j = 1,2, \ldots, N_y$, and $k = 1,2, \ldots, r+1$, and $\tilde{F}(x,y_{jk})$ is the H_0^1-projection of $F(x,y_{jk})$ onto $P_{r+1}(x_{i-1},x_i) \cap H_0^1(x_{i-1},x_i)$. Applying the Peano Kernel Theorem [11] and a local inverse assumption on F yields the desired result.

LEMMA 3.4. Let $\hat{U} = (\hat{U}^1,\hat{U}^2)$ satisfy (3.3). Then

$$|||(\hat{U}-u)^1|||_x + |||(\hat{U}-u)^2|||_y \leq C\, h^{r+2}.$$

PROOF. We have

$$|||(\hat{U}-u)^1|||_x^2 = \sum_i \sum_j h_{i-1} h'_{j-1} \sum_k \sum_\ell ((\hat{U}-u)^1(\bar{x}_{i\ell},y_{jk}))^2$$

$$= \sum_i \sum_j h_{i-1} h'_{j-1} \sum_k \sum_\ell [\hat{U}(\bar{x}_{i\ell},y_{jk})-\hat{U}_{jk}^1(\bar{x}_{i\ell})+\hat{U}_{jk}^1(\bar{x}_{i\ell})$$

$$-u^1(\bar{x}_{i\ell},y_{jk})]^2 ,$$

where $\hat{U}_{jk}^1$ is the L_2-projection of u_{jk}^1 onto $\mathring{M}_0^{r+1}(\Delta_x)$ and u_{jk}^1 is the restriction of u^1 to $\{(x,y_{jk})\,|\,x \in I\}$. Thus, by one-dimensional approximation theory,

$$\| \hat{U}^1-u^1 \|_x^2 \leq 2 \sum_i \sum_j \sum_k \sum_\ell h_{i-1}h'_{j-1}\Big\{\Big[\hat{U}^1(\bar{x}_{i\ell},y_{jk})-\hat{U}_{jk}(\bar{x}_{i\ell})\Big]^2 + \Big[\hat{U}^1_{jk}(\bar{x}_{i\ell})-u^1_{jk}(\bar{x}_{i\ell})\Big]^2\Big\}$$

$$\leq C \sum_j \sum_k \sum_i \sum_\ell h_{i-1}h'_{j-1}\Big[\hat{U}^1(\bar{x}_{i\ell},y_{jk})-\hat{U}^1_{jk}(\bar{x}_{i\ell})\Big]^2 + C\,h^{2(r+2)}.$$

Thus, it suffices to bound

$$T = \sum_j \sum_k \sum_i \sum_\ell h_{i-1}h'_{j-1}\Big[\hat{U}^1(\bar{x}_{i\ell},y_{jk})-\hat{U}^1_{jk}(\bar{x}_{i\ell})\Big]^2$$

Now,

$$T \leq C \sum_j \sum_k \sum_i \sum_\ell h_{i-1}h'_{j-1}A_k B_\ell\Big[\hat{U}^1(\bar{x}_{i\ell},y_{jk})-\hat{U}_{jk}(\bar{x}_{i\ell})\Big]^2,$$

where A_k (B_ℓ) is the kth (the ℓth) weight at the kth (the ℓth) Gauss point for the (r+1)-point ((r+2)-point) Gauss quadrature formula, or

$$T \leq C \sum_j \sum_k h'_{j-1}A_k \int_0^1 (\hat{U}^1(x,y_{jk})-\hat{U}^1_{jk}(x))^2 dx.$$

Now, by the definition of $\hat{U}^1$, we have

$$\begin{aligned} 0 &= (\hat{U}^1-u^1, v_i^x\, w_{jk}^y) \\ &= \int_0^1 \int_{y_{j-1}}^{y_j} (\hat{U}^1-u^1)(x,y)v_i^x(x)w_{jk}^y(y)\,dx\,dy \\ &= \int_0^1 \Big[A_k h'_{j-1}(\hat{U}^1-u^1)(x,y_{jk})-E_j(x)\Big]v_i^x(x)\,dx, \end{aligned}$$

where $E_j(x)$ is the error term in the (r+1)-point Gauss quadrature formula for u^1. Thus,

$$A_k h'_{j-1} \int_0^1 (\hat{U}^1 - u^1)(x, y_{jk}) v_i^x(x)\,dx = -\int_0^1 E_j(x) v_i^x(x)\,dx,$$

or

$$A_k h'_{j-1} \int_0^1 \left[\hat{U}^1(x, y_{jk}) - \hat{U}^1_{jk}(x)\right] v_i^x(x)\,dx = -\int_0^1 E_j(x) v_i^x(x)\,dx.$$

By replacing $v_i^x(x)$ by $\hat{U}^1(x, y_{jk}) - \hat{U}^1_{jk}(x)$ we obtain

$$A_k h'_{j-1} \int_0^1 (\hat{U}^1(x, y_{jk}) - \hat{U}^1_{jk}(x))^2 dx = -\int_0^1 E_j(x)(\hat{U}^1(x, y_{jk}) - \hat{U}^1_{jk}(x))\,dx.$$

Thus,

$$T \leqslant C \sum_j \sum_k h'_{j-1} A_k \int_0^1 (\hat{U}^1(x, y_{jk}) - \hat{U}^1_{jk}(x))^2 dx$$

$$\leqslant C \sum_j \sum_k \left[-\int_0^1 E_j(x)(\hat{U}^1(x, y_{jk}) - \hat{U}^1_{jk}(x))\,dx\right]$$

$$\leqslant C \sum_j \sum_k \| E_j \|_{L_2(I)} \| \hat{U}^1(\cdot, y_{jk}) - \hat{U}^1_{jk} \|_{L_2(I)},$$

so it suffices to bound

$$E_j(x) = \int_{y_{j-1}}^{y_j} -u^1(x, y) w^y_{jk}(y)\,dy + A_k h'_{j-1} u^1(x, y_{jk}).$$

Since

$$L(f) \equiv \int_{y_{j-1}}^{y_j} f(x, y) w^y_{jk}(y)\,dy - A_k h'_{j-1} f(x, y_{jk}) = 0$$

for all $f(x,y)$ such that f is a polynomial of degree $\leqslant r+1$ in x, then by the Peano Kernel Theorem,

$$E_j(x) = \int_{y_{j-1}}^{y_j} \frac{\partial^{r+1}}{\partial\tau^{r+2}}(-u^1)(x,\tau)L_y\left(\frac{(y-\tau)_+^{r+1}}{(r+1)!}\right)d\tau$$

$$= \int_{y_{j-1}}^{y_j} \frac{\partial^{r+2}}{\partial\tau^{r+2}}(-u^1)(x,\tau)\left[\int_{y_{j-1}}^{y_j}\frac{(y-\tau)_+^{r+1}}{(r+1)!}w_{jk}^y(y)dy - A_k h'_{j-1}\frac{(y_{jk}-\tau)_+^{r+1}}{(r+1)!}\right]d\tau.$$

Hence,

$$\| E_j\|^2_{L_2(I)} \leqslant h'^{2(r+2)}_{j-1}\, h'_{j-1}\int_0^1\int_{y_{j-1}}^{y_j}\left(\frac{\partial^{r+2}}{\partial\tau^{r+2}}u^1(x,\tau)\right)^2 d\tau dx$$

$$\leqslant C\, h'^{2(r+2)}_{j-1}\, h'_{j-1}\left\|\frac{\partial^{r+2}u^1}{\partial y^{r+2}}\right\|^2_{L_2(I\times[y_{j-1},y_j])}.$$

Thus,

$$T \leqslant C\, h^{r+2}\left(\left[\sum_j\sum_k h'^{\frac12}_{j-1}\| \hat{U}^1(\cdot,y_{jk})-\hat{U}^1_{jk}\|_{L_2(I)}\right]^2\right)^{\frac12}\left\|\frac{\partial^{r+2}u^1}{\partial y^{r+2}}\right\|_{L_2(\Omega)}$$

$$\leqslant C\, h^{r+2}\left(\sum_j\sum_k h'_{j-1}\| \hat{U}^1(\cdot,y_{jk})-\hat{U}^1_{jk}\|^2_{L_2(I)}\right)^{\frac12}.$$

On the other hand,

$$T \geqslant C\sum_j\sum_k h'_{j-1}\| \hat{U}^1(\cdot,y_{jk})-\hat{U}^1_{jk}\|^2_{L_2(I)}.$$

Thus,

$$\left[\sum_j\sum_k h'_{j-1}\| \hat{U}^1(\cdot,y_{jk})-\hat{U}^1_{jk}\|^2_{L_2(I)}\right]^{\frac12} \leqslant C\, h^{r+2},$$

or

$$T \leqslant C\, h^{2(r+2)}.$$

Hence, $||| \hat{U}^1-u^1 |||_x \leqslant C\, h^{r+2}.$

LEMMA 3.5. Assume that $(P-\hat{P},1) = 0$. Then

$$\| P-\hat{P} \|_{L_2(\Omega)} \leq C\, h^{r+2}.$$

PROOF. Consider the boundary value problem

$$\operatorname{div} \psi \equiv -\nabla\cdot a\nabla\phi = P-\hat{P}, \quad x \in \Omega ,$$

$$a\nabla\phi\cdot\nu = 0 \quad \text{on} \quad \partial\Omega .$$

Then for $\tilde{\psi} \in V_h^r$, the H_0^1-projection of ψ onto V_h^r, we have

$$\begin{aligned}
\| P-\hat{P} \|_{L_2(\Omega)}^2 &= (P-\hat{P},\ \operatorname{div} \psi) \\
&= (P-\hat{P},\ \operatorname{div}(\psi-\tilde{\psi})) + (P-\hat{P},\ \operatorname{div} \tilde{\psi}) \\
&= (P-\hat{P},\ \operatorname{div}(\psi-\tilde{\psi})) + (a^{-1}(U-\hat{U}_a),\tilde{\psi}) \\
&= (a^{-1}(U-\hat{U}_a),\tilde{\psi}) .
\end{aligned}$$

Thus, by elliptic regularity, we have

$$\begin{aligned}
\| P-\hat{P} \|_{L_2(\Omega)}^2 &\leq \| a^{-1}(U-\hat{U}_a) \|_{L_2(\Omega)} [\| \tilde{\psi}-\psi \|_{L_2(\Omega)} + \| \psi \|_{L_2(\Omega)} \\
&\leq C\ \| U-\hat{U}_a \|_{L_2(\Omega)} \| P-\hat{P} \|_{L_2(\Omega)} .
\end{aligned}$$

Hence,

$$\| P-\hat{P} \|_{L_2(\Omega)} \leq C\ \| U-\hat{U}_a \|_{L_2(\Omega)} \leq C\, h^{r+2}$$

by Lemma 3.2.

LEMMA 3.6. There exists a constant $C > 0$, independent of h, such that

$$|||\, p-\hat{P}\, ||| \leq C\, h^{r+2} .$$

PROOF. The proof of this result closely follows that of Lemma 3.4.

PROOF OF THEOREM 3.1. To prove the theorem we observe that for $|||u^1-U^1|||_x$ it suffices to estimate $|||u^1-\hat{U}_a^1|||_x$ and $|||\hat{U}_a^1-U^1|||_x$ by the triangle inequality. For $|||u^1-\hat{U}_a^1|||_x$ it suffices to estimate $|||u^1-\hat{U}^1|||_x$ and $\| \hat{U}^1-\hat{U}_a^1 \|_{L_2(\Omega)}$. Also for $|||\hat{U}_a^1-U^1|||_x$ it suffices to estimate $\| U^1-\hat{U}^1 \|_{L_2(\Omega)}$ and $\| \hat{U}^1-\hat{U}_a^1 \|_{L_2(\Omega)}$ by (3.8). The estimate of $|||u^1-U^1|||_x$ follows from Lemmas 3.1, 3.2 and 3.4. We obtain the estimate for $|||u^2-U^2|||_y$ similarly. The estimate of $|||p-P|||$ follows from Lemmas 3.4 and 3.6.

The above superconvergence results can be used to derive the following optimal L_∞ rates of convergence for two-dimensional problems. These estimates require the use of an inverse assumption; hence a quasi-uniformity assumption on the mesh must be assumed.

THEOREM 3.2. Let $(u;p)$ satisfy (2.1) - (2.2) and assume $p \in W_3^{r+2}(\Omega)$ and $u^i \in W_\infty^{r+2}(\Omega)$ for $i = 1,2$. Further assume that the mesh defined by the tensor product partitions Δ_x and Δ_y is quasi-uniform and that $(p-P,1) = 0$. Let $(U;P)$ satisfy (2.3) - (2.4). Then there exists a constant $C > 0$, independent of h, such that

$$\| p-P \|_{L_\infty(\Omega)} + \| u^1-U^1 \|_{L_\infty(\Omega)} + \| u^2-U^2 \|_{L_\infty(\Omega)} \leq C\, h^{r+1} .$$

PROOF. One can easily verify that

$$\| p-\hat{P} \|_{L_\infty(\Omega)} + \| u-\hat{U} \|_{L_\infty(\Omega)} \leq C\, h^{r+1} ,$$

From (3.8), Lemmas 3.1 and 3.2, we deduce that

$$\| U-\hat{U} \|_{L_2(\Omega)} \leq C\, h^{r+2} .$$

By an inverse assumption, we obtain

$$\| U-\hat{U} \|_{L_\infty(\Omega)} \leq C\, h^{r+1} .$$

From Lemma 3.5 and an inverse assumption we have

$$\| P-\hat{P} \|_{L_\infty(\Omega)} \leq C\, h^{r+1} .$$

The proof of theorem is now immediate.

4. SUPRA-CONVERGENCE FOR THE FIVE-POINT BLOCK-CENTERED FINITE DIFFERENCE METHOD

For completeness we first present a result [10] that shows that the block-centered finite difference method is a mixed finite element method $(r = 0)$ with special numerical quadrature formulas.

Set $x_{-1} = -x_1$, $y_{-1} = -y_1$, $x_{N_x+1} = 1 + h_{N_x-1}$, and $y_{N_y+1} = 1 + h_{N_y-1}$. Let $x_{i+\frac{1}{2}} = (x_i + x_{i+1})/2$ and $y_{j+\frac{1}{2}} = (y_j + y_{j+1})/2$ and let $a_{n,m} = a(x_n, y_m)$. Set $h_{i-\frac{1}{2}} = x_{i+\frac{1}{2}} - x_{i-\frac{1}{2}}$ and $h'_{j-\frac{1}{2}} = y_{j+\frac{1}{2}} - y_{j-\frac{1}{2}}$. The block-centered finite difference approximation to (1.1)-(1.3) with $a^1 = a^2 = a$ is

$$
\begin{aligned}
-\frac{1}{h_i}\Bigg[& a_{i+1,j+\frac{1}{2}}\left(\frac{P_{i+\frac{3}{2},j+\frac{1}{2}} - P_{i+\frac{1}{2},j+\frac{1}{2}}}{h_{i+\frac{1}{2}}}\right) \\
& -a_{i,j+\frac{1}{2}}\left(\frac{P_{i+\frac{1}{2},j+\frac{1}{2}} - P_{i-\frac{1}{2},j+\frac{1}{2}}}{h_{i-\frac{1}{2}}}\right)\Bigg] \qquad (4.1) \\
& - \text{similar terms in } y \text{ direction} \\
& = f_{i+\frac{1}{2},j+\frac{1}{2}}, \qquad \begin{array}{l} i = 0,1,\ \dots\ , N_x-1, \\ j = 0,1,\ \dots\ , N_y-1, \end{array}
\end{aligned}
$$

with boundary reflection conditions

$$
\begin{aligned}
P_{-\frac{1}{2},j+\frac{1}{2}} &= P_{\frac{1}{2},j+\frac{1}{2}}, \\
P_{N_x+\frac{1}{2},j+\frac{1}{2}} &= P_{N_x-\frac{1}{2},j+\frac{1}{2}}, \\
P_{i+\frac{1}{2},-\frac{1}{2}} &= P_{i+\frac{1}{2},\frac{1}{2}}, \qquad (4.2) \\
P_{i+\frac{1}{2},N_y+\frac{1}{2}} &= P_{i+\frac{1}{2},N_y-\frac{1}{2}}.
\end{aligned}
$$

Multiplying (4.1) by $h_i h'_j$, the area of the cell, we obtain

$$h'_j((U^1)_{i+1,j+\frac{1}{2}} - (U^1)_{i,j+\frac{1}{2}})$$
$$+ h_i((U^2)_{i+\frac{1}{2},j+1} - (U^2)_{i+\frac{1}{2},j}) \tag{4.3}$$
$$= h'_j\ h_i\ f_{i+\frac{1}{2},j+\frac{1}{2}},$$

where U^1 and U^2 are velocity components defined by

$$(U^1)_{i+1,j+\frac{1}{2}} = -a_{i+1,j+\frac{1}{2}}\left(\frac{P_{i+\frac{3}{2},j+\frac{1}{2}} - P_{i+\frac{1}{2},j+\frac{1}{2}}}{h_{i+\frac{1}{2}}}\right), \tag{4.4a}$$

$$(U^2)_{i+\frac{1}{2},j+1} = -a_{i+\frac{1}{2},j+1}\left(\frac{P_{i+\frac{1}{2},j+\frac{3}{2}} - P_{i+\frac{1}{2},j+\frac{1}{2}}}{h'_{j+\frac{1}{2}}}\right). \tag{4.4b}$$

Let U be the unique function in V_h^0 satisfying (4.3)-(4.4), and let P be the piecewise-constant function with cell values $P_{i+\frac{1}{2},j+\frac{1}{2}}$. We show that $(U;P) \in V_h^0 \times D_h^0$ satisfies (2.3)-(2.4) if certain numerical quadrature formulas are used. However, we first note that by reflection

$$(U^1)_{0,j+\frac{1}{2}} = (U^1)_{N_x,j+\frac{1}{2}} = 0,$$
$$(U^2)_{i+\frac{1}{2},0} = (U^2)_{i+\frac{1}{2},N_y} = 0;$$

the boundary conditions are satisfied.

Now the left-hand side of (4.3) is equal to

$$h'_j h_i \frac{\partial}{\partial x}(U^1) + h_i h'_j \frac{\partial}{\partial y}(U^2) = \int_{x_i}^{x_{i+1}} \int_{y_j}^{y_{j+1}} \nabla\cdot U\ dx,$$

and the right-hand side is the midpoint-rule integral of f over the cell. Thus, U satisfies

$$(\nabla\cdot U,w) = (f,w)_{Mx\ My}, \qquad w \in D_h^0, \tag{4.5}$$

where $Mx\ My$ denotes the midpoint-rule quadrature in both directions. Equation (4.4a) implies that

$$\tfrac{1}{2}(h_i+h_{i+1})h'_j \frac{1}{a_{i+1,j+\frac{1}{2}}} (U^1)_{i+1,j+\frac{1}{2}}$$

$$- h'_j(P_{i+\frac{1}{2},j+\frac{1}{2}} - P_{i+\frac{3}{2},j+\frac{1}{2}})$$

$$= 0,$$

which can be rewritten as

$$(a^{-1}U^1, v^x_{i+1} w^y_{j+1})_{Tx\ My} - (P, \frac{\partial}{\partial x}(v^x_{i+1} w^y_{j+1})) = 0 , \qquad (4.6)$$

where v^x_{i+1} is the C^0, piecewise linear basis function satisfying $v^x_{i+1}(x_k) = \delta_{i+1,k}, w^y_{j+1} = w^y_{j+1,1}$ is the piecewise constant basis function, and Tx My denotes trapezoidal integration in the x-direction tensored with the midpoint rule in the y-direction.

Similarly,

$$(a^{-1}U^2, w^x_{i+1} v^y_{j+1})_{Mx\ Ty} - (P, \frac{\partial}{\partial y}(w^x_{i+1} v^y_{j+1})) = 0. \qquad (4.7)$$

Combining (4.6) and (4.7), we deduce that

$$(a^{-1}U^1, v^1)_{Tx\ My} + (a^{-1}U^2, v^2)_{Mx\ Ty} - (P, \nabla\cdot v) = 0, \qquad (4.8)$$

$$v = (v^1, v^2) \in V^0_h .$$

By (4.5) and (4.8), the block-centered pressure and associated velocity satisfy the mixed method equations (2.3)-(2.4), if the indicated quadrature formulas are used.

For existence of (U;P) we assume that

$$\mathcal{R}_h = \sum_{i=0}^{N_x-1} \sum_{j=0}^{N_y-1} f_{i+\frac{1}{2},j+\frac{1}{2}} h_i h'_j = 0;$$

otherwise we modify the right-hand side of (4.1) by subtracting the term $\mathcal{R}_h/N_x N_y$.

In this section we modify the definition of the x and y semi-norms of Section 2. For a function $f:\Omega \to \mathbb{R}$, we define

$$|||f|||^2_x = \sum_{i=0}^{N_x} \sum_{j=0}^{N_y-1} h_{i-\frac{1}{2}}\, h'_j\, f^2(x_i, y_{j+\frac{1}{2}})$$

and

$$|||f|||_y^2 = \sum_{j=0}^{N_y} \sum_{i=0}^{N_x-1} h'_{j-\frac{1}{2}} h_i f^2(x_{i+\frac{1}{2}}, y_j).$$

The main result of this section is the following superconvergence result.

THEOREM 4.1. Let $(u;p)$ satisfy (1.1)-(1.3) and assume $p \in W_3^4(\Omega)$ and $u^i \in W_3^3(\Omega)$, $i = 1,2$. Let $(U;P)$ satisfy (4.1), (4.2) and (4.4). Further assume that $(p-P,1) = 0$. Then there exists a constant $C > 0$, independent of h, such that

$$|||p-P||| + |||u^1-U^1|||_x + |||u^2-U^2|||_y \leq C h^2.$$

As in the proof of Theorem 3.1, we construct two projections, $(\hat{U};\hat{P})$ and $(\tilde{U};\tilde{P})$ in $V_h^0 \times D_h^0$. We have

$$(a^{-1}\hat{U}^1, v^1)_{Tx\ My} + (a^{-1}\hat{U}^2, v^2)_{Mx\ Ty} - (\hat{P}, \nabla \cdot v) = 0, \qquad (4.9)$$

$$v = (v^1, v^2) \in V_h^0$$

and

$$(\text{div } \tilde{U}, w) = (f,w)_{Mx\ My}, \qquad w \in D_h^0. \qquad (4.10)$$

Note that (4.9) implies that $\hat{U}$ satisfies (4.4a) and (4.4b) with U and P replaced by $\hat{U}$ and $\hat{P}$ respectively.

We now briefly describe the construction of $(\hat{U};\hat{P})$. For convenience of notation, we set $z_i = x_{i-\frac{1}{2}}$ and $w_j = y_{j-\frac{1}{2}}$. Let Δ_z (Δ_w) be the partition defined by the z_i's $(w_j$'s). Denote by $p^L \in M_0^1(\Delta_z) \otimes M_0^1(\Delta_w)$ the piecewise bilinear interpolant satisfying

$$p^L(z_i, w_j) = p(z_i, w_j).$$

Set $\varepsilon^L = p - p^L$. Define $p^{Q_x} \in M_0^2(\Delta_z) \otimes M_0^1(\Delta_w)$ by

$$p^{Q_x}(z_i, w_j) = 0$$

and

$$p^{Q_x}(x_i, w_j) = \varepsilon^L(x_i, w_j).$$

Similarly, we define $p^{Q_y} \in M_0^1(\Delta_z) \otimes M_0^2(\Delta_w)$ by

$$p^{Q_y}(z_i, w_j) = 0$$

and

$$p^{Q_y}(z_i, y_j) = \varepsilon^L(z_i, y_j).$$

Set

$$p^Q = p^L + p^{Q_x} + p^{Q_y}.$$

We now construct $\hat{P}_{i-\frac{1}{2},j-\frac{1}{2}}$. Let

$$\hat{P}_{i-\frac{1}{2},j-\frac{1}{2}} = \hat{P}^L_{i-\frac{1}{2},j-\frac{1}{2}} + \hat{P}^{Q_x}_{i-\frac{1}{2},j-\frac{1}{2}} + \hat{P}^{Q_y}_{i-\frac{1}{2},j-\frac{1}{2}},$$

where

$$\hat{P}^L_{i-\frac{1}{2},j-\frac{1}{2}} = p^L(z_i, w_j) \qquad i = 1, \ldots, N_x,\ j = 1, \ldots, N_y,$$

$$\hat{P}^{Q_x}_{i+\frac{1}{2},j-\frac{1}{2}} = \begin{cases} 0 & i=0,\ j=1, \ldots, N_y, \\ \hat{P}^{Q_x}_{i-\frac{1}{2},j-\frac{1}{2}} + h_{i-\frac{1}{2}} \dfrac{\partial}{\partial x} p^{Q_x}(z_{i+1}, w_j), & i=1, \ldots, N_x-1,\ j=1, \ldots, N_y, \end{cases}$$

and

$$\hat{P}^{Q_y}_{i-\frac{1}{2},j+\frac{1}{2}} = \begin{cases} 0 & i=1, \ldots, N_x,\ j=0, \\ \hat{P}^{Q_y}_{i-\frac{1}{2},j-\frac{1}{2}} + h'_{j-\frac{1}{2}} \dfrac{\partial}{\partial y} p^{Q_y}(z_i, w_{j+1}), & i=1, \ldots, N_x,\ j=1, \ldots, N_y-1 \end{cases}.$$

The following lemma is established in Weiser and Wheeler [12].

LEMMA 4.1. Let $p \in W_3^4(\Omega)$. Then there exists a constant $C > 0$, independent of h, such that

$$|p_{i+\frac{1}{2},j+\frac{1}{2}} - \hat{P}_{i+\frac{1}{2},j+\frac{1}{2}}| \leqslant C\, h^2,$$

$$|u^1_{i+1,j+\frac{1}{2}} - \hat{U}^1_{i+1,j+\frac{1}{2}}| \leqslant C\, h^2,$$

and

$$|u^2_{i+\frac{1}{2},j+1} - \hat{U}^2_{i+\frac{1}{2},j+1}| \leqslant C\, h^2,$$

where $\hat{U}$ is defined by (4.4) with P replaced by $\hat{P}$.

To construct $\tilde{U} = (\tilde{U}^1, \tilde{U}^2) \in V_h^0$ satisfying (4.10) we consider the three-point block-centered finite difference scheme in one dimension. We define $\tilde{P}^1 \in D_h^0$ by

$$-\frac{1}{h_i}\left[a_{i+1,j+\frac{1}{2}}\left(\frac{\tilde{P}_{i+\frac{3}{2},j+\frac{1}{2}} - \tilde{P}_{i+\frac{1}{2},j+\frac{1}{2}}}{h_{i+\frac{1}{2}}}\right) - a_{i,j+\frac{1}{2}}\left(\frac{\tilde{P}_{i+\frac{1}{2},j+\frac{1}{2}} - \tilde{P}_{i-\frac{1}{2},j+\frac{1}{2}}}{h_{i-\frac{1}{2}}}\right)\right]$$

$$= \left(f + \frac{\partial}{\partial y}\left(a(x,y)\frac{\partial p}{\partial y}\right)\right)(x_{i+\frac{1}{2}}, y_{j+\frac{1}{2}})$$

and $\tilde{U}^1 \in V_h^0$ by (4.4a) with P replaced by $\tilde{P}$. Similarly, $(\tilde{U}^2, \tilde{P}^2)$ are defined. One can easily verify that (4.10) holds.

We state without proof the following lemma which can be found in [12].

LEMMA 4.2. Let $(u;p)$ satisfy the hypotheses of Theorem 4.1. Then there exists a constant $C>0$, independent of h, such that

$$|||\, p-\tilde{P}^1 \,||| + |||\, p-\tilde{P}^2 \,||| + |||u^1-\tilde{U}^1|||_x + |||u^2-\tilde{U}^2|||_y \leqslant C\, h^2.$$

PROOF OF THEOREM 4.1. Subtracting (4.9) from (4.8) and setting $v = U-\tilde{U}$, we have

$$(a^{-1}(U-\hat{U})^1, (U-\tilde{U})^1)_{Tx\ My} + (a^{-1}(U-\hat{U})^2, (U-\tilde{U})^2)_{Mx\ Ty} = 0. \tag{4.11}$$

Thus by (4.10) and Lemmas 4.1-4.2,

$$|||\, U^1-\hat{U}^1 \,|||_x + |||\, U^2-\hat{U}^2 \,|||_y \leqslant C\,(|||\, \hat{U}^1-\tilde{U}^1 \,|||_x + |||\, \hat{U}^2-\tilde{U}^2 \,|||_y) \leqslant C\, h^2,$$

or

$$|||U^1-u^1|||_x + |||U^2-u^2|||_y \leqslant C\,h^2.$$

The estimate of $|||p-P|||$ follows from estimating $|||P-\hat{P}|||$ as in Lemma 3.5.

5. NUMERICAL RESULTS FOR THE FIVE-POINT BLOCK-CENTERED FINITE DIFFERENCE METHOD

In this section, we present sample numerical results demonstrating the second-order convergence predicted by Theorem 4.1 $(r=0)$. The block-centered finite difference method (4.1)-(4.2) was used to approximate the pressure p and velocities $-(1+xy)p_x$ and $-p_y$ for the model problem

$$-((1+xy)p_x)_x - p_{yy} = f(x,y) \quad \text{in } \Omega,$$

$$-\nabla p \cdot \nu = 0 \qquad \text{on } \partial\Omega,$$

where f was chosen so that

$$p(x,y) = \cos(\pi x)\cos(\pi y) + y^2(1-y)^2.$$

The model problem was chosen so that p had a relatively simple analytic form with no obvious symmetries in x or y.

The finite difference method was used with two sets of grids with $N_x = N_y = M$. The first set of grids was chosen so that

$$h_i = \begin{cases} h, & i \text{ even} \\ h/\alpha, & i \text{ odd} \end{cases}$$

with h_j' chosen by the same rule for some constant $\alpha \geqslant 1$. Note that the case $\alpha = 1$ is the usual uniform grid. The second set of grids was chosen by the rule

$$\bar{h}_i = 1 + \text{rnd}_i \times (\beta-1)$$

$$h_i = \bar{h}_i \Big/ \sum_{m=1}^{M} \bar{h}_m$$

with h_j' chosen by the same rule, where each rnd_i was a random number between 0 and 1 for some constant $\beta \geqslant 1$. Note that the maximum neighboring gridblock length ratio was α for the first set of grids, and β for the second set of grids.

The finite difference method was run in single precision on a Cray 1, using a direct solver for the resulting linear systems.

TABLE 5.1

	M	$\bar{h}^{-1}$	ε	$\bar{\varepsilon}$	ε_1	$\bar{\varepsilon}_1$	ε_2	$\bar{\varepsilon}_2$
$\alpha=1$	5	5.0	.017	.43	.029	.73	.027	.68
	10	10.0	.0042	.42	.0073	.73	.0067	.67
	15	15.0	.0019	.42	.0032	.73	.0030	.67
	20	20.0	.0010	.42	.0018	.73	.0017	.67
	25	25.0	.00067	.42	.0012	.72	.0011	.67
	30	30.0	.00046	.42	.00081	.72	.00075	.67
	35	35.0	.00034	.42	.00059	.72	.00055	.67
	40	40.0	.00026	.42	.00045	.72	.00042	.67
$\alpha=3$	5	3.7	.038	.51	.066	.89	.061	.82
	10	6.7	.011	.47	.022	1.00	.021	.93
	15	10.3	.0045	.48	.0091	.97	.0083	.88
	20	13.3	.0026	.47	.0057	1.01	.0052	.92
	25	17.0	.0016	.47	.0034	.99	.0031	.89
	30	20.0	.0012	.47	.0025	1.01	.0023	.91
	35	23.7	.00084	.47	.0018	.99	.0016	.90
	40	26.7	.00066	.47	.0014	1.01	.0013	.91
$\alpha=5$	5	3.4	.043	.50	.075	.86	.069	.80
	10	6.0	.013	.47	.027	.98	.025	.91
	15	9.4	.0054	.48	.011	.94	.0097	.86
	20	12.0	.0033	.47	.0069	.99	.0063	.90
	25	15.4	.0020	.47	.0041	.96	.0037	.87
	30	18.0	.0015	.47	.0031	.99	.0028	.90
	35	21.4	.0010	.47	.0021	.97	.0019	.88
	40	24.0	.00082	.47	.0017	.99	.0016	.90
$\beta=3$	5	3.6	.018	.23	.034	.45	.035	.45
	10	6.9	.0070	.33	.011	.50	.014	.64
	15	9.5	.0033	.30	.0083	.75	.0053	.48
	20	13.7	.0019	.37	.0040	.74	.0029	.55
	25	15.4	.0012	.28	.0019	.45	.0028	.66
	30	19.4	.00090	.34	.0015	.57	.0018	.68
	35	23.1	.00055	.29	.0012	.61	.0011	.57
	40	26.3	.00047	.32	.00093	.64	.00089	.61
$\beta=5$	5	3.4	.018	.21	.037	.43	.038	.44
	10	6.1	.0089	.33	.012	.44	.022	.84
	15	8.7	.0034	.26	.0094	.72	.0078	.59
	20	13.5	.0019	.34	.0043	.77	.0034	.62
	25	14.8	.0013	.28	.0030	.65	.0024	.54
	30	17.1	.00090	.26	.0026	.74	.0013	.38
	35	20.8	.00075	.33	.0017	.75	.0014	.62
	40	24.0	.00055	.31	.0012	.68	.00093	.54

The results are presented in Table 5.1, where

$$\bar{h} = \max_{i\,j}(h_i, h'_j)$$

$$\varepsilon = |||\, p - P \,|||$$

$$\bar{\varepsilon} = \varepsilon\, \bar{h}^{-2}$$

$$\varepsilon_1 = |||\, u^1 - U^1 \,|||_x$$

$$\bar{\varepsilon}_1 = \varepsilon_1\, \bar{h}^{-2}$$

$$\varepsilon_2 = |||\, u^2 - U^2 \,|||_y$$

$$\bar{\varepsilon}_2 = \varepsilon_2\, \bar{h}^{-2}\ .$$

Second order convergence is apparent for both pressures and velocities in each case, with $\bar{\varepsilon}$, $\bar{\varepsilon}_1$, and $\bar{\varepsilon}_2$ behaving roughly as constants as M increases. For a fixed $\bar{h}$, the nonuniform meshes resulted in errors no larger than uniform meshes. For fixed M, however, the nonuniform meshes generally resulted in errors 2 to 3 times larger than the errors for the uniform meshes. Thus, for fixed $\bar{h}$, there was no accuracy penalty for using nonuniform meshes, but there was a computer cost penalty since more gridblocks were required.

ACKNOWLEDGEMENT

The authors wish to thank Exxon Production Research Company for supporting this work.

REFERENCES

1. BREZZI, F., On the existence, uniqueness and approximation of saddle point problems arising from Lagrangian multiplier, *RAIRO, Anal. Numer*, 2, 129-151 (1974).

2. DOUBLAS, J. JR., DUPONT, T. and WAHLBIN, L., Optimal L_∞ error estimates for Galerkin approximation to solutions of two-point boundary value problems, *Math. of Comp.*, 29, 130, 475-483 (1975).

3. DUPONT, T., L_p-boundedness of the L_2-projection into continuous piecewise polynomial spaces over an interval, in preparation.

4. EWING, R. E. and WHEELER, M. F., Computational aspects of mixed finite element methods, in: R. S. Stepleman (Ed.), *Numerical Methods for Scientific Computing*, North-Holland, New York, to appear.

5. EWING, R. E., KOEBBI, J. V., GONZALEZ, R., and WHEELER, M.F., Computing accurate velocities for fluid flow in porous media, *Proceedings of TICOM Conference*, Austin, Texas (1983).

6. FALK, R. S. and OSBORN, J. E., Error estimates for mixed methods, *RAIRO Anal. Numer.* 14, 249-277 (1980).

7. KREISS, H. O., MANTEUFFEL, T. A., SWARTZ, B. K., WENDROFF, B., and WHITE, A. B., Supra-convergent schemes on irregular grids, to appear.

8. MANTEUFFEL, T. A. and WHITE, A. B., Private communication.

9. RAVIART, P. A. and THOMAS, J. M., A mixed finite element method for 2nd order elliptic problems, in: *MATHEMATICAL ASPECTS OF THE FINITE ELEMENT METHOD, Lecture Notes in Mathematics*, Springer-Verlag, Heidelberg (1977).

10. RUSSELL, T. F. and WHEELER, M. F., Finite element and finite difference methods for continuous flows in porous media, in: R. E. Ewing (Ed.), *Mathematics of Reservoir Simulation*, SIAM Publications, Philadelphia (1984).

11. SARD, A., *Linear Approximation*, American Mathematical Society, Providence, Rhode Island (1963).

12. WEISER, A. and WHEELER, M. F., On convergence of five-point block-centered finite differences for elliptic problems, to appear.

13. WHEELER, M. F., An optimal L_∞ error estimate for Galerkin approximations to solutions of two-point boundary value problems, *SIAM Jour. Num. Anal.*, 10, 914-917 (1973).

ON OPTIMIZATION ASPECTS OF A CFD FINITE ELEMENT PENALTY ALGORITHM

A. J. Baker

University of Tennessee, Knoxville, TN USA

I. INTRODUCTION

Finite element theory provides a formal basis for construction of discrete and/or semi-discrete approximate solutions to differential equation descriptions in mechanics, in particular fluid dynamics. The classical theory is founded within the calculus of variational boundary value problems [1,2], and the earliest fluid mechanics applications employed direct extensions, cf. [3-4]. Shortly thereafter, the connection with a weak (Galerkin weighted residuals) statement was established, essentially ending the search for pseudo-variational principles [5] and yielding the resultant extension to a certain class of non-linear fluid mechanics problems, cf. [6-8].

In the same time period, it was recognized that the unaltered weak statement formulation was inflexible for broad based application in computational fluid dynamics (CFD). The concept of constraining the extremization of a variational boundary value statement [9] led in a natural way to construction of a penalty form of the weak statement. This provided a particularly useful theoretical construction for the incompressible Navier-Stokes problem class in primitive variables [10] including a three-dimensional description [11]. The basic concept of the incompressibility constraint has been extended considerably and reduced to practice for a three-dimensional turbulent subsonic aerodynamic flow CFD algorithm [12].

Coincident with development of incompressible penalty algorithms was the realization that the unmodified weak statement appeared inadequate as the problem class characteristc Reynolds (and Peclet) number became large. The artificial viscosity concepts developed by the finite difference community [13,14] were reexamined, and a number of modifications developed and implemented within finite element algorithm statements. Several of these can be reconstructed as an augmentation of the basic weak statement by an appropriately defined penalty term. Included herein at least is the Petrov-Galerkin formulation for the incompressible Navier-Stokes equations [10] and the compressible Euler equations [15], the artificial density formulation for shocked transonic potential flows [16], and the modified penalty-

ISBN 0-12-747255-X

Galerkin formulation for the unsteady Euler and Navier-Stokes equation [17]. Within this penalty interpretation framework could also be included the Taylor-Galerkin algorithm [18] and the generalized Galerkin method [19].

It seems fair to conclude therefore, that penalized weak statements of the conservation law systems for fluid mechanics may emerge as a basic theoretical error extremization statement for a broad range of problem classes in computational fluid dynamics. With this as a precept, the issue immediately turns to accuracy and efficiency of the developed algorithms, in particular the use of large time steps within an implicit integration procedure, use of boundary conforming discretizations, and definition and use of solution adaptive meshes. This paper develops and examines these issues for the penalty-Galerkin formulation of a finite element algorithm for the high speed Euler and/or Navier Stokes equations.

2. PROBLEM STATEMENT

2.1 Conservation Law System

In nondimensional conservation form, using Cartesian tensor summation notation, the Navier-Stokes equation system for a compressible, viscous, heat-conducting fluid is

$$L(\rho) = \frac{\partial \rho}{\partial t} + \frac{\partial}{\partial x_j}\left[\rho u_j\right] = 0 \tag{1}$$

$$L(\rho u_i) = \frac{\partial(\rho u_i)}{\partial t} + \frac{\partial}{\partial x_j}\left[u_j \rho u_i + p\delta_{ij} - \sigma_{ij}\right] = 0 \tag{2}$$

$$L(\rho e) = \frac{\partial(\rho e)}{\partial t} + \frac{\partial}{\partial x_j}\left[u_j \rho e + u_j p - \sigma_{ij} u_i - q_j\right] = 0 \tag{3}$$

In equations 1-3, ρ is density, ρu_i the momentum vector, p the pressure, and e the mass specific total energy. For a polytropic gas, $p = (\gamma - 1)\rho\varepsilon$, and the equation of state is

$$p = (\gamma - 1)\left[\rho e - \tfrac{1}{2}\rho u_j u_j\right] \tag{4}$$

The Stokes stress tensor σ_{ij}, heat flux vector q_j, and specific internal energy ε are defined as

$$\sigma_{ij} = \frac{\mu}{Re}\left[\frac{\partial u_i}{\partial x_j} + \frac{\partial u_j}{\partial x_i}\right] - \frac{2\mu}{3Re}\frac{\partial u_k}{\partial x_k}\delta_{ij} \tag{5}$$

$$q_j = -\kappa \frac{\partial e}{\partial x_j} \tag{6}$$

$$\varepsilon = e - \tfrac{1}{2}u_i u_i \tag{7}$$

with μ the absolute viscosity, κ the coefficient of heat conductivity, and δ_{ij} the Kronecker delta. The Euler equations are contained within equations 1-4 for equations 5-6 vanishing identically.

2.2 The Penalty-Galerkin Solution Algorithm

The Navier-Stokes equation system 1-7 describes the three-dimensional evolution of the array of state points defining the corresponding flow field. For the state vector of dependent variables $\underline{q}(\underline{x},t)$ with scalar components q_α, $1 \leq \alpha \leq p$, where p equals the total degrees of freedom of the system, let $\underline{L}(\underline{q}(\underline{x},t)) = \underline{0}$ represent equations 1-7 defined on the solution domain $\Omega = R^n \times t$. Let $\ell(\underline{q}) = \underline{0}$ represent the corresponding constraint statement for $\underline{q}$, including normal derivatives, on the boundary $\partial\Omega = \partial R \times t$ of Ω.

The finite element procedure expands on the formal statement of **approximation** to the (exact) solution $\underline{q}(\underline{x},t)$. The approximate solution $\underline{q}^h(\underline{x},t)$ is computed on a discretization formed as the union of non-overlapping subdomains Ω^h, ie., $\Omega \approx \cup\Omega^h$, hence also $\partial\Omega \equiv \cup\partial\Omega^h$, where superscript "h" denotes mesh measure and emphasizes that the **quality** of $\underline{q}^h(\underline{x},t)$ depends in a fundamental way on the choice of Ω^h. Typically, the discretization of Ω is the semi-discretization $\Omega^h \equiv R^n_e \times t$, wherein "space" and "time" are assumed separable. The union of the n-dimensional, non-overlapping finite element domains R^n_e forms the mesh of measure h on R^n, and $\underline{q}^h(\underline{x},t)$ is constructed thereon as the semi-discrete approximation,

$$\underline{q}^h(\underline{x},t) \equiv \bigcup_e \underline{q}^e(\underline{x},t) \quad (8) \qquad\qquad \underline{q}^e(\underline{x},t) \equiv \underline{N}_k(\underline{\eta})^T\underline{Q}(t)_e \quad (9)$$

Hence, $\underline{q}^h(\underline{x},t)$ is the union of semi-discrete approximations $\underline{q}^e(\underline{x},t)$, each of which is defined on an element domain $R^n_e \times t$ by the matrix inner product of the (element-independent) cardinal basis set $\underline{N}_k(\underline{\eta})$ and the undetermined time- and element-dependent expansion coefficient set $\underline{Q}(t)_e$. Elements of $\underline{N}_k$ are typically chosen from a finite dimensional subspace of the Hilbert space H^m_0 containing all functions whose m^{th} derivatives are square integrable on R^n and which satisfy the boundary conditions.

Substitution of $\underline{q}^h(\underline{x},t)$ into $\underline{L}(\cdot)$ and $\underline{\ell}(\cdot)$ yields both non-homogeneous, hence measures of the (semi-)discrete approximation error (distribution) $\underline{e}^h(\underline{x},t)$. The penalty-Galerkin finite element algorithm is the formal statement of constraints to be enforced on the error distribution $\underline{e}^h(\underline{x},t)$. The Galerkin statement requires that $\underline{L}(\underline{q}^h)$ and $\underline{\ell}(\underline{q}^h)$ be orthogonal to the elements of the approximation subspace $\underline{N}_k$. The penalty statement requires that the semi-discrete error expressed by the substantial derivative $L^c(\cdot) \equiv \partial(\cdot)/\partial t + u_i\, \partial(\cdot)/\partial x_i$ be orthogonal to the gradient of the subspace $\underline{N}_k$. The penalty Galerkin statement is the linear combination of these expressions, i.e.,

$$\int_{R^n} \underline{N}_k \underline{L}(q^h_\alpha) + \underline{\beta}_\alpha \cdot \int_{R^n} \nabla \underline{N}_k\, \underline{L}^c(q^h_\alpha) + \lambda_\alpha \int_{\partial R} \underline{N}_k\, \underline{\ell}(q^h_\alpha) \equiv \underline{0} \quad (10)$$

In equation 10, λ_α is a scalar multiplier defined to cancel appropriate expressions in the first term associated with imposition of non-homogeneous Neumann boundary conditions. The penalty term is multiplied by the vector constant $\underline{\beta}_\alpha$, with functional form [17, Ch.8],

$$\underline{\beta}_\alpha \equiv h_e \left[\underline{\nu}^1_\alpha \delta_t + \text{sgn}(\underline{u})\, \underline{\nu}^2_\alpha \delta_x \right] \tag{11}$$

where h_e is the measure of the element R^n_e, $\underline{\nu}^i_\alpha$ are constants dependent upon q_α, and δ_t and δ_x are Kronecker-type identifiers that limit application to the first and second terms respectively in $L^c(\cdot)$.

When evaluated, equation 10 yields the system of ordinary differential equations

$$\underline{\underline{A}} \frac{d\underline{Q}}{dt} + \underline{B}(\underline{Q}(t)) + \underline{b} = \underline{0} \tag{12}$$

where $\underline{A}$ is a non-diagonal matrix, the column matrix $\underline{B}(\underline{Q}(t))$ is a complicated non-linear function of the discrete approximation $\underline{Q}$, and $\underline{b}$ is a column matrix of constants (data). Equation 12 is of use in evaluating the Taylor series,

$$\underline{F} \equiv \underline{Q}_{j+1} - \underline{Q}_j - \Delta t \left.\frac{d\underline{Q}}{dt}\right|_{j+\theta} - \tfrac{1}{2}\Delta t^2 \left.\frac{d^2\underline{Q}}{dt^2}\right|_{j+\theta} + \ldots \equiv \underline{0} \tag{13}$$

which defines the expansion coefficient set $\underline{Q}$ at time $t_{j+1} = t_j + \Delta t$, in terms of the known data $\underline{Q}_j$ and the derivative evaluated within the interval, $0 \le \Theta \le 1$. For the Navier-Stokes CFD problem class, equation 13 is strongly non-linear for $\Theta > 0$, which is reduced to a problem in linear algebra using an appropriate algorithm, for example, the Newton (-Raphson) procedure,

$$\underline{\underline{J}}^{\,p}_{j+1}\, \delta\underline{Q}^{\,p+1}_{j+1} = -\underline{F}^{\,p}_{j+1} \tag{14}$$

where p is the iteration index,

$$\underline{Q}(t_j + \Delta t)^{p+1}_{j+1} \equiv \underline{Q}^{\,p+1}_{j+1} \equiv \underline{Q}^{\,p}_{j+1} + \delta\underline{Q}^{\,p+1}_{j+1} \tag{15}$$

and the Jacobian of the Newton algorithm is defined as

$$\underline{\underline{J}} \equiv \frac{\partial \underline{F}}{\partial \underline{Q}} \tag{16}$$

3. THEORETICAL ANALYSIS

Theoretical statements of accuracy and convergence are typically derived as asymptotic error estimates in Sobolev norms $\|\cdot\|_{H^m}$. For linear elliptic or parabolic equations L( ), the statement is of the form [1,2],

$$\|e^h(n\Delta t)\|_{H^m(\Omega)} \le C_1 h^\gamma \|\,\text{data}\,\|_{H^p(\Omega)} + C_2 \Delta t^\alpha \|\,\text{data}\,\|_{H^\alpha(\Omega_0)} \tag{17}$$

where C_1 and C_2 are constants, $\gamma = \min(k+1-m, \mu)$ where $\mu \ge 0$ depends principally on the quality of the discretization Ω^h, and $\alpha > 0$ is related to Θ and the specific form of equation 13. Results of numerical experiments [8,20] indicate that the form of equation 17 is appropriate for $1 \leq k \leq 2$ in equation 9, for the important (and non-linear) laminar and turbulent boundary layer flow CFD problem class.

For a linear scalar hyperbolic equation L( ), equation 17 is modified to the form [2],

$$\| e^h(n\Delta t) \|_{H^1(\Omega)} \leq C_1 h^{k+1} \| \text{data} \|_{H^p(\Omega)} + C_2 \Delta t^{\alpha} \| \text{data} \|_{H^{\alpha}(\Omega_0)} + C_3 \, h \int_0^{n\Delta t} \| q(\tau) \|_{H^{k+1}} d\tau \qquad (18)$$

where k is defined in equation 9. In equations 17-18, $\|$ data $\|$ denotes the appropriate order Sobolev norm of domain and boundary data including that associated with interpolation of the initial condition specification onto $\underline{Q}(t=0)$. The results of numerical experiments [21] for a linear convection problem confirm the dominance of the k-independent term in equation 18, for $1 \leq k \leq 2$ in equation 9.

The real interest in the broad CFD problem class is for flows in which the non-linearity in equation 2 dominates, which can yield generation of non-smooth solutions (from smooth initial data), e.g., shocks and contact discontinuities. In such situations, one anticipates that only $\| \cdot \|_{H^p}$, $p \leq 0$, may exist for the analytical solution. However, semi-discrete approximate solutions remain definable for $k \geq 1$ in equation 9, which presents an interesting analysis requirement for the mathematician. Of course, artificial viscosity is (typically) embedded, implicitly or explicitly, within an algorithm definition statement, eg., equation 10, such that the non-linear hyperbolic character of $\underline{L}(\cdot)$ becomes subjugated to an elliptic operator in the discrete approximation.

4. OPTIMIZATION ASPECTS

4.1 The Penalty Parameter Set $\underline{\beta}_{\alpha}$

The original analysis of the penalty-Galerkin (pG) formulation [22], developed for a linear scalar transport equation, derived an optimum estimate for β based upon a Fourier stability analysis and extremization of the algorithm order of accuracy. Referring to equation 11, the original scalar form was $\beta \equiv \nu h_e$, and the optimum determination yielded $\nu_o = (15)^{-\frac{1}{2}}$. Numerical experimentation with linear and non-linear (non-smooth) example problems determined that the form of equation 11, with ν^1 and ν^2 distinct, was preferred for the Euler equation system. The resultant Fourier stability analysis, for a scalar linear equation, determines the optimal order of accuracy for the pG (k=1) algorithm to be expressed by the constraint [23],

$$\nu^2 = \frac{(d)^2[1/180 - (\nu^1)^4] + (\nu^1)^2}{(d)^2[\nu^1/12 - (\nu^1)^3] + \nu^1} \qquad (19)$$

where $d \equiv 2\pi h/\lambda$ and λ is the Fourier mode wavelength. Figure 1 is a graph of equation 19 for $\lambda = Nh$, $2 \le N \le 22$, with N as a parameter. All curves coalesce at $\nu^1 = \nu_o = \nu^2$, in agreement with the original analysis. Sixth order accuracy can be achieved only for $\nu^1 > 0$, and for any value of ν^1, ν^2 ranges over an order of magnitude as a function of N with the extreme level required for the smallest N. Setting $\nu^1 \equiv 0$ and repeating the Fourier analysis confirms the k=1 pG algorithm is second order accurate, while it is fourth order accurate for $\nu^1 \equiv 0 \equiv \nu^2$.

The Riemann shock tube specification of Sod [24], the inviscid solution of which yields a strong shock, a contact discontinuity and a rarefaction wave interspersed by a high and low temperature plateau, respectively, was examined using the pG algorithm to numerically refine the estimate of equation 19 for the one-dimensional Euler equation system, equations 1-4. Figure 2a is the graph of the k=1, pG discrete solution for momentum for ν^1 and ν^2 defined at the locus of the curves in Figure 1, ie., for $1 \le \alpha \le 3$,

$$\nu^1_\alpha \equiv \nu_o(1, 0, 1), \qquad \nu^2_\alpha \equiv \nu_o(1, 1, 1) \tag{20}$$

Figure 2b is the comparison k=1 pG algorithm solution for the definition,

$$\nu^1_\alpha \equiv \nu_o(3/8, 0, 1/4), \qquad \nu^2_\alpha \equiv \nu_o(3/4, 2, 1) \tag{21}$$

and Figure 2c graphs the companion k=2 pG algorithm solution for the definition

$$\nu^1_\alpha \equiv \nu_o(0, 0, 0), \qquad \nu^2_\alpha \equiv \nu_o(1/4, 3/4, 1/2) \tag{22}$$

The excessive Gibbs phenomena behind the shock, for the definition in equation 20, is effectively eliminated by the refinements of equations 21-22 for the k=1 and k=2 pG algorithms respectively. The shock is crisply defined, at the expense of additional smearing of the contact discontinuity, although the interspersed temperature plateaus are rather adequately defined, Figures 3a-b. For the definition in equation 21, the k=1 pG algorithm is operating just to the left of the coalescence at ν_o, Figure 1, wherein the optimal order-accurate range for ν^2 is $1/2 \le \nu^2/\nu_o \le 3/2$. This tends to indicate the appropriateness of the linearized k=1 pG algorithm stability analysis in estimating the constraint set for the non-linear Euler equation system. The companion stability analysis for the k=2 pG algorithm is not completed, hence, equation 22 is the result of "pure numerical optimization." These levels of ν^2 are about half the k=1 values, which is in proportion to the doubling of h for a k=2 discretization employing the same number of nodes. That $\nu^1_\alpha \equiv 0$ is "optimal" is not well understood, except in the light of the numerical experience [17, Ch.4] that the basic k=2 Galerkin formulation enjoys an improved phase accuracy distribution in comparison to the k=1 implementation.

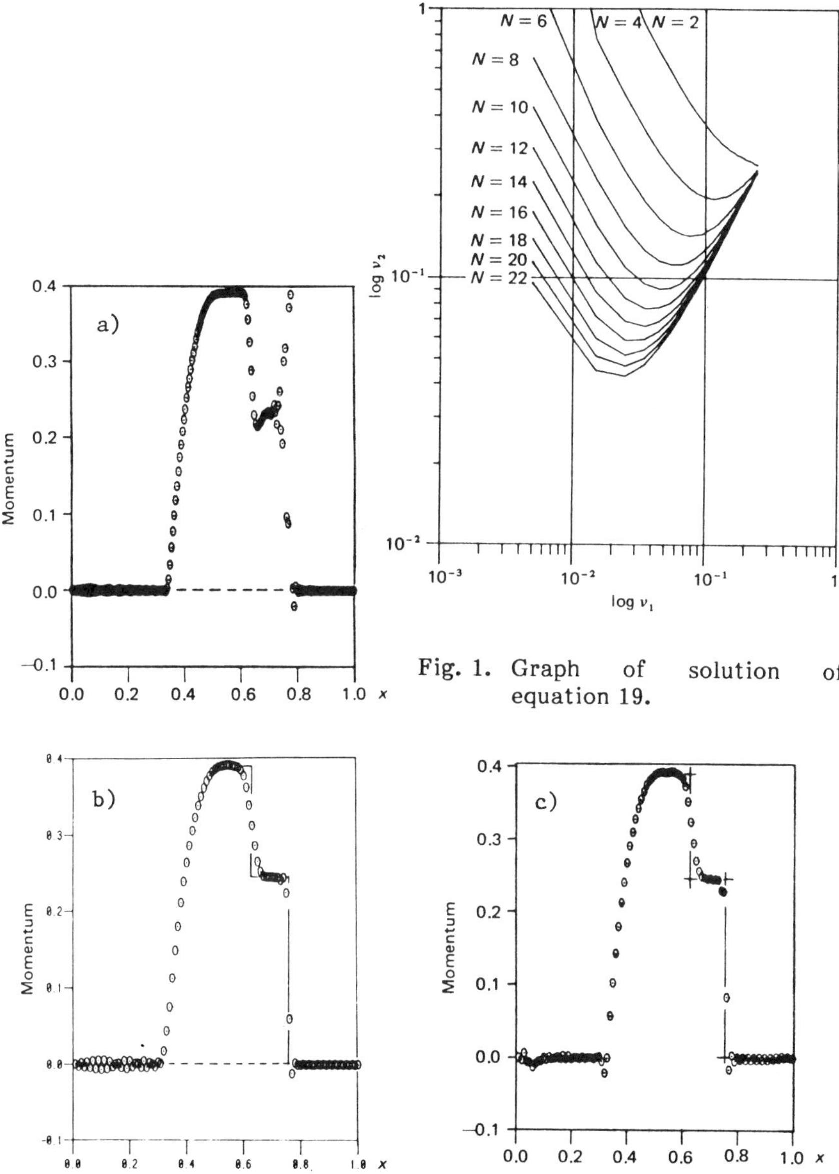

Fig. 1. Graph of solution of equation 19.

Fig. 2. Momentum discrete approximate solution, Riemann shock tube, pG finite element algorithm, t = 0.142s;

a) k=1, M=200, $\nu_{\alpha}^{1} = \nu_{0}(1, 0, 1)$, $\nu_{\alpha}^{2} = \nu_{0}(1, 1, 1)$

b) k=1, M=100, $\nu_{\alpha}^{1} = \nu_{0}(3/8, 0, 1/4)$, $\nu_{\alpha}^{2} = \nu_{0}(3/4, 2, 1)$

c) k=2, M=50, $\nu_{\alpha}^{1} = \nu_{0}(0, 0, 0)$, $\nu_{\alpha}^{2} = \nu_{0}(1/4, 3/4, 1/2)$

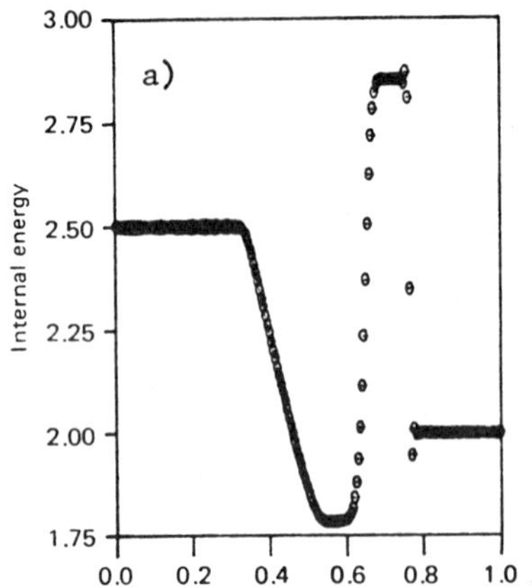

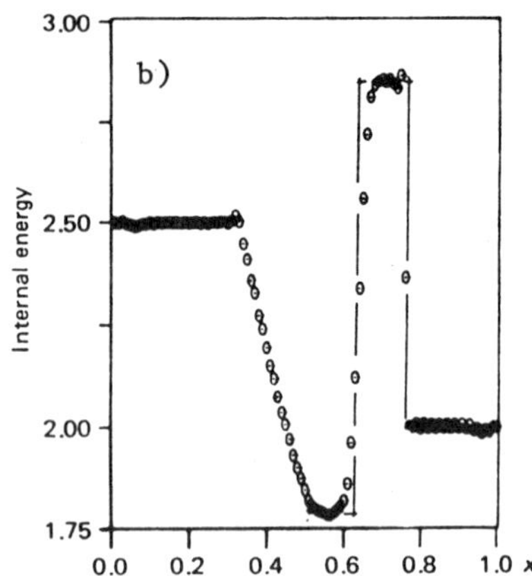

Fig. 3. Temperature distribution, Riemann shock tube, pG finite element algorithm, $t_f = 0.142s$,

a) $k=1$, $M=100$, ν_α^i optimal.

b) $k=2$, $M=50$, ν_α^i optimal.

The shock Mach number of the Sod specification is $M_S \approx 1.6$, and the cited pG algorithm results compare quite favorably (in the "eyeball norm") with those of the better finite difference algorithms as well as the Taylor-Galerkin algorithm [18]. A slightly more difficult test, with a shock Mach number $M_S \approx 2$, is obtained by imparting an initial momentum to the fluid in the high pressure chamber and using the initial conditions $\rho = 0.445$, $m = 0.311$ and $g = 8.928$ on $0 \leq x < 1/2$, and $\rho = 0.5$, $m = 0$ and $g = 1.473$ on $1/2 \leq x \leq 1$. This initial condition specification yields a sharp elevation in the density level, between the shock and contact discontinuity, in graphic similarity to the temperature distribution of the Sod definition, see Figure 3. Figure 4 graphs the k=1 pG algorithm discrete approximate solution of the dependent variable set, at $t_f = 0.15$, as obtained using the definitions in equation 21. Both the shock and contact discontinuity are unacceptably diffused, which prompted a reexecution in which the values of ν^2 (only) were halved, ie., $\nu_\alpha^2 = \nu_0(3/8, 1, 1/2)$. Figure 5 graphs the resultant k=1 pG algorithm discrete solution in which the representation of the shock is quite acceptable. However, the contact discontinuity is still rather diffused, such that the associated pressure still exhibits a local bump.

The results of these Riemann shock tube specifications have confirmed that the pG algorithm can produce discrete approximate solutions to the Euler equation system that are of acceptable accuracy. A criticism could and should be leveled at the arbitrariness represented by the penalty parameter set ν_α^i. In defense, the range of ν_α^i over which numerical experimentation has yielded estimates distinct from ν_0 is quite small on the scale of the data in Figure 1. An improved theoretical analysis could cast both the functional form of the penalty constraint, and its associated scalar parameter set, on firmer ground.

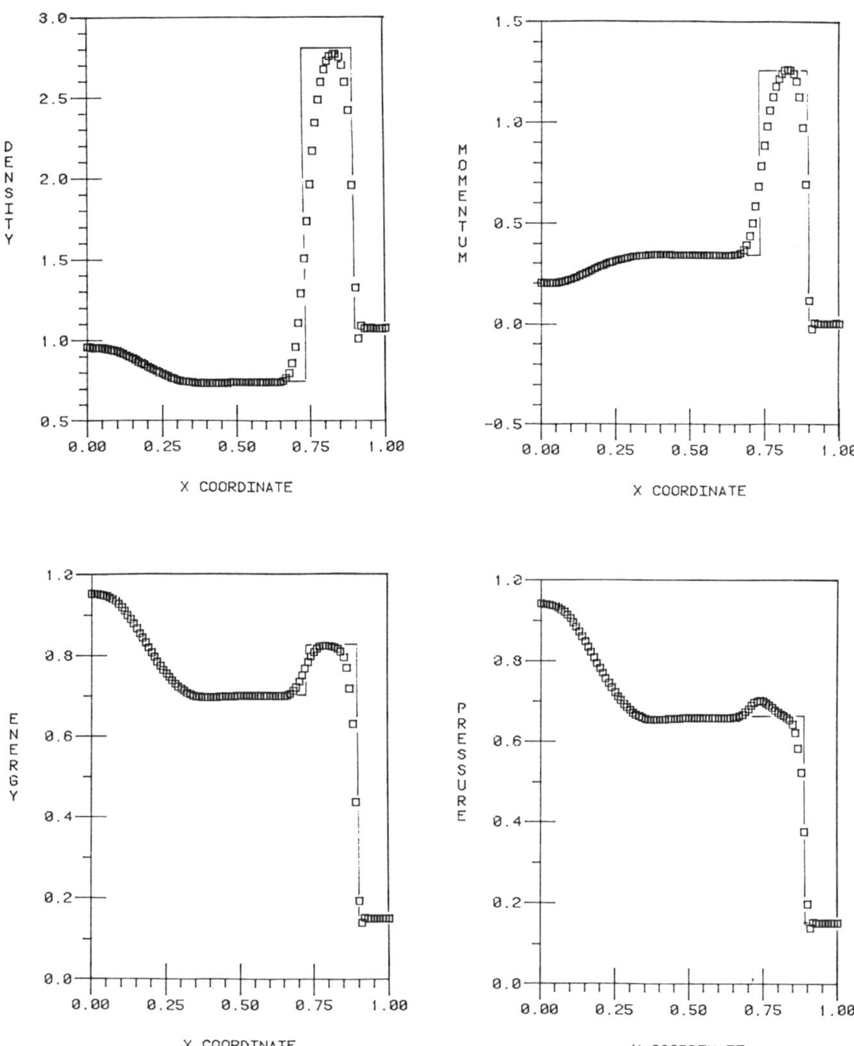

Fig. 4. Euler discrete approximate solution, Riemann shock tube no. 2, pG algorithm, $t_f = 0.15s$, ν_α^i optimal.

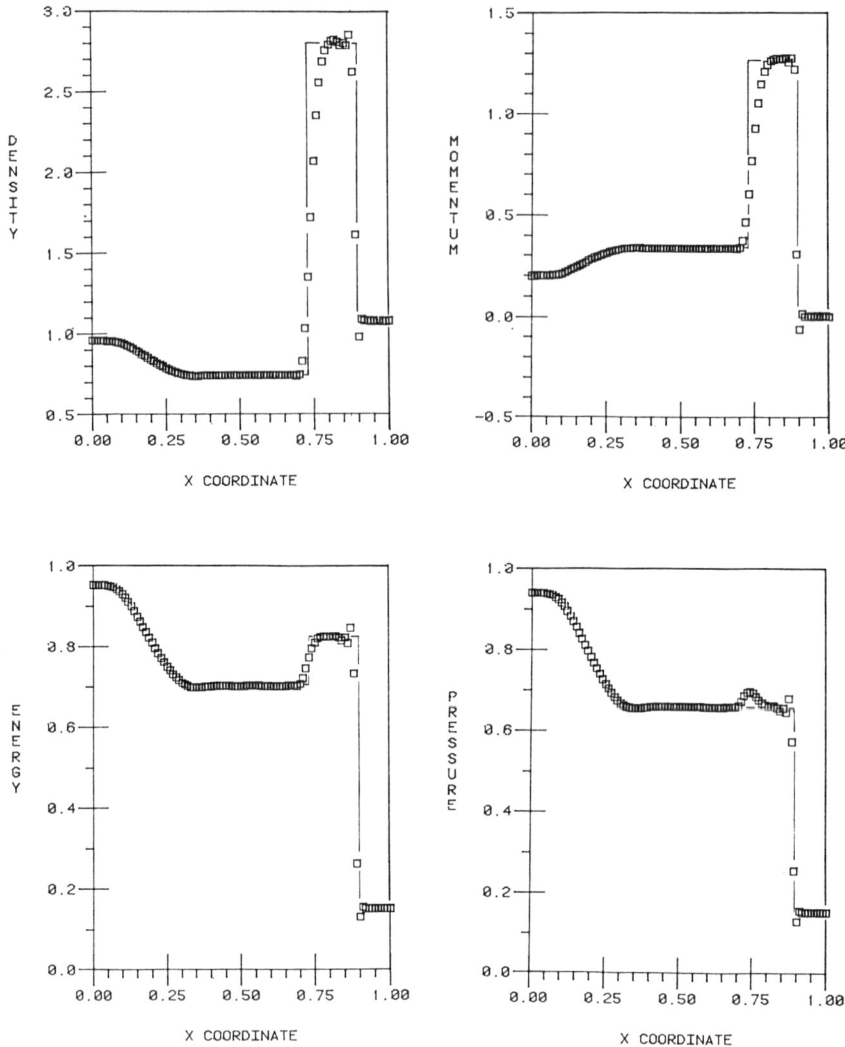

Fig. 5. Euler discrete approximate solution, Riemann shock tube no. 2, pG algorithm, t_f = 0.15s, ν_α^i $\frac{1}{2}$(optimal).

4.2 Boundary Conditions

The deLaval nozzle geometry, with mixed subsonic-supersonic flow regimes including off-design operation, provides an excellent test configuration to confirm consistent boundary condition specifications as well as use of large Courant numbers. The associated Euler equation specification for quasi-one-dimensional flow is contained within equations 1-4 by replacement of the convection velocity (u) with the product $u\ell nA$, where A(x) is the flow cross-sectional area distribution [17, Ch.8]. A characteristic analysis confirms that three of the four dependent variables at the inlet are eligible for a Dirichlet constraint, while only one may be so specified at the outlet. From the laboratory physics of the associated flow, the exit pressure is typically defined and the inlet approximates a stagnation chamber. Further, the flow is choked when the throat Mach number reaches unity, hence thereafter the mass flow rate is independent of the exit pressure and independent of the shock Mach number during off-design operation. Finally, a shock in the diverging flow section can be driven back to the throat, by raising the exit pressure, and once the shock is swallowed, the mass flow rate is controlled by the exit pressure level.

Any numerical algorithm for quasi-one-dimensional inviscid flow should faithfully meet these requirements, and the pG finite element algorithm has been thoroughly confirmed for consistency. A duct of variable cross-section was defined such that the exit Mach number for supersonic on-design operation was $M_e = 1.5$. The exit chamber is specified of uniform cross-section such that vanishing normal derivative boundary conditions could be defined for the non-Dirichlet constrained variables. The converging section upstream of the throat is extended to a local Mach number of 0.10, and a uniform cross-section inlet region was defined over 10 elements upstream of this point as an approximation to a "stagnation" reservoir. At the inlet end of this reservoir, a vanishing normal boundary condition is applied to the single non-Dirichlet constrained variable.

The pG finite element algorithm meets each consistency requirement for a range of on- and off-design specifications. The exit Dirichlet boundary condition is defined on pressure, and a vanishing derivative is applied to the remaining three variables for flow at arbitrary Mach number distributions. If the Mach number equals or exceeds unity anywhere in the nozzle, Dirichlet inlet boundary conditions are defined for density, momentum and pressure. As a consequence of equation 4, of necessity the energy at inlet is also a Dirichlet constraint. The inlet-Dirichlet boundary condition for momentum (only) is changed to vanishing derivative (within the computer program) if at any time there exists no point of sonic or region of supersonic flow.

Figure 6 summarizes the results of the pG algorithm prediction of Mach number distribution for a range of numerical experiments. Curve 1 shows the Mach number distribution for the exit pressure p_e defined

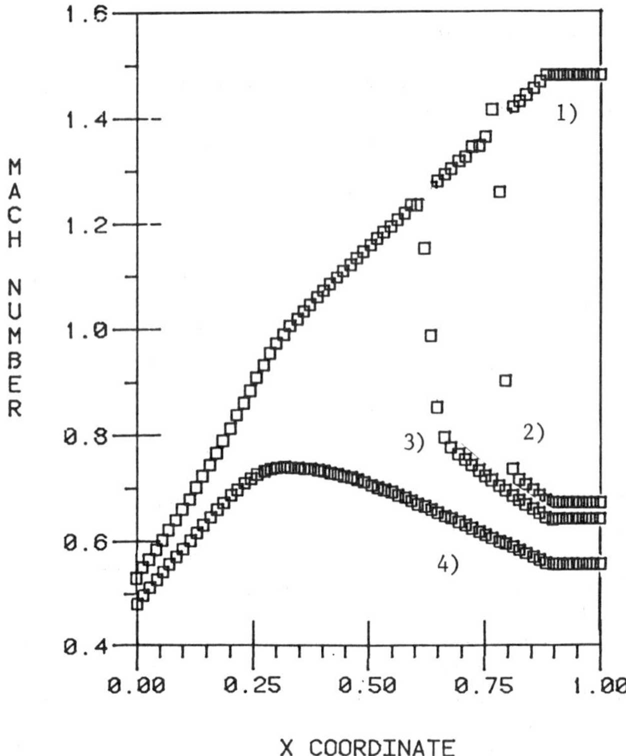

Fig. 6. Mach number discrete approximate solution, deLaval nozzle, pG algorithm, ν_α^i optimal, steady-state solution,

1) supersonic on-design, $M_e = 1.5$
2) supersonic, off-design, $M_s = 1.4$
3) transitional, off-design, $M_s = 1.28$
4) sonic-subsonic, $M_e = 0.55$.

for a supersonic expansion on-design. Curve 2 graphs the pG discrete solution for an off-design exit pressure specification that requires a shock to occur at $M_s = 1.4$. Using the Riemann test specification for ν_α^i, equation 21, this steady-state solution predicts the correct Mach number, exhibits negligible Gibbs phenomena, and smears the shock over about three elements. This solution served as the initial condition specification for a test to evaluate upstream repositioning of the shock by increasing p_e (only) at the last downstream node, such that the new steady-state shock Mach number should be $M_s = 1.28$. The solution transient is essentially completed in a time interval $\Delta t \approx 0.1$ sec, during which the (implicit) pG algorithm extremum Courant number C_m ranged $5 \leq C_m \leq 50$. The steady-state solution was pursued to $t_f = 0.5$

sec, at which time $C_m > 10{,}000$. Curve 3 graphs the resultant steady-state solution, for which $M_s = 1.25$ is a close approximation to design with the shock smeared over three elements. For the companion experiment, curve 4 graphs the smooth solution for on-design flow with a sonic throat, but subsonic expansion to a higher exit pressure. From this specification, the exit pressure was then lowered (at the last node), to correspond to the off-design flow with $M_s = 1.28$, which produces the requirement for the pG algorithm to predict a non-smooth solution from smooth initial data. The resulting steady-state discrete solution is identical to that graphed as curve 3, the solution transient spanned $\Delta t \approx 0.1$, and $C_m > 10{,}000$ at $t_f = 0.5$.

Figure 7 is a composite graph of the pG algorithm steady-state discrete solutions for momentum for this test range. Since $M_t = 1$ for all variations, the flow is choked, hence the momentum is correctly predicted as insensitive to the various exit pressure definitions, curve 1. Some modest Gibbs phenomena is generated at the shock, for both $M_s = 1.4$ and $M_s = 1.28$. The final test is initialized by the condition corresponding to curve 4, Figure 6, but the exit pressure is raised rather than lowered. The result is a purely subsonic solution with reduced mass flow rate. The inlet momentum boundary condition is switched to Neumann, and the pG solution again proceeds rapidly to the new steady state, for which the Mach number distribution and (reduced) mass flow rate are in essentially exact agreement with the isentropic analytical solution.

4.3 Efficiency

Data have already been cited indicating the implicit integration implementation of the pG algorithm exhibits certain efficiency aspects for predictions going to steady-state. The Newton algorithm specification, equations 13-16, admits generation of specifiably time-accurate solutions, as well as rapid time-inaccurate transit towards a steady-state including use of local integration time step distributions. The functional form of equation 13, upon insertion of equation 12 and rearrangement is [17, Ch.8],

$$\underline{F} = \underline{\underline{A}}(\cdot)\underline{\delta Q} + \Delta t\,\underline{B}(\cdot) = \underline{0} \tag{23}$$

where $\underline{\underline{A}}$ is a square matrix, and $\underline{\delta Q}$ and $\underline{B}$ are column matrices. Both $\underline{\underline{A}}(\cdot)$ and $\underline{B}(\cdot)$ involve $\underline{Q}_{j+1}$ and $\underline{Q}_j$, and recalling equation 15, the solution to equation 14 yields the p+1st estimate of $\underline{\delta Q}$ at t_{j+1}. Although a Taylor series could be used to estimate the p=0 array $\underline{Q}^0_{j+1}$, the simple expedient is to assume

$$\underline{Q}^0_{j+1} \equiv \underline{Q}_j \tag{24}$$

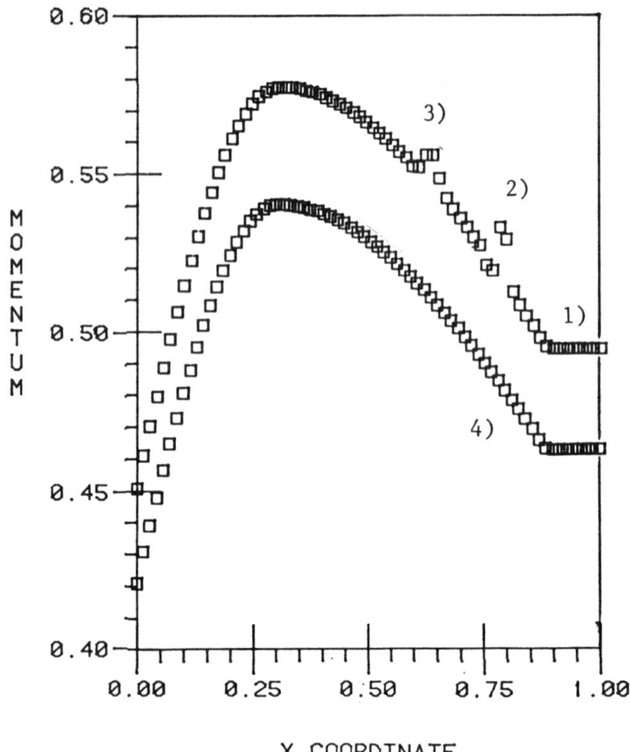

Fig. 7. Momentum discrete approximate solution, deLaval nozzle, pG algorithm, ν_{α}^{i} optimal, steady-state solution,
1) supersonic on-design, $M_e = 1.5$
2) supersonic, off-design, $M_s = 1.4$
3) transitional, off-design, $M_s = 1.28$
4) sonic-subsonic, $M_e = 0.55$.

By definition then, $\underline{\delta Q}_{j+1}^{1} \equiv \underline{0}$ and $\underline{F}$ reduces to a linear multiple of $\Delta t \equiv \Delta t_{j+1} - t_j$. The corresponding functional dependence in the Jacobian, equation 16, must also be evaluated using $\underline{Q}_{j+1}^{o} = \underline{Q}_j$ (and $\Delta t_j \equiv t_j - t_{j-1}$). Combining, the functional form of equation 14 for the first iteration (p=0) is

$$\underline{\underline{J}}(\underline{Q}_j, \Delta t_j)\, \underline{\delta Q}_{j+1}^{1} = -\Delta t_{j+1} \underline{B}(\underline{Q}_j) \tag{25}$$

Hence,

$$\left(\frac{1}{\Delta t_{j+1}}\right) \underline{\delta Q}_{j+1}^{1} = -\underline{\underline{J}}^{-1} \underline{B} \tag{26}$$

and the right side is a known (column) matrix. Hence, each element of $\underline{\delta Q}_{j+1}$ can be individually scaled for "optimal size" by defining the corresponding $\underline{\Delta t}_{j+1}$ distribution. Provided only the p=0 step of each Newton iteration is taken, then equations 25-26 can be cycled to yield a rapid (time-inaccurate) progression to steady-state, provided the solution process does not diverge, ie., there exists some upper bound on Δt_{j+1}. This also eliminates the need to evaluate the first term in equation 23, yielding an associated savings in computational effort.

If more than the p=0 iteration step is used, then equation 26 remains valid to estimate a nominal (scalar) value for $(\Delta t_{j+1})_{max}$. Then, the Newton algorithm is exercised in its normal fashion yielding a (time-) accurate solution $\underline{Q}_{j+1}^{p+1}$, for which $|\underline{\delta Q}_{j+1}^{p+1}|_{max} < \varepsilon$, where ε is the convergence requirement. Dependent upon the accuracy of equation 24, the Newton iteration procedure should yield a quadratic rate of convergence to ε. The time accurate solutions for the Riemann problems, obtained marching at a fixed constant Δt, have documented a quadratic (and even better) iterative convergence, after the first few time steps (to homogenize the step initial conditions), for $0.1 \leq |\underline{\delta Q}_{j+1}^{1}|_{max} \leq 1.0$, see Figure 8. In all situations, the extremum $\underline{\delta Q}$ occurs at the shock as it traces through the mesh.

Also graphed in Figure 8 is the typical convergence history for the supersonic off-design deLaval nozzle experiment. These data predict a convergence rate somewhat smaller than quadratic, but were obtained using an "optimization" procedure [25] wherein Δt_{j+1} was adjusted at the beginning of each iteration to otain as large a Courant number as practical. In this instance, except as approaching steady-state, the definition in equation 24 is inaccurate and should be replaced with a Taylor series representation through several upstream time stations.

These issues regarding efficient time-step selection and robust iterative convergence rates carry over to the multi-dimensional problem definition for the implicit pG algorithm. There, the construction of a "suitable" approximation to the Newton algorithm Jacobian is required, and one has been developed employing tensor matrix products [23]. Rewriting the pG algorithm in a generalized coordinates description [17, Ch.8], and using the kth degree isoparametric tensor product basis set for equation 9, permits a ready restatement of the Newton Jacobian, equation 16, as

$$\underline{\underline{J(F)}} = \underline{\underline{J}}_1 \otimes \underline{\underline{J}}_2 \otimes \underline{\underline{J}}_3 \tag{27}$$

Each component $\underline{\underline{J}}_\eta$ is block tri- or penta-diagonal, dependent upon k, and constructed from its definition assuming interpolation and differentiation are one-dimensional. Equation 14 then takes the form

$$\underline{\underline{J}}_1 \otimes \underline{\underline{J}}_2 \otimes \underline{\underline{J}}_3 \; \underline{\delta Q}_{j+1}^{p+1} = -\underline{F}_{j+1}^{p} \tag{28}$$

Defining,

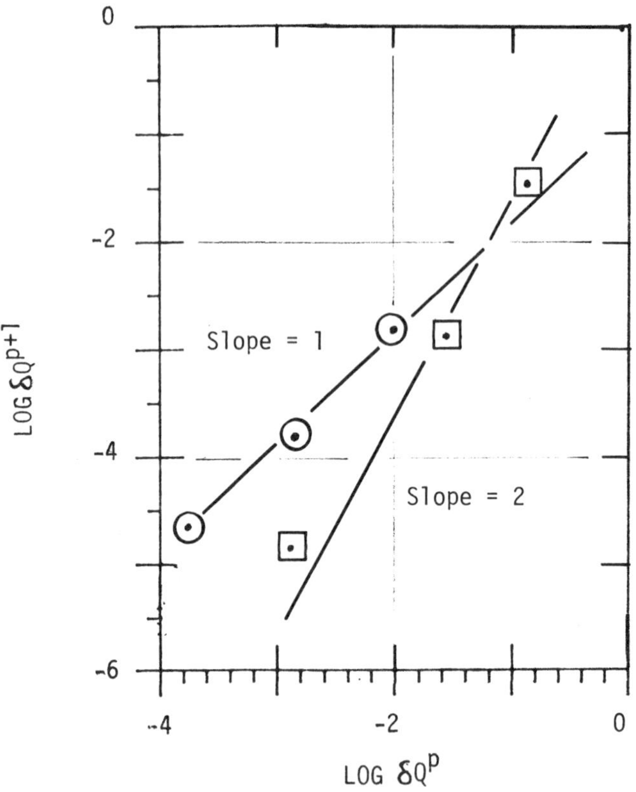

Fig. 8. Graph of iterative convergence rate, ⊡ Riemann shock tube, ⊙ de Laval nozzle.

$$\begin{aligned} \underline{P1}_{j+1}^{p+1} &\equiv \underline{\underline{J}}_2 \otimes \underline{\underline{J}}_3 \; \underline{\delta Q}_{j+1}^{p+1} \\ \underline{P2}_{j+1}^{p+1} &\equiv \underline{\underline{J}}_3 \; \underline{\delta Q}_{j+1}^{p+1} \end{aligned} \tag{29}$$

the operational approximation to equation 28 becomes the sequence

$$\begin{aligned} \underline{\underline{J}}_1 \, \underline{P1}_{j+1}^{p+1} &= - \underline{F}_{j+1}^{p} \\ \underline{\underline{J}}_2 \, \underline{P2}_{j+1}^{p+1} &= \underline{P1}_{j+1}^{p+1} \\ \underline{\underline{J}}_3 \, \underline{\delta Q}_{j+1}^{p+1} &= \underline{P2}_{j+1}^{p+1} \end{aligned} \tag{30}$$

The definition and use of this form of approximate factorization, where permitted by appropriately regular discretizations Ω^h, essentially removes the storage requirement of the Newton Jacobian as a consideration of consequence in programming the pG algorithm.

Also, for the multi-dimensional flow description, an aspect of efficiency and versatility accrues to retention of the viscous stress tensor and heat flux vector as dependent variables, ie., nodal degrees of freedom. A non-penalized Galerkin statement is suitable to cast the corresponding discrete approximation statements $\underline{F}$, cf. [17, pg. 432]. Furthermore, a stress-strain rate closure law other than equation 5 could be defined, including for example replacement by a time averaged form for turbulent flow that requires solution of an additional set of partial differential equations, e.g., the k-ε two equation closure model.

Combining the generalized coordinates description for the pG algorithm with this decision yields an efficient formulational structure. For example, for equation 2, the first term in the pG constraint statement, equation 10, yields for $\underline{m} \equiv \rho\underline{u}$,

$$\begin{aligned}\int_{R^n} \underline{N}_k\, \underline{L}(\underline{m})d\underline{x} = S_e \Big[& \underline{DET}_e^T\, \underline{\underline{M}}3000\, \underline{MI}'_e \\ & - \underline{UBARK}_e^T \left(\underline{\underline{M}}30K0 - \hat{n}_K\, \underline{\underline{N}}3000 \right) \underline{MI}_e \\ & - \left(\underline{ETAKL}_e^T\, \underline{\underline{M}}30K0 - \hat{n}_L\, \underline{DET}_e^T\, \underline{\underline{N}}3000 \right) \left(\underline{P}_e\, \delta_{IL} - \underline{SIGIL}_e \right) \Big]\end{aligned} \quad (31)$$

In equation 31, S_e is the assembly operator, the indices K and L obey the tensor summation rule, $1 \leq (K,L) \leq n$, and I is the free index for the scalar components of $\underline{m}$. The element matrix $\underline{M}30K0$ is the hypermatrix equivalent of $\partial/\partial\eta_k$ (transformed), contracted as indicated with corresponding nodal distributions of the scalar components $\underline{ETAKL}_e$ of the generalized coordinate transformation. The element matrices $\underline{M}3000$ and $\underline{N}3000$ are hypermatrix equivalents of interpolation on R_e^n of variable quadratic products, $\hat{n}_L$ is the outwards pointing unit normal vector for $\partial R_e \cap \partial R$, and δ_{IL} is the discrete Kronecker delta. Finally, $\underline{MI}_e$, $\underline{P}_e$ and $\underline{SIGIL}_e$ are nodal values of momentum, pressure and stress tensor, while $\underline{UBARK}_e$ contains nodal values of the velocity resolution into contravariant convection components parallel to the $\underline{\eta}$ coordinate system. Viewing equation 31, only three distinct element matrices are required, and inclusion of the penalty term adds only two more.

Combining these various aspects, inviscid and viscous ($Re = 10^5$) flow solutions are reported [23] for a two-dimensional simulation of the Sod definition of the Riemann shock tube. The two-dimensional inviscid solutions are identical with the one-dimensional solution on the corresponding discretization Ω^h. For the two-dimensional viscous simulation, no-slip and cold wall boundary conditions replace the vanishing Neumann constraints, and the mesh in the transverse direction was refined to permit capture of the wall region gradients. Figure 9 graphs in perspective the resulting pG algorithm solutions for discrete pressure, momentum and principal shear stress. Away from the walls, the viscous and inviscid solutions are identical, while sharp gradients in shear stress and momentum are predicted near the walls by the viscous solution. This problem definition employed a 32 x 20 rectangular mesh and specified 10 dependent variables/node, plus 5 pieces of metric data, yielding ~10,000 nodal degrees of freedom. The viscous and inviscid solutions were generated implicitly using the identical time step, the same convergence histories were recorded (90 total iterations, 3 iterations/time step), and the viscous solution was accomplished using only 70K words of central memory.

4.4 Adaptive Mesh

Another efficiency aspect accrues to the generalized coordinates formulation of the pG finite element CFD algorithm as regards inclusion of a solution adaptive discretization procedure. Extremization of a constrained variational principle provides a robust theoretical framework [26] for an elliptic grid generation procedure, and numerical results abound for boundary-conforming mesh generation, cf. [27]. For the pG algorithm, the fundamental modification occurs in the definition of $\underline{q}^h(\underline{x},t)$, equation 8, wherein the (matrix) elements of the $k\underline{th}$ degree cardinal basis $\underline{N}_k$ become functions of time yielding equation 9 in the form

$$\underline{q}^e(\underline{x},t) \equiv \underline{N}_k(\underline{\eta},t)^T \underline{Q}(t)_e \tag{32}$$

The time derivative terms in both $\underline{L}(\cdot)$ and $\underline{L}^c(\cdot)$, equation 10, then become,

$$\frac{\partial}{\partial t}(\underline{q}^e(\underline{x},t)) = \underline{N}_k(\cdot)^T \frac{d\underline{Q}_e}{dt} + \frac{\partial}{\partial t}(\underline{N}_k(\cdot)^T)\,\underline{Q}_e \tag{33}$$

which generates appropriate additional contributions in the algorithm statement equation 10.

Since $\underline{N}_k(\cdot)$ expresses the isoparametric coordinate transformation, by definition the intrinsic $\underline{\eta}$ coordinate system is invariant yielding the time-dependence embedded within the Lagrangian coordinates $\underline{XI}$, $1 \leq I \leq n$, of the nodes of the mesh on R^n. Denoting the elements of $\underline{\dot{XI}}_e$ as the corresponding Lagrangian velocity, the last term in equation 33 can be expressed as,

$$\frac{\partial}{\partial t}(\underline{N}_k(\cdot)^T)\underline{Q}_e = \underline{XI}^T \underline{\underline{N}}_p\, \underline{Q}_e \tag{34}$$

where the elements of the matrix $\underline{\underline{N}}_p(\cdot)$ are determined upon definition of k in equation 9. Equation 34 is then employed in the generalized coordinates implementation of equation 10 in the same manner as all other terms. Considering the expansion for $\underline{L}(\underline{m})$ for example, a non-vanishing Langrangian grid velocity distribution becomes an augmentation of the convection velocity field $\underline{\bar{u}}_k$, with nodal values $\underline{UBARK}_e$, in the form

$$\int_{R_e^n} \underline{N}_k \underline{L}(\underline{m}) d\underline{x} = \mathop{S}_e \Big[\cdot - \underline{UBARK}_e^T (\underline{\underline{M}}30K0 + \cdot)\underline{MI}_e + \underline{XK}_e^T (\underline{\underline{P}}300K + \cdot)\underline{MI}_e - \ldots \Big] \tag{35}$$

where $\underline{\underline{P}}300K$ is an additional element-independent hypermatrix.

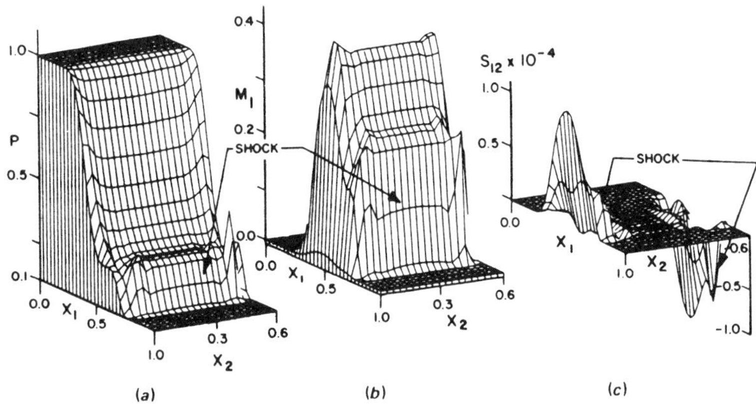

Fig. 9. Discrete approximate solution, two-dimensional viscous Riemann shock tube, pG algorithm, ν_α^i optimal, Re = 10^5, t_f = 0.142s, a) pressure, b) momentum c) shear stress.

The formulation is easy to complete for the one-dimensional restriction for equations 1-4 and for definition of the linear (k=1) tensor product basis in equation 9. The matrix elements of $\underline{N}_1$ are $\frac{1}{2}(1 \mp \eta/h_e)$, where $h_e \equiv X_R - X_L$ is the element measure and X_R and X_L are the corresponding (time-dependent) right and left node coordinates of R_e^1. Then,

$$\frac{\partial}{\partial t}\underline{N}_1(\cdot) = \frac{\partial}{\partial h_e}\left[\underline{N}_1(\cdot)\right]\frac{\partial h_e}{\partial t}$$

$$= \frac{1}{2he}(\pm\ \eta/h_e)(\dot{X}_R - \dot{X}_L)_e \tag{36}$$

where $\dot{X}_R$ and $\dot{X}_L$ are the Lagrangian velocities of the nodes of R_e^1. Hence, the specific form of equation 34 is,

$$X_e^T \underline{\underline{N}}_p Q_e = \frac{1}{2he}\dot{\underline{X}}_e^T (\mp 1)(\pm\ \eta/h_e)^T \underline{Q}_e$$

$$= \frac{1}{2h_e}\dot{\underline{X}}_e^T \begin{bmatrix} -\eta/h_e & \eta/h_e \\ \eta/h_e & -h/he \end{bmatrix} \underline{Q}_e \tag{37}$$

which defines the matrix elements of $\underline{\underline{N}}_p$. The additional term in equation 35, defining the impact of moving the nodal coordinate distribution within the first term of equation 10, is then computed using transposition of scalars and rearranging to yield

$$\int_{R_e^1} \underline{N}_1 \underline{X}_e^T \underline{\underline{N}}_p \underline{Q}_e \, dx = \frac{1}{4h_e} \dot{\underline{X}}_e^T \int \begin{Bmatrix} -1 \\ +1 \end{Bmatrix} (1 \mp \eta/h_e)(\pm \eta/h_e)^T d\eta \, \underline{Q}_e$$

$$= \frac{1}{6} \dot{\underline{X}}_e^T \begin{bmatrix} \begin{Bmatrix} -1 \\ +1 \end{Bmatrix} & \begin{Bmatrix} +1 \\ -1 \end{Bmatrix} \\ \begin{Bmatrix} -2 \\ +2 \end{Bmatrix} & \begin{Bmatrix} +2 \\ -2 \end{Bmatrix} \end{bmatrix} \underline{Q}_e \tag{38}$$

Thus, equation 38 defines the corresponding elements of $\underline{\underline{P}}3001$, equation 35.

The use of adaptive meshes for finite element algorithm prediction of representative sample problems is reported [28-30], with the latter results generated using the pG algorithm construction. Figure 10 summarizes the results for a transient scalar convection problem, the definition for which is obtained from equation 1 by setting $u \equiv U_0$ a constant. The initial condition is an expanded sine wave, and the exact solution is translation of the initial condition parallel to the x axis with velocity U_0, Figure 10a). Figure 10b) graphs the non-penalty Galerkin algorithm prediction, obtained on a uniform stationary discretization with nodal coordinate distribution as illustrated. The lagging short wavelength phase error has substantially distorted the solution as the consequence of using (the relatively large) Courant number $C = 0.5$. Figure 10c) graphs the non-penalty Galerkin algorithm prediction on a moving mesh which dynamically adapts a non-uniform discretization clustering regions of refined grid according to the solution energy semi-norm distribution. The nodal coordinates are noted, and the resultant solution exhibits excessive Gibbs phenomena. Figure 10d) is the pG algorithm prediction as obtained, for $\nu^1 = \nu_0 = \nu^2$, on a dynamically adaptive mesh with indicated node coordinate distribution. This solution is completely free of shortwave oscillations and represent an improvement (in the "eyeball norm") over the reference fixed mesh case. Of greater potential utility, the adaptive mesh solution has acquired a leading phase error component, compare Figure10b), which indicates the potential exists to optimize this fundamental aspect.

5. SUMMARY AND CONCLUSIONS

A penalty-Galerkin finite element algorithm, for construction of discrete approximate solutions to the high speed Euler and Navier-Stokes equations, has been examined for attributes of optimization. The results highlighted in this paper may be interpreted as confirming the contention that the pG algorithm, in a generalized coordinates description, appears to exhibit many of the required efficiency and versatility aspects. Of course, there remains a great deal of research and numerical evaluation to be completed before this assertion can be

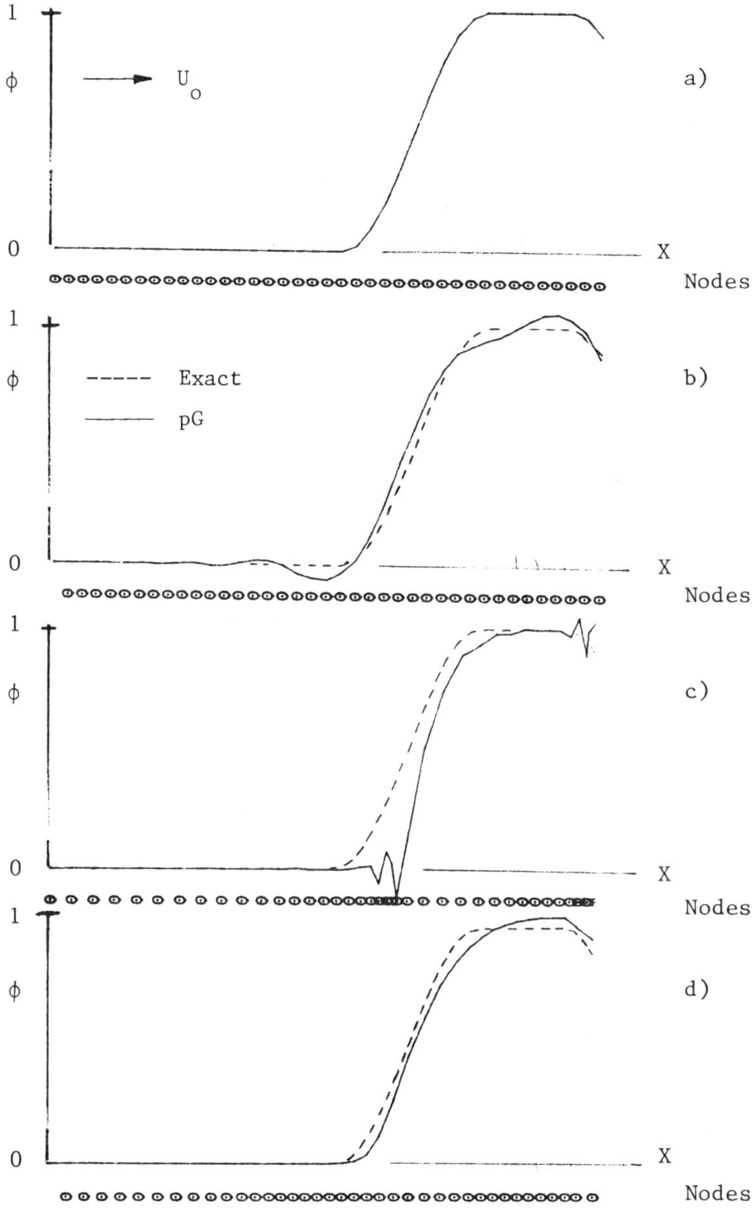

Fig. 10. Discrete approximate solution, sine wave convection, adaptive grid pG algorithm,
a) exact solution
b) fixed mesh, $C = 0.5$, $\nu^i = 0$
c) moving mesh, $C = 0.5$, $\nu^i = 0$
d) moving mesh, $C = 0.5$, $\nu^i - \nu_o$

stated with unblemished confidence. The results presented herein appear sufficient to pursue this course of analysis.

6. ACKNOWLEDGMENTS

The ongoing research and engineering projects in CFD algorithm analysis are coordinated through the Computational Fluid Dynamics Laboratory at The University of Tennessee, the financial support for which, from the private and university sectors, is gratefully acknowledged. The significant contribution of Dr. M. O. Soliman is of fundamental importance, and I wish to acknowledge as well as the important work of Mr. George Moore. Finally, the generosity of the University Computer Center, with whose support this work could not be completed, is thoroughly appreciated.

7. REFERENCES

1. STRANG, F., and FIX, G.J., ***An Analysis of the Finite Element Method,*** Prentice-Hall, New Jersey (1973).
2. ODEN, J.T., and REDDY, J.N., ***An Introduction to the Mathematical Theory of Finite Elements,*** John Wiley, New York (1976).
3. ZIENKIEWICZ, O.C., and CHEUNG, Y.K., Finite Elements in the Solution of Field Problems, ***The Engineer,*** 507-510 (1965).
4. DeVRIES, G., and NORRIE, D.A., The Application of the Finite Element Technique to Potential Flow Problems, ***Trans. A.S.M.E., J. Appl. Mech.,*** 798-802 (1971).
5. FINLAYSON, B.A., ***The Method of Weighted Residuals and Variational Principles,*** Academic Press, New York (1972).
6. BAKER, A.J., Finite Element Solution Algorithm for Viscous Incompressible Fluid Dynamics, ***Int. J. Num Mtd Engr.,*** 6, 89-101 (1973).
7. POPINSKI, Z. and BAKER, A.J., An Implicit Finite Algorithm for the Boundary Layer Equations, ***J. Comp. Phys.,*** 21, 55-84 (1976).
8. SOLIMAN, M.O., and BAKER, A.J., Accuracy and Convergence of a Finite Element Algorithm for Turbulent Boundary Layer Flow, ***Comp. Mtd. Appl. Mech. Engr.,*** 28, 81-102 (1981).
9. CAREY, G.F., and ODEN, J.T., ***Finite Elements: A Second Course, Vol. II,*** Prentice Hall, Englewood Cliffs (1983).
10. HUGHES, T.R.J., LIU, W.K., and BROOKS, A., Review of Finite Element Analysis of Incompressible Viscous Flows by the Penalty Function Formulation, ***J. Comp. Phys.,*** 30, 1-60 (1979).
11. REDDY, J.N., Penalty Finite Element Analysis of 3-D Navier-Stokes Equations, ***Comp. Mtd. Appl. Mech. Engr.,*** 35, 87-106 (1982).
12. BAKER, A.J., and ORZECHOWSKI, J.A., An Interaction Algorithm for Three-Dimensional Turbulent Subsonic Aerodynamic Juncture Region Flow, ***AIAA J.,*** 21 524-533 (1983).

13. von NEUMANN, J., and RICHTMYER, R.D., A Method for the Numerical Calculation of Hydrodynamic Shocks, *J. Appl. Phys.*, 21, 232-237 (1950).
14. ROACHE, P.J., *Computational Fluid Dynamics,* Hermosa Publishers, Albuquerque (1972).
15. HUGHES, T.R.J., and TEZDUYAR, T.E., Finite Element Methods for the Compressible Euler Equations, in Carey and Oden (eds), *Proc 5th Int. Sym. Finite Elements and Flow Problems,* Univ. Texas, Austin, 397-400 (1984).
16. HABASHI, W.G., and HAFEZ, M.M., Finite Element Solutions of Transonic Flow Problems, *AIAA J.,* 20, 1368-1376 (1982).
17. Baker, A.J., *Finite Element Computational Fluid Mechanics,* McGraw-Hill/Hemisphere, New York (1983).
18. SELMIN, V., DONEA, J., and QUARTAPELLE, L., Taylor-Galerkin Method for Non-Linear Hyperbolic Equations, presented at *Int. Conf. Num. Mtd. for Transient and Coupled Problems,* Venice, Italy, July 9-13 (1984).
19. MORTON, K.W., and PARROTT, A.K., Generalized Galerkin Methods for First-Order Hyperbolic Equations, *J. Comp. Phys.,* 36, 249-270 (1980).
20. SOLIMAN, M.O., and BAKER, A.J., Accuracy and Convergence of a Finite Element Algorithm for Laminar Boundary Layer Flow, *Comp. and Fluids,* 9, 43-62 (1981).
21. BAKER, A.J., and SOLIMAN, M.O., Utility of a Finite Element Solution Algorithm for Initial-Value Problems, *J. Comp. Phys.,* 32, 289-324 (1979)
22. RAYMOND, W.H., and GARDER, A., Selective Damping in a Galerkin Method for Solving Wave Problems with Variable Grids, *Monthly Weather Rev.,* 104, 1583-1590 (1976).
23. BAKER, A.J., and SOLIMAN, M.O., A Finite Element Algorithm For Computational Fluid Dynamics, *AIAA J.,* 21, 816-827 (1983).
24. SOD, G.A., A Survey of Several Finite Difference Methods for Systems of Non-Linear Hyperbolic Conservation Laws, *J. Comp. Phys.,* 27, 1-31 (1978).
25. MOORE, G.E., *An Analysis of Factors Affecting Convergence and Efficiency of an Implicit Finite Element Algorithm for Solution of the Quasi One Dimensional Euler Equations,* M.Sc. Thesis, University of Tennessee (1984).
26. SALTZMAN, J., and BRACKBILL, J., Applications and Generalizations of Variational Methods for Generating Adaptive Meshes, in J.F. Thompson (ed.), *Numerical Grid Generation,* North Holland Press, 865-884 (1982).
27. THOMPSON, J.F. (ed.), *Numerical Grid Generation,* North Holland Publishers, Amsterdam (1982).
28 GELINAS, R.J., DOSS, S.K., and MILLER, K., The Moving Finite Element Method: Applications to General Partial Differential Equations with Multiple Large Gradients, *J. Comp. Phys.,* 40, 202-249 (1981).

29. DIAZ, A.R., KIKUCHI, N., and TAYLOR, J.E., A Method of Grid Optimization For Finite Element Methods, *J. Comp. Mtd. Appl. Mech & Engr.*, 41, 29-45 (1983).
30. BAKER, A.J., and SOLIMAN, M.O., On A Solution Adaptive Mesh Algorithm for a Generalized Coordinates Finite Element CFD Algorithm, presented at *5th Int. Sym. Fin. El. Mtd in Flow Problems*, Univ. of Texas, Austin (1984).

FINITE ELEMENTS FOR NONLINEAR INTEGRO-DIFFERENTIAL EQUATIONS AND THEIR INTEGRATION IN TIME

J. M. Sanz-Serna and I. Christie

Departamento de Ecuaciones Funcionales, Facultad de Ciencias, Universidad de Valladolid, Valladolid, Spain.
Department of Mathematics, WestVirginia University, Morgantown, WV 26506, USA.

1. INTRODUCTION

It is well known that the familiar wave equation $w_{tt} = w_{xx}$ constitutes only a first approximation in the study of the transversal vibration of a string [1]. The dimensionless nonlinear integro-differential equation

$$w_{tt} = (1+\int_0^\pi w_x^2\,dx)\, w_{xx}, \quad 0<x<\pi, \quad t>0 \tag{1.1}$$

provides a better model in that it takes into account the increase in tension resulting from the extension of the string [2]. Similar nonlinear integro-differential corrections have been suggested in the study of beams and plates [3,4].

In this paper we consider a Galerkin method for the solution of (1.1) rewritten as the first order system

$$u_t = (1+\int_0^\pi v^2\,dx)\, v_x, \qquad v_t = u_x, \tag{1.2a}$$

which arises from the substitutions $u = w_t$, $v = w_x$.

Along with equations (1.2a) we have the boundary conditions

$$u(0,t) = u(\pi,t) = 0, \quad t > 0 \tag{1.2b}$$

(corresponding to the string being fixed at both ends) and the initial conditions

$$u(x,0),\ v(x,0) \text{ given } 0<x<\pi . \tag{1.2c}$$

ISBN 0-12-747255-X

For (1.2a) subject to (1.2b) the law of *conservation of energy* holds under the form [2]

$$\int_0^\pi u^2 dx + \int_0^\pi v^2 dx + \tfrac{1}{2} \left(\int_0^\pi v^2 dx\right)^2 = \text{constant}. \tag{1.3}$$

A semidiscrete (continuous time) Galerkin method for the numerical treatment of (1.2) is described in Section 2. The semi-discrete solutions are shown to satisfy the conservation law (1.3). In Section 3 we consider several schemes for the time integration of the semidiscrete equations. Some of these methods are linearly implicit, while others are fully implicit and include a modified Crank-Nicolson technique whereby (1.3) can be exactly conserved after the time integration. The last Section is devoted to comparative numerical experiments. The emphasis lies in the nonlinear stability aspects of the time integration and in the role of conservation laws in connection with that stability.

2. A GALERKIN METHOD

Following [5], we divide the interval $[0,\pi]$ into N elements of equal length $h=\pi/N$ by means of a mesh $x_i=ih$, $i=0,1,\dots,N$. We consider the space S^h of real, continuous, piecewise linear functions in $[0,\pi]$ and the subspace S^h_o consisting of the functions in S^h that vanish at $x=0$ and $x=\pi$.

The Galerkin approximation of u and v is a mapping $t \to (U(t), V(t))$; $U(t)\in S^h_o$, $V(t)\in S^h$ defined by

$$(U_t,\phi) = \left(\left(1+\int_0^\pi V^2 dx\right) V_x,\phi\right), \quad \phi\in S^h_o \tag{2.1}$$

$$(V_t,\psi) = (U_x,\psi), \quad \psi\in S^h \tag{2.2}$$

$$U(0)-u(\cdot,0), \quad V(0)-v(\cdot,0) \quad \text{'small'} \tag{2.3}$$

We denote by $(\cdot,\cdot)$, $\|\cdot\|$ the inner product and norm in the space $L^2[0,\pi]$ respectively. It has been proved in [5] that (2.1)-(2.3) possesses a unique solution for $0\le t\le\infty$ and that furthermore the following a priori L^2-estimate holds

$$\| U(t) \|^2 + \| V(t) \|^2 + \tfrac{1}{2}\| V(t) \|^4 = \text{constant} . \tag{2.4}$$

Notice that, after (1.3), this estimate establishes the law of conservation of energy. We have also proved in [5] (by means of the energy method) that the method (2.1)-(2.3) possesses optimal *second order of convergence* in the L^2 norm.

When the usual basis functions $\phi_i\in S^h$, $i=0,1,\dots,N$, $\phi_i(x_i)=1$, $\phi_i(x_j)=0$, $i\neq j$ are introduced, the approximations $U(t)$, $V(t)$ can be written in the form $U(t) = \sum_1^{N-1} U_i(t)\phi_i$, $V(t) = \sum_0^{N} V_i(t)\phi_i$, where

$U_i(t)$, $V_i(t)$ are the nodal values of $U(t)$, $V(t)$ respectively. Conditions (2.1), (2.2) are readily seen to be equivalent to the following system of ordinary differential equations for the nodal values

$$\begin{aligned} M\,(d/dt)\underline{U} &= (1+h\underline{V}^T N\underline{V})\,Q\underline{V} \\ N\,(d/dt)\underline{V} &= -Q^T\underline{V} \end{aligned} \tag{2.5}$$

where $\underline{U} = [U_1, U_2, \ldots, U_{N-1}]^T$, $\underline{V} = [V_0, V_1, \ldots, V_N]^T$ and M, N, Q, are suitable tridiagonal matrices with dimensions $(N-1)\times(N-1)$, $(N+1)\times(N+1)$ and $(N-1)\times(N+1)$ respectively [5].

The conservation law (2.4) may now be rewritten as

$$h\underline{U}^T M\underline{U} + h\underline{V}^T N\underline{V} + \tfrac{1}{2}\,h^2(\underline{V}^T N V)^2 = \text{constant}. \tag{2.6}$$

3. INTEGRATION IN TIME

In order to advance in time the solution of (2.5) we first consider the *trapezoidal rule* (Crank-Nicolson)

$$\begin{aligned} M(\underline{U}^{n+1}-\underline{U}^n) &= (k/2)\,[G(\underline{V}^{n+1})Q\underline{V}^{n+1}+G(\underline{V}^n)Q\underline{V}^n], \\ N(\underline{V}^{n+1}-\underline{V}^n) &= -(k/2)\,[Q^T\underline{U}^{n+1}+Q^T\underline{U}^n], \end{aligned} \tag{3.1}$$

here k is the stepsize in time and $\underline{U}^n$, $\underline{V}^n$ are the approximations of $\underline{U}(nk)$, $\underline{V}(nk)$ respectively. In this section we use the notation $G(\underline{V}) = 1+h\underline{V}^T N\underline{V}$.

An alternative to (3.1) is provided by the closely related *implicit mid-point rule*

$$\begin{aligned} M(\underline{U}^{n+1}-\underline{U}^n) &= k\,G(\tfrac{1}{2}(\underline{V}^{n+1}+\underline{V}^n))\,Q(\tfrac{1}{2}(\underline{V}^{n+1}+\underline{V}^n)), \\ N(\underline{V}^{n+1}-\underline{V}^n) &= -k\,Q^T(\tfrac{1}{2}(\underline{U}^{n+1}+\underline{U}^n)). \end{aligned} \tag{3.2}$$

We refer to [6] for a discussion of the relation between (3.1) and (3.2). The approximations generated by (3.1) or (3.2) *do not* possess the conservation law (2.6) enjoyed by the theoretical and time-continuous solutions. It has often been argued (see [7, 8] for extensive references) that the exact conservation of invariants of motion is a feature that discretizations of nonlinear problems should have. In order to obtain a scheme which conserves (2.6) we follow a technique originally introduced by Strauss and Vázquez (see [9] for recent references). The *modified Crank-Nicolson rule* which conserves (2.6) is given by [5].

$$M(\underline{U}^{n+1}-\underline{U}^{n}) = (k/4)\ [G(\underline{V}^{n+1})+G(\underline{V}^{n})]\ Q(\underline{V}^{n+1}+\underline{V}^{n}),$$
$$N(\underline{V}^{n+1}-\underline{V}^{n}) =-(k/2)\ Q^{T}(\underline{U}^{n+1}+\underline{U}^{n}). \tag{3.3}$$

Care should be exercised in the implementation of the schemes (3.1)-(3.3) as the system of nonlinear equations for the unknowns $\underline{U}^{n+1}$, $\underline{V}^{n+1}$ is not *banded* due to the presence of the 'nonlocal' term $G(\underline{V}^{n+1}) = 1+h\underline{V}^{T}N\underline{V}$. In particular the use of Newton's method is not recommended as the relevant Jacobian matrix is not sparse. The following predictor-corrector technique is useful. (We restrict the presentation to (3.1); the cases (3.2), (3.3)are similar.) Once $\underline{U}$, $\underline{V}$ are known at level n, we consider the *predictor* stage

$$N(\underline{V}_{(0)}-\underline{V}^{n}) =-k\ Q^{T}\underline{U}^{n} \tag{3.4a}$$

and the *corrector* stages s=1,2,...

$$M(\underline{U}_{(s)}-\underline{U}^{n}) = (k/2)\ [G(\underline{V}_{(s-1)})Q\underline{U}_{(s)}+G(\underline{V}^{n})QV^{n}],$$
$$N(\underline{V}_{(s)}-\underline{V}^{n}) =-(k/2)\ [Q^{T}\underline{U}_{(s)}+Q^{T}\underline{U}^{n}]. \tag{3.4b}$$

Formulae (3.4b) are applied until a value of s is found for which the iterates s, s-1 are identical within the accuracy of the machine. Then $\underline{U}^{n+1}$, $\underline{V}^{n+1}$ are taken to be $\underline{U}_{(s)}$, $\underline{V}_{(s)}$ respectively.

The matrix N can be factorized once and for all and therefore an application of the predictor demands a forward and a backward substitution on an N+1 vector. Each corrector stage requires the solution of a linear system in 2N unknowns, whose matrix changes with n and s. This matrix is block-tridiagonal with 2×2 blocks, provided that the unknowns are taken in the order V_0, U_1, V_1, U_2, V_2,..., V_N.

Clearly the accuracy and properties of the algorithm (3.4) are identical [10, p.86] with those of (3.1). Some computational effort can be saved by allowing only one application of the corrector per time step. In that case the algorithm is only linearly implicit and not identical with the fully implicit scheme (3.1)[10].

In order to identify the methods being used we shall employ the symbol TR to refer to the trapezoidal rule (3.1), the symbol MP to refer to the mid-point method (3.2) and the symbol mC to refer to the modified Crank-Nicolson scheme (3.3). We shall also consider the linearly implicit methods lTR, lMP, lmC obtained from TR, MP mC respectively by allowing one iteration per step of the corresponding corrector.

4. NUMERICAL EXPERIMENTS

We applied the six methods outlined above to the problem given by the system (1.2) with initial conditions $u(x,0) = 0$, $v(x,0) = 0.25 \cos x$, whose exact solution is known [5]. In the table we have displayed a set of typical results corresponding to the L^2 error in $\underline{U}$ and the energy (2.6), when N=20, k=0.1 and t=200 (2000 time steps). For this values of N, k the time-integration errors dominate over those arising from the discretization in space.

method	error	energy
exact	-	0.1025515
TR	0.068	0.1025518
MP	0.074	0.1025513
mC	0.072	0.1025515
lTR	0.059	0.1022890
lMP	0.062	0.1025566
lmC	0.062	0.1024229

Note that even though mC is the only method for which the conservation of energy is theoretically guaranteed, the use of the implicit methods MP and TR has also resulted in practice in a conservation of energy to six decimal places. Furthermore the three implicit methods which have conserved the energy are more inaccurate that those methods which have not done so. (The most accurate scheme is in fact that which has worst conserved the energy.) No stability problem arose with any of the six methods even in very long runs. Thus *exact conservation does not always benefit the accuracy of the scheme nor is necessary to guarantee stability.* (This conclusion is also reached in [11] in a different context.)

We also conclude, in agreement with [12], that the good *stability* properties of *implicit* methods are not lost when they are replaced by their *linearly implicit* counterparts. (However, note that there is a danger in treating some terms in an *explicit* way such as the predictor corrector of [13], cf. [11]).

Finally we observe in the table that each linearly implicit scheme has produced a smaller error than its (fully) implicit counterpart. In our experience this behaviour is highly problem dependent.

ACKNOWLEDGEMENTS

J.M. S. is thankful to 'Comisión Asesora' for financial suport.

REFERENCES

1. ANTMAN, S.S., The equations for large vibrations of strings. *Amer. Math. Monthly*, 87, 359-370 (1980).
2. DICKEY, R.W., Infinite systems of nonlinear oscillation equations related to the string. *Proc. Amer. Math. Soc.*, 23, 459-468 (1969).
3. WOINOWSKY-KRIEGER, S., The effect of axial forces on the vibrations of hinged bars. *J. Appl. Mech.*, 17, 35-36 (1950).
4. EISLEY, J.G., Nonlinear vibrations of beams and rectangular plates. *Z. Agnew. Math. Phys.*, 15, 167-175 (1964).
5. CHRISTIE, I. and SANZ-SERNA, J.M., A galerkin method for a nonlinear integro-differential wave system. *Comput. Meths. Appl. Mech. Eng.* (to appear).
6. DAHLQUIST, G., On one-leg multistep methods, *SIAM J. Numer. Anal.*, 20, 1130-1138 (1983).
7. SANZ-SERNA, J.M., An explicit finite difference scheme with exact conservation properties. *J. Comp. Phys.*, 47, 199-210 (1982).
8. SANZ-SERNA, J.M. and MANORANJAN, V.S., A method for the integration in time of certain partial differential equations. *J. Comp. Phys.*, 52, 273-289 (1983).
9. SANZ-SERNA, J.M., Methods for the numerical solution of the nonlinear Schroedinger equation. *Math. Comp.* (to appear).
10. LAMBERT, J.D., *Computational Methods in Ordinary Differential Equations*, Wiley, London (1973).
11. SANZ-SERNA, J.M. and VERWER, J.G., Conservative and nonconservative schemes for the solution of the nonlinear Schroedinger equation, in preparation.
12. SANZ-SERNA, J.M., Linearly implicit variable coefficient methods of Lambert-Sigurdsson type. *IMA J. Numer. Anal.*, 1, 39-45 (1981).
13. GRIFFITHS, D.F., MITCHELL, A.R. and MORRIS, J.Ll., A numerical study of the nonlinear Schroedinger equation. *Comp. Meths. Appl. Mech. Eng.* (to appear).

MOVING FINITE ELEMENT METHODS FOR THE SOLUTION OF EVOLUTIONARY EQUATIONS IN ONE AND TWO DIMENSIONS

A.J. Wathen, M.J. Baines and K.W. Morton+

University of Reading, Whiteknights, Reading, Berks.
now at *+Oxford University Computing Lab., 8-11 Keble Road, Oxford.*

1. INTRODUCTION

The Moving Finite Element Method (MFE) was introduced by Miller & Miller [4] in 1981. The method seeks an approximate solution of a time-dependent partial differential equation as an expansion in terms of local basis functions, where nodal positions are allowed to be time-dependent. The essential feature is that equations for the movement of the nodes (as well as those for the evolution of the solution generally) are found by minimising the L_2 norm of the residual of the differential equation over time derivatives in a simultaneous way.

Suppose that the differential equation is

$$u_t = L(u) \tag{1.1}$$

and that u is approximated by the expansion

$$U = \sum_{i=1}^{N} a_i(t)\ \alpha_i(x,\underline{s}(t)) \tag{1.2}$$

where $\underline{s}(t)$ is the vector of nodal positions s_i, α_i is a local basis function defined in terms of $\underline{s}$ and $a_i(t)$ are coefficients. (We restrict the argument to one dimension but it extends to any number of dimensions - see [5] and §9). Differentiation of U with respect to time gives

$$U_t = \sum_{i=1}^{N} (\dot{a}_i\alpha_i + a_i\dot{\alpha}_i) \tag{1.3}$$

$$= \sum_{i=1}^{N} \left\{\dot{a}_i\alpha_i + a_i \sum_{j=1}^{N} \dot{s}_j \frac{\partial\alpha_i}{\partial s_j}\right\}$$

$$= \sum_{i=1}^{N} (\dot{a}_i\alpha_i + \dot{s}_i\beta_i) \tag{1.4}$$

ISBN 0-12-747255-X

where the dot denotes time differentiation and $\beta_i = \sum_{\ell=1}^{N} a_\ell \frac{\partial \alpha_\ell}{\partial s_i}$.

(1.5)

In the case of piecewise linear basis functions α_i is piecewise linear with support on just two elements: the functions $\frac{\partial \alpha_i}{\partial s_j}$ can easily be constructed and shown to be also piecewise linear with the same support as α_i. However, whereas α_i is continuous and independent of U, the function β_i is discontinuous at the node i and moreover its slope depends on the slope of the solution U. As such it might seem an unlikely candidate as a basis function, but note the similarity with test functions in the streamline diffusion method. To find the coefficients $\dot{a}_i$ and $\dot{s}_i$ the L_2 norm of the residual

$$\| U_t - L(U) \|_2 \tag{1.6}$$

is minimised over $\dot{a}_i$ and $\dot{s}_i$, (i=1,2,...,N). This leads to the equations

$$\begin{aligned} \langle \alpha_i, U_t - L(U) \rangle &= 0 \\ \langle \beta_i, U_t - L(U) \rangle &= 0 \end{aligned} \tag{1.7}$$

(i=1,2,...,N) which in turn provide the MFE matrix system

$$A(\underline{y})\dot{\underline{y}} = \underline{g}(\underline{y}) \tag{1.8}$$

Here $\underline{y} = (a_1, s_1; a_2, s_2; \ldots; a_N, s_N)^T$, $A(\underline{y})$ is 2x2 block tri-diagonal with blocks

$$A_{ij} = \begin{pmatrix} \langle \alpha_i, \alpha_j \rangle & \langle \alpha_i, \beta_j \rangle \\ \langle \beta_i, \alpha_j \rangle & \langle \beta_i, \beta_j \rangle \end{pmatrix} \qquad j-1 \leqq i \leqq j+1 \tag{1.9}$$

and $\underline{g}(\underline{y})$ arises from the term L(U) in (1.7) above.

Exact solution of the non-linear system of equations (1.8) provides the variation of $\underline{y}$ and hence of a_i, s_i (i=1,2,...,N) with time. The nodal parameters and positions both adjust so as to effect the minimisation of (1.6). Moreover, as we shall see, the solution in certain important cases carries the best L_2 fit to the exact solution over the variables a_i, s_i (i=1,2, ...,N). These useful properties are, however, offset by special difficulties which occur when A becomes singular: this may be caused by element-folding (node overtaking) or "parallelism" (the presence of collinear nodes). To avoid these difficulties Miller introduced penalty functions which prevent such singularities occurring. As a result the matrix A becomes non-singular but rather badly conditioned and the system (1.8) is awkward to solve. Our approach here, in contrast, is to focus attention on these singularities and to take special action when they arise.

In practice $\underline{y}$ is found from (1.8) by some time-stepping scheme. This approximation undermines the exact minimisation of the residual (1.6) and best fit properties, which strictly hold only for the semi-discrete method.

Initial conditions may be introduced by an L_2 projection of the initial data. Boundary conditions (of various types) are easily incorporated by appropriately constraining boundary nodes: for Dirichlet conditions both a and s can be set, for Neumann s only is set.

2. DECOMPOSITION OF A

A useful decomposition of A is found as follows. Number the elements (in 1-D) as k=1,2,...,n=N+1, where the element k=i+1 is that between nodes i and $i+1$. Instead of the expansion (1.4), write

$$U_t = \sum_{k=1}^{n} \sum_{\nu=1}^{2} \dot{w}_{k\nu}(t)\ \phi_{k\nu}(x,\underline{s}(t)) \tag{2.1}$$

where the basis functions ϕ_{k1}, ϕ_{k2} are non-zero only on element k and are respectively those parts of α_i and α_{i+1} on that element. The two sets of basis functions $\{\alpha_i,\beta_i\}$ and $\{\phi_{k\nu}\}$ span the same space. From (1.4) and (2.1)

$$U_t = \underline{\alpha}^T\dot{\underline{y}} = \underline{\phi}^T\dot{\underline{w}} \tag{2.2}$$

where $\underline{\alpha} = (\alpha_1,\beta_1;\alpha_2,\beta_2;\ldots;\alpha_N,\beta_N)^T$, $\underline{\phi} = (\phi_{11},\phi_{12};\phi_{21},\phi_{22},\ldots;\phi_{n1},\phi_{n2})$.

Consider now $\|U_t\|_2^2$, which provides the left hand side of (1.8). From (1.4) and (2.1) again

$$\dot{\underline{y}}^T\underline{\alpha}\ \underline{\alpha}^T\dot{\underline{y}} = \dot{\underline{w}}^T\underline{\phi}\ \underline{\phi}^T\dot{\underline{w}} \tag{2.3}$$

or $$\dot{\underline{y}}^T A\ \dot{\underline{y}} = \dot{\underline{w}}^T C\dot{\underline{w}}\ . \tag{2.4}$$

Here C is a 2x2 block diagonal matrix with blocks

$$C_k = \begin{bmatrix} \langle\phi_{k1},\phi_{k1}\rangle & \langle\phi_{k1},\phi_{k2}\rangle \\ \langle\phi_{k2},\phi_{k1}\rangle & \langle\phi_{k2},\phi_{k2}\rangle \end{bmatrix} = \frac{1}{6}(s_{i+1} - s_i)\begin{bmatrix} 2 & 1 \\ 1 & 2 \end{bmatrix} \tag{2.5}$$

We shall refer to the expansion (2.1) as an elementwise description, while (1.4) is a nodewise description. The relationship between $\dot{\underline{y}}$ and $\dot{\underline{w}}$ emerges from (2.2).

It may readily be shown (by direct differentiation or from [3]) that β_i is a linear sum of the ϕ's that make up α_i. Using the element-node numbering described above

$$\left.\begin{aligned} \alpha_i &= \phi_{k-1\,2} + \phi_{k\,1} \\ \beta_i &= -m_{i-\frac{1}{2}}\,\phi_{k-1\,2} - m_{i+\frac{1}{2}}\,\phi_{k\,1}, \end{aligned}\right\} \tag{2.6}$$

where

$$m_{i+\frac{1}{2}} = \frac{a_{i+1} - a_i}{s_{i+1} - s_i} \tag{2.7}$$

is the slope of U in the element k (between the nodes i and $i+1$). It follows from (2.2) that

$$\dot{\underline{w}} = M\dot{\underline{y}} \quad , \tag{2.8}$$

where M is a 2x2 block diagonal matrix with blocks

$$M_i = \begin{bmatrix} 1 & -m_{i-\frac{1}{2}} \\ 1 & -m_{i+\frac{1}{2}} \end{bmatrix} . \tag{2.9}$$

Note that the 2x2 blocks of M are staggered with respect to the 2x2 blocks of C. Comparison of (2.4) and (2.8) shows that

$$A = M^T CM \quad . \tag{2.10}$$

(In higher dimensions the decomposition (2.10) also holds: M is now rectangular in general, although it may be shown to be similar to a block rectangular 'diagonal' matrix (see [5])).

One advantage of (2.10) is that it immediately enables the sources of singularity of A to be picked out. There are only two such sources, namely, collinearity of nodes (termed "parallelism") and node overtaking, corresponding to singularity of M and C, respectively. (A full discussion is given in [5]). We presently discuss each of these in turn, but first consider the conditioning and inversion of A in the absence of singularities.

3. NON-SINGULAR A

The matrix A occurring in (1.8) turns out, in the absence of singularities, to be exceptionally well conditioned in the sense that the corresponding Jacobi iteration matrix

$$J = I - D^{-1}A \tag{3.1}$$

(where D consists of the diagonal blocks of A) has only two distinct eigenvalues $\pm\frac{1}{2}$ (together possibly with zeros corresponding to boundary terms). This is true for any number of nodes and any solution U. The result follows directly from the property

$$A_{ii} = 2(A_{i-1\ i} + A_{i+1\ i}) \tag{3.2}$$

of the submatrices A_{ij} of (1.9), a consequence of the finite element assembly in one dimension, and certain orthogonality relations. Full details are given in [5] and [6].

As a result of the above property J is convergent and the corresponding SOR iteration for A (with easily calculated optimal parameter $8-4\sqrt{3}$) converges with a convergence factor better than 0.072. We have in fact chosen to use a conjugate gradient method [1] and obtain convergence in only two or three iterations.

In higher dimensions the spectrum of J disperses but remains tightly bounded. The general result is that for d dimensions it lies in the interval

$$[-\tfrac{1}{2}d,\tfrac{1}{2}]$$

for any U and any mesh.

Explicit inversion of A is possible in one dimension since M and C are both 2x2 block diagonal and $A^{-1}=M^{-1}C^{-1}M^{-T}$ (see (2.5) and (2.9) but this result does not extend to higher dimensions.

4. TREATMENT OF PARALLELISM

Rather than introduce penalty functions to prevent parallelism when M becomes singular, we have adopted the following alternative procedure. If the slopes of U in successive elements are equal then the basis function β_i becomes a multiple of α_i (see 2.6) and the two equations of (1.7) become identical. The rank of A is reduced by one and A becomes singular. However, the null space of A is trivially that spanned by by the single vector

$$\underline{u} = \{0,0;0,0;\ldots;m,1;\ldots;0,0\}^T \tag{4.1}$$

where m is the common slope. To obtain the full set of solutions of (1.8) we need only one particular solution to which we merely add any multiple of $\underline{u}$.

One particular solution of (1.8) may be obtained by replacing the second equation of (1.7) by an arbitrary equation: we choose it to be

$$\dot{s}_i = 0 \tag{4.2}$$

(i.e. we seek the particular solution for which the node i is fixed). The matrix A becomes modified such that each element in the 2i'th row and column is zero, except on the diagonal where it may be set to 1. This gives a new system of equations (c.f. (1.8))

$$A^*(\underline{y}^*)\dot{\underline{y}}^* = \underline{g}^*\underline{y}^* \quad . \tag{4.3}$$

When (4.3) has been solved for the particular solution $\dot{\underline{y}}^*$, the general solution of (1.8) is

$$\dot{\underline{y}} = \dot{\underline{y}}^* + c\underline{u} \quad , \tag{4.4}$$

where c is any constant which can be chosen to satisfy some external condition, for example that s_i lies midway between s_{i-1} and s_{i+1}. Whatever c is chosen the residual (1.6) is unaffected.

The procedure can be extended to any number of parallel nodes. Moreover the largest and smallest eigenvalues of J remain $\pm\frac{1}{2}$ when A is modified to A^* however many parallel nodes are present (see [6]). There is a well defined corresponding procedure in higher dimensions.

5. NODE OVERTAKING IN HYPERBOLIC PROBLEMS

Again, instead of introducing penalty functions to prevent node overtaking when C becomes singular, we introduce - for hyperbolic problems - a procedure which recognises such node overtaking as indicative of the formation of a shock. When C becomes singular $s_i = s_{i+1}$, say, (c.f. (2.5)) and two equations of the set of equations (1.7) are lost (c.f. (1.8) and (2.10)). In that case the MFE system (1.8) decouples into two disjoint subsystems, the matrix of each of which is non-singular in general. To solve the two systems requires extra conditions, in effect internal boundary conditions, which may be taken to be the jump condition for the hyperbolic equation, namely,

$$s_i = s_{i+1} = \text{Lim} \int_{s_L}^{s_R} L(u)dx/(a_R - a_L) \tag{5.1}$$

as $a_L \to a_i$, $s_L \to s_i$ from the left, $a_R \to a_{i+1}$, $s_R \to s_{i+1}$ from the right (with $s_i = s_{i+1}$). The solution will then maintain the shock and move it at the appropriate speed.

When a third node runs into a shock that has already formed, a composite form of (5.1) can be used to give the new shock speed and there is a redundant node which may be deleted: this is consistent with the notion of loss of information associated with a shock. In higher dimensions a similar procedure may be followed although the details are more intricate.

6. BEST FIT PROPERTY

For the equation

$$u_t + a(u)u_x = 0 \tag{6.1}$$

with piecewise linear initial data $U(x,0)$, the exact solution after time Δt is

$$u(\xi,\Delta t) = U(x,0) , \qquad \xi = x + a(U(x,0))\Delta t. \tag{6.2}$$

The best L_2 fit $U(\xi,\Delta t)$ to $u(\xi,\Delta t)$ from piecewise linear functions with free nodes is given by

$$\begin{aligned} \langle U(\xi,\Delta t) - u(\xi,\Delta t), \ \alpha_i(\xi,\underline{s}(\Delta t))\rangle &= 0 \\ \langle U(\xi,\Delta t) - u(\xi,\Delta t), \ \beta_i(\xi,\underline{s}(\Delta t))\rangle &= 0 \end{aligned} \tag{6.3}$$

$(i=1,2,\ldots,N)$(c.f. (1.7)). Substituting (6.2) into (6.3) and taking the limit as $\Delta t \to 0$ can be shown to yield the MFE equations (1.7) for the equation (6.1). Thus for this equation the semi-discrete MFE equations (1.7) describe the evolution of the best L_2 fit to the exact solution.

7. TIME STEPPING

In the Miller MFE method a stiff solver is used for the set of ordinary differential equations (1.8) to cope with the stiffness arising from the penalty functions he uses. In contrast we have found that simple explicit Euler time stepping is always adequate for hyperbolic problems. The reason for this is associated with the following argument.

For the equation (6.1) with $a(u) = u$ it can readily be shown [5] that the MFE method with piecewise linears gives the ordinary differential equations for the characteristics, which in this case are straight lines. It follows that in this simple case a first order Euler solver is exact with arbitrary time step (at least until shocks form). For more general equations a piecewise linear initial function will not retain that form and/or the characteristics will be curved. Large time steps will then incur errors but when small time steps are taken the solution will approximately follow the characteristics. From this point of view the method is like a Lagrangian method although, since the nodal amplitudes are also subject to change, not precisely so. It is clear from the above example that the usual restrictions on Eulerian time steps are not appropriate here and that accuracy (rather than stability) is the main criterion to be satisfied. The only absolute limit on time steps is imposed by the occurrence of shocks.

We give below some examples showing the power of the method and this completes the description of the method for hyperbolic problems.

8. DIFFUSION PROBLEMS

For diffusion problems (those for which MFE has been most widely used hitherto) there are extra difficulties associated with representation of the second derivatives in $L(u)$ and with node overtaking (when the arguments of §5 above do not apply).

The use of piecewise linear basis functions leads to difficulties with the calculation of the inner product

$$\langle \beta_i, U_{xx} \rangle \tag{8.1}$$

since β_i is discontinuous at s_i and U_{xx} has the character of a delta function at that point. Miller [4] evaluates (8.1) using a limiting procedure (δ-mollification) but we have approached the problem by focussing attention on u_{xx}. For solutions in which large curvature is important, for example in the steep fronts occurring in convection diffusion problems, a piecewise linear approximation U is inadequate: we have therefore sought to recover the character of the solution by replacing U locally by a smoother 'recovered' function.

It has been shown by Johnson [2] that Miller's δ-mollification method is equivalent to recovering U (for the purposes of evaluating (8.1)) by a local Hermite cubic whose slopes at the nodes are averages of the slopes in the neighbouring elements. Another way of dealing with the difficulty is to carry out for each element a quadratic recovery of U_x using three values, the value at the midpoint and the mean values at the end points.

The other difficulty is that of node overtaking, prevalent except when time steps are very small. Since shocks are not admissable solutions either a new interpretation of node overtaking is required or its avoidance altogether. Methods based on overturning manifolds, including the merging of nodes, for overcoming time step limitations are under consideration. Another approach is to exploit implicit (rather than explicit) time stepping in this particular case.

9. NUMERICAL RESULTS

The four frames in Fig. 1 show the initial data and MFE solution for the inviscid Burgers equation

$$u_t + uu_x = 0 \tag{9.1}$$

after 0,5,10 and 15 explicit Euler time steps with $\Delta t = 0.1$. Note the effect of the parallelism algorithm described in §4. In Fig. 2 we show the MFE solution of the two dimensional Buckley-Leverett-like equation

$$u_t + \frac{\partial}{\partial x}\left(\frac{4u^2}{u^2 + \frac{3}{4}(1-u)^2}\right) + \frac{\partial}{\partial y}\left(\frac{4u^2}{u^2 + \frac{3}{4}(1-u)^2}\right) = 0 \ , \tag{9.2}$$

with data

$$u(x,y,0) = \frac{0.1}{0.1 + \sqrt{x^2+y^2}} , \quad u(0,0,t) = 1, \quad \frac{\partial u}{\partial n} = 0 \quad (9.3)$$

on the boundary, after 18 explicit Euler time steps with Δt = 0.025. The initial mesh was regular, all elements being isosceles or right angled triangles. The inset shows the distorted mesh after 18 time steps and the main picture is an isoplot view of the MFE solution. The regular mesh of spot heights is purely an artifact of the graph plotting. This solution required 2-3 seconds for each time step (irrespective of the distortion of the mesh) on the NORD 500 mini computer at Reading.

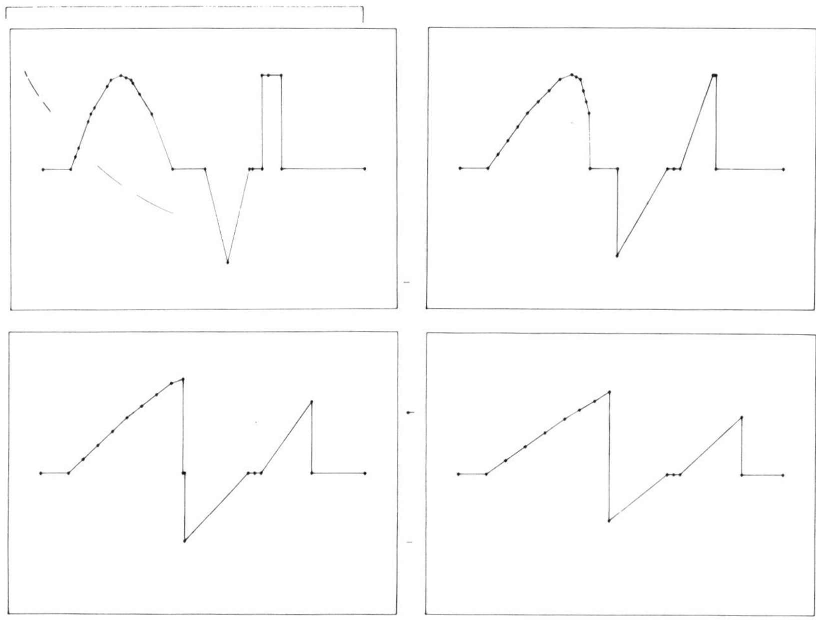

Fig. 2

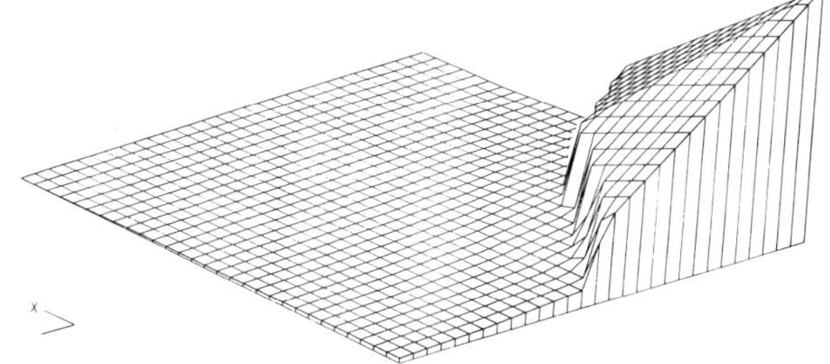

10. CONCLUSION

We have seen that many of the original difficulties of the MFE method are overcome when penalty functions are avoided, particularly in the case of hyperbolic problems. Space prohibits discussion of the treatment of systems of equations where, provided that each component of the system is given its own finite element basis, the algebraic structure of the scalar case is preserved. Further work will concern questions of accuracy, in particular criteria for the introducing or deletion of nodes: both of these are easy to implement using the ideas in this paper. Our work in higher dimensions already shows the power of the method but needs to be taken further, particularly for practical applications.

ACKNOWLEDGEMENTS

One of us (AJW) acknowledges the support of the SERC and British Gas via a CASE Studentship.

REFERENCES

1. CONCUS, P., GOLUB, G.H. & O'LEARY, D.P., A Generalised Conjugate Gradient Method for the Numerical Solution of Elliptic Partial Differential Equations. *Sparse Matrix Computations*, Academic Press, (1975).

2. JOHNSON, I.W. The Moving Finite Element Method for the Viscous Burgers' Equation, *Univ. of Reading, Num. Anal. Report 3/84*, (1984).

3. LYNCH, D.R., Unified Approach to Simulation on Deforming Elements with Application to Phase Change Problems. *J. Comput. Phys.* 47, 387-411 (1982).

4. MILLER, K., & MILLER, R.N. Moving Finite Elements, Part I. *SIAM J. Numer. Anal.* 18, 1019-1032 (1981).

5. WATHEN, A.J., *Ph.D. Thesis, Univ. of Reading* (1984).

6. WATHEN, A.J., & BAINES, M.J. On the Structure of the Moving Finite Element Equations. *Univ. of Reading, Num. Anal. Report 5/83*, (1983).

ON THE CONVERGENCE OF A FINITE ELEMENT METHOD FOR SOLVING A NONLINEAR FREE BOUNDARY PROBLEM RELATED TO MHD EQUILIBRIA

G. Caloz and J. Rappaz

Ecole Polytechnique Fédérale de Lausanne, Switzerland

1. INTRODUCTION

Let Ω be a bounded and connected domain of $\mathbb{R}^2$ with a Lipschitzian boundary $\partial\Omega$ and let I be a positive number. We seek real positive numbers d, λ and a function $u \in H^2(\Omega)$ satisfying:

$$-\Delta u = \lambda u^+ \quad \text{in} \quad \Omega \,, \tag{1.1}$$

$$u = -d \quad \text{on} \quad \partial\Omega \,, \tag{1.2}$$

$$-\int_{\partial\Omega} \frac{\partial u}{\partial n}\, d\sigma = I \,, \tag{1.3}$$

where Δ is the Laplacian operator, u^+ is the function defined by $u^+(x) = u(x)$ if $u(x) \geqslant o$ and $u^+(x) = o$ if $u(x) < o$, $\frac{\partial u}{\partial n}$ is the outward normal derivative of u on $\partial\Omega$.

Problem (1.1) to (1.3) is directly related to ideal MHD equilibria in torus (see [17]). Several authors have treated this problem or similar ones. First Temam [17] establishes the existence of solutions, while Berestycki and Brezis [2] prove the existence of solutions when the right-hand side term of (1.1) has a more general form. A unicity result has been obtained by Temam [18] and Puel [13]. An example of non uniqueness in Problem (1.1) to (1.3) has been given by Schaeffer [15]. Kinderlehrer and Spruck [11] study the shape and smoothness of the plasma domain $\Omega_p = \{x \in \Omega : u(x) > o\}$, when u is a solution of (1.1) to (1.3) for fixed numbers d and λ. For the mathematical and numerical study of this kind of problem, we mention furthermore the work of Sermange [16] and Cipolatti [6]. In [10], Kikuchi constructs a path of solutions of (1.1) to (1.2), with $d = 1$ and λ close to the first eigenvalue λ_1 of the operator $-\Delta$ with homogeneous boundary condition, by using an iterative method based on the principle of contracting mappings. We remark here that if we set $w = u/d$, with $d > o$, then the pair (λ, w) satisfies (1.1) to (1.2) in which we set $d = 1$.

With another point of view, Rappaz [14] proves Kikuchi's results by using a variant of the implicit function theorem stated

ISBN 0-12-747255-X

by Girault and Raviart in [8]; the advantage of this technique is that one can apply it for analysing the approximation of Problem (1.1) to (1.2) with $d = 1$ and λ close to λ_1 by a finite element method, and the error analysis of the discrete problem appears immediately.

In the works of Kikuchi [10] and Rappaz [14], the solutions (λ,u) of Problem (1.1) to (1.2) with $d = 1$ are parametrized by a parameter ε which has no physical meaning. In this paper, we consider the full problem (1.1) to (1.3) and we show, by using the same technique as in [14], that it is possible to construct a path of solutions (λ,u) parametrized by d, when λ is included between the two first eigenvalues λ_1,λ_2 of the operator $-\Delta$ with homogeneous boundary condition. Note that d is a physical parameter which represents a magnetic potential difference between $\partial\Omega_p$ and $\partial\Omega$. Then, we show that the finite element method for approximating Problem (1.1) to (1.3) with polynomials of degree two is convergent in a mesh dependent norm. The main difficulty of this step is that the normalization equation (1.3) is ill-posed in the space $H^1(\Omega)$ containing the finite element spaces which we use; that is the reason why we work in a space H_h^2 which depends on the mesh, such that $H^2(\Omega) \subset H_h^2 \subset H^1(\Omega)$. Finally we give an efficient algorithm to compute the solutions of (1.1) to (1.3) as d is fixed and we state the convergence when d is sufficiently small. We mention that some partial results have been stated without proof in [4].

The organization of the paper is as follows. In Section 2 we state all the theoretical results concerning Problem (1.1) to (1.3) and its approximation by a finite element method. In Section 3 we give a sketch of the proofs concerning the results of Section 2; the complete proofs shall be published in the thesis of the first author [3].

2. THEORETICAL RESULTS

Throughout the paper we use the Sobolev spaces $H^m(\Omega)$ (with integer $m \geq o$) provided with the norm $\|\cdot\|_{m,\Omega}$; we denote $H_o^1(\Omega) = \{f \in H^1(\Omega) : f = o \text{ on } \partial\Omega\}$. For the sake of simplicity we assume that Ω is a convex polygonal domain in the plane.

Let $\lambda_1 < \lambda_2$ be the two first eigenvalues of the classical eigenproblem: find $(\lambda,\varphi) \in \mathbb{R} \times H_o^1(\Omega)$, $\varphi \neq o$ such that

$$-\Delta\varphi = \lambda\varphi \quad \text{in} \quad \Omega \,. \tag{2.1}$$

λ_1 is of multiplicity 1 and we choose the corresponding eigenvector φ_1 such that (1.3) is satisfied. Our first result is:

THEOREM 1. *Assume that* (d_o,λ_o,u_o) *is a solution of Problem* (1.1) *to* (1.3) *which satisfies:*

$$\lambda_1 \leqslant \lambda_o < \lambda_2 , \tag{2.2}$$
$$\text{measure } \{x \in \Omega : u_o(x) = o\} = o . \tag{2.3}$$

Then there exist $\varepsilon > o$ *and two continuous mappings*

$$\lambda : d \in (d_o-\varepsilon,\ d_o+\varepsilon) \to \lambda(d) \in \mathbb{R} , \tag{2.4}$$
$$u : d \in (d_o-\varepsilon,\ d_o+\varepsilon) \to u(d) \in H^2(\Omega), \tag{2.5}$$

such that for $d \in (d_o-\varepsilon,\ d_o+\varepsilon)$, *the triple* $(d,\lambda(d),u(d))$ *is a solution of Problem* (1.1) *to* (1.3) *with* $\lambda(d_o) = \lambda_o$ *and* $u(d_o) = u_o$.

The existence and the uniqueness of (d_o,λ_o,u_o) satisfying (2.2) can be obtained by using the results of Puel [13]; notice that (o,λ_1,φ_1) is a solution of Problem (1.1) to (1.3). Hypothesis (2.3) is discussed in the work of Kinderlehrer and Spruck [11].

In order to give a finite element approximation of Problem (1.1) to (1.3), we introduce, for $h > o$, a triangulation τ_h of Ω by triangles K of diameter less or equal to h. We assume the family of triangulations $\{\tau_h\}$ satisfies the minimal angle condition and is quasi-uniform (see [5]). For the integer $k \geqslant 1$ we define the space

$$V_{h,k} = \{v \in H^1(\Omega) : v/_K \in P_k , \quad \forall K \in \tau_h\},$$

where P_k is the space of polynomials of degree k or less. In the following we set $\overset{o}{V}_{h,k} = V_{h,k} \cap \overset{1}{H}_o(\Omega)$. The finite element approximation of Problem (1.1) to (1.3) consists of finding positive constants d, λ and a function $u_h \in V_{h,k}$ satisfying:

$$\int_\Omega \nabla u_h \nabla v_h dx = \lambda \int_\Omega \overset{+}{u_h} v_h dx , \quad \forall v_h \in \overset{o}{V}_{h,k} , \tag{2.6}$$
$$u_h = - d \qquad \text{on } \partial\Omega , \tag{2.7}$$
$$- \int_{\partial\Omega} \frac{\partial u_h}{\partial n}\, d\sigma = I . \tag{2.8}$$

We note the $V_{h,k}$ is not a subspace of $H^2(\Omega)$ and, before we state our main result, we introduce a mesh dependent space as in Babuska et al [1]. Let $\Gamma_h = (\bigcup_{K\in\tau_h} \partial K) - \partial\Omega$ where ∂K is the boundary of the triangle $K \in \tau_h$.
We define $H_h^2 = \{v \in H^1(\Omega) : v/_K \in H^2(K) \text{ for all } K \in \tau_h\}$ and on H_h^2 we choose the norm $\|v\|_{2,h} = (\sum_{K\in\tau_h} \|v\|_{2,K}^2 + h^{-1} \int_{\Gamma_h} |J\, \frac{\partial v}{\partial n}|^2 d\sigma)^{\frac{1}{2}}$, where $J\, \frac{\partial v}{\partial n}$ is the jump of the normal derivative of v at the boundary of two adjacent triangles of τ_h. In fact we have $H^2(\Omega) \subset H_h^2 \subset H^1(\Omega)$ and if $v \in H^2(\Omega)$ then $\|v\|_{2,\Omega} = \|v\|_{2,h}$. Our main result is:

THEOREM 2. *Assume the hypotheses of Theorem* 1 *and suppose* $k = 2$. *Then there exist two positive numbers* h_o, $\bar{\varepsilon} < \varepsilon$ *and, for* $h \leqslant h_o$, *two continuous mappings*

$$\lambda_h : d \in (d_o-\bar{\varepsilon},\ d_o+\bar{\varepsilon}) \to \lambda_h(d) \in \mathbb{R} , \tag{2.9}$$

$$u_h : d \in (d_o-\bar{\varepsilon}, d_o+\bar{\varepsilon}) \to u_h(d) \in V_{h,2}, \tag{2.10}$$

such that for $d \in (d_o-\bar{\varepsilon}, d_o+\bar{\varepsilon})$, *the triple* $(d,\lambda_h(d),u_h(d))$ *is a solution of Problem* (2.6) *to* (2.8). *Moreover, if* $u(.)$ *is continuous from* $(d_o-\bar{\varepsilon}, d_o+\bar{\varepsilon})$ *into* $H^3(\Omega)$ *we have the error estimate*

$$|\lambda(d)-\lambda_h(d)| + \| u(d)-u_h(d)\|_{2,h} \leq Ch, \tag{2.11}$$

for all $d \in (d_o-\bar{\varepsilon}, d_o+\bar{\varepsilon})$, $h \leq h_o$, *where* C *is a constant independent of* h *and* d.

In the case $k=1$, it is not possible to prove the convergence in the norm $\|\cdot\|_{2,h}$ of a solution branch of (2.6) to (2.8) to the exact branch given by Theorem 1. However, it seems, by using an other technique, it is possible to prove a similar result as in Theorem 2 if we replace the norm $\|\cdot\|_{2,h}$ by $\|\cdot\|_{1,\Omega}$.

Finally we want to give a simple algorithm to compute, for fixed d, the quantities $\lambda_h(d)$ and $u_h(d)$. Our iteration scheme is very close to an inverse power method for eigenvalue problems proposed by Lackner [12]; it can be sketched as follows:

(i) fix d and choose a starting function $u^{(o)}$ in $V_{h,k}$,

(ii) determine, for $n=o,1,2,\ldots$, $w^{(n+1)} \in \overset{\circ}{V}_{h,k}$, $\lambda^{(n+1)}$, $u^{(n+1)} \in V_{h,k}$, by:

$$\int_\Omega \nabla w^{(n+1)} \nabla v dx = \int_\Omega u^{(n)+} v\, dx, \quad \forall v \in \overset{\circ}{V}_{h,k}, \tag{2.12}$$

$$\lambda^{(n+1)} = -I/\int_{\partial\Omega} \frac{\partial w^{(n+1)}}{\partial n} d\sigma, \tag{2.13}$$

$$u^{(n+1)} = \lambda^{(n+1)} w^{(n+1)} - d. \tag{2.14}$$

We remark that in (2.12), $w^{(n+1)}$ is the solution of a finite dimensional linear problem with a fixed right member.

THEOREM 3. *Assume the hypotheses of Theorems* 1 *and* 2 *and suppose that* d_o *is small enough. Then, if* $h \leq h_o$ *is small enough and if* $u^{(o)}$ *is chosen sufficiently close to* $u_h(d)$ *in the norm* $\|\cdot\|_{2,h}$, *we have:*

$$\lim_{n\to\infty}(|\lambda_h(d)-\lambda^{(n)}| + \| u_h(d)-u^{(n)}\|_{2,h}) = o, \tag{2.15}$$

where $(\lambda_h(d),u_h(d))$ *is given in Theorem* 2 *and* $(\lambda^{(n)},u^{(n)})$ *in* (2.12) *to* (2.14).

3. PROOF

In order to prove Theorem 1, we rewrite Problem (1.1) to (1.3). If we set $w=u+d$, we have $w \in H^2(\Omega) \cap H^1_0(\Omega)$ and

$$-\Delta w = \lambda(w-d)^+ \quad \text{in } \Omega, \tag{3.1}$$

$$\int_{\partial\Omega} \frac{\partial w}{\partial n} d\sigma + I = o . \tag{3.2}$$

Let us consider the continuous linear operator T from $L^2(\Omega)$ into $H_o^1(\Omega)$ defined by: for $f \in L^2(\Omega)$, $v = Tf$ is such that

$$- \Delta v = f \quad \text{in } \Omega , \tag{3.3}$$
$$v = o \quad \text{on } \partial\Omega . \tag{3.4}$$

It is well known that the range of T is included in $H^2(\Omega)$ when Ω is a convex polygonal domain (see [9] for instance), and (3.1) is equivalent to $w = \lambda T(w-d)^+$. If we set $V = H^2(\Omega) \cap H_o^1(\Omega)$ and define the mapping $S: \mathbb{R} \times \mathbb{R} \times V \to \mathbb{R} \times V$ by:

$$S(d,\lambda,w) = (I + \int_{\partial\Omega} \frac{\partial w}{\partial n} d\sigma \; ; \; w - \lambda T(w-d)^+), \tag{3.5}$$

Problem (3.1) - (3.2) is strictly equivalent to finding $(d,\lambda,w) \in \mathbb{R} \times \mathbb{R} \times V$ such that $S(d,\lambda,w) = o$.

Clearly $S(d_o,\lambda_o,w_o) = o$ with $w_o = u_o + d_o$. Theorem 1 is proved if we show there exist $\varepsilon > o$ and two continuous mappings $\lambda: d \in (d_o-\varepsilon,d_o+\varepsilon) \to \lambda(d) \in \mathbb{R}$, $w: d \in (d_o-\varepsilon,d_o+\varepsilon) \to w(d) \in V$ such that

$$S(d,\lambda(d),w(d)) = o , \tag{3.6}$$
$$\lambda(d_o) = \lambda_o, w(d_o) = w_o ; \tag{3.7}$$

the function $u(\cdot)$ is given by $u(d) = w(d) - d$.

For proving (3.6) - (3.7) we use a variant of the implicit function theorem stated in [8,14] for a class of "almost C^1" functions because S is not a C^1-mapping. To this end we introduce the set $\Omega_p = \{x \in \Omega: u_o(x) > o\} = \{x \in \Omega: w_o(x) > d_o\}$ and the function χ such that $\chi(x) = 1$ if $x \in \Omega_p$, $\chi(x) = o$ if $x \notin \Omega_p$. Under Hypothesis (2.3), we easily verify that the mapping $v \in H^1(\Omega) \to v^+ \in L^2(\Omega)$ is Frechet differentiable at the point u_o; the derivative being the linear operator $v \in H^1(\Omega) \to \chi v \in L^2(\Omega)$. Consequently, the derivative B of S at the point (d_o,λ_o,w_o), with respect to λ and w, exists and is given by

$$B(\lambda,w) = (\int_{\partial\Omega} \frac{\partial w}{\partial n} d\sigma ; \; w - \lambda_o T\chi w - \lambda T(w_o-d_o)^+). \tag{3.8}$$

LEMMA 1. *Under the hypotheses of Theorem* 1, *the operator* B *is an isomorphism of* $\mathbb{R} \times V$.

Proof: The operator B can be written as a sum of the identity operator in $\mathbb{R} \times V$ with a compact operator from $\mathbb{R} \times V$ into itself; consequently B is a Fredholm operator of index zero and it is sufficient to verify the injectivity of B to prove Lemma 1.

Let $(\lambda,w) \in \mathbb{R} \times V$ be such that $B(\lambda,w) = o$, i.e.:

$$\int_{\partial\Omega} \frac{\partial w}{\partial n} d\sigma = o , \tag{3.9}$$

$$w = \lambda_o T\chi w + \lambda T(w_o - d_o)^+ . \tag{3.10}$$

By replacing (3.10) in (3.9) and using a Green formula, we obtain:

$$o = \int_{\partial\Omega} \frac{\partial}{\partial n} (\lambda_o T\chi w + \lambda T(w_o - d_o)^+) d\sigma = \lambda_o \int_\Omega \chi w dx + \lambda \int_\Omega (w_o - d_o)^+ dx . \tag{3.11}$$

If we denote by $a(.,.)$ the Dirichlet form and by $(.,.)_o$ the L^2-scalar product, we obtain by (3.10) and (3.11):

$$\begin{aligned} a(w,w_o) &= \lambda_o(\chi w, w_o)_o + \lambda((w_o - d_o)^+, w_o)_o = \\ &= \lambda_o(\chi w, w_o - d_o)_o + \lambda((w_o - d_o)^+, (w_o - d_o))_o = \\ &= \lambda_o(w, (w_o - d_o)^+)_o + \lambda((w_o - d_o)^+, (w_o - d_o)^+)_o . \end{aligned} \tag{3.12}$$

Since $S(d_o, \lambda_o, w_o) = o$, we have $a(w_o, w) = \lambda_o((w_o - d_o)^+, w)_o$ and (3.12) becomes:

$$\lambda((w_o - d_o)^+, (w_o - d_o)^+)_o = o . \tag{3.13}$$

From $S(d_o, \lambda_o, w_o) = o$ we deduce that $\lambda_o \int_\Omega (w_o - d_o)^+ = I$; consequently (3.13) implies $\lambda = o$. Relation (3.10) is reduced to

$$w = \lambda_o T\chi w . \tag{3.14}$$

It remains to prove that (3.9) and (3.14) imply $w = o$. We first remark that, for $g \in H_0^1(\Omega)$, we have $a((T - T\chi)g, g) = ((1-\chi)g, g)_o = \int_{\Omega - \Omega_p} g^2 dx \geq o$. So, T and $T\chi$ are selfadjoint compact linear operators of $H_0^1(\Omega)$ provided with the scalar product $a(.,.)$ and $(T - T\chi)$ is positive semi-definite. The characteristic values of T are smaller than those of $T\chi$ (see [7] p.909). From Hypothesis (2.2) it follows that λ_o may be either the smallest characteristic value of $T\chi$ or a regular value of $T\chi$. In the second case, (3.14) implies $w = o$. In the first case, i.e. λ_o is the smallest characteristic value of $T\chi$, we show that if w is solution of (3.14), then either $w(x) \geq o$, $\forall x \in \Omega_p$ or $w(x) \leq o$, $\forall x \in \Omega_p$. By using (3.9) and (3.14) we obtain: $o = \int_{\partial\Omega} \frac{\partial w}{\partial n} d\sigma = \int_\Omega \lambda_o \chi w dx$ which means that $\chi w = o$ and so $w = o$. ∎

Proof of Theorem 1: Clearly if S was a C^1-mapping from $\mathbb{R} \times \mathbb{R} \times V$ into $\mathbb{R} \times V$, the relations (3.6), (3.7) would be a direct consequence of Lemma 1 and the classical implicit function theorem, and Theorem 1 would be proved. Since S is not a C^1-mapping, we modify the above arguments by using the variant of the implicit function theorem given in [14, Lemma 2]. With the technique used in the proof of [14, Theorem 1] we can complete the proof of Theorem 1. ∎

Before proving Theorem 2 we introduce the linear operator T_h

from $L^2(\Omega)$ into $V_{h,k}^o$ defined by: for $f \in L^2(\Omega)$, $v_h = T_h f$ is such that

$$\int_\Omega \nabla v_h \nabla s_h dx = \int_\Omega f s_h dx , \quad \forall s_h \in V_{h,k}^o . \tag{3.15}$$

By setting $w_h = u_h + d$ in Problem (2.6) to (2.8), we see that $w_h \in V_{h,k}^o$ and

$$w_h = \lambda T_h (w_h - d)^+ , \tag{3.16}$$

$$\int_{\partial\Omega} \frac{\partial w_h}{\partial n} d\sigma + I = o . \tag{3.17}$$

We define the mapping S_h: $\mathbb{R}\times\mathbb{R}\times H_h^2 \to \mathbb{R}\times H_h^2$ by :

$$S_h(d,\lambda,w) = (I + \int_{\partial\Omega} \frac{\partial w}{\partial n} d\sigma \; ; \; w - \lambda T_h (w-d)^+) \; ; \tag{3.18}$$

(3.16) - (3.17) is strictly equivalent to find (d,λ,w_h) in $\mathbb{R}\times\mathbb{R}\times H_h^2$ such that $S_h(d,\lambda,w_h) = o$. In fact if $S_h(d,\lambda,w_h) = o$, then w_h belongs to the range of T_h, i.e. $w_h \in V_{h,k}^o$. In order to compare S_h to S we first establish the

LEMMA 2. *Assume that* $k = 2$. *Then we have*

$$\lim_{h\to o} \sup_{f\in H^1(\Omega)} \frac{\| (T-T_h) f \|_{2,h}}{\| f \|_{1,\Omega}} = o . \tag{3.19}$$

Moreover, if f *is such that* $Tf \in H^3(\Omega)$, *we have the estimate*

$$\| (T-T_h) f \|_{2,h} \leq ch \| Tf \|_{3,\Omega} , \tag{3.20}$$

where c *is independent of* h.

Proof: Let Π_h: $v \in H_o^1(\Omega) \to \Pi_h v \in V_{h,2}^o$ be the projector defined by

$$\int_\Omega \nabla (v - \Pi_h v) \nabla v_h dx = o , \quad \forall v_h \in V_{h,2}^o , \quad v \in H_o^1(\Omega) . \tag{3.21}$$

Clearly we have $T_h = \Pi_h T$ and, by using the same technique as in [1], we can verify the estimate:

$$\| v - \Pi_h v \|_{2,h} \leq ch^{\ell-2} \| v \|_{\ell,\Omega} \tag{3.22}$$

for all $v \in H^r(\Omega) \cap H_o^1(\Omega)$, where $r \geq 2$, $2 \leq \ell \leq \min(r,3)$. Error estimate (3.20) is a direct consequence of (3.22) with $r = \ell = 3$.

To prove (3.19) we assume that it is false. Since T is a compact operator from $H^1(\Omega)$ into $H^2(\Omega)$, there exist $\varepsilon > o$, $g \in V$, $f_h \in H^1(\Omega)$ with $\| f_h \|_{1,\Omega} = 1$ such that

$$\| (I - \Pi_h) T f_h \|_{2,h} \geq \varepsilon \tag{3.23}$$

and $\quad \| g - T f_h \|_{2,\Omega} \to o \quad$ when $\quad h \to o$. (3.24)

From (3.23) we have

$$\varepsilon \leqslant \| g - Tf_h \|_{2,h} + \| g - \Pi_h g \|_{2,h} + \| \Pi_h (g - Tf_h) \|_{2,h} .$$

By using (3.22) with $r = \ell = 2$, together with (3.24), we obtain, for $h \leqslant h_o$ small enough,

$$\| g - \Pi_h g \|_{2,h} \geqslant \frac{\varepsilon}{2} . \tag{3.25}$$

Inequality (3.25) contradicts (3.22) with $r = \ell = 3$ and the fact that $H^3(\Omega)$ is dense in $H^2(\Omega)$. ■

Now we remark that the mapping S in (3.5) can be extended as an operator from $\mathbb{R} \times \mathbb{R} \times H_h^2$ into $\mathbb{R} \times H_h^2$. A direct consequence of Lemma 2 is:

LEMMA 3. *Assume that* $k = 2$. *Then for all bounded subsets* Λ *of* $\mathbb{R} \times \mathbb{R} \times H_h^2$ *we have*

$$\lim_{h \to o} \sup_{(d,\lambda,w) \in \Lambda} ||| S(d,\lambda,w) - S_h(d,\lambda,w) ||| = o , \tag{3.26}$$

where $||| (\delta, v) ||| = (\delta^2 + \| v \|_{2,h}^2)^{\frac{1}{2}}$. ■

LEMMA 4. *Under the hypotheses of Theorem* 1, *the operator* B *given in* (3.8) *is an isomorphism of* $\mathbb{R} \times H_h^2$, *the inverse of which is uniformly bounded with respect to* h.

Proof: By applying a Green formula on each triangle $K \in \tau_h$, it is not difficult to prove the existence of a constant c (independent of h) such that

$$\left| \int_{\partial\Omega} \frac{\partial v}{\partial n} d\sigma \right| \leqslant c \, \| v \|_{2,h} , \quad \text{for all } v \in H_h^2 . \tag{3.27}$$

In particular, $B: \mathbb{R} \times H_h^2 \to \mathbb{R} \times H_h^2$ is bounded. As in the proof of Lemma 1, we show that B is an injective Fredholm operator of index zero. An explicit calculus using (3.27) shows that B^{-1} is uniformly bounded with respect to h. ■

Proof of Theorem 2: We use Lemmas 3, 4 and the arguments of the proof of [14, Theorem 2] to show the existence of $\bar{\varepsilon} < \varepsilon$ and of a continuous mapping (λ_h, w_h) defined on $(d_o - \bar{\varepsilon}, d_o + \bar{\varepsilon})$ with values in a neighborhood of (λ_o, w_o) such that $S_h(d, \lambda_h(d), w_h(d)) = o$, for all d with $|d - d_o| < \bar{\varepsilon}$. If $u_h(d) = w_h(d) - d$, the triple $(d, \lambda_h(d), u_h(d))$ is a solution of Problem (2.6) to (2.8). The error estimate (2.11) is a consequence of the implicit function theorem given in [14, Lemma 2] and of (3.20). ■

Proof of Theorem 3: If we define the mapping

$$F(d,u) = -I(\int_{\partial\Omega} \frac{\partial}{\partial n}(Tu^+)d\sigma)^{-1} Tu^+ - d \qquad (3.28)$$

and if $F_h(d,u)$ is given by (3.28) when we replace T by T_h, then it is easy to see that the algorithm (2.12) to (2.14) can be put on the form:

$$u^{(n+1)} = F_h(d,u^{(n)}). \qquad (3.29)$$

We verify that φ_1 is a fix point of F(o,.), i.e. $F(o,\varphi_1) = \varphi_1$, and that the spectral radius of the derivative $D_uF(o,\varphi_1)$ of F with respect to u at the point (o,φ_1) is equal to $\lambda_1/\lambda_2 < 1$. By using an argument of continuity together with (3.19) and a contracting theorem, we prove that for h and d small enough, the iterative scheme (3.29) is convergent when $u^{(o)}$ is chosen sufficiently close to φ_1. ∎

REFERENCES

[1] BABUSKA I., OSBORN J., PITKÄRANTA J., Analysis of mixed methods using mesh dependent norms. *Math. Comp.*35, 1039-1062 (1980).

[2] BERESTYCKI H., BREZIS H., Sur certains problèmes de frontière libre. *C.R. Acad.Sc.Paris* 283, Série A, 1091-1094 (1976).

[3] CALOZ G., Thesis in preparation.

[4] CALOZ G., RAPPAZ J., On the numerical approximation of a free boundary problem related to MHD equilibria. *Comp.Phys. Com.* (to appear).

[5] CIARLET P.G., *The finite element method for elliptic problems*. Noth-Holland-Studies in mathematics and its applications (1978).

[6] CIPOLATTI R., Considérations sur un problème non linéaire: l'équilibre d'un plasma confiné. *Thèse 3e cycle, Université Paris XI* (1982).

[7] DUNFORD N., SCHWARTZ J.T., *Linear operators Part II.* (3rd printing). Interscience publishers (1963).

[8] GIRAULT V., RAVIART P.A., An analysis of upwind schemes for the Navier-Stokes equations. *SIAM J. Numer. Anal.*19, 312-333 (1982).

[9] GRISVARD P., Behavior of the solutions of an elliptic boundary value problem in a polygonal or polyhedral domain. pp.207-274 of B. Hubbard (Ed.), *Numerical Solution of Partial Differential Equations*. Acad. Press (1976).

[10] KIKUCHI F., Construction of a path of MHD equilibrium solutions by an iterative method. *Institute of Space and Aeronautical Science University of Tokyo* 574, 97-111 (1979).

[11] KINDERLEHRER D., SPRUCK J., The shape and smoothness of stable plasma configurations. *Ann. Scuola N. Sup. Pisa* 5, Série IV, 131-148 (1978).

[12] LACKNER K., Computation of ideal MHD equilibria. *Comp.Phys. Com.* 12, 33-44 (1976).

[13] PUEL J.P., A free boundary, nonlinear eigenvalue problem. pp. 400-410 of de la Penha, Medeiros (Eds.), *Contemporary Developments in Continuum Mechanics and Partial Differential Equations*. North-Holland (1978).

[14] RAPPAZ J., Approximation of a nondifferentiable nonlinear problem related to MHD equilibria. *Numer. Math.*(to appear).

[15] SCHAEFFER D.G., Non-uniqueness in the equilibrium shape of a confined plasma. *Comm. in PDE* 2, 587-600 (1977).

[16] SERMANGE M., *Etude mathématique et numérique de problèmes aux limites non linéaires intervenant en physique des plasmas*. Thèse d'Etat, Université Paris XI (1982).

[17] TEMAM R., A nonlinear eigenvalue problem: the shape of equilibrium of a confined plasma. *Arch. Rat. Mech. Anal.*60, 51-73 (1975).

[18] TEMAM R., Remarks on a free boundary value problem arising in plasma physics. *Comm. in PDE* 2, 563-585 (1977).

ON THE NORM OF THE SOBOLEV IMBEDDING OF $H^2(G)$ INTO $C(G)$ FOR SQUARE DOMAINS IN $\mathbb{R}^2$

J. T. Marti

ETH-Zentrum, CH-8092 Zürich, Switzerland

1. INTRODUCTION

Let G be the square domain $(0,s)^2$ in $\mathbb{R}^n$, $\| \ \|_{\infty,G}$ the supremum norm on G and $\| \ \|_{k,G}$ the Sobolev norm, given by

$$\| f \|_{k,G} := \left[\sum_{|p| \leqq k} \int_G D^p f(x)^2 dx \right]^{1/2}, \qquad f \in C^2(G) ,$$

where $C^k(G)$ is the vector space of p-times continuously partially differentiable real functions on G for $|p| \leqq k$. The Sobolev space $H^k(G)$ is the completion of $\{f \in C^k(G) : \| f \|_{k,G} < \infty\}$ with respect to $\| \ \|_{k,G}$. It is well-known that $H^2(G)$, and thus all $H^k(G)$ for $k \geqq 2$, may be imbedded in the Banach space $C(G)$ of continuous real functions f on G with finite $\| f \|_{\infty,G}$ in the sense that there is a bounded linear operator S, the Sobolev imbedding operator, from $H^2(G)$ into $C(G)$ such that $Sf = f$ for f in $H^2(G)$ [1, Theorems 3.16 and 5.4]. If c denotes the norm of S then c is obviously the least constant in the Sobolev inequality

ISBN 0-12-747255-X

$$\| f \|_{\infty,G} \leqq c \| f \|_{2,G} , \qquad f \in H^2(G) . \tag{1}$$

In the one-dimensional case, where $G = (0,s)$, it is known [3] that $c = \tanh^{-1/2} s$. This result may be obtained by the finite element method or by a closed evaluation of a variational method.

In this paper, explicit upper and lower bounds for c are derived, showing that c tends to $1/s$ for s close to zero and that c is in the interval $[2/5 , 35/2]$ for $s \geqq 1$. These bounds become poor for large values of s . It is conjectured that c tends to a value in a neighbourhood of 0.45 when s increases to infinity.

Much better lower bounds may be obtained by the method of finite elements. Since conforming finite elements corresponding to approximations in $H^2(G)$ are of class C^1 we choose plate elements for a uniform partition of G into a number of n^2 squares. Then, excellent lower bounds c_n for c can be computed directly from the elements of the upper triangular matrix $\underline{R}$ of the Cholesky decomposition $\underline{R}^T\underline{R}$ of the global matrix $\underline{A}$ of the relation $\| f \|^2_{2,G} = y^T A y$, $y \in \mathbb{R}^{4m}$, where $m := (n+1)^2$ and y is the vector of partial derivatives of order 0 and 1 of f at the corresponding knots (vertices). In Theorem 1 we can show that the lower bounds c_n indeed converge to the least constant c of (1). The three cases considered numerically are $s = 0.1$, 1 and 10 . The observed convergence rates are $O(h^p)$, where $p \approx 4$, 3 and 1 and the values of c are $10.0250\ldots$, $1.24796\ldots$ and $0.915\ldots$ respectively. Up to now there is no proof for a certain convergence rate for $\lim_n c_n = c$ given in Theorem 1. This would only guarantee that at least some digits of the numbers given above are true. Therefore, from a strictly logical point of view, though a convergence proof is available in Theorem 1, the values indicated above are conjectured values for the desired constants c only.

2. UPPER AND LOWER BOUNDS FOR c

In order to obtain upper bounds for the least constant c in the Sobolev inequality (1) we integrate Taylor's formula over G, apply Schwarz's inequality, Fubini's theorem, substitute $z = x + t(y-x)$ and again use Schwarz's inequality. This yields $y = x + t^{-1}(z-x)$ and

$$\begin{aligned}
s^2 f(x) &= \sum_{|p|\leqq 1} \int_G D^p f(y)(x-y)^p dy \\
&\quad + \sum_{|p|=2} \int_G \left[\int_0^1 tD^p f(x+t(y-x))(x-y)^p dt \right] dy \\
&\leqq \sum_{|p|\leqq 1} \| D^p f \|_{0,G} \left[\int_G |(x-y)^p|^2 dy \right]^{1/2} \\
&\quad + \sum_{|p|=2} \int_0^1 \left[\int_G |D^p f(x+t(y-x))(x-y)^p| dy \right] tdt \\
&= \sum_{|p|\leqq 1} a_p \| D^p f \|_{0,G} \\
&\quad + \sum_{|p|=2} \int_0^1 \left[\int_{x+t(-x+G)} |D^p f(z)| \, |t^{-2}(z-x)^p| dz \right] t^{-1} dt \\
&\leqq \sum_{|p|\leqq 1} a_p \| D^p f \|_{0,G} \\
&\quad + \sum_{|p|=2} \| D^p f \|_{0,G} \int_0^1 \left[\int_{x+t(-x+G)} |t^{-2}(z-x)^p|^2 dz \right]^{1/2} t^{-1} dt \\
&= \sum_{|p|\leqq 1} a_p \| D^p f \|_{0,G} + \sum_{|p|=2} \| D^p f \|_{0,G} \left[\int_G |(x-y)^p|^2 dy \right]^{1/2} \\
&\leqq \sum_{|p|\leqq 2} a_p \| D^p f \|_{0,G} , \quad x \in G ,
\end{aligned}$$

where the positive constants a_p are given by $a_p^2 = s^2$, $s^4/3$,

$s^4/3$, $s^6/9$, $s^6/5$, $s^6/5$ for $p = (0,0)$, $(0,1)$, $(1,0)$, $(1,1)$, $(0,2)$ and $(2,0)$ respectively. The inequality of Schwarz for the sums $\sum_{|p|\leqq 2}$ finally implies

$$f(x) \leqq s^{-2}\left(\sum_{|p|\leqq 2} a_p^2\right)^{1/2} \| f \|_{m,G} .$$

Therefore,

$$c \leqq (s^{-2} + 2/3 + 23s^2/45)^{1/2} \tag{2}$$

and for $s \geqq 1$ one has $c \leqq (98/45)^{1/2}s \simeq 1.47573s$.

The special case where f in (1) is a nonzero constant function immediately yields $c \geqq 1/s$. A more sophisticated lower bound for the constant c in (1) can be obtained by the use of the special trial function $f(x) = \exp(-3^{-1/4}(x_1+x_2))$, $x \in G$. An elementary computation then gives

$$c \geqq d/(1-\exp(-3^{-1/4}s))$$

for $d := (2+2\cdot 3^{1/2})^{-1/2} \simeq 0.427800$. In the special case $s = 1$ one has $c \geqq 0.803747$ and for s tending to 0 or ∞ lower bounds of $1/s$ or d , respectively, are obtained. Comparing the upper and lower bounds for s close to zero yields the result that the least constant c in (1) tends to $1/s$ when s tends to zero.

3. SQUARE PLATE ELEMENTS FOR THE FINITE ELEMENT COMPUTATIONS

The following two-dimensional standard form functions may be used for the construction of $C^1(G)$-functions (see [2], Theorem 2.2.14, or [4], p. 126, the Bogner-Fox-Schmidt rectangle) spanning a 4m-dimensional subspace V_n of $H^2(G)$:

$$hpq(x) := g_{p_1q_1}(x_1)g_{p_2q_2}(x_2) ,$$

$$p,q \leqq (1,1) , \quad x = (x_1,x_2)^T \in (0,1)^2 ,$$

$$g_{00}(t) := 1 - 3t^2 + 2t^3 , \quad g_{01}(t) := g_{00}(1-t) ,$$

$$g_{10}(t) := t - 2t^2 + t^3 , \quad g_{11}(t) := -g_{10}(1-t) , \quad 0 \leqq t \leqq 1 ,$$

where the multi-indices p denote the order of the partial derivatives and the multi-indices q, interpreted as a vector in $\mathbb{R}^2$, denote a vertex of the unit square $(0,1)^2$. An element f_n of V_n is represented by a 4m-vector $y = \{y_{4k-3+2p_1+p_2}\}$ $(1 \leqq k \leqq m,\ 0 \leqq p \leqq (1,1))$ whose elements are partial derivatives $D^p f_n$ of f_n of order p at the knots located at the grid points of a uniform partition of G into a number of n^2 squares:

$$f_n((j-1+x_1)h , s-(i-x_2)h) = \sum_{0 \leqq p,q \leqq 1} y_{4k(i,j,q)-3+2p_1+p_2} h_{pq}(x) ,$$

$$1 \leqq i,j \leqq n , \quad x \in (0,1)^2 ,$$

where $h := s/n$ is the edge length of the plate elements and $k(i,j,q)$ denotes the number of a knot with $p = 0$ at the vertex of a finite element located in row i and column j corresponding to the vertex q of the unit square $(0,1)^2$. With this notation one has

$$\| f_n \|^2_{2,G} = \sum_{|r| \leqq 2} h^{-2|r|} \sum_{1 \leqq i,j,i',j' \leqq n} \sum_{0 \leqq p,q,p',q' \leqq (1,1)}$$

$$y_{k(i,j,q)+2p_1+p_2}\ y_{k(i',j',q')+2p'_1+p'_2}$$

$$\int_0^h \int_0^h (D^r h_{pq})(nx_1,nx_2)(D^r h_{p'q'})(nx_1,nx_2)\,dx_1 dx_2 .$$

The abbreviation

$$(b_{2q_1+q_2+1,2q_1'+q_2'+1})_{2p_2+p_1+1,2p_2'+p_1'+1}$$

$$:= \sum_{|r|\leqq 2} h^{-2|r|} \int_{(0,1)^2} D^r h_{pq}(x) D^r h_{p'q'}(x)dx \ ,$$

$$0 \leqq p,p' \leqq (0,1) \ , \qquad 0 \leqq q \leqq q' \leqq (0,1) \ ,$$

simplifies the above expression to

$$\| f_n \|^2_{2,G} = \underline{y}^T \underline{A} \underline{y} \ ,$$

where $\underline{A}$ is a symmetric tridiagonal $n+1$ by $n+1$ block band matrix of the form

$$\underline{A} = \begin{bmatrix} \underline{B} & \underline{D}^T & 0 & \cdot \\ \underline{D} & \underline{B}+\underline{C} & \cdot & 0 \\ 0 & \cdot & \underline{B}+\underline{C} & \underline{D}^T \\ \cdot & 0 & \underline{D} & \underline{C} \end{bmatrix} .$$

The matrices $\underline{B}$, $\underline{C}$ and $\underline{D}$ occurring in $\underline{A}$ are themselves tridiagonal block matrices, given by the elements b_{ij} of an upper triangular 4 by 4 block matrix whose elements consist of 4 by 4 matrices,

$$\underline{B} = \begin{bmatrix} b_{22} & b_{24} & 0 & \cdot \\ b_{24}^T & b_{22}+b_{44} & \cdot & 0 \\ 0 & \cdot & b_{22}+b_{44} & b_{24} \\ \cdot & 0 & b_{24}^T & b_{44} \end{bmatrix}$$

$$\underline{C} = \begin{bmatrix} b_{11} & b_{13} & 0 & \cdot \\ b_{13}^T & b_{11}+b_{33} & \cdot & 0 \\ 0 & \cdot & b_{11}+b_{33} & b_{13} \\ \cdot & 0 & b_{13}^T & b_{33} \end{bmatrix}$$

$$\underline{D} = \begin{bmatrix} b_{12} & b_{14} & 0 & \cdot \\ b_{23}^T & b_{12}+b_{34} & \cdot & 0 \\ 0 & \cdot & b_{12}+b_{34} & b_{14} \\ \cdot & 0 & b_{23}^T & b_{34} \end{bmatrix}$$

The effective size of the matrices is $4m \times 4m$ for $\underline{A}$ and $4(n+1) \times 4(n+1)$ for $\underline{B}$, $\underline{C}$ and $\underline{D}$. Lower bounds c_n for the least constant c in (1) may now be obtained by

$$\begin{aligned} c_n &:= \sup\left\{y_{4k-3} / \| f_n \|_{2,G} \;:\; 0 \neq y \in \mathbb{R}^{4m},\ 1 \leqq k \leqq m\right\} \\ &= \max\left\{c_{nk} \;:\; 1 \leqq k \leqq m\right\} , \end{aligned} \tag{3}$$

where $\{e_1, \ldots, e_{4m}\}$ is the standard basis of $\mathbb{R}^{4m}$,

$$c_{nk} := \left(y_{4k-3}^2 / \underline{y}^T \underline{A}\, \underline{y}\right)^{1/2}$$

and $\underline{y}$, depending on k, is the solution of

$$\underline{A}\, \underline{y} = e_{4k-3} .$$

Indeed, the above equation for $\underline{y}$ may be obtained by the application of Lagrange's method of multipliers for the evaluation of the maximum of $\| f_n \|^2_{2,G}$ for all f_n in V_n under the restriction that $y_{4k-3} = 1$. It is clear that the matrix $\underline{A}$ is positive definite and thus invertible and that c_{nk} is the square root of the $(4k-3)$th diagonal element of $\underline{A}^{-1}$. Hence a standard computation yields the numbers $c_{n1}, \ldots, c_{n,k}$ and, therefore, c_n from the elements of the upper triangular matrix $\underline{R}$ of the Cholesky decomposition $\underline{R}^T\underline{R}$ of $\underline{A}$. From the computational point of view this means that one Cholesky decomposition simultaneously gives the solutions c_{nk} of a number of m variational problems.

4. THE CONVERGENCE OF THE LOWER BOUNDS c_n TO THE LEAST CONSTANT c OF (1)

THEOREM 1 *If* c *is the least constant in the Sobolev inequality (1), i.e. the norm of the Sobolev imbedding operator* S, *then the numbers* c_n *given by (3) satisfy*

$$\lim_n c_n = c \quad .$$

Proof. First of all we observe that by (2)

$$c \leq c' := s + 1/s \quad .$$

For every $\varepsilon > 0$ there is an f in $H^2(G)$ such that $\| f \|_{2,G} = 1$ and

$$\| f \|_{\infty,G} \geq c - \varepsilon/6 \quad .$$

Since the restrictions to G of the elements of $C^3(\mathbb{R}^2)$ are dense in $H^2(G)$ (see e.g. [1], Theorem 3.18) there is a g in $C^3(\mathbb{R}^2)$ such that

$$\| f-g \|_{2,G} \leqq \varepsilon/(6c') \quad .$$

Let now $X(n)$ $(n \in \mathbb{N})$ be the set of knots of the elements of V_n. In view of the uniform continuity of g on the closure of G there is an n_o such that

$$\| g \|_{\infty,G} \leqq \| g \|_{\infty,X(n)} + \varepsilon/6 , \qquad n \geqq n_o$$

Next, by [2, Theorems 2.2.15 and 6.1.6] there is an $n_1 \geqq n_o$ such that the finite element interpolating functions g_n of g in V_n satisfy

$$\| g_n - g \|_{2,G} \leqq \varepsilon/(6c') , \qquad n \geqq n_1 .$$

Due to (1) and the triangle inequality we then have

$$\frac{\| g_n \|_{\infty,X(n)}}{\| g_n \|_{2,G}} \geqq \frac{\| g \|_{\infty,X(n)} - \| g_n - g \|_{\infty,X(n)}}{\| g \|_{2,G} + \| g_n - g \|_{2,G}}$$

$$\geqq \frac{\| g \|_{\infty,G} - \varepsilon/6 - \| g_n - g \|_{\infty,G}}{\| g \|_{2,G} + \| g_n - g \|_{2,G}}$$

$$\geqq \frac{\| f \|_{\infty,G} - \| f-g \|_{\infty,G} - \varepsilon/6 - \| g_n - g \|_{\infty,G}}{1 + \| f-g \|_{2,G} + \| g_n - g \|_{2,G}}$$

$$\geqq \frac{\| f \|_{\infty,G} - c' \| f-g \|_{2,G} - \varepsilon/6 - c' \| g_n - g \|_{2,G}}{1 + \varepsilon/(3c')}$$

$$\geqq \frac{c - 2\varepsilon/3}{1 + \varepsilon/(3c')} = c - \frac{c\varepsilon/(3c') + 2\varepsilon/3}{1 + \varepsilon/(3c')} \geqq c - \varepsilon, \quad n \geqq n_1 .$$

Let $k(n)$ in $[1,m]$ be such that $|g_n(4k(n)-3)| = \| g_n \|_{\infty,X(n)}$. The construction of $c_{nk(n)}$ in the last section now yields

$$|g_n(4k(n)-3| \, / \, \| g_n \|_{2,G} \leqq c_{nk(n)} \quad .$$

Therefore, (3) and the above inequalities finally imply

$$\begin{aligned} c - \varepsilon &\leq \| g_n \|_{\infty, X(n)} / \| g_n \|_{2,G} \\ &= |g_n(4k(n)-3)| / \| g_n \|_{2,G} \leq c_{nk(n)} \leq c_n \\ &= \sup \left\{ \| f_n \|_{X(n),\infty} / \| f_n \|_{2,G} : 0 \neq \underline{y} \in \mathbb{R}^{4m} \right\} \\ &\leq \sup \left\{ \| f_n \|_{G,\infty} / \| f_n \|_{2,G} : 0 \neq \underline{y} \in \mathbb{R}^{4m} \right\} \leq c . \quad \square \end{aligned}$$

5. NUMERICAL RESULTS

The numerical results for $s = 0.1$, 1 and 10 are presented in the following Table 1 where the values for $n = \infty$ are extrapolated by Aitken's method.

TABLE 1

Numerical lower bounds c_n for the least constant c in the Sobolev inequality (1)

n	s = 0.1	1	10
1	10.02500	1.24632	0.847
2	10.02500	1.24780	0.901
4	10.02500	1.24794	0.909
8	10.02496	1.24796	0.914
∞	10.02470	1.24796	0.915

The observed convergence rates for $s = 0.1$, 1 and 10 are of the order h^p for $p \approx 4$, 3 and 1, respectively.

REFERENCES

1. ADAMS, R.A., *Sobolev Spaces*. Academic Press, New York (1975).
2. CIARLET, P.G., *The Finite Element Method for Elliptic Problems*. North-Holland, Amsterdam (1978).
3. MARTI, J.T., Evaluation of the least constant in Sobolev's inequality for $H^1(0,s)$. *SIAM J. Numer. Anal.* 20, 1239-1242 (1983).
4. SCHWARZ, H.R., *Methode der finiten Elemente*. Teubner, Stuttgart (1980).

GROUPS, SYMMETRY, AND VARIATIONAL BOUNDARY-VALUE PROBLEMS

A. Bossavit

Electricité de France, Clamart, France

1. INTRODUCTION

Group theoretic concepts are common tools in modern physics, but rarely appear in finite-element literature. The present paper purports to direct attention towards this topic, and to show how geometrical symmetry in boundary-value problems can be exploited to reduce the size of the computations.

Suppose one has to solve the Poisson problem

$$- \Delta u = f, \quad u_{|\Gamma} = 0, \tag{1}$$

on a domain Ω which has some symmetry (Fig. 1). This symmetry (which is not in general shared by f) is aptly defined by the group of isometries for which Ω is globally invariant. There is a fundamental domain, or symmetry cell C, such that

$$\overline{\Omega} = \cup\{g \in G \mid g(\overline{C})\}. \tag{2}$$

One would like to be able to solve (1) by solving a collection of problems of the same kind, but on C instead of Ω. As each reduced problem is of smaller size, this would result in appreciable savings. Such a procedure would be attractive too for the eigenvalue problem

$$- \Delta u = \lambda u, \quad u_{|\Gamma} = 0. \tag{3}$$

Investigations into this reveal an interesting situation. When G is abelian, a suitable procedure is easily found. (Consider a symmetry plane; take symmetric and antisymmetric parts of f; this results in two subproblems on half of Ω, a Neumann one and a Dirichlet one; repeat the procedure for each of these, using the next symmetry element; the case of cyclic symmetries is no problem provided G is abelian: boundary conditions with complex multipliers--roots of 1-- appear.)

On the other hand, when G is not commutative, the above procedure fails. (Conversations with practising engineers in vibration analysis show that they are well aware of the difficulty,

ISBN 0-12-747255-X

and do not use the smaller possible symmetry cell, for fear of missing some modes, or because of the uneasy feeling that subtle traps might exist [8].) Nevertheless, it is still possible to achieve the desired purpose [3, 6]. For instance, in the case of Fig. 1 (group D_3, non abelian), one has to solve 6 problems on C: one with Dirichlet conditions on Ω, one with Neumann conditions, and two sets of problems coupled together by boundary conditions of an unusual kind. This result is obtained only at the price of quite obscure computations if one stays at the elementary level (without group theory).

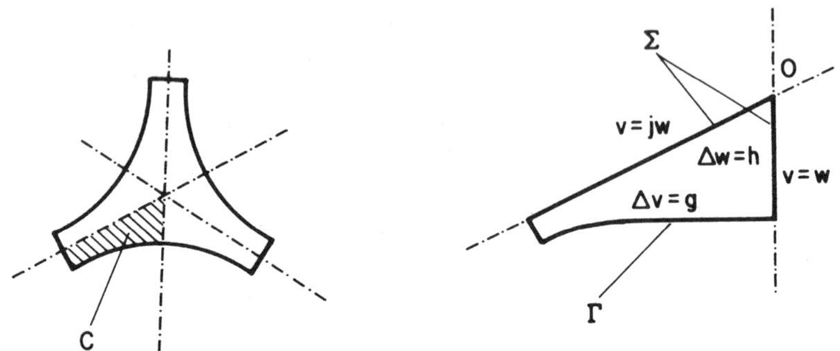

FIG. 1. Domain with dihedral symmetry D_3, with symmetry cell, and boundary conditions for one of the reduced coupled problems.

Only group representation theory allows one to fully understand what goes on. As shown in [5], the commutative case is a generalisation of classical Fourier analysis. What is needed is a general procedure of non-abelian harmonic analysis, and classical representation theory [7] contains the necessary tools for this. Along with an exposition of the relevant parts of the theory, we shall concentrate on the problem of how to derive the right boundary conditions in Fig. 1: there is a simple rule to find them, once the irreducible representations (irreps) of G are known (and they are, for all common groups). The occurence of coupled problems is due to the existence of irreps of degree 2 or more for non-abelian groups.

2. COMMUTATIVE HARMONIC ANALYSIS OF BVP's

2.1 Definitions

The ingredients we need are: a domain Ω with regular boundary Γ in $\mathbb{R}^d$ ($d = 2$ or 3), a Hilbert space V of functions on Ω (e.g. $H^1_0(\Omega)$), $C(\bar{\Omega})$ being dense in V, a scalar product $a(u, v)$ on V, like for instance

$$a(u, v) = \int_{\Omega} \operatorname{grad} u \cdot \operatorname{grad} \bar{v} \quad , \tag{4}$$

an antilinear functional $L: V \to \mathbb{C}$, and a group G of isometries such that

$$g(\Omega) = \Omega \qquad \forall\, g \in G \quad . \tag{5}$$

The object of study is the variational equation

$$a(u, v) = L(v) \qquad \forall\, v \in V \tag{6}$$

i.e. the "weak" form of some boudary value problem, for instance (1). We define, first on continuous functions, then on V by linear continuous extension, unitary operators U_g in $L(V)$:

$$(U_g u)(x) = u(g^{-1}x) \qquad \forall \quad x \in \Omega. \tag{7}$$

The U_g form a group, which is homeomorphic to G. The basic assumption is known as *equivariance*:

$$a(u, v) = a(U_g u, U_g v) \qquad \forall\, u, v \in V, \quad \forall\, g \in G. \tag{8}$$

(Note that G may have to be smaller than the group of *all* isometries for which (5) hold, if (8) is to be satisfied; (8) contains symmetry of the domain and of the material properties.)

Definition 1: Let X be a finite dimensional vector space. A representation (of *degree* d = dim(X)) is a mapping $\rho: G \to GL(X)$ which is a group homomorphism, i.e.

$$\rho(1) = 1, \quad \rho(gh) = \rho(g)\,\rho(h) \qquad \forall \quad g, h \in G \; . \tag{9}$$

One can always assume (by a suitable choice of the norm in X) that all $\rho(g)$ are unitary, and we shall do so.

Definition 2: A representation is irreducible if no other subspace of X than $\{0\}$ or X is such that

$$\rho(g)\; Y \subset Y \qquad \forall\, g \in G \tag{10}$$

If ρ is reducible, such a proper subspace exists, and by unitarity, its orthogonal complement is also invariant. So ρ is the *sum*, in an obvious sense, of its restrictions over two smaller subspaces, and the process can be carried forward until a sum of *irreducible* representations is obtained. So irreps are the building-blocks from which all representations are made.

Theorem 1: All irreps are of degree 1 iff the group is abelian.

For the proof, see [7]. Irreps in this case are complex-valued functions on G. One can endow them with a group structure, and prove (Pontryagin's theorem) that the group of irreps of their group is G itself. In particular, if n is the order of G, there are exactly n irreps.

If G is cyclic, $G = \{1, g, \ldots, g^{n-1}\}$, the irreps are, due to (9),

$$\rho_\nu(g^k) = \exp(2\,i\,\pi k\,\nu/n), \qquad \nu = 0, 1, \ldots, n-1. \tag{11}$$

An abelian group is a direct product of cyclic groups. As irreps of direct products are products of the factor irreps, the irreps of an abelian group are easily computed. The procedure described in the introduction is in fact equivalent to a step by step computation of the irreps of G.

2.2 *Decomposition of (6) into sub-problems*

Let ρ_ν be an irrep of G, and define

$$P_\nu = \int_G \overline{\rho_\nu(g)}\, U_g \, dg\,, \qquad \nu = 1, 2, \ldots, n. \tag{12}$$

where the summation sign means the average value of the integrand on G. The central result we need [7] is:

Theorem 2: The P_ν are orthogonal projectors. Their images $V_\nu = P_\nu V$ are mutually orthogonal, and

$$V = \oplus\, V_\nu\,, \tag{13}$$

the sum being taken over all irreps ρ_ν of G, $\nu = 1, \ldots, n$. More,

$$U_g P_\nu = \rho_\nu(g)\, P_\nu \qquad \forall\, \nu, \;\; \forall\, g \in G. \tag{14}$$

i.e. U_g is homothetic on each subspace.

Equivariance (8) and property (14) combine to give

$$a(P_\nu u, v) = a(u, P_\nu v) \qquad \forall\, u, v \in V, \; \nu = 1, \ldots, n,$$

and thus, if u is solution of (6), one has

$$a(P_\nu u, v) = L(v) \qquad \forall\, v \in V_\nu, \quad \nu = 1, \ldots, n. \tag{15}$$

So we have n separate variational problems on "smaller" subspaces. *A "block-diagonalisation" of the initial problem has been achieved.* As far as abstract variational problems are concerned, we may stop here. But the interesting question is of course, *if (6) is a BVP like e.g. (1), how can (15) be interpreted as a BVP on the symmetry-cell C, and what are the boundary conditions on the part of ∂C which is not part of Γ?*

Definition 3: A symmetry cell, relative to (Ω, G), is any connected open part C of Ω, with regular boundary, such that

$$\Omega \subset \mathrm{cl}\Big(\bigcup_{g \in G} g(C)\Big) \qquad \text{and} \qquad g(C) \cap h(C) = \emptyset \quad \text{if } g \neq h.$$

The "new boundary", on which we need boundary conditions, is $\Sigma = \partial C - \Gamma$.

Proposition 1: For a.e. $x \in \Sigma$ (relative to the measure on Σ), there exists a unique $g \in G$, $g \neq 1$, such that $gx \in \Sigma$.

Therefore, one can define, for u continuous on Ω, and a.e. on Σ,

$$\gamma_\nu u(x) = u(x) - \rho_\nu(g)\, u(g\,x), \qquad \nu = 1, \ldots, n, \tag{16}$$

g being for each x the only element of G referred to in Prop. 1, and, by linear continuous extension, obtain a continous operator γ_ν on V, similar to the trace operator.

Proposition 2: Let i be the restriction mapping on cl(C). There is an isomorphism $\tilde{i} : V_\nu \to \tilde{V}_\nu$, between V_ν and

$$\tilde{V}_\nu = i(V_\nu) = \{\tilde{v} \in i(V) \mid \gamma_\nu \tilde{v} = 0\}. \tag{17}$$

The inverse mapping (from $\tilde{V}_\nu$ onto V_ν) is denoted by j_ν.

The proof comes from (14) and (16). By this result, we can associate with (15) boundary value problems *on C:*

Proposition 3: Define $a_\nu(u, v) = a(j_\nu u, j_\nu v)$ and $L_\nu(v) = L(j_\nu v)$. Then, if u is solution of (6), its projections $P_\nu u = u_\nu$ are solutions of

$$\begin{aligned} &a_\nu(u_\nu, \tilde{v}) = L_\nu(\tilde{v}) \qquad \forall\, v \in \tilde{V}_\nu \\ &u_\nu \in \tilde{V}_\nu \quad (\text{i.e. } \gamma_\nu u_\nu = 0) \end{aligned} \tag{18}$$

In the case of our model problem (a is given by (4)), one checks (using invariance of the Lebesgue measure under the isometries), that

$$a_\nu(u, v) = n \int_C \mathrm{grad}\, u \cdot \mathrm{grad}\, \overline{v} \tag{19}$$

and $L_\nu(v) = n \int_C f_\nu \overline{v}$, with

$$f_\nu(x) = \int_G \rho_\nu(g)\, f(gx)\, dg \tag{20}$$

To summarize, *we found a recipe to make use of symmetry* in problems like (1) when the symmetry-group is abelian:

1) Find all irreducible representations of the group,
2) Choose a symmetry cell C,
3) Form left-hand sides by formula (20),
4) Work out boundary constraints (16) on the "new boundary",
5) Solve the sub-problems

$$-\Delta \tilde{u}_\nu = f_\nu \quad \text{on } C, \qquad \tilde{u}_\nu = 0 \text{ on } \Gamma,\ \gamma_\nu \tilde{u}_\nu = 0 \text{ on } \Sigma, \tag{21}$$

6) Synthesize u, by using the formula

$$u(gx) = \sum_\nu \overline{\rho_\nu(g)}\, \tilde{u}_\nu(x) \tag{22}$$

The similarity of (22) with *Fourier synthesis* (just as (20) is *Fourier analysis*) was expected.

3. NON-COMMUTATIVE HARMONIC ANALYSIS OF BVP's

3.1 Modifications to the previous theory

The irreps may now be of degree 2 or more, thus $\rho_\nu(g)$ may be a matrix. Let us denote by ρ_ν^{ij} the matrix entries, and define

$$P_\nu^{ij} = d_\nu \int_G \overline{\rho_\nu^{ij}(g)}\, U_g\, dg, \qquad i,j = 1, \ldots, d_\nu \tag{23}$$

where d_ν is the degree of the ν-irrep. Now the P^{ij} still are orthogonal projectors, but only the P^{ii}'s exist in the decomposition of V [7]:

$$V = \bigoplus_{\nu,\, i} P^{ii}\, V = \bigoplus V_\nu^i \tag{24}$$

But it is still necessary to compute all projections u_ν^{ij} of the solution u, though only some of them figure in the synthesis formula:

$$u = \sum_{\nu,\, i} u_\nu^{ii} \tag{25}$$

This is so because if one tries to find how some element in V_ν^{ii}

behaves under the action of the operators U_g--which is how we found boundary conditions previously--formula (14) should be replaced by:

$$
\begin{aligned}
U_g P_\nu^{ij} &= d_\nu \int_G \overline{\rho_\nu^{ij}(h)}\, U_{gh}\, dh \\
&= d_\nu \sum_k \overline{\rho_\nu^{ik}(g^{-1})} \int_G \rho_\nu^{kj}(gh)\, U_{gh}\, dh \\
&= \sum_k \rho_\nu^{ki}(g)\, P_\nu^{kj}
\end{aligned}
\tag{26}
$$

And so, *all* the projections of u are required. Their number is still n, the order of the group, for

$$
n = \sum_\nu d_\nu^2 \quad , \tag{27}
$$

a standard result in representation theory.

Our previous recipe goes unchanged, but step 4, which relies on (26), is more delicate, and we shall explain it on a specific example.

3.2 The model problem in the case of Fig. 1 (D_3 symmetry)

Let us call s the symmetry with respect to the vertical axis in Fig. 1, and r the rotation around 0 by an angle of 2 /3. The group generated by s and r has six elements and is noted D_3. As any text-book, e.g. [7] can tell us, it has 3 irreps, two of degree 1,

group elements	1	r	r^2	s	sr	sr^2
ρ_1	1	1	1	1	1	1
ρ_2	1	1	1	-1	-1	-1

and one of degree 2 ($j = (1 + i\sqrt{3})/2$) :

	1	r	r^2	s	sr	sr^2
ρ_3:	$\begin{vmatrix} 1 & 0 \\ 0 & 1 \end{vmatrix}$	$\begin{vmatrix} j & 0 \\ 0 & j^2 \end{vmatrix}$	$\begin{vmatrix} j^2 & 0 \\ 0 & j \end{vmatrix}$	$\begin{vmatrix} 0 & 1 \\ 1 & 0 \end{vmatrix}$	$\begin{vmatrix} 0 & j^2 \\ j & 0 \end{vmatrix}$	$\begin{vmatrix} 0 & j \\ j^2 & 0 \end{vmatrix}$

There is no trouble with the two subproblems corresponding to ρ_1 (Neumann conditions on Σ), and ρ_2 (Dirichlet conditions). Let us drop index 3 and call u^{ij}, i and j = 1 or 2, the four projections of u corresponding to ρ_3.

Relation (26) may be written more conveniently:

$$u^{ij}(gx) = \sum_k \rho^{ki}(g)\, u^{kj}(x) \tag{28}$$

Applying this to g = r and g = s, we get, in matrix form,

$$\begin{vmatrix} u^{11} & u^{12} \\ u^{21} & u^{22} \end{vmatrix} (x) = \begin{vmatrix} j & 0 \\ 0 & j^2 \end{vmatrix} \begin{vmatrix} u^{11} & u^{12} \\ u^{21} & u^{22} \end{vmatrix} (rx)$$

$$\begin{vmatrix} (id.) \end{vmatrix} (sx) = \begin{vmatrix} 0 & 1 \\ 1 & 0 \end{vmatrix} \begin{vmatrix} (id.) \end{vmatrix} (x)$$

and therefore (cf. Fig. 2)

$$u^{1\cdot} = u^{2\cdot} \text{ on } S, \tag{29}$$

$$u^{1\cdot} = j\,u^{2\cdot} \text{ on } R. \tag{30}$$

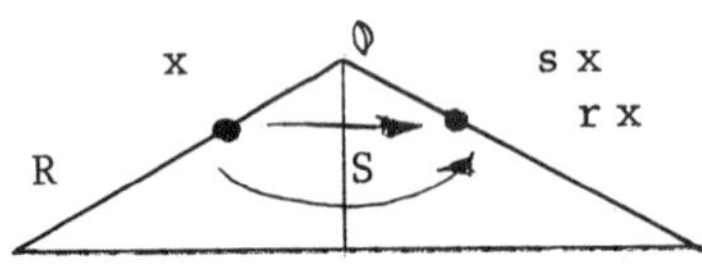

FIG. 2. Sketch of boundary constraints for (1), Fig. 1.

So we have two sets of two problems "with coupling at the boundary" to solve on the symmetry-cell.

REFERENCES

(Very few papers address the topic, as was already noticed in [4] in 1973!)

1. THOMAS, D.L., Dynamics of Rotationally Periodic Structures. *Int. J. Num. Meth. Engng.*, 14, 81-102 (1979).

2. FAESSLER, A., *Applications of Group Theory to the Method of Finite Elements ... (Thesis)*. ETH Zürich (1976).

3. FRICKER, A.J., POTTER, S., WHISTON, G.S., Series of CERL Reports, ca. 1980.

4. GLOCKNER, P.G., Symmetry in Structural Mechanics. *J. Struct. Div. ASCE*, ST1, 71-89 (1973).

5. MACKEY, G.W., Harmonic Anlysis as the Exploitation of Symmetry. *Bull. AMS*, 3, 1, 543-698 (1980).

6. OHAYON, R., VALID, R., Structures à Symétrie Cyclique, in P. Lascaux (Ed.), *Les méthodes Numériques de l'Ingénieur*, Pluralis, Paris (1983).

7. SERRE, J.P., *Représentations linéaires des groupes finis*. Hermann, Paris (1978).

8. WILLIAMS, F.H., A Warning on the Use of Symmetry ... *Int. J. Num. Meth. Engng.*, 12, 379-83 (1978).

NUMERICAL PROBLEMS IN 3D FINITE ELEMENT ANALYSIS BASED ON DEGENERATED ELEMENTS

J. Altenbach, H. Berger and U. Gabbert

Technische Hochschule "Otto von Guericke", Magdeburg, DDR

1. INTRODUCTION

Three dimensional finite element analysis is both expensive and time consuming. In order to be undertaken effectively, for structures of any shape without excessively fine meshes, it requires a library of suitable elements. Such a library is available in the program COSAR, [1], developed at the T.H. Magdeburg. COSAR contains the basic 20-node brick element and a number of degenerated elements which are derived from it, Fig.1. Use of these elements enables most shapes to be partitioned, [2]-[4], but produces certain effects in the stress analysis which need special treatment both theoretically and numerically.

2. STRESS ANALYSIS

The shape functions G_L for the degenerated 20 node brick elements contain *correcting* terms $\Delta G_{i\alpha}$, in addition to those of the standard element, which ensure compatibility over the triangular faces. These are shown in Table 1, whilst their derivatives are set out in Table 2. The standard stress analysis procedure is

$$u_i(\xi_1,\xi_2,\xi_3) = G_L(\xi_1,\xi_2,\xi_3)u_{iL}, \quad i = 1,2,3, \; L = 1,2,\ldots,N, \tag{2.1}$$

$$\sigma_{ij}(\xi_1,\xi_2,\xi_3) = E_{ijrs}\frac{1}{2}\left(\frac{\partial G_L}{\partial x_s} u_{rL} + \frac{\partial G_L}{\partial x_r} u_{sL}\right), \; i,j,r,s = 1,2,3, \; L = 1,2,\ldots,N, \tag{2.2}$$

$$x_i(\xi_1,\xi_2,\xi_3) = G_L(\xi_1,\xi_2,\xi_3)x_{iL}, \quad i = 1,2,3, \; L = 1,2,\ldots,N, \tag{2.3}$$

and derivatives of shape functions in global coordinates x_i are expressed in terms of derivatives in local coordinates ξ_i, so that

ISBN 0-12-747255-X

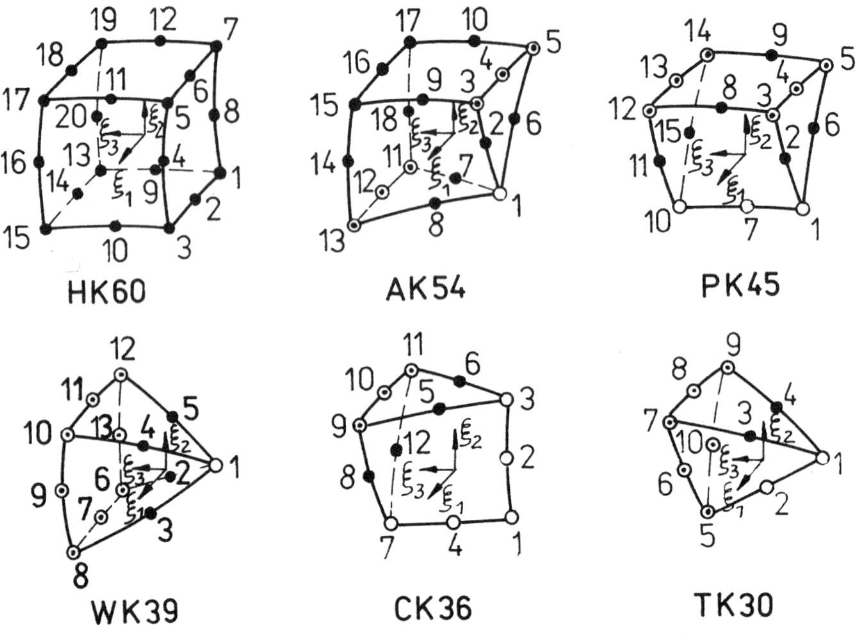

FIG. 1. COSAR 3D Element Library
● standard node ○ degenerated node
⊙ degenerated node with modified shape functions (additional terms)

$$\frac{\partial G_L}{\partial \xi_i} = \frac{\partial x_s}{\partial \xi_i} \frac{\partial G_L}{\partial x_s} = J_{is} \frac{\partial G_L}{\partial x_s} ,$$

$$\frac{\partial G_L}{\partial x_s} = \frac{\partial \xi_i}{\partial x_s} \frac{\partial G_L}{\partial \xi_i} = J^{-1}_{si} \frac{\partial G_L}{\partial \xi_i} ,$$

where $J^{-1}_{si} = \bar{J}^{-1}_{si}/\det J$. For a degenerated element such as the prism of Fig. 2, this enables stress components to be derived at all points $P(\xi_{ip})$ of the element, except the degenerated nodes where the determinant of the Jacobian is singular. At such points D, using the L'Hospital rule, we define C_s as

$$\lim_{\xi_K \to \xi_{KD}} \frac{\partial G_L(\xi_i)}{\partial x_s} = \lim_{\xi_K \to \xi_{KD}} \frac{\bar{J}^{-1}_{si} \frac{\partial G_L}{\partial \xi_i}}{\det J} = \left(\frac{\frac{\partial}{\partial \xi_K}\left[\bar{J}^{-1}_{si} \frac{\partial G_L}{\partial \xi_i}\right]}{\frac{\partial}{\partial \xi_K}\left[\det J\right]} \right)_{\xi_{KD}} = C_s ,$$

where ξ_1 is the non-uniquely defined coordinate of D and $\xi_2 = \xi_K$

TABLE 1

Shape functions G_L *and additional terms* $\Delta G_{i\alpha}$ *of COSAR 3D elements*

Element	Shape functions
HK60	Corner nodes $G_L = \frac{1}{8}(1+\xi_{1L}\xi_1)(1+\xi_{2L}\xi_2)(1+\xi_{3L}\xi_3)(\xi_{1L}\xi_1+\xi_{2L}\xi_2+\xi_{3L}\xi_3-2)$ L=1,3,... midside nodes $G_L = \frac{1}{4}(1+\xi_{1L}\xi_1)(1+\xi_{2L}\xi_2)(1+\xi_{3L}\xi_3)[1-(\xi_1\xi_{2L}\xi_{3L})^2-(\xi_{1L}\xi_2\xi_{3L})^2-(\xi_{1L}\xi_{2L}\xi_3)^2]$ L=2,3,... $\Delta G_{i\alpha} = \frac{1}{16}[1+(-1)^{\alpha}\xi_i](1-\xi_j^2)(1-\xi_k^2)$; ξ_{iL} local coordinates for the nodes L
AK54	$G_1^A = G_1+G_2+G_3$; $G_3^A = G_5+\Delta G_{31}$; $G_4^A = G_6-2\Delta G_{31}$; $G_5^A = G_7+\Delta G_{31}$; $G_{11}^A = G_{13}+\Delta G_{21}$; $G_{12}^A = G_{14}-2\Delta G_{21}$; $G_{13}^A = G_{15}+\Delta G_{21}$; $G_L^A = G_{L+2}$ (L = 2,6,7,8,9,10,14,15,16,17,18)
PK54	$G_1^P = G_1+G_2+G_3$; $G_3^P = G_5+\Delta G_{31}$; $G_4^P = G_6-2\Delta G_{31}$; $G_5^P = G_7+\Delta G_{31}$ $G_7^P = G_9+G_{10}$; $G_{10}^P = G_{13}+G_{14}+G_{15}$; $G_{12}^P = G_{17}+\Delta G_{32}$; $G_{13}^P = G_{18}-2\Delta G_{32}$; $G_{14}^P = G_{19}+\Delta G_{32}$; $G_L^P = G_{L+2}$ (L = 2,6) $G_L^P = G_{L+3}$ (L=8,9); $G_L^P = G_{L+5}$ (L=11,15)
WK39	$G_1^W = \sum_{L=1}^{8} G_L$; $G_L^W = G_{L+7}$ (L = 2,3,4,5); $G_6^W = G_{13}+\Delta G_{11}+G_{21}$; $G_7^W = G_{14}-2\Delta G_{21}$; $G_8^W = G_{15}+\Delta G_{21}+\Delta G_{12}$; $G_9^W = G_{16}-2\Delta G_{12}$; $G_{10}^W = G_{17}+\Delta G_{12}+\Delta G_{22}$; $G_{11}^W = G_{18}-2\Delta G_{22}$; $G_{12}^W = G_{19}+\Delta G_{22}+\Delta G_{11}$; $G_{13}^W = G_{20}-2\Delta G_{11}$
CK36	$G_1^C = G_1+G_2+G_3$; $G_2^C = G_4+G_8$; $G_3^C = G_5+G_6+G_7$; $G_4^C = G_9+G_{10}$; $G_7^C = G_{13}+G_{14}+G_{15}$; $G_9^C = G_{17}+\Delta G_{22}+\Delta G_{32}$ $G_{10}^C = G_{18}-2(\Delta G_{22}+\Delta G_{32})$; $G_{11}^C = G_{19}+\Delta G_{22}+\Delta G_{32}$ $G_L^C = G_{L+6}$ (L = 5,6); $G_L^C = C_{L+8}$ (L = 8,12)
TK30	$G_1^T = \sum_{L=1}^{8} G_L$; $G_2^T = G_9+G_{10}$; $G_5^T = G_{13}+G_{14}+G_{15}+\Delta G_{12}+\Delta G_{11}$ $G_6^T = G_{16}-2\Delta G_{12}$; $G_7^T = G_{17}+\Delta G_{12}+\Delta G_{22}+0.5(1+\xi_3)\Delta G_{32}$ $G_8^T = G_{18}-2\Delta G_{22}-(1+\xi_3)\Delta G_{32}$ $G_9^T = G_{19}+\Delta G_{22}+\Delta G_{11}+0.5(1+\xi_3)\Delta G_{32}$; $G_{10}^T = G_{20}-2\Delta G_{11}$ $G_L^T = G_{L+8}$ (L = 3,4)

TABLE 2

Derivatives of the shape-functions G_L and the additional term $\Delta G_{i\alpha}$ in local-coordinates

Corner nodes

$$\frac{\partial G_L}{\partial \xi_i} = \frac{1}{8}\,\xi_{iL}(1+\xi_{jL}\xi_j)(1+\xi_{kL}\xi_k)\cdot(2\xi_{iL}\xi_i+\xi_{jL}\xi_j+\xi_{kL}\xi_k-1)$$

midside nodes

$$\frac{\partial G_L}{\partial \xi_i} = \frac{1}{4}\,(1+\xi_{jL}\xi_j)(1+\xi_{kL}\xi_k)\cdot\left\{\xi_{iL}[1-(\xi_j\xi_{kL})^2-(\xi_{jL}\xi_k)^2] - 2\xi_i(\xi_{jL}\xi_{kL})^2\right\}$$

$$\frac{\partial(\Delta G_{i\alpha})}{\partial \xi_i} = \frac{1}{16}\,(-1)^\alpha(1-\xi_j^2)(1-\xi_k^2)$$

$$\frac{\partial(\Delta G_{i\alpha})}{\partial \xi_j} = -\frac{1}{8}\,[1+(-1)^\alpha\xi_i]\xi_j(1-\xi_k^2)$$

$$\frac{\partial(\Delta G_{i\alpha})}{\partial \xi_k} = -\frac{1}{8}\,[1+(-1)^\alpha\xi_i](1-\xi_j^2)\xi_k$$

$$\alpha = 1,2\ ; \quad i,j,k = 1,2,3\ , \quad i \neq j \neq k\ .$$

TABLE 3

Optimal points for stress-analysis - 20 nodes Hexahedron

Isoparametric approximation second order for the displacements u_i

$$u_i(\xi_1,\xi_2,\xi_3)=[1,\xi_1,\xi_2,\xi_3,\xi_1^2,\xi_1\xi_2,\xi_2^2,\xi_2\xi_3,\xi_3^2,\xi_3\xi_1,\xi_1^2\xi_2,\xi_1\xi_2^2,\xi_2^2\xi_3,\xi_2\xi_3^2,\xi_3^2\xi_1,\xi_3\xi_1^2,\xi_1\xi_2\xi_3,\xi_1^2\xi_2\xi_3,\xi_1\xi_2^2\xi_3,\xi_1\xi_2\xi_3^2]\underline{a}_i=\underline{s}^T(\xi_1,\xi_2,\xi_3)\underline{a}_i\ .$$

Complete polynomial third order approximation for the displacements u_i

$$\bar{u}_i(\xi_1,\xi_2,\xi_3)=[1,\xi_1,\xi_2,\xi_3,\xi_1^2,\xi_1\xi_2,\xi_2^2,\xi_2\xi_3,\xi_3^2,\xi_3\xi_1,\xi_1^2\xi_2,\xi_1\xi_2^2,\xi_2^2\xi_3,\xi_2\xi_3^2,\xi_3^2\xi_1,\xi_3\xi_1^2,\xi_1\xi_2\xi_3,\xi_1^3,\xi_2^3,\xi_3^3]\underline{b}_i = \underline{p}\,(\xi_1,\xi_2,\xi_3)\underline{b}_i$$

Adjunct nodes-displacements-vectors

$$\underline{v}_i^T = [u_{i1},u_{i2},u_{i3},\dots,u_{iL},\dots,u_{i20}]\ ; \quad \underline{v}_i = \underline{A}\,\underline{a}_i$$

$$\underline{\bar{v}}_i^T = [\bar{u}_{i1},\bar{u}_{i2},\bar{u}_{i3},\dots,\bar{u}_{iL},\dots,\bar{u}_{i20}]\ ; \quad \underline{\bar{v}}_i = \underline{B}\,\underline{b}_i$$

$$\underline{v}_i \approx \underline{\bar{v}}_i \Rightarrow \underline{a}_i = \underline{A}^{-1}\underline{B}\,\underline{b}_i\ ;$$

$$\frac{\partial u_i}{\partial \underline{\xi}} = \frac{\partial \underline{s}^T}{\partial \underline{\xi}}\,\underline{a}_i\ ; \quad \frac{\partial \bar{u}_i}{\partial \underline{\xi}} = \frac{\partial \underline{p}^T}{\partial \underline{\xi}}\,\underline{b}_i \Rightarrow \boxed{\frac{\partial \underline{s}^T}{\partial \underline{\xi}}\,\underline{A}^{-1}\underline{B}\,\underline{b}_i = \frac{\partial \underline{p}^T}{\partial \underline{\xi}}\,\underline{b}_i}$$

valid for all components of $\frac{\partial u_i}{\partial \underline{\xi}}$ if $\underline{\xi}^T = \left[\frac{1}{\sqrt{3}}\,,\ \frac{1}{\sqrt{3}}\,,\ \frac{1}{\sqrt{3}}\right]$

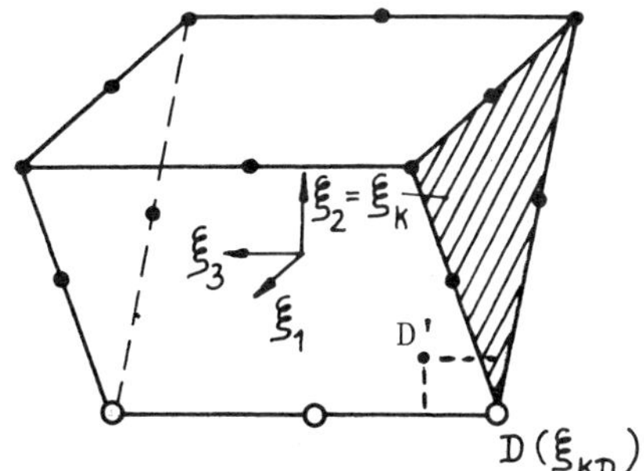

FIG. 2. D = D(ξ_K = −1)degenerated node
D' = D'(0, $\xi_{2D} + \varepsilon, \xi_{3D} + \varepsilon$) (coordinates local to D)

is the variable coordinate in the face containing D. This technique demands that a special algorithm be implemented to calculate C_s for each specific degenerated element. This process can be avoided if for degenerated elements the analysis is carried out at points D' = $\{0, \xi_{2D} + \varepsilon, \xi_{3D} + \varepsilon\}$ near to D for suitable ε. Numerical tests indicate that 0.001 is a suitable value for ε.

3. ELEMENT DISTORSION AND ACCURACY OF STRESS ANALYSIS

The use of degenerated elements is demonstrated in Fig. 3 for the case of a beam loaded in pure bending $F_{\sigma 11} = \{(\sigma_{11exact} - \sigma_{11apr}) / \sigma_{11exact}\} \cdot 100\%$. Results so obtained are shown in Fig. 4 to be superior to those derived from distorted brick elements; the exact σ_{11} values are ±210 N/mm^2 and the ε-displacement values are < 0.5%

4. OPTIMAL POINTS FOR DEGENERATED ELEMENTS

An attempt has been made to translate the concept of *optimal points*, [5]-[7], to the degenerated elements. Table 3 contains the relevant information for the 20 node brick and Table 4 the generalisation to the degenerated elements of COSAR. The most important result in the degenerate case is that there are no points at which all the derivatives of the components of the displacement vector are optimal, Table 5. For brick elements the derivatives each have value unity on two faces ξ_i = C; in the degenerated elements this occurs only for special lines or points, and all *optimal points* for $\partial u_i / \partial \xi$ are on the element surfaces, [8]. The results of the numerical tests of Fig. 5 permit the following conclusion: for higher accuracy the stress components for degenerated elements should be calculated at the 2 × 2 × 2 Gauss points. For degenerated elements there is no superconvergence, but in all the cases tested there is a significant improvement in accuracy.

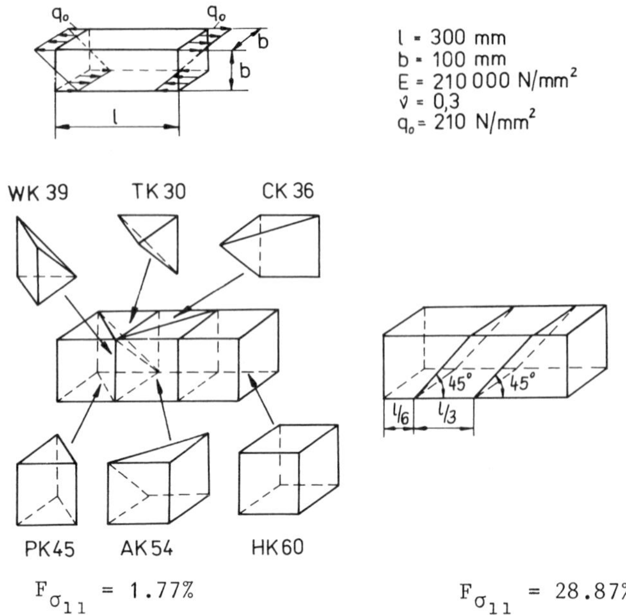

FIG. 3a Beam - partition of degenerated elements

FIG. 3b Beam - partition of distorted brick elements

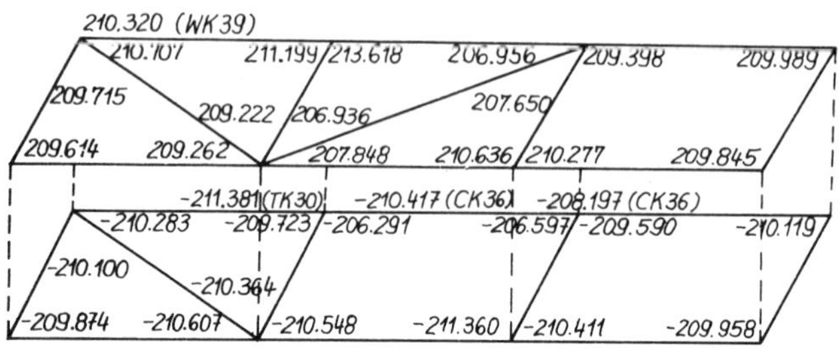

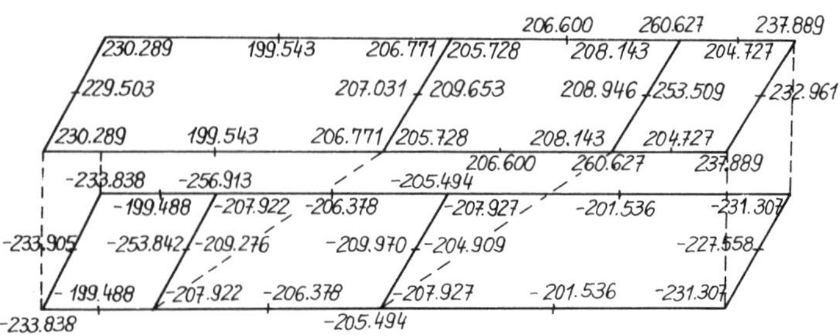

FIG. 4. Bending stresses σ_{11} at corner nodes,
top: degenerated elements, bottom: distorted brick elements

TABLE 4

Generalisation of the optimal points concept

$$u_i(\xi_1,\xi_2,\xi_3) = [G_1,G_2,\ldots,G_N]\underline{v}_i \quad ; \quad G_L \text{ - shape-functions}$$

$$\underline{v}_i = \underline{I}\,\underline{v}_i$$

$$\underline{v}_i \approx \overline{\underline{v}}_i \Rightarrow \underline{v}_i = \underline{B}\,\underline{b}_i$$

$$\overline{\underline{v}}_i = \underline{B}\,\underline{b}_i$$

$$\frac{\partial u_i}{\partial \underline{\xi}} = \left[\frac{\partial G_1}{\partial \underline{\xi}} ;\ldots; \frac{\partial G_N}{\partial \underline{\xi}}\right]\underline{v}_i = \left[\frac{\partial G_1}{\partial \underline{\xi}} ;\ldots; \frac{\partial G_N}{\partial \underline{\xi}}\right]\underline{B}\,\underline{b}_i \quad \Rightarrow$$

$$\frac{\partial u_i}{\partial \underline{\xi}} = \frac{\partial \underline{p}}{\partial \underline{\xi}}\,\underline{b}_i$$

$$\frac{\partial G_1}{\partial \underline{\xi}} ;\ldots; \frac{\partial G_N}{\partial \underline{\xi}}\;\underline{B}\,\underline{b}_i = \frac{\partial \underline{p}^T}{\partial \underline{\xi}}\,\underline{b}_i$$

valid only in different points for the components of $\frac{\partial u_i}{\partial \underline{\xi}}$.

TABLE 5

Faces, lines and points of highest accuracy for the components of the gradient of displacements $u_i(\xi_1,\xi_2,\xi_3)$

$-1 \leqq c \leqq +1$, $i = 1, 2, 3$

	$\frac{\partial u_i}{\partial \xi_1}$			$\frac{\partial u_i}{\partial \xi_2}$			$\frac{\partial u_i}{\partial \xi_3}$		
	ξ_1	ξ_2	ξ_3	ξ_1	ξ_2	ξ_3	ξ_1	ξ_2	ξ_3
HK60	$\pm\frac{1}{\sqrt{3}}$	c	c	c	$\pm\frac{1}{\sqrt{3}}$	c	c	c	$\pm\frac{1}{\sqrt{3}}$
AK54	$\pm\frac{1}{\sqrt{3}}$	1	c	±1	$\pm\frac{1}{\sqrt{3}}$	c	±1	c	$\pm\frac{1}{\sqrt{3}}$
	$\pm\frac{1}{\sqrt{3}}$	c	1	c	$\pm\frac{1}{\sqrt{3}}$	1	c	1	$\pm\frac{1}{\sqrt{3}}$
PK45	$\pm\frac{1}{\sqrt{3}}$	1	c	±1	$\pm\frac{1}{\sqrt{3}}$	c	±1	c	$\pm\frac{1}{\sqrt{3}}$
							c	1	$\pm\frac{1}{\sqrt{3}}$
WK39	$\pm\frac{1}{\sqrt{3}}$	c	1	c	$\pm\frac{1}{\sqrt{3}}$	1	±1	±1	$\pm\frac{1}{\sqrt{3}}$
CK36	$\pm\frac{1}{\sqrt{3}}$	1	1	±1	$\pm\frac{1}{\sqrt{3}}$	c	±1	c	$\pm\frac{1}{\sqrt{3}}$
TK30	$\pm\frac{1}{\sqrt{3}}$	1	1	±1	$\pm\frac{1}{\sqrt{3}}$	1	±1	1	$\pm\frac{1}{\sqrt{3}}$
							c	−1	$\pm\frac{1}{\sqrt{3}}$

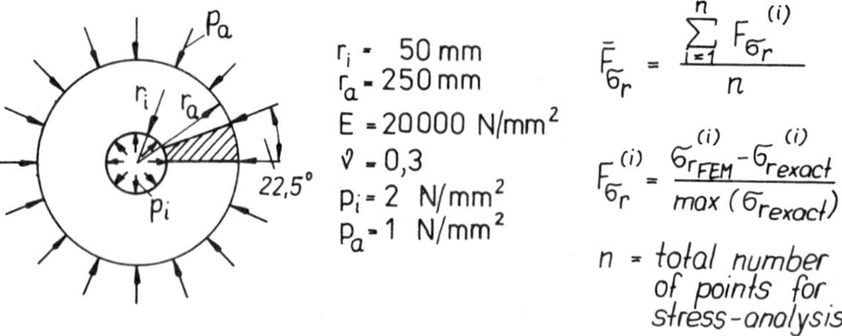

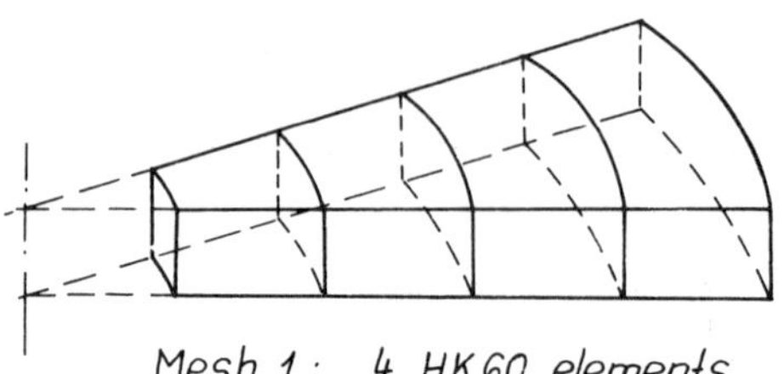

Mesh 1: 4 HK60 elements

Average error $\bar{F}_{\sigma_r}$ for σ_r

nodal point : 2,478

3×3×3 Gausspoints : 1,092

2×2×2 Gausspoints : 0,051

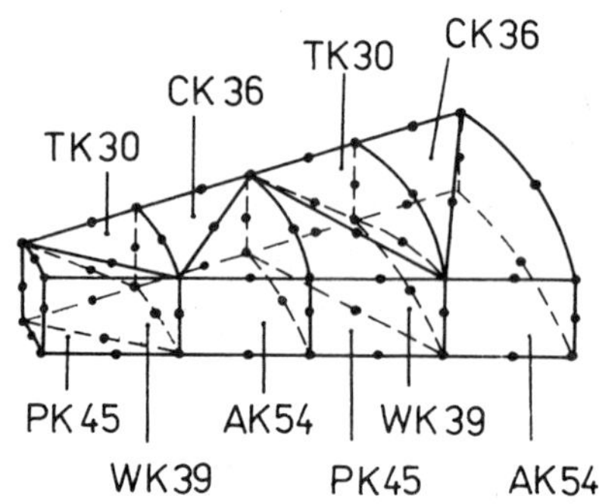

Mesh 2: 10 degenerated element

nodalpoints : 1,367

3×3×3 Gausspoint : 0,677

2×2×2 Gausspoint : 0,492

FIG. 5. Average error for meshes with and without degenerated elements

5. HIGHER ACCURACY OF STRESSES AT ELEMENT NODES

Methods have been proposed [9] and implemented in COSAR for increasing accuracy of local and global stress fields and for having higher accuracy at element nodes. Extrapolation formulas from Gauss points to element nodes are effective for improving nodal stresses. Linear, bilinear and quadratic extrapolation schemes have been tested for the 20 node brick element and transferred to the degenerated element family of COSAR. Special treatment was again necessary for the degenerated nodal points. Smoothing algorithms based on weighted integral methods, the least squares method and conjugate stress analysis have also been implemented. There are significant similarities between the different methods which facilitate their implementation, [7]. From numerical evidence it can be said that for all 3D elements in COSAR the linear extrapolation formula using $2 \times 2 \times 2$ points is the most effective from the viewpoint of expense and accuracy.

REFERENCES

1. ALTENBACH, J. and SACHAROV, A., *Finite Elemente in der Festkörpermechanik*. VEB Fachbuchverlag, Leipzig (1982).
2. NEWTON, R.E., Degeneration of brick-type isoparametric elements. *Int. J. Numer. Meth. Eng.* 7, 579-581 (1973).
3. IRONS, B.M., A technique for degenerating brick-type isoparametric elements using hierarchical midside nodes. *Int. J. Numer. Meth. Eng.* 8, 203-209 (1974).
4. ZIENKIEWICZ, O.C., *The Finite Element Method, Third Edition*. McGraw Hill (1977).
5. ZLAMAL, M., Superconvergence and reduced integration in finite element method. *Math. Comp.* 32, 663-685 (1978).
6. BARLOW, J., Optimal stress locations in finite element models. *Int. J. Numer. Meth. Eng.* 10, 243-251 (1976).
7. BERGER, H., Beitrag zur Spannungsberechnung mit Hilfe der Methode der finiten Elemente auf der Grundlage von dreidimensionalen Verschiebungselementen. *Dissertation A, TH Magdeburg* (1982).
8. BERGER, H. and ALTENBACH, J., Optimale Punkte für die Spannungsberechnung bei finiten 3D-Verschiebungselementen. *Technische Mechanik* 4, 24-32 (1983).
9. BERGER, H. and ALTENBACH, J., Berechnung verbesserter Spannungswerte für dreidimensionale finite Elemente. *Technische Mechanik* 5 (1984), (in press).

POSTBUCKLING AND IMPERFECTION SENSITIVITY OF ELASTIC STRUCTURES

H. Beem, U. Eckstein, R. Harte, R.K. Jürcke
W.B. Krätzig and U. Wittek

Ruhr-Universität Bochum, Bochum, BRD

1. INTRODUCTION

The *finite element analysis* of thin structures involves consideration of buckling and postbuckling behaviour. *Curved shell structures*, which are able to sustain heavy loads, particularly demand careful design and efficient computational methods in order that decrease of sustainable load due to buckling may be calculated.

Figure 1.1 indicates the variety of possible buckling phenomena. Generally two basic types can be distinguished:

- *bifurcation buckling*,
- *snap-through buckling*.

In this paper we want to discuss the application of finite element methods to bifurcation problems. In this context we will consider a method for detecting neutral equilibrium points and for computing the *nonlinear postbuckling path* of equilibrium, whether that be stable or unstable. In addition to this a postbuckling analysis will be instigated to obtain results for structures with small shape-deviations, generally called *imperfections*. The classification of bifurcation problems into asymmetric and symmetric cases is shown in Figure 1.2.

ISBN 0-12-747255-X

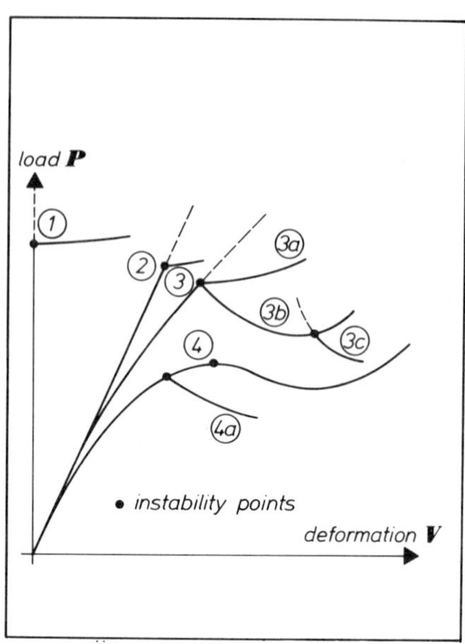

FIG. 1.1 Buckling phenomena

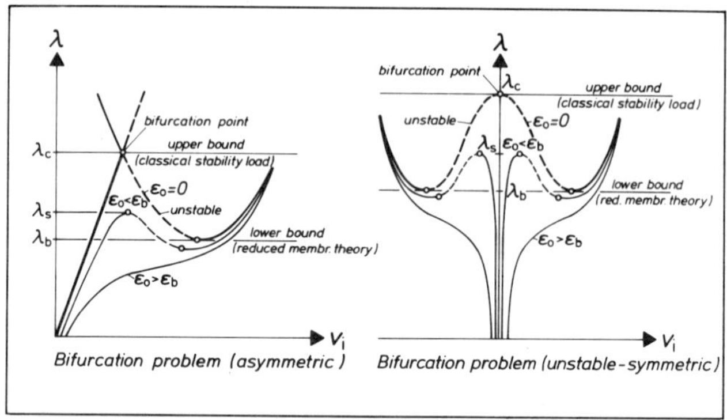

FIG. 1.2 Critical buckling loads of perfect and imperfect structures

2. PHYSICAL STRUCTURE OF GEOMETRICALLY NONLINEAR THEORIES

Shell structures are taken as an example for highly geometrically nonlinear problems. The total potential energy π as the basic functional for finite elements is written as [10]:

$$\pi = \pi_i + \pi_e = {}^1/_2 \iint_{\overset{\circ}{F}} (n^{(\alpha\beta)} \alpha_{(\alpha\beta)} + m^{(\alpha\beta)} \omega_{(\alpha\beta)}\, d\overset{\circ}{F} + $$
$$+ \iint_{\overset{\circ}{F}} (\overset{\circ}{p}{}^{\alpha} v_{\alpha} + \overset{\circ}{p}{}^{3} v_3)\, d\overset{\circ}{F} - \int_{C_t} \overset{\circ}{n}_t v_t + \overset{\circ}{n}_u v_u + \overset{\circ}{n}_3 v_3 + \overset{\circ}{m}_t \omega_t) d\overset{\circ}{s} \tag{2.1}$$

where the first integral denotes the potential energy of the internal forces and moments, the second integral denotes the potential energy of external middle surface forces, and the third one represents the energy of the external boundary forces. All integrals are related to the undeformed structure ($d\overset{\circ}{F}$, $d\overset{\circ}{s}$).

The stress tensor $n^{(\alpha\beta)}$ and the moment tensor $m^{(\alpha\beta)}$ can easily be replaced by the 1st and 2nd strain tensor using a Hookean constitutive relationship. Within the framework of a Kirchhoff-Love-hypothesis the strain tensor can be expressed in terms of the middle surface displacements v_α (in-plane) and v_3 (normal), e.g. by

$$\alpha_{(\alpha\beta)} = \frac{1}{2} (v_{\alpha}|_{\beta} + v_{\beta}|_{\alpha} - 2v_3\, \overset{\circ}{b}_{\alpha\beta} + v_3|_{\alpha}\, v_3|_{\beta}) ,$$
$$\omega_{(\alpha\beta)} = -v_3|_{\alpha\beta} , \tag{2.2}$$

where $\overset{\circ}{b}_{\alpha\beta}$ denotes the curvature tensor of the undeformed middle surface $\overset{\circ}{F}$ [10].

This well known equation of the Donnell/Marguerre shell theory represents a simple, but effective nonlinear theory which covers small rotations for shallow structures. Transforming the tensor variables into a matrix notation we end up with the more common form of the total potential energy [16]:

$$\pi = \frac{1}{2} \int \underline{\varepsilon}^T\, \underline{E}\, \underline{\varepsilon}\, d\overset{\circ}{F} - \int \underline{u}^T\, \underline{\overset{\circ}{p}}\, d\overset{\circ}{F} - \int \underline{r}^T\, \underline{\overset{\circ}{t}}\, ds , \tag{2.3}$$

where t^o indicates the stress along the boundary and $\underline{r}$ denotes the kinematic boundary quantities.

The strain matrix ε is associated with the displacement vector $\underline{u}$ via a differential operator relation:

$$\underline{\varepsilon} = \underline{D}_k\, \underline{u} = (\underline{D}_{kL} + \underline{D}_{kN}\, (\underline{u}))\, \underline{u} . \tag{2.4}$$

$\underline{D}_{kL}$ and $\underline{D}_{kN}$ denote linear and nonlinear differential operators, whereby $\underline{D}_{kN}$ itself is a linear functional in $\underline{u}$ and its derivatives. The operators can easily be derived for the Donnell/Marguerre theory from equation (2.2) [11].

3. FINITE ELEMENT IMPLEMENTATION

In order to calculate nonlinear equilibrium points a Total Lagrange formulation is adopted and three different states of deformation are defined [7].

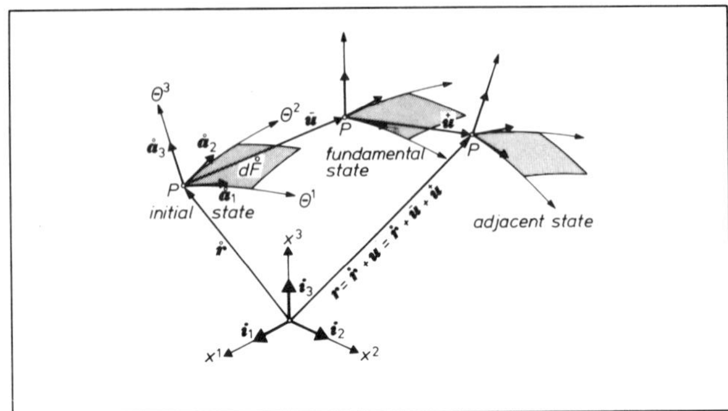

FIG. 3.1 Different states of deformation

Any displacement field shall be decomposed according to $\underline{u} = \bar{\underline{u}} + \overset{+}{\underline{u}}$, where $\overset{+}{\underline{u}}$ represents the variation of the displacement $\bar{\underline{u}}$.

For imperfect structures an additional pre-state $(\overset{i}{\ldots})$ is used, with $\overset{i}{\underline{u}}$ leading to the initial state [5]. Applying equation (2.4) to the decomposition of $\underline{u}$ leads to several strain terms [6]:

$$\varepsilon = \bar{\varepsilon} + \bar{\bar{\varepsilon}} + \overset{+}{\varepsilon} + \overset{-+}{\varepsilon} + \overset{+-}{\varepsilon} + \overset{++}{\varepsilon} \tag{3.1}$$

where $\quad \tilde{\varepsilon} = \underline{D}_{kL} \cdot \tilde{\underline{u}} \quad$ and $\quad \tilde{\hat{\varepsilon}} = \underline{D}_{kN}\,(\tilde{\underline{u}}) \cdot \hat{\underline{u}}$ (~,^ := −, +) .

Substituting these strains into the total potential energy (2.3) we obtain an incremental formulation of the energy. Postulating the fundamental state $\bar{\underline{u}}$ to be an equilibrium state we can write

$$\delta\pi = \pi'[\underline{u}]\ \delta\overset{+}{\underline{u}} = 0 \tag{3.2}$$

where $\pi'\delta\overset{+}{\underline{u}}$ denotes a Gateaux variation of the functional π [2, 12].

The discretization of the displacement field $\bar{\underline{u}}$ and $\overset{+}{\underline{u}}$ within finite elements, the application of equation (3.2) in the element and the summation to global discrete displacements $\bar{\underline{V}}$, $\overset{+}{\underline{V}}$ leads to the tangent stiffness relation in figure 3.2 [11]. This stiffness matrix is the sum of the linear elastic stiffness ($\underline{K}_e$), linear and nonlinear geometric stiffness ($\underline{K}_{gL}$, $\underline{K}_{gN}$), and linear and nonlinear initial displacement stiffness ($\underline{K}_{uL}$, $\underline{K}_{uN}$). The right side of the equation is the central equation

Discretization (Displacement modell):

$$\bar{\boldsymbol{u}}^{p} = \boldsymbol{\Omega}^{p}\bar{\boldsymbol{v}}^{p}, \quad \bar{\boldsymbol{v}} = \boldsymbol{a}\bar{\boldsymbol{V}}$$

$$\overset{+}{\boldsymbol{u}}^{p} = \boldsymbol{\Omega}^{p}\overset{+}{\boldsymbol{v}}^{p}, \quad \overset{+}{\boldsymbol{v}} = \boldsymbol{a}\overset{+}{\boldsymbol{V}}$$

Tangent stiffness relation:

$$\boldsymbol{K}_{T}\overset{+}{\boldsymbol{V}} = [\boldsymbol{K}'_{e} + \boldsymbol{K}'_{gL} + \boldsymbol{K}'_{gN} + \boldsymbol{K}'_{uL} + \boldsymbol{K}_{uN}]\overset{+}{\boldsymbol{V}} = \boldsymbol{P} - \boldsymbol{P}_{i} = \overset{+}{\boldsymbol{P}}$$

FIG. 3.2 Basic discretized relation

for an iterative method to equalize internal equilibrium forces ($\underline{P}_i$) and applied external loads ($\underline{P}$).

The tensor formulation allows an efficient way of programing [7], and an effective double curved shell element family (fig. 2.3) was developed and implemented into the program used [1]. For nonlinear calculations excellent results have been obtained with the conforming triangular element NACS 54.

	NONLINEAR ARBITRARILY CURVED SHELL ELEMENTS				
	Δ-ELEMENTS				□-ELEMEN.
	NACS 27	NACS 36	NACS 54	NACS 63	NACS 48
Basic functions	$u_\lambda : p_k^3$ $u_3 : p_k^3$	$u_\lambda : p_k^3$ $u_3 : p_k^5$	$u_\lambda : p_k^5$ $u_\lambda : p_k^5$	$u_\lambda : p^5$ $u_3 : p^5$	$u_\lambda : p_b^3$ $u_3 : p_b^3$
	p_k^3 cubical plynomial (TOCHER, HOLAND, ZIENKIEWICZ)		p_k^5 quintic polynomial (COWPER) p^5 quintic polynomial (BELL)		p_b^3 bicubic polynomial
Degrees of Freedom	v_λ $v_{\lambda,\alpha}$ v_3 $v_{3,\alpha}$	v_λ $v_{\lambda,\alpha}$ v_3 $v_{3,\alpha}$ $v_{3\alpha\beta}$	v_λ $v_{\lambda,\alpha}$ $v_{\lambda,\alpha\beta}$ v_3 $v_{3,\alpha}$ $v_{3\alpha\beta}$	v_λ $v_{\lambda,\alpha}$ $v_{\lambda\alpha\beta}$ v_3 $v_{3,\alpha}$ $v_{3,\alpha\beta}$ $v_{\lambda,\nu}$ $v_{3,\nu}$	v_λ $v_{\lambda,\alpha}$ $v_{\lambda,12}$ v_3 $v_{3,\alpha}$ $v_{3,12}$
Edges/Nodes/DOF	3/3/27	3/3/36	3/3/54	3/6/63	4/4/48
No. of integration-points	12	21	21	21	16

FIG. 3.3 Geom. nonlinear Shell-Elements NACS

4. COMPUTATIONAL ALGORITHMS

Algorithms for nonlinear response in finite element computation are usually based on combined incremental and iterative techniques. The scalar load factor λ is incremental stepwise, and the linearisation error is corrected by equilibrium [5]. In order to jump over singularity points within the deflection path a sophisticated Riks/Wempner/Wessel linear arc-length method is used [4, 5, 13] and described in figure 4.1.

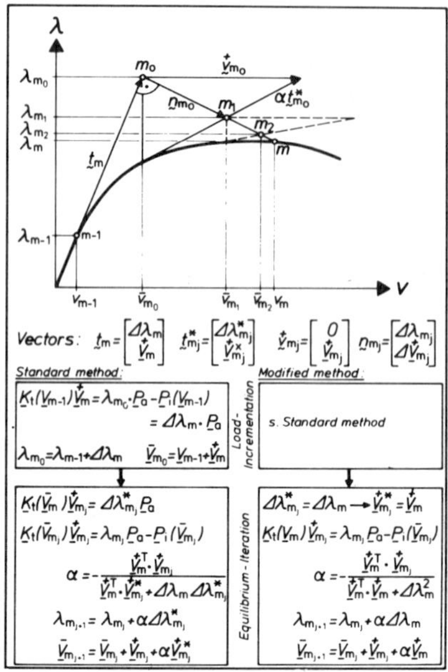

FIG. 4.1 Riks/Wempner/Wessel Algorithm

The detection of bifurcation points (in general instability points), can be expressed mathematically by solving [2]:

$$\pi'' \; [\underline{u}, \lambda_c] \; \underline{u}_1 \; \delta \, \underline{u} = 0 \, , \tag{4.1}$$

where λ_c denotes the lowest eigenvalue and $\underline{u}_1$ the corresponding eigenmode. In matrix formulation we end up with the quadratic eigenvalue problem of figure 4.2.

$$\boldsymbol{u} = \lambda \bar{\boldsymbol{u}} + \overset{+}{\boldsymbol{u}}, \quad \boldsymbol{V} = \lambda \bar{\boldsymbol{V}} + \overset{+}{\boldsymbol{V}}$$

with $\bar{\boldsymbol{u}}, \bar{\boldsymbol{V}}$: *displacement field at point of neutral equilibrium*

$$\bar{\boldsymbol{u}} = \bar{\boldsymbol{u}}_c \, , \bar{\boldsymbol{V}} = \bar{\boldsymbol{V}}_c \text{ for } \lambda = \lambda_c = 1$$

$$\boldsymbol{K}_T \overset{+}{\boldsymbol{V}} = [\boldsymbol{K}_e + \lambda(\boldsymbol{K}_{gL} + \boldsymbol{K}_{uL}) + \lambda^2(\boldsymbol{K}_{gN} + \boldsymbol{K}_{uN})] \overset{+}{\boldsymbol{V}} = 0$$

Linearization for $\lambda = \lambda_c = 1$

$$[\boldsymbol{K}_e + \lambda(\boldsymbol{K}_{gL} + \boldsymbol{K}_{uL} + \boldsymbol{K}_{gN} + \boldsymbol{K}_{uN})] \overset{+}{\boldsymbol{V}} = 0$$

FIG. 4.2 Quadratic eigenvalue Problem

Several states of linearization of this equation are achieved by neglecting $\underline{K}_{uN}$, K_{gN}, $\underline{K}_{uL}$. Following a nonlinear prebuckling path two methods are applied in practice for detecting bifurcation points:

- a check is kept on the diagonal elements of a Gaussian decomposition matrix D, corresponding to the stability coefficients defined in [15],
- the linearized eigenvalue problem is solved until $|1-\lambda| < \delta$, where δ is the approximation factor ($\delta \leq 0.05$).

The last method, although being the more complicated at first glance, can help to steer the load incrementation more efficiently in the prebuckling path of deformation.

5. APPLICATION TO POSTBUCKLING THEORY

Postbuckling analysis dates back to the pioneering work of Koiter [9] and has been adapted and reformulated by several authors e.g. Budiansky, Hutchinson [2,3,8]. Today finite element methods serve as a new tool for applying basic ideas of this theory.

The main interesting assumption is the asymptotic polynomial expansion of the load factor λ and the displacement $\underline{u}$ at the bifurcation point:

$$\lambda = \lambda_c + \xi \lambda_1 + \xi^2 \lambda_2 + \dots \tag{5.1}$$

$$\underline{u} = \lambda \underline{u}_o + \xi \underline{u}_1 + \xi^2 \underline{u}_2 + \dots \tag{5.2}$$

where λ_c is the lowest eigenvalue, $\underline{u}_1$ the corresponding eigenmode, $\underline{u}_2$ another orthogonalized displacement and ξ a perturbation factor valid near the bifurcation point.

The first practically useful result of this theory, of course self-evident in the eigenvalue problem, is the application of $\xi\underline{u}_1$ as a perturbation of the prebuckling deformation to enter the secondary, mostly unstable, postbuckling path. The factor ξ here has to be big enough for the equilibrium iteration to be convergent to a postbuckling point and small enough to remain suitably near to the buckling point.

The second result is the ability to calculate the λ_1 and λ_2 parameters, representing respectively the tangent and the curvature of the postbuckling path. These parameters and also the additional displacement field $\underline{u}_2$ are found by putting the asymptotic expansion (5.1),(5.2) into a Taylor expansion of the instability equation (4.1) with respect to λ and ordering the terms according to power terms of ξ.

The outstanding advantage of Koiter's theory is the estimation of the imperfection influence on the load carrying capacity.

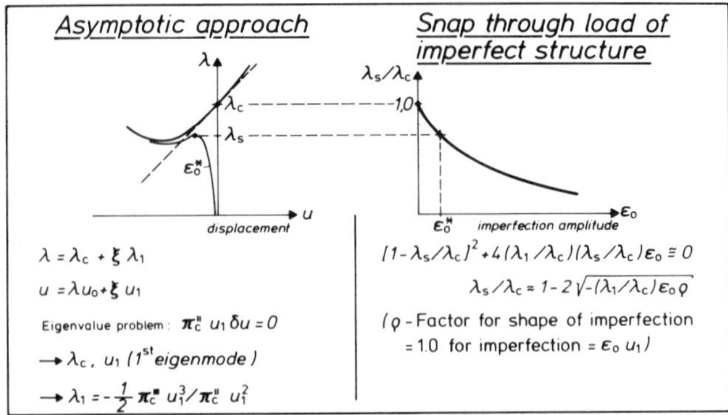

FIG. 5.1 Postbuckling Analysis - Asymmetric Case

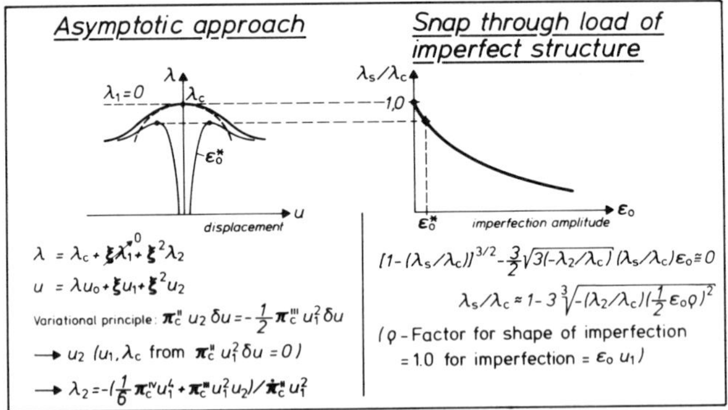

FIG. 5.2 Postbuckling Analysis - Symmetric Case

The last figures contain the basic variational functionals and formulae for postbuckling analysis. The relation between initial imperfections and the load reduction factor is plotted by using Koiter's formulae and presented for the asymmetric (fig. 5.1) and the symmetric cases (fig. 5.2).

The approximation fits well for eigenmode like imperfections; for arbitrary ones the practical calculation of ρ is still a problem. The treatment of multi-mode bifurcation has also not been solved for finite element application more generally, although theoretical results exist [14].

It should be emphasized that the finite element calculation of λ_1 and λ_2 costs less than 1.5 the eigenvalue computation λ_c and consists of nearly straight linear algorithms.

6. EXAMPLES

6.1 *Shallow Cylindrical Segment Axially Loaded*

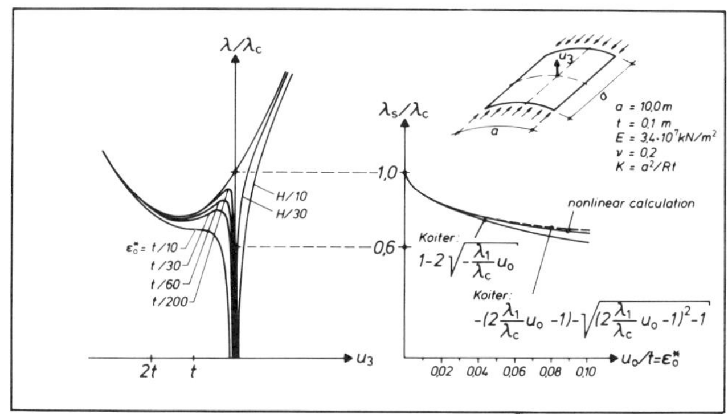

FIG. 6.1 Postbuckling of shallow cylindrical shell

Figure 6.1 shows the typical asymmetric bifurcation case. The chosen structure itself is extremely imperfection sensitive. The most dangerous imperfection shape is the 1st eigenmode. The approximation of imperfect snap-through load by Koiter's formulae is rather sufficient.

6.2 *Square Plate With Inplane Load*

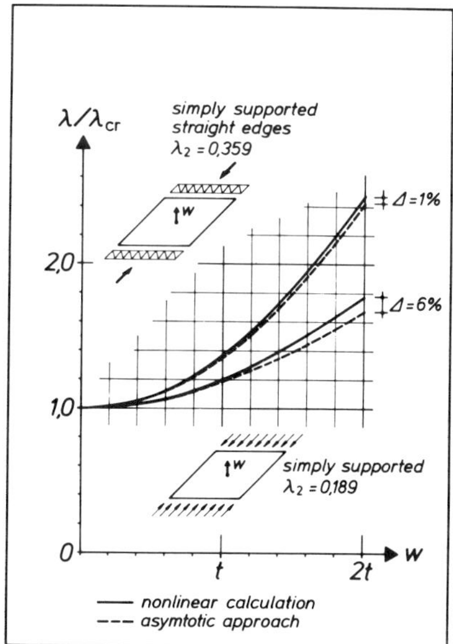

FIG. 6.2 Postbuckling behaviour of a square plate

The square plate is a well known example for stable symmetric bifurcation. In-plane and normal deformations are decoupled. After buckling a stress concentration at the edges causes the stable reaction. Therefore boundary conditions influence decisively the postbuckling behaviour. The estimation of the midpoint normal displacement by the asymptotic approach fits well as shown.

6.3 *Cylindrical Shell With Radial Pressure*

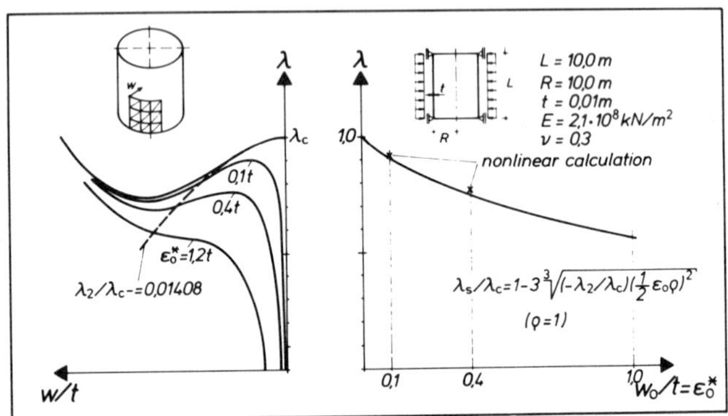

FIG. 6.3 Postbuckling of cylindrical shell

Of great practical interest are shells representing the unstable symmetric bifurcation case. Again figure 6.3 points out the ability of nonlinear calculation of perfect and imperfect structures and in addition the efficiency of finite element postbuckling analysis.

ACKNOWLEDGEMENT

The authors gratefully like to acknowledge the support by the "Deutsche Forschungsgemeinschaft" (DFG) for this work.

7. REFERENCES

1. BEEM, H., LECHTLEITNER, B., *FEMAS - Finite Element Moduln allgemeiner Strukturen*. 2nd Edition, Ruhr-Universität Bochum, KIB III, (1984).
2. BUDIANSKY, B., Theory of Buckling and Post-Buckling Behavior of Elastic Structures. *Advances Appl. Mech.* 14 (1974).
3. BUDIANSKY, B., Dynamic Buckling of Elastic Structures, Criteria and Estimates, in: *Dynamic Stability of Structures*. ed. Herrmann, Pergamon Press, (1967).

4. CRISFIELD, M.A., An arc-length method including line search and accelerations. *Int. J. Numer. Meth. Eng.* 19, 1269-1289 (1983).
5. ECKSTEIN, U., *Nichtlineare Stabilitätsberechnung elastischer Schalentragwerke.* Ruhr-Universität Bochum, TWM 83-3 (1983).
6. ECKSTEIN, U., HARTE, R., KRÄTZIG, W.B. and WITTEK, U., Solution strategies for linear and nonlinear instability phenomena for arbitrarily curved thin shell structures. *7th Int. Conf. on Struct. Mech. in Reactor Techn.* Chicago (1983).
7. HARTE, R., *Doppelt gekrümmte finite Dreieckselemente für die lineare und geometrisch nichtlineare Berechnung allgemeiner Flächentragwerke.* Ruhr-Universität Bochum, TWM 82-10 (1982).
8. HUTCHINSON, J.W., Lecture Notes on the Postbuckling Behaviour of Structures. *St. Venant Session, Udine Italy* (1974).
9. KOITER, W.T., On the Stability of Elastic Equilibrium. *University of Delft, Diss. 1945.* NASA TT F-10 (1967).
10. KRÄTZIG, W.B., BASAT, Y. and WITTEK, U., Nonlinear behaviour and elastic stability of shells - theoretical concept - numerical computations - results. *Results, Proc. Coll. Buckling of Shells, Stuttgart* (1982).
11. KRÄTZIG, W.B., WITTEK, U. and BASAR, Y., Buckling of general shell-theory and numerical analysis. In *Collapse*, J.M.T. Thompson/ G.W. Hunt (eds.). Cambridge University Press (1983).
12. RALL, L.B., *Computational Solution of Nonlinear Operator Equations.* John Wiley & Sons, London (1969).
13. RAMM, E., Strategies for tracing nonlinear response near limit points. *Proc. of Europe-U.S. Workshop*, Ruhr-Universität Bochum (1980).
14. STUMPF, H., *Unified Operator Description, Nonlinear Buckling and Postbuckling Analysis of Thin Elastic Shells.* Ruhr-Universität Bochum, IfM 34 (1982).
15. THOMPSON, A general theory for the equilibrium and stability of discrete conservative systems. *J. Appl. Math. Phys, ZAMP* 20, 797-846 (1969).
16. ZIENKIEWICZ, O.C., *The Finite Element Method, 3rd edition.* McGraw-Hill, London (1977).

AN ANNULAR FINIE ELEMENT IN POLAR COORDINATES FOR PLATE BENDING PROBLEMS

A.B. Sabir and H.G.V. Der Avanessian

Department of Civil and Structural Engineering, University College, Cardiff, Wales

1. INTRODUCTION

The finite element analysis of three dimensional plated structures requires the use of generalised finite element stiffness matrices which include the effect of out-of-plane bending as well as deformation within the plane of the plated elements. For example in box girders the upper deck and the lower flange require to be modelled by such generalised finite elements. In the case of the curved box girder shown in Fig. 1

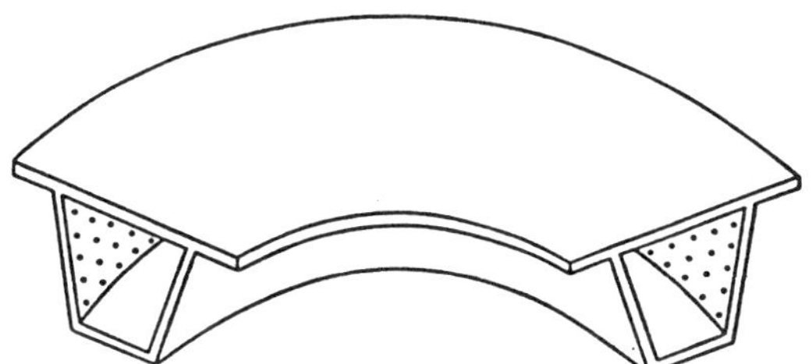

FIG.1 A curved box girder

the upper and lower decks are more conveniently modelled by annular elements bound by two concentric curves and two straight radial edges. The derivation of the stiffness matrices for such elements requires the use of polar coordinates and if the analysis is confined to displacements which are small when compared with the dimensions of the box girder, the inplane and out-of-plane deformations are regarded as independent of each other and the generalised stiffness matrix can be written as given by the following matrix equation

$$\begin{bmatrix} p_i \\ p_o \end{bmatrix} = \begin{bmatrix} K_i & 0 \\ 0 & K_o \end{bmatrix} \begin{bmatrix} \delta_i \\ \delta_o \end{bmatrix} \qquad (1)$$

where p is the load vector, K is the stiffness matrix and δ is

ISBN 0-12-747255-X

the generalised nodal displacements. Suffices i and 0 refer to inplane and out-of-plane deformations respectively. A new class of finite elements for inplane deformation both in cartesian and polar coordinates are developed in references [1,2]. The shape functions for such inplane elements are based on simple independent strain rather than displacement functions. Their convergence performance is examined and compared with the more traditionally accepted displacement elements by analysing several types of problems. In the present paper we shall confine our attention to the available stiffness matrices for out-of-plane deformation K_0 and propose a new element having a displacement function which satisfies the exact representation of rigid body modes of deformation and possesses only the essential external nodal degrees of freedom.

2. AVAILABLE SECTOR ELEMENTS FOR PLATE BENDING

A limited number of elements for out-of-plane deformations have been developed in a system of polar coordinates. Three such elements are discussed and a brief description for each is given below.

Olson and Lindburg [3] developed an annular segmental plate bending element with four corner nodes each having three external nodal degrees of freedom. The displacement function for the lateral deflection, w, was taken to consist of the same polynomial terms as for the well-known nonconforming twelve degrees rectangular plate bending element [4]. The nodal degrees of freedom used were the vertical deflection w, and the two rotations $\partial w/\partial r$ and $(1/r)(\partial w/\partial\Theta)$.

Sawko and Merriman [5] were guided by the form of the analytical solution to the differential equation of bending of circular plates in polar coordinates and suggested the following shape function

$$
\begin{aligned}
w = {} & A_1+A_2r+A_3r^2+A_4r^3+A_5\cos\Theta+A_6r\cos\Theta+A_7r^2\cos\Theta+A_8r^3\cos\Theta \\
& +A_9\sin\Theta+A_{10}r\sin\Theta+A_{11}r^2\sin\Theta+A_{12}r^3\sin\Theta+A_{13}\sin2\Theta \\
& +A_{14}r\sin2\Theta+A_{15}r^2\sin2\Theta+A_{16}{}^3\sin2\Theta \qquad (2)
\end{aligned}
$$

The above expression contains sixteen independent terms and requires not only the essential three nodal degrees of freedom mentioned earlier, but an additional one which is taken to be $(1/r)\ \partial^2w/\partial r\partial\Theta$. This corresponds to the additional degree of freedom $\partial^2w/\partial x\partial y$ used for the conforming plate bending rectangular element [6]. The stiffness matrix was calculated using a numerical integration technique. This element satisfies the requirement of strain-free rigid body movements and the general elasticity compatibility equations.

Raju, Kirshna and Rao [7] used the natural mode technique to derive a stiffness matrix for the annular element. The displacement distributionswere split into nine straining or natural modes and three rigid modes. This element is nonconforming , inter-element compatibility of vertical deflection w is maintained on the straight edges of the element but is violated on the curved sides.

3. THE PROPOSED ELEMENT

Consider an element bound by two radial straight edges and two concentric circular arcs as shown in Fig. 2. The strain-displacement relationships in polar coordinates can be written in the following form, see Ref. [8].

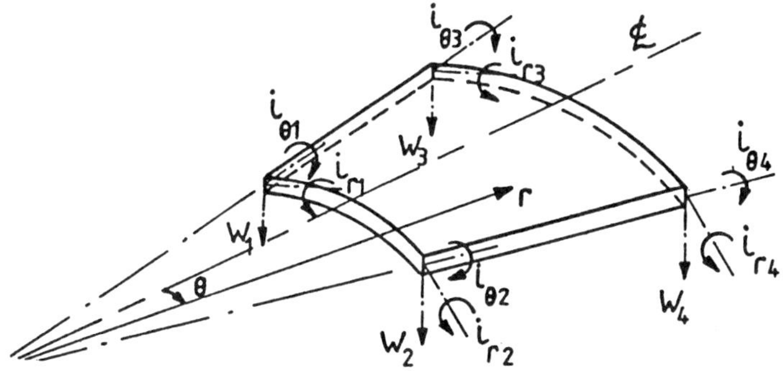

FIG. 2 Annular element in polar coordinates

$$
\begin{aligned}
\chi_r &= -\frac{\partial^2 w}{\partial r^2} \\
\chi_\Theta &= -\frac{1}{r}\frac{\partial w}{\partial r} - \frac{1}{r^2}\frac{\partial^2 w}{\partial \Theta^2} \\
\psi &= 2\left(\frac{1}{r}\frac{\partial^2 w}{\partial r \partial \Theta} - \frac{1}{r^2}\frac{\partial w}{\partial \Theta}\right)
\end{aligned}
\qquad (3)
$$

where χ_r and χ_Θ are the bending curvatures in the radial and tangential directions and ψ is the twisting curvature of the mid-surface. If the shape function for the element is to satisfy the exact representation of the strain free rigid body modes of deformations, the terms corresponding to this type of deformation can be obtained by integrating equations (3) when χ_r, χ_Θ and ψ are made equal to zero, see Ref. [9]. The resulting function for the lateral displacement w is

$$w_R = a_1 + a_2 r\cos\Theta + a_3 r\sin\Theta \qquad (4)$$

Three rigid body modes are defined by (4), the constant term corresponds to a translational mode and the terms $r\cos\Theta$ and

$r\sin\Theta$ are of rotational modes.

In order to avoid the difficulties associated with interal and non-geometric degrees of freedom such as $(1/r)(\partial^2/\partial r\partial\Theta)$ and also to keep the element as simple as possible the proposed element is allowed to have only the essential external degrees of freedom $(w, \partial w/\partial r, (1/r)(\partial w/\partial\Theta))$ for each of the corner nodes. Hence the shape function should contain 12 independent constants only. Having used three for the representation of the rigid body modes, the other nine independent terms were chosen to be polynomial in form and were selected from the shape function given in Ref. [3].This selection was made by omitting the three polynomial terms which would have approximately satisfied the strain free rigid body modes of deformation when the element is small. Hence

$$w = a_1+a_2 r\cos\Theta+a_3 r\sin\Theta+a_4\Theta+a_5 r^2+a_6\Theta^2+a_7 r^2\Theta+a_8 r\Theta^2+a_9 r^3 + a_{10}\Theta^3+a_{11}r^3\Theta+a_{12}r\Theta^3 \quad (5)$$

4. PROBLEMS CONSIDERED

A comprehensive test of the performance of the available elements, references [3], [5] and [7], and the proposed element is made by applying them to the solution of the variety of problems quoted in the finite element publications on sector elements. The problems of circular plates with holes loaded by line and point lateral loads causing rotationally symmetric and unsymmetrical deformations are first considered. A more practical problem of a curved bridge deck is then considered.

4.1 Circular plates with central holes

The NODAL solution routine reference [10]is used to analyse two such plates. The first deforming in a rotationally symmetric manner. The plate is simply supported around its outer circumference and loaded by a uniformly distributed line load around the edge of the hole. The dimensions used for this problem are the same as in reference [5]. A convergence test was first carried out for the lateral deflection at the edge of the hole and the results obtained from the proposed element and those of refs. [3] and [5] were satisfactory. For a mesh consisting of two elements in each direction all the above elements gave a result which differed by less than 0.2%. A sample of results for the stresses at the middle of the elements is given in Table 1. This table gives the variation of the radial bending stress resultant when a quarter of the circular plate is divided into a uniform mesh of eight elements in each direction. The results are in good agreement with the analytical solution, reference [8]. The performance of element (3) appears to be slightly better than the other two giving a difference of about 0.7% in the maximum stress occurring while this difference is about 1.6% and 2.6% for both the proposed and element (5).

radius	Radial moment (M_r) for the 8x8 mesh Nm/m			
	The proposed element	Ref [3]	Ref [5]	Exact Ref [8]
$r_o - 0.5\beta$	0.7741	0.8059	0.7657	0.8149
$- 1.5\beta$	2.2831	2.3176	2.2418	2.3320
$- 2.5\beta$	3.5743	3.6122	3.4816	3.6315
$- 3.5\beta$	4.5551	4.5968	4.3671	4.6234
$- 4.5\beta$	5.0934	5.1407	5.0401	5.1770
$- 5.5\beta$	5.0000	5.0523	4.9011	5.1040
$- 6.5\beta$	4.0000	4.0535	3.9522	4.1282
$- 7.5\beta$	1.6526	1.7227	1.6510	1.8313

Table 1 Variation of the radial bending stress resultant where r_o is the outer radius and $\beta = (r_o - r_i)/8$

The second problem considered is that of a plate deforming unsymmetrically. The plate is loaded by a point lateral load at its outer edge and held rigidly at the edge of the hole. The dimensions are as given in Fig. 3. Table 2 gives the results for the lateral deflection under the point load p for the previously mentioned three elements together with those of Ref. [7]. The proposed and the element of Ref. [3] give more satisfactory

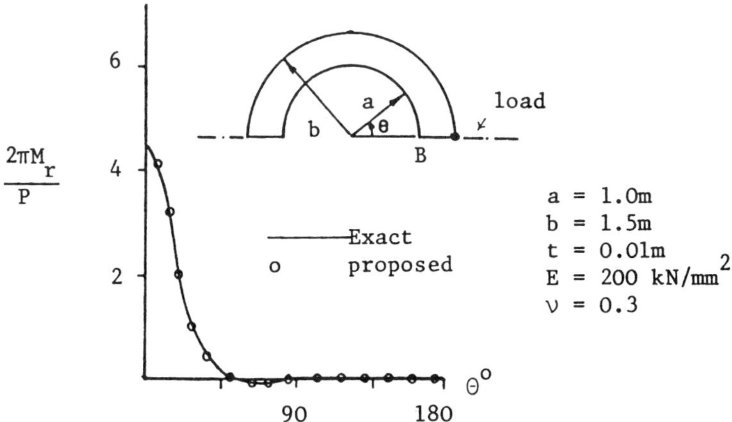

FIG. 3 Variation of radial bending stress resultant

results than the other two elements. However, the results of Refs. [5] and [7] appear to converge monotonically.

Mesh ($r \times \Theta$)	Deflection WD/p			
	The proposed element	Ref [3]	Ref [5]	Ref [7]
1 x 6	0.05086	0.05089	0.04691	0.04955
2 x 8	0.05223	0.05225	0.04885	0.05031
2 x 12	0.05137	0.05137	0.04992	0.05046
4 x 24	0.05088	0.05088	0.05054	-
8 x 24	0.05086	0.05086	0.05054	-
Exact value = 0.05072				

Table 2 Convergence of lateral deflection under the applied point load $p.D = Et^3/12(1-\nu^2)$

The mid element radial bending stress resultant along the first radial row of elements next to the applied load is given in a non dimensional form of $2\pi M_r/P$ in Table 3. The results are for the case when half the plate is divided into a mesh of 8x24 elements. The results obtained from the proposed and the elements of reference [5] are in agreement while those given by the element of reference [3] appear to be in error. To obtain the stresses at the edge where the plate is held rigidly a graphical method of extrapolation is used and the results from the proposed element are shown in Fig. 3. This figure shows the satisfactory agreement of the results for the stresses when compared with the analytical solution. A point of detail which warranted a special attention is the calculation of the radial bending stress at point B (Fig. 3). A more refined mesh was used to obtain an extrapolated finite element result of 4.50 instead of the analytical value of 4.45.

Element No.	Ref [5]	Ref [3]	The proposed element
1	-0.210	-94.78	-0.225
2	-0.635	-43.21	-0.605
3	-1.002	-20.32	-0.942
4	-1.355	- 9.79	-1.291
5	-1.745	- 4.90	-1.684
6	-2.219	- 2.54	-2.160
7	-2.838	- 1.11	-2.776
8	-3.707	+ 2.11	-3.638

Table 3 Variation of radial bending stress resultant along first row of elements adjacent to P

4.2 *Annular segments in plate bending*

An annular segment subjected to a point lateral load acting either at the centre of the outer or the inner curved edges is considered. The plate is simply supported along the radial edges and free along the curved edges. The dimensions used are those given by Meyer and Scordelis [11]. Table 4 gives the results for the lateral deflection at the centre of the plate when the load is acting at the centre of the outer edge.

Mesh	The proposed element	Ref [3]	Ref [5]
4 x 4	0.33460	0.33460	0.33537
6 x 6	0.33476	0.33476	0.34048
8 x 8	0.33479	0.33486	failed
8 x 16	0.33480	0.33488	"

Table 4 Convergence of lateral deflection at the centre of plate

This table shows that both the proposed and the element of Ref. [3] give similar results while the element of Ref. [5] appears to be numerically unstable for meshes greater than (6x6). The radial bending stresses at the centre of the elements were then calculated when half the plate is divided into a mesh of 8x8 elements. A disagreement in the results given by the proposed and the element of Ref. [3] was observed. Table 5 gives a sample of such results for the radial bending stress resultant along the central line of the first radial row of elements adjacent to

Element No.	The proposed element	Ref [3]	Modified Ref [3]
1	-0.02857	-0.064408	-0.02855
2	-0.06902	-0.31418	-0.06916
3	-0.08508	-0.24996	-0.085083
4	-0.08903	-0.19188	-0.089029
5	-0.08427	-0.15133	-0.084274
6	-0.07154	-0.11572	-0.071544
7	-0.05024	-0.07006	-0.050241
8	-0.01853	-0.02479	-0.018534

Table 5 The variation of the radial bending stress resultant. Element 1 is adjacent to the point load

mid span. In Ref. [3] the element stiffness matrix is calculated on the assumption that $\Theta=0$ lies on the left hand side straight edge of the element and the results given in the third column of Table 5 were calculated on this basis. The authors calculated this matrix by taking $\Theta=0$ to coincide with the centre of the element and the results obtained from this modified stiffness matrix are given in the fourth column of Table 5. This column shows results which are in agreement with those obtained from the proposed element. The results given in the third column of Table 3 can also be improved if this procedure is used. Because of restriction to space in the present paper the reader is referred to reference [12] where a more detailed discussion of this problem together with the case when the load is applied at the centre of the inner edge is given. The finite element results for the stresses due to the radial moment at the centre of the elements as obtained by the proposed element were used together with a graphical method of interpolation to obtain the stresses along the central radial line of the plate. Figure 4 shows a satisfactory agreement of these results when compared

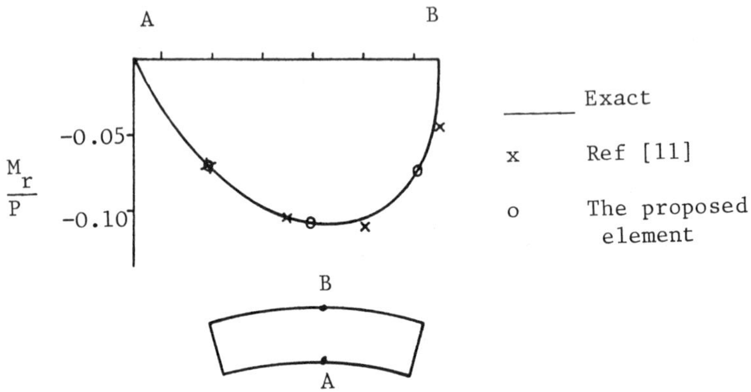

FIG. 4 Variation of bending stress resultant along AB

with the closed form elasticity solution of the plate equations for this special boundary value problem. This closed form solution is referred to as exact. This figure also shows the results obtained by Meyer and Scordelis [11] using a finite strip method of analysis.

CONCLUSION

The performance of a proposed simple annular element is compared with existing elements by applying them to a variety of bending problems. The proposed element satisfies the exact representation of rigid body modes of deformations. In this way it has an advantage over the element given in Ref. [3] in that it requires less computations for the calculation of the stiffness matrix.

References

1. SABIR, A.B. A New Class of Finite Elements for Plane Elasticity Problems. *Proc. of 7th Int. Conference on Structural Mechanics in Reactor Technology, Seminar on the Computational Aspects of Finite Elements*. Chicago USA, Aug. (1983).

2. SABIR, A.B. Finite Elements in Polar Coordinates for Plane Elasticity Problems. *In the Press*.

3. OLSON, M.D. and LINDBURG, G.M. Annular and Circular Sector Finite Elements for Plate Bending. *Int.J.of Mech.Sci.* 12, (1970).

4. ZIENKIEWICZ, O.C. and CHEUNG, Y.K. The Finite Element Method for Analysis of Elastic Isotropic and Orthotropic Slabs. *Proc.Inst.Civ.Eng.,* 28, 471-88, (1964).

5. SAWKO, F. and MERRIMAN, P.A. An Annular Segment Finite Element for Plate Bending. *Int.J. for Num.Methods in Eng.* 3, (1971).

6. WALKER, A.C. Rayleigh-Ritz Method for Plate Flexure. *J.of the Mech.Eng.Div.Proc. ASCE.* 93, (1967).

7. RAJU, I.S., KRISHNA, A.V. and RAO, A.K. Sector Elements for Matrix Displacement Analysis. *Int.J. for Num.Methods in Eng.* 6, (1973).

8. TIMOSHENKO, S.P. and GOODIER, J.N. *Theory of Elasticity.* McGraw Hill, New York, (1971).

9. ASHWELL, D.G., SABIR, A.B. and ROBERTS, T.M., Further Studies in the Application on Curved Finite Elements to Circular Arches. *Int.J.Mech.Sci.* 13, *507-517.* (1971).

10. SABIR, A.B. The Nodal Solution Routine for the Large Number of Linear Simultaneous Equations in the Finite Element Analysis of Plates and Shells. *Chapter 5, Finite Element for Thin Shells and Curved Members.* John Wiley & Sons (1976).

11. MEYER, C. and SCORDELIS, A.C. Analysis of Curved Folded Plate Structures. *J. of The Structural Div. ASCE,* (1971).

12. DER AVANESSIAN, H.G.V. Analysis of Curved Box Girders. Ph.D. Thesis, University of Wales, Cardiff,(1984).

ON THE CONVERGENCE OF A FOUR-NODE PLATE BENDING ELEMENT BASED ON MINDLIN/REISSNER PLATE THEORY AND A MIXED INTERPOLATION

K.J. Bathe * and F. Brezzi †

**Massachusetts Institute of Technology, Cambridge, U.S.A.*
†University of Pavia, Pavia, Italy

1. INTRODUCTION

The aim of this paper is to analyze, from a theoretical point of view, a four-node element for Mindlin/Reissner plates recently introduced by Bathe and Dvorkin [3]. We recall that the Mindlin/Reissner formulation for "moderately thick" plates amounts to assuming that the "in-plane" displacements u_1 and u_2 have the form [2]

$$u_1(x,y,z) = -z\beta_1(x,y); \quad u_2(x,y,z) = -z\beta_2(x,y) \tag{1.1}$$

In the Mindlin/Reissner theory the unknowns are $\beta_1(x,y)$, $\beta_2(x,y)$ and u_3 which is assumed as

$$u_3(x,y,z) = w(x,y) \tag{1.2}$$

The unknowns $\underline{\beta}$ and w are defined in $\Omega \subseteq \mathbb{R}^2$, and the undeformed configuration of the plate occupies the region $\Omega\times]-t/2,t/2[$ (hence, t is the plate thickness). The corresponding strain field is therefore

$$\varepsilon_{11} = -z\partial\beta_1/\partial x \qquad \varepsilon_{22} = -z\partial\beta_2/\partial y \qquad \varepsilon_{33} = 0$$

$$2\varepsilon_{12} = -z(\partial\beta_1/\partial y + \partial\beta_2/\partial x) \quad 2\varepsilon_{13} = \partial w/\partial x-\beta_1 \quad 2\varepsilon_{23} = \partial w/\partial y - \beta_2 \tag{1.3}$$

THE MATHEMATICS OF FINITE
ELEMENTS AND APPLICATIONS V

ISBN 0-12-747255-X

and the corresponding stress field is

$$\sigma_{11} = (\varepsilon_{11}+\nu\varepsilon_{22})\ E/(1-\nu^2) \quad \sigma_{22} = (\varepsilon_{22}+\nu\varepsilon_{11})\ E/(1-\nu^2)$$

$$\sigma_{ij} = E\ \varepsilon_{ij}/(1+\nu) \quad i,j = 1,2,3;\ i\neq j \tag{1.4}$$

where E is the Young's modulus and ν is the Poisson's ratio. If $p(x,y)$ is the transverse loading per unit area, the total potential energy is given by

$$\Pi = \frac{1}{2} t^3 a\ (\underline{\beta},\underline{\beta}) + \frac{1}{2}\lambda t \int_\Omega |\underline{\nabla} w-\underline{\beta}|^2\ d\Omega - \int_\Omega pwd\Omega \tag{1.5}$$

where

$$a(\underline{\beta},\underline{\beta}) = \frac{E}{12(1-\nu^2)}\int_\Omega [(\partial\beta_1/\partial x+\nu\partial\beta_2/\partial y)\partial\beta_1/\partial x + (\nu\partial\beta_1/\partial x \tag{1.6}$$

$$+\partial\beta_2/\partial y)\ \partial\beta_2/\partial y + 2^{-1}(1-\nu)(\partial\beta_1/\partial y+\partial\beta_2/\partial x)^2]dxdy$$

$$\lambda = Ek/(2(1+\nu)) \tag{1.7}$$

and k is a constant to account for the actual nonuniformity of the shearing stresses (namely, since $\sigma_{13}=\sigma_{23}=0$ on the lower and upper surfacesof the plate, we must "correct" (1.3) which assumes ε_{13} and ε_{23} to be constant with z; for more details, see e.g. [2]).

We assume here, for the sake of simplicity(†) that our plate is clamped along the entire boundary $\partial\Omega$; hence we look for $(\underline{\beta}, w)$ in the space

$$V = \{(\underline{\eta},\zeta)\ |\ \underline{\eta}\in(H_0^1\ (\Omega))^2,\quad \zeta\in H_0^1\ (\Omega)\} \tag{1.8}$$

The well-known Korn's inequality (see e.g. [6]) states that

$$\exists\ c>0 \text{ such that } \forall\ \underline{\eta}\in(H_0^1\ (\Omega))^2$$
$$a(\underline{\eta},\underline{\eta}) \geq c\ \|\underline{\eta}\|_1^2 \tag{1.9}$$

This implies that, from the mathematical point of view, for any $t>0$ the functional Π in (1.5) is strictly convex and hence has a unique minimiser in V. However, it is also well known that many finite element discretizations of (1.5) fail when the

(†)Although best tractable for mathematical analysis, in actual computations, the clamped plate problem usually yields the worst convergence properties.

thickness t of the plate is "too small". In order to understand this phenomenon, we thought it necessary to construct a sequence of problems $(P_t)_{t>0}$ such that the corresponding solutions $(\underline{\beta}(t), w(t))$ stay bounded for $t\to 0$. Such a sequence is meant as a test for the numerical discretizations and our analysis will concentrate on the behaviour of the discretization in [3] when applied to the sequence. For a similar analysis on several beam elements we refer to the very good paper by Arnold [1]. Unfortunately, our (two-dimensional) case is much harder, so that the generality of the results of [1] on beams is (at this time) out of reach for plates.

An outline of the paper is as follows. In Section 2 we introduce the sequence of problems (P_t) and we study the behaviour of the corresponding solutions for $t\to 0$. In Section 3 we recall the definition of the element in [3] and we show that, when it is applied to the sequence of problems (P_t), the corresponding discrete solutions satisfy

$$||\underline{\beta}(t)-\underline{\beta}_h(t)||_1 + ||w(t)-w_h(t)||_1 < ch(||\underline{\beta}(t)||_3+||\underline{\gamma}(t)||_2) \quad (1.10)$$

with c independent of h and t. We also have the (non-optimal) result

$$t||\underline{\gamma}(t)-\underline{\gamma}_h(t)||_0 \lesssim ch \quad (1.11)$$

where $\underline{\gamma} = t^{-2}(\underline{\nabla}w-\underline{\beta})$ and $\underline{\gamma}_h \simeq t^{-2}(\underline{\nabla}w_h-\underline{\beta}_h)$ are the continuous and discrete "shear strains". For the sake of simplicity, the analysis of Section 3 is carried out on the particular case of a rectangular Ω and for a decomposition into rectangles. The analysis of more general cases (using the element in shell problems [7]) is not a straight-forward generalization of the present case.

2. THE SEQUENCE OF PROBLEMS $(P_t)_{t>0}$.

We construct our test sequence in the following way. We assume that Ω, E, ν are kept constant, while the pressure $p_t(x,y)$ varies as

$$p_t(x,y) \equiv t^3 f(x,y) \quad (2.1)$$

where f(x,y) is a given function, say, in $L^2(\Omega)$, which obviously does not change with t. By dividing the potential energy in (1.5) by t^3 we have therefore the following sequence of problems $(P_t)_{t>0}$

$$\left.\begin{array}{l}\text{Minimize, for } (\underline{\beta},w)\in V \text{ the functional}\\ \frac{1}{2}a(\underline{\beta},\underline{\beta}) + \frac{\lambda}{2}t^{-2}||\underline{\nabla}w-\underline{\beta}||_0^2 - (f,w)\end{array}\right\} \quad (2.2)$$

where, as usual, (,) denotes the $L^2(\Omega)$ inner product.

Again Korn's inequality (1.9) ensures that (2.2) has a unique solution, $(\underline{\beta}(t), w(t))$, for all $t>0$. It is also an easy matter to check that (say, for $0<t<t_0$)

$$\|\underline{\beta}(t)\|_1^2 + \|w(t)\|_1^2 + t^2\|\underline{\gamma}(t)\|_0^2 \leq \text{ c indep. of t} \tag{2.3}$$

where $\underline{\gamma}(t)$ is defined by

$$\underline{\gamma}(t) := t^{-2}(\underline{\nabla}w(t)-\underline{\beta}(t)) \tag{2.4}$$

In order to analyze the behaviour of the solution for $t\to 0$ it is convenient to consider the Euler equations associated with (2.2)

$$\left.\begin{array}{l}\text{find } (\underline{\beta}(t), w(t)) \in V \text{ such that} \\ a(\underline{\beta}(t),\underline{\eta}) + \lambda(\underline{\gamma}(t),\underline{\nabla}\zeta-\underline{\eta}) = (f,\zeta) \;\forall\; (\underline{\eta},\zeta) \quad V \\ \underline{\gamma}(t) = t^{-2}(\underline{\nabla}w(t)-\underline{\beta}(t))\end{array}\right\} \tag{2.5}$$

It is clear that in order to study the behaviour of the solutions for $t\to 0$ we need estimates on $\underline{\gamma}(t)$ rather than on $t\underline{\gamma}(t)$ as in (2.3). For this we introduce the space

$$H_0(\text{rot};\Omega) = \{\underline{\chi} \mid \underline{\chi} \in (L^2)^2,\ \text{rot}\underline{\chi} \in L^2(\Omega),\ \underline{\chi}\cdot\underline{\tau} = 0\} \tag{2.6}$$

where

$$\text{rot}(\chi_1,\chi_2) := \partial\chi_1/\partial y - \partial\chi_2/\partial x \tag{2.7}$$

$$\underline{\tau} = \text{counterclockwise tangent unit vector to } \partial\Omega \tag{2.8}$$

$$\|\underline{\chi}\|^2_{H_0(\text{rot};\Omega)} := \|\underline{\chi}\|_0^2 + \|\text{rot}\underline{\chi}\|_0^2 =: \|\underline{\chi}\|_H^2 \tag{2.9}$$

and the space

$$\Gamma := (H_0(\text{rot};\Omega))' = \text{dual space of } H_0(\text{rot};\Omega) \tag{2.10}$$

Theorem 2.1 *We have*

$$\|\underline{\gamma}(t)\|_\Gamma \leq c \text{ independent of } t \tag{2.11}$$

Proof. From (2.10) we have that there exists $\underline{\chi} \in H_0(\text{rot};\Omega)$ such that

$$(\underline{\gamma}(t),\underline{\chi}) = \|\underline{\gamma}(t)\|_{\Gamma}^{2} = \|\underline{\chi}\|_{H}^{2} \tag{2.12}$$

We shall show that it is possible to find $(\underline{\eta},\zeta) \in V$ such that

$$\underline{\chi} = \underline{\nabla}\zeta - \underline{\eta} \tag{2.13}$$

$$\|\underline{\eta}\|_1 + \|\zeta\|_1 \le c\|\underline{\chi}\|_H \tag{2.14}$$

with c independent of χ. If this is true, then from (2.3), (2.5) and (2.12)-(2.14) we get

$$\begin{aligned}\|\underline{\gamma}\|_{\Gamma}\ \|\underline{\chi}\|_H &= (\underline{\gamma},\underline{\chi}) = (f,\zeta) - a(\underline{\beta},\underline{\eta}) \\ &\le c(\|\zeta\|_1 + \|\underline{\eta}\|_1) \le c\ \|\underline{\chi}\|_H\end{aligned} \tag{2.15}$$

which implies (2.11). Hence we have just to show (2.13), (2.14). We first choose $\theta \in (H_0^1(\Omega))^2$ such that

$$\operatorname{div} \underline{\theta} = -\operatorname{rot} \underline{\chi} \tag{2.16}$$

$$\|\underline{\theta}\|_1 \le c\ \|\operatorname{rot} \underline{\chi}\|_0 \tag{2.17}$$

This is always possible (see e.g. [9]) since rot $\underline{\chi}$ has zero mean value on Ω. Then we set

$$\underline{\eta} \equiv (\eta_1,\eta_2) := (-\theta_2,\theta_1) \tag{2.18}$$

so that from (2.16) - (2.18) we have

$$\operatorname{rot} \underline{\eta} = -\operatorname{rot} \underline{\chi} \tag{2.19}$$

$$\|\underline{\eta}\|_1 \le c\|\operatorname{rot} \underline{\chi}\|_0 \le c\|\underline{\chi}\|_H \tag{2.20}$$

Now we choose ζ as the unique solution of: $\Delta\zeta = \operatorname{div} \underline{\gamma} + \operatorname{div} \underline{\eta}$, $\zeta \in H_0^1(\Omega)$. Clearly

$$\begin{aligned}\|\zeta\|_1 &\le c(\|\operatorname{div}\underline{\chi} + \operatorname{div} \underline{\eta}\|_{-1}) \le c(\|\underline{\chi}\|_0 + \|\underline{\eta}\|_0) \\ &\le c\|\underline{\chi}\|_H\end{aligned} \tag{2.21}$$

We have now

$$\left.\begin{aligned}&\operatorname{div}(\underline{\nabla}\zeta-\underline{\eta}) = \operatorname{div}\underline{\chi};\ \operatorname{rot}(\underline{\nabla}\zeta-\underline{\eta}) = -\operatorname{rot}\underline{\eta} = \operatorname{rot}\underline{\chi} \\ &(\underline{\nabla}\zeta-\underline{\eta})\cdot\underline{\tau} = \underline{\chi}\cdot\underline{\tau} = 0 \text{ on } \partial\Omega\end{aligned}\right\} \tag{2.22}$$

which easily imply (2.13). On the other hand (2.14) follows from (2.20) (2.21).

□

We are now able to analyze the behaviour of $\underline{\beta}(t)$, $w(t)$ for $t\to 0$.

Theorem 2.2 We have, for $t\to 0$,

$$\left.\begin{array}{l}\underline{\beta}(t) \to \underline{\beta}_0,\ w(t) \to w_0 \text{ in } H_0^1(\Omega) \\ \underline{\gamma}(t) \to \underline{\gamma}_0 \text{ in } \Gamma\end{array}\right\} \tag{2.23}$$

where $\underline{\beta}_0$, w_0, $\underline{\gamma}_0$ satisfy:

$$a(\underline{\beta}_0,\underline{\eta}) + \lambda < \underline{\gamma}_0, \underline{\nabla}\zeta - \underline{\eta} > = (f,\zeta) \ \forall\ (\underline{\eta},\zeta) \in V \tag{2.24}$$

$$\underline{\nabla} w_0 = \underline{\beta}_0 \tag{2.25}$$

$$E\,\Delta^2 w_0 = 12(1-\nu^2) f \tag{2.26}$$

Proof. (2.23) is obvious from (2.3) and (2.11). Then (2.24), (2.25) follow immediately passing to the limit in (2.5). Now, taking $\zeta=0$ in (2.24) we have

$$-A\ \underline{\beta}_0 = \lambda\ \underline{\gamma}_0 \tag{2.27}$$

where $-A$ is the (2-d) linear elasticity operator (associated with the bilinear form $a(\ ,\ )$). Taking instead $\underline{\eta}=0$ in (2.24) we get

$$\operatorname{div} \underline{\gamma}_0 = -f \tag{2.28}$$

and (2.26) follows from (2.25), (2.27), (2.28).

□

Remark. Let $w_K(t)$ be the solution of the Kirchhoff model

$$\frac{E\,t^3}{12(1-\nu^2)}\,\Delta^2 w_K(t) = p_t = t^3 f \tag{2.29}$$

corresponding to the same values of E, ν, and load p_t of our test sequence of Mindlin plates. Then theorem 2.2 states that the Mindlin solution $w(t)$ converges to the Kirchhoff solution w_K (which actually is independent of t). Note also that this is independent of the choice of the correction factor k in (1.7). For additional results on this convergence question see e.g. Destuynder [5].

Remark. It can be shown that the space Γ defined in (2.10) could also be written as

$$\Gamma = \{\underline{\kappa} \mid \underline{\kappa} \in (H^{-1}(\Omega))^2,\ \operatorname{div} \underline{\kappa} \in H^{-1}(\Omega)\} \tag{2.30}$$

It seems natural to guess that optimal error estimates for $\underline{\gamma}(t)$ should be given in the space Γ. However, for the case analyzed in the next section we were not able to do so. The task is much easier in the one-dimensional case where the space corresponding to Γ is just L^2.

Remark. The main reason for proving theorem 2.2 is to show that the sequence $(\underline{\beta}(t), w(t))$ does not tend to zero when $t \to 0$. The test would not be serious otherwise.

3. CONVERGENCE OF THE FOUR-NODE ELEMENT

We assume now that Ω is a rectangle and that we are given a sequence $\{T_h\}_h$ of uniform decompositions of Ω into rectangles R; if h_1 and h_2 are the lengths of the edges of R, we set

$$|h| = \max_i h_i \tag{3.1}$$

and we assume as usual that there exist two constants c_1, $c_2 > 0$ such that

$$c_2 |h| \leq h_1/h_2 \leq c_1 \; |h| \tag{3.2}$$

for all T_h.

We define, for all T_h,

$$Q_h = \{\phi | \phi \in H_0^1(\Omega);\ \phi_{|R} \in Q_1 \ \forall\ R \in T_h\} \tag{3.3}$$

where Q_1 is the space of bilinear functions. We also introduce

$$\Gamma_h = \{\underline{\chi} | \underline{\chi} \in H_0(\mathrm{rot};\Omega),\ \underline{\chi}_{|R} \in (Q_{01},\ Q_{10})\ \forall\ R \in T_h\} \tag{3.4}$$

where Q_{01} = span $\{1,y\}$ and Q_{10} = span $\{x,1\}$. Note that the condition $\underline{\chi} \in H_0(\mathrm{rot};\Omega)$ means in this case that $\underline{\chi} \cdot \underline{\tau} = 0$ on $\partial\Omega$, χ_1 is continuous on the horizontal edges, and that χ_2 is continuous on the vertical edges. Hence the degrees of freedom in Γ_h are: values of χ_1 on the internal horizontal edges and values of χ_2 on the internal vertical edges.

For any vector $\underline{\eta}_h \in (Q_h)^2$ we define now its interpolant $\overset{*}{\underline{\eta}}_h$ in Γ_h by the formulas

$$\left.\begin{array}{l} \overset{*}{\eta}_{h,1} = \eta_{h,1} \text{ at the midpoints of horizontal edges} \\ \overset{*}{\eta}_{h,2} = \eta_{h,2} \text{ at the midpoints of vertical edges} \end{array}\right\} \tag{3.5}$$

We finally set

$$V_h = (Q_h)^2 \times Q_h \tag{3.6}$$

Applying the four-node element introduced in [3] to each problem (2.2) we have the following sequence of problems:

$$\left.\begin{array}{l}\text{Minimize, for } (\underline{\beta}_h, w_h) \in V_h \text{ the functional} \\ \frac{1}{2}\, a(\underline{\beta}_h, \underline{\beta}_h) + \frac{\lambda t^{-2}}{2} \,||\, \underline{\nabla}\, w_h - \overset{*}{\underline{\beta}}_h ||_0^2 - (f,w)\end{array}\right\} \tag{3.7}$$

It is easy to check that each problem (3.7), for $t>0$, has a unique solution $(\underline{\beta}_h(t), w_h(t))$. The dependence on t will not be made explicit in the notation when unnecessary. Our aim is to estimate the error $||\underline{\beta}(t) - \underline{\beta}_h(t)||_1 + ||w(t) - w_h(t)||_1$ uniformly in t. As in the continuous problem, it will be convenient to introduce

$$\underline{\gamma}_h(t) = t^{-2} \, (\underline{\nabla} w_h(t) - \overset{*}{\underline{\beta}}_h(t)) \tag{3.8}$$

and to write the Euler equations of (3.7) in the form

$$\left.\begin{array}{l}\text{find } (\underline{\beta}_h, w_h) \in V_h \text{ such that} \\ a(\underline{\beta}_h, \underline{\eta}_h) + \lambda\, (\underline{\gamma}_h, \underline{\nabla}\zeta_h - \overset{*}{\underline{\eta}}_h) = (f, \zeta_h) \;\; \forall (\underline{\eta}_h, \zeta_h) \in V_h \\ \underline{\gamma}_h = t^{-2}\, (\underline{\nabla} w_h - \overset{*}{\underline{\beta}}_h)\end{array}\right\} \tag{3.9}$$

The following lemma summarizes a few properties that can be proved by simple algebraic manipulations or standard numerical integration techniques.

<u>Lemma 3.1</u> <u>We have, for all</u> $(\underline{\eta}_h, w_h)$ <u>in</u> V_h <u>and for all</u> $\underline{\chi}_h \in \Gamma_h$:

$$\underline{\nabla} w_h \in \Gamma_h \tag{3.10}$$

$$(\underline{\chi}_h, \underline{\eta}_h) = (\underline{\chi}_h, \overset{*}{\underline{\eta}}_h) \tag{3.11}$$

$$\text{rot}\, \overset{*}{\underline{\eta}}_h|_R = \frac{1}{|R|}\int_R \text{rot}\, \underline{\eta}_h \, dxdy \quad \forall\, R \in T_h \tag{3.12}$$

□

The following lemma provides some asymptotic estimates that will be used later on.

<u>Lemma 3.2</u> <u>We have</u>

$$||\zeta - \zeta^I||_{1,R} \le c|h|(||\partial^2\zeta/\partial x^2||_{0,R}^2 + ||\partial^2\zeta/\partial y^2||_{0,R}^2)^{1/2} \quad \forall\, R \in T_h \tag{3.13}$$

<u>for</u> ζ <u>smooth and</u> ζ^I = <u>interpolant of</u> ζ <u>in</u> Q_h

$$\left.\begin{array}{l} \|\underline{\eta}_h - \overset{*}{\underline{\eta}}_h\|_0 \leq c|h|\ \|\underline{\eta}_h\|_1 \\ \text{for } \underline{\eta}_h \in (Q_h)^2 \end{array}\right\} \tag{3.14}$$

Proof. The inequality (3.13) is well known (see e.g. [4] [11]). The proof of (3.14) is an easy exercise.

□

We now give the main theorem.

Theorem 3.1 Let $(\underline{\beta}(t), w(t))$ and $(\underline{\beta}_h(t), w_h(t))$ be the solutions of (2.5) and (3.9) respectively. We have

$$\begin{aligned} &\|\underline{\beta}(t)-\underline{\beta}_h(t)\|_1 + \|w(t)-w_h(t)\|_1 + t\|\underline{\gamma}(t)-\underline{\gamma}_h(t)\|_0 \\ &\leq c|h|\ (\|\underline{\beta}(t)\|_3 + \|\underline{\gamma}(t)\|_2) \end{aligned} \tag{3.15}$$

with c independent of h and t, and $\underline{\gamma}(t)$, $\underline{\gamma}_h(t)$ given in (2.5) and (3.9).

Proof. Let $\underline{\beta}_I \in (Q_h)^2$ be such that

$$\|\underline{\beta} - \underline{\beta}_I\|_1 \leq c|h|\ \|\underline{\beta}\|_3 \tag{3.16}$$

Other requirements on $\underline{\beta}_I$ will be given later on. Using the Korn's inequality (1.9) we have

$$\begin{aligned} c\|\underline{\beta}-\underline{\beta}_h\|_1^2 &\leq a(\underline{\beta}-\underline{\beta}_h, \underline{\beta}-\underline{\beta}_h) \\ &= a(\underline{\beta}-\underline{\beta}_h, \underline{\beta}-\underline{\beta}_I) + a(\underline{\beta}-\underline{\beta}_h, \underline{\beta}_I-\underline{\beta}_h) \\ &\leq c|h|\ \|\underline{\beta}\|_3 + a(\underline{\beta}-\underline{\beta}_h, \underline{\beta}_I-\underline{\beta}_h) \end{aligned} \tag{3.17}$$

On the other hand, from (2.5) and (3.9) we obtain

$$\begin{aligned} a(\underline{\beta}-\underline{\beta}_h, \underline{\beta}_I-\underline{\beta}_h) &= \lambda(\underline{\gamma}, \underline{\beta}_I-\underline{\beta}_h) - \lambda(\underline{\gamma}_h, \overset{*}{\underline{\beta}}_I-\overset{*}{\underline{\beta}}_h) \\ &\leq \lambda(\underline{\gamma}-\underline{\gamma}_h, \overset{*}{\underline{\beta}}_I-\overset{*}{\underline{\beta}}_h) + \|\underline{\gamma}\|_0\ c|h|\ \|\underline{\beta}_I-\underline{\beta}_h\|_1 \end{aligned} \tag{3.18}$$

where we have used (3.14) for $\underline{\eta}_h = \underline{\beta}_I - \underline{\beta}_h$. Now we have

$$\begin{aligned} \lambda(\underline{\gamma}-\underline{\gamma}_h, \overset{*}{\underline{\beta}}_I-\overset{*}{\underline{\beta}}_h) &= \lambda t^2(\underline{\gamma}-\underline{\gamma}_h, \underline{\gamma}_h-\underline{\gamma}) - \lambda(\underline{\gamma}-\underline{\gamma}_h, t^2\underline{\gamma}_h + \overset{*}{\underline{\beta}}_h) \\ &\quad + \lambda(\underline{\gamma}-\underline{\gamma}_h, t^2\underline{\gamma} + \overset{*}{\underline{\beta}}_I) \end{aligned} \tag{3.19}$$

Note now that from (2.5) and (3.9) we have

$$(\underline{\gamma}-\underline{\gamma}_h, \underline{\nabla\zeta}_h) = 0 \qquad \forall\ \zeta_h \in Q_h \tag{3.20}$$

so that, since $t^2\underline{\gamma}_h + \overset{*}{\underline{\beta}}_h = \underline{\nabla w}_h$, (3.19) becomes

$$\lambda(\underline{\gamma}-\underline{\gamma}_h, \overset{*}{\underline{\beta}}_I-\underline{\beta}_h) = -\lambda t^2 \|\underline{\gamma}-\underline{\gamma}_h\|_0^2 + \lambda(\underline{\gamma}-\underline{\gamma}_h, t^2\underline{\gamma}+\overset{*}{\underline{\beta}}_I)$$

Combining (3.17), (3.18), (3.20) we have

$$c\|\underline{\beta}-\underline{\beta}_h\|_1^2 + \lambda t^2\|\underline{\gamma}-\underline{\gamma}_h\|_0^2 \leq c|h| \, (\|\underline{\beta}\|_3 \;\; \|\underline{\beta}-\underline{\beta}_h\|_1 + \|\underline{\gamma}\|_0 \;\; \|\underline{\beta}_I-\underline{\beta}_h\|_1) + \lambda(\underline{\gamma}-\underline{\gamma}_h, t^2\underline{\gamma}+\overset{*}{\underline{\beta}}_I) \tag{3.21}$$

Here we need more help from $\overset{*}{\underline{\beta}}_I$. More precisely we require that

$$\operatorname{rot}\overset{*}{\underline{\beta}}_I\Big|_R = -\frac{1}{|R|}\int_R t^2 \operatorname{rot}\underline{\gamma} = \frac{-1}{|R|}\int_R \operatorname{rot}\underline{\beta} \tag{3.22}$$

for all R in T_h. The existence of a $\underline{\beta}_I$ satisfying both (3.16) and (3.22) is proved in lemma 3.3 below. Next we choose $\underline{q} \in (H_0^1(\Omega))^2$ such that

$$\operatorname{rot}\underline{q}\,\Big|_R = t^2 \operatorname{rot}\underline{\gamma} + \operatorname{rot}\overset{*}{\underline{\beta}}_I \tag{3.23}$$

It follows from (3.22) that $\underline{q}$ may be chosen such that

$$\|\underline{q}\|_1 \leq c|h|t^2 \, \|\operatorname{rot}\underline{\gamma}\|_1 \tag{3.24}$$

(see (2.19), (2.20) for a similar argument, plus the standard bound $\|\operatorname{rot}\underline{\gamma} - \frac{1}{R}\int_R \operatorname{rot}\underline{\gamma}\|_0 \leq c|h| \, \|\operatorname{rot}\underline{\gamma}\|_1$). Hence $t^2\underline{\gamma}+\overset{*}{\underline{\beta}}_I-\underline{q}$ has zero rotation, so that we may write

$$t^2\underline{\gamma} + \overset{*}{\underline{\beta}}_I - \underline{q} = \underline{\nabla\zeta} \tag{3.25}$$

for some $\zeta \in H_0^1(\Omega)$. Using (3.25) and (3.20) we have

$$\begin{aligned}\lambda(\underline{\gamma}-\underline{\gamma}_h, t^2\underline{\gamma}+\overset{*}{\underline{\beta}}_I) &= \lambda(\underline{\gamma}-\underline{\gamma}_h, \underline{q} + \underline{\nabla\zeta})\\ &= \lambda(\underline{\gamma}-\underline{\gamma}_h, \underline{q}) + \lambda(\underline{\gamma}-\underline{\gamma}_h, \underline{\nabla\zeta}-\underline{\nabla\zeta}^I)\end{aligned} \tag{3.26}$$

We have finally from (3.13) and (3.25)

$$\begin{aligned}\|\zeta-\zeta_I\|_{1,R} &\leq c|h| \, (\|\partial^2\zeta/\partial x^2\|_{0,R}^2 + \|\partial^2\zeta/\partial y^2\|_{0,R}^2)^{1/2}\\ &\leq c|h| \, (t^2\|\underline{\gamma}\|_1 + \|\underline{q}\|_1)\end{aligned} \tag{3.27}$$

and collecting (3.26), (3.24) and (3.27) we get

$$\lambda(\underline{\gamma}-\underline{\gamma}_h, t^2\underline{\gamma}-\overset{*}{\underline{\beta}}_I) \leq c|h|t^2 \, \|\underline{\gamma}-\underline{\gamma}_h\|_0 \;\; \|\underline{\gamma}\|_2 \tag{3.28}$$

Combining now (3.21) and (3.28) we obtain

$$||\underline{\beta}-\underline{\beta}_h||_1 + \lambda t^2 \, ||\underline{\gamma}-\underline{\gamma}_h||_0^2 \le c|h| \, \{||\underline{\beta}||_3 \, ||\underline{\beta}-\underline{\beta}_h||_1 + ||\underline{\gamma}||_0 \, ||\underline{\beta}_I-\underline{\beta}_h||_1 + t^2 ||\underline{\gamma}-\underline{\gamma}_h||_0 \, ||\underline{\gamma}||_2\} \tag{3.29}$$

and hence easily (for fixed λ)

$$||\underline{\beta}-\underline{\beta}_h||_1 + t||\underline{\gamma}-\underline{\gamma}_h||_0 \le c|h| \, (||\underline{\beta}||_3 + ||\underline{\gamma}||_2) \tag{3.30}$$

Finally

$$\underline{\nabla}w-\underline{\nabla}w_h = \underline{\beta}-\overset{*}{\underline{\beta}}_h + t^2\underline{\gamma} - t^2\underline{\gamma}_h \tag{3.31}$$

and (3.15) follows from (3.30), (3.31) using again (3.14), now with $\underline{\eta}_h = \underline{\beta}_h$. ±

The proof of theorem 3.1 used the existence of a function $\underline{\beta}_I$ satisfying (3.16) and (3.22). The proof of existence of $\underline{\beta}_I$ is the object of the following final lemma.

<u>Lemma 3.3</u> <u>For any</u> $\underline{\beta} \in (H^3 \cap H_0^1)^2$ <u>there exists</u> $\underline{\beta}_I \in (Q_h)^2$ <u>such that</u> (3.16) <u>and</u> (3.22) <u>are satisfied</u>.

<u>Proof</u>. Let us first set $\theta_1 := -\beta_2$ and $\theta_2 := \beta_1$ so that $\mathrm{rot}\underline{\beta} = \mathrm{div}\underline{\theta}$. Next consider the auxiliary problem

$$\left.\begin{aligned} -\Delta\underline{\bar{\theta}} + \underline{\nabla}p &= -\Delta\underline{\theta} \\ \mathrm{div}\underline{\bar{\theta}} &= \mathrm{div}\underline{\theta} \end{aligned}\right\} \tag{3.32}$$

which obviously has the unique solution $\underline{\bar{\theta}} = \underline{\theta}$ and $p = 0$. Next consider its finite element approximation

$$\begin{aligned} &\int \underline{\nabla\theta}_h \cdot \underline{\nabla\eta}_h - \int p_h \, \mathrm{div}\underline{\eta}_h = \int \underline{\nabla\theta}\cdot\underline{\nabla\eta}_h \quad \forall\, \underline{\eta}_h \in (Q_h)^2 \\ &\int_R \mathrm{div}\underline{\theta}_h = \int_R \mathrm{div}\underline{\theta} \quad \forall\, R \in \mathcal{T}_h \end{aligned} \tag{3.33}$$

where $\underline{\theta}_h$ is sought in $(Q_h)^2$ and p_h among piecewise constants. Following [8], [10] (and their notations) we obtain that (3.33) has a unique solution, which satisfies

$$||\pi p_h||_0 + h||p_h||_0 \le c \, ||\underline{\theta}-\underline{\theta}_h||_1 \tag{3.34}$$

where π is a projection operator that filters out the checker board modes by blocks of four elements. Hence we have, for all $\underline{\theta}_I \in (Q_h)^2$,

$$||\underline{\nabla\theta}-\underline{\nabla\theta}_h||_0^2 = \int(\underline{\nabla\theta}-\underline{\nabla\theta}_h)\cdot(\underline{\nabla\theta}-\underline{\nabla\theta}_I) + \int(\underline{\nabla\theta}-\underline{\nabla\theta}_h)\cdot(\underline{\nabla\theta}_I-\underline{\nabla\theta}_h) \tag{3.35}$$

On the other hand

$$\int(\underline{\nabla\theta}-\underline{\nabla\theta}_h)\cdot(\underline{\nabla\theta}_I-\underline{\nabla\theta}_h) = \int p_h \operatorname{div}(\underline{\theta}_I-\underline{\theta}_h) \tag{3.36}$$

$$= \int p_h \operatorname{div}(\underline{\theta}_I-\underline{\theta}) = \int \pi p_h \operatorname{div}(\underline{\theta}_I-\underline{\theta}) + \int (I-\pi)p_h \operatorname{div}(\underline{\theta}_I-\underline{\theta})$$

If we choose now $\underline{\theta}_I$ to be the interpolant of $\underline{\theta}$ by blocks of four elements ([8]) we have $\int (I-\pi p_h) \operatorname{div}\underline{\theta}_I=0$. Then

$$\int \pi p_h \operatorname{div}(\underline{\theta}_I-\underline{\theta}) \leq c||\underline{\theta}-\underline{\theta}_h||_1 \; ||\underline{\theta}_I-\underline{\theta}||_1 \tag{3.37}$$

$$\int (I-\pi)p_h \operatorname{div}\underline{\theta} \leq c||p_h||_0 \; |h|^2 \; ||\operatorname{div}\underline{\theta}||_2 \tag{3.38}$$

because of the shape of $(I-\pi)p_h$. Combining (3.35)-(3.38) we have

$$||\underline{\theta}-\underline{\theta}_h||_1 \leq c|h| \; ||\underline{\theta}||_3 \tag{3.39}$$

Now we rotate back, setting

$$\beta_{I,1} := \theta_{h,2} \; , \; \beta_{I,2} = -\theta_{h,1} \tag{3.40}$$

Therefore (3.39) implies (3.16) and (3.33) with (3.12) gives (3.22).

Remark. In the proof of theorem 3.1, we discard at several points, information (for instance an estimate of order $|h|t$ would be enough in (3.28) instead of the $|h|t^2$). This is due to two reasons. Firstly, our estimate is not optimal. The optimal estimate to be expected should be, for instance

$$||\underline{\beta}-\underline{\beta}_h||_1 + ||w-w_h||_1 + ||\underline{\gamma}-\underline{\gamma}_h||_\Gamma \leq c|h|$$

for $\underline{\beta}$, w, $\underline{\gamma}$ smooth enough. Secondly, since the estimate is already non-optimal, we did not endeavor to reduce the regularity required on $\underline{\beta}$ and $\underline{\gamma}$. An improvement of (3.15) in this respect should also be possible.

Remark. We used the assumption that the mesh T_h is uniform only in proof of lemma 3.3 (namely in (3.38)). Actually a general rectangular mesh can be allowed, provided that each rectangle is then split into sixteen equal subrectangles, see [10] for more details.

4. CONCLUSIONS

We analyzed from the mathematical point of view the finite element discretization proposed in [3] for Mindlin plates. At least for particular cases (like a uniform rectangular mesh) we proved that the element is uniformly stable with respect to the thickness parameter t and that it converges with optimal rate $O(|h|)$ in H^1, uniformly in t. We did not prove uniform stability of the "shear strains" $\underline{\gamma}_h = t^{-2}(\underline{\nabla} w_h - \underline{\beta}_h^*)$ nor uniform convergence. Actually, we have $||\underline{\gamma}-\underline{\gamma}_h||_0 \leq c|h|t^{-1}$ which is basically unsatisfactory. Probably a filtering procedure should be applied to $\underline{\gamma}_h$ in order to have L^2 stability and an optimal rate of convergence uniformly in t.

REFERENCES

[1] D.N. ARNOLD, *Num. Math.*, 37, 405-421 (1981).

[2] K.J. BATHE, *Finite Element Procedures in Engineering Analysis*. Prentice-Hall, Englewood Cliffs, New Jersey (1982).

[3] K.J. BATHE and E.N. DVORKIN, A Four-Node Plate Bending Element Based on Mindlin/Reissner Plate Theory and a Mixed Interpolation, *Int. J. Num. Meth. in Eng.*, in press.

[4] P.G. CIARLET, *The Finite Element Method for Elliptic Problems*. North Holland, Amsterdam (1978).

[5] P. DESTUYNDER, *Thèse d'état*. Paris (1980).

[6] G. DUVAUT and J.L. LIONS, *LES INÉQUATIONS EN MECHANIQUE ET EN PHYSIQUE*. Dunod, Paris (1972).

[7] E.N. DVORKIN and K.J. BATHE, A Continuum Mechanics Based Four-Node Shell Element for General Nonlinear Analysis, *Engineering Computations*, 1, 77-88 (1984).

[8] C. JOHNSON and J. PITKÄRANTA, *Math. Comp.* 38, 375-400 (1982).

[9] O. LADYSHENSKAYA, *The Mathematical Theory of Viscous Incompressible Flow*, Gordon and Breach, New York (1969).

[10] J. PITKÄRANTA and R. STENBERG, Error Bounds for the Approximation of the Stokes Problems using Bilinear/Constant Elements on Irregular Quadrilateral Meshes, *Rep. Mat. A222*, Helsinki Univ., (1984).

[11] G. STRANG and G. FIX, *An Analysis of the Finite Element Method*. Prentice-Hall, Englewood Cliffs, NJ (1973).

FORMULATION AND FINITE ELEMENT ANALYSIS OF A GENERAL CLASS OF ROLLING CONTACT PROBLEMS WITH FINITE ELASTIC DEFORMATIONS

J. Tinsley Oden

and

E. B. Becker, T. L. Lin, and L. Demkowicz

The University of Texas at Austin, Austin, Texas U.S.A.

1. INTRODUCTION

The rolling contact problem in solid mechanics involves the description of the behavior of a deformable cylindrical body rolling at steady state while in contact with another body. The great importance of this problem in engineering is clear, as rolling contact arises naturally in the analysis of pneumatic tires, train tires, gears, bearings, and many mechanisms and machine parts. Contact problems involving finite deformations may be encountered in the study of rubber rollers in printers and presses, pneumatic tires, elasometric gears, seals, and gaskets.

Despite its importance, surprisingly few papers can be found on the general problem of rolling contact with finite elastic deformations. The bulk of the literature on rolling contact focuses on the very special case of infinitesimal deformations and generally makes use of the assumption that the contact pressures are Hertzian (i.e., determined by the classical Hertz solution). For a survey of literature and methods for problems of this type, see Kalker [1,2]. Batra [3,4] has considered static contact problems in finite elasticity and Padavon and Zeid [5,6] studied formulations of the general rolling contact problem and gave some numerical solutions to cases involving linearly elastic materials and frictionless contact. A rather complete account of the subject of contact problems in elasticity and numerical methods for solving them is given in the forthcoming treatise of Kikuchi and Oden [7].

In the present paper, a general variational principle is presented which governs the problem of finite elastic deformations of a cylindrical body rolling at a prescribed

ISBN 0-12-747255-X

angular velocity ω on a rough surface and subjected to internal pressures, axle loads, and frictional forces. The problem class includes models of the motion of rubber pneumatic tires. The variational principle features a highly nonlinear variational inequality which is characterized by non-convex energy potentials, non-conservative pressure loadings, non-differentiable frictional terms, unilateral constraints, possibly multiple solutions, and nonlinear resonance phenomena at critical angular velocities. A regularized form of this principle is also derived, in which an exterior penalty method is used to accommodate the unilateral constraint and a regularization of the friction terms is introduced to produce a differentiable approximation of friction effects. These regularizations lead to a nonlinear variational equality-characterization of the rolling contact problem.

Following our discussion of these new variational principles, we describe a special finite-element approximation of the regularized problem and details of algorithms for handling the friction, inertia, unilateral constraint terms as well as constraints arising from possible material incompressibility. In this part of our study, we confine ourselves to two-dimensional cases. We then describe algorithms for solving the systems of highly nonlinear equations that arise in this analysis. These algorithms are of the Riks-Keller type [8,9]; they are continuation (arc length) methods which advance solutions along tangents to solution paths and which feature Newton-Raphson iterations to keep the approximate solution on or near paths. Variants of these methods, such as those due to Crisfield [10], which have special features designed to accelerate convergence rates can also be used. These methods are capable of handling turning points, bifurcations, and multiple branching. We conclude the study with a discussion of numerical results obtained by analyzing several representative examples.

2. KINEMATICS OF ROLLING CONTACT

The physical situation we wish to consider is as follows: a deformable cylinder is rolling on a foundation at a constant velocity relative to the foundation and is in a state of steady motion. The cylinder can be attached to a rigid cylindrical axle of radius R_1, and this axle spins at a constant angular acceleration ω. An observer riding on the axle will see the same geometry of the deformed cylinder at all times t, as indicated in Fig. 1. The situation is also equivalent to a cylinder spinning about a fixed axis 0 and in contact with a moving belt, as indicated in the figure. Throughout, we shall place no limitations on magnitudes of deformation measures.

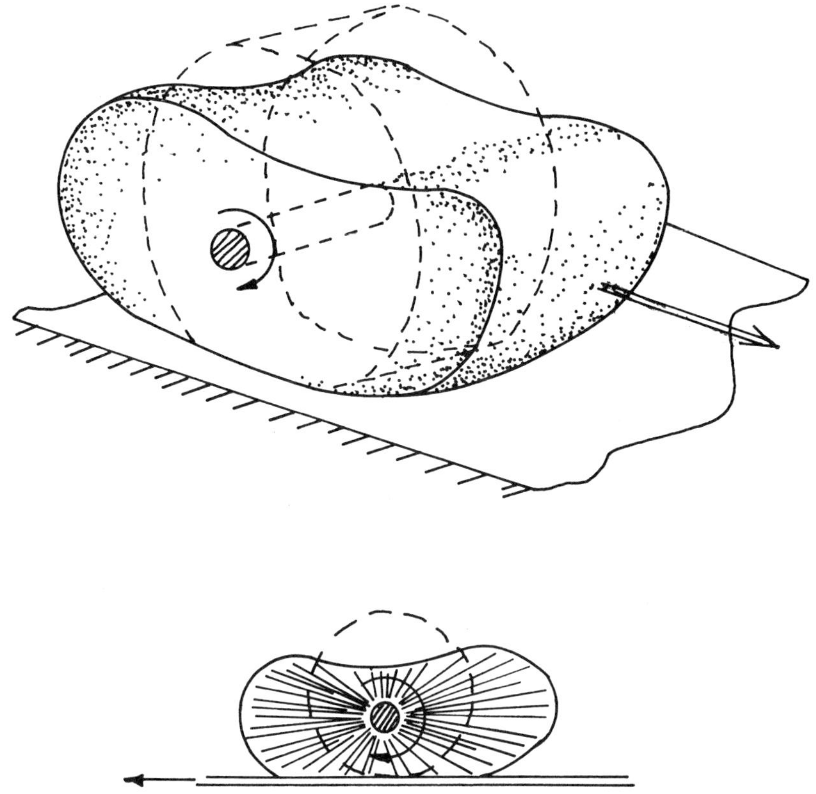

FIG. 1. A general class of rolling contact problems: a) a highly deformable cylinder rolls at a constant velocity over a rigid foundation or, b) a fixed deformable wheel spins about its axis as it contacts a belt moving relative to that axis at a constant velocity.

The description of deformation of a body involves the comparision of geometries in two or more of its configurations. Since the rolling cylinder is available to us only in its current deformed state, it is natural to compare this geometry with that of a rigid cylinder spinning about its axis at a constant angular velocity ω. Towards this end, we establish the following coordinate frames:

R, Θ, Z = material coordinates of a particle X in the rigid spinning cylinder; these are particle labels which coincide with the cylindrical-polar coordinates of X at a time $\tau = 0$.

r, θ, z = reference coordinates; these are the coordinates of particles in the rigid spinning cylinder at an arbitrary time $t \geq 0$, and are related to the material coordinates as follows:

$$r = R, \ \theta = \Theta + \omega t \ , \ z = Z \tag{1}$$

x_1, x_2, x_3 = spatial coordinates; these are the Cartesian coordinates of points in space relative to a moving coordinate frame with origin fixed on the axes of the moving cyclinder, x_1 parallel to the foundation, x_3 directed along the axes of the cylinder, and x_2 normal to the foundation.

These coordinates and the geometry of the rigid cylinder are shown in Fig. 2a. The reference coordinates are thus dependent on time t, but this dependence is only formal since the shape of the deformed cylinder is the same at all times.

The motion of the cylinder is defined by a differentiable and invertible map $\underset{\sim}{\chi}$ that takes the reference configuration C_t of the cylinder at time t (the places occupied by particles of the rigid cylinder at time t) onto the current (deformed) configuration at the same time t. In particular the motion carries points (r, θ, z) in the reference configuration to positions x_i, $i=1,2,3$, relative to the spatial frame of reference, given by

$$x_i = \chi_i(r, \theta, z) \tag{2}$$

Thus, time enters the description of motion only implicitly as $\theta = \Theta + \omega t$.

Various kinematical quantities can be computed in terms of the motion $\underset{\sim}{\chi}$. The deformation gradient tensor $\underset{\sim}{F}$ is given by

$$\underset{\sim}{F} = \nabla \underset{\sim}{\chi}$$

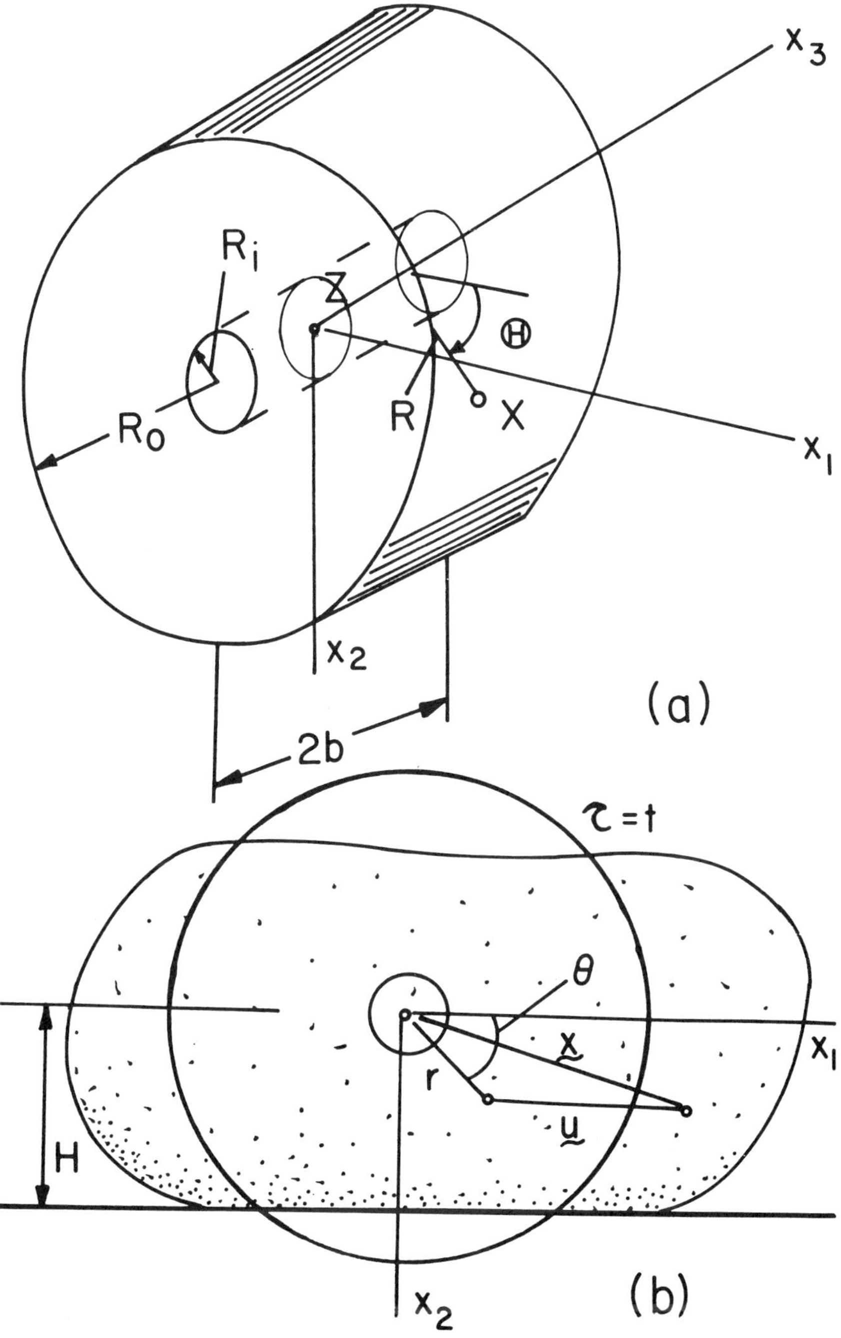

Fig. 2. Geometry of a) rigid and b) deformed cylinder.

or

$$F^{i}_{,j}(r, \theta, z) = \left[\partial \chi_i/\partial r, \ \partial \chi_i/\partial \theta, \ \partial \chi_i/\partial z\right] \tag{3}$$

and the right- and left- Cauchy-Green deformation tensors $\underset{\sim}{C}$ and $\underset{\sim}{B}$ are given, respectively, by

$$\underset{\sim}{C} = \underset{\sim}{F}^T \underset{\sim}{F} \ , \ \underset{\sim}{B} = \underset{\sim}{F} \underset{\sim}{F}^T \tag{4}$$

If $\underset{\sim}{r}$ denotes the position vector of the particle with reference coordinates (r, θ, z) at time t, the vector

$$\underset{\sim}{u} = \underset{\sim}{\chi}(r, \theta, z) - \underset{\sim}{r} \tag{5}$$

is the displacement vector of this particle. The reference velocity vector is given by

$$v_i = \frac{\partial \chi_i(r, \theta, z)}{\partial t} = \omega \frac{\partial \chi_i}{\partial \theta} \tag{6}$$

and the acceleration vector is given by

$$a_i = \frac{\partial v_i(r,\theta,z)}{\partial t} = \omega^2 \frac{\partial^2 \chi_i}{\partial \theta^2} \tag{7}$$

Note that since the reference coordinates are time dependent,

$$\frac{\partial u_i}{\partial t} = v_i - \frac{\partial r_i}{\partial t} = \omega\left(\frac{\partial \chi_i}{\partial \theta} - \frac{\partial r_i}{\partial \theta}\right) \tag{8}$$

where r_i are the Cartesian components of $\underset{\sim}{r}$.

3. A BOUNDARY-VALUE PROBLEM FOR ROLLING CONTACT

We shall now establish equations and inequalities governing a class of rolling contact problems for highly deformable materials. We shall adopt a referential ("material") description of the problem in the sense that quantities (vectors, tensors, densities, etc.) are referred to the reference configuration and are functions of the referential coordinates (r, θ, z) . The behavior of the general body in rolling contact is governed by the following conditions:

1. *Linear Momentum*

Linear momentum is balanced at each particle if and only if,

$$\text{Div } \underset{\sim}{S} + \rho_0 \underset{\sim}{b} = \rho_0 \ddot{\underset{\sim}{x}} \tag{9}$$

where Div denotes the divergence operator referred to the reference configuration, superimposed dots indicated time rates, and

$\underset{\sim}{S}$ = the Piola-Kirchhoff stress tensor
ρ_0 = the mass density in the reference configuration
$\underset{\sim}{b}$ = the body force per unit mass in the reference configuration
$\ddot{\underset{\sim}{x}}$ = the acceleration vector $\left[\ddot{x}_i = \omega^2 \frac{\partial^2 \chi_i}{\partial \theta^2}\text{, by (7)}\right]$

The balance of angular momentum requires that $\underset{\sim}{S}\,\underset{\sim}{F}^T = \underset{\sim}{F}\,\underset{\sim}{S}^T$, so that $\underset{\sim}{S}$ is, in general, not symmetric.

2. *Constitutive Equations*

While it is not necessary at this point to identify the constitution of the material of which the cylinder is composed, we generally have in mind an elastic material for which the Cauchy stress $\underset{\sim}{T}$ is given by

$$\left.\begin{aligned} \underset{\sim}{T} &= \underset{\sim}{F}(\underset{\sim}{C}) = \underset{\sim}{F}(\nabla \underset{\sim}{x}^T \nabla \underset{\sim}{x}) \\ &\text{or} \\ \underset{\sim}{S} &= \det \underset{\sim}{F}\, \underset{\sim}{F}(\underset{\sim}{C}) \underset{\sim}{F}^{-T} \end{aligned}\right\} \tag{10}$$

where $\underset{\sim}{F}$ is a smooth response functional.

In most of the applications, we assume that the material is an isotropic hyperelastic material characterized by a stored energy function W, representing the strain energy per unit volume in the reference configuration, which is given as a differential function of the principal invariants of the deformation tensor $\underset{\sim}{C}$:

$$W = W\,(I, II, III) \tag{11}$$

$$\left.\begin{aligned} I &= I_{\underset{\sim}{C}} = \operatorname{tr} \underset{\sim}{C} \qquad (\text{tr} = \text{trace}) \\ II &= II_{\underset{\sim}{C}} = \frac{1}{2}(\operatorname{tr} \underset{\sim}{C})^2 - \frac{1}{2} \operatorname{tr} \underset{\sim}{C}^2 \\ III &= III_{\underset{\sim}{C}} = \det \underset{\sim}{C} = \det \underset{\sim}{F}^T \underset{\sim}{F} = (\det \underset{\sim}{F})^2 \end{aligned}\right\} \tag{12}$$

In this case,

$$\underset{\sim}{S} = \frac{\partial W}{\partial \nabla \underset{\sim}{\chi}} = \frac{\partial W}{\partial \underset{\sim}{F}} \tag{13}$$

or

$$\underset{\sim}{S} = \frac{\partial W}{\partial I} \cdot \frac{\partial I}{\partial \underset{\sim}{\nabla} \underset{\sim}{\chi}} + \frac{\partial W}{\partial II} \cdot \frac{\partial II}{\partial \underset{\sim}{\nabla} \underset{\sim}{\chi}} + \frac{\partial W}{\partial III} \cdot \frac{\partial III}{\partial \underset{\sim}{\nabla} \underset{\sim}{\chi}} \tag{14}$$

If the material is incompressible (which is often assumed to be the case when analyzing many rubber-like materials), all motions are subject to the incompressibility constraint,

$$\det \underset{\sim}{F} = 1 \tag{15}$$

Then W reduces to a function of only I, II and determines $\underset{\sim}{S}$ only to within an arbitrary hydrostatic pressure p:

$$W = \hat{W}\,(I, II) \; ; \; \underset{\sim}{S} = \frac{\partial \hat{W}}{\partial \underset{\sim}{\nabla} \underset{\sim}{\chi}} + p\,\frac{\partial III}{\partial \underset{\sim}{\nabla} \underset{\sim}{\chi}} \tag{16}$$

The possibility of non-homogeneous materials presents no special difficulties: we can assume that W is also an explicit function of (r, Ⓗ, z) and that its "form" (functional dependence on the invariants) can change from point to point.

3. *Unilateral Contact Condition*

The unilateral constraint or contact condition, expressing the fact that the motion of the cylinder in the x_2-direction is constrained by the presence of the rigid roadway is simply

$$\chi_2(r, \theta, z) \le H \qquad (r, \theta, z) \in \Gamma \tag{17}$$

where H is the distance from the axis of the deformed cylinder to the foundation (Fig. 2b) and Γ is the exterior surface of the cylinder. If $\underset{\sim}{\sigma}$ denotes the stress vector,

$$\underset{\sim}{\sigma} = \underset{\sim}{S}\,\underset{\sim}{n} \tag{18}$$

where $\underset{\sim}{n}$ is a unit exterior normal to the surface of the rigid spinning reference cylinder, then we also must have on Γ,

$$\chi_2 < H \Rightarrow \underset{\sim}{\sigma} = \underset{\sim}{0} \; , \quad \chi_2 = H \Rightarrow \underset{\sim}{\sigma} \cdot \underset{\sim}{i}_2 \le 0 \tag{19}$$

where $\underset{\sim}{i}_i$ is a set of orthonormal basis vectors tangent to x_i.

4. *Friction/Slip Conditions*

If the cylinder rolls relative to the foundation, friction is developed on the contact surface. We shall first describe effects with a classical model, although more general friction laws can be easily accommodated. Let

v_0 = the magnitude of the velocity of foundation (roadway or belt) relative to the axis of the cylinder (with no slipping, $v_0=R_0\omega$ for rigid bodies).

σ_n = $\underset{\sim}{\sigma} \cdot \underset{\sim}{i}_2$ = contact pressure = the normal component of the stress vector $\underset{\sim}{\sigma}$ on the contact surface

$\underset{\sim}{\sigma}_T$ = the tangential component of $\underset{\sim}{\sigma}$ on Γ .

$\underset{\sim}{w}_T$ = the slip velocity, defined by

$$\underset{\sim}{w}_T = \dot{\underset{\sim}{\chi}} - v_0 \underset{\sim}{i}_1 \tag{20}$$

The slip-stick conditions on the contact surface are then, everywhere on Γ ,

$$\left.\begin{aligned}
&|\underset{\sim}{\sigma}_T| < \nu\, |\sigma_n| \Rightarrow \underset{\sim}{w}_T = \underset{\sim}{0} \\
&\qquad\qquad [\text{or, equivalently,} \\
&\omega \frac{\partial \chi_1}{\partial \theta} - v_0 = 0 \text{ and} \\
&\frac{\partial \chi_3}{\partial \theta} = 0] \\
&|\underset{\sim}{\sigma}_T| = \nu |\sigma_n| \Rightarrow \text{there exists } \lambda \in \mathbb{R}, \\
&\qquad\qquad \lambda > 0, \text{ such that} \\
&\underset{\sim}{w}_T = -\lambda \underset{\sim}{\sigma}_T
\end{aligned}\right\} \tag{21}$$

Here ν is the coefficient of friction. It can be demonstrated that the above conditions are equivalent to

$$\underset{\sim}{\sigma}_T \cdot \underset{\sim}{w}_T + \nu|\sigma_n|\; |\underset{\sim}{w}_T| = 0 \tag{22}$$

5. *Other Boundary Conditions*

If the cylinder is fixed to the rigid spinning axle of radius R_i, then

$$\chi_i(R_i, \theta, z) = (R_i, \theta, z) \text{ on } \Gamma_0 \tag{23}$$

where

$$\Gamma_0 = \{(r, \theta, z)|\; r = R_i,\; |z| \le b\}. \tag{24}$$

Alternatively, the cylinder could be hollow with the interior surface Γ subjected to a prescribed ("inflation") pressure p_0. Then, if $\underset{\sim}{n}$ denotes a unit normal to the material surface Γ_0 in the current (deformed) configuration, we must have the traction condition

$$\underset{\sim}{S} \cdot \underset{\sim}{n}_0 = - p_0(\det \underset{\sim}{F})\underset{\sim}{F}^{-T}\underset{\sim}{n}_0 \qquad \text{on } \Gamma_0 \tag{25}$$

where $\underset{\sim}{n}_0$ is a unit vector normal to Γ_0 in the reference configuration.

The five sets of conditions above characterize the general rolling contact problem: If $\underset{\sim}{S}$ is given as a function of $\nabla\chi$, we seek a motion χ that simultaneously satisfies all of these conditions for given ρ_0, $\underset{\sim}{b}$, ω, v_0, and H.

4. A VARIATIONAL PRINCIPLE

We shall now set the stage for the development of a variational principle governing the general class of rolling contact problems described in the preceding section. Different sets of trial and test functions are called for: the set V of admissible motions,

$$V = \{\underset{\sim}{\chi} = \underset{\sim}{\chi}(r, \theta, z) \;\Big|\; \underset{\sim}{\chi} - \underset{\sim}{r} = 0 \quad \text{on} \quad \Gamma_0\} \tag{26}$$

the set K of admissible motion with constraints and the set $\overline{K}$ of mixed velocity-motions,

$$K = \{\underset{\sim}{\chi} \in V \;\Big|\; \chi^2 \leq H \text{ on } \Gamma_0\}$$

$$\overline{K} = \{\underset{\sim}{m} = (\omega \frac{\partial\chi_1}{\partial\theta}, \chi_2, \omega \frac{\partial\chi_3}{\partial\theta}) \text{ for some } \underset{\sim}{\chi} \in K \;\Big|\; m_1 = -\omega m_2,\ m_3 = 0 \quad \text{on} \quad \Gamma_0\} \tag{27}$$

In the case of incompressible materials, we add to the definition of K and $\overline{K}$ the incompressibility constraint,

$$\det \nabla\underset{\sim}{\underset{\sim}{\chi}} - 1 = 0 \qquad \text{a.e. in } \Omega \tag{28}$$

In general, V and K will be subsets of an appropriate space of functions, the character of which will depend upon the form of the constitutive equations. For example, let us define

$$\underset{\sim}{S}(\nabla\underset{\sim}{\chi}) \equiv (\det \underset{\sim}{F})\underset{\sim}{F}(\nabla\underset{\sim}{\chi}^T\nabla\underset{\sim}{\chi})\underset{\sim}{F}^{-T}$$

Then, the condition,

$$\underset{\sim}{\chi} \in K : \left|\int_\Omega \underset{\sim}{S}(\nabla\underset{\sim}{\chi}) : \nabla\underset{\sim}{\chi} dV_0\right| < \infty$$

where Ω is the interior of the rigid reference cylinder and dV_0 is an element of volume in Ω, will, in many important cases, imply that

$$K \subset (W^{1,p}(\Omega))^3, \quad p \geq 2$$

where $W^{1,p}(\Omega)$ is the Sobolev space of order (1,p); i.e., the space of integrable functions defined on Ω distributional derivatives of first order in $L^p(\Omega)$. However, we will then have

$$\overline{K} \subset W^{2,p}(\Omega) \times W^{1,p}(\Omega) \times W^{2,p}(\Omega)$$

since additional L^p-differentiability in the x_1- and x_3-directions is required of functions in $\overline{K}$. In other words, the *necessity of anisotropic function spaces, containing functions with different orders of differentiability in different directions, appears to arise naturally in the class of rolling contact problems under consideration.*

We next introduce a collection of nonlinear forms on V x K:

$$\left.\begin{aligned}
&A : K \times \overline{K} \to \mathbb{R} \\
&A(\underset{\sim}{\chi}, \underset{\sim}{v}) = \int_\Omega \underset{\sim}{S}(\nabla\underset{\sim}{\chi}) : \nabla\underset{\sim}{v}\, dV_0 \\
&j : K \times \overline{K} \to \mathbb{R} \\
&j(\underset{\sim}{\chi}, \underset{\sim}{v}) = \int_\Gamma \nu|\underset{\sim}{S}(\nabla\underset{\sim}{\chi})\underset{\sim}{n}_0 \cdot \underset{\sim}{i}_2||\underset{\sim}{v}_T - v_0\underset{\sim}{i}_1|dA_0 \\
&B : K \times \overline{K} \to \mathbb{R} \\
&B(\underset{\sim}{\chi}, \underset{\sim}{v}) = \int_\Omega \rho_0 \frac{\partial\chi_i}{\partial\theta} \cdot \frac{\partial v_i}{\partial\theta}\, dV_0
\end{aligned}\right\} \quad (29)$$

We also introduce the functional,

$$f : \overline{K} \to \mathbb{R}, \; f(\underset{\sim}{\nu}) = \int_\Omega \rho_0\underset{\sim}{b} \cdot \underset{\sim}{\nu}\, dV_0 \quad (30)$$

The functionals in (29) represent virtual work (or power) contributions: $A(\cdot,\cdot)$ is the virtual work performed by the

Piola-Krichhoff stresses on the deformation gradients, $j(\cdot,\cdot)$ is a virtual work (power) contribution from frictional forces on the contact surfaces, $B(\cdot,\cdot)$ is the virtual power of inertia forces, and $f(\cdot)$ is the virtual work due to external forces. Note that the strain energy

$$U(\chi) = \int_0^1 A(s\chi, \chi)ds = \int_0^1 W(\nabla\chi)dV_0$$

is not generally a convex functional of χ and that $j(\cdot,\cdot)$ is not differentiable (in the Gâteaux sense).

The variational problem is now put forth as follows: Find $\chi \in K$ with $(\dot{\chi}_1=\omega\delta_\theta\chi_1,\ \dot{\chi}_2=\chi_2,\ \dot{\chi}_3=\omega\delta_\theta\chi_3,\ \dot{\chi}\in \bar{K})$ such that

$$\left.\begin{aligned} &A(\chi,\ v - \chi) + j(\chi,\ v) - j(\chi,\chi) \\ &\quad \geq \omega^2 B(\chi,\ v - \chi) + f(v - \chi) \\ &\qquad \forall\ v \in \bar{K} \end{aligned}\right\} \tag{31}$$

It remains only to show that this variational inequality is equivalent in some sense to the boundary-value problem of rolling contact described earlier. This is accomplished by standard agreements given, for example, in Duvant and Lions [11] and the details shall not be furnished here. The result is summarized in the following theorem:

Theorem 1. (Equivalence). Let the motion χ be a solution to the general rolling contact problem; i.e., χ satisfies (9) with S given as a function of $\nabla\chi$ by (10) and satisfies (17), (19), (22), and (23) [or (25)]. Then χ is a solution of the variational inequality (31). Conversely, let χ be any sufficiently smooth solution of (31). Then χ is also a solution of the general rolling contact problem given in Section 3. □

5. REGULARIZATION

We now consider a regularization of the variational problem which consists of three components:

1. *Regularization of the Friction Functional*

The friction functional $j(\cdot,\cdot)$ is replaced by a differentiable regularization $j_\varepsilon(\cdot,\cdot)$, depending on a real parameter $\varepsilon > 0$, which approximates $j(\cdot,\cdot)$ artibrarily closely as ε tends to zero.

If $w = \dot{\chi}_1 - v_0$ is the slip velocity and σ_T is the frictional stress on the contact surface, the power developed by σ_T on w is not differentiable at $w = 0$. Similar regularization techniques for friction problems have been used by Campos et al. [12], Oden and Pires [13, 14] and Martins and Oden [15]. We replace the friction law with a differentiable form, dependent on ε, of the form

$$|\underset{\sim}{\sigma}^*_T| = \phi(\varepsilon, w, \sigma_n)$$

$$= \begin{cases} -\nu|\sigma_n|w/\varepsilon & \text{if } |w| \leq \varepsilon \\ -\nu|\sigma_n|\operatorname{sgn}(w) & \text{if } |w| \geq \varepsilon \end{cases}$$

Then

$j(\underset{\sim}{\chi}, \underset{\sim}{v}) - j(\underset{\sim}{\chi}, \underset{\sim}{\chi})$ is replaced by

$$j_\varepsilon(\underset{\sim}{\chi}, \underset{\sim}{v}) = \int_{\Gamma_c} \nu|\sigma_n(\underset{\sim}{\chi})|\begin{Bmatrix} w/\varepsilon \\ w/|w| \end{Bmatrix} v_1 \, ds \tag{32}$$

where $\dot{\chi}^1 = -x^2 \dfrac{\partial\chi_1}{\partial X_1} + X_1 \dfrac{\partial\chi_1}{\partial X_2}$

2. *The Unilateral Constraint*

The unilateral constraint $\chi_2 - H \leq 0$ is represented by the equality

$$(\chi_2 - H)_+ = (\phi)_+ \quad \{(\phi)_+ = \max(0, \phi)\}$$

and this constraint is relaxed by the introduction to the energy the exterior penalty functional,

$$P_\delta(\underset{\sim}{\chi}) = \frac{1}{2\delta} \int_\Gamma (\chi_2 - H)_+^2 \, ds \tag{33}$$

where δ is an arbitrary positive number. The corresponding work term is

$$\langle P_\delta'(\underset{\sim}{\chi}), \underset{\sim}{v}\rangle = \delta^{-1} \int_\Gamma (\chi_2 - H)_+ v_2 \, ds \tag{34}$$

Physically, the quantity $\delta^{-1}(\chi_2 - H)_+$ represents a normal contact force acting on Γ. This can, in fact, be used in computation as an approximation of the normal stress component σ_n on the contact surface.

3. *Incompressibility*

The incompressibility constraint, det $\underset{\sim}{\nabla}\underset{\sim}{\chi} = 1$, is relaxed by introducing a penalty functional,

$$I_{\varepsilon_0}(\underset{\sim}{\chi}) = \frac{1}{2\varepsilon_0} \int_{\Omega} (\det \underset{\sim}{\nabla}\underset{\sim}{\chi}-1)^2 dV_0 \tag{35}$$

for arbitrary $\varepsilon_0 > 0$. The corresponding virtual work is

$$<I'_{\varepsilon_0}(\underset{\sim}{\chi}), \underset{\sim}{v}> = \varepsilon_0^{-1} \int_{\Omega_0} (\det \underset{\sim}{\nabla}\underset{\sim}{\chi}-1) \cdot \text{adj } \underset{\sim}{\nabla}\underset{\sim}{\chi} \cdot \underset{\sim}{\nabla}\underset{\sim}{v} dV_0 \tag{36}$$

where adj $\underset{\sim}{\nabla}\underset{\sim}{\chi}$ (the adjucate of $\underset{\sim}{\nabla}\underset{\sim}{\chi}$) is the transpose of cofactors of $\nabla\chi$.

The regularized form of the variational problem is as follows:

Find $\underset{\sim}{\chi}_{\varepsilon} \in V$ ($\underset{\sim}{\varepsilon} = (\varepsilon, \delta, \varepsilon_0)$ such that

$$\left. \begin{aligned} &A(\underset{\sim}{\chi}_{\varepsilon}, \underset{\sim}{v}) + j_{\varepsilon}(\underset{\sim}{\chi}_{\varepsilon}, \underset{\sim}{v}) + <P'_{\delta}(\underset{\sim}{\chi}_{\varepsilon}), \underset{\sim}{v}> \\ &\qquad + <I'_{\varepsilon_0}(\underset{\sim}{\chi}_{\varepsilon}), \underset{\sim}{v}> = f(\underset{\sim}{v}) \qquad \forall \underset{\sim}{v} \in V \end{aligned} \right\} \tag{37}$$

6. FINITE ELEMENT MODELS

6.1 The Discrete Problem

The finite element approximation of the relaxed variational boundary-value problem (37) begins with the introduction of a partition of Ω into a mesh of E finite elements and the representation of admissible motions over each element by polynomial shape functions $\psi_N(r, \theta, z)$. In this way, we construct a family of finite-dimensional supspaces $\{V^h\}$ of V, h being the mesh parameter, with

$$\left. \begin{aligned} V^h = \{&\underset{\sim}{\chi}^h = (\chi_i^h)_{1\le i\le 3} \in V | \\ &\chi_i^h \in C^0(\bar{\Omega}) \;;\; \chi_i^h|_{\Omega_e} \\ &= \sum_{N=1}^{N_e} \chi_i^N \psi_N(r, \theta, z) \;,\; 1 \le e \le N_e \} \end{aligned} \right\} \tag{38}$$

Here Ω_e is a typical element, χ_i^N is the value of χ^h at node N of element Ω_e, and N_e is the number of nodes in element Ω_e.

Formally, the discrete problem can be stated as follows:

Find $\underset{\sim}{\chi}_{\varepsilon}^h \in V^h$ such that

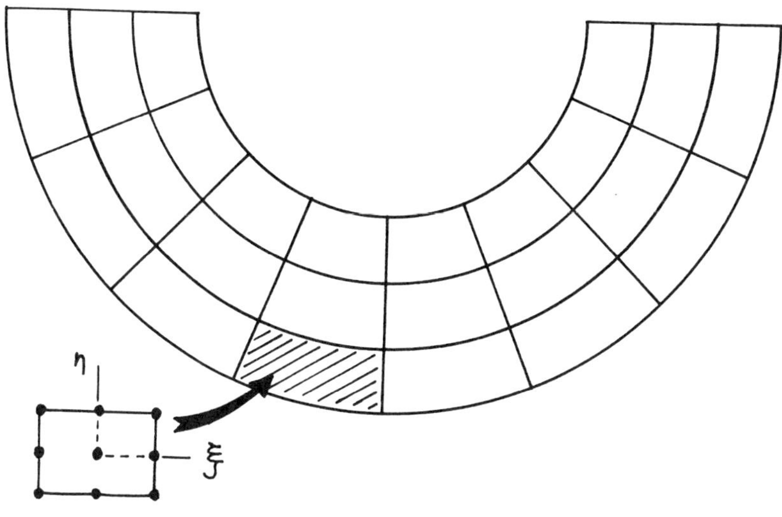

FIG. 3. A finite element mesh of 9-node biquadratic elements

$$A(\chi^h_\varepsilon, v^h) + j^h_\varepsilon(\chi^h_\varepsilon, v^h) + \langle P'_\delta(\chi^h_\varepsilon), v^h\rangle$$
$$+ \langle I'_{\varepsilon_0}(\chi^h_\varepsilon), v^h\rangle = f(v^h) \qquad \forall v^h \in V^h \tag{39}$$

6.2 *Elements and Stiffnesses*

In the present paper, we focus on two-dimensional approximations which employ 9-node biquadratic (Q_2) elements or eight-node isoparametric (I8) elements. All integrals in (39) are evaluated using Gaussian quadrature; e.g. for a mesh of E elements, we use

$$\left.\begin{aligned} A(\chi^h, v^h) &= \sum_{e=1}^{E} A^h_e(\chi^h, v^h) \\ A^h_e(\chi^h_\varepsilon, \psi_{N,i}) &= \sum_{j=1}^{G} W^e_j \frac{\partial W(\nabla\chi^h_\varepsilon)}{\partial \chi^N_k}(\xi^e_j)\psi_{N,i}(\xi^e_j) \\ j^h_\varepsilon(\chi^h_\varepsilon, \psi_N) &= \sum_{e=1}^{E'}\sum_{j=1}^{G'} Q^e_j \frac{\nu}{\delta}(\chi^h_\varepsilon{}^2 - H)_+ \begin{Bmatrix} w^h/\varepsilon \\ w^h/|w^h| \end{Bmatrix} \psi_N(\xi^e_j) \end{aligned}\right\} \tag{40}$$

etc. Here W^e_j and Q^e_j are the quadrature weights, ξ^e_j are the quadrature points (with W^e_j Gaussian quadrature weights and Q^e_j corresponding to Simpson's rule on the boundary). For Q_2-elements, we typically take G = 9 = 3 x 3. In the expression for j^h_ε, the normal stress σ_n is approximated by $\delta^{-1}(\chi^h_{\varepsilon_2} - H)_+(\xi^e_j)$ at the quadrature points. The bracket { } indicates that the upper entry is to be used when the slip velocity at a quadrature point is less than or equal to ε (corresponding to a non-slip situation) and the lower entry to a slip, for which this entry is greater than or equal to ε.

In the calculation of $\langle I'_{\varepsilon_0}(\chi^h_\varepsilon), v^h\rangle$, we typically use either "underintegration", by which we mean that a quadrature rule of order G = 2x2 is used or a perturbed Lagrangian element. The former corresponds to a local hydrostatic pressure approximation by discontinuous bilinear elements; the latter is the Q_2/P_1 element in which a Q_2-approximation of the motion χ_ε is used with a P_1 (piecewise linear) approximation of the hydrostatic pressure. The stability and convergence of this element for linear Stokes problems was established by Oden and Jacquotte [16]; the Q_2/P_1-element was first proposed and used in Oden [17] and was successfully applied to finite element approximations in finite elasticity by Miller [18].

7. ALGORITHMS

7.1 Introductory Comments

The finite element approximation of the regularized form (37) of the governing variational equality leads to a highly nonlinear system of algebraic equations of the form

$$F(\tilde{x}, \rho) = 0 \tag{41}$$

where

$\tilde{F}$ = an N-vector of nonlinear equilibrium (momentum) equations governing the discrete model

$\tilde{x} = (x_1, x_2, \ldots, x_N)^T$ = a vector of N degrees of freedom representing nodal values of the motion $\tilde{\chi}$.

ρ = a collection of critical parameters such as:

H - the distance of the axis of the cylinder to the foundation

ω^2 - the squared angular velocity

p_0 - the internal (inflation) pressure

In addition, the system will also depend upon the regularization and penalty parameters $\varepsilon, \delta, \varepsilon_0$. In general, the system may have multiple solutions for sufficiently large H and it may also exhibit reasonance-type behavior for certain critical frequencies ω^2. In the present investigation, we shall describe computational procedures for solving (41) for fixed ω^2 and p_0 which are built around a so-called continuation methodology in which ρ = H is treated as unknown and in which we attempt to trace families of solutions (x, ρ) of (41) along paths in N+1-dimensional χ-, ρ- space. Such paths are described by parametric equations of the type

$$\tilde{x} = \tilde{x}(s), \ \rho = \rho(s) \qquad s \in [0, 1] \tag{42}$$

where s is a real parameter, often regarded as an arc-length of solution paths of (42).

The general class of continuation methods to which our method belongs is known as a Riks or Riks-Wempner type method, in reference to early versions of these schemes proposed by Riks [8] (and Wempner). Among important features of these methods are that 1) they can accommodate decreases in load parameters, such as ρ, at points on solution paths (a feature which allows one to determine limit points and which is not generally present in traditional incremental loading schemes in which ρ is regarded as a parameter), and 2) they provide a convenient framework for the calculation of bifurcation points and multiple branches of solution paths. Several variants of these types of continuation schemes have been developed, and we mention in this regard the works of Keller [9], Rheinboldt [19,20], Crisfield [10, 21], Ramm [22], Padovan [5, 23] and Endo et al [24].

The methods employed here involve the solution of a nonlinear system of ordinary differential equations by corrector-predictor methods in which the predictor establishes the tangent hyperplane to the solution path and computes an approximate solution in it; the corrector modifies the solution by constrained Newton-Raphson or quasi-Newton schemes. Details are given in the following subsections.

7.2 A Riks-Keller-Type Scheme

Returning to (41), if (42) holds, then at every point on the solution path,

$$F_i(\underset{\sim}{x}(s), \rho(s)) = 0 \qquad s \in [0, 1] \tag{43}$$

$$\left.\begin{aligned} J_{ij}(\underset{\sim}{x}, \rho)\dot{x}_j + g_i(\underset{\sim}{x}, \rho)\,\dot{\rho} &= 0 \\ \dot{x}_j\,\dot{x}_j + \dot{\rho}^2 &= 1 \end{aligned}\right\} \tag{44}$$

where

$$\left.\begin{aligned} J_{ij}(\underset{\sim}{x}, \rho) &= \frac{\partial F_i(\underset{\sim}{x}, \rho)}{\partial x_j} \\ g_i(\underset{\sim}{x}, \rho) &= \frac{\partial g_i(\underset{\sim}{x}, \rho)}{\partial \rho} \\ \dot{x}_j &= \frac{dx_j}{ds} \\ \dot{\rho} &= \frac{d\rho}{ds} \end{aligned}\right\} \tag{45}$$

Equations (43) and (44) hold on the path $\Gamma = \{(x, \rho) |\ \underset{\sim}{x} = \underset{\sim}{x}(s), \rho = \rho(s), 0 \leq s \leq 1\}$ in N+1-dimensional space. Repeated indices are summed from 1 to N. The second member of (44) is, of course, the definition of arc length of Γ.

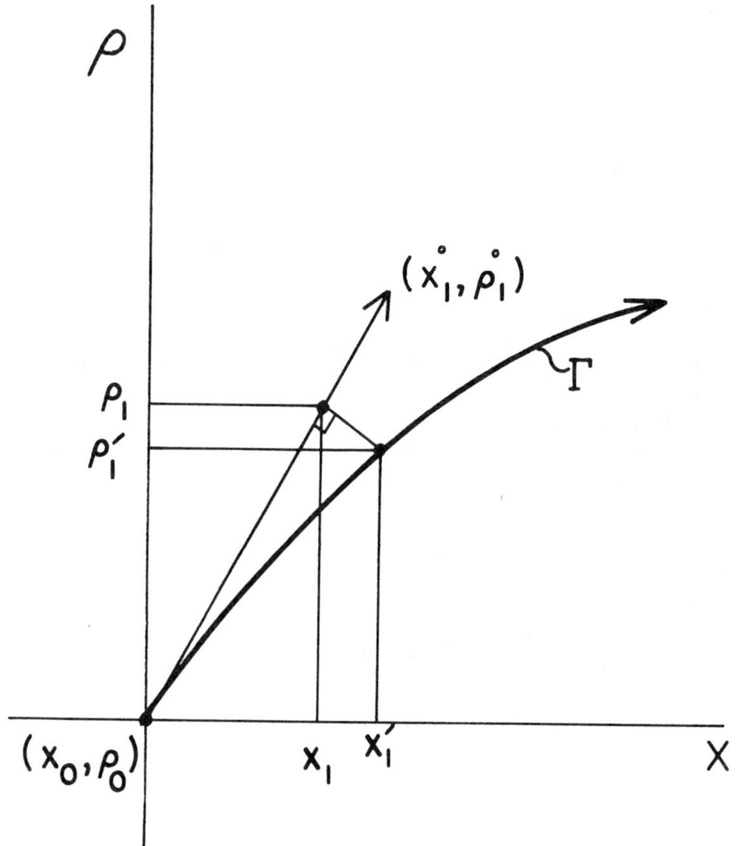

FIG. 4. Tangent plane to solution path

It is easily verified that (44) is equivalent to the system

$$J_{ij}(\underset{\sim}{x}, \rho)v_j = g_i(\underset{\sim}{x}, \rho)$$

$$\dot{\rho}^2 = \frac{1}{1+v_i v_i} \tag{46}$$

$$\dot{x}_j = -\dot{\rho}v_j$$

It is clear that if the linear system $(46)_1$ is solved for v_j, $\dot{\rho}$ and $\dot{x}_j$ are then determined immediately from $(46)_2$. Equations (46) (or (44)) are sufficient to determine directions $(\dot{x}_j, \dot{\rho})$ which define a hyperplane tangent to Γ. These can be used to define a linear extrapolation of the solution into this plane. It then makes sense to correct this approximate solution so that a point on the solution path Γ is obtained. One algorithm for such a procedure is given as follows:

Step 1 (tangent Hyperplane)

With initial data $\underset{\sim}{x}_0$, ρ_0, $(\Delta\rho)_0$, compute

$$J_{ij}(\underset{\sim}{x}_0, \rho_0)v_j^{(1)} = g_i(\underset{\sim}{x}_0, \rho_0)$$

$$\dot{\rho}_1 = 1 \Big/ \sqrt{1 + v_j^{(1)} v_j^{(1)}}$$

$$\dot{x}_{1j} = -\dot{\rho}_1 v_j^{(1)}$$

Step 2 (Extrapolation)

Compute

$$(\Delta s)_1 = (\Delta\rho)_0/\dot{\rho}_1$$

$$\underset{\sim}{x}_1 = \underset{\sim}{x}_0 + \dot{\underset{\sim}{x}}_1(\Delta s)_1$$

$$\rho_1 = \rho_0 + \dot{\rho}_1(\Delta s)_1$$

Step 3 (Correction)

$$J_{ij}^{(r)} w_j^{(r)} = -F_i^{(r)}$$

$$J_{ij}^{(r)} v_j^{(r)} = g_i^{(r)}$$

$$(\Delta\rho)^{(r)} = \frac{-\dot{\underset{\sim}{x}}_1 \cdot \underset{\sim}{w}^{(r)}}{\dot{\rho}_1 - \dot{\underset{\sim}{x}}_1 \cdot \underset{\sim}{v}^{(r)}}$$

$$(\Delta \underset{\sim}{x})^{(r)} = \underset{\sim}{w}^{(r)} - \underset{\sim}{v}^{(r)} (\Delta\rho)^{(r)}$$

with

$$J_{ij}^{(r)} = J_{ij}(\underset{\sim}{x}^r, \rho^r), \quad F_i^{(r)} = F_i(\underset{\sim}{x}^r, \rho^r)$$

$$\underset{\sim}{x}^r = \underset{\sim}{x}_1 + \sum_{k=1}^{r} (\Delta \underset{\sim}{x})^{(k)}$$

$$\rho^r = \rho_1 + \sum_{k=1}^{r} (\Delta\rho)^{(k)}$$

$$(\Delta \underset{\sim}{x})^k = \underset{\sim}{x}^{k+1} - \underset{\sim}{x}^k; \quad (\Delta\rho)^k = \rho^{k+1} - \rho^k$$

$$1 \le i, j \le N, \quad 1 \le r \le R$$

Set

$$\underset{\sim}{x}_1' = \underset{\sim}{x}_1 + \sum_{r=1}^{R} \Delta \underset{\sim}{x}^r = \underset{\sim}{x}^R$$

$$\rho_1' = \rho_1 + \sum_{r=1}^{R} (\Delta\rho)^{(r)} = \rho^R$$

Step 4

Return to Step 1 and continue the process with $(\underset{\sim}{x}_0, \rho_0)$ replaced by $(\underset{\sim}{x}_1', \rho_1')$.

Remarks:

1. In Step 1, it is, in general, inappropriate to set $(\underset{\sim}{x}_0, \rho_0) = (0, 0)$ since this is not a point on the solution path. The starting point $(\underset{\sim}{x}_0, \rho_0)$ is computed by specifying a small initial value of ρ_0 and computing $\underset{\sim}{x}_0$ by Newton's method.

2. The procedure in Step 2 was advocated by Keller [9] and has the attractive feature of preserving the symmetry and bandwidth of J_{ij} (when symmetry exists) as opposed to treating the full system (46) at once.

3. The algorithm given in Step 3 is also a Keller-type scheme, similar in structure to that of Step 2, and is equivalent to the constrained Newton-Raphson scheme,

$$\left.\begin{aligned} J_{ij}(\underset{\sim}{x}^r, \rho^r)\Delta x_j^r + g_i(\underset{\sim}{x}^r, \rho^r)\Delta\rho^r \\ = -F_i(\underset{\sim}{x}^r, \rho_i^r) \\ \dot{\underset{\sim}{x}}_1 \cdot \Delta\underset{\sim}{x}^r + \rho_1(\Delta\rho)^{(r)} = 0 \end{aligned}\right\} \quad (47)$$

4. In (47), the second equation is a constraint on the Newton-Raphson process which forces the scheme to progress toward the solution path Γ in a direction normal to the tangent plane. This is indicated graphically in Fig. 4. Some acceleration of this iterative process can be realized by using alternative constraints which make the iterative scheme follow a spherical path (Crisfield [10]), or an elliptical path (Padovan [23]), and such variants are easily implemented.

5. If a Cholesky-Gauss decomposition of the jacobean matrix J_{ij} can be constructed of the form

$$\underset{\sim}{J} = \underset{\sim}{L}\,\underset{\sim}{D}\,\underset{\sim}{L}^T, \quad \underset{\sim}{D} = \text{diag}\,\{\lambda_i\}$$

the rank of $\underset{\sim}{J}$ can be monitored by following sign changes of the pivots $\{\lambda_i\}$. If, at a point on Γ where sign λ_i changes, we have

$$|\dot{\rho}_m^{(r)}| < \text{tol}$$

where "tol" is a small preassigned tolerance, we assume that a limit point has been reached (which, occasionally, may not be the case), and we set

$$\dot{\rho}_m = (\text{sign}\,\lambda_i)\ 1\ /\ (1 + v_i v_i)^{\frac{1}{2}}$$

At such singular points, the orientation of tangent planes to branches through these points is determined by a linear eigenvalue analysis. Once the vectors $(\dot{\underset{\sim}{x}}, \dot{\rho})$ defining hyperplanes tangent to solution paths through equilibrium points are determined, the procedure given here can be used to continue the solution along any such path. Thus, the method can be used to determine bifurcation points, limit points, and to trace multiple branches.

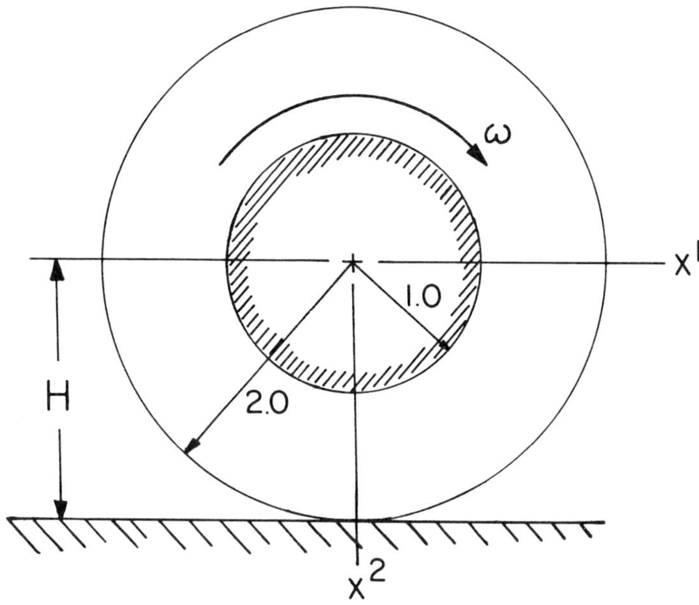

FIG. 5 Rotating cylinder

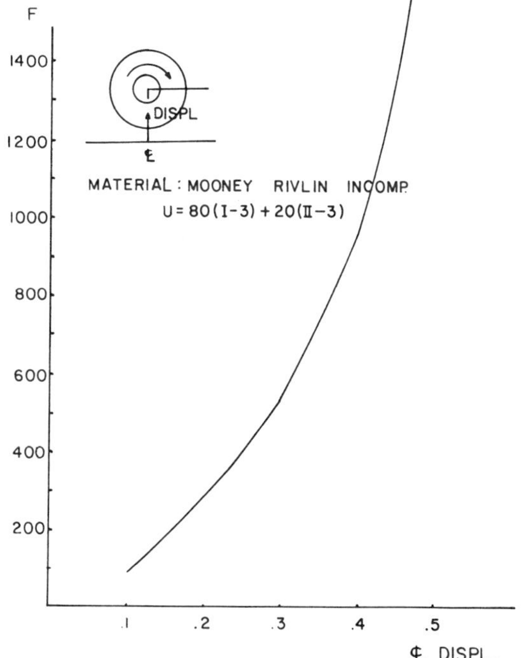

FIG. 6 Axle load versus centre displacement

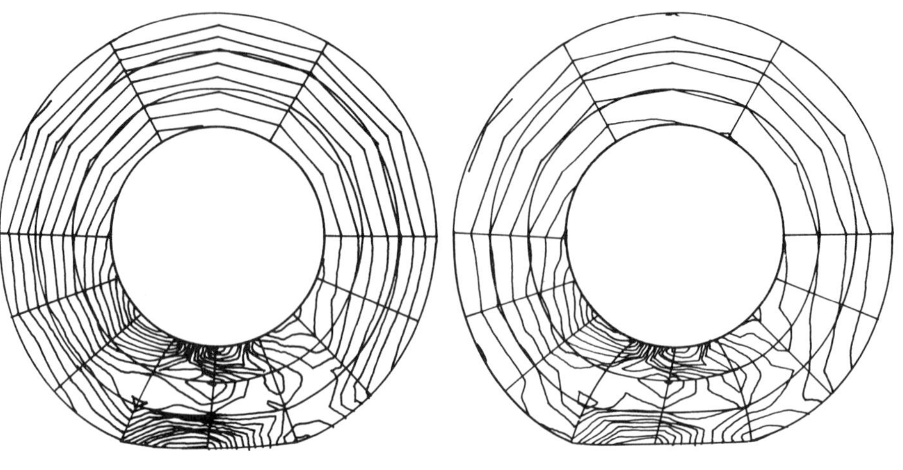

FIG. 7a Stress contours for rolling cylinder, Step 2

FIG. 7b Step 3

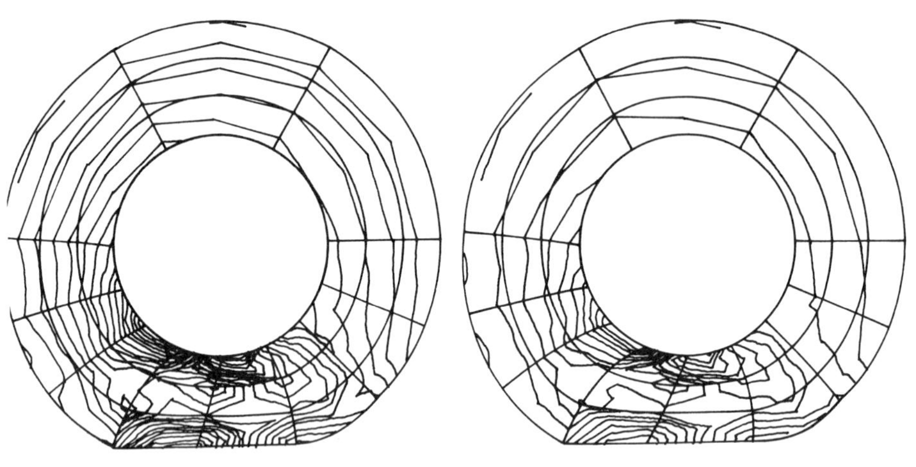

FIG. 7c Step 4

FIG. 7d Step 5

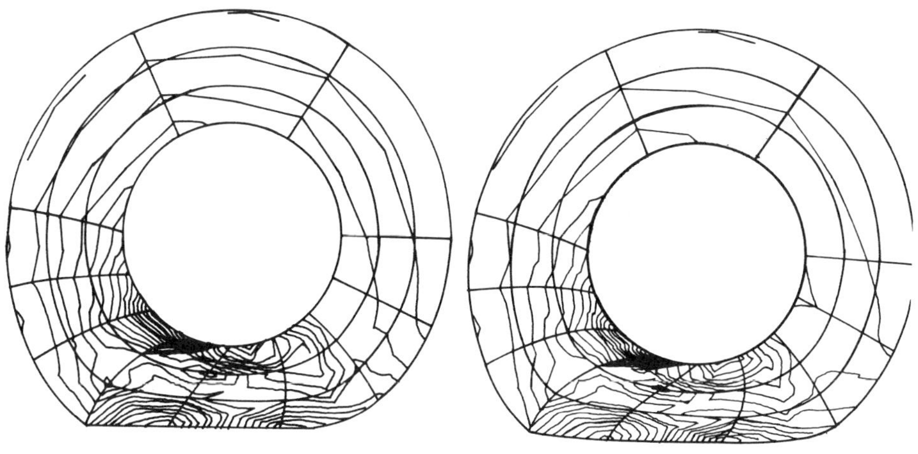

FIG. 7e Stress contours for rolling cylinder, Step 6

FIG. 7f Step 7

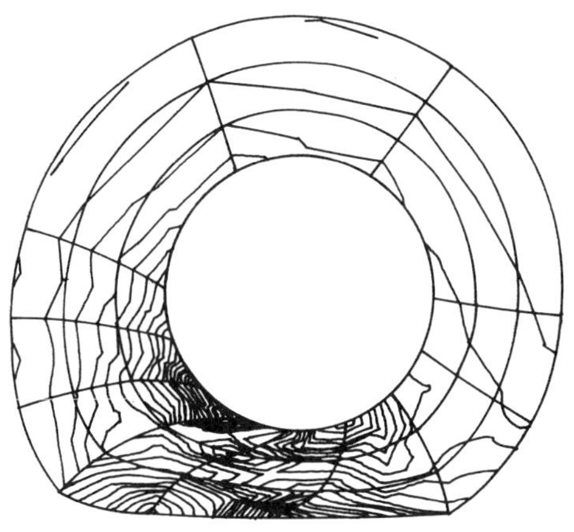

FIG. 7g Step 8

8. NUMERICAL RESULTS

We shall now present the results of several numerical experiments run using the methods and algorithms discussed previously. As a model problem, we consider the rolling contact problem illustrated in Fig. 5. The problem is one of finite plane strain of a rubber cylinder composed of a Mooney-Rivlin material with

$$W = C_1(I_{\tilde{C}}-3) + C_2(II_{\tilde{C}}-3)$$

with material constants $C_1 = 80$ psi, $C_2 = 20$ psi. The ground velocity $v_0 = 0$, the coefficient of friction is taken to be $\nu = 0.3$, the referential mass density $\rho_0 = 1$, and the initial outside radius is $R_0 = 2.0$ inches. The cylinder is spinning at an angular velocity $\omega = 5$ rps.

As a loading parameter, we take the distance H from the axis of the cylinder to the roadway, and consider increments such that H = 1.9, 1.82, 1.73, 1.71, 1.65, 1.61, 1.58, 1.55, 1.54, and 1.53.

Computed results are given in Figs. 6 and 7. In Fig. 6, the progressive contact pressure profiles for various values of H are shown. The deformed cylinder geometries for various values of H are indicated in Figs. 7a-g. Note that very large strains are realized in this example.

ACKNOWLEDGMENT

This work was completed during the course of a research project in the National Tire Modelling Program supported by NASA under Contract NAS 1-17359, Computational Mechanics Co. The encouragement of Mr. John Tanner of Langley Research Center is gratefully acknowledged.

REFERENCES

1. KALKER, J.J., The Computation of Three-Dimensional Rolling Contact with Dry Friction, *Int. J. Num. Meth. Eng.*, 14, 1293-1307 (1979,1).

2. KALKER, J.J., Survey of Wheel -- Rail Rolling Contact Theory, *Veh. Sys. Dyn.*, 5, 317-358 (1979,2).

3. BATRA, R.C., Quasistatic Indentation of a Rubber-Covered Roll by a Rigid Roll, *I.J.N.M.E.*, 17, 1823-1833 (1981).

4. BATRA, R.C., Rubber Covered Rolls -- The Nonlinear Elastic Problem, *J. Appl. Mech.*, 47, 82-86 (1980).

5. PADOVAN, J. and ZEID, I., Finite Element Analysis of Steadily Moving Contact Fields, *Comput. & Struct.* (18)2, 111-200 (1984).

6. ZEID, I. and PADOVAN, J., Large and Small Deformation Finite Element Analysis of Rolling Contact, Ph.D. thesis, Univ. of Akron (1980).

7. KIKUCHI, N. and ODEN, J.T., *Contact Problems in Elasticity*, SIAM Publishers, Philadelphia (in press).

8. RIKS, E., The Application of Newton's Method to Problem of Elastic Stability, *J. Appl. Mech.* 39, 1060-1066 (1972).

9. KELLER, H.B., Practical Procedures in Path Following Near Limit Point, in Glowinski and Lions (eds.), *Computing Methods in Applied Sciences and Engineering*, North-Holland Pub. Inc. (1982).

10. CRISFIELD, M.A., A Fast Incremental/Iterative Solution Procedure that Handles 'Snap-Through', *Comput. & Struct.* 13, 55-62 (1981).

11. DUVAUT, G. and Lions, J., *Inequalities in Mechanics and Physics*, Springer-Verlag, Berlin, Heidelberg, N.Y. (1972).

12. CAMPOS, L.T., ODEN, J.T., and KIKUCHI, N., A Numerical Analysis of a Class of Contact Problems with Frictions in Elastostatics, *Comp. Meth. in Appl. Mech. & Engrg.* 34, 821-845 (1982).

13. ODEN, J.T. and PIRES, E., Nonlocal Friction in Contact Problems in Place Elasticity, *Finite Elements: Special Problems in Solid Mechanics, V*, Prentice-Hall, Chapter 5 (1984).

14. ODEN, J.T. and PIRES, E., Nonlocal and Nonlinear Friction Laws and Variational Principles for Contact Problems in Elasticity," *J. Appl. Mech.* 50(1), 67-76 (1983).

15. MARTIN, J.A.C. and ODEN, J.T., A Numerical Analysis of a Class of Problems in Elastodynamics with Friction, *TICOM Report* 84-2, Univ. of Texas, Austin (1983).

16. ODEN, J.T. and JACQUOTTE, O., Stability of Some Mixed Finite Element Methods for Stokesian Flows, *Comput. Meth. in Appl. and Engrg.* (1984) (to be published).

17. ODEN, J.T., Penalty Methods for Constrained Problems in Nonlinear Elasticity, in Carlson and Shield (eds.), *Finite Elasticity*, Presented at the Lehigh Symposium in Philadelphia (1980).

18. MILLER, T.H., A Finite Element Study of Instabilities in Rubber Elasticity, Ph.D. Thesis, Univ. of Texas, Austin (1982).

19. RHEINBOLDT, W.C., *Methods for Solving Systems of Nonlinear Equations*, SIAM Publications, Philadelphia (1974).

20. ORTEGA, J.M. and RHEINBOLDT, W.C., *Iterative Solution of Nonlinear Equations in Several Variables*, Academic Press, New York (1970).

21. CRISFIELD, M.A., An Arc-length Method Including Line Searches and Accelerations, *I.J.N.M.E.* 19(9), 1269-1289 (1983).

22. RAMM, E., Strategies for Tracing the Nonlinear Response Near Limit Points, in Wunderlich, Stein, and Bathe (eds.), *Nonlinear Finite Element Analysis in Structural Mechanics*, Springer-Verlag (1980).

23. PADOVAN, J. and ARCHAGA, T., Formal Convergence Characteristics of Elliptically Constraint Incremental Newton-Raphson Algorithms, *Int. J. Eng. Sci.* 20(10), 1077-1097 (1982).

24. ENDO, T., ODEN, J.T., BECKER, E.B. and MILLER, T., A Numerical Analysis of Contact and Limit-point Behavior in a Class of Problems of Finite Elastic Deformation, *Comput. & Struct.* 18(5), 899-910 (1984).

INDIRECT OPTIMIZATION ALGORITHMS FOR NONLINEAR CONTACT PROBLEMS

D. Bischoff

Universität Hannover, Hannover, Germany

1. INTRODUCTION

Finite element programs for bilateral problems in nonlinear elasticity of solids essentially use the minimum properties of the solutions. Here quasi Newton methods (especially the BFGS method) have proven to be effective [15].

Unilateral problems however lead to constrained minimization problems or variational inequalities, respectively, the solution of which causes difficulties with conventional finite element programs. With the help of so-called indirect optimization methods constrained minimization problems however can be described equivalently by unconstrained problems or can be approximated, respectively [9].

For that purpose multiplier methods are especially suited [8] since they avoid the known difficulties of the Lagrange- and penalty-methods. If one uses the generalization of the Uzawa algorithm developed by Grossmann and Kaplan [10] (in particular for the form of Wierzbicki [17]) stopping criterion be given for the solution of subproblems and the convergence rate can be controlled by an adaptive increase of the penalty parameter.

The BFGS method could be used again to solve the subproblems. A limited memory version should be applied to avoid a restriction in the number of update steps.

2. MINIMUM PRINCIPLE FOR CONTACT PROBLEMS IN ELASTICITY

We consider hyperelastic materials, i.e. elastic materials possessing a stored energy function. For such materials a typical boundary value problem states:

We look for a vector field $u_0 : \Omega \to R^n$, which minimizes the

ISBN 0-12-747255-X

integral:

$$\pi(u) := \int_\Omega f(x,u(x),\nabla u(x))\, d\omega + \int_\Gamma h(x,u(x))\, d\gamma \tag{2.1}$$

in a suitable class of functions V.

In general V is a Sobolev space $W^{m,p}(\Omega)$ ($m\in N$, $m\geq 1$, $1<p<\infty$) which guarantees the existence of the integral and contains the essential boundary conditions and constraints.

Ω is a nonempty, constrained, open subset of R^n, $n=1,2,3$, whose boundary is Γ and the integrand f has the form

$$f(x,u,\nabla u) = W(x,\nabla u) + B(x,u)$$

whereby W equals the stored energy function and B is a body force potential.

W is given by

$$W(u) = \frac{1}{2}\sigma\varepsilon$$

whereby σ is the second Piola Kirchhoff stress tensor and ε is the Green strain tensor.

Statements concerning the existence and uniqueness of solutions of (2.1) depend on the solution spaces V and the convexity of π. For the threedimensional case this was extensively examined by Ball [3] and in the case of bars and shells by Antmann in e.g. [1].

Regarding contact problems the deformation of the body is unilaterally constrained by a rigid obstacle whose surface is described for example by the function

$$g(x) := \Psi(x_1,x_2) - x_3.$$

If one assumes that $\Gamma_m \subset \Gamma$ is the possible contact region the unilateral constraint can be indicated by the inequality

$$\phi(u)(x) := \Psi(x_1+u_1(x),x_2+u_2(x)) - (x_3+u_3(x)) \leq 0 \quad \text{for all } x\in\Gamma_m$$

by means of the function g and the deformation field u.

If additional boundary conditions $u=\bar{u}$ on $\Gamma_u \subset \Gamma$ are prescribed one gets

$$V = \{u\in W^{m,p}(\Omega) \mid \gamma_0 u=\bar{u} \text{ on } \Gamma_u,\ \phi(\gamma_0 u) \leq 0 \text{ on } \Gamma_m\}$$

as function space whereby $\gamma_0 u$ is the trace of u on Γ.

If π is convex and V a convex closed subset of a reflexive Banach space the convex functional analysis [7] yields information concerning existence and uniqueness of solutions.

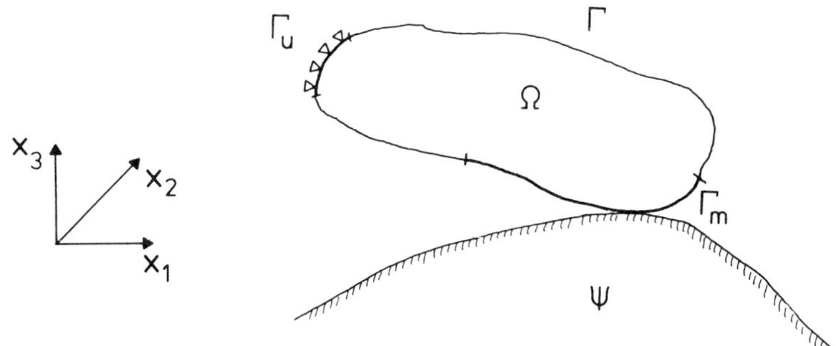

FIG. 1. Body Ω in contact with an obstacle described by Ψ.

3. DISCRETIZATION

With the help of finite elements the deformations $u \in V$ are replaced by linear combinations of shape functions $U_h = \underline{u}_h \, \underline{N}(x)$ and thus the infinite dimensional space V is replaced by finite dimensional spaces V_h (h characterizes the meshwidth and by the shape functions $\{N_i(x)\}$ a basis of V_h is given).

To assure the convergence of the solution U_h determined in those subspaces against the exact solution u_o conforming "elements" are assumed (i.e. $V_h \subset V$). Error estimates for such approximations are found in [6], [12].

If one integrates (2.1) numerically the minimization problem transforms into

$$\pi_h(\underline{u}_h) = \min \{\pi_h(\underline{v}_h) \mid \underline{v}_h \in K_h\} \tag{3.1}$$

whereby

$$K_h := \{\underline{v}_h \in R^N \mid \underline{\phi}_h(\underline{v}_h) \leq \underline{0}, \ \ \underline{v}_h = \underline{\bar{u}}_h\}.$$

$$\bar{U}_h(x) = \underline{\bar{u}}_h \, \underline{N}(x) \quad \text{on } \Gamma_u$$

approximates the boundary conditions on Γ_u and

$$\underline{\phi}_h(\underline{v}_h)\underline{N}(x) \leq 0$$

approximates the constraints on Γ_m.

Assuming that π_h is a convex and differentiable function and K_h is convex and closed, then (3.1) is equivalent to the variational inequality

$$\underline{\nabla\pi}_h(\underline{u}_h)^T(\underline{v}_h-\underline{u}_h) \geq 0 \quad \text{for all } \underline{v}_h \in K_h.$$

This inequality transforms into the equation

$$\underline{\nabla\pi}_h(\underline{u}_h)^T \underline{v}_h = 0 \quad \text{for all } \underline{v}_h \in K_h$$

if K_h is assumed to be a linear space and can thus be solved by the usual finite element program systems.

4. DESCENT METHODS FOR MINIMIZATION PROBLEMS

4.1 Direct Methods

To determine the minimum point $\underline{u}_h$ of π_h one uses in general descent methods in which case one calculates the next iteration point $\underline{u}_h^{k+1} \in K_h$ in such a way that one arrives at

$$\pi_h(\underline{u}_h^{k+1}) < \pi_h(\underline{u}_h^{k}).$$

One thus gets $\underline{u}_h^{k+1}$ by proceeding from $\underline{u}_h^{k}$ along a feasible descent direction $\underline{d}^k$ for a certain length determinded by ρ^k, so that a sufficient descent is guaranteed, i.e.

$$\underline{u}_h^{k+1} := \underline{u}_h^{k} + \rho^k \underline{d}^k$$

If K_h is not an affine space the determination of the descent direction $\underline{d}^k$ generally causes difficulties.

4.2 Indirect Methods

For this reason one frequently resorts to auxiliary problems for which the descent direction is easily to be determined. Hereby one can consider either equivalent problems for which the constraints can be handled more easily (e.g. dual problems) or one approximates the original problem by a sequence of problems without constraints (e.g. penalty problems).

The following paragraph serves to depict the shortcomings of the methods as well as the way for overcoming them by a combination of both methods.

5. INDIRECT OPTIMIZATION METHODS

Original problem (primal problem)

For fixed h in (3.1) $\underline{u}_h$ will be replaced by $\underline{w}^*$ and $\underline{v}_h$ by $\underline{w}$. Thus we look for $\underline{w} \in K_h$ with

$$\pi_h(\underline{w}^*) = \min \{\pi_h(\underline{w}) \mid \underline{w} \in K_h\} \tag{5.1}$$

5.1 Lagrange Multipliers

Let $L(\underline{w},\underline{\lambda}) := \pi_h(\underline{w}) + \underline{\lambda}^T \underline{\phi_h}(\underline{w}) \quad \underline{w} \in R^N,\ \underline{\lambda} \in R^M$

Because of

$$\sup_{\underline{\lambda} \in R^M_+} L(\underline{w},\underline{\lambda}) = \begin{cases} \pi_h(\underline{w}) & \text{for } \underline{\phi}_h(\underline{w}) \leq \underline{0} \\ +\infty & \text{else} \end{cases} =: \tilde{L}(\underline{w})$$

the original problem is equivalent to

$$\inf_{\underline{w} \in R^N} \left(\sup_{\underline{\lambda} \in R^M_+} L(\underline{w},\underline{\lambda}) \right) = \inf_{\underline{w} \in R^N} \tilde{L}(\underline{w}).$$

One gets as dual problem

$$\beta = \sup_{\underline{\lambda} \in R^M_+} \underset{\sim}{L}(\underline{\lambda}) \tag{5.2}$$

whereby

$$\underset{\sim}{L}(\underline{\lambda}) := \inf_{\underline{w} \in R^N} L(\underline{w},\underline{\lambda}).$$

Predictions concerning the existence of solutions $\underline{\lambda}^*$ of (5.2) and the connection with $\underline{w}^*$ are stated by the strong duality theorem (e.g. [11]).

An algorithm for the solution of (5.2) originates from Arrow, Hurwitz and Uzawa [2] and works as follows.

Step 1: Choice of a starting value $\underline{\lambda}^0 \in R^M_+$; $k := 0$

Step 2: Determine $\underset{\sim}{L}(\underline{\lambda}^k)$ (i.e. determine $\underline{w}^k \in R^N$ in a way that $L(\underline{w}^k,\underline{\lambda}^k) \leq L(\underline{w},\underline{\lambda}^k)\ \forall\ \underline{w} \in R^N$).

Step 3: Stop if $\underline{\phi}_h(\underline{w}^k) \leq \underline{0}$, else choose a suitable $\rho^k \in R_+$,

compute $\tilde{\underline{\lambda}}^k := \underline{\lambda}^k + \rho^k \underline{\phi}_h(\underline{w}^k)$,

$\underline{\lambda}^{k+1} := \underline{\max}\ \{\underline{0}, \tilde{\underline{\lambda}}^k\}$.

Step 4: $k := k+1$, goto step 2.

This algorithm describes an ascent method for problem (5.2) whose direction equals the gradient direction. As a consequence the convergence rate is at most linear and the algorithm is not very effective. Another disadvantage is the fact that ρ^k is to be determined adaptively at big expense only and should therefore be available as a global constant ρ (possibilities to determine ρ are found in [9]).

5.2 Penalty Methods

For the original problem with constraints one constructs the following approximate problems:

If $\alpha : R^N \to R_+$ is convex and continuous and $\alpha(\underline{w})=0 \Leftrightarrow \underline{w} \in K_h$

$$(\text{e.g. } \alpha(\underline{w}) := \sum_{i=1}^{M} (\max\{(\phi_h(\underline{w}))_i, 0\})^2)$$

and if

$$\pi_{hr}(\underline{w}) := \pi_h(\underline{w}) + r \cdot \alpha(\underline{w}),\ \underline{w} \in R^N,\ r \in R_+$$

then one can consider the following penalty problem:

We look for $\underline{w}_r \in R^N$ with

$$\pi_{hr}(\underline{w}_r) = \min\{\pi_{hr}(\underline{w}) \mid \underline{w} \in R^N\} \tag{5.4}$$

in which no constraints appear.

This results in $\underline{w}_r \to \underline{w}^*$ if $r \to \infty$ (see [5]). Convergence for $\alpha \in C^1$ can generally only be observed if $r \to \infty$ but in this case the problems become numerically unstable.

5.3 Multiplier Methods

The difficulties mentioned under *5.1* and *5.2* are avoided if one uses the dual problems developed by Grossmann and Kaplan [10] as substitute problems. If

$$K_b : R^N \times Y \times R^M \to \bar{R},\quad Y \in R^{M+K},\quad K \geq 0$$

whereby

$$K_b(\underline{w},\underline{y},\underline{v}) := \begin{cases} \pi_h(\underline{w}) + b(\underline{y},\underline{v}) & \text{if } \underline{\phi_h}(\underline{w}) \leq \underline{v} \\ +\infty & \text{else} \end{cases}$$

and if

$$L_b(\underline{w},\underline{y}) := \inf_{\underline{v} \in R^M} K_b(\underline{w},\underline{y},\underline{v})$$

whereby $b: Y \times R^M \to R$ has the following properties

(i) $b(\underline{y},\cdot)$ is strictly convex and continuous on R^M for all $\underline{y} \in Y$

(ii) $\lim_{\|\underline{v}\| \to \infty} (\underline{a}^T\underline{v} + b(\underline{y},\underline{v})) = +\infty \quad \forall\ \underline{a} \in R^M,\ \underline{y} \in Y$

(iii) $b(\underline{y},\underline{0}) \geq b(\underline{y},\underline{v}) - \underline{\nabla}_v b(\underline{y},\underline{v})^T\underline{v} + \delta(\underline{v}) \quad \forall\ \underline{v} \in R^M,\ \underline{y} \in Y$

whereby $\delta: R_+ \to R_+$ with $\delta(0) = 0$, $\delta(s) > 0\ \forall\ s>0$ and δ monotonically increasing as dual problem one gets:

$$\underset{\sim}{\Lambda}(\underline{y}^*) := \max \{\underset{\sim}{\Lambda}(\underline{y}) \mid \underline{y} \in Y\} \tag{5.5}$$

whereby $\underset{\sim}{\Lambda}(\underline{y}) := \underset{\sim b}{L}(\underline{y}) - b(\underline{y},\underline{0})$

and $\underset{\sim b}{L}(\underline{y}) := \min \{L_b(\underline{w},\underline{y}) \mid \underline{w} \in R^N\}$.

For the solution the following algorithm can be utilized [11] which respresents a generalization of Uzawa's algorithm:

Step 1: Choice of a starting value $\underline{y}^0 \varepsilon\ Y$, $k := 0$.

Step 2: Determine $\underset{\sim b}{L}(\underline{y}^k)$
(i.e. determine $\underline{w}^k := \arg\min \{L_b(\underline{w},\underline{y}^k) \mid \underline{w} \in R^N\}$),
determine $\underline{v}^k := \arg\min \{b(\underline{y}^k,\underline{v}) \mid \underline{v} \geq \underline{\phi}_h(\underline{w}^k)\}$.

Step 3: Stop for $\underline{v}^k = \underline{0}$, else (5.6)
determine $\underline{y}^{k+1}$ with
$\underline{\nabla_v} b(\underline{y}^{k+1},\underline{0}) = \underline{\nabla_v} b(\underline{y}^k,\underline{v}^k)$.

Step 4: $k := k+1$, goto step 2.

Remark: The determination of $\underline{v}^k$ and $\underline{y}^{k+1}$ should be done explicitly, if possible.

Example: For $Y = R^M$ and

$$b_1(\underline{y},\underline{v}) := \frac{r}{2}\| y + \underline{v} \|^2, \quad r > 0 \text{ (Wierzbicki, [17])}$$

one gets $(j = 1,2,\ldots,M)$

$$v_j^k = \max \{(\phi_h(\underline{w}^k))_j, -y_j^k\}$$

$$(\nabla_v b(\underline{y},\underline{v}))_j = r(\underline{y}_j + \underline{v}_j)$$

$$y_j^{k+1} = \max \{0, y_j^k + (\phi_h(\underline{w}^k))_j\} = y_j^k + \max \{(\phi_h(\underline{w}^k))_j, -y_j^k\}$$

An acceleration of the method is obtained if one increases r successively for it proves that

$$\underline{\nabla^2 \underset{\sim}{\Lambda}_{b_1}}(y) \to r\underline{I} \quad \text{(see [14])},$$

i.e. the above-mentioned method represents an approximation of the Newton method of the dual problem (5.5) for the case of big r.

5.4 Simple Demonstration Example

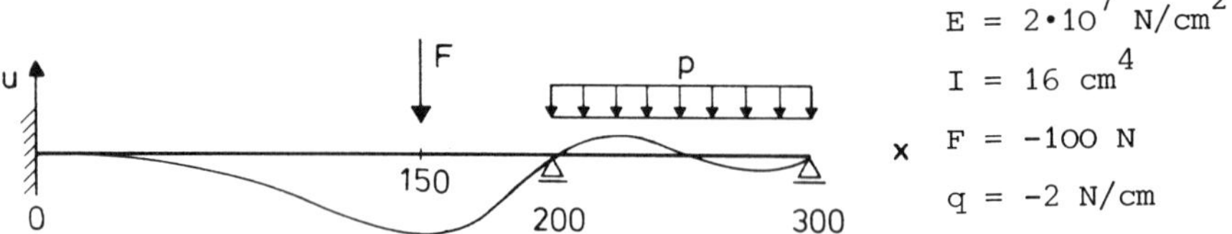

FIG. 2. Continuous beam

One gets

$$u(0) = 0 \qquad u'(0) = 0$$
$$u(150) = -0.00604248 \qquad u'(150) = 0.000113525$$
$$u(200) = 0 \qquad u'(200) = 0.000019531$$
$$u(300) = 0 \qquad u'(300) = 0.000120443$$

If the solution is computed iteratively one gets with different values of the starting parameter r_o:

TABLE 1

r_o	Lagrange	Penalty	Wierzbicki
10^5		$26/65/10^{14}$	$15/49/10^{10}$
10^8	100/300 *)	$14/46/10^{13}$	$6/37/10^{10}$
10^{10}		$10/26/10^{13}$	$3/18/10^{11}$

*) Iteration stopped; two accurate digits were obtained
Listed are: Number of update steps for $\underline{y}^k$ /
Number of minimization steps to determine $\underline{w}^k(\underline{y}^k)$/
Final value of the penalty parameter

6. REMARKS CONCERNING THE NUMERICAL IMPLEMENTATION

If one solves a contact problem with the algorithm (5.6) a few pointers are to be observed when implementing it with a finite element program.

6.1 Approximate Solution of Subproblems

The accuracy with which $\tilde{L}_b(\underline{y}^k)$ is determined should depend on the accuracy with which the contact region has been approximated up to this point.

If π_h is strictly convex, i.e. if a constant $\gamma > 0$ exists with

$$\pi_h(\underline{w}) \geq \pi_h(\underline{w}) + \underline{\nabla\pi_h}(\underline{\tilde{w}})^T(\underline{w} - \underline{\tilde{w}}) + \frac{\gamma}{2}\| \underline{w} - \underline{\tilde{w}} \|^2$$

$$\text{for all } \underline{w},\ \underline{\tilde{w}} \in R^N$$

one can stubstitute step 2 in the algorithm (5.6) by ([10])

Step 2': Determine $\underline{w}_\varepsilon{}^k \in R^N$ with

$$\|\nabla_w L_b(\underline{w}_\varepsilon{}^k, \underline{y}_\varepsilon{}^k)\|^2 \leq \min \{2\gamma(L(\underline{w}_\varepsilon{}^k, \underline{\lambda}_\varepsilon{}^k) - R(\underline{w}_\varepsilon{}^k, \underline{y}_\varepsilon{}^k)), \varepsilon_k\}$$

whereby L is the Lagrange function belonging to the original problem

$$R(\underline{w},\underline{y}) := L_b(\underline{w},\underline{y}) - b(\underline{y},\underline{O})$$

$$\underline{\lambda_\varepsilon}^k := \underline{\nabla_v} b(\underline{y_\varepsilon}^k, \underline{v_\varepsilon}^k)$$

$$\underline{v_\varepsilon}^k := \arg\min \{b(\underline{y}_\varepsilon{}^k, \underline{v}) \mid \underline{v} \geq \underline{\phi_h}(\underline{w}_\varepsilon{}^k)$$

$\{\varepsilon_k\}$ is a prescribed positive sequence tending to zero.

Example: For $b_1(\underline{y},\underline{v}) = \frac{r}{2} \| \underline{y+v} \|^2$ one gets the stopping criterion:

$$\|\nabla_{\underline{w}} L_b(\underline{w}^k, \underline{y}^k)\|^2 \leq \min \{\gamma \cdot r \, \|\max\{\underline{\phi_h}(\underline{w}_\varepsilon{}^k), -\underline{y}_\varepsilon{}^k\}\|^2, \varepsilon_k\}$$

since

$$L(\underline{w}_\varepsilon{}^k, \underline{\lambda}_\varepsilon^k) - R(\underline{w}_\varepsilon{}^k, \underline{y}_\varepsilon{}^k) =$$

$$= b(\underline{y}_\varepsilon{}^k, 0) - b(\underline{y}_\varepsilon{}^k, \underline{v}_\varepsilon{}^k) + \underline{\nabla_v} b(\underline{y}_\varepsilon{}^k, \underline{v}_\varepsilon{}^k)^T \underline{v}_\varepsilon{}^k$$

$$= \frac{r}{2} \|\underline{\max} \{\underline{\phi_h}(\underline{w}_\varepsilon{}^k), -\underline{y}_\varepsilon{}^k\}\|^2$$

6.2 *Estimation of the Convergence Rate*

The following general statement [10] can be used for the estimation of the convergence rate:

Suppose that

$\underline{\tilde{w}} \in R^N$ with $\phi_h(\underline{\tilde{w}}) < \underline{O}$ exists,

$\chi(\underline{v}) := \min \{\pi_h(\underline{w}) \mid \underline{\phi_h}(w) \leq \underline{v}\}$ is twice differentiable in a neighbourhood of $\underline{O}$,

$\{\underline{y}^k\}$ and $\{\underline{v}^k\}$ are the sequences produced by the algorithm (5.6), $\underline{B}_k$ are symmetrical and positive definite matrices with

$$\nabla_{\underline{v}} b(\underline{y}^{k+1}, \underline{v}^{k+1}) - \nabla_{\underline{v}} b(\underline{y}^{k+1}, \underline{0}) = \underline{B}_k \underline{v}^{k+1};$$

then for every $\alpha > \|\underline{\nabla}^2 \chi(\underline{0})\|$ there exists a $k_o \in N$ with

$$\|\underline{v}^{k+1}\| \leq \frac{\alpha}{\alpha+\beta_k} \|\underline{v}\|^k \quad \text{for all } k \geq k_o$$

whereby β_k is the smallest eigenvalue of $\underline{B}_k$.

Example: If one chooses $r = y_o$ as a variable in the method of Wierzbicki, i.e.

$$Y = [r_o, \infty[\times R^M \quad \text{and}$$

$$b_2(\underline{y}, \underline{v}) := y_o \sum_{i=1}^{M} (y_i - v_i)2$$

this results in

$$y_i^{k+1} = \frac{y_o}{y_o^{k+1}} \max \{0, y_i^k - (\phi_h(\underline{w}^k))_i\}, \; i=1,2,\ldots,M$$

and

$$\nabla_{\underline{v}} b(\underline{y}^{k+1}, \underline{v}^{k+1}) - \underline{\nabla}_v b(\underline{y}^{k+1}, \underline{0}) = 2y_o^{k+1} \underline{I}\, \underline{v}^{k+1};$$

i.e. $\beta_k = 2y_o^{k+1}$ is the smallest eigenvalue of $\underline{B}_k$.

Thus the above estimate yields

$$\|\underline{v}^{k+1}\| \leq \frac{\alpha}{2y_o^{k+1} + \alpha} \|\underline{v}^k\| \text{ for all } k \geq k_o.$$

Therefore, if

$$\|\underline{y}_o^{k+1}\| := \max \{r_o, \frac{c}{\|\underline{v}_k\|}\}, \; c > 0, \; k = 0,1,\ldots$$

is chosen, $\underline{v}^k$ converges to 0 at least quadratically

6.3 Examples

6.3.1 For the demonstration example one gets:

TABLE 2

r	Wierzbicki	adaptive control	approx. minimum	direct method *)
10^8	$6/37/10^{10}$	$3/19/10^8$	$2/10/10^3$	10

*) projected quasi-Newton method after Bertsekas [4].

6.3.2 As a nontrivial example we report the computation of the pressure distribution in the contact region of a car tyre. The geometrically nonlinear calculation was done with an isoparametric four node element. With a linear material law the calculation showed a very good correspondence with data determined by experiments [13].

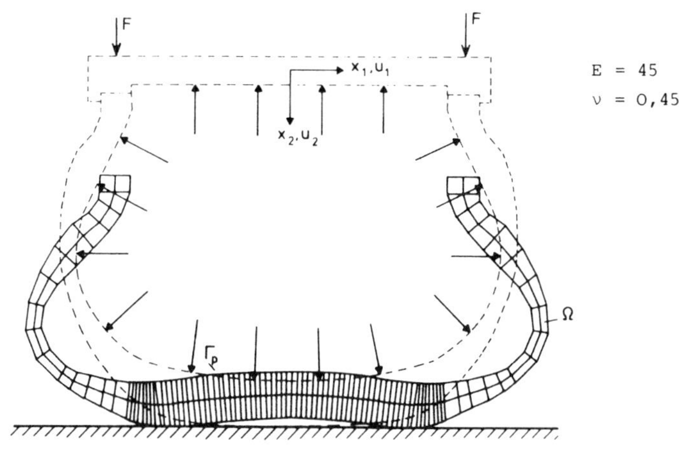

FIG. 3. Contact between a torus segment and a rigid plane (with interior pressure)

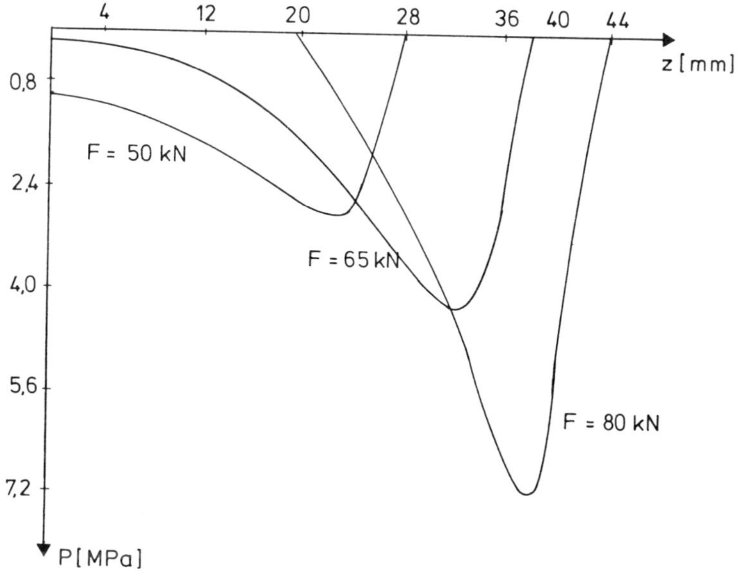

FIG. 4. Pressure distribution in the contact region (p=0,1)

For the determination of $L_{\tilde{b}}(\underline{y}^k)$ the BFGS method was used. For large problems a preconditioned version with limited memory (e.g.[16]) can be utilized for the optimal use of the available storage.

ACKNOWLEDGEMENT

The support of the Stiftung Volkswagenwerk is gratefully acknowledged.

REFERENCES

1. ANTMAN, S.S., Monotonicity and invertibility conditions in one-dimensional nonlinear elasticity. pp. 57-92 of R.W. Dickey (Ed.), *Nonlinear Elasticity*, Academic Press, New York (1973).
2. ARROW, K.J., HURWICZ, H., UZAWA, H., *Studies in linear and nonlinear programming*. Stanford University Press, Stanford (1958).
3. BALL, J.M., Convexity conditions and existence theorems in nonlinear elasticity. *Arch. Rational Mech. Anal.* 63, 337-403 (1976/77).
4. BERTSEKAS, D.P., *Constrained optimization and Lagrange multiplier methods*. Academic Press, New York (1982).
5. BISCHOFF, D., *Penalty-Verfahren in reflexiven Banach-Räumen*. Diss., Universität Hannover, Hannover (1982).
6. BREZZI, F., HAGER, W.W., RAVIART, P.A., Error estimates for the finite element solution of variational inequalities. Part I. Primal theory. *Numer. Math.* 28, 431-443 (1977).
7. EKELAND, I., TEMAN, R., *Convex analysis and variational problems*. North Holland, Amsterdam (1976).
8. FORTIN, M., GLOWINSKI, R., *Augmented Lagrangian methods: applications to the numerical solution of boundary value problems*. North Holland, Amsterdam (1983).
9. GLOWINSKI, R., LIONS, J.L., TREMOLIERES, R., *Numerical analysis of variational inequalities*. North Holland, Amsterdam (1981).
10. GROSSMANN, C., KAPLAN, A.A., *Strafmethoden und modifizierte Lagrangefunktionen in der nichtlinearen Optimierung*. B.G. Teubner, Leipzig (1979).
11. GROSSMANN, C., KLEINMICHEL, H., *Verfahren der nichtlinearen Optimierung*. B.G. Teubner, Leipzig (1976).
12. HASLINGER, J., On numerical solution of a variational inequality of the 4th order by finite element method. *Aplikace Matematiky* 23, 334-345 (1978).
13. JACOBI, W., *Das geometrisch nichtlineare Kontaktproblem "elastischer Körper - starres Hindernis" bei verformungsabhängiger Belastung als restringiertes Minimalproblem*. Mitt. des Inst. f. Statik d. Univ. Hannover Nr. 29, Hannover (1983).

14. LUENBURGER, D.G., *Introduction to linear and nonlinear programming*. Addison-Wesley, Reading (1973).
15. MATTHIES, H., STRANG, G., The solution of nonlinear finite element equations. *Int. J. for Num. Meths. in Eng.* 14, 1613-1626 (1979).
16. NOCEDAL, J., Updating quasi-Newton matrices with limited storage. *Math. of Comput.* 35, 773-782 (1980).
17. WIERZBICKI, A.P., A penalty-shifting method in constrained static optimization and its convergence properties. *Arch. Automat. Telemech.* 16, 395-416 (1971).

NONCONVEX UNILATERAL CONTACT PROBLEMS AND APPROXIMATION

P.D.Panagiotopoulos

School of Technology, Aristotle University Thessaloniki, Greece and
Institute of Technical Mechanics RWTH Aachen, F.R.G.

1. INTRODUCTION

In several contact problems reaction-displacement diagrams may arise which are of nonmonotone nature and may involve discontinuities as, for instance, happens in the case of unilateral contact of a deformable body with a granular support. If these diagrams were monotone then, as is well known, the problem could be formulated in terms of variational inequalities [1][2]. In the nonmonotone case we obtain, due to the lack of monotonicity, hemivariational inequalities. The mechanical theory of nonconvex problems has recently been initiated in [3] (see also [4][5]), where numerous examples of mechanical problems having as a variational formulation a hemivariational inequality, are given. The arising mathematical problem has been studied for the coercive case in [6][7] and for the semicoercive case in [8] concerning the existence of the solution. Note that contrasting with the monotone (or convex) case, which is closely connected to the notion of subdifferential of convex functionals, the nonmonotone (or nonconvex) case is connected to the notion of generalized gradient introduced in 1975 by F.H.Clarke [9][10] for nonconvex functionals. Here we will formulate and study the nonconvex unilateral contact problem of linear elastic plates.

2. FORMULATION OF THE NONCONVEX UNILATERAL CONTACT BOUNDARY CONDITION.

Let us consider an open bounded connected subset Ω of R^3 which is occupied by a linear elastic body in its undeformed state. The boundary Γ of Ω is assumed to be "sufficiently regular". On $\Gamma_S \subset \Gamma$ we assume that nonconvex unilateral contact boundary conditions hold. They have the form

ISBN 0-12-747255-X

$$\text{if} \quad u_N<0 \quad \text{then} \quad S_N=0, \quad S_{T_i}=C_{T_i}, \quad i=1,2,3 \tag{2.1}$$

$$\text{if} \quad u_N\geqq 0 \quad \text{then} \quad S_N+k(u_N)=0, \quad S_{T_i}=C_{T_i}, \quad i=1,2,3 \tag{2.2}$$

Here S_N (resp. u_N) denotes the projection of the boundary stress resultant $S=\{S_i\}$ (resp. the displacement $u=\{u_i\}$) onto the outward unit normal vector to Γ, $S_T=\{S_{T_i}\}$ (resp. $u_T=\{u_{T_i}\}$) is the respective tangential component and $C_T=\{C_{T_i}\}$ is a given function. $k=k(u_N)$ is a nonmonotone function which idealizes the action of the support on the body. Note that the law (2.1), (2.2) is of the form of the well-known Winkler law (idealization of the contact action by one-dimensional springs). The graph $\{u_N, k(u_N)\}$ may include a finite number of finite jumps (vertical segments) which correspond to local fracture effects. Eqs. (2.1) hold in the regions of Γ_S which are not in contact with the support, whereas eqs.(2.2) hold in the regions of contact. Further we write eqs. (2.1)(2.2) compactly as

$$-S_N\in\hat{b}(u_N), \quad S_{T_i}=C_{T_i} \tag{2.3}$$

where $\hat{b}:\mathbb{R}\to\mathcal{P}(\mathbb{R})$ is generally a nonmonotone multivalued function. We assume that with respect to $\hat{b}$ a measurable selection $b:\mathbb{R}\to\mathbb{R}$ can be determined such that $b(\xi)\in\hat{b}(\xi)$ $\forall\xi\in\mathbb{R}$ (This is obvious in (2.1),(2.2)) Moreover let $b\in L^\infty_{loc}(\mathbb{R})$. For any $\mu>0$ and $\xi\in\mathbb{R}$ we define

$$\bar{b}_\mu(\xi)=\underset{|\xi_1-\xi|<\mu}{\text{ess sup}}\, b(\xi_1) \quad \text{and} \quad \underline{b}_\mu(\xi)=\underset{|\xi_1-\xi|<\mu}{\text{ess inf}}\, b(\xi_1) \tag{2.4}$$

and by taking the limits as $\mu\to 0$ we obtain the functions $\bar{b}(\xi)$, $\underline{b}(\xi)$ Let us further assume that $\hat{b}(\xi)=[\bar{b}(\xi), \underline{b}(\xi)]$. (This assumption is verified in (2.1),(2.2). If $b(\xi_{\pm 0})$ exists at every $\xi\in\mathbb{R}$ then a locally Lipschitz function $\varphi:\mathbb{R}\to\mathbb{R}$ can be determined such that [11]

$$\hat{b}(\xi)=\bar{\partial}\varphi(\xi) , \tag{2.5}$$

where $\bar{\partial}$ denotes the generalized gradient in the sense of Clarke. Here φ is such that

$$\varphi^0(\xi,\bar{\xi})=\underset{\substack{h\to 0\\ \lambda\to 0_+}}{\limsup}\, \frac{1}{\lambda}\int_{\xi+h}^{\xi+h+\lambda\bar{z}} b(\xi)d\xi \tag{2.6}$$

where $\varphi^0(.,.)$ is the directional derivative in the sense of Clarke. Thus eqs (2.3) take the form

$$-S_N\in\bar{\partial}\varphi(u_N), \quad S_{T_i}=C_{T_i} \quad \text{on } \Gamma_S \tag{2.7}$$

which is equivalent to the hemivariational inequality

$$\varphi^{o}(u_N,v_N-u_N)\geqq -S_N(v_N-u_N) \quad \forall v_N\in R,\quad S_{T_i}=C_{T_i} \text{ on } \Gamma_S. \tag{2.8}$$

In the framework of the Kirchhoff plate theory $\Omega=\Omega_1\times(-\frac{h}{2},+\frac{h}{2})$, where Ω_1 is an open bounded connected subset of R^2 and h is the thickness of the plate. We denote by z,J the vertical displacements of the plate and we assume, for simplicity's sake, that on the boundary Γ_1 of Ω_1, J=0 and $\partial J/\partial n=0$ (for sign conventions cf. [2]). Let $\underline{\underline{f}}$ be the external vertical loading. We assume that $f=\bar{f}+f$ where $\bar{f}\in L^2(\Omega_1)$ is given and $\bar{f}$ is the unilateral contact reaction i.e.

$$-\bar{f}\in\hat{b}(J) \text{ on } \Omega_2\subseteq\Omega_1\ (\bar{\Omega}_2\cap\Gamma_1=\emptyset),\ \bar{f}=0 \text{ on } \Omega_1-\Omega_2 \tag{2.9}$$

Due to the mechanical model used, we do not consider any tangential forces.

Let $Z=\overset{o}{H}{}^2(\Omega)$. Applying the method of [3] we obtain by means of (2.8) the following variational problem. We denote by $\alpha(.,.)$ the bilinear form of the elastic energy of the plate and by (.,.) the $L^2(\Omega_1)$ - inner product.

Problem 1: Find J∈Z which satisfies the hemivariational inequality

$$\alpha(J,z-J)+\int_{\Omega_2}\varphi^{o}(J,z-J)d\Omega\geqq(\bar{\bar{f}},z-J)\quad \forall z\in Z \tag{2.10}$$

Note that for φ convex $\bar{\partial}\varphi=\partial\varphi$ (subdifferential) and therefore (2.10) reduces to the well-known variational inequality of convex unilateral contact problems.

3. EXISTENCE OF THE SOLUTION OF PROBLEM 1.

We apply the regularization method. Let $\rho\in C_{\infty}^{o}(-1,+1)$ be a mollifier (see e.g. [12]) such that

$$\rho\geqq 0 \quad \text{and} \quad \int_{-\infty}^{+\infty}\rho(\xi)d\xi=1 \tag{3.1}$$

and let b_ε be the convolution $\rho_\varepsilon * b$ where

$$\rho_\varepsilon(\xi)=\frac{1}{\varepsilon}\rho(\frac{\xi}{\varepsilon})\quad \varepsilon>0 \tag{3.2}$$

We assume that

$$\liminf_{\xi\to\infty} b(\xi)=b(+\infty)>b(-\infty)=\limsup_{\xi\to-\infty} b(\xi) \tag{3.3}$$

and we now consider the following problem:

Problem 1_ε: Find $J_\varepsilon\in Z$ such that

$$\alpha(J_\varepsilon, z) + \int_{\Omega_2} b_\varepsilon(J_\varepsilon) z d\Omega = (\overline{\overline{f}}, z) \quad \forall z \in Z \qquad (3.4)$$

Note that $b_\varepsilon(J_\varepsilon(x)) = \frac{d\varphi_\varepsilon}{dJ}(J_\varepsilon(x))$. Problem 1_ε is the regularized form of the problem 1.

Proposition 1: Suppose that (3.3) holds. Then problem 1 has a solution.
Proof: We consider a Galerkin basis $\{w_i\}$ of Z and the n-dimensional subspace W_n of Z, and we assume that $\varepsilon = \varepsilon(n)$ such that $\varepsilon \to 0$ if and only if $n \to \infty$. Note that $Z \subset L^\infty(\Omega_1)$. Let $J_{\varepsilon n}$ be a solution of the problem $1_{\varepsilon n}$: find $J_{\varepsilon n} \in W_n$ such that

$$\alpha(J_{\varepsilon n}, z_n) + \int_{\Omega_2} b_\varepsilon(J_{\varepsilon n}) z_n d\Omega = (\overline{\overline{f}}, z_n) \quad \forall z_n \in W_n . \qquad (3.5)$$

We prove [6][7], that $J_{\varepsilon n}$ exists and is bounded in Z: thus as $\varepsilon \to 0$ $J_{\varepsilon n} \to J$ weakly in Z, strongly in $L^2(\Omega_1)$ and a.e in Ω_1; that $\{b_\varepsilon(J_{\varepsilon n})\}$ is weakly precompact in $L^1(\Omega_1)$; and thus $b_\varepsilon(J_{\varepsilon n}) \to \chi$ weakly in $L^1(\Omega_1)$ where $\chi \in \hat{Z} \cap L^1(\Omega_1)$. To complete the proof we have to show that $\chi \in \hat{b}(J)$ a.e on Ω_2. Since $J_{\varepsilon n} \to J$ a.e in Ω_1 we may find for $\alpha > 0$ $\omega \subset \Omega_1$, with mes $\omega < \alpha$, such that $J_{\varepsilon n} \to J$ uniformly in $\Omega_1 - \omega$, $J \in L^\infty(\Omega_1 - \omega)$. Thus for any $\mu > 0$ we can determine $n_o > 2/\mu$ and ε_o such that for $n > n_o$ and $\varepsilon < \varepsilon_o$, $|J_{\varepsilon n}(x) - J(x)| < \mu/2$ $\forall x \in \Omega_1 - \omega$; $b_\varepsilon(J_{\varepsilon n}(x))$ is an average value of $\xi \to b(\xi)$ for $n > n_o$ and $\varepsilon < \varepsilon_o$ and for $\xi \in (-\mu/2 + J_{\varepsilon n}(x), \mu/2 + J_{\varepsilon n}(x))$, $x \in \Omega_1 - \omega$. Since the range of ξ is a subset of $[-\mu + J(x), \mu + J(x)]$ we find that

$$b_\varepsilon(J_{\varepsilon n}(x)) \in [\underline{b}_\mu(J(x)), \bar{b}_\mu(J(x))] . \qquad (3.6)$$

Thus for $e > 0$ a.e. in $\Omega_1 - \omega$, $e \in L^\infty(\Omega_1 - \omega)$, we may write that

$$\int_{\Omega_2 - \omega} e\, b_\varepsilon(J_{\varepsilon n}(x)) d\Omega \in [\int_{\Omega_2 - \omega} e\, \underline{b}_\mu(u(x)) d\Omega, \int_{\Omega_2 - \omega} e\, \bar{b}_\mu(u(x)) d\Omega] \qquad (3.7)$$

Thus by taking the limit as $\varepsilon \to o$ we obtain

$$\chi \in \hat{b}(J(x)) \text{ a.e. on } \Omega_2 - \omega \quad . \qquad (3.8)$$

The proof is completed by considering α as small as possible q.e.d. No uniqueness result can be proved.

4. STRONG CONVERGENCE

On the assumptions of prop. 1 the following proposition holds (cf. also [13]).

Proposition 2: Let there exist a constant $c > o$ and $q > 1$ such that

$$|b(\xi)| \leqq c(1+|\xi|^q) \quad \forall \xi \in \mathbb{R} \tag{4.1}$$

Then as $\varepsilon \to 0$, $J_{\varepsilon n} \to J$ strongly.

Proof: Let J (resp. $J_{\varepsilon n}$) be one of the solutions of problem 1 (resp. problem $1_{\varepsilon n}$). The coerciveness of $\alpha(.,.)$ implies that

$$c\| J-J_{\varepsilon n}\|_Z^2 \leqq (f,J-z_n) + \int_{\Omega_2} \varphi^o(J,J_{\varepsilon n}-J)d\Omega + \int_{\Omega_2} b_\varepsilon(J_{\varepsilon n})(z_n-J_{\varepsilon n})d\Omega +$$

$$+ \alpha(J_{\varepsilon n},z_n-J) \quad \forall z_n \in W_n . \tag{4.2}$$

Prop. 1 implies that as $\varepsilon \to 0$, $J_{\varepsilon n} \to J$ weakly in Z. Using the compact imbedding $\overset{o}{H}{}^2(\Omega_1) \subset L^q(\Omega_1)$ for $1<q<\infty$ we find that $J_{\varepsilon n} \to J$ strongly in $L^q(\Omega_1)$. The density of W_n as $n \to \infty$ in Z implies that a sequence $\{\bar{z}_n\}$, $z_n \in W_n$, exists such that $\bar{z}_n \to J$ strongly in Z. We put in (4.2) $z_n = \bar{z}_n$ and we show that any term in the right-hand side of (4.2) tends to zero. Indeed, the definition of the generalized gradient and (4.1) imply that

$$\left|\int_{\Omega_2} \varphi^o(J,J_{\varepsilon n}-J)d\Omega\right| \leqq c \int_{\Omega_2} (1+|J|^q)|J_{\varepsilon n}-J|\, d\Omega \to 0 \tag{4.3}$$

since $J_{\varepsilon n} \to J$ strongly in $L^q(\Omega_1)$, $q>1$. b_ε which is defined as the regularization of b and which determines $\varphi_\varepsilon^o(.,.)$ implies the same estimation as b does, i.e.

$$|b_\varepsilon(\xi)| \leqq c(1+|\xi|^q) \quad \forall \xi \in \mathbb{R}. \tag{4.4}$$

Thus

$$\left|\int_{\Omega_\alpha} b_\varepsilon(J_{\varepsilon n})(\bar{z}_n-J_{\varepsilon n})d\Omega\right| \leqq c\int_{\Omega_2} (1+|J_{\varepsilon n}|^q)|\bar{z}_n-J_{\varepsilon n}|d\Omega \leqq$$

$$\leqq c \int_{\Omega_2} (1+|J_{\varepsilon n}|^q)(|\bar{z}_n-J| + |J_{\varepsilon n}-J|)d\Omega \to 0 \tag{4.5}$$

and the proof is completed q.e.d.

An existence result concerning the semicoercive case is given in [8] (cf. also [14]). Note that for the semicoercive problem only the strong convergence of the seminorms can be shown analogously to the proof of prop. 2.

5. A NOTE ON THE DYNAMIC PROBLEMS

For the dynamic problem we consider a reaction-displacement law of the form $(J' = \partial J/\partial t)$

$$-\bar{f} \in \hat{b}(J') \text{ in } \Omega_2 \subset \Omega_1, \ \bar{\Omega}_2 \cap \Gamma = 0, \ \bar{f}=0 \text{ in } \Omega_1 - \Omega_2 \tag{5.5}$$

and we pose the following problem
Problem 1′: Find $J:[0,T]\to Z$ with $J'(t)\in Z$ and $J''(t)\in L^2(\Omega_1)$ such as to satisfy for a.e. $t\in[0,T]$ the hemivariational inequality

$$(\rho J'',z-J')+\alpha(J,z-J')+\int_{\Omega_2}\varphi^0(J',z-J')d\Omega\geq(\bar{\bar{f}},z-J')\quad\forall z\in Z \qquad (5.2)$$

and the initial conditions

$$J=0 \quad \text{and} \quad J'=0 \quad \text{at } t=0 \qquad (5.3)$$

If φ is convex, i.e. if b is monotone increasing, (5.2) reduces to the well-known dynamic variational inequality of plate theory [2]. The following proposition holds.
Proposition 3: Let (3.3) hold and assume that

$$\bar{\bar{f}}_i,\ \bar{\bar{f}}_i'\in L^2(0T\times\Omega_1). \qquad (5.4)$$

Then problem 1′ has at least one solution such that

$$J,\ J'\in L^\infty(0T,\overset{o}{H}{}^2(\Omega)),\ J''\in L^\infty(0T,L^2(\Omega)). \qquad (5.5)$$

The proof proceeds by the standard technique of time-discretization. At every time step the existence of the solution is guaranteed by prop. 1. Similarly, the regularization procedure can be applied. Analogously to problem 1′ we can formulate a hemivariational inequality with respect to the variation z-J if instead of (5.1), (2.9) holds (cf. also [15]) Summarizing it can be said that both in the static and the dynamic problems the existence proofs are constructive and supply a method for approximating a solution of the problem.

REFERENCES

1. FICHERA,G. Boundary Value Problems in Elasticity with Unilateral Constraints, p.p. 150-245, of S. Flügge (Ed.) *Encyclopedia of Physics VIa/2*. Springer-Verlag, Berlin, Heidelberg, New York (1972).
2. DUVAUT,G. and LIONS,J.L., *Les inéquations en Mécanique et en Physique*. Dunod, Paris (1972).
3. PANAGIOTOPOULOS P.D., Non-convex Energy Functions. Hemivariational Inequalities and Substationarity Principles. *Acta Mechanica* 48, 111-130 (1983).
4. PANAGIOTOPOULOS P.D., Nonconvex Superpotentials in the sense of F.H.Clarke and Applications. *Mech.Res.Comm.* 8, 335-340 (1981)
5. PANAGIOTOPOULOS P.D. and BANIOTOPOULOS C.C., A Hemivariational Inequality and Substationarity approach to the interface problem: Theory and prospects of applications. *Eng.*

Anal. 1, 20-31 (1984.
6. PANAGIOTOPOULOS P.D., Nonconvex Problems of Semipermeable Media and Related Topics. To appear in *ZAMM*, 64 (1984).
7. PANAGIOTOPOULOS P.D., Une generalization non-convexe de la notion du sur-potentiel et ses applications. *C.R. Acad.Sc. Paris* 296 II 1105-1108 (1983).
8. PANAGIOTOPOULOS P.D., Hemivariational Inequalities. Existence and Approximation Results. *Proc. 2nd CISM Conf. on Unilateral Problems in Structural Analysis, Ravello Italy 1983.* Springer Verlag,Wien 1984.
9. CLARKE F.H., Generalized gradients and applications. *Trans. A.M.S.* 205, 247-262 (1975).
10. CLARKE F.H., *Optimization and Nonsmooth Analysis.* Wiley-Interscience, New York (1983).
11. CHANG K.C., Variational Methods for Nondifferentiable Functionals and their Applications to Partial Differential Equations. *J. Math. Anal. Appl.* 80, 102-129 (1981).
12. ODEN J.T, and REDDY J.N., *An Introduction to the Mathematical Theory of Finite Elements.* Wiley-Interscience, New York(1976).
13. HASLINGER J., and PANAGIOTOPOULOS P.D., On the Approximation of the Solution of Certain Hemivariational Inequalities. To appear.
14. McKENNA P.J. and RAUCH J., Strongly Nonlinear Perturbations of Nonnegative Boundary Value Problems with Kernel. *J.Dif.Eq.* 28, 253-269 (1978).
15. RAUCH J., Discontinuous Semilinear Differential Equations and Multiple Valued Maps. *Proc.A.M.S.* 64, 277-282 (1977).

ON OPTIMAL SHAPE DESIGN OF AN ELASTIC BODY ON A RIGID FOUNDATION

J. Haslinger

Charles University, Prague, Czechoslovakia

P. Neittaanmäki

Lappeenranta University of Technology, Finland

T. Tiihonen

University of Jyväskylä, Finland

1. INTRODUCTION

During the last ten years optimal shape design problems in connection with the finite element method have become very popular and very important from the practical point of view. A relatively large number of papers have been devoted to finding an optimal shape in problems which are governed by variational equations of elliptic type (see [7] and references therein). On the other hand, only few papers analyze problems in which the state variational equation is replaced by an elliptic inequality (see [1-4], for example). One of the most important applications of variational inequalities in elasticity is the so called contact problem.

In this contribution we firstly study the existence of the optimal shape for an elastic body unilaterally supported by a rigid foundation. Secondly, we apply finite elements to these problems and analyze the relation of the continuous and the discrete problem. Finally we present some numerical results.

ISBN 0-12-747255-X

2. STATEMENT OF THE OPTIMAL SHAPE DESIGN PROBLEM

Let us consider a two-dimensional elastic body $\Omega = \Omega(\alpha) \subset \mathbb{R}^2$ having the following geometrical structure.

$$\Omega(\alpha) = \{(x_1,x_2) \in \mathbb{R}^2 | a \leq x_1 \leq b, \quad 0 \leq \alpha(x_1) \leq x_2 \leq \gamma\},$$

$a,b,\gamma \geq 0$ given constants and $\alpha \in C^{0,1}([a,b])$ a function;

$$\partial\Omega(\alpha) = \overline{\Gamma}_D \cup \overline{\Gamma}_F \cup \overline{\Gamma}_C(\alpha),$$

where the shape of the contact surface is defined by a control parameter α from the set $\mathcal{U}_{ad}$ of admissible controls,

$$\mathcal{U}_{ad} = \{\alpha \in C^{0,1}([a,b]) | 0 \leq \alpha(x_1) \leq \gamma, \ \alpha(a) = A, \quad \alpha(b) = B,$$
$$\left|\frac{d}{dx_1}\alpha\right| \leq C_1, \quad \text{meas}\ (\Omega(\alpha)) = C_2\}, \tag{2.1}$$

A, B, C_1, C_2 are given positive constants (see FIG. 1.)

Suppose that the body $\Omega(\alpha)$ is unilaterally supported by a rigid frictionless foundation (here by the set $\{(x_1,x_2) \in \mathbb{R}^2 | x_2 \leq 0\}$) and subjected to a body force $F = (F_1,F_2)$ and to a surface traction $P = (P_1,P_2)$ on Γ_F.

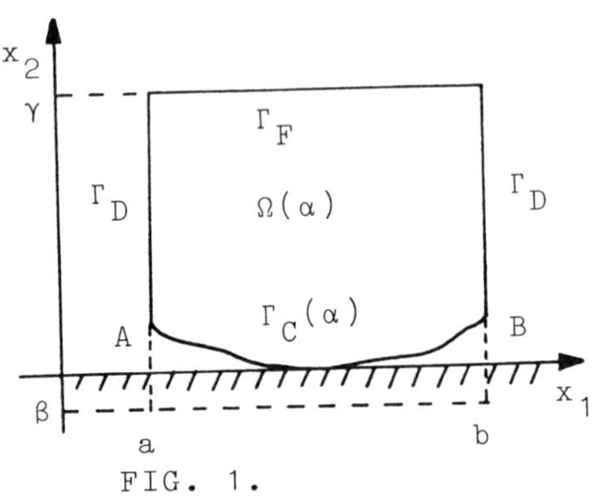

FIG. 1.

In the classical formulation of contact problems one looks for a displacement field $u = u(\alpha) = (u_1(\alpha),u_2(\alpha))$ satisfying the equilibrium equations (the dependence of u on α is emphasized by writing $u = u(\alpha)$)

$$\frac{\partial}{\partial x_j}\tau_{ij} + F_i = 0 \quad \text{in}\ \Omega(\alpha), \quad i = 1,2 \tag{2.2}$$

where the stress tensor $\tau(u) = \{\tau_{ij}(u)\}_{i,j=1}^{2}$ is related to the strain tensor $\varepsilon(u) = \{\varepsilon_{ij}(u)\}_{i,j=1}^{2}$ by means of the linear Hooke's law:

$$\tau_{ij}(u) = C_{ijkl}\, \varepsilon_{kl}(u), \quad \varepsilon_{kl}(u) = \frac{1}{2}\left\{\frac{\partial u_k}{\partial x_l} + \frac{\partial u_l}{\partial x_k}\right\}$$

with elasticity coefficients C_{ijkl} satisfying the usual symmetry and ellipticity conditions. The following boundary conditions will be assumed (cf FIG. 1.):

$$u_i = 0 \quad \text{on } \Gamma_D, \ i = 1,2 \tag{2.3}$$

$$\tau_{ij}(u)n_j = P_i \quad \text{on } \Gamma_F, \ i = 1,2 \tag{2.4}$$

$$u_2(x_1,\alpha(x_1)) \geq -\alpha(x_1) \ \forall\, x_1 \in [a,b] \tag{2.5}$$

$$T_1(u) = 0 \quad \text{on } \Gamma_C(\alpha) \tag{2.6}$$

$$T_2(u) \leq 0, \ (u_2 + \alpha)T_2(u) = 0 \quad \text{on } \Gamma_C(\alpha), \tag{2.7}$$

where the standard notations and conventions of elasticity are used ([6]).

In order to give the variational inequality formulation of (2.2) - (2.7) we introduce a Hilbert space $V(\alpha)$ of virtual displacements

$$V(\alpha) \equiv V(\Omega(\alpha)) = \{v \in [H^1(\Omega(\alpha))]^2 \mid v_i = 0 \text{ on } \Gamma_D, \ i = 1,2\} \tag{2.8}$$

and its closed convex subset $K(\alpha)$ of admissible displacements:

$$K(\alpha) \equiv K(\Omega(\alpha)) = \{v \in V(\alpha) \mid v_2(x_1,\alpha(x_1)) \geq -\alpha(x_1) \ \forall\, x_1 \in [a,b]\} \tag{2.9}$$

The variational form of (2.2) - (2.7) reads as follows:

Find $u = u(\alpha) \in K(\alpha)$ such that

$$(\mathcal{P}) \quad (\tau(u),\varepsilon(v-u))_{0,\Omega(\alpha)} \geq \langle L, v-u\rangle_\alpha \ \forall\, v \in K(\alpha),$$

where L is the given distribution of external forces,

$$\langle L,\xi\rangle_\alpha = \int_{\Omega(\alpha)} F_i\xi_i \, dx + \int_{\Gamma_F} P_i\xi_i \, ds \tag{2.10}$$

with $F \in [L^2(\Omega_\beta)]^2$, $P \in [L^2(\Gamma_F)]^2$, $\Omega_\beta = (a,b) \times (\beta,\gamma)$

and $(\tau(u),\varepsilon(v))_{0,\Omega(\alpha)} = \int_{\Omega(\alpha)} \tau_{ij}(u)\varepsilon_{ij}(v) \, dx.$

We state the shape optimization problem.

Problem ($\mathbb{P}$). *Find* $\alpha^* \in U_{ad}$ *such that*

$$E(u(\alpha^*);\alpha^*) = \min_{\alpha \in U_{ad}} E(u(\alpha);\alpha), \tag{2.11}$$

where $E\colon K(\alpha) \times U_{ad} \to \mathbb{R}$ *is total potential energy functional,*

$$E(u(\alpha);\alpha) = \frac{1}{2}(\tau(u(\alpha)),\varepsilon(u(\alpha)))_{0,\Omega(\alpha)} - \langle L,u(\alpha)\rangle_\alpha$$

in which $u(\alpha)$ *is the solution of the state inequality* (2.9).

It holds [4]

Theorem 1.1. Let U_{ad} *be given by* (2.1). *Then there exists at least one solution* α^* *of Problem* ($\mathbb{P}$).

To avoid the difficulties with non-differentiability of the mapping $\alpha \to u(\alpha)$, the following approach will be used. Instead of the state variational inequality (2.9) a family of penalized problems will be introduced: Find $u_\varepsilon = u_\varepsilon(\alpha) \in V(\alpha)$ such that

$$(P_\varepsilon) \quad (\tau(u_\varepsilon),\varepsilon(v))_{0,\Omega(\alpha)} + \frac{1}{\varepsilon}\int_a^b ([u_{2\varepsilon}+\alpha]^-)^2 \, v \, dx_1 = \langle L,v\rangle_\alpha, \quad \forall\, v \in V(\alpha),$$

where $\varepsilon \to 0+$ is a penalty parameter. The penalized optimal shape design problem ($\mathbb{P}_\varepsilon$) now reads as follows:

Problem ($\mathbb{P}_\varepsilon$). *Find* $\alpha_\varepsilon^* \in U_{ad}$ *such that*

$$E(u_\varepsilon(\alpha_\varepsilon^*);\alpha_\varepsilon^*) = \min_{\alpha \in U_{ad}} E(u_\varepsilon(\alpha);\alpha), \tag{2.12}$$

where $u_\varepsilon(\alpha) \in V(\alpha)$ *is the solution of* (P_ε).

The solvability of $(\mathbb{P}_\varepsilon)$ and the relation between $(\mathbb{P})$ and $(\mathbb{P}_\varepsilon)$ is given in the following theorem. Let us choose a sequence of positive numbers $\{\varepsilon_k\}$ such that $\varepsilon_k \to 0$ as $k \to \infty$. We have (cf [2]).

Theorem 2.2. For every $\varepsilon_k > 0$ *Problem* $(\mathbb{P}_{\varepsilon_k})$ *has at least one solution* $\alpha_k^* \in \mathcal{U}_{ad}$. *Moreover, there exists a subsequence* $\{\alpha_{k_j}^*, u_{k_j}(\alpha_{k_j}^*)\} \subset \{\alpha_k^*, u_k(\alpha_k^*)\}$ *and elements*

$$\{\alpha^*, u(\alpha^*)\} \in \mathcal{U}_{ad} \times K(\alpha^*) \text{ such that}$$

$$\alpha_{k_j}^* \to \alpha^* \quad \text{in} \quad C^0([a,b]), \text{ for } \quad j \to \infty$$

$$u_{k_j}(\alpha_{k_j}^*) \to u(\alpha^*) \text{ (weakly) in } \quad H^1(G_m(\alpha^*)), \text{ for } \quad j \to \infty$$

and for any m, where $G_m(\alpha^*) = (a,b) \times (\alpha^* + 1/m, \gamma)$ *and where* α^* *is the solution of Problem* $(\mathbb{P})$ *and* $u(\alpha^*)$ *is the corresponding state, solving the state problem (P).*

3. APPROXIMATION BY FINITE ELEMENTS

Approximation of the problem $(\mathbb{P})$ will be defined starting from $(\mathbb{P}_\varepsilon)$ by means of finite elements. We suppose that $\mathcal{U}_{ad}$ is replaced by the set

$$\mathcal{U}_{ad}^h = \{\alpha \in C([a,b]) \,|\, \alpha|_{[a_{i-1},a_i]} \in P_1([a_{i-1},a_i])\} \cap \mathcal{U}_{ad} \qquad (3.1)$$

where $a = a_0 < a_1 < \ldots < a_N = b$ is the partition of $[a,b]$, P_1 denotes the set of *linear functions*. For any $\alpha_h \in \mathcal{U}_{ad}^h$ we define $\Omega(\alpha_h) = (a,b) \times (\alpha_h, \gamma)$, i.e., the variable part of the boundary $\Gamma_C(\alpha)$ is now approximated by a piecewise linear arc $\Gamma_C(\alpha_h)$. Let $\mathcal{T}_h$ be a triangulation of $\Omega(\alpha_h)$. To any $\mathcal{T}_h$ a finite dimensional space $V_h(\alpha_h) \subset V(\Omega(\alpha_h))$ will be associated:

$$V_h(\alpha_h) = \{v_h \in [C(\overline{\Omega(\alpha_h)})]^2 \,|\, v_h|_{T_i} \in [P_1(T_i)]^2,\ v_h = 0 \text{ on } \Gamma_u\}$$

The approximation of $(\mathbb{P})$ is now defined as follows

Problem $(\mathbb{P}_{\varepsilon(h)})_h$. *Find* $\alpha^*_{\varepsilon h} \in \mathcal{U}^h_{ad}$ *such that*

$$E_h(u_{\varepsilon h}(\alpha^*_{\varepsilon h});\alpha^*_{\varepsilon h}) = \min_{\alpha_h \in \mathcal{U}^h_{ad}} E_h(u_{\varepsilon h}(\alpha_h), \alpha_h)$$

where $u_{\varepsilon h}(\alpha_h) \in V_h(\alpha_h)$ *is the solution of*

$$(\tau(u_{\varepsilon h}), \varepsilon(v_h))_{0,\Omega(\alpha_h)} + \frac{1}{\varepsilon(h)} \int_a^b ([u_{2\varepsilon h} + \alpha_h]^-)^2 v_h \, dx_1$$

$$= \langle L, v_h \rangle_{\alpha_h} \quad \forall\, v_h \in V_h(\Omega(\alpha_h)) \qquad (3.2)$$

where $\varepsilon = \varepsilon(h) \to 0+$ *iff* $h \to 0+$

The solvability of the Problem $(\mathbb{P}_{\varepsilon(h)})_h$ can be proven with compactness arguments. Assuming the regularity of the triangulation $\mathcal{T}_h$ one can prove.

Theorem 3.1. Let $\alpha^*_h \in \mathcal{U}^h_{ad}$ *be a solution of Problem* $(\mathbb{P}_{\varepsilon(h)})_h$ *and let* $u_{\varepsilon h}$ *be the corresponding solution of the state equation* (3.2). *Then there exists a subsequence* $\{\alpha^*_{h_j}\} \subset \{\alpha^*_h\}$, *an element* $\alpha^* \in \mathcal{U}_{ad}$ *and* $\mathbf{y}(\alpha^*) \in K(\alpha^*)$ *such that*

$$\alpha^*_{h_j} \to \alpha^* \qquad \text{in } C^0([a,b]), \text{ for } h_j \to 0+$$

$$u_{\varepsilon h_j}(\alpha^*_{h_j}) \rightharpoonup u(\alpha^*) \quad \text{in } H^1(G_m(\alpha)), \text{ for } h_j \to 0+,$$

and for any $m \geq m_0$, *where* α^* *is the solution of Problem* $(\mathbb{P})$, $u(\alpha^*)$ *the corresponding state and* $G_m(\alpha^*)$ *is defined as in Theorem 2.2.*

The proof will be presented in [5] (cf [3]).

4. NUMERICAL RESULTS

We consider two types of test examples: The undeformed body is in contact at point $(a+b)/2$ (Example 1) and at points a and b (Example 2).

Example 1. The parameters for $\mathcal{U}^h_{ad}$ (corresponding to (2.1) and (3.1)) are: $a=0$, $b=8$, $A=B=.05$, $\gamma=1$, $C_1=.05$, $C_2=7.8$. Let the body force $F\equiv 0$ and let $P_2=10^5$ for $2\leq x_1 \leq 6$, $=0$ elsewhere, $P_1\equiv 0$. The Young's modulus $=2.2 \times 10^6$ and the Poisson ratio $=.28$.

Due to the symmetry with respect to $x_1=4$ only the half of the body is discretized (128 triangular elements, 9x9 grid). In FIG. 2 we see the initial design $\Gamma_C(\alpha_h)$ before (•) and after (x) deformation, the nodes in contact are denoted by (⊗)

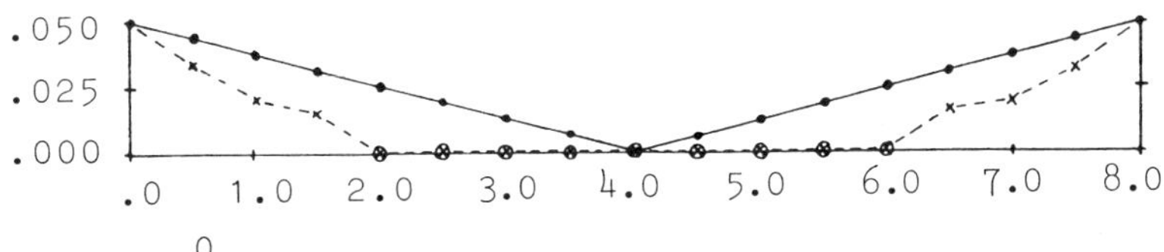

FIG. 2. $E^0 = -9.13$

In order to find a better design for $\Gamma_C(\alpha_h)$ we have applied gradient algorithm (minimization routine VE01A of Harwell Subroutine Library). The gradient of cost functional was computed applying material derivatives or nodal derivatives ([3]). In FIG. 3 we see the situation after 20 iterations.

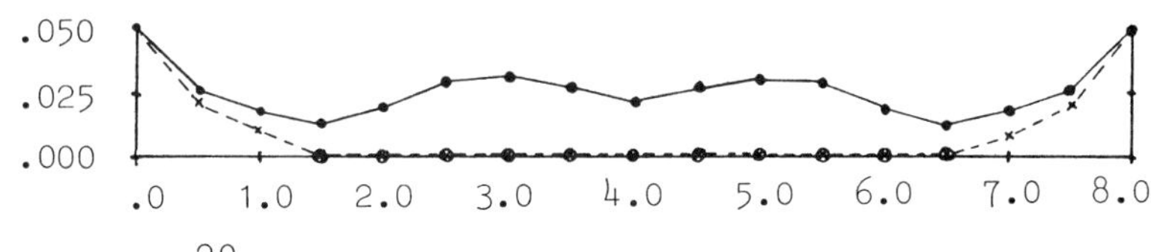

FIG. 3. $E^{20} = -14.80$

We find that the value of cost functional is reduced from -9.13 to -14.80. After deformation the body is in contact from $x_1=1.5$ to $x_1=6.5$

Example 2. The parameters A and B are chosen to be =0; the others being the same as in Example 1 FIG. 4 shows the initial situation and FIG. 5 the result after 10 iterations.

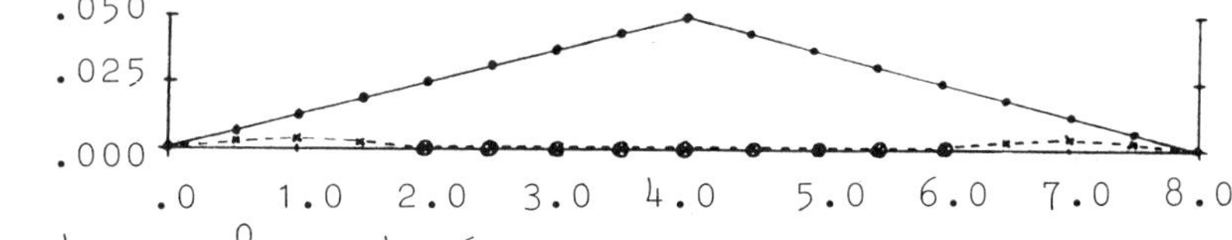

FIG. 4. $E^0 = -24.16$

.050
.025
.000
.0 1.0 2.0 3.0 4.0 5.0 6.0 7.0 8.0

FIG.5. $E^{10} = -25.62$

We note that the change in the value of the cost functional is not significant although the contact zone expands to include $\Gamma_C(\alpha_h)$ completely.

For further numerical experiments we refer to [5].

REFERENCES

1. BENEDICT, R.L. and TAYLOR, J.E., Optimal design for elastic bodies in contact. pp. 1553-1569 of E.J. Haug and J. Cea (Ed.), *Optimization of Distributed Parameter Structures*, Vol. II. Nato Advanced Study Institute Series, Series E, no 49, Sijthoff & Noordhoff, Alphen aan den Rijn (1981).

2. HASLINGER, J. and NEITTAANMÄKI, P., Penalty method in design optimization of systems governed by a unilateral boundary value problem. *Ann. Fac. Sci. Tolouse*, Vol. V, 199-216 (1983).

3. HASLINGER, J. and NEITTAANMÄKI, P., On optimal shape design of systems governed by mixed Dirichlet - Signorini boundary value problem. *Math. Meth. Appl. Sci.* (to appear).

4. HASLINGER, J. and NEITTAANMÄKI, P., On the existence of optimal shape in contact problems, University of Jyväskylä, *Dept. Math. Preprint* 23, 1983.

5. HASLINGER, J., NEITTAANMÄKI, P. and TIIHONEN, T., to appear.

6. HLÁVAČEK, I., HASLINGER, J., NEČAS, J. and LOVÍŠEK, J., Numerical solution of variational inequalities (in Slovak), *ALFA, SNTL* (1982), English translation, to appear.

7. PIRONNEAU, O., *Optimal shape design for elliptic systems* Springer Series in Comput. Physics, Springer Verlag, New York, (1984).

NODAL METHODS FOR THE TRANSPORT EQUATION

Pierre Lesaint

Universite de Franche-Comté, 2503 Besancon, France

1. INTRODUCTION

During the last few years, various new methods have been derived for solving the discrete-ordinates transport problems. Among them is the so called discontinuous finite element method defined by W.H. REED and T.R. HILL [5], mathematically studied by P. LESAINT and P.A. RAVIART [3] and more recently by C. JOHNSON [2]. This method is always stable and rather accurate, and can be used for rectangular (x,y), cylindrical and spherical geometry, with either triangular or convex quadrilateral meshes. Characteristic methods have been extensively studied and used (see for example [1] and the references of this paper). More recently modal methods have been defined for rectangular meshes [6]. The aim of this paper is to sketch a mathematical study of these methods and to show how they can be extended to triangular meshes. For detailed proofs, we refer to a forthcoming paper [4]. Essentially, we get an error of order $h^{1/2}$ in the maximum norm and of order h in the mean square norm, when using a constant approximation of the solution along each side of the mesh cell.

2. NODAL METHODS ON RECTANGLES

Consider the discrete-ordinates transport in (x,y) geometry for one energy group, and for the angular direction (μ,ν)

$$\mu \frac{\partial \phi}{\partial x} + \nu \frac{\partial \phi}{\partial y} + \sigma\phi = S \quad \text{in} \quad \Omega =]0,1[\times]0,1[\tag{2.1}$$

$$\phi = 0 \qquad \text{on } \partial_-\Omega = \{(x,y) \ \partial\Omega \ ; \ \mu n_x + \nu n_y < 0\} \tag{2.2}$$

where $\vec{n} = (n_x, n_y)$ denotes the outer normal on $\partial\Omega$. We assume in

THE MATHEMATICS OF FINITE ELEMENTS AND APPLICATIONS V

ISBN 0-12-747255-X

the sequel that $\mu > 0$, $\nu > 0$. The method will be defined for positive, piece wise constant cross section σ. For approximating the flux ϕ on a mesh $K = [a,b] \times [c,d]$, we first define three distincts representations of ϕ. The first one $\phi_x(x,y)$ (resp. second one $\phi_y(x,y)$) is a polynomial with degree $\leqslant k$ with respect to the variable x (resp. y). These polynomials will be expanded in term of their k+1 first moments of order $\leqslant k$, for example, if k=1.

$$\frac{1}{\Delta y}\int_c^d \phi_y(x,y)dy \quad , \quad \frac{12}{\Delta y^2}\int_c^d (y-\frac{1}{2}(c+d))\,\phi_y(x,y)dy,$$

where $\Delta y = d-c$, and more generally $\int_c^d \phi_y(x,y)\,w(y)dy$ where the functions w are suitably chosen polynomials with degree $\leqslant k$. The third representation $\Pi\phi$ of ϕ is a polynomial with degree $l \geqslant 0$.

Multiplying equation (2.1) by w(y) and integrating with respect to y, we define the following system of differential equations in x.

$$\mu\frac{d}{dx}\left(\int_c^d \phi_y(x,y)\,w(y)dy\right) + \sigma\int_c^d \phi_y(x,y)\,w(y)dy =$$

$$\int_c^d \left(\bar{S}w + \nu\phi_y(x,y)\frac{dw}{dy} - \nu\frac{d}{dy}(\phi_x(x,y)w)\right)dy \qquad (2.3)$$

for all polynomials w with degree $\leqslant k$. We also write for all polynomials v with degree $\leqslant k$.

$$\nu\frac{d}{dy}\int_a^b \phi_x(x,y)\,v(x)dx + \sigma\int_a^b \phi_x(x,y)\,v(x)dx =$$

$$\int_a^b \left(\bar{S}v + \mu\phi_x(x,y)\frac{dv}{dx} - \mu\frac{d}{dx}(\phi_y(x,y)v)\right)dx \qquad (2.4)$$

we define $\bar{S}$ as a projection of S on the space P(l) of the polynomials with degree $\leqslant l$.

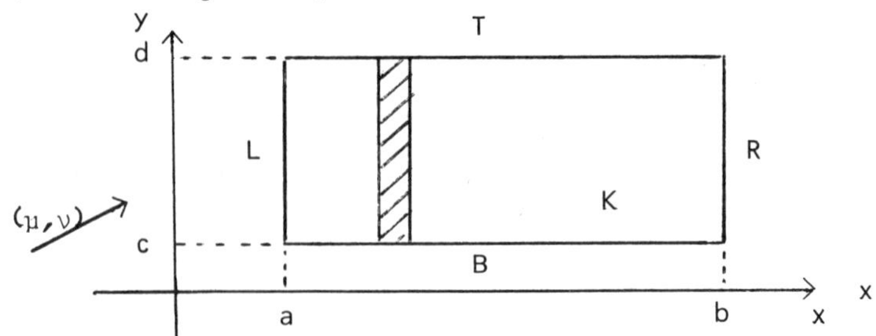

FIG. 1. The rectangular mesh K. The differential equation (2.3) is being integrated from the left to the right.

Relations (2.3) and (2.4) yield a system of 2k+2 linear differential equations allowing the calculation of the k+1 first moments of the functions $\phi_x(x,d)$ and $\phi_y(b,y)$, assuming that the functions $\phi_x(x,c)$ and $\phi_y(a,y)$ are already known from previous calculations.

To compute the third representation $\Pi\phi$ of the flux ϕ, to be used, for example when solving an eigenvalue problem, to construct the source term for the next iteration, we average the balance equation (2.1) as follows

$$\int_a^b \int_c^d (\Pi\phi)\ (-\mu \frac{\partial\Psi}{\partial x} - \nu \frac{\partial\Psi}{\partial y} + \sigma\Psi) - S\Psi\ )dx\ dy +$$
$$\nu \int_a^b ((\phi_x\Psi)(x,d) - (\phi_x\Psi)(x,c))dx + \mu\int_c^d ((\phi_y\Psi)(b,y) - (\phi_y\Psi)(a,y))dy = 0 \quad (2.5)$$

for all Ψ belonging to the space P(l).

For k=l=0, k=l-1=0, k=l=1, we get respectively the constant (C.C.), the constant linear (C.L.) and the linear linear (L.L.4) schemes [6]. In the case of the linear linear scheme, further approximation of the fluxes ϕ_x and ϕ_y are done inside the element K, permitting to decouple the system of differential equations [7].

3. STABILITY AND ERROR BOUNDS FOR THE C.C. AND C.L. SCHEMES

The only differences between these two schemes are the approximation $\bar{S}$ of the source term S and the definition $\Pi\phi$ of the flux inside the meshes, so that the stability properties are identically the same for both schemes.

We let $\phi_T = \frac{1}{\Delta x}\int_a^b \phi_x(x,d)dx,\quad \phi_R = \frac{1}{\Delta y}\int_c^d \phi_y(b,y)dy,$

with similar definitions for ϕ_B and ϕ_L. We get the following system of two equations with two unknowns ϕ_T and ϕ_R :

$$\phi_T + \frac{1-e^{-\beta}}{\alpha}\ \phi_R = e^{-\beta}\ \phi_B + \frac{1-e^{-\beta}}{\alpha}\ \phi_L + \frac{1-e^{-\beta}}{\sigma}\ \bar{S} \quad (3.1)$$

$$\phi_R + \frac{1-e^{-\alpha}}{\beta}\ \phi_T = e^{-\alpha}\ \phi_L + \frac{1-e^{-\alpha}}{\beta}\ \phi_B + \frac{1-e^{-\alpha}}{\sigma}\ \bar{S} \quad (3.2)$$

where $\alpha = \frac{\sigma\Delta x}{\mu}$, $\beta = \frac{\sigma\Delta y}{\nu}$. We let the matrices

$$\underline{A} = \begin{bmatrix} 1 & \frac{1}{\alpha}(1-e^{-\beta}) \\ \frac{1}{\beta}(1-e^{-\alpha}) & 1 \end{bmatrix} \qquad \underline{B} = \begin{bmatrix} e^{-\beta} & \frac{1}{\alpha}(1-e^{-\beta}) \\ \frac{1}{\beta}(1-e^{-\alpha}) & e^{-\alpha} \end{bmatrix}$$

We have the following properties : det A > 0 and the eigen values of the matrix $\underline{P} = \underline{A}^{-1}\,\underline{B}$ are real, of opposite signs and have absolute values ≤ 1, which means that an Von Neumann stability condition is guaranteed. For small values of α and β, with $\alpha \neq \beta$, the diagonal entries of the matrix P are of opposite sign, which indicates that negative fluxes may occur.

For $\alpha = \beta$ and for small values of α, we have

$$\phi_T = O(1)\ \ \phi_L + O(h)\ \phi_B$$

$$\phi_R = O(1)\ \ \phi_B + O(h)\ \phi_L$$

with h = max $(\Delta x, \Delta y)$, which indicates that the method behaves nearly like a characteristic method.

For $\alpha = \beta$, the matrix $\underline{P}$ is symmetric, with spectral radius ≤ 1, and we have stability for the euclidean norm. For $\alpha \neq \beta$, the matrix $\underline{P}$ is no longer symmetric and the spectral radius of $\underline{P}^t\underline{P}$ may be bigger than one since the sum s of the eigenvalues satisfies for small values of α and β :

$$s \# 2 + 4\ \left(\frac{\alpha-\beta}{\alpha+\beta}\right)^2 ,$$

so that stability in the euclidean norm is no longer true.

We let $\underline{Q} = \begin{bmatrix} 1 & 1 \\ 1 & -\frac{\alpha}{\beta} \end{bmatrix}$, $\underline{Y} = \underline{Q}^{-1} \begin{bmatrix} \phi_T \\ \phi_R \end{bmatrix}$ $\underline{X} = \underline{Q}^{-1} \begin{bmatrix} \phi_B \\ \phi_L \end{bmatrix}$

We have

$$\underline{Y} = \underline{Q}^{-1}\ \underline{P}\ \underline{Q} + \text{source terms}$$

For small values of α and β, we can show that the norm of the matrix $\underline{Q}^{-1}\,\underline{P}\,\underline{Q}$, induced by the maximum norm of vectors in R^2 is smaller than $1-Ch$ for some constant C.

4. ERROR BOUNDS FOR THE C.C. AND C.L. SCHEMES

Let $S_y(x) = \frac{1}{\Delta y}\int_c^d S(x,y)\,dy$, $\overline{\phi}_H = \frac{1}{\Delta x}\int_a^b \phi(x,d)\,dx$,

with similar definitions for $S_x(y)$, $\overline{\phi}_B$, $\overline{\phi}_R$ and $\overline{\phi}_L$. We the have

$$\begin{bmatrix} \phi_H - \overline{\phi}_H \\ \phi_R - \overline{\phi}_R \end{bmatrix} = \underline{P} \begin{bmatrix} \phi_B - \overline{\phi}_B \\ \phi_L - \overline{\phi}_L \end{bmatrix} + \underline{A}^{-1} \begin{bmatrix} E_y \\ E_x \end{bmatrix} \tag{4.1}$$

where $E_y = \int_c^d \frac{1}{\nu}(\overline{S} - S_x(y))\,e^{-\frac{\sigma}{\nu}(d-y)}\,dy$

$$+ \frac{\mu}{\nu\Delta y}\int_c^d (\phi(b,y)-\overline{\phi}_R - (\ (a,y)-\overline{\phi}_L))\,e^{-\frac{\sigma}{\nu}(d-y)}\,dy,$$

with a similar definition for E_x.

Using an extension of Bramble and Hilbert Lemma [3], we show that E_x and E_y are bounded by $O(h^2)$. After summation over all the meshes, we get the following.

Theorem 4.1. - For h sufficiently small, we have

$$\max (|\phi_H - \overline{\phi}_H| + |\phi_R - \overline{\phi}_R|) = O(h^{1/2}),$$

the maximum being taken on all the meshes of the domain.

Let $\tilde{\phi}$ denote the mean value of the flux ϕ on the rectangle K we then have for small values of α and β :

$$\Pi\phi - \phi = - \frac{\alpha}{\alpha+\beta} (\overline{\phi}_B - \phi_B) - \frac{\beta}{\alpha+\beta} (\overline{\phi}_L - \phi_L) +$$

$$\frac{1}{\beta \det A} (E_y - \frac{1-e^{\beta}}{\alpha} E_x) + \frac{1}{\alpha \det A} (E_x - \frac{1-e^{-\alpha}}{\beta} E_y)$$

and we get the following.

Theorem 4.2. - For h sufficiently small, we have

$$\left(\Delta x \ \Delta y \sum_{K \subset \Omega} (\Pi - \tilde{\phi})^2\right)^{1/2} = O(h).$$

Remark 4.1. - For the C.L. method and when $\alpha = \beta$ we get for both theorems 4.1 an 4.2 an error of order $h^{3/2}$.

Remark 4.2. - The results of both theorems are pessimistic since the numerical experiments show a better order of convergence.

5. NODAL METHODS ON TRIANGLES

Let K be the triangle A B C, with two sides C A and C B determined by characteristic direction (μ,ν). Define the mapping F_K by

$$x = \frac{1}{2} (\xi-\eta)x_A + \frac{1}{2} (\xi+\eta)x_B + (1-\xi)x_C \tag{5.1}$$

$$y = \frac{1}{2} (\xi-\eta)y_A + \frac{1}{2} (\xi+\eta)y_B + (1-\xi)y_C \tag{5.2}$$

The triangle K is the image of the reference triangle K by the mapping F_K, as shown on figure 2.

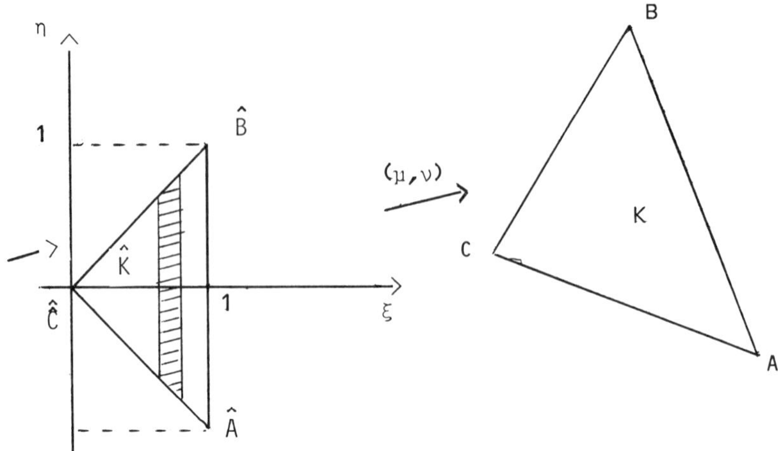

FIG. 2. Triangles $\hat{K}$ and K.

To any function v defined on K, we associate $\hat{v} = v_o F_K^{-1}$. Using this coordinate transformation, equation (2.1) may be written as :

$$z \frac{\partial\hat{\phi}}{\partial\xi} + t \frac{\partial\hat{\phi}}{\partial\eta} + \sigma \hat{\phi} = s \tag{5.3}$$

where $\tilde{\sigma} = \hat{\sigma} J_K$, $\tilde{S} = \hat{S} J_K$, $J_K = \dfrac{\text{area } K}{\text{area } \hat{K}}$.

It is easy to see that $z > 0$ so that the sides $\hat{C}\hat{A}$ and $\hat{C}\hat{B}$ of triangle $\hat{K}$ are determined by the characteristic direction (z,t). Integrating with respect to η, we have the following differential equation :

$$z \frac{d}{d\xi}\left(\int_{-\xi}^{\xi} \hat{\phi} d\eta\right) + \tilde{\sigma}\int_{-\xi}^{\xi} \hat{\phi}\, d\eta = \int_{-\xi}^{\xi} \tilde{S} d\eta + (z+t)\hat{\phi}(\xi,-\xi) + (z-t)\hat{\phi}(\xi,\xi) \tag{5.4}$$

The fluxes on each side and the source term $\tilde{S}$ are approximated by their mean values and we get the scheme

$$\hat{\phi}(\hat{A}\,\hat{B}) = \frac{1}{\tilde{\sigma}}\left(1-e^{-\frac{\tilde{\sigma}}{z}}\right)\left((z+t)\hat{\phi}(\hat{C}\,\hat{A}) + (z-t)\hat{\phi}(\hat{C}\,\hat{B})\right) + \left(\frac{1}{\tilde{\sigma}} - \frac{z}{(\tilde{\sigma})^2} + \frac{z}{(\tilde{\sigma})^2} e^{-\frac{\tilde{\sigma}}{z}}\right) \tilde{S}$$

A similar procedure is derived when only one side of the triangle is determined by the characteristic direction. We then have to solve two differential equations.

The method is always stable and we have the same error bounds [4] as for the rectangle case.

REFERENCES.

1. R.E. ALCOUFFE, E.W. LARSEN - A review of characteristic methods used to solve the linear transport equation. Proceedings of the International Topical meeting on Advances in Mathematical Methods for the solution of nuclear engineering problems. Vol. 1, pp. 3-17, München, April 1981.

2. C. JOHNSON - Finite element methods for convection diffusion problems. Presented at the Fifth International Symposium on Computing Methods in Engineering and Applied Sciences, INRIA, Versailles, Décembre 14-18, 1981.

3. P. LESAINT, P.A. RAVIART - On a finite element method for solving the neutron transport equation Mathematical aspects of finite elements in partial differential equations, pp. 89-123, Academic Press, 1974.

4. P. LESAINT - Triangular nodal methods for the transport equation. To appear.

5. W.H. REED, T.R. HILL - Triangular mesh methods for the neutron transport equation. Proc. Amer. Nucl. Soc., 1973.

6. W.F. WALTERS, R.D. O'DELL - Nodal methods for discrete ordinates transport problems in (x,y) geometry. Proceedings of the International Topical meeting an Advances in Mathematical methods for the solution of nuclear engineering problems, Vol. 1, pp. 115-130, München, April 27-29, 1981.

7. W.F. WALTERS, R.D. O'DELL - A comparison of linear nodal, linear discontinuous and diamond schemes for solving the transport equation in (x,y) geometry, LA UR 81-2142 Los Alamos Scientific Laboratory.

LEAST SQUARES FORMULATION OF EXTREMUM PRINCIPLES AND WEIGHTED RESIDUAL METHODS USED IN FINITE ELEMENT CODES FOR SOLVING THE BOLTZMANN EQUATION FOR NEUTRON TRANSPORT

R. T. Ackroyd

UKAEA, Northern Division, Risley, Warrington, Cheshire, England.

1. INTRODUCTION

Deterministic solutions of reactor physics and shielding problems are based on the solution of the one-group Boltzmann equation [1]

$$\underline{\Omega}.\nabla\phi_o + \Sigma\phi_o = S \tag{1.1}$$

for the angular flux $\phi_o(\underline{r},\underline{\Omega})$, at $\underline{r}$ for the direction $\underline{\Omega}$, induced by the distributed source $S(\underline{r},\underline{\Omega})$ and by surface sources. The operator Σ is defined in terms of the removal cross-section $\sigma(\underline{r})$ and the differential scattering cross-section $\sigma_s(\underline{r},\underline{\Omega}.\underline{\Omega}')$ by

$$\Sigma\phi_o = \sigma\phi_o - \int_{\Omega'} \sigma_s(\underline{r},\underline{\Omega}.\underline{\Omega}')\phi_o(\underline{r},\underline{\Omega}')d\Omega'$$

with the integration over all directions. Finite element methods for neutron transport are being developed to provide faster solutions than the stochastic Monte Carlo method, the current procedure for problems on systems with complex geometries.

The parity equations [2]

$$\begin{aligned} \underline{\Omega}.\nabla\phi_o^- + C\phi_o^+ &= S^+ \\ \underline{\Omega}.\nabla\phi_o^+ + G^{-1}\phi_o^- &= S^- \end{aligned} \tag{1.2}$$

with $u^{\pm} = [u(\underline{r},\underline{\Omega}) \pm u(\underline{r},-\underline{\Omega})]/2$, are equivalent to the first-order Boltzmann equation (1.1). The operators C, G and their inverses are positive definite and self-adjoint, and they are related to Σ by

$$\Sigma u = Cu^+ + G^{-1}u^- \quad \text{and} \quad \Sigma^{-1}u = C^{-1}u^+ + Gu^- \tag{1.3}$$

ISBN 0-12-747255-X

The parity equations yield the second-order Boltzmann equations

$$- \underline{\Omega}.\nabla[G\underline{\Omega}.\nabla\phi_o^+] + C\phi_o^+ = S^+ - \underline{\Omega}.\nabla[GS^-] \quad (1.4)$$

$$- \underline{\Omega}.\nabla[C^{-1}\underline{\Omega}.\nabla\phi_o^-] + G^{-1}\phi_o^- = S^- - \underline{\Omega}.\nabla[C^{-1}S^+] \quad (1.5)$$

The even-parity equation (1.4) is the one most frequently used for finite element methods based on variational or weighted residual methods. Table 1 lists the boundary conditions of the various forms of the Boltzmann equation for the exterior surfaces and interfaces of the system in Fig 1 of volume V partitioned by surfaces S_j into sub-regions V_i. In general there is no way known to satisfy the boundary conditions for S_b and S_s, consequently emphasis is placed on methods which minimise the errors incurred by approximate solutions.

Least squares, variational and weighted residual methods have been employed for the finite element methods of solving one form or other of the Boltzmann equation.

The majority of workers use a finite element representation for the spatial representation of the solution; and employ either an expansion in spherical harmonics, or the piece-wise linear interpolation of the discrete ordinates method, to describe the dependence of the solution on neutron direction. Extremum principles provide for the simple, but exacting, edge-cell problem of Fig 2 a good solution with computable precise bounds for the reaction rate in a locality of the system. A straightforward weighted residual method, however, provides unsatisfactory solutions of Fig 3 for the total flux

$$\Phi_o = \int_\Omega \phi_o(\underline{r},\underline{\Omega})\,d\Omega$$

The generalised least squares method reviewed here has the merit of encompassing a number of classical variational and weighted residual methods and leads to extremum principles and associated weighted residual methods which demand no satisfaction of boundary and interface conditions. Principles are obtained [3] for all three forms of the Boltzmann equation from a parent principle according to the scheme of Table 2. Complementary principles can be used with a bi-variational method to obtain upper and lower bounds on a local characteristic of an approximate solution. Some realistic applications [4], [5] are shown in the poster section and in the proceedings of an international seminar [6].

TABLE 1

Comparison of Boundary Conditions for Transport Equations

Equation \ Surface	Bare S_b	S_s with source	Perfect reflector S_{pr}	Interface $S_i \cap S_j$
		$\mathbf{n}$ outward normal for exterior surfaces		$\mathbf{n}_i$ outward normal to S_i
First order	$\phi_0 = 0$ for $\mathbf{\Omega}\cdot\mathbf{n} < 0$	$\phi_0 = T(\mathbf{r},\mathbf{\Omega})$ for $\mathbf{\Omega}\cdot\mathbf{n} < 0$	$\phi_0(\mathbf{r},\mathbf{\Omega}) = \phi_0(\mathbf{r},\mathbf{\Omega}^x)$ for all $\mathbf{\Omega}\cdot\mathbf{n} = -\mathbf{\Omega}^x\cdot\mathbf{n} = 0$ where $\mathbf{\Omega}^x$ is the reflected direction to $\mathbf{\Omega}$	ϕ_0 continuous for $\mathbf{\Omega}\cdot\mathbf{n}_i \neq 0$
Second order	$\phi_0^+ + G[S^- - \mathbf{\Omega}\cdot\nabla\phi_0^+] = 0$ for $\mathbf{\Omega}\cdot\mathbf{n} < 0.$	$\phi_0^+ + G[S^- - \mathbf{\Omega}\cdot\nabla\phi_0^+] = T(\mathbf{r},\mathbf{\Omega})$ for $\mathbf{\Omega}\cdot\mathbf{n} < 0.$	$\phi_0^+(\mathbf{r},\mathbf{\Omega}) = \phi_0^+(\mathbf{r},\mathbf{\Omega}^x)$ $G[S^-(\mathbf{r},\mathbf{\Omega}) - \mathbf{\Omega}\cdot\nabla\phi_0^+(\mathbf{r},\mathbf{\Omega}^x)$ $-S^-(\mathbf{r},\mathbf{\Omega}^x) + \mathbf{\Omega}^x\cdot\nabla\phi_0^+(\mathbf{r},\mathbf{\Omega}^x)] = 0$	ϕ_0^+ and $G[\mathbf{\Omega}\cdot\nabla\phi_0^+ - S^-]$ continuous for $\mathbf{\Omega}\cdot\mathbf{n}_i \neq 0$
Even-parity	$\phi_0^+ - G[S^- - \mathbf{\Omega}\cdot\nabla\phi_0^+] = 0$ for $\mathbf{\Omega}\cdot\mathbf{n} > 0.$	$\phi_0^+ - G[S^- - \mathbf{\Omega}\cdot\nabla\phi_0^+] = T(\mathbf{r},-\mathbf{\Omega})$ for $\mathbf{\Omega}\cdot\mathbf{n} > 0.$		
Odd-parity	$\phi_0^- + C^{-1}[S^+ - \mathbf{\Omega}\cdot\nabla\phi_0^-] = 0$ for $\mathbf{\Omega}\cdot\mathbf{n} < 0.$ $\phi_0^- - C^{-1}[S^+ - \mathbf{\Omega}\cdot\nabla\phi_0^-] = 0$ for $\mathbf{\Omega}\cdot\mathbf{n} > 0.$	$\phi_0^- + C^{-1}[S^+ - \mathbf{\Omega}\cdot\nabla\phi_0^-] = T(\mathbf{r},\mathbf{\Omega})$ for $\mathbf{\Omega}\cdot\mathbf{n} < 0.$ $\phi_0^- - C^{-1}[S^+ - \mathbf{\Omega}\cdot\nabla\phi_0^-] = -T(\mathbf{r},-\mathbf{\Omega})$ for $\mathbf{\Omega}\cdot\mathbf{n} > 0.$	$\phi_0^-(\mathbf{r},\mathbf{\Omega}) = \phi_0^-(\mathbf{r},\mathbf{\Omega}^x)$ $C^{-1}[S^+(\mathbf{r},\mathbf{\Omega}) - \mathbf{\Omega}\cdot\nabla\phi_0^-(\mathbf{r},\mathbf{\Omega})$ $-S^+(\mathbf{r},\mathbf{\Omega}^x) + \mathbf{\Omega}^x\cdot\nabla\phi_0^-(\mathbf{r},\mathbf{\Omega}^x)] = 0$	ϕ_0^- and $C^{-1}[\mathbf{\Omega}\cdot\nabla\phi_0^- - S^+$ continuous for $\mathbf{\Omega}\cdot\mathbf{n}_i \neq 0$

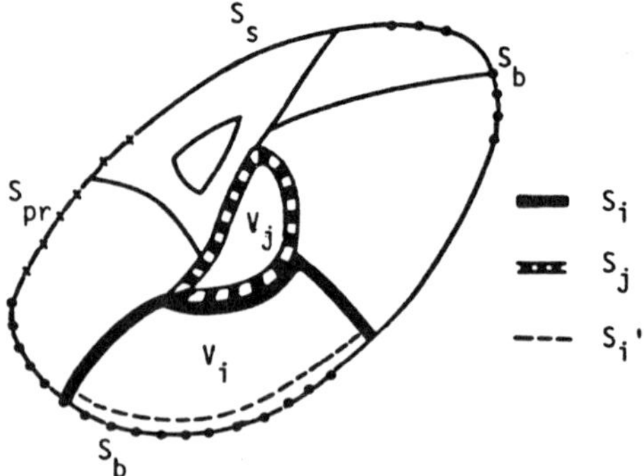

FIG 1. System specification.

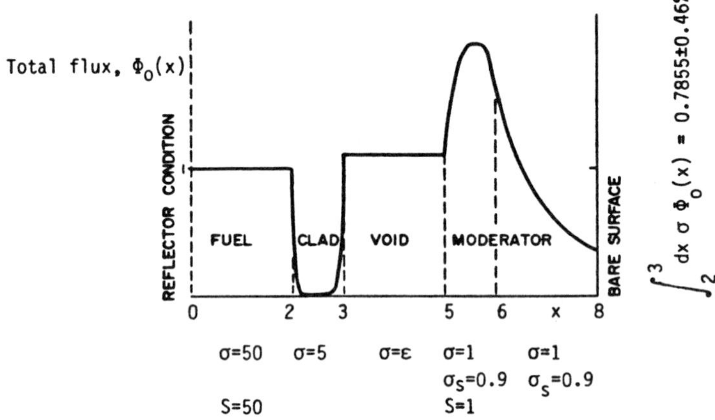

FIG 2. Edge-cell problem solved by extremum principles.

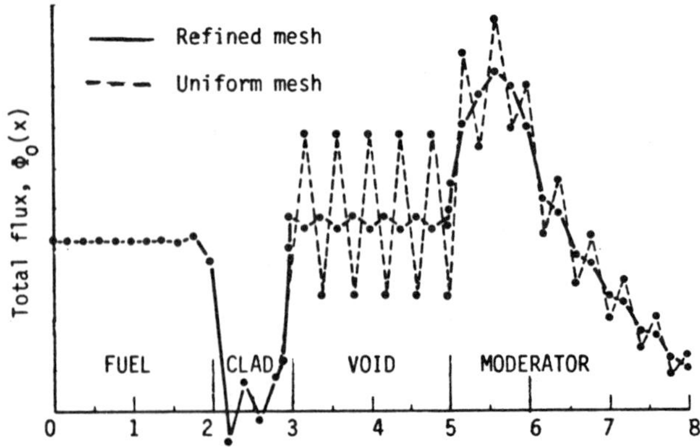

FIG 3. Edge-cell problem solved by a weighted-residual method.

TABLE 2

$K^{\pm}$ Principle Family Tree

Principle	Type	Derivation	Trial functions	Equation	Boundary conditions on trial functions	Use
K^{+-}	Maximum	Least squares	ϕ^+ and ϕ^- for ϕ_0^+ and ϕ_0^-	Transport parity pair	None anywhere, but weights $w^+(\mathbf{r},\mathbf{\Omega})$ and $w^-(\mathbf{r},\mathbf{\Omega})$ assigned to S_{pr} and interfaces	Global approximation
R	Maximum	From K^{+-}	ϕ for ϕ_0	First order	None anywhere. Weight constraint $w^+ = w^-$	Global approximation
$\tilde{R}$	Maximum	Least squares or R	ϕ	First order	Continuity of ϕ across interfaces Reflection condition for S_{pr}	Global approximation
K^+	Maximum	$\phi^- = \phi_0^-$ in K^{+-}	ϕ^+	Second order for ϕ_0^+	Continuity of ϕ^+ across interfaces Reflection condition for S_{pr}	Global approximation (Bi-variational bounds)
K^-	Maximum	$\phi^+ = \phi_0^+$ in K^{+-}	ϕ^-	Second order for ϕ_0^-	Continuity of ϕ^- across interfaces Reflection condition on S_{pr}	Global approximation (Bi-variational bounds)
E^+	Minimum	$\phi^- = G(S^- - \mathbf{\Omega}\cdot\nabla\phi^+)$ in K^{+-}	ϕ^+	Second order for ϕ_0^+	Continuity of ϕ^+ and $G(S^- - \mathbf{\Omega}\cdot\nabla\phi^+)$ across interfaces Reflection condition for ϕ^+ on S_{pr}	Global approximation Local approximation
$\alpha - K^-$	Minimum	$\phi^+ = \phi_0^+$ in K^{+-}	ϕ^-	Second order for ϕ_0^-	As for K^- $2\alpha = \int_V [\langle C^{-1}S^+, S^+\rangle + \langle GS^-, S^-\rangle]\, dV$	Global approximation (Bi-variational bounds)
$K^+ - \alpha$	Maximum	$\phi^- = \phi_0^-$ in K^{+-}	ϕ^+	Second order for ϕ_0^+	As for K^+ $+2\int_{S_e}\int_{\mathbf{\Omega}\cdot\mathbf{n}<0} \lvert\mathbf{\Omega}\cdot\mathbf{n}\rvert\, T^2\, d\Omega\, dS$	Global approximation (Bi-variational bounds) Local approximation

2. GENERALISED LEAST SQUARES METHOD

The method treats the approximation $\phi^{\pm}$ for $\phi_o^{\pm}$ as the exact solution of an approximating system, which differs from the given system by the presence of additional surface and volume sources. The approximating system is forced as close as possible to the given system by minimising a functional

$$W^{+-}(\phi^+,\phi^-) = \left(I + J\right) + \left(I_b + I_s\right) + \left(I_f + I_{pr}\right) \tag{2.1}$$

The first, second and third groupings of terms are measures of the additional volume sources, the additional sources on the exterior surfaces S_b and S_s; and the additional sources on the interfaces $U(S_i \cap S_j)$ and the perfect reflector S_{pr} [2]. The third group is weighted with arbitrary positive weights $w^{\pm}(\underline{r},\underline{\Omega})$ and a positive parameter λ and it caters for discontinuities across interfaces and on reflectors

$$I + J = \int_V\int_\Omega \left\{ \begin{array}{l} [\underline{\Omega}.\nabla\phi^+ + G^{-1}\phi^- - S^-]G[\underline{\Omega}.\nabla\phi^+ + G^{-1}\phi^- - S^-] \\ + [\underline{\Omega}.\nabla\phi^- + C\phi^+ - S^+]C^{-1}[\underline{\Omega}.\nabla\phi^- + C\phi^+ - S^+] \end{array} \right\} d\Omega dV \tag{2.2}$$

$$\begin{aligned} I_b + I_s = & \int_{\underline{\Omega}.\underline{n}>0} |\underline{\Omega}.\underline{n}| \left\{ \int_{S_b} [\phi^+ - \phi^-]^2 dS + \int_{S_s} [\phi^+ - \phi^- - T(\underline{r},-\underline{\Omega})]^2 dS \right\} d\Omega \\ & + \int_{\underline{\Omega}.\underline{n}<0} |\underline{\Omega}.\underline{n}| \left\{ \int_{S_b} [\phi^+ + \phi^-]^2 dS + \int_{S_s} [\phi^+ + \phi^- - T(\underline{r},\underline{\Omega})]^2 dS \right\} d\Omega \end{aligned} \tag{2.3}$$

$$I_f + I_{pr} = \lambda \int_\Omega |\underline{\Omega}.\underline{n}| \left\{ \begin{array}{l} \displaystyle\int_{U(S_i \cap S_j)} \left(\begin{array}{l} w^+(\underline{r},\underline{\Omega})[\phi^+(\underline{r}+0,\underline{\Omega}) - \phi^+(\underline{r}-0,\underline{\Omega})]^2 \\ + w^-(\underline{r},\underline{\Omega})[\phi^-(\underline{r}+0,\underline{\Omega}) - \phi^-(\underline{r}-0,\underline{\Omega})]^2 \end{array} \right) dS \\ + \displaystyle\int_{S_{pr}} \left(\begin{array}{l} w^+(\underline{r},\underline{\Omega})[\phi^+(\underline{r},\underline{\Omega}) - \phi^+(\underline{r},\underline{\Omega}^x)]^2 \\ + w^-(r,\underline{\Omega})[\phi^-(r,\underline{\Omega}) - \phi^-(\underline{r},\underline{\Omega}^x)]^2 \end{array} \right) dS \end{array} \right\} d\Omega \tag{2.4}$$

On expanding the (I + J) term, the identity

$$K^{+-}(\phi^+,\phi^-) + W^{+-}(\phi^+,\phi^-) = \int_V\int_\Omega [S^+C^{-1}S^+ + S^-GS^-]d\Omega dV$$
$$+ 2\int_{S_s}\int_{\underline{\Omega}.\underline{n}<0} |\underline{\Omega}.\underline{n}|T^2 d\Omega dS = 2\alpha \qquad (2.5)$$

is obtained, giving the maximum principle

$$K^{+-}(\phi^+,\phi^-) \leqq 2\alpha \quad \text{with equality iff} \quad \phi^\pm = \phi_o^\pm \qquad (2.6)$$

3. CLASSICAL EXTREMUM PRINCIPLES

Classical principles for either ϕ^+ or ϕ^- can be obtained, as in [1], by making particular choices of ϕ^- and ϕ^+ respectively, and imposing the appropriate boundary conditions listed in Table 1. For example, the pairs ϕ^+ with ϕ_o^- and ϕ^+ with $\phi^- = G[S^- - \underline{\Omega}\nabla\phi^+]$ in the K^{+-} principle lead to the complementary principles

$$K^+(\phi^+) \leqq F^+(\phi_o^+,\phi_o^+) \leqq E^+(\phi^+) \qquad (3.1)$$

where the bi-linear functional

$$F^+(\phi^+,\psi^+) = \int_V\int_\Omega \left\{\underline{\Omega}.\nabla\phi^+ G\underline{\Omega}.\nabla\psi^+ + \phi^+C\psi^+\right\} d\Omega dV + \int_{S_b \cup S_s}\int_\Omega |\underline{\Omega}.\underline{n}|\phi^+\psi^+ d\Omega dS \qquad (3.2)$$

This functional is, by the way, fundamental for establishing upper and lower bounds for a local characteristic of an approximate solution [7].

The weighted terms $(I_f + I_{pr})$ in K^{+-} principle serve to tie together the various parts of a system with the exterior surfaces S_b and S_s, but these terms vanish for the classical principles. How then is the tying together accomplished for the classical principles?

The operator weights G and C^{-1} in the (I + J) terms give rise to the term $\mathrm{div}[\underline{\Omega}\phi^+\phi^-]$ in the expansion of $W^+(\phi^+,\phi^-)$ used in deriving the K^{+-} principle. This term in conjunction with the interface conditions imposed on the trial function of a classical principle provides the necessary tie via the divergence theorem.

4. NON-CLASSICAL PRINCIPLES

4.1 Boundary Free Principles

The K^{+-} maximum principle gives rise [3] to a maximum principle K^{-} for the first-order Boltzmann equation (1) when the weight constraint of Table 2 is used. This unusual circumstance is discussed further in Section 4.2.

The edge-cell problem provides [8] a severe test for the K^{+-} principle, which permits the use of different orders of spherical harmonic expansions for the trial functions $\phi^{\pm}$ in the various regions. The errors in the total flux for P_7 and $P_{1,7,3}$ spherical harmonic expansions are shown in Fig 4. They are indistinguishable for $x < 5.5$. The errors have been determined by comparison with the analytical solution of Garcia and Siewert [9]. Wilson and Williams [10] have applied the K^{+-} principle to a number of benchmark problems with similar success.

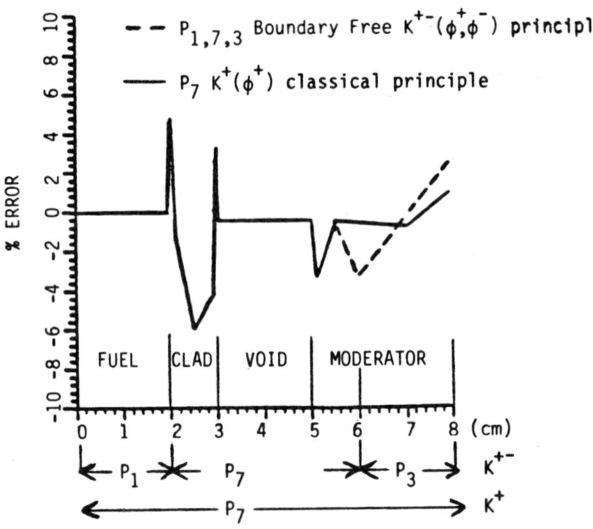

FIG 4. Errors in total flux for classical and boundary-free extremum principles for the edge-cell problem.

To ease the computational problem of two trial functions in the K^{+-} principle, one can be eliminated by choosing for example $\phi^{-} = G(S^{-} - \underline{\Omega}.\nabla\phi^{+})$, so that ϕ^{-} and ϕ^{+} are related as ϕ_{o}^{-} is to ϕ_{o}^{+}.

The way in which the optimised K^{+-} approximations $\hat{\phi}_{\lambda}^{\pm}$ for $\phi_{o}^{\pm}$ converge with the relative weighting λ is as follows. If

$$I_r(\hat{\phi}_\lambda^+,\hat{\phi}_\lambda^-) = I(\hat{\phi}_\lambda^+,\hat{\phi}_\lambda^-) + J(\hat{\phi}_\lambda^+,\hat{\phi}_\lambda^-) + I_b(\hat{\phi}_\lambda^+,\hat{\phi}_\lambda^-) + I_s(\hat{\phi}_\lambda^+,\hat{\phi}_\lambda^-) \quad (4.1)$$

$$\lambda I_d(\hat{\phi}_\lambda^+\ \hat{\phi}_\lambda^-) = I_f(\hat{\phi}_\lambda^+,\hat{\phi}_\lambda^-) + I_{pr}(\hat{\phi}_\lambda^+,\hat{\phi}_\lambda^-) \quad (4.2)$$

$$\left.\begin{array}{l} I_r(\hat{\phi}_\lambda^+,\hat{\phi}_\lambda^-) > I_r(\hat{\phi}_c^+,\hat{\phi}_c^-) - \dfrac{A}{\lambda}\ ,\ \text{A a constant} \\[2ex] I_d(\hat{\phi}_\lambda^+,\hat{\phi}_\lambda^-) < A/\lambda^2 \end{array}\right\} \quad (4.3)$$

where $\phi_c^\pm$ are the optimised approximations given by the classical K^+ and K^- principles. The trends of I_d with λ is shown for a shielding problem in Fig 5.

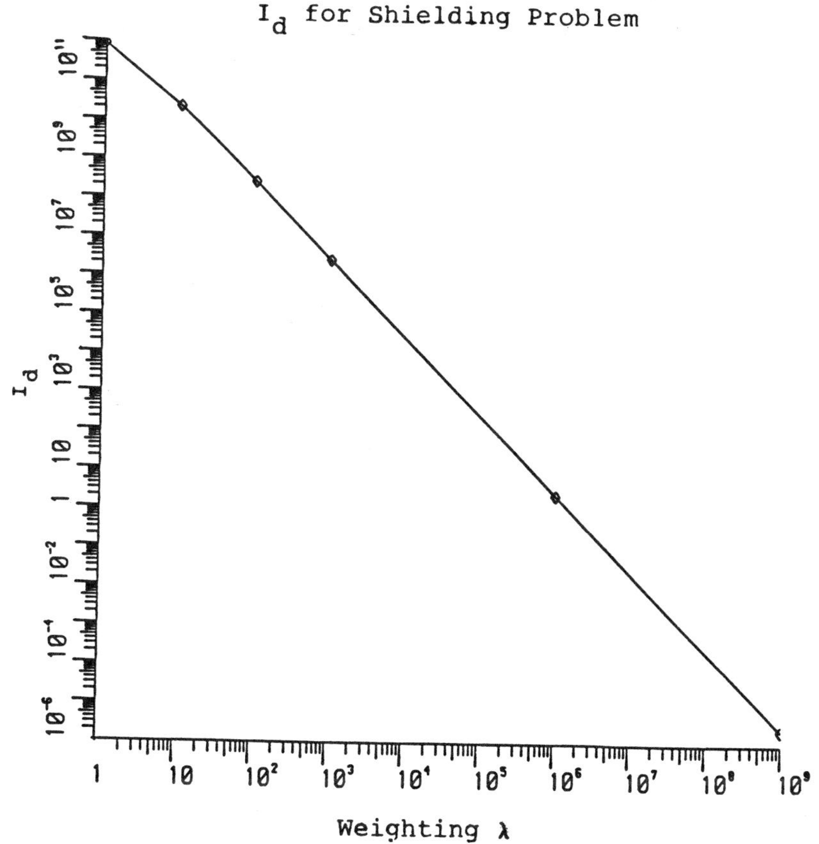

FIG 5. Trend of I_d with relative weighting λ for a three-moment spherical harmonic expansion.

4.2 First Order Principles

The $\overline{K}(\phi)$ and $\tilde{K}(\phi)$ principles can be derived directly by the least squares method or from the $K^{+-}(\phi^+,\phi^-)$ principle as in [3] and [11]. Here an outline is given of how the boundary-tied $\tilde{K}(\phi)$ principle fits in with Euler-Lagrange theory. Similar remarks apply to the $\overline{K}(\phi)$ principle.

The Euler-Lagrange equation for the $\tilde{K}$ functional is of the second-order, because the functional is quadratic in $\underline{\Omega}.\nabla\phi$. By taking even and odd parity components of the Euler-Lagrange equation and its associated boundary conditions the second-order Boltzmann equations and their associated boundary conditions are obtained, which imply the first-order equation and its boundary conditions. The fact that the Euler-Lagrange equation of $\tilde{K}(\phi)$ is a disguised form of the first-order Boltzmann equation can be demonstrated directly [11], using the properties of the operator Σ. Thus the statement [12] that there is no variational principle for the first-order Boltzmann equation is refuted.

5. GEOMETRICAL INTERPRETATION OF EXTREMUM PRINCIPLES

The K^{+-} principle and its derivatives can be given a geometrical interpretation with the aid of a suitable Hilbert space as described in [1], [2] and [11]. The projection theorem shows that approximating solution vector $\underline{\Phi}$ and the exact solution vector $\underline{\Phi}_o$ satisfy, when $\underline{\Phi}$ is optimised by trial functions $\phi^{\pm}$

$$(\underline{\Phi}-\underline{\Phi}_o).\underline{\Phi} = 0 \tag{5.1}$$

and
$$\underline{\Phi}^2 + (\underline{\Phi}-\underline{\Phi}_o)^2 = \underline{\Phi}_o^2 = 2\alpha \text{ (known)} \tag{5.2}$$

$$K^{+-}(\phi^+,\phi^-) = \underline{\Phi}^2 \tag{5.3}$$

The orthogonality of $\underline{\Phi}-\underline{\Phi}_o$ to every vector in the subspace used to specify $\underline{\Phi}$ gives rise to weighted residual equations for $\phi^{\pm}$, which can be cast into a variety of Galerkin and Petrov-Galerkin forms. For example, with functions ϕ_ℓ^+ as a basis for ϕ^+ and $\phi^- = \phi_o^-$ the Galerkin equations

$$0 = \int_V \int_\Omega \phi_\ell^+ \left(-\underline{\Omega}.\nabla[G\underline{\Omega}.\nabla\phi^+] + C\phi^+ - S^+ + \underline{\Omega}.\nabla GS^-\right) d\Omega dV$$
$$+ \text{ boundary terms} \tag{5.4}$$

are obtained for the second-order Boltzmann equation. On the other hand, if $\phi^- = G[S^- - \underline{\Omega}.\nabla\phi^+]$, the Petrov-Galerkin equations

$$0 = \int_V \int_\Omega \left(\phi_\ell + \Sigma^{-1}\underline{\Omega}.\nabla\phi_\ell\right)\left(\underline{\Omega}.\nabla\phi + \Sigma\phi - S\right) d\Omega dV + \text{boundary terms} \quad (5.5)$$

are obtained with $\phi_\ell = \phi_\ell^+ + \phi_\ell^-$ for the first-order Boltzmann equation. Fletcher [13] using the formulation (5.4) has excellent results, whereas the use of the intuitive weight ϕ_ℓ^+ only for the first-order equation gives unsatisfactory results.

6. BOUNDS FOR A LOCAL CHARACTERISTIC

Complementary principles such as K^+ and E^+ or Davis's principles can be used [7] to provide precise upper and lower bounds for a local characteristic of the exact solution, such as the reaction rate in an arbitrarily small neighbourhood. Here the system is assumed to have a bare surface. The bounds are obtained by considering an associated system, physically the same as the given system, but with source S_a replacing the given source S_o of the given system. Associated with K^+ and E^+ for a system, with boundary S_b only is the functional (3.2), for which

$$F^+(\phi_o^+, \phi_a^+) = \int_V \int_\Omega \left\{ \phi_o^+ S_a^+ + \underline{\Omega}.\nabla\phi^+ G S_a^- \right\} d\Omega dV \quad (6.1)$$

If, for example, S_a^+ is equal to a cross-section inside a neighbourhood and zero elsewhere, while S_a^- is zero everywhere, then $F^+(\phi_o^+, \phi_a^+)$ is the reaction rate in the neighbourhood.

Bounds on $F^+(\phi_o^+, \phi_a^+)$ can be placed by considering two families of problems having sources $bS_o \pm 1/b\ S_a$, with b a parameter. The span of the bounds can be minimised with respect to b using a bi-variational method, similar to that of Barnes & Robinson [14] but giving at its best bounds of half the span of Barnes-Robinson bounds. The way in which the bounds for capture rate in an absorbing slab are reduced by increasing the numbers of elements and spherical harmonic moments is shown in Figs 6 & 7.

7. ANALOGIES WITH DIFFUSION THEORY

The diffusion equation

$$\text{div}\underline{\phi}_o + \Sigma\phi_o = S \quad \text{with} \quad \underline{\phi}_o = -D\nabla\phi_o \quad (7.1)$$

with D and Σ continuous functions of $\underline{r}$ is often a fair approximation to the Boltzmann equation. For a system of volume V_1, interface S_1 and bare surface S_2 the boundary conditions are

$$\phi_o \text{ and } \underline{\phi}_o.\underline{n} \text{ continuous across } S_1 \quad (7.2)$$

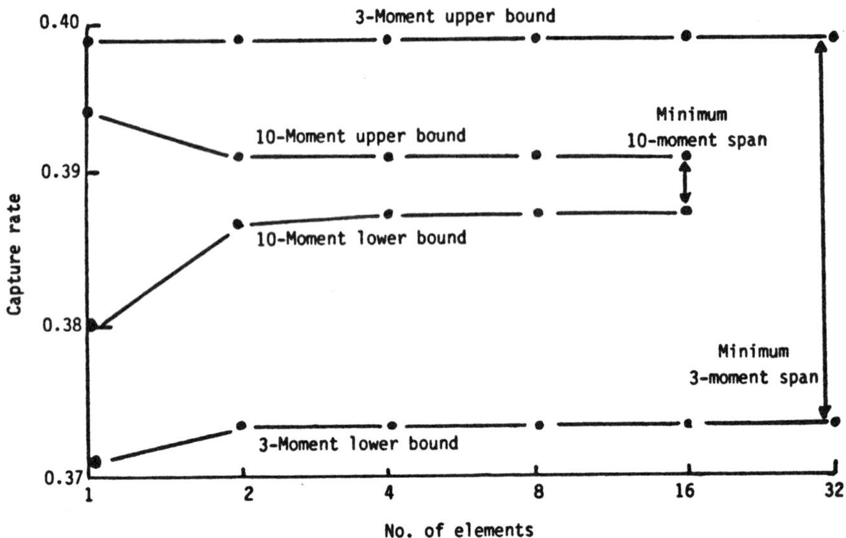

FIG 6. Convergence of bounds on capture rate with mesh refinement.

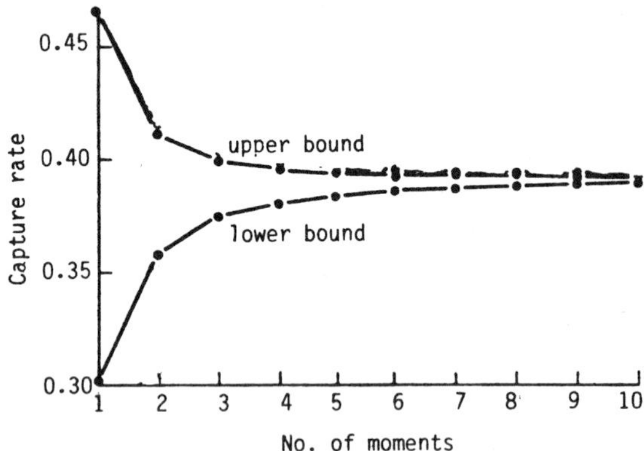

FIG 7. Convergence of bounds on capture rate with spherical harmonic moments.

$$\phi_o - \alpha^2 \underline{n}.\phi_o = 0 \quad \text{on } S_2 \tag{7.3}$$

A least squares treatment of this classical problem [15], analogous to the procedure for the Boltzmann equation, starts with the positive definite functional

$$\check{H}(\phi,\underline{\phi}) = \int_V \left\{[\text{div}\underline{\phi}+\Sigma\phi-S]\Sigma^{-1}[\text{div}\underline{\phi}+\Sigma\phi-S]+[\underline{\phi}+D\nabla\phi].D^{-1}[\underline{\phi}+D\nabla\phi]\right\} dV$$

$$+\int_{S_1} \left\{w\left([\phi]_{\underline{r}-o}^{\underline{r}+o}\right)^2 + w'\left([\underline{n}.\underline{\phi}]_{\underline{r}-o}^{\underline{r}+o}\right)^2\right\} dS$$

$$+\int_{S_2} \frac{1}{\alpha^2}\,[\phi-\alpha^2\underline{n}.\underline{\phi}]^2\, dS \tag{7.4}$$

with w and w' arbitrary positive weights.

The identity

$$\check{H}(\phi,\underline{\phi}) + \hat{H}(\phi,\underline{\phi}) = \int_V S\Sigma^{-1}S\, dV \tag{7.5}$$

gives the maximum principle

$$\hat{H}(\phi,\underline{\phi}) \leqq \int_V S\Sigma^{-1}S\, dV \tag{7.6}$$

for the trial function ϕ and the trial vector $\underline{\phi}$ with equality iff $\phi = \phi_o$ and $\underline{\phi} = -D\nabla\phi_o$.

By way of example, the classical variational principle that

$$\int_V [2\phi S-D(\nabla\phi)^2-\Sigma\phi^2]dV + \int_{S_2} \frac{\phi^2}{\alpha}\, dS$$

is maximised by ϕ_o is obtained by using the trial pair ϕ and $\underline{\phi}_o$, with ϕ continuous across S_1 as the essential boundary condition. A complementary principle is obtained with the trial pair ϕ_o and $\underline{\phi}$ with $\underline{n}.\underline{\phi}$ continuous across S_1, viz.

$$\int_V \left\{S\Sigma^{-1}S - 2\Sigma^{-1}\text{div}\underline{\phi} + \Sigma^{-1}(\text{div}\underline{\phi})^2 + D^{-1}\underline{\phi}^2\right\} dV + \int_{S_2} \alpha^2(\underline{n}.\underline{\phi})^2 dS$$

is minimised by $\underline{\phi}_o$.

If $\underline{\phi} = D\nabla\phi$, then the essential boundary condition for the complementary minimum principle is $D\underline{n}.\nabla\phi$ continuous across S_1, whereas the essential boundary condition for the maximum principle is ϕ continuous across S_1. Thus the complementary principles have essential boundary conditions which are complementary.

8. GENERAL REMARKS

The finite element method based on the least squares and penalty function methods encompass the established procedures and some new ones. The accuracy of the approximate solution can be gauged precisely, both locally and globally, because extremum principles are employed. Codes based on extremum principles also have the merit of providing checks on each other. For example, a code written for anisotropic scattering can be checked for a very extreme form of anisotropic scattering by a code written for isotropic scattering [16]. When the scattering cross-section has an isotropic term plus delta function terms, for backward and forward scattering, the isotropic version of the $K^+(\phi^+)$ principle with modified boundary conditions can be used to solve problems on this extreme case of anisotropic scattering [17].

REFERENCES

1. ACKROYD, R.T., The Why and How of Finite Elements. *Ann. nucl. Energy* 8, 539-566 (1981).

2. ACKROYD, R.T., Completely Boundary-free Minimum and Maximum Principles for Neutron Transport and their Least Squares and Galerkin Equivalents. *Ann. nucl. Energy* 9, 95-124 (1982).

3. ACKROYD, R.T., A Finite Element Method for Neutron Transport - VII. *Ann. nucl. Energy* 10, 243-261 (1983).

4. FLETCHER, J.K., A Finite Element Solution of the Three Dimensional Neutron Transport Equation. *Poster Session of MAFELAP Conference* (1984).

5. ISSA, J.G., QUAH, C.S. & GODDARD, A.J.H., Finite Element Method Applied to Neutron Transport Problems. *Poster Session of MAFELAP Conference* (1984).

6. WILLIAMS, M.M.R. & GODDARD, A.J.H., (Eds), Seminar on *Finite Elements in Radiation Physics*. Ann. nucl. Energy 8, 539-721 (1981).

7. ACKROYD, R.T. & SPLAWSKI, B.A., A Finite Element Method for Neutron Transport - VI. *Ann. nucl. Energy* 9, 315-330 (1982)

8. ACKROYD, R.T. et al., Some Benchmark Shielding Problems Solved by the Finite Element Method. *Sixth International Conference on Radiation Shielding*, Tokyo, Japan (1983).

9. GARCIA, R.D.M. & SIEWERT, C.E., A Multi-region Calculation in the Theory of Neutron Diffusion. *Nucl. Sci. Eng.* 75, 53-56 (1980).

10. WILSON, E. & WILLIAMS, M.M.R., Private Communication.

11. ACKROYD, R.T., Least-Squares Derivation of Extremum and Weighted-Residual Methods for Equations of Reactor Physics - I. *Ann. nucl. Energy* 10, 65-99 (1983).

12. DUDERSTADT, J.R. & MARTIN, W.R., *Transport Theory*. Wiley-Interscience, New York (1979).

13. FLETCHER, J.K., A Solution of the Multi-group Transport Equation using a Weighted-Residual Technique. *Ann. nucl. Energy* 8, 647-656 (1981).

14. BARNSLEY, M.F. & ROBINSON, P.D., Bi-variational Bounds. *Proc. R. Soc. Lond., A.* 338, 527-533 (1974).

15. ACKROYD, R.T., Least-Squares Derivation of Extremum and Weighted-Residual Methods for Equations of Reactor Physics - II. (To be published).

16. ACKROYD, R.T. & WILLIAMS, M.M.R., An Extended Variational Principle for an Albedo Boundary Condition. (To appear in *Ann. nucl. Energy*).

17. WILLIAMS, M.M.R. & WOOD, J., Private Communication.

A-POSTERIORI ERROR ESTIMATION, ADAPTIVE MESH REFINEMENT AND MULTIGRID METHODS USING HIERARCHICAL FINITE ELEMENT BASES

A.W. Craig, J.Z. Zhu, O.C. Zienkiewicz

University College of Swansea, Wales, U.K.

1. INTRODUCTION

The finite element method today is probably the most popular method for the solution of systems of partial differential equations arising from physical problems. However, despite this popularity there are still several fundamental questions about the method which have only recently been addressed.

Firstly, and perhaps most importantly, comes the question of a-posteriori error estimation. The theory of finite elements is based on rigourous, asymptotic, a-priori error estimation. The success of the method is largely based on the fact that for a wide range of problems this analysis guarantees us the best solution from our chosen finite element subspace. We can therefore use the finite element method with confidence, even on relatively coarse meshes. However, this still begs the question 'How good is the 'best' solution?'. To answer this question we need to have a theory of a-posteriori error analysis. Such a theory would aim to provide an estimate of the error in the solution of a problem as a function of the approximate solution itself. Thus the finite element user could decide on the basis of the program output whether in fact the solution was adequate, or whether it was necessary to increase the dimension of the approximating subspace. Of course, in practical situations this is precisely what happens at the moment, but the decision is based on a mixture of guesswork and experience. We shall demonstrate, in this paper, how this heuristic element can be eliminated.

One question naturally follows from what we have said, and that is the question of how to refine the mesh, and indeed how to design a suitable mesh in the first place. Again for most finite-element users this is a heurisitic procedure. The experienced engineer will place a fine mesh in areas where there are expected to be, for example, large stresses, and will place a coarse mesh where the solution is very smooth. Such a process is not only inefficient in terms of creating a good finite

THE MATHEMATICS OF FINITE
ELEMENTS AND APPLICATIONS V

ISBN 0-12-747255-X

element solution space for any given number of degrees of freedom, it is also wasteful of the analyst's time, a high proportion of it being spent typing (seemingly) endless lists of data into a computer, followed by the equally frustrating task of attempting to find the errors in the data which cause the program to fail. It is our thesis that we should only have to design a mesh which is adequate to resolve the geometry of the problem and then the computer should automatically refine the mesh in an intelligent manner (intelligent meaning refining the mesh in areas where it needs to be refined and leaving it alone otherwise). It is not difficult to see that if we have a code which can do this, coupled with an a-posteriori error estimation facility, then all the user need do is to specify the simplest mesh possible for the problem, together with a tolerance within which the solution is required, that the computer will then do best the work which it is designed to do. The key to these problems is the use of hierarchical finite element bases.

2. HIERARCHICAL SUBSPACES

In order to clarify what follows we shall begin with a definition of hierarchical bases. Let us assume that V is some Hilbert space with a(.,.) and (.,.) defined as inner products on that space, and let us consider the following problem:

Find $u \in V$ such that

$$a(u,v) = (f,v) \qquad \forall v \in V \tag{2.1}$$

Under the standard conditions the Lax-Milgram lemma holds and we have a unique solution to the problem. The standard Galerkin approximation is to introduce a finite dimensional subspace $V_h \subset V$ and to pose the problem:

find $u_h \in V_h$ such that

$$a(u_h, v_h) = (f, v_h) \qquad \forall v_h \in V_h \tag{2.2}$$

We have a basis for the space V_h such that

$$\underset{i=1,n}{\text{span}} \{\Phi_i\} = V_h \tag{2.3}$$

(where Φ_i are our finite element basis functions). Thus taking $v_h = \Phi_j$, $j=1,n$ we obtain a matrix equation in the coefficients of these basis functions.

Definition

We shall describe a basis $\{\Phi_i\}_{i=1,n}$ of a space V_h, to be hierarchical if we can write

$$V_h = V_1 \oplus V_2 \tag{2.4}$$

where for some k, $1<k<n$

$$V_1 = \underset{i=1,k}{\text{span}} \{\Phi_i\}$$
$$V_2 = \underset{i=k+1,n}{\text{span}} \{\Phi_i\} \tag{2.5}$$

and V_1 is a conforming finite element subspace of V. (Of course the bases representing V_1 and V_2 may in themselves be hierarchical.)

Remark 1

The term 'hierarchical' is not immediately apparent from this definition, it comes from the fact that if we have

$$V_h = \underset{i=1,n}{\oplus} V_i \tag{2.6}$$

then

$$V_1 \subset V_1 \oplus V_2 \subset V_1 \oplus V_2 \oplus V_3 \subset \ldots\ldots \subset V_h \tag{2.7}$$

Remark 2

As we are concerned with these hierarchies (2.7) the subspace V_2 need have no worthwhile approximation properties on its own, as we are now interested in $V_1 + V_2$. It will also be seen later that we may wish to use non-conforming subspaces in some applications.

We shall be concerned, in this paper, with the bases produced by hierarchical h- and p-refinement, although it is equally possible to have the spaces V_1 and V_2 based on entirely different meshes, or to have the functions in V_2 to be global functions introduced to model singularities in the solution.

3. ADAPTIVE REFINEMENT

The first advantage of using hierarchical bases in an adaptive refinement scheme appears when we consider the structure of the resulting matrix equations. Suppose we applied a finite element discretisation over a space $V_1 \subset V$ to problem (2.1), the resulting equations are: find $u_1 \in V_1$ such that

$$a(u_1, v_1) = (f, v_1) \quad \forall v_1 \in V_1 \tag{3.1}$$

or in matrix form

$$K_1 \underline{u}_1 = \underline{F}_1 \tag{3.2}$$

where $\underline{u}_1$ is the vector of coefficients of the function u_1 expressed in terms of the basis.

If we then refine our mesh hierarchically by introducing an auxilliary space V_2, the problem becomes: find $u_2 \in V_1 \oplus V_2$ such that

$$a(u_2, v_2) = (f, v_2) \quad \forall v_2 \in V_1 \oplus V_2 \tag{3.3}$$

or again in matrix form

$$\begin{bmatrix} K_1 & K_{1,2} \\ \\ K_{2,1} & K_2 \end{bmatrix} \underline{u}_2 = \begin{bmatrix} \underline{F}_1 \\ \\ \underline{F}_2 \end{bmatrix} \qquad (3.4)$$

where the matrix K_1 and vector F_1 appearing in equation (3.4) are precisely those appearing in equation (3.2).

Definition

We may define a correction indicator as a measure of the decrease in the error obtained by augmenting the finite element space.

It is on the basis of these correction indicators that our self-adaptive procedure refines the mesh. The correction indicator which we use is the one introduced by Peano et al [1]. If we assume that we have solved an n-dimensional subspace, obtained a solution u_h, and then introduce a new basis function Φ_{n+1} then this has the form

$$\eta^2 = \frac{[(f,\Phi_{n+1}) - a(u_h, \Phi_{n+1})]^2}{a(\Phi_{n+1}, \Phi_{n+1})} \quad . \qquad (3.5)$$

This is obtained by assuming that the introduction of the new function does not alter the coefficients of the original basis functions, solving the extra line introduced into the equations and calculating the energy norm of the new function. In fact it is possible to show that this is equivalent to

$$\eta^2 = \frac{a^2(e,\Phi_{n+1})}{a(\Phi_{n+1}, \Phi_{n+1})} \qquad (3.6)$$

where e is the discretisation error. Peano used this quantity as an error estimator, whereas in our context it is used as a correction indicator. Having calculated these indicators there are obviously many refinement strategies open, in all of them attention turning to the highest indicators, and for details of how we may proceed the reader is referred to Zienkiewicz and Craig [2].

This quantity (3.6) has several drawbacks as an estimator, the main one being that it is possible for the error to be orthogonal to Φ_{n+1} in a(.,.). While this would tell us (correctly) not to add the function Φ_{n+1}, it would also tell us that the error is zero which may not be the case. However, we have

$$\|e\|_E^2 \geq \frac{a^2(e, v)}{a(v, v)} \qquad (3.7)$$

(in particular when v= Φ_{n+1}), so we may define a more useful estimator in the following way: considering equation (3.1) we may have that

$$V_1 = \underset{i=1,n}{\text{span}}\{\Phi_i\} \tag{3.8}$$

and

$$V_2 = \underset{i=n+1,\ n+m}{\text{span}} [\Phi_1] \qquad m \geq 1 \tag{3.9}$$

then let

$$\eta^2 = \max_{v \in V_2} \frac{a^2(e,v)}{a(v,v)} \tag{3.10}$$

(when m = 1, this is just (3.6)). It is therefore simple to show that

$$\eta^2 \to a(e,e) \text{ as } \dim(V_2) \to \infty \tag{3.11}$$

This quantity is cheap to calculate for low values at m, and in fact it is possible to improve further on this estimate by introducing a measure of the degree of orthogonality between V_1 and V_2 (see Zienkiewicz & Craig [3]).

The following table shows the result of applying the adaptive procedures to a linear elasticity problem, a gravity dam with upstream loading. We present the relative accuracy of the solution, and the number of degrees of freedom used against refinement level.

TABLE 1

Accuracy against refinement level, $\frac{\|e\|}{\|u\|} \times 100$

(figure in brackets refers to number of degrees of freedom used).

Solution Number		Complete Refinement	Adaptive Refinement
	1 (Linear)	14.5% (118)	14.5% (118)
	2 (Quadratic)	5.5% (370)	6.0% (222)
	3 (Cubic)	4.5% (622)	5.0% (305)
	4 (Quartic)	2.0% (978)	2.5% (354)

As can be seen from the table the strategy of complete refinement needs almost three times as many degrees of freedom to obtain comparable accuracy with adaptive refinement. For further examples see [3]. We have used, in our program, higher order basis functions to construct V_2, however this is obviously not necessary, although it saves some computation. It may be more desirable to project into other forms of functions. This alternative is under investigation.

4. MULTIGRID METHODS

As mentioned above we are interested in solving the resulting adaptive equations efficiently. To this end we have been working with multigrid/multilevel algorithms. However the results presented here are valid for any multigrid algorithm posed in terms of hierarchical basis functions, so in this sense, this section can stand alone from the rest of the paper.

For a comprehensive introduction to the basic ideas of multigrid the reader is referred to Brandt [4]. We shall take the algorithm of Bank & Dupont [5] as our model. The algorithm (in our notation for hierarchical bases) is as follows; $\exists$ a sequence of subspaces of V such that

$$V_1 \subset V_1 \oplus V_2 \subset V_1 \oplus V_2 \oplus V_3 \subset \ldots \subset \bigoplus_{i=1,k} V_i \subset V \tag{4.1}$$

and we wish to find the function

$$u_k \in \bigoplus_{i=1,k} V_i \; : \; a(u_k, v) = G(v) \quad \forall v \bigoplus_{i=1,k} V_i \tag{4.2}$$

we choose parameters p,m and on initial guess $u^\circ \in \bigoplus_{i=1,k} V_i$, then

for i=1,m

$$\text{(i)} \quad (u^i - u^{i-1}, v) = \Lambda_k^{-1}[G(v) - a(u^{i-1}, v)] \quad \forall v \in \bigoplus_{i=1,k} V_i \tag{4.3}$$

(or if k=1 solve directly)

(ii) obtain $q \in \bigoplus_{i=1,k-1} V_i$ by applying the scheme p times to the equation

$$a(q,v) = G(v) - a(u^m, v) \quad \forall v \in \bigoplus_{i=1,k-1} V_i \tag{4.4}$$

$$\text{(iii)} \quad u^{m+1} = u^m + q \tag{4.5}$$

where Λ_k is the maximum eigenvalue associated with space $\bigoplus_{i=1,k} V_i$.

There are several difficulties associated with this algorithm when stated in terms of standard bases. Firstly, the matrices corresponding to a(.,.) in equations (4.3) and (4.4) are completely different, that is we must calculate and store the discretisation of the differential operator on each mesh which we use. As we have seen in equation (3.4), this not the case for hierarchical bases as the matrix takes a nested form. The same remarks are also true for the data.

More important however is the calculation of the residual term $G(v)-a(u^m,v)$ in equation (4.4), as u^m and v are in terms of different bases we must calculate the nodal residuals on the fine mesh and perform a weighted transfer to the coarse mesh, thus we need to introduce a restriction operator at each level of approximation. Similarly, to perform the operation in equation (4.5) we need first to interpolate q onto the finer mesh before we can add the solutions pointwise. Such a procedure is time consuming, but for example, direct residual injection is very cheap. This has been tested in the framework of the finite difference method [4] but does not give good results. However let us re-examine the algorithm in terms of hierarchical bases. Writing out the iteration applied to the second level equation (4.4) in full we obtain:

$$(q^i - q^{i-1}, v) = \Lambda_{k-1}^{-1}(G(v) - a(u^m, v) - a(q^{i-1}, v)) \qquad \forall\, v \varepsilon \oplus_{i=1,k-1} V_i \tag{4.6}$$

where u^m is obtained from the previous iteration. In principle we could update u^m directly instead of obtaining the correction vector q however, in practice, for standard bases, this obviously requires the use of the interpolation operator at each step of the iteration. But for hierarchical bases we may write,

$$u^m = \tilde{u}^m + \bar{u}^m \tag{4.7}$$

where

$$\tilde{u}^m \ \varepsilon \ \oplus_{i=1,k-1} V_i \tag{4.8}$$

$$\bar{u}^m \ \varepsilon V_k$$

Now the addition

$$\tilde{u}^{m+i} = \tilde{u}^m + q^i \tag{4.9}$$

may be performed by pointwise addition of the vectors of coefficients, and we may rewrite (4.6) as

$$(\tilde{u}^{m+i} - \tilde{u}^{m+i-1}, v) = \Lambda_{k-1}^{-1}(G(v) - a(\bar{u}^m, v) - a(\tilde{u}^{m+i-1}, v)) \qquad \forall\ v \varepsilon \oplus_{i=1,k-1} V_i \tag{4.10}$$

in other words, we update the coefficients of the lower order basis functions directly, and leave those corresponding to V_k unchanged. This not only removes the problems of restriction and prolongation, but also does away with the need to calculate and store vectors corresponding to q. For example if we had a matrix equation posed on $V_1 + V_2$,

$$\begin{bmatrix} k_{11} & k_{12} \\ k_{21} & k_{22} \end{bmatrix} \begin{bmatrix} \underline{u}_1 \\ \underline{u}_2 \end{bmatrix} = \begin{bmatrix} \underline{f}_1 \\ \underline{f}_2 \end{bmatrix} \tag{4.11}$$

the algorithm would then simply state;

(i) choose $\begin{bmatrix} \underline{u}_1^\circ \\ \underline{u}_2^\circ \end{bmatrix}$

(ii) perform m iterations on a relaxed Jacobi algorithm on

equations (4.1) to obtain $\begin{bmatrix} \underline{u}_1^m \\ \underline{u}_2^m \end{bmatrix}$

(iii) solve directly

$k_{11}\underline{u}_1^{m+1} = \underline{f}_1 - k_{12}\underline{u}_2^m$

(iv) let $\begin{bmatrix} \underline{u}_1^{m+1} \\ \underline{u}_2^m \end{bmatrix} = \begin{bmatrix} \underline{u}_1^\circ \\ \underline{u}_2^\circ \end{bmatrix}$ and return to (ii)

that is, a simple block iteration scheme.

The implementation of these algorithms is very simple for p-type hierarchic elements, and can be done in a straight-forward manner. There are some problems however with h-type elements, due to the integration of the inner products of high order with low order elements (the supports of the basis functions may differ greatly), however it is of course possible to define a linear mapping on an element level, between the standard and hierarchical subspaces. This mapping is in fact just another form of the restriction mapping used in standard multigrid procedures. Therefore it is possible to pose our problems in terms of standard elements, then map the element stiffness matrices obtained into the corresponding hierarchical ones, and then proceed as before. We emphasize that this transformation need only be performed once, at the beginning of the program, as opposed to many times in a standard multigrid procedure.

References

1. PEANO, A., FANELLI, M., RICCIONI, R. and SARDELLA, L., Self-Adaptive Convergence at the Crack Tip of a Dam Buttress. *Int. Conf. Num. Meth. Frac. Mech.*, Swansea (1979).

2. ZIENKIEWICZ, O.C. and CRAIG, A.W., *Adaptive Mesh Refinement and A-Posteriori Error Estimation for the P-Version of the Finite Element Method. Adaptive Computational Methods for Partial Differential Equations*, (Ed. I. Babuska, J. Chandra, J.F. Flaherty), SIAM Philadelphia, (1987).

3. ZIENKIEWICZ, O.C. and CRAIG, A.W., Keynote Lecture on Adaptive-Multigrid Procedures. *Int. Conf. of Accuracy Estimates & Adaptive Refinement Procedures for Finite Element Computations (ARFEC)*, Lisbon (1984).

4. BRANDT, A., Multi-level Adaptive Solutions to Boundary-Value Problems, *Math. Comp.* 31, 333-390 (1977).

5. BANK, R., and DUPONT, T., An Optimal Order Process for Solving Elliptic Finite Element Equations, *Math. Comp.* 36 (1981).

DYNAMIC IMPLEMENTATION OF THE h-VERSION OF THE FINITE ELEMENT METHOD

M. C. Rivara

University of Chile, Santiago, Chile
Faculty of Physical and Mathematical Sciences

1. INTRODUCTION

This paper is concerned with the idea that the combination of fully irregular and conforming triangulations, appropriate multigrid methods, and selfadaptivity of the mesh is the most flexible and effective way of exploiting the intrinsic features of the h-version of the finite element method in 2-dimensions.

In recent years, adaptivity for finite elements has been widely justified and developed by Babuska and Rheinboldt [1-3], and mainly applied to rectangular meshes. A first step on the implementation of adaptive multigrid software over triangular grids has been accomplished by Bank and Sherman [5] by using regular local mesh refinement. The implementation of multigrid algorithms and adaptivity of the mesh in the framework of fully irregular and conforming triangulations have been recently studied and discussed by Rivara [7-11]. Such a new approach depends on several critical related items: the mesh refinement algorithms, the local error estimators used to estimate the global error; the multigrid algorithm to be used in the adaptive context; a data structure that efficiently manages the adaptive generation of the grids and the multigrid algorithm; and the overall solution process. It is the purpose here to summarize some of the findings of [7-11], mainly those concerning the new conforming mesh refinement algorithms for triangulations, the molecular list data structure developed to manage sequences of fully irregular conforming triangulations, and the new dynamic multigrid algorithm. The main characteristics of the experimental EXPPES package are also discussed; and numerical results, showing that optimal rates of convergence have been achieved for singular solutions in adaptive manner are presented.

THE MATHEMATICS OF FINITE ELEMENTS AND APPLICATIONS V

ISBN 0-12-747255-X

2. CONFORMING AND 1/2-NON-CONFORMING TRIANGULATIONS.

In [10], two families of conforming triangulations have been considered: the usual family of *quasi-uniform triangulations* and the family of *admissible non-quasi-uniform triangulations* τ^* which satisfies the smoothness condition (a) and the non-degeneracy condition (b). (a) For any $\tau \epsilon \tau^*$ and for any pair of non-disjoint triangles t_1, $t_2 \epsilon \tau$ (with respective diameter h_1 and h_2)

$$\frac{\min(h_1,h_2)}{\max(h_1,h_2)} \geq \delta > 0,$$

(b) If α_t is the minimum interior angle of triangle t, then

$$\min_{t \epsilon \tau} \alpha_t \geq \alpha_o > 0, \ \forall \tau$$

To discuss the conforming mesh refinement algorithms, the concept of 1/2-non-conforming triangulation needs to be introduced. Thus we shall say that a triangle t(a,b,c) in τ, with vertices a,b,c is a 1/2-non-conforming triangle in τ with the 1/2-non-conforming point d, where d is the midpoint of side $\overline{ac}$, if there exist two triangles t(a,d,e), t(e,d,c) in τ (see Fig.1) such that $t(a,d,e) \cap t(e,d,c) = \overline{de}$, $t(a,d,e) \cap t(a,b,c) = \overline{ad}$ and $t(e,d,c) \cap t(a,b,c) = \overline{dc}$. Finally, τ will be a 1/2 non-conforming triangulation if the triangles of τ are either conforming or 1/2-non-conforming in τ.

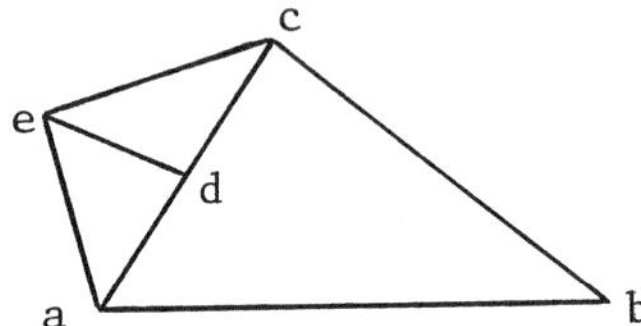

FIG.1. Triangle t(a,b,c) is a 1/2-non-conforming triangle in this triangulation.

3. CONFORMING MESH REFINEMENT ALGORITHMS

The conforming mesh refinement algorithms are simple and effective tools for the selective refinement of conforming triangular grids. These algorithms, based on the bisection of triangles by the longest side, were recently developed by Rivara [7,8] and they have been used in the implementation of the EXPDES package [9-11]. For any given non-degenerate and conforming triangulation τ and for any submesh of refinement $V \subset \tau$ (the subset of triangles to be refined), these algorithms produce a new refined, conforming triangulation τ^* by using a finite number of intermediate 1/2-non-conforming triangulations If we denote by $\tilde{\tau}_i$ the ith 1/2-non-conforming triangulation generated throughout the process, and by R_i the subset of its 1/2-non-conforming triangles, the basic algorithm can be schematically described by the simple procedure:

0 Input $\{\tau, V\}$
1 Obtain $\tilde{\tau}_1$ by bisecting all the triangles of V by the longest side. Set $i \leftarrow 1$
2 While $\text{card}(R_i) > 0$ do
Construct $\tilde{\tau}_{i+1}$ as follows
For any $t \varepsilon R_i$ with the 1/2-non-conforming point P, do
Bisect t by the longest side. Let Q be the intersection point generated in this way. If $P \neq Q$, then join them
Set $i \leftarrow i+1$
3 $\tau^* = \tilde{\tau}_{i+1}$

The main result concerning these algorithms can be summarized as follows [7,8]. For any initial, non-degenerate and conforming triangulation, the arbitrary and iterative application of these algorithms always produce nested triangulations that belong to the same family of admissible non-quasi-uniform triangulations τ^*, where the constants that define the family only depend on the geometrical characteristics of the initial triangulation. In other words, all the triangulations generated by these algorithms are guaranteed to be nested, conforming, non-degenerate and smooth. These are desirable features both from the point of view of the finite element method and the multigrid algorithms [2,5,9]. For a more extensive discussion of these algorithms and their properties see [7,8,10].

4. MOLECULAR LIST DATA STRUCTURE FOR ADAPTIVE MULTIGRID SOFTWARE OVER FULLY IRREGULAR AND CONFORMING TRIANGULATIONS

The implementation of adaptive multigrid finite element software over fully irregular and conforming triangulations critically depends on the data structure to be used to manage the adaptive generation of the grids and the dynamic multigrid algorithm. In [9,10], such a critical problem has been solved effectively by introducing the concept of *molecular structure* associated with any conforming triangulation. In this approach, each conforming triangulation is represented as an ordered set of ordered list associated with the vertices or nodes of the triangulation. More specifically, let τ be any non-degenerate conforming triangulation defined by the set $N(\tau)$ of its nodes or vertices. Then, for each node $j \varepsilon N(\tau)$, we shall define the molecule j as the polygonal region whose vertices are nodes of τ and whose interior only contains the node $j \varepsilon N(\tau)$. We shall also define the integer valued function n such that n(j) will be the number of nodes of τ lying on the boundary of molecule j. Then the molecular List associated with the molecule j will be the ordered List $M(j) = (j_1, j_2, \ldots, j_{n(j)})$, where the j_i are the vertices of τ lying on the boundary of the molecule j ordered in counter clockwise sense. If j is an interior node, there are n(j) possible representations of this molecule. Conversely, if j is a boundary node, the molecular list M(j) is unique and j_1 must also be a boundary node. To illustrate these concepts, consider the triangulation of Fig.2, where, on the one hand, for the interior node 7, n(7) = 6, and M(7) = (1,2,3,4,5,6) is a possible representation of the molecular structure of this node; and on the other hand, for the boundary node 6, n(6) = 3, and M(6) = (1,7,5) is unique.

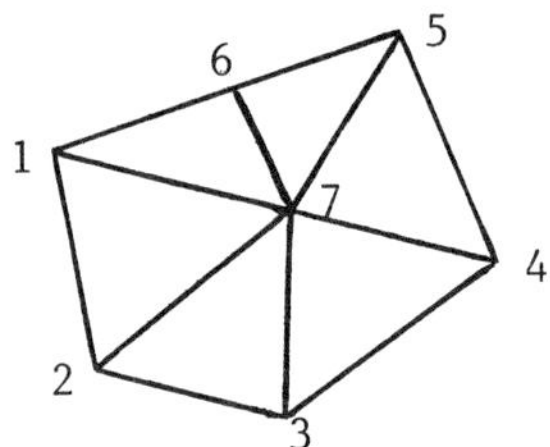

FIG.2. Molecule associated with node 7.

Thus, the conforming triangulation τ can be represented by the ordered List $L(\tau) = (M(1), M(2), \ldots, M(N))$, where N is the number of elements of $N(\tau)$. The triangles of τ can be easily generated by traversing the molecules sequentially, in such a way that each triangle T(i,j,k) is considered only once, that is for the molecule associated with its smallest numbered vertex. The order of the molecules is indeed used in our software each time that the triangles of a given triangulation need to be traversed, e.g. in the generation of the (non-symmetric) stiffness matrices, in the global conforming mesh refinement algorithm, and in the computation of the error estimators.

The data structure we have developed for adaptive multigrid software is indeed based on the basic molecular List structure discussed above. Consequently, each grid will be represented by a set of molecular Lists. However, since for highly singular solutions, the adaptive procedure can generate large sequences of only locally different triangulations (see also section 5), we have developed a compact data structure where each molecular List appears only once. Such a compact data structure can be adequately implemented as a set of vectors whose elements can contain pointers to other vectors, and its most distinguished features are the following. For linear finite elements, this data structure allows the management of the non-zero, non-diagonal entries of the associated sequence of stiffness matrices in a compact real array that uses the same pointers needed to manage the sequence of molecules (the diagonal elements can be easily managed in a separate vector); as a consequence, a reduced amount of data storage is needed to manage the sequence of triangulations and stiffness matrices involved in the multigrid method. Moreover, since the machine storage increases linearly with the number of vertices, the compact molecular structure compares advantageously with those that classical finite element software use with direct solvers of sparse linear systems. It should be also pointed out here that, when a small number of vertices is adaptively created between two consecutive grids k and k+1, all the information concerning the molecular List associated with the part of the triangulation k+1 that is different from triangulation k, is stored in a contiguous block of data. This property has been used to develop an adaptive multigrid algorithm whose main features will be discussed in the next section. For a more detailed discussion of the data structure and the implementation of the different algorithms, see references [10,11].

5. THE DYNAMIC MULTIGRID ALGORITHM

As was stated earlier in the preceding section, adaptivity in the presence of severe singularities can produce a sequence of consecutive finite element spaces of slowly increasing dimension, and consequently, the efficiency of the classical full multigrid algorithm [6] can be seriously affected. To deal with this problem, an adaptive multigrid algorithm, which indeed corresponds to a modified full multigrid algorithm, was developed in [10]. The main features of this algorithm are the following two:

(i) The finite element problem at the current level k+1 is solved by using a dynamic multigrid iteration, only if the number of nodes adaptively created between levels k and k+1 is big enough with respect to the number of nodes of triangulation k. Otherwise, a fixed number of iterative sweeps are performed over an associated local subproblem in order to obtain an approximation of the finite element solution at level k+1.

(ii) The dynamic multigrid iteration essentially corresponds to a V-cycle of the classical multigrid algorithm [6,10], where local or global transfer of the residuals and iterative sweeps are performed between consecutive grids depending on the number of nodes adaptively generated between the two levels.

From the point of view of the implementation, this algorithm exploits the useful properties of the compact molecular data structure described in section 4. A detailed formulation of this algorithm based on the concepts of local subproblems and submeshes is given in reference [10].

6. THE EXPDES PACKAGE AND NUMERICAL RESULTS

The EXPDES package [10,11] combines adaptivity together with all the techniques described in the preceding sections over linear finite element spaces. The selfadaptive procedure is essentially based on the ideas of Babuska and Rheinboldt [1-3]: once a finite element solution is computed over the current grid, a local error estimator of the error (developed in [10] for triangles) is used to estimate the global error. Then a mesh equilibration strategy of the error is used to predict which elements should be refined in the next grid.

The EXPDES package constitutes a general, easy to use, flexible software that provides reliable information about the quality of the numerical results computed, and makes most of initial boring tasks and difficult decisions involved in the use of classical finite element software. This software can be used in selfadaptive or user-adaptive form. In both cases the user only needs to define a coarse initial triangulation of the region. All the remaining triangulations are automatically generated by the computer, and they are always nested, conforming, non-degenerate and smooth. Numerical experiments performed with singular problems [9,10] have shown that, in selfadaptive way, optimal sequences of finite element solutions (with respect to the involved number of degress of freedom) are generated.

To illustrate the behaviour of the EXPDES package, we shall consider the Laplace problem with Dirichlet boundary conditions over the L-shaped region formed as the union of three unitary squares [1,9,10]. The solution of this problem is the function $u=r^{2/3} \sin(2\theta/3)$, which is singular at the reentrant corner. Consequently for this problem, the asymptotic rate of convergence for linear elements and quasi-uniform triangulations diminishes to 1/3. However, it is known that for an adequate non-quasi-uniform triangulation, the maximum rate of convergence 1/2 can again be achieved [4,12].

To solve this problem we have used both the selfadaptive procedure with the dynamic multigrid algorithm and a uniform-refinement procedure that generates a sequence of quasi-uniform triangulations. The convergence behaviour of this problem is shown in Fig.3, where the slope of the curves represents the numerical rate of convergence achieved. Here one can see that, on one hand, the numerical rate of convergence for the quasi-uniform sequence is in complete agreement with the theory, and on the other hand, the selfadaptive procedure has achieved the maximum rate of convergence for linear elements. In this sense we say that the adaptive sequence of finite element problems generated is optimal. In order to illustrate the quality of the triangulations generated, we have included in Fig. 4 the last triangulation adaptively generated.

Finally, with respect to the error estimates computed, it

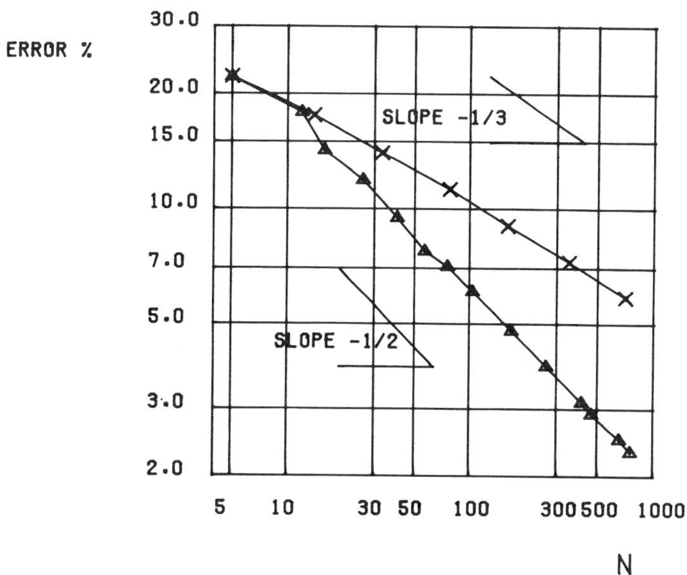

FIG. 3. Convergence behaviour for the L-shaped problem both for the adaptive (Δ) and quasi-uniform (x) procedures.

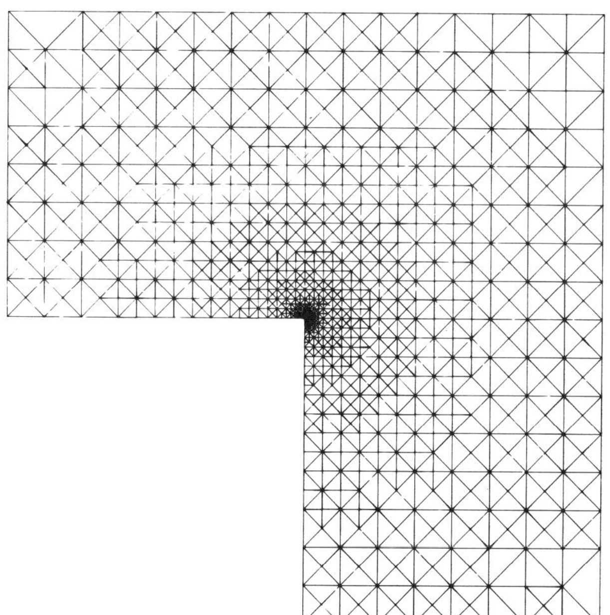

FIG. 4.

should be pointed out here that those adaptively generated were better than those obtained for the quasi-uniform sequence of triangulations. However, in both cases the ratio estimated error/exact error tends quickly to 1 when the error diminishes.

AKNOWLEDGEMENTS

This research was performed at the Catholic University of Leuven.

REFERENCES

1. BABUSKA, I., The selfadaptive approach in the finite element method. pp. 125-142 of J.R. Whiteman (Ed.), *The Mathematics of Finite Elements and Applications*. Academic Press, London (1976).
2. BABUSKA, I. and RHEINBOLDT, W.C., Error estimates for adaptive finite element computations, *SIAM J. Numer. Anal.* 15, 736-754 (1978).
3. BABUSKA, I. and RHEINBOLDT, W.C., Reliable error estimation and mesh adaptation for the finite element method, pp.67-108 of J.T. Oden (Ed.), *Computational Methods in Nonlinear Mechanics*, North Holland (1980).
4. BABUSKA, I. and SZABO, B., On the rates of convergence of the finite element method, *Int. J. Numer. Math. Engrg.* 18,323-341 (1982).
5. BANK, R.E. and SHERMAN A.H., The use of adaptive grid refinement for badly behaved elliptic partial differential equations, pp. 18-24 of *Computers in Simulation XXII*, North Holland (1980).
6. HACKBUSCH, W., Multi-grid convergence theory, pp 177-219 of W. Hackbusch and U. Trottenberg (Eds), *Multigrid Methods*, Lecture Notes in Math. 960, Springer-Verlag (1982).
7. RIVARA, M.C., Algorithms for refining triangular grids suitable for adaptive and multigrid techniques, *Int. J. Numer. Meth. Engrg.*, to appear.
8. RIVARA, M.C., Mesh refinement processes based on the generalized bisection of simplices, *SIAM J. on Numer. Anal*, to appear.
9. RIVARA, M.C., Design and data structure of a fully adaptive multigrid finite element software, *ACM Trans. on Math. Software*, to appear.
10. RIVARA, M.C., *Adaptive multigrid software for the finite element method*, Doctoral dissertation, K.U. Leuven, Belgium (1984).
11. RIVARA, M.C., EXPDES Users'Manual, Dept. Computer Science, K. U. Leuven (1984).
12. WHITEMAN, J.R., Finite elements for singularities in two and three dimensions, pp. 37-55 of J.R. Whiteman (Ed.), *The Mathematics of Finite Elements and Applications*, Ac. Press London (1982).

ABSTRACTS OF POSTER SESSIONS

1. A RASTER SCAN ALGORITHM FOR GENERATING HIDDEN ISOPARAMETRIC SURFACES

J.E. Akin*, J.M. Montgomery* and W.H. Gray**
*Department of Mechanical Engineering and Materials Science, Rice University, Houston, Texas 77001, U.S.A.
**Oak Ridge National Laboratory, Oak Ridge, Tennessee, U.S.A.

Finite element analyses and computer aided design systems must often present graphical representations of three-dimensional isoparametric surfaces. Various algorithms have been developed to describe color surfaces and to determine visibility. A procedure for using isoparametric interpolation to define such surfaces is presented. It is an extension of an earlier algorithm for defining contour lines. When applied on a raster scan color graphics device (or a line printer) the procedure allows the true representation of isoparametric surfaces. Surfaces can be presented in different colors. The intersection of surfaces is automatically included in the procedure for determining visibility. The algorithm can also be easily extended to allow color contour shading of the surfaces to represent function values. The procedure will be compared to other methods, such as that in MOVIE. BYU. Sample results will be illustrated.

[1] Akin, J.E. and Gray, W.H., Contouring on isoparametric surfaces, I.J.N.M.E., 11, 1893-1894, 1977.

2. APPLICATION OF A DIRECT ELIMINATION PROCEDURE IN THE BOUNDARY ELEMENT ANALYSIS OF MULTI-REGION PROBLEMS

M.C. Au
Department of Civil Engineering, Carleton University, Ottawa, Ontario, K15 5B6, Canada

The classical procedures in boundary element analysis of field problems give rise to systems of equations in which the matrices are always fully populated. From a computational point of view this characteristic is often viewed as a disadvantage of a boundary element formulation when compared with finite difference or finite element schemes. The procedure of setting up a boundary element matrix however permits the elimination of elements in the

ISBN 0-12-747255-X

matrix during its formulation. Such a procedure will yield a system of equations within an upper triangular matrix instead of the fully populated version. Such a technique can also be easily implemented in the generalized solution of a multi-region problem. This paper illustrates the application of the elimination procedure for the solution of certain multi-region problems in elastostatics and potential theory. The efficiency of the elimination scheme is illustrated by appeal to several examples in which the computational requirements for both the classical scheme and the elimination scheme are compared.

3. A FULLY INTERACTIVE MULTIGRID FINITE ELEMENT STRESS PACKAGE

K.E. Barrett, D.M. Butterfield, S. Ellis and J.H. Tabor
Department of Mathematics, Coventry (Lanchester) Polytechnic, Coventry, CV1 5FB, England

As computer speeds and storage have dramatically increased over recent years, it has now become possible to analyse quite significantly sized stress problems, typically containing a 1000 or so nodal degrees of freedom in real time. A complete package which takes the user from an input stage, where a few parameters are used to define the outline shape of the region to be stressed, through to the stage where stress contour plots can be given for a 256 degree of freedom problem, was demonstrated via a video film. The film was produced via a BBC microcomputer (acting as a graphics terminal) linked to a Harris 500 machine and lasted for fifteen minutes. The total CPU time used was under one minute on the Harris time sharing system and the filming was deliberately slowed down to enable a commentary to be superimposed. The processor time was small as the mesh generation part of the package uses a multigrid solution technique, an order N technique, to solve for the x and y coordinates of the N nodal points in the grid. The discrete form of the stress equations is also solved by multigrid methods, so that again the processing time is proportional to the number of nodes. Timings and details of the general performance of the package on the ACT Sirius and Apricot microcomputers was also presented.

4. TREATMENT OF A RE-ENTRANT VERTEX IN A THREE-DIMENSIONAL POISSON PROBLEM

A.E. Beagles and J.R. Whiteman
Department of Mathematics and Statistics, Brunel University Uxbridge, Middlesex, England

We consider the solution, u of the following problem based on the Laplacian, Δ,

$$-\Delta u = f \quad \text{in } \Omega$$
$$u = 0 \quad \text{on } \partial\Omega$$

where $\Omega \subset \mathbb{R}^3$ is an open polyhedral domain with boundary $\partial\Omega$ and $f \in L_2(\Omega)$. We introduce a technique for obtaining estimates of the singular behaviour of u in the neighbourhood of a vertex in $\partial\Omega$. It is known that the behaviour of u near a vertex is given in spherical polar coordinates centred on the vertex by

$$u(r,\theta,\phi) = r^{\alpha}Q(\theta,\phi) + \omega(r,\theta,\phi)$$

where $\omega \in H^2(\Omega)$, $\alpha = -\frac{1}{2} + \sqrt{\frac{1}{4} + \lambda^2}$ and λ, Q are respectively the first eigenvalue and eigenfunction of the Laplace-Beltrami eigenproblem on the subset of the surface of the unit sphere $(r = 1)$ cut off by the planes defining the vertex, with the boundary conditions of the original problem imposed on these planes. We obtain a weak form of the eigenproblem and use the finite element method in the θ,ϕ plane to obtain an upper bound on λ and hence on α. Applying our technique to the re-entrant vertex defined by three orthogonal planes we obtain an improvement to the upper bounds for this problem published to date.

5. THE HOMOGENIZATION METHOD APPLIED TO FREE SURFACE SEEPAGE IN LAYERED POROUS MEDIA

J.C. Bruch, Jr. and A. Shechter
Department of Mechanical and Environmental Engineering, University of California, Santa Barbara, California 93106, U.S.A.

The two dimensional problem of steady state seepage from a rectangular channel through a layered porous medium underlain by a drain at a finite depth is considered. The layered porous medium consists of alternating homogeneous, isotropic layers having hydraulic conductivities, k_1 and k_2, respectively. The solution approach used the Baiocchi method (which assumed saturated flow) and transformation on a problem having the hydraulic conductivity a function of y only which yields a minimum problem that can be solved by means of the finite element method in a successive over-relaxation scheme with projection. In applying this approach to the layered medium problem a limitation on the application of the Baiocchi method becomes evident. This is discussed along with some numerical results, showing that even though the positiveness of the transformed solution function is maintained, zones of unsaturated flow can occur, which violates the assumption of saturated flow in the method. The method of homogenization is then applied to this problem and it is shown numerically that as the number of layers increases the solution approaches the homogeneous case. Also in the application of the homogenization method the results are consistent with the theory, i.e. a totally saturated flowfield.

Acknowledgement
The research leading to this paper was supported by the University of California, Water Resources Center, as a part of Water Resources Center Project UCAL-WRC-W-607.

6. A SIMPLE TEST EXAMPLE FOR ERROR BOUNDS FOR THE FINITE ELEMENT METHOD

L. Collatz
Institut für Angewandte Mathematik, Universität Hamburg, Bundesstrasse 55, 2000 Hamburg, West Germany

A torsion problem for a beam with U-cross-section is considered. This contains two essential singularities, one at each reentrant corner. The cross-section can be divided into two finite elements meeting at an interface. By the use of approximation methods, monotonicity and the jump condition at the interface, an inclusion can be obtained. Numerical results are presented.

7. BODY ORIENTED COORDINATES APPLIED TO THE FINITE-ELEMENT METHOD

W.A. Cook
Technical Engineering Support Group, Los Alamos National Laboratory, Los Alamos, New Mexico 87545, U.S.A.

The objective of this research is to increase the accuracy of the finite-element method using coordinate systems that are intrinsic to the shape of the body being analyzed. Present finite elements use Cartesian coordinates and are more accurate for rectangular problems than for nonrectangular problems.

The feasibility of this research was checked by developing a finite element using cylindrical coordinates (CYL) and comparing its solutions of cylindrical problems with the solutions of a finite element using Cartesian coordinates (CART). The CYL solutions were the most accurate.

The body coordinate element (BCE), uses a coordinate system developed from linearly blended interpolation formulas. Each element is defined by one constant and one linear body coordinate identical to rectangular elements being defined with Cartesian coordinates. The displacements are approximated and the stiffness and loads formulated using these body coordinates. The BCE has been used to solve several problems. It reduces to CART for rectangular problems and to CYL for circular problems. Four problems have been solved where the body coordinates are nonorthogonal. In two of these, BCE is more accurate than CART and in the other two, CART is more accurate than BCE.

8. CHARACTERISTIC IMPEDANCE OF TRANSMISSION LINES USING DUAL FINITE-ELEMENTS

P. Daly
Department of Electrical and Electronic Engineering, University of Leeds, Leeds, LS2 9JT, England.

Microwave transmission lines presently under development in miniature integrated circuit technology normally use guiding structure consisting of a uniform line with rectangular cross-section within which there may be several conductors and/or dielectrics. By a standard formulation of an energy variational principle, both upper and lower bounds to the characteristic impedance may be set up. These bounds are obtained in turn by using electric and magnetic field potentials which are duals, obtained by interchanging bounding electric and magnetic walls. The solution of the matrix equations resulting from a finite-element formulation of these expressions leads to estimates which are at least an order of magnitude more accurate than would be produced by either bound itself.

By defining an enclosing rectangular structure within which the wanted structure is defined, it is simple to carry out the majority of the tiresome pre-processing tasks required to generate the mesh, to label points and to define the finite elements. A simple code is used to characterise each point so as to facilitate the generation of a reduced problem matrix. Since the mesh generation is essentially a topological rather than geometric problem, the location of all the points and elements is a separate question determined at a later stage usually by means of scaling factors. Owing to the identical nature of the variational expressions for both the original problem and its dual, no programming changes are called for in taking upper or lower bounds. These changes are contained in the potential codes and dielectric constants for the dual problems.

Finally the solution of the finite-element matrix equations consists of a surface integration of the field and a grid of potential values all of which are stored in a special data file used by dedicated contour plotting routines. Examples are presented of computations for upper and lower bounds to the characteristic impedance of particular microwave integrated circuits.

9. ON CHEBYSHEV PSEUDOSPECTRAL CALCULATIONS WITH FINITE ELEMENT PRECONDITIONING

M. Deville* and E. Mund**
**Unité de Mecanique Appliquée, Université Catholique de Louvain, Louvain-la-Neuve, Belgium*
***Service de Métrologie Nucléaire, Université Libre de Bruxelles, Bruxelles, Belgium*

There has been a great interest in recent times, in the use of preconditioning techniques for iterative solutions of the algebraic system of equations resulting from the application of spectral (or pseudo-spectral) methods to elliptic boundary value problems. Spectral methods are widely used in fluid mechanics. When they can be used, they are extremely accurate and yield typically an error of the order of computer round-off. Their main drawbacks however lie in the fact that they lead to non-sparse, non-symmetric, ill-conditioned systems of algebraic equations and that they are difficult to implement in the case of complicated geometries. The aim of this paper is to show how finite elements may be used to alleviate the first difficulty while keeping the full power of the spectral method. Pathways for the second difficulty are also suggested.

Let $Lu = f$ represent a boundary value problem, the pseudo-spectral approximation of which leads to a $N \times N$ algebraic system $L_N u_N = f_N$. Denoting by L_{ap}, the finite element matrix on the same discretization and applying Stetter's IDec principle, the basic equation may be cast into the iterative form

$$L_{ap} u_N^{(k+1)} = L_{ap} u_N^{(k)} - \alpha (L_N u_N^{(k)} - f_N) \qquad k = 0,1,\ldots$$

where α is an optimal acceleration parameter. This may be viewed as a succession of finite element-type problems with different right-hand sides, the residues $L_N u_N^{(k)} - f_N$ being evaluated using fast direct and inverse discrete Chebyshev transforms. Numerical experiments on Poisson's equation with bilinear elements indicate a sharp decrease in error level: 10^{-13} in no more than 10 iterations with 16×16 Chebyshev grids.

10. THE IMPLEMENTATION OF FINITE ELEMENT METHODS ON THE ICL DAP PARALLEL PROCESSOR

P.G. Ducksbury

Numerical Optimisation Centre, Hatfield Polytechnic, P.O. Box 109, Hatfield, Hertfordshire, AL10 9AB, England

This paper will describe the work which is currently being undertaken into the solution of partial differential equations by finite element methods using the ICL Distributed Array Processor (DAP) computer.

The DAP is a large single instruction multiple data (SIMD) parallel processing machine which has a connected grid of 64×64 processors. This forms an extremely efficient machine for the solution of PDE's by the two methods to be described below. Each of the DAP's 4096 processors has the capability to store and process its own finite element.

Two problems will be used to illustrate the work being carried out. The first of these is the two-dimensional heat conduction equation for which the solution method follows the Galerkin finite element approach which leads to a set of linear simultaneous equations. The solution is then completed by a linear conjugate gradient algorithm.

The important point to note here is that the matrix of linear equations A is normally obtained as the summation over the element matrices

$$A = \sum_e A^e \ .$$

However, A^e will be non-zero only in those row/columns corresponding to variables in the eth element. It therefore becomes possible to employ the conjugate gradient algorithm without having to assemble the matrix A. It is this storing of the element matrices together with the accessing of the local variables in the eth element which maps very efficiently onto the DAP's hardware, and this will be illustrated.

The natural extension to this work was to look at the solution of nonlinear problems. The method of solution for this type of problem was to employ the finite element least squares approach and to solve the resulting optimisation problem with a nonlinear version of conjugate gradient algorithms or by a truncated Newton approach.

For the above two problems numerical results will be quoted for both parallel and sequential implementations.

11. OPTIMAL INTERPOLATION SCHEMES FOR QUADRILATERAL PLATE BENDING ELEMENTS

S. Ellis, K.E. Barrett and J.H. Tabor
Department of Mathematics, Coventry (Lanchester) Polytechnic, Coventry, CV1 5FB, England

Despite recent advances in the understanding of 'shear locking' phenomena in beam and plate elements a simple plate bending element which behaves well in all mesh configurations has not been forthcoming.

Here reduced integration schemes are shown to be more accurate than those involving different order polynomials for displacement and rotations, and to be equivalent to interpolation schemes based on a least squares techniques.

The optimal integration rules for plate elements are known to be dependent on the element alignment and hence their use in general mesh configurations is not possible.

However the least squares technique permits the formulation of quadrilateral plate elements which are independent of element alignment. This is illustrated by the derivation of a four node quadrilateral element which does not lock and achieves good accuracy in all mesh configurations and plate thicknesses.

12. SIMULATION OF THE ELASTOPLASTIC INDENTATION OF A SPHERE AGAINST A HALFSPACE USING THE FINITE ELEMENT METHOD

N. Endahl
Department of Mechanical Engineering, Linköping University, S-581 83 Linköping, Sweden

Permanent deformations of bearing raceway surfaces caused by static overloading can be of a great importance for the lifetime of the bearing. To find accurate dimensioning criteria, Lundberg and Palmgren undertook an experimental investigation in 1943. In the last few years other authors also have performed similar investigations with nearly the same results. In the seventies and during the last years numerical calculations were performed for the indentations of rigid bodies against elastoplastic halfspaces. These calculations leave, however, no information about the total permanent deformation, which is the primary variable among dimensioning rules used today. The total permanent deformation is composed of the residual deformations both in the sphere and the halfspace respectively and their mutual ratio is not constant, but is instead a function of the actual load level.

Residual deformations can with relative simplicity be measured, but on the other hand residual stresses and strains cause a great deal of difficulty, especially when one is leaving the surface and is asking for the state in the interior of the bodies. Another factor to be taken into account is the friction between the bodies. This influences the residual deformations as pointed out, e.g. by Johnson with the experimental investigations.

The material properties in presented calculations were assumed to be elastic linearly isotropic strain hardening together with the von Mises equivalent stress and associated flow rule. The reason for choosing such a simple material model is that the experimental knowledge of the material characteristics is not so well known and is also rather sensitive for heat transfer, both quenching and annealing. A parametric study of the influence from the strain hardening parameter is also very easy to perform with this material description.

The emphasis of the presented investigation is a numerical simulation of the indentation problem and comparisons with existing experimental results. Possibilities for improvement of the existing design criteria have also been investigated. The paper presents results of finite element analysis where both the material and contact nonlinearities are taken into account.

13. FINITE ELEMENT ANALYSIS OF SINGULAR DIFFUSION EQUATIONS

R. England and I. Rudomin
IIMAS, Universidad Nacional Autónoma de Mexico

This study is concerned with the nonlinear diffusion equation

$$\frac{\partial u}{\partial t} = \nabla\cdot[p(x,u,t)\nabla u] - q(x,u,t)u + f(x,u,t)$$

with Dirichlet boundary conditions on at least part of the boundary. Such an equation can have a singularity along any curve where $p = 0$, and in particular along part of the boundary if $p(x,0,t) = 0$. The leading term of the solution may then be proportional to $\sqrt{r}$, where r is the distance from the singular curve.

Three basic methods have been reported for treating point singularities in linear elliptic equations: local refinement; special non-polynomial basis functions; isoparametric "quarter-point" elements. These methods are adapted to treat line singularities, and are tested on a number of simple problems in both one and two space dimensions.

Results in one dimension are entirely compatible with theoretical expectations. For all problems, with or without a singular solution, local refinement gives better results than a standard mesh. "Quarter-point" elements give significantly better results when the solution is singular, but are counter productive otherwise. The special basis functions used are essentially equivalent to "quarter-point" elements, although more subject to perturbation by rounding errors.

In two dimensions, there are many ways of performing mesh refinement, or of choosing special basis functions. With the particular choice taken, first results do not entirely confirm theoretical expectations. It may be that the choice is inappropriate, but it seems likely that, because of limited computing facilities, it has not been possible to refine the mesh sufficiently to ascertain the asymptotic behaviour. Details of the test problems and results are given.

14. A FINITE ELEMENT SOLUTION OF THE THREE-DIMENSIONAL NEUTRON TRANSPORT EQUATION

J.K. Fletcher
UKAEA Northern Division, Risley Nuclear Power Development Establishment, Warrington, Cheshire, WA3 6AT, England

A solution of the three-dimensional multi-group neutron transport equation is described. The flux $\psi(\underline{r},\underline{\Omega})$ at position $\underline{r}$ in the direction of unit vector $\underline{\Omega}$ is expanded as a series of unnormalised spherical harmonics and takes the form

$$\psi(\underline{r},\underline{\Omega}) = \sum_{\ell=0}^{N} (2\ell+1) \sum_{m=0}^{\ell} P_\ell^m(\cos\theta)(\psi_{\ell m}(\underline{r})\cos(m\phi) + \gamma_{\ell m}(\underline{r})\sin(m\phi))$$

with θ and ϕ the axial and azimuthal angles of $\underline{\Omega}$ respectively. Using the orthogonality properties of the expansion functions and the recurrence relations of the associated Legendre polynomials, $P_\ell^m(\cos\theta)$, a set of first order differential equations for the coefficients or moments of the series, $\psi_{\ell m}(\underline{r})$, $\gamma_{\ell m}(\underline{r})$ follows. On elimination of odd ℓ terms a second order system results and this is solved by a weighted residual approach. To minimise computer fast store requirements a semi-iterative algorithm is employed which calculates a pseudo-source term so that $\psi_{\ell m}(\underline{r})$ and $\gamma_{\ell m}(\underline{r})$ are evaluated individually in order of ascending ℓ,m.

Results for various test problems are shown and compared with finite difference solutions, particularly for the case of wide variation in the values of $\psi(\underline{r},\underline{\Omega})$ as might occur in radiation shielding studies.

15. ON THE CONVERGENCE OF MIXED FINITE ELEMENT METHODS FOR THE TRANSIENT RESPONSE OF THE TIMOSHENKO BEAM

T. Geveci
National Research Institute for Mathematical Sciences, CSIR, P.O. Box 395, Pretoria 0001, South Africa

The equations corresponding to the transient response of the *Timoshenko beam* can be expressed as

$$(D_t^2\phi_d(t),\psi) + \frac{1}{d^2}(D_t^2 w_d(t),v) + (D_x\phi_d(t),D_x\psi) + \frac{1}{d^2}(\phi_d(t) - D_x w_d(t),\psi - D_x v) = 0$$

for $t>0$, ψ, $v \in H_0^1(0,1)$. $\phi_d(t) \in H_0^1(0,1)$ is the rotation of the vertical fibres, $w_d(t) \in H_0^1(0,1)$ is the vertical displacement, d is the thickness of the beam which is identified with the interval [0,1] and is clamped at the end-points, $(\cdot,\cdot)$ denotes the $L^2(0,1)$-inner product. Appropriate initial data are provided.

If S_h^r is the space of continuous functions which vanish at the end-points and which reduce to a polynomial of degree $\leq r$ on each subinterval of a partition of [0,1] with mesh size h, and $Q_h^r = \{v' + c : v \in S_h^r,\ c \in \mathbb{R}\}$, a certain mixed finite element formulation requires the determination of $(\phi_{dh}(t),w_{dh}(t),\xi_{dh}(t)) \in S_h^r \times S_h^r \times Q_h^r$ for $t>0$ such that

$$(D_t^2\phi_{dh}(t),\psi_h) + \frac{1}{d^2}(D_t^2 w_{dh}(t),v_h) + (D_x\phi_{dh}(t),D_x\psi_h) + (\xi_{dh},\psi_h - D_x v_h) = 0\ ,$$

$$(\phi_{dh}(t) - D_x w_{dh}(t),\eta_h) - d^2(\xi_{dh}(t),\eta_h) = 0$$

for all $(\psi_h,v_h,\eta_h) \in S_h^r \times S_h^r \times Q_h^r$. Initial data are appropriately chosen.

D.N. Arnold's uniform estimates (in the thickness d) for the static case (Num. Mat. 37 (1981), 405-421) are utilized in order to obtain error estimates for $\| \phi_{dh}(t) - \phi_d(t) \|_{H_0^1}$ and $\| w_{dh}(t) - w_d(t) \|_{H_0^1}$ in the dynamic case.

16. THE NAG/SERC FINITE ELEMENT LIBRARY

C. Greenough, C.R.I. Emson and I.M. Smith
Rutherford Appleton Laboratory, Chilton, Didcot, Oxfordshire OX11 0QX, England

Until recently finite element techniques were used almost exclusively in structural engineering problems. But now there is a growing awareness that they have great potential in many other fields outside structural analysis. As a result the variety of practitioners of the method has grown dramatically and it is now being applied to almost all areas of numerical modelling. Coupled with this growth in application many theoreticians are now engaged in the analysis of the finite element method providing the mathematical foundations of uniqueness and convergence.

As a result of the new diversity of interest, the applications of finite element analysis have outgrown the existing computer software used to exploit it. Most of the existing software is aimed towards the solution of stressing problems and has usually been in the form of large software packages. There have been very few systems written for other application areas and even less software for general use.

The NAG/SERC Finite Element Library is an attempt to make the techniques available to a much larger population. The flexibility needed to accomplish this can only really be provided in the context of a software library.

The Library is in two Levels: Level 0, which consists of subroutines to perform most of the basic operations required in a finite element analysis, and Level 1, a set of example programs in various application areas. These programs provide an illustration of the programming philosophy used in the Library and can be used as a starting point from which the user can develop a program to meet his own particular requirements.

The software and documentation of the Finite Element Library provide a firm software base upon which research and teaching programmes can be founded. In the research environment the highly portable and modular nature of the Library provides an excellent vehicle for collaboration in algorithm and software development. This also encourages software exchange between co-workers and helps to avoid the wastage of duplication.

The structure and modularity of the Library programs make them ideal for teaching since many of the facets of finite element analysis are made transparent. This can also enable students to make modifications to the software and hence gain valuable experience in finite element programming.

17. GALERKIN OR COLLOCATION FOR SECOND KIND INTEGRAL EQUATIONS?

S. Joe
School of Mathematics, University of New South Wales, P.O. Box 1, Kensington, N.S.W. 2033, Australia

Consider the numerical solution of the integral equation

$$y(t_1,t_2)=1+i\lambda\int_0^1\int_0^1 \log\sqrt{(t_1-s_1)^2+(t_2-s_2)^2}\,y(s_1,s_2)\,ds_1 ds_2, \qquad t_1,t_2 \in [0,1]\ , \tag{*}$$

where λ is a real scalar and $i^2 = -1$. This equation arises in the problem of determining the distribution of an alternating current in an infinitely long square conducting bar.

Two methods which could be used to solve (*) are the Galerkin and collocation methods with say piecewise bilinear approximating functions. However, it is not clear which of these two methods should be used. At first glance, it might appear that the collocation method should be preferred, as both approximations have the same order of convergence, but the Galerkin method is computationally more demanding, since it requires the evaluation of 4-dimensional integrals, as compared to the double integrals required in the collocation method.

However, the Galerkin method has a compensating advantage when we consider the iterated variants of the two methods. It is known that with piecewise bilinear approximating functions, the iterated Galerkin solution has an order of convergence twice that of the Galerkin solution itself, whereas the iterated collocation solution exhibits no superconvergence. Since, in principle, the iteration step does not require much more work, it is possible that the Galerkin scheme may require less computational work than the collocation scheme to achieve a certain level of accuracy.

To study more closely the iterated variants of the two methods, we consider the situation in one dimension. We see that under suitable conditions, the iterated collocation solution may also exhibit superconvergence, but the conditions under which it does so are more restricted than for the iterated Galerkin method. In particular, the smoothness conditions on the true solution are much stronger for the iterated collocation method than for the iterated Galerkin method, if superconvergence is to be achieved.

We also give a simple class of examples which shows that the smoothness conditions on the solution for the iterated collocation method are essentially necessary. Thus in practical situations, where the exact solution typically has only a limited number of derivatives, the results suggest that the order of convergence for the iterated collocation method in one dimension (and in higher dimensions) may be much inferior to that for the iterated Galerkin method.

18. VARIATIONAL INEQUALITIES IN PLASTICITY - DUAL FINITE ELEMENT APPROACH

Z. Kestránek
Computer Centre, CKD Praha, 190 02 Praha 7, Czechoslovakia

In this paper we study incremental finite element methods for finding approximate solutions of quasi-static plasticity problems. We propose another variant of the incremental finite element method, starting from the formulation of the quasi-static problem in terms of stresses and hardening parameters only.

Whereas in the mixed method the stresses and hardening parameters are approximated by piecewise constant functions and the displacement by piecewise linear functions, we employ piecewise linear functions for both the stresses and hardening parameters.

The stress approximations consist of Watwood-Hartz equilibriated triangular elements. The finite element method will produce approximations to the stresses successively at a finite number of time levels. At each time level one has to solve a constrained nonlinear optimization problem. We also discuss the Lagrange multiplier method with slack variables for solving this problem. With the particular choice of finite element spaces the optimization problem can easily be solved.

Numerical experiments are presented to show the convergence behaviour of the proposed method.

19. A THREE-DIMENSIONAL FINITE ELEMENT ANALYSIS OF BULGING DURING THE CONTINUOUS CASTING OF STEEL SLABS AND BLOOMS

B.M. Leckenby*, B.A. Lewis* and B. Barber**
* *Department of Mathematics and Computer Science, Sunderland Polytechnic, Sunderland, England.*
** *British Steel Corporation, Teesside Laboratories, England*

The continuous casting process consists of pouring liquid steel into a water cooled mould such that a thin shell of solidifying steel containing a liquid core can be withdrawn through a series of guide rolls, with water sprays positioned between them, until complete solidification takes place. Due to the ferrostatic pressure of liquid steel, the solidifying strand bulges between the support rolls, the amount of bulging depending upon many factors, including shell thickness, temperature, ferrostatic pressure and the section size being cast. In wide slab casting this bulging can lead to quality defects in the steel product and to operational problems with the casting machine.

Excessive bulging results in high strains at the solid/liquid interface which will, if they exceed certain limits, lead to internal cracking. Strand bulging also increases the resistance to withdrawal and can result in the drive rolls slipping due to excessive withdrawal resistance.

It is therefore necessary to be able to calculate inter roll bulging for use in machine design, where it is important to determine the distribution of roll spacing throughout the casting machine. It is also important to be able to relate product quality to the stresses and strains in the strand such that the effect of changing operating parameters such as casting speed, on quality, can be determined.

Two dimensional mathematical models already exist for wide slabs. These models perform calculations on a longitudinal section along the centre of the slab in the direction of slab movement. In these models it is assumed that the roll pitch is much smaller than the slab width, such that the deformation at the midface of the slab may be considered independent of width; it is also assumed that the section is in a state of plane stress. A new three dimensional finite element model, using 20 noded brick elements, has been developed. This model uses the theory of visco-plasticity and the creep law of Myazawa and Schwerdtfeger to calculate the bulging over one roll spacing. Using this model it is now possible to consider both slabs and blooms and to analyse sections in planes normal to the direction of slab movement.

Results are presented comparing the recent three-dimensional work with earlier two-dimensional models.

20. SUPERCONVERGENT RECOVERY OF THE GRADIENT FROM FINITE ELEMENT APPROXIMATIONS ON LINEAR TRIANGLES

N. Levine
Department of Mathematics, University of Reading, Whiteknights, Reading, RG6 2AX, England

For quadrilateral elements, gradient superconvergence has been well established since Veryard improved the accuracy of the gradients of biquadratic Galerkin approximations by sampling them at the second order Gauss points in each element. Such points of exceptional accuracy of derivatives are known as "stress points" and their existence is an example of the phenomenon of "superconvergence"; we associate these terms here with the sampling or recovery of gradients to an order of accuracy higher than is globally possible.

Stress points are locations where the derivative of the polynomial which dominates the error expansion coincides with its approximation (the derivative of a lower degree polynomial). This property is at the heart of superconvergence, for it leads us directly to the stress points of the "unknown" functions's interpolant. Further, the gradient of the interpolant is at all points a superconvergent approximation to the gradient of the finite element solution. Therefore the stress points and superconvergence properties of these two functions are identical.

In this paper we consider piecewise linear approximations on triangular elements and apply the above ideas in a way which is to some extent related to the work of Zlámal on quadrilaterals. We average the (piecewise constant) gradient between two neighbouring elements and show that this "recovered" gradient is a superconvergent approximation to the true gradient at the midpoint of their common edge.

Related schemes recover the gradient at all other points in the problem domain. In particular, the scheme can be modified to estimate normal derivatives at a Dirichlet boundary. All of the recovery schemes are simple, local and explicit.

We show that the triangulation may be greatly distorted but must be smooth in a certain sense. In addition, exactly six elements must meet at each internal node. We give numerical demonstrations of the necessity of these conditions. On general domains they may not be possible with straightforward meshes. We discuss the choice between non-straightforward meshes and the restriction of superconvergence to certain subdomains.

21. DIFFERENT FINITE ELEMENT APPROXIMATIONS FOR REVERSE-BIASED SEMICONDUCTOR DEVICES

L.D. Marini
Istituto di Analisi Numerica del C.N.R, 27100 Pavia, Italy

We study the approximations of variational inequalities related to the analysis of reverse biased semiconductor devices. We also compare the numerical results obtained by using various types of finite element discretizations. All the methods analyzed were first order accurate (in energy norm) from the theoretical point of view. However, on "real life" cases (i.e. with rapidly changing coefficients) the mixed (Raviart-Thomas) approach appears to be decidedly superior in the approximation of the electric field even for piecewise constants coefficients (that is when the famous "homogenizing effect" does not come into play). The reasons for such behaviour still need to be clarified at present.

22. STABILITY OF ROTATING SYSTEMS USING NONLINEAR FINITE ELEMENT METHODS

H.G. Matthies and C. Nath
Germanifcher Lloyd Aktiengesellschaft, P.O. Box 11 16 06, D-2000 Hamburg 11, West Germany

For systems with periodic solutions, such as mechanical systems with rotating parts, it is often important to establish the conditions under which the periodic solution is stable under small perturbations. The method presented for dynamic stability analysis is based on the theory of periodic differential equations. To be specific, the results will be phrased in terms of a mechanical system, but the theory and numerical approximation are equally applicable to any other dynamical system. The stability of periodic solutions, which will be investigated, has to be distinguished from, and is more subtle than, the phenomenon of resonance, in which case most probably a periodic solution will not exist at all.

Since the systems to be considered are described by a large number of equations, a full stability analysis seems infeasible and an approximate procedure must be introduced. A Galerkin approximation is proposed with the natural vibration modes as basis vectors, and the stability analysis is performed for the reduced system.

The theory is applied to the nonlinear finite element model of the wind energy convertor GROWIAN. (Grosse Windenergieanlage). Due to lack of experience in the construction and operation of such very large wind energy conversion systems global dynamic stability analysis emerges as one of the main issues, in addition to the stress and vibration analyses.

23. THE APPLICATION OF BOUNDARY ELEMENT METHOD TO EXTERNAL FLOW PROBLEMS

T.T. Mohamad
Department of Mechanical Engineering, Coventry (Lanchester) Polytechnic, Coventry, CV1 5FB, England

The use of the direct approach of the Boundary Element Method (BEM) has been extended to solve external flow problems about a body of an arbitrary shape. The problem is derived and tested for two cases: uniform flow, and circulatory flow.

The motivation for carrying out this work was that these types of flow provide simple test examples of fixed domain problems governed by Laplace's equation for which analytical solutions are available. Moreover, the direct approach of the BEM is computationally flexible and can deal with a superimposed vortex of a given strength and location. The emphasis on using the BEM was to provide improved accuracy of solutions accompanied by drastic reduction in computer storage, data preparation and processing. Furthermore, a distinct physical advantage is achieved by requiring that only the boundary itself need be discretised analytically integrating the undisturbed flow domain. Thus, it is no longer needed that fictitious boundaries at semi-infinity be assumed, as is the usual procedure in other domain methods.

Comparative tests were conducted to study both computational cost and accuracy with discretization. Tests showed that this method can accurately predict the behaviour of uniform flow and that with circulation past a solid boundary of an arbitrary shape.

24. SAFE - STRESS ANALYSIS USING FINITE ELEMENTS

A.O. Moscardini, B.A. Lewis and M. Cross
Department of Mathematics and Computer Science, Sunderland Polytechnic, Sunderland, England

The continued use of the finite element method for stress analysis in complex structures has created the need for specific computer software which will not only provide the appropriate numerical algorithms to solve problems but will also simplify the large data input and present the results in a comprehensible form. SAFE is an acronym for Stress-linked Analysis with Finite Elements and is a system of closely linked computer codes which satisfy the following criteria: a) data input is minimal, b) user flexibility, c) the programs are completely interactive, d) automatic generation of output, e) low response time in a time sharing environment.

The codes are completely interactive, written in BASIC and run on both a VAX 11750 and a PDP 1140. Three, four, six or eight noded two dimensional elements can be generated to cover complex-shapes which may contain "holes". GINO graphics on the VAX enables instant graphics to be displayed on a VDU, line-printer or as hardcopy. SAFE can be used by the experienced engineer or as a teaching aid to undergraduates. This versatility is one of its greatest assets and its performance has been most successful to date.

25. ON THE QUASI-COUPLED SIGNORINI PROBLEM WITH FRICTION IN LINEAR THERMO-ELASTICITY

J. Nedoma
General Computing Centre, Czechoslovak Academy of Science, P.O. Box 5, 182 07 Praha 8, Czechoslovakia

Let $G \in R^2$ be the open domain with a Lipschitz boundary $\partial G = \overline{\Gamma}_\tau \cup \overline{\Gamma}_\alpha \cup \overline{\Gamma}_u$. Let $u \in W^1 = [H^1(G)]^2$ be the displacement vector, $T \in H^1(G)$ the temperature. It is required that the following problem be solved: find a vector function $u \in W^1$ and a scalar function $T \in H^1(G)$ satisfying

$$(c_{ijkl}e_{kl}(u)_{,j} + F_i = 0, \quad -(H_{ij}(x)T_{,j})_{,i} + \rho c_e v_j T_{,j} = Q \text{ in } G , \tag{1}$$

$$\tau_{ij}n_j = P_{0i} , \quad T = T_p \ (\text{or } H_{ij}T_{,i}n_j = q_0) \qquad \text{on } \Gamma_\tau , \tag{2}$$

$$u_i = u_{0i} , \quad T = T_1 \ (\text{or } H_{ij}T_{,i}n_j = 0) \qquad \text{on } \Gamma_u , \tag{3}$$

$$u_n \leq 0 , \quad \tau_n(u) \leq 0 , \quad u_n\tau_n(u) = 0 \qquad \text{on } \Gamma_\alpha , \tag{4a}$$

$$|\tau_t(u)| \leq F|\tau_n(u)| , \quad |u_t|(|\tau_t(u)| - F|\tau_n(u)|) = 0 ,$$

$$(F\tau_n(u))(x) < 0 \Rightarrow \exists\theta \geq 0 , \quad u_t(x) = -(\theta\tau_t)(x) ,$$

and

$$T \leq T_2 \ , \ q \leq 0 \ , \ (T - T_2)q = 0 \ , \qquad (4b)$$

where u_t, τ_n, u_t, τ_t are respectively the normal and tangential displacements and stress, F is the coefficient of friction in the sense of Coulomb's law, θ is a non-negative function on Γ_α, q is the heat flow, $F_i = f_i - (\beta_{ij}(T-T_0))_{,j} \in L^2(G)$, $P_{0i} \in H^{-1/2}(\Gamma_\tau)$, $Q \in L^2(\overline{G})$ are the body and surface forces and the heat sources, T_p, T_1, T_2 the given temperatures, $\rho \in C(\overline{G}), c_e \in C(\overline{G})$, $H_{ij} \in C^1(\overline{G})$, $c_{ijkl}(x) \in C^1(\overline{G})$ are respectively the density, the specific heat, the thermal conductivity and the elastic constants, and where

$$0<a_0 \leq c_{ijkl}(x)\xi_{kl}\xi_{ij}\|\xi\|^{-2} \leq A_0 < +\infty, \forall\ x \in G, \xi=(\xi_{ij}) \in R^4, i,j=1,2,$$
$$a_0, A_0 = \text{const.} > 0$$

$$H_{ij}(x)\zeta_i\zeta_j \geq c\|\xi\|^2 \ , \quad \forall\ x \in G \ , \ c = \text{const.} > 0 \ .$$

Theorem: Let G be the domain defined above, $c_{ijkl}(x), H_{ij}(x)$ satisfy the above conditions, $u_0 \in W^1$ be such that $u_0|_{\overline{\partial G \setminus \Gamma u}} = 0$. Let $F \in [L^2(\overline{G})]^2$, $P \in [H^{-1/2}(\partial G)]^2$ be such that $[P,u] = 0$ for every $w \in [H^{1/2}(\partial G)]^2$ with $w|_{\overline{\partial G \setminus \Gamma_\tau}} = 0$, (where $[.,.]$ is the scalar product in $[L^2(\partial G)]^2$), $Q \in L^2(\overline{G})$, $T_r \in H^1(\partial G)$, $r = p$ or 1, $s = \tau$ or u. $T_r|_{\overline{\partial G \setminus \Gamma_s}} = 0$, and let $F \in C^{0,1}(\Gamma_\alpha)$ have a compact support in Γ_α. Then $\|F\|_\infty < \sqrt{a_0/2A_0} < \sqrt[4]{\mu/(\lambda+2\mu)}$ and there exists at least one solution of the Signorini problem with friction for the quasi-coupled case in thermoelasticity.

The finite element method was used to solve the problem numerically and the rate of convergence is proved to be O(h), provided the exact solution is sufficiently regular. As the regularity hypotheses are not fulfilled in general, the convergence of finite element approximation to the exact solution without any regularity assumption is justified for one class of problems. The problem discussed is also a model problem from geodynamics, based on the plate tectonic hypothesis and the thermoelasticity. It is closely connected with an in situ exploration of the stress field in the collision of rift zones of the lithosphere.

26. APPLICATION OF FINITE ELEMENT METHOD FOR SOLUTION OF ELECTROANALYTICAL PROBLEMS

M. Penczek and Z. Sojek
Department of Chemistry, Warsaw University, 1 Pasteura, 02-093 Warsaw, Poland

Usually in electroanalytical problems a general set of two parabolic equations

$$\frac{\partial}{\partial t} C_0(x,t) = D_0 \frac{\partial^2}{\partial x^2} C\ (x,t) \ ,$$

$$\frac{\partial}{\partial t} C_R(y,t) = D_R \frac{\partial^2}{\partial x^2} C_R(y,t) \ ,$$

where C_i are concentrations, i = 0, R-species, with the various boundary conditions, determined by the experimental parameters, must be solved. For many cases analytical solutions have not been obtained, and so numerical methods have been applied.

Seeking good stability and accuracy of the solution, we have introduced the FEM into electroanalytical chemistry. Our model was determined by the following boundary conditions:

$$D_0 \frac{\partial}{\partial x} C_0(0,t) + D_R \frac{\partial}{\partial y} C_R(0,t) = 0$$

$$C_0(0,t) + F(t)\ C_R(0,t) = 0$$

$$\frac{\partial}{\partial x} C_0(1,t) = 0 \quad \frac{\partial}{\partial y} C_R(s,t) = 0 \ .$$

Both 0 and R could diffuse either in finite or infinite space. Continuous and discontinuous F(t) functions were employed. Particularly good, low cost results were obtained with irregular space increments in the piecewise Hermite approximation.

27. FELTRAN

C.S. Quad, J. Issa, A.J.H. Goddard and R.T. Ackroyd
Department of Mechanical Engineering, Imperial College of Science and Technology, London, SW7 2BX, England

The principles used in the finite element code FELTRAN for solving the energy dependent Boltzman equation for neutron transport are described. Comparisons with other methods are made for some difficult problems. Solutions for the energy dependent angular flow use a finite element representation for the spatial dependence and a spherical harmonic expanision for the directional dependence. For two-dimensional and finite cylindrical systems rectangular and arbitrary triangular elements are used. For three-dimensional systems of arbitrary cross-section the elements are triangular prisms.

The code FELTRAN is written in modular form to permit easy development. The finite element equations are solved by direct elimination. An out-of-core solver is used for large problems.

Comparison of FELTRAN solutions with results for other deterministic methods and the Monte-Carlo method are given. FELTRAN gives very accurate solutions with reasonable computer times. In particular the method is free from the spurious oscillations of the ray-effect, which can be troublesome with other determinative methods. Coarse mesh calculations, using either quadratic or cubic elements, provide very good approximations.

28. AN ADAPTIVE FINITE ELEMENT APPROACH TO A PLATE PROBLEM

E. Rank
Institut für Bauingenieurwesen I, Technische Universität München, Arcisstrasse 21, D-8000 München, West Germany

A general problem in finite element analysis is the design of optimal FE-meshes, which should be fine enough to produce sufficient accuracy. At the same time too large a number of elements has to be avoided in view of computer time and storage requirements. Therefore an optimal mesh should be graded according to a uniform error distribution. As the error can hardly be predicted by the data, the finite element solution itself should provide some information on the magnitude and distribution of the error. These a-posteriori-estimates have been put on a reliable mathematical foundation for a class of bilinear forms by Babuska et al. in the energy norm. Computer error estimators for the magnitude and error indicators for the distribution of the error have been obtained for some one-dimensional problems and for the problem of plain strain and plane stress.

In this paper the general theory of Babuska is applied to plate-problems according to Mindlin's plate theory. In contrast to Kirchoff's theory shear forces are considered. Thus, better results near corners are obtained and, furthermore, thick plates can be treated. For quadrilateral elements an error indicator is derived, which is composed of element residuals and of the jumps of bending moments and shear forces across element edges.

Some numerical examples show the efficiency of the indicators. Finally, it is demonstrated that the indicators can be used to control an adaptive refinement of the finite element mesh. Thus, nearly optimal meshes are constructed by the FE-code itself, which minimize the effort for data preparation and give best accuracy for a given number of degrees of freedom.

29. ACCURATE TWO-DIMENSIONAL FINITE ELEMENT MODELLING OF CHARGE COUPLED DEVICES

M. Razaz
Department of Electronic and Electrical Engineering, University of Birmingham, P.O. Box 363, B15 2TT, England

A two-dimensional transient algorithm using the finite element method is presented, which can simulate accurately the dynamic behaviour of a charge coupled device (CCD). Although in the past a number of approximate 2D models for CCD's have been published they all suffer from one or more fundamental shortcomings such as (i) assuming instantaneous distribution of majority carriers in response to the potentials applied to the electrodes, and ignoring the time dependent redistribution process of both minority and majority carriers within the device, and their combined effect on the potential distribution and (ii) assuming that minority carriers are precisely at, and move along, the semiconductor-oxide interface. The model presented in this paper uses none of the above approximations, as it solves, in two space dimensions, the exact transient system of coupled nonlinear partial differential equations describing the dynamic behaviour of a CCD. These are the Poisson equation for the potential distribution and the two continuity equations for the dynamics of majority and minority carriers within the device. A fast and self-consistent algorithm is used iteratively to solve these equations subject to appropriate device initial and boundary conditions. Numerical results for charge transfer in a surface channel CCD are also presented.

30. A FINITE ELEMENT METHOD FOR SOLVING A PACKED BATCH DISTILLATION PROBLEM

A Said, L. Pibouleau and S. Domenech
Institut du Genie Chimique, L.A. C.N.R.S. n°192,
Chemin de la Loge, 31078 Toulouse Cedex, France

The purpose of this poster is to solve the mathematical modeling steady-state equations of a packed batch distillation column, in which only the axial dispersion phenomena are taken into account.

The differential equation for the vapour phase, with regard to the level Z in the column, can be stated as follows:

$$- V \frac{\partial Y_i}{\partial Z} + D_v h_v C_v \frac{\partial Y_i}{\partial Z^2} + k V_i a(Y_i^* - Y_i) = 0 \ .$$

The boundary conditions are:

$$V Y_{ei} = \left| V Y_i - D_v h_v C_v \frac{\partial Y_i}{\partial Z} \right|_{Z=0} , \quad \left| \frac{\partial Y_i}{\partial Z} \right|_{Z=Z} = 0 \ .$$

For the liquid phase, the differential equation, and the boundary conditions are similar. For the reboiler and the condenser, we have the classical material balance equations.

The discretized equations are obtained by the method of weighted residuals, using the Galerkin criterion. By making use of integration by parts and inserting the boundary conditions, we obtain a system of nonlinear algebraic equations, which are solved by the Broyden method.

For the binary mixture cyclohexane-toluene, the numerical results are compared with the experimental results. The finite element method gives very good results, and converges more rapidly to the experimental results, than the difference method. From this example, it appears that the finite element method is potentially a very powerful tool for solving problems of packed column simulation.

31. SPACES OF PIECEWISE POLYNOMIALS IN TWO VARIABLES

L.L. Schumaker
Center for Approximation Theory, Texas A & M University, College Station, Texas 77843, U.S.A.

Recently there has been considerable effort to create a theory of multivariate splines analogous to the well-known and well-developed univariate theory. Considerable progress has been made on several fronts: 1) identifying the dimension of spaces of piecewise polynomials on partitions of the plane, 2) constructing locally supported basis elements (B-splines), 3) discussing their spanning and other algebraic properties, and 4) determining their approximation power. This presentation focuses on area 1) where there are already many surprises and a variety of open questions.

32. BOUNDARY ELEMENT METHODS FOR CRACK PROBLEMS IN THE PRESENCE OF BODY FORCES

R.N.L. Smith and J.C. Mason
The Royal Military College of Science, Mathematics Branch, Shrivenham, Swindon, Wiltshire, SN6 8LA, England.

A boundary element model, which incorporates subregions and quarter-point elements at crack tips, has been developed and successfully applied to a number of two-dimensional symmetric and asymmetric curved crack problems. The model is here extended to include certain types of body forces.

The use of subregions, with suitable continuity conditions specified across partitions, gives added flexibility to the boundary element method (BEM). In particular, it allows internal data for stresses and displacements to be obtained on subregion partitions without further integration, while permitting the accurate modelling of the geometry of cracks or holes by placing them along partitions.

The use of quarter-point and traction-singular quarter-point elements at crack tips provides an efficient model of the known limiting behaviour of stresses and displacements. It also permits accurate values of stress intensity factors to be obtained for crack problems based on a computed solution for the displacement near the crack tip.

If body forces are present, difficulties may arise in the BEM. Elements may then be required in the interior of the problem domain, which would offset one of the usual advantages of the BEM over other methods. However, certain types of body force problems may be solved by adopting a new fundamental solution, in which the effects of body forces are obtained by boundary integration. Examples of body forces which permit such an approach are those due to gravitation, rotation, and thermal stress.

The use of the new fundamental solution in combination with BE subregions has not previously been reported. We therefore consider the implementation of such an approach and give numerical results for some sample problems. In particular, we determine stress intensity factors for cracks in rotating regions and carry out a stress and fracture analysis of a thermal stress problem. The new combination of techniques is shown to be effective, and this therefore extends the range and versatility of the BEM.

33. A COMPUTATIONAL STUDY OF BRANCHING IN NONLINEAR WATER WAVES

E.F. Toro
Department of Aerodynamics, College of Aeronautics, Cranfield Institute of Technology, Cranfield, Bedford, MK43 0AL, England.

Free surface gravity waves have been the subject of several investigations using finite and boundary elements, perturbation expansions with the aid of Padé approximants, and results have usually been presented in terms of the relationship between the amplitude A and the non-dimensionalised wave velocity for prescribed values of the fluid depth.

Here we present a numerical study of waves in terms of a bifurcation problem. Computations are carried out using a new Ritz-Kantorovich algorithm based on a variational formulation with functional $J = J_{Q,\lambda}(h(x),\psi(x,y))$. Q and λ are the parameters of the problem and denote the discharge and the domain length respectively, and are prescribed. The quantities $h(x)$ and $\psi(x,y)$ govern the position of the free surface and the internal flow field (to be computed).

Results are displayed through a bifurcation diagram with bifurcation branches (λ,A) for several chosen values of Q. A transition wavelength λ_t is computed which separates a region of deep-water waves from other waves. For branches in this region bifurcation is to the left. Branches on the shallow-water region bifurcate to the right and tend asymptotically to a value A_{max}. Branches in between these two regions bifurcate to the right and have a turning point near the stagnation level.

34. THEORY AND PRACTICE OF EFFICIENT CAD-IMPLEMENTATIONS IN STRUCTURAL ANALYSIS

Ch. Troeder, J. Böhm and A. Spielvogel
Institut für Maschinenelemente und Maschinengestaltung, RWTH, Schinkelstrasse 8, D-5100 Aachen, West Germany.

The linkage of the design-process to the calculation-phase will be illustrated by the example of a structural analysis with FEM and with BEM. The geometric data resulting from the drawing are extracted out of the CAD-system and transmitted to a program, which is installed on a host computer.

The demands on the data-communication and the structural preprocessing such as mesh generation and boundary conditions input are summarized. The transformation of these demands into a program system is demonstrated by some examples. Also the optimized usage concerning quick response time and easy handling of FEM and BEM in connection with CAD-systems will be outlined.

35. STEADY STATE TEMPERATURE DISTRIBUTION IN THE HUMAN TORSO

E.H. Twizell[*] and P. Smith[†]

[*]*Department of Mathematics and Statistics, Brunel University, Uxbridge, Middlesex, UB8 3PH England.*

[†]*Department of Mathematics and Computer Studies, Sunderland Polytechnic, Green Terrace, Sunderland, SR1 3SD, England.*

The mathematical modelling of heat flow in the human body is important in aviation medicine and space science and in certain industrial processes.

In this paper, a cross-section of the torso is approximated by a circular core surrounded by a circular layer of muscle and insulating circular layers of fat and skin. The region occupied by the cross-section of the torso is discretized into a mesh of triangles and the boundary of the torso, that is, the skin surface, is consequently approximated by a polygon.

The elliptic partial differential equation, together with the associated boundary conditions, are expressed in equivalent variational form. Linear basis functions are used and the resulting integral is minimized over the region bounded by the approximating polygon.

This simple model, which employs the most elementary finite element technique, is seen to give very good results for two numerical experiments.

Index

Italic numbers denote pages on which diagrams appear.

C

E

G

H

N

Q

R

U

V

W

Y